Plant Genetics and Genomics: Crops and Models

Volume 9

Series Editor
Richard A. Jorgensen

For further volumes:
http://www.springer.com/series/7397

Renate Schmidt · Ian Bancroft
Editors

Genetics and Genomics of the Brassicaceae

Springer

Editors
Renate Schmidt
Leibniz-Institut für
Pflanzengenetik und
Kulturpflanzenforschung (IPK)
Corrensstr.3
06466 Gatersleben
Germany
schmidtr@ipk-gatersleben.de

Ian Bancroft
John Innes Institute
Norwich Research Park
NR4 7UH Colney, Norwich
United Kingdom
ian.bancroft@bbsrc.ac.uk

ISBN 978-1-4419-7117-3 e-ISBN 978-1-4419-7118-0
DOI 10.1007/978-1-4419-7118-0
Springer New York Dordrecht Heidelberg London

Printed on acid-free paper

Springer is part of Springer Science+Business Media (www.springer.com)

Preface

The angiosperm family Brassicaceae is commonly termed the mustard family or, because of their characteristic flowers consisting of four petals in the form of a Greek cross, the Cruciferae. It comprises 338 genera assigned to 25 tribes and includes the widely studied species *Arabidopsis thaliana* (L.) Heynh. (thale cress) of the Camelineae and a diverse array of cultivated types within the Brassiceae, including oilseed rape, mustards, leafy vegetables, root vegetables, and cole (stem vegetable) crops.

The initial focus for the application of genomic approaches in the Brassicaceae was *A. thaliana*, which had been selected by the late 1980s as a "model" species in which to study plant biology at the molecular level. This was the first plant species for which a genome sequencing program was launched in the mid-1990s by the *Arabidopsis* Genome Initiative, culminating in 2000 in the landmark publication of an analysis of its complete genome sequence. This resource has facilitated the unprecedented expansion in our understanding of plant biology over the last decade.

Genomic and comparative genomic analyses have been applied to a number of species within the Brassicaceae, revealing much about genome evolution in plants, particularly after the publication of *Arabidopsis* genome sequence. Such studies revealed, for example, that the ancestral karyotype for the Brassicaceae was probably $n=8$, and that numerous chromosomal rearrangements and a reduction in chromosome number shaped the genome of *A. thaliana*. Angiosperms have a propensity to undergo chromosome doubling, or polyploidization. Such events are followed by a process of "diploidization," during which genomes stabilize and gene copy number is reduced. The Brassicaceae presents an excellent opportunity to study these processes. The genome sequence of *A. thaliana* provides evidence for as many as three polyploidization events, the last of which occurred near the origin of the Brassicaceae, and is anticipated to be present throughout the family. A distinctive feature of the tribe Brassiceae is extensive subsequent genome triplication, indicative of a hexaploidy event. In addition, several species, particularly within the Brassiceae, are recently formed allotetraploids, e.g., *Brassica napus* ($n=19$) was formed by hybridization of *Brassica rapa* ($n=10$) and *B. oleracea* ($n=9$). Our increasing understanding of these processes is crucial for the interpretation of data from comparative genomic analyses.

The most extensive genomic resources have been developed for the tribe Camelineae, principally, though not exclusively, for *A. thaliana*. These include 130 Mb genome sequence of *A. thaliana*, 1.5 m *A. thaliana* ESTs, multiple commercially available *A. thaliana* microarrays, publicly available gene knockout lines, resource centers for plant lines and DNA stocks, and public databases.

Genomic resources are rapidly developing for the tribe Brassiceae, mainly driven by the economic importance of the *Brassica* species crops. These include ongoing genome sequencing of the *B. rapa* genespace, 0.8 m ESTs from *Brassica* species, a commercially available *Brassica* microarray, many linkage maps (some being integrated), mutagenized populations, resource centers for distribution of BAC libraries, public databases, and a steering committee to coordinate research efforts.

Genomic resources are being developed for a few species from other tribes, which are being studied primarily because of particular attributes, e.g., *Thlaspi caerulescens* which is being studied for metal hyperaccumulation, *Arabis alpina* for perenniality, *Thellungiella halophila* for salt tolerance, and *Boechera* sp. for apomixis.

Contents

1 **Phylogeny, Genome, and Karyotype Evolution of Crucifers (Brassicaceae)** 1
Martin A. Lysak and Marcus A. Koch

2 **Brassicaceae in Agriculture** 33
Suzanne I. Warwick

3 **The Non-coding Landscape of the Genome of *Arabidopsis thaliana*** 67
Georg Haberer, Yu Wang, and Klaus F.X. Mayer

4 **Natural Variation in *Arabidopsis thaliana*** 123
Maarten Koornneef, Matthieu Reymond, and Carlos Alonso-Blanco

5 **Chasing Ghosts: Comparative Mapping in the Brassicaceae** . . . 153
Isobel Parkin

6 **Comparative Genome Analysis at the Sequence Level in the Brassicaceae** 171
Chris Town, Renate Schmidt, and Ian Bancroft

7 **Structural and Functional Evolution of Resynthesized Polyploids** 195
J. Chris Pires and Robert T. Gaeta

8 **Genetics of *Brassica rapa* L.** 215
Nirala Ramchiary and Yong Pyo Lim

9 **The Genetics of *Brassica oleracea*** 261
Carlos F. Quiros and Mark W. Farnham

10 **The Genetics of *Brassica napus*** 291
Federico L. Iniguez-Luy and Maria L. Federico

11 **Genetics of *Brassica juncea*** 323
Akshay K. Pradhan and Deepak Pental

12 ***Arabidopsis lyrata* Genetics** 347
Outi Savolainen and Helmi Kuittinen

13 **The Genetics of *Capsella*** 373
Günter Theißen

14 **Self-Incompatibility in the Brassicaceae** 389
June B. Nasrallah

15 **Sequencing the Gene Space of *Brassica rapa*** 413
Jeong-Hwan Mun and Beom-Seok Park

16 **Germplasm and Molecular Resources** 437
Emma M. Knee, Luz Rivero, Deborah Crist, Erich Grotewold, and Randy Scholl

17 **Resources for Metabolomics** 469
Christoph Böttcher, Edda von Roepenack-Lahaye, and Dierk Scheel

18 **Transformation Technology in the Brassicaceae** 505
Penny A.C. Sparrow, Cassandra M.P. Goldsack, and Lars Østergaard

19 **Resources for Reverse Genetics Approaches in *Arabidopsis thaliana*** 527
Bekir Ülker and Bernd Weisshaar

20 **Resources for Reverse Genetics Approaches in *Brassica* Species** 561
Thomas Wood, Pauline Stephenson, and Lars Østergaard

21 **Bioinformatics Resources for *Arabidopsis thaliana*** 585
Neil Graham and Sean May

22 **Bioinformatics Resources for the *Brassica* Species** 597
Martin Trick

23 **Perspectives on Genetics and Genomics of the Brassicaceae** 617
Renate Schmidt and Ian Bancroft

Index 633

Contributors

Carlos Alonso-Blanco Departamento de Genética Molecular de Plantas, Centro Nacional de Biotecnologia (CNB) Consejo Superior de Investigaciones Científicas (CSIC), 28049 Madrid, Spain, calonso@cnb.csic.es

Ian Bancroft John Innes Centre, Norwich Research Park, Norwich NR4 7UH, UK, ian.bancroft@bbsrc.ac.uk

Christoph Böttcher Department of Stress and Developmental Biology, Leibniz Institute of Plant Biochemistry, Weinberg 3, 06120 Halle/Saale, Germany, cboettch@ipb-halle.de

Deborah Crist Department of Plant Cellular and Molecular Biology, Arabidopsis Biological Resource Center, The Plant Biotechnology Center, The Ohio State University, Columbus, OH 43210, USA, abrc@arabidopsis.org

Mark W. Farnham USDA-ARS-U.S. Vegetable Laboratory, Charleston, SC 29414, USA, mark.farnham@ARS.USDA.GOV

Maria L. Federico Genomics and Bioinformatics Unit (UGB), Agri aquaculture Nutritional Genomic Center (CGNA), INIA-Carillanca Km 10 Camino Cajon-Vilcun, PO Box 58D, Temuco, Chile, maria.federico@cgna.cl

Robert T. Gaeta Division of Biological Sciences, University of Missouri, Columbia, MO 65211, USA, gaetar@missouri.edu

Cassandra M.P. Goldsack John Innes Centre, Norwich Research Park, Norwich, NR4 7UH, UK, cassie.goldsack@bbsrc.ac.uk

Neil Graham Plant and Crop Sciences Division, Nottingham Arabidopsis Stock Centre, University of Nottingham, Loughborough, LE12 5RD UK, neil@arabidopsis.org.uk

Erich Grotewold Department of Plant Cellular and Molecular Biology, Arabidopsis Biological Resource Center, The Plant Biotechnology Center, The Ohio State University, Columbus, OH 43210, USA, abrc@arabidopsis.org

Georg Haberer MIPS, Institute for Bioinformatics and Systems Biology, Helmholtz Center Munich, 85764 Neuherberg, Germany, g.haberer@helmholtz-muenchen.de

Federico L. Iniguez-Luy Genomics and Bioinformatics Unit (UGB), Agri aquaculture Nutritional Genomic Center (CGNA), INIA-Carillanca Km 10 Camino Cajon-Vilcun, PO Box 58D, Temuco, Chile, federico.iniguez@cgna.cl

Emma M. Knee Department of Plant Cellular and Molecular Biology, Arabidopsis Biological Resource Center, The Plant Biotechnology Center, The Ohio State University, Columbus, OH 43210, USA, abrc@arabidopsis.org

Marcus A. Koch Heidelberg Institute of Plant Sciences, Biodiversity and Plant Systematics, University of Heidelberg, Heidelberg, Germany, mkoch@hip.uni-hd.de

Maarten Koornneef Department of Plant Breeding and Genetics, Max Planck Institute for Plant Breeding Research, 50829 Cologne, Germany, koornnee@mpipz.mpg.de

Helmi Kuittinen Department of Biology and Biocenter, University of Oulu, Oulu FIN-90014, Finland, helmi.kuittinen@oulu.fi

Yong Pyo Lim Department of Horticulture, Chungnam National University, Daejeon 305-764, South Korea, yplim@cnu.ac.kr

Martin A. Lysak Department of Functional Genomics and Proteomics, Institute of Experimental Biology, Masaryk University, Brno, Czech Republic, lysak@sci.muni.cz

Sean May Plant and Crop Sciences Division, Nottingham Arabidopsis Stock Centre, University of Nottingham, Loughborough, LE12 5RD UK, sean@arabidopsis.org.uk

Klaus F.X. Mayer MIPS, Institute for Bioinformatics and Systems Biology, Helmholtz Center Munich, 85764 Neuherberg, Germany, k.mayer@helmholtz-muenchen.de

Jeong-Hwan Mun Genomics Division, Department of Agricultural Biotechnology, National Academy of Agricultural Science, Suwon 441-707, Korea, munjh@rda.go.kr

June B. Nasrallah Department of Plant Biology, Cornell University, Ithaca, NY 4853, USA, jbn2@cornell.edu

Lars Østergaard John Innes Centre, Norwich, UK, lars.ostergaard@bbsrc.ac.uk

Beom-Seok Park Genomics Division, Department of Agricultural Biotechnology, National Academy of Agricultural Science, Suwon 441-707, Korea, pbeom@korea.kr

Isobel Parkin Agriculture and Agri-Food Canada, Saskatoon Research Centre, Saskatoon, SK, Canada S7N 0X2, isobel.parkin@agr.gc.ca

Deepak Pental Department of Genetics, Centre for Genetic Manipulation of Crop Plants, University of Delhi South Campus, New Delhi 110021, India, dpental@gmail.com

J. Chris Pires Division of Biological Sciences, 371 Life Sciences Center, University of Missouri, Columbia, MO 65211, USA, piresjc@missouri.edu

Akshay K. Pradhan Department of Genetics, Centre for Genetic Manipulation of Crop Plants, University of Delhi South Campus, New Delhi 110021, India, pradhancgmcp@gmail.com

Carlos F. Quiros Department of Plant Sciences, University of California, Davis, CA 95616, USA, cfquiros@ucdavis.edu

Nirala Ramchiary Department of Horticulture, Chungnam National University, Daejeon 305-764, South Korea, nrudsc@gmail.com

Matthieu Reymond Department of Plant Breeding and Genetics, Max Planck Institute for Plant Breeding Research, 50829 Cologne, Germany; Institut Jean-Pierre Bourgin, UMR1318 INRA-AgroParisTech, INRA Centre de Versailles-Grignon, Route de St-Cyr (RD10), 78026 Versailles Cedex, France, reymond@mpipz.mpg.de

Luz Rivero Department of Plant Cellular and Molecular Biology, Arabidopsis Biological Resource Center, The Plant Biotechnology Center, The Ohio State University, Columbus, OH 43210, USA, abrc@arabidopsis.org

Outi Savolainen Department of Biology and Biocenter, University of Oulu, Oulu FIN-90014, Finland, outi.savolainen@oulu.fi

Dierk Scheel Department of Stress and Developmental Biology, Leibniz Institute of Plant Biochemistry, Weinberg 3, 06120 Halle/Saale, Germany, dscheel@ipb-halle.de

Renate Schmidt Leibniz Institute of Plant Genetics and Crop Plant Research (IPK), 06466 Gatersleben, Germany, schmidtr@ipk-gatersleben.de

Randy Scholl Department of Plant Cellular and Molecular Biology, Arabidopsis Biological Resource Center, The Plant Biotechnology Center, The Ohio State University, Columbus, OH 43210, USA, scholl.1@osu.edu; abrc@arabidopsis.org

Penny A.C. Sparrow John Innes Centre, Norwich Research Park, Norwich NR4 7UH, UK, penelope.sparrow@bbsrc.ac.uk

Pauline Stephenson John Innes Centre, Norwich, UK, pauline.stephenson@bbsrc.ac.uk

Günter Theißen Department of Genetics, Friedrich Schiller University Jena, Jena D-07743, Germany, guenter.theissen@uni-jena.de

Chris Town J. Craig Venter Institute, 9704 Medical Center Drive, Rockville, MD 20850, USA, cdtown@jcvi.org

Martin Trick Computational and Systems Biology Department, John Innes Centre, Norwich Research Park, Norwich NR4 7UH, UK, martin.trick@bbsrc.ac.uk

Bekir Ülker Plant Molecular Engineering Group, IZMB (Institute of Cellular and Molecular Botany), University of Bonn, D-53115 Bonn, Germany, ulker@uni-bonn.de

Edda von Roepenack-Lahaye Department of Stress and Developmental Biology, Leibniz Institute of Plant Biochemistry, Weinberg 3, 06120 Halle/Saale, Germany, vonroep@ipb-halle.de

Yu Wang MIPS, Institute for Bioinformatics and Systems Biology, Helmholtz Center Munich, 85764 Neuherberg, Germany, yu.wang@helmholtz-muenchen.de

Suzanne I. Warwick Eastern Cereal and Oilseed Research Centre, Agriculture and Agri-Food Canada, Central Experimental Farm, Ottawa, ON, Canada K1A OC6, warwicks@agr.gc.ca

Bernd Weisshaar Faculty of Biology, Bielefeld University, D-33594, Bielefeld, Germany, bernd.weisshaar@uni-bielefeld.de

Thomas Wood John Innes Centre, Norwich, UK, tom.wood@bbsrc.ac.uk

Chapter 1
Phylogeny, Genome, and Karyotype Evolution of Crucifers (Brassicaceae)

Martin A. Lysak and Marcus A. Koch

Abstract Brassicaceae (crucifers or the mustard family) is a large plant family comprising over 330 genera and about 3,700 species, including several important crop plants (e.g. *Brassica* species), ornamentals as well as model organisms in the plant sciences (e.g. *Arabidopsis thaliana*). In recent years, the wealth of *Arabidopsis* and *Brassica* genomic resources along with newly established tools and techniques fostered the unprecedented progress in phylogenetics and genomics of crucifers. Multigene phylogenetic analyses paved the way for a new infrafamiliar classification based on phylogenetically circumscribed genera and tribes. A new generation of comparative genetic, cytogenetic, and genomic studies as well as whole-genome sequencing projects unveil general principles of karyotype and genome evolution in Brassicaceae.

Keywords Brassicaceae · Cruciferae · Phylogeny · Tribal classification · Genome and karyotype evolution · Chromosomes · Genome size · Whole-genome duplication · Polyploidy · Genome collinearity

Abbreviations

ACK	Ancestral crucifer karyotype
AK	Ancestral chromosome of the ACK
ancGS	Ancestral genome size
APG	Angiosperm phylogeny group
CCP	Comparative chromosome painting
cpDNA	Chloroplast DNA
DAPI	4′, 6-diamidino-2-phenylindole
FISH	Fluorescence in situ hybridization
gDNA	Genomic DNA
GISH	Genomic in situ hybridization

M.A. Lysak (✉)
Department of Functional Genomics and Proteomics, Institute of Experimental Biology, Masaryk University, Brno, Czech Republic
e-mail: lysak@sci.muni.cz

R. Schmidt, I. Bancroft (eds.), *Genetics and Genomics of the Brassicaceae*, Plant Genetics and Genomics: Crops and Models 9, DOI 10.1007/978-1-4419-7118-0_1,

GS	Genome size
ITS	Internal transcribed spacer within rDNA
NOR	Nucleolar organizing region
mtDNA	Mitochondrial DNA
mya	Million years ago
PCK	Proto-Calepineae karyotype
rDNA	Ribosomal DNA
RFLP	Restriction fragment length polymorphism
RGC	Rare genomic change
WGD	Whole-genome duplication

Contents

1.1 General Introduction 2
1.2 Phylogenetic Position of Brassicaceae and Recognition of Infrafamiliar Taxa . . . 5
1.3 Genome and Chromosomal Evolution 10
1.3.1 Prehistory of Crucifer Genomes: Whole-Genome Duplications and the Age of the Family 10
1.3.2 Genome Size Variation 12
1.3.3 Chromosomes and Chromosome Number Variation 13
1.3.4 Hybridization and Polyploidy 15
1.3.5 Genome and Chromosome Collinearity 18
1.3.6 Revealing Chromosome Homeology Through Comparative Chromosome Painting 18
1.3.7 Ancestral Crucifer Karyotype (ACK, n=8) 20
1.3.8 Overview of Karyotype Evolution in Brassicaceae 22
References 24

1.1 General Introduction

The Brassicaceae (Cruciferae or mustard family) is a large plant family with approximately 338 genera and 3,709 species (see Warwick et al. 2006b) and is of special interest as it include many crop plants (*Brassica oleracea*, *Brassica napus*, *Armoracia rusticana*, and many more), ornamentals (*Aubrieta*, *Iberis*, *Lunaria*, *Arabis*, *Draba* and others) as well as model organisms in the plant sciences (*Arabidopsis thaliana*, *Arabidopsis lyrata*, *Arabidopsis halleri*, *B. napus*, *Capsella rubella*, *Thellungiella halophila*, *Arabis alpina*, and few others). The family shows a worldwide distribution, except Antarctica (Fig. 1.1). Most of the taxa are found in temperate regions of the Northern Hemisphere. However, numerous genera are also found in the Southern Hemisphere (such as *Draba*, *Lepidium*, and *Cardamine*), and some of them are even endemic to southern regions (e.g., South African genera:

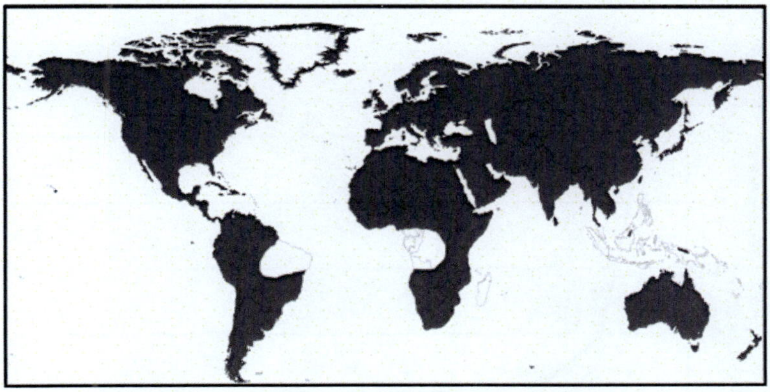

Fig. 1.1 Worldwide distribution of the Brassicaceae

Brachycarpea, *Chamira*, *Schlechteria, Silicularia*, all subsumed by Al-Shehbaz and Mummenhoff 2005 under a broadly defined genus *Heliophila*). In the tropics, the distribution of Brassicaceae is limited to mountainous and alpine regions. *A. alpina* represents the classical example of a plant with a worldwide Northern Hemispheric distribution in mountainous, alpine and arctic habitats including the East African high mountains in Kenya, Tanzania and Ethiopia (Koch et al. 2006). The worldwide distribution of Brassicaceae provides an excellent basis to perform various evolutionary, biogeographic or phylogeographic studies on various taxonomic levels (Koch and Kiefer 2006). However, species diversity is not distributed equally and the most important diversification centers are found in the Irano-Turanian region (ca. 150 genera and ca. 900 species with 530 endemics) and the Mediterranean region (ca. 113 genera and ca. 630 species with 290 endemics). The Saharo-Sindian region (ca. 65 genera and 180 species with 62 endemics) and North America (ca. 99 genera and 778 species with 600 endemics) show a significant reduction in species diversity (Hedge 1976, Al-Shehbaz 1984, Rollins 1993, Appel and Al-Shehbaz 2002). This reduction of species diversity is continued in the Southern Hemisphere (South America with 40 genera and 340 species; Southern Africa with 15 genera and at least 100 species; and Australia and New Zealand with 19 genera and 114 species) (Allan 1961, Marais 1970, Hewson 1982, Al-Shehbaz 1984, Appel and Al-Shehbaz 2002). This overall distribution pattern might provide some evidence for the origin of the family in the Irano-Turanian region (Franzke et al. 2009). It should be noted that the taxonomic circumscription of many taxa, on the generic as well as species level, is still provisional. This is obvious, when we have a closer look on the number of species within a given genus (Fig. 1.2) with 62 or 72% of the genera with three species or less and five species or less, respectively.

Species of Brassicaceae can be easily distinguished morphologically from species of any other family of vascular plants, even if closely related, due to its highly conserved and fairly uniform flower architecture. The flowers are bilaterally symmetrical and consist of four, almost always free sepals in two whorls, free petals (though sometimes lacking), often six, free, tetradynamous stamens (outer

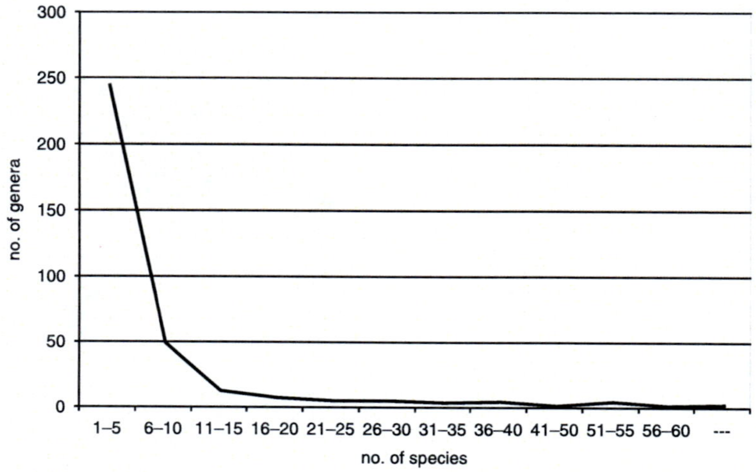

Fig. 1.2 Distribution of species number within crucifer genera

two shorter than inner four; though many *Lepidium* species have four or only two stamens), and a bicarpellate ovary. Only few species and genera show an asymmetrical perianth (e.g., *Iberis* or *Teesdalia*) or divided petals (*Berteroa*). In contrast, fruits of the Brassicaceae exhibit enormous diversity in size, shape, and structure. Therefore, the fruits are used as the most important diagnostic characters for the delimitation and identification of taxa at various ranks. The fruits are often dehiscent, two-valved capsules divided longitudinally by a false septum into two loculi, though in many genera the fruits are indehiscent and/or the septum is totally lacking. Some groups are characterized by angustiseptate fruits (compressed at a right angle to the septum), as in the members of the tribe Lepidieae sensu Schulz (1936), while in the majority of the family the fruits are either latiseptate (compressed parallel to the septum) or not flattened (terete or angled). On the basis of length/width ratio, the fruits have been traditionally divided into those with silicles (length less than three times the width) or with siliques (length more than three times the width), but such division, though useful diagnostically, is arbitrary and has no phylogenetic implications (Al-Shehbaz 1984, Appel and Al-Shehbaz 2002, Al-Shehbaz et al. 2006). Additional important taxonomic characters include embryologic features (position of cotyledons in relation to radicle), nectary glands, trichomes, as well as growth form, chromosome numbers, and seed-coat anatomy and surface (Appel and Al-Shehbaz 2002).

The history of systematics, phylogenetics, and evolutionary research in Brassicaceae family can be divided into three periods. The first period provided us with comprehensive taxon descriptions and several classification systems proposed from the early nineteenth to the mid-twentieth century. The most notable among them are those of De Candolle (1821), Prantl (1891), Hayek (1911), Schulz

(1936), and Janchen (1942). According to these systems, Brassicaceae was artificially divided into 4–19 tribes and 20–30 subtribes. A second period cumulated after a summary of knowledge was obtained about this family more than 30 years ago (Vaughan et al. 1976), followed by a contribution of Tsunoda et al. (1980) that dealt with the biology and breeding of *Brassica* crops and their wild allies. This period was also characterized with a series of papers describing species and their taxonomy and re-defining various tribes and subtribes (e.g. Al-Shehbaz 1984, 1985, 1988a, b, c, for review see also Koch 2003, Koch et al. 2003). The third and most dynamic period of significant taxonomic changes started in the early 1990s, initially based on isozymes but continuously increasing the amount of utilized DNA data (e.g. Koch 2003, Koch et al. 2003). Since then molecular biology and DNA techniques have revolutionized plant systematics and phylogenetics, and because of the selection of *A. thaliana* (*Arabidopsis* thereafter) as the most prominent model plant system, crucifer species belong to an intensely studied plant group.

What are the most important milestone accomplishments in crucifer phylogenetics during the past two decades? In principle, and aside from the wealth of knowledge and resources for *A. thaliana* and other model species, they are at least four: (i) achieving a new infrafamiliar classification based on phylogenetically circumscribed genera and tribes, (ii) phylogenetic circumscription of the order Brassicales and the determination of Cleomaceae as the closest and sister family to Brassicaceae, (iii) unravelling general principles of crucifer evolution by exploring species- or genus-specific evolutionary histories, and (iv) detailed information on karyotype and genome evolution across the family including genetic and cytogenetic maps as well as whole-genome DNA sequence data sets. In this chapter, these four principal achievements were divided into two main sections: Phylogenetic position of Brassicaceae and recognition of infrafamiliar taxa (Section 1.2) and genome and chromosomal evolution (Section 1.3).

1.2 Phylogenetic Position of Brassicaceae and Recognition of Infrafamiliar Taxa

In the most recent comprehensive angiosperm phylogeny (Stevens 2001) the order Brassicales (extended order Capparales) comprises 17 families, 398 genera, and approximately 4,450 species. Overall Brassicales contains approximately 2.2% of the eudicot diversity (Magallön et al. 1999) with its earliest fossil known from the Turonian [89.5 million years ago (mya)]. The age of a stem and crown group was estimated to be 90–85 and 71–69 mya, respectively (Wikström et al. 2001).

Based on strictly morphological studies, Judd et al. (1994) indicated that Brassicaceae is nested within the paraphyletic Capparaceae (including Cleomaceae) and suggested their union into one family, Brassicaceae s.l. However, molecular studies (Hall et al. 2002, 2004, Schranz and Mitchell-Olds 2006) clearly demonstrated that Brassicaceae is sister to Cleomaceae and both are sister to Capparaceae. As a result, three families are currently recognized.

The history of tribal classification systems is long and well summarized in various reviews (e.g., Appel and Al-Shehbaz 2002, Koch 2003, Koch et al. 2003, Mitchell-Olds et al. 2005, Al-Shehbaz et al. 2006) and does not need to be repeated here. Prior to 2005 as summarized in Koch and Mummenhoff (2006), the most important conclusion reached in phylogenetic studies was that except for the Brassiceae, the other tribes were artificially delimited and did not reflect the phylogenetic relationships of their component genera. The other exception was thought to be the tribe Lepidieae (e.g., Zunk et al. 1999), but that too was shown to be artificially circumscribed (Al-Shehbaz et al. 2006).

Of the 49 infrafamiliar taxa (19 tribes and 30 subtribes) previously recognized by Schulz (1936), only nine tribes (Alysseae, Arabideae, Brassiceae, Euclidieae, Heliophileae, Hesperideae, Lepidieae, Schizopetaleae, and Sisymbrieae) were maintained in the re-circumscription of Brassicaceae by Al-Shehbaz et al. (2006). In addition, 16 tribes were either newly described or re-established. This tribal classification followed the first comprehensive phylogeny of Brassicaceae based on the plastidic gene *ndh*F (Beilstein et al. 2006). The study identified three major, significantly supported clades (Lineage I–III; Fig. 1.3). The authors have recently confirmed these data with an extended data set utilizing also nuclear phytochrome A sequence data (Beilstein et al. 2008). A subsequent, ITS-based phylogeny (Bailey et al. 2006) provided substantial support to the new tribal system. In a more recent analysis focusing primarily on the evolution of plastidic *trn*F pseudogene in the mustard family, a supernetwork was reconstructed based on nuclear alcohol dehydrogenase (*adh*), chalcone synthase (*chs*), internal transcribed spacer of nuclear ribosomal DNA (ITS), and plastidic maturase (*mat*K) sequence data (Koch et al. 2007). The supertree was largely in congruence with the corresponding *trn*LF derived phylogeny, and all three major lineages identified by Beilstein et al. (2006) were confirmed. The supertree approach clearly demonstrated that there is a substantial conflicting "phylogenetic signal" at the deeper nodes of the family tree resulting in virtually unresolved phylogenetic trees at the generic level. Some results were contradictory, such as the ancestral position of the Cochlearieae (Koch et al. 2007), which was not revealed by the *ndh*F (Beilstein et al. 2006, 2008) or ITS data (Bailey et al. 2006). Remarkably, a phylogenetic study using the mitochondrial *nad*4 intron (Franzke et al. 2009) is highly congruent with the ITS- and *ndh*F-based studies. In summary, most of the tribes recognized by Al-Shehbaz et al. (2006) are clearly delimited, however, much less significant support is available for the relationships between the various tribes.

Despite the use of multi-gene phylogenies, the lack of resolution in the skeletal backbone of the family's phylogenetic tree is not yet understood, which could be due to one of the following hypotheses. First, early radiation events were quite rapid and were characterized by low levels of genetic variation separating the different lineages. Second, reticulate evolution (e.g., as found in the tribe Brassiceae) resulted in conflicting gene trees that do not reflect species phylogenies. The mitochondrial *nad*4 intron data presented by Franzke et al. (2009) perhaps favor the former hypothesis. Koch et al. (2007), who found that the micro-structural evolutionary changes may be useful for inferring early events of divergence, also

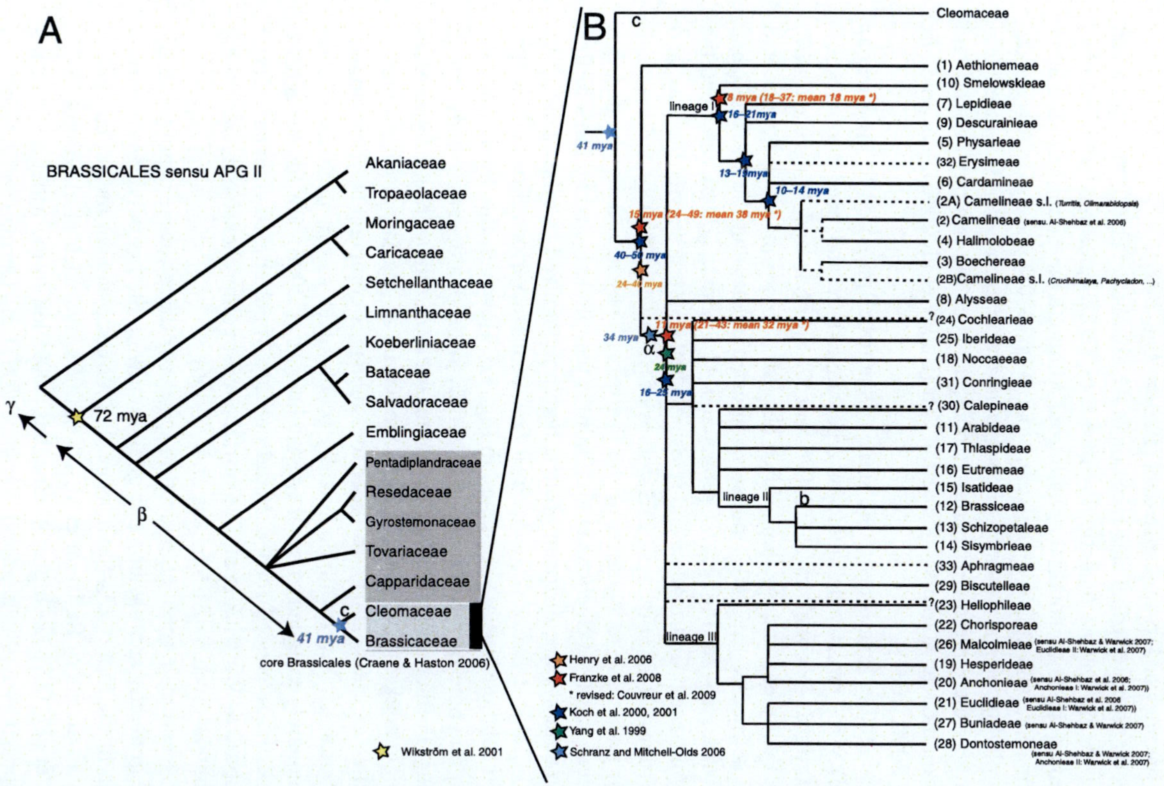

Fig. 1.3 Synopsis of phylogenetic hypothesis, divergence time estimates, and genome duplication events (see also Fig. 1.4). **a** Brassicales according to APG II (2003). **b** Brassicaceae: combination from various sources of tribal relationships in the Brassicaceae (for details refer to the text). Lineages I–III are described in Beilstein et al. (2006), (2008). Koch et al. (2007) used different numbers and we suggest to use Beilstein's version only in order to avoid future confusion. *Dashed lines* indicate uncertain phylogenetic position. However, it should be kept in mind that this synopsis is not derived from one single phylogenetic analysis. All branches are not drawn to scale

favored this scenario. In fact, the two structural rearrangements in the *trn*L-F region alone (Koch et al. 2007) indicate ancient patterns of divergence and their early occurrence is supported by phylogenetic analysis of that region even while excluding the micro-structural mutations (Koch et al. 2007) and is further supported from analyses of the nuclear ITS sequence data (Bailey et al. 2006).

The recently proposed tribal classification of Al-Shehbaz et al. (2006) recognized 25 tribes (see below). More recently, Franzke et al. (2009) presented a family phylogeny based on the mitochondrial *nad*4 intron. Although the sampling in the latter study was smaller, both cpDNA (Beilstein et al. 2006, 2008) and mtDNA (Franzke et al. 2009) phylogenies are mostly in congruence with each other. However, it is still unclear why there are major inconsistencies between the two phylogenies and those based on nuclear markers, such as the ITS (Bailey et al. 2006), *adh*, and *chs* (Koch et al. 2000, 2001).

Additional studies have shown that some of the tribes proposed by Al-Shehbaz et al. (2006) were broadly delimited or are paraphyletic and need further splitting. For example, Warwick et al. (2007) and Al-Shehbaz and Warwick (2007) showed the Euclidieae and Anchonieae to consist of more than one lineage and were newly defined as the Malcolmieae, Dontostemoneae, and the re-established Buniadeae (tribes 26–28, Fig. 1.3). German and Al-Shehbaz (2008) described the new Aphragmeae and Conringieae and re-established Calepineae, Biscutelleae, and Erysimeae (tribes 29–33, Fig. 1.3). ITS studies of Bailey et al. (2006) and Warwick et al. (2010) justify the recognition of the last tribe. They also demonstrated that the Camelineae sensu Al-Shehbaz et al. (2006) is paraphyletic and requires further division into at least three different groups, herein recognized as tribes 2, 34 (2A), and 35 (2B) (Table 1.1, Fig. 1.3). An overview of these various tribes and a synopsis of their relationships are presented in Fig. 1.3. However, this figure does not represent an outcome of a comprehensive family-wide phylogenetic study which is still lacking. Furthermore, it should be emphasized that phylogenetic hypotheses based on single markers (e.g., plastidic, mitochondrial or nuclear) possess a limited value (Koch et al. 2001, 2007). In order to establish a comprehensive phylogeny of the entire family, several taxonomically problematic genera need to be sampled and adequately assigned to tribes (Al-Shehbaz et al. 2006). For further generic delimitations, the interested reader should consult Appel and Al-Shehbaz (2002) and the database of Warwick et al. (2006b); the most comprehensive modern tribal classification was given by Al-Shehbaz et al. (2006).

Although a complete tribal classification system of Brassicaceae is not yet available, we are gradually approaching this goal. Following the first phylogenetic tribal classification of the family (Al-Shehbaz et al. 2006), subsequent molecular studies (e.g., Bailey et al. 2006, Beilstein et al. 2006, 2008, Warwick et al. 2006a, 2007, 2008, Koch et al. 2007, Koch and Al-Shehbaz 2009) led to the tribal adjustments recently proposed by Al-Shehbaz and Warwick (2007) and German and Al-Shehbaz (2008). Table 1.1 summarizes and updates our knowledge of the tribal placement of nearly two-thirds (62.7%) of the 338 genera and 87.6% of the 3,709 species compiled by Warwick et al. (2006b).

Table 1.1 Overview on tribes, genera, and species of Brassicaceae analyzed for their phylogenetic position using molecular markers (Note: The family comprises approximately 338 genera and 3,709 species in total, Warwick et al. 2006b)

	No. of Genera	No. of Species	References
1. Aethionemeae	1	45	Koch and Al-Shehbaz (2009)
2. Camelineae	7	35	Koch and Al-Shehbaz (2009)
3. Boechereae	7	118	Al-Shehbaz et al. (2006)
4. Halimolobeae	5	39	Bailey et al. (2007)
5. Physarieae	7	133	Koch and Al-Shehbaz (2009)
6. Cardamineae	9	333	Koch and Al-Shehbaz (2009)
7. Lepidieae	4	235	Koch and Al-Shehbaz (2009)
8. Alysseae	15	283	Koch and Al-Shehbaz (2009), Warwick et al. (2008)
9. Desurainieae	6	57	Al-Shehbaz et al. (2006)
10. Smelowskieae	1	25	Al-Shehbaz et al. (2006)
11. Arabideae	8	470	Koch and Al-Shehbaz (2009)
12. Brassiceae	46	230	Al-Shehbaz et al. (2006)
13. Schizopetaleae s.l.	28	230	Al-Shehbaz et al. (2006)
14. Sisymbrieae	1	40	Al-Shehbaz et al. (2006)
15. Isatideae	2	65	Koch and Al-Shehbaz (2009)
16. Eutremeae	1	26	Warwick and Al-Shehbaz (2006)
17. Thlaspideae	7	27	Al-Shehbaz et al. (2006)
18. Noccaeeae	3	90	Koch and Al-Shehbaz (2009)
19. Hesperideae	1	45	Al-Shehbaz et al. (2006)
20. Anchonieae	8	68	Al-Shehbaz and Warwick (2007)
21. Euclidieae	13	115	Al-Shehbaz and Warwick (2007)
22. Chorisporeae	3	47	Al-Shehbaz and Warwick (2007)
23. Heliophileae	1	80	Al-Shehbaz et al. (2006)
24. Cochlearieae	1	21	Al-Shehbaz et al. (2006)
25. Iberideae	1	27	Al-Shehbaz et al. (2006)
26. Malcolmieae	8	37	Al-Shehbaz and Warwick (2007)
27. Buniadeae	1	3	Al-Shehbaz and Warwick (2007)
28. Dontostemoneae	3	28	Al-Shehbaz and Warwick (2007)
29. Biscutelleae	1	53	German and Al-Shehbaz (2008)
30. Calepineae	3	8	German and Al-Shehbaz (2008)
31. Conringieae	2	9	German and Al-Shehbaz (2008)
32. Erysimeae	1	180	German and Al-Shehbaz (2008)
33. Aphragmeae	1	11	German and Al-Shehbaz (2008)
34. Unnamed (Camelineae 2A)	2	5	Koch and Al-Shehbaz (2009)
35. Unnamed (Camelineae 2B)	3	20	Koch and Al-Shehbaz (2009)
Total	212	3,249	

The following paper added a significant number of unassigned genera to existing datasets and as a consequence bringing the total number of tribes to 44 [WARWICK S.I., MUMMENHOFF K., SAUDER C.A., KOCH M.A., AL-SHEBAZ I.A. (2010) Closing the gaps: Phylogenetic relationships in the Brassicaceae based on DNA sequence data of nuclear ribosomal ITS region. Pl. Syst. Evol. 285 (3-4): 209-232.]

An ongoing comprehensive phylogenetic study of the family on the generic level (Warwick et al. 2010) aims to cover more than 95% of all currently recognized genera. The major difficulty is obtaining adequate material for molecular studies on species of numerous monospecific or oligospecific genera (see Fig. 1.2). Many of these are known only from the collections of their type species. As most of the larger crucifer genera (e.g., *Cardamine, Draba, Erysimum, Heliophila, Lepidium, Rorippa*) are reasonably well-surveyed molecularly and were shown to be largely monophyletic, it is the smaller and medium-sized genera (especially of the tribes Brassiceae and Schizopetaleae) that need further studies. We assume that many of these genera will be merged with others, and that the total number of genera might be reduced substantially.

1.3 Genome and Chromosomal Evolution

1.3.1 Prehistory of Crucifer Genomes: Whole-Genome Duplications and the Age of the Family

One of the fundamental recent findings of crucifer genomics is the importance of ancient whole-genome duplications (WGDs) which influenced literally every structural and functional aspect of Brassicaceae genomes. The sequencing of the *Arabidopsis* genome revealed that at least 60% of its genome is segmentally duplicated (AGI 2000). Over the past several years a general consensus has been reached on the origin of duplicated segments in *Arabidopsis* through ancient tetraploidization events α, β, and γ (Blanc et al. 2003, Bowers et al. 2003, De Bodt et al. 2005, Henry et al. 2006, Maere et al. 2005, Simillion et al. 2002). Nevertheless, the concept of the three WGDs has been challenged by the analysis of whole-genome shotgun sequence of grapevine (*Vitis vinifera*). Jaillon et al. (2007) showed that the grapevine genome, consisting of three ancestral genomes, has a paleohexaploid origin. The comparison with other plant genome sequences available (*Arabidopsis*, poplar, and rice) showed that the paleohexaploidy event must have occurred before the radiation of Eurosids and probably after the separation of dicots and monocots. In *Arabidopsis*, the ancient triplication was equated with the WGD event γ, which was followed by subsequent tetraploidization events β and α (Jaillon et al. 2007). As the two latter paleotetraploidy events were not identified in Caricaceae (papaya, *Carica papaya*; Ming et al. 2008, Tang et al. 2008), belonging to the Brassicales as *Arabidopsis*, it can be assumed that the two paleopolyploid events are specific for the core Brassicales including Brassicaceae (cf. Hall et al. 2004). Furthermore, an independent paleohexaploid WGD detected in Cleomaceae (Schranz and Mitchell-Olds 2006), a sister family of Brassicaceae, is suggesting that the α WGD event post-dates the Cleomaceae–Brassicaceae split. In addition, the absence of two paralogs of arginine decarboxylase gene (*Adc*) in the genus *Aethionema* (Aethionemeae) as compared to the remaining Brassicaceae species (Galloway et al. 1998, Schranz and Mitchell-Olds 2006) pinpoints the α event to be specific for "core Brassicaceae." Further research is needed to resolve the position of β and α WGDs within Brassicales and Brassicaceae, respectively. The position of the four WGDs within a phylogenetic framework and their dating are shown in Fig. 1.4a, b, respectively.

Comparative genetic mapping using *Arabidopsis* markers has shown that genomes of *Brassica* species with diploid-like chromosome numbers were triplicated due to a hexaploidy event (Lagercrantz 1998, Parkin et al. 2005). The whole-genome triplication has also been unambiguously evidenced cytogenetically by comparative chromosome painting (CCP) using *Arabidopsis* BAC contigs in several *Brassica* species and the tribe Brassiceae (Lysak et al. 2005, 2007, Ziolkowski et al. 2006). Therefore, such a relatively recent genome duplication which can be uncovered by CCP and comparative genetic mapping, unlike the three ancient WGDs, is classified here as a mesopolyploid WGD event. Ongoing CCP analyses in a wide range of Brassicaceae species indicate that mesopolyploid WGDs were relatively common within different crucifer lineages (Mandáková et al. 2010 and unpublished data). Paleopolyploid and mesopolyploid WGDs were obscured by chromosomal and genetic diploidization processes including chromosome rearrangements, leading to chromosome number reduction, genome size downsizing, and diploid-like inheritance. Diploid-like genomes give rise to neopolyploid cytotypes and species via autopolyploidy and hybridization-driven allopolyploidy. Therefore, as in other angiosperm lineages, the evolution of Brassicaceae genomes is characterized by cyclic rounds of WGDs followed by diploidization (Fig. 1.4b).

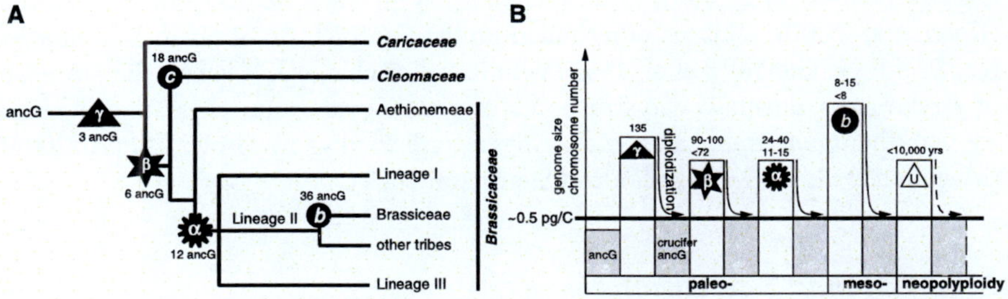

Fig. 1.4 **a** Discussed whole-genome duplication (WGD) events mapped onto the schematic phylogeny of the Brassicales. Besides γ and β events, lineage-specific tetraploidization α (core Brassicaceae) and whole-genome triplications *c* (Cleomaceae) and *b* (Brassiceae) are shown. The assumed number of ancestral genomes (ancG) multiplied by WGD events is also given. Modified after Beilstein et al. (2006), Lysak et al. (2006), Schranz and Mitchell-Olds (2006), Jaillon et al. (2007), Ming et al. (2008), and Tang et al. (2008). **b** Model of cyclic genome evolution in Brassicaceae. Multiple ancient and more recent WGDs (γ, β, α, and *b*; the U event refers to the origin of *Brassica* allopolyploids of the U's *triangle*) increasing genome size and chromosome number are counteracted by the genome diploidization, reducing genome size and chromosome number. For simplicity, the model assumes that polyploid genomes are being diploidized toward the reconstructed ancestral crucifer genome of ~0.5 pg/C (Lysak et al. 2009). The size of the ancestral angiosperm genome triplicated by the γ event was chosen arbitrarily. Rough time estimates of the WGDs are given in million years ago following Rana et al. (2004), Lysak et al. (2005), Henry et al. (2006), Franzke et al. (2008), and Ming et al. (2008)

Divergence time estimates (Fig. 1.3) are still controversial. The usage of K_S values, as presented by Schranz and Mitchell-Olds (2006) and Maere et al. (2005), might be more reliable since they do not make assumptions about molecular clocks. K_S values represent the synonymous substitution rate that can be inferred for protein-coding genes only. It is the calculation of a rate considering only synonymous mutations not resulting in amino acid changes. Schranz and Mitchell-Olds (2006) estimated the time of very early radiation of Brassicaceae as 34 mya. This was based on a genome-wide (referring to *A. thaliana*) estimated K_S average (K_S = 0.67) reflecting the most recent (α) WGD event and using *A. thaliana* as a reference (Simillion et al. 2002, Bowers et al. 2003, De Bodt et al. 2005). Genome-wide comparison of K_S values from Cleomaceae and Brassicaceae suggests that the corresponding mean K_S value is 0.82, which indicates 41 mya as a divergence time estimate for these two families. This range of divergence fits well the assumption that the split between Caricaceae and the core Brassicales had occurred 72 mya (Wikström et al. 2001). It is noteworthy that the divergence time estimates were obtained by different methods (calculating synonymous mutation rates and running simulation models, Koch et al. 2000; calculating genome-wide K_S values, Schranz and Mitchell-Olds 2006; or using relaxed molecular clock approaches, Franzke et al. 2009). Despite a relatively large variation, it is remarkable to see that the order of magnitude is similar. This is even the case for the divergence data provided by Franzke et al. (2009) showing much more recent estimates and which have been, however, re-calculated by the same authors providing much older estimates (Couvreur et al. 2009). A critical aspect is dating and calibration of the phylogenetic trees as the Brassicaceae taxa are poorly represented as macrofossils (Appel and Al-Shehbaz 2002). Some studies relied partly on macrofossils from the Miocene (Koch et al. 2000), whereas other relied on the Turonian fossil taxon *Dressiantha* (ca. 89.5 mya; Gandolfo et al. 1998) with its set of characters such as the presence of a gynophore, unequal petal size, or a bicarpellate gynoecium suggesting an affinity with the order Brassicales sensu Angiosperm Phylogeny Group (APGII 2003) (Magallon et al. 1999, Magallon and Sanderson 2001).

1.3.2 Genome Size Variation

Multiple whole-genome duplications and triplications have necessarily increased genome size (GS) of ancestral Brassicaceae species. Nevertheless this tendency is not apparent when GS of the extant crucifer taxa are surveyed. Despite the 16-fold variation in GS across the family, all crucifer species analyzed so far are classified as having very small ($\leq$ 1.4 pg) or small ($\leq$ 3.5 pg/C; Leitch et al. 1998) genomes, with a mean *C* value of 0.63 pg (Lysak et al. 2009). The smallest *C* values have been estimated for *A. thaliana* (0.16 pg/C; Camelineae) and some *Sphaerocardamum* species (0.15–0.16 pg/C; Halimolobeae). On the opposite pole, *Bunias orientalis* has the largest genome (2.43 pg/C; Buniadeae), followed by high GS values in some species of *Physaria* (2.23–2.34 pg/C; Physarieae) and *Matthiola* (2.11–2.29 pg/C; Anchonieae) (Lysak et al. 2009).

Two recent studies (Oyama et al. 2008, Lysak et al. 2009) attempted to disentangle the evolution of GS in Brassicaceae using available phylogenetic frameworks. Both studies revealed increases as well as decreases in GS across the analyzed species sets without apparent evolutionary directionality. Lysak et al. (2009) reconstructed a theoretical ancestral genome size (ancGS) as 0.5 pg/C and tested modes of GS evolution along phylogenetic lineages based on five gene markers. When comparing the *C* values, 50% of crucifer taxa analyzed showed a GS decrease and increase, respectively, and GS evolution shows no dominant tendency. Overall GS in the extant species as compared to the ancGS remained relatively stable across evolutionary time (Fig. 1.4b) and increases were generally moderate, with significant increases found only in the Anchonieae, Buniadeae, and Physarieae. Hence, despite dynamic processes having the potential to increase GS (e.g., transposable element amplification and WGDs), mechanisms eliminating accumulated DNA and/or suppressing DNA amplification must be active in Brassicaceae. Although possible mechanisms of GS downsizing were suggested [e.g., illegitimate recombination: Devos et al. (2002) or gross chromosome rearrangements: Lysak and Lexer (2006), Gaeta et al. (2007)], the modes of GS stasis in Brassicaceae are still poorly understood.

Particularly intriguing is genome evolution in crucifer species with large C values (2.0 pg/C), large chromosomes, and low, diploid-like chromosome number ($2n = 8$–14). This group of taxa thus far represents two *Bunias* ($2n = 14$), one *Physaria* ($2n = 8$), and three *Matthiola* ($2n = 14$) species (Lysak et al. 2009). To resolve this, it was proposed that the diploid-like number and large chromosome size may have arisen from ancient polyploidy events followed by diploidization involving chromosome number reduction without extensive DNA loss, or via retrotransposon amplification equally increasing size of all chromosomes as well as GS of the entire genome (Lysak et al. 2009). Although the polyploidization–diploidization scenario seems to be more conceivable, the issue cannot be resolved at present.

1.3.3 Chromosomes and Chromosome Number Variation

Mitotic chromosomes of crucifer species are generally very small, spanning only a few micrometers in size (e.g., 1.5–2.8 μm in *A. thaliana*, Koornneef et al. 2003, 2–5 μm in *Brassica*, Cheng et al. 1995; 1.2–5.8 μm in *Boechera*, Kantama et al. 2007). As noted before, larger chromosomes seem to be frequently observed in species with low chromosome numbers ($n = 4$–7) and large genome sizes (e.g., *Bunias*, *Matthiola*, *Physaria*; Manton 1932, Lysak et al. 2009), although large chromosomes were also reported in *Menonvillea* with $2n = 22$ (Manton 1932).

Chromosome morphology and structure of most of the crucifer species are closely linked to the discrete distribution of repetitive DNA elements along a longitudinal chromosome axis. The highest percentage of dispersed and tandem repeats including ribosomal DNA repeats constitutes distinct heterochromatic arrays (Fig. 1.5). Euchromatic chromosome regions largely coincide with gene-rich sequences, though they are not completely depleted of dispersed repeats. Differential staining techniques reveal pericentromere regions as a prominent heterochromatic component of crucifer chromosomes. Both dispersed and satellite repeats compose interstitial and terminal heterochromatic knobs. Similar knobs known as nucleolar organizing regions (NORs) are formed by tandem arrays of 45S ribosomal DNA (rDNA). Distinct chromosome organization also determines the characteristic chromatin organization within interphase nuclei with distinct heterochromatic chromocenters comprising pericentromeres, knobs, and NORs, interspersed by euchromatic chromosome territories (Fransz et al. 2002, Koornneef et al. 2003, Pecinka et al. 2004).

Chromosome numbers, recently compiled by Warwick and Al-Shehbaz (2006), are known for ca. 70% of genera and 40% of crucifer species. Chromosome numbers in Brassicaceae vary over 32-fold, with the lowest chromosome number of $n = 4$ found only in *Physaria* (Physarieae) and *Stenopetalum* (Camelineae); five chromosome pairs ($n = 5$) were observed in *A. thaliana* (Camelineae), one *Matthiola* (Anchonieae), five *Stenopetalum* and 21 *Physaria* species (Warwick and Al-Shehbaz 2006). The highest chromosome numbers have been reported by Montgomery (1955), Easterly (1963), and Harriman (1965) in North American polyploid *Cardamine* (formerly *Dentaria*) species (*Cardamine angustata*, *Cardamine concatenata*, *Cardamine diphylla*, *Cardamine dissecta*, and *Cardamine maxima*). Among these polyploid species the highest counts were obtained in *Cardamine concatenata* (*Delphinula laciniata*; $2n = \pm 240$ by Montgomery 1955 and $2n = 256$ by Easterly 1963) and *C. diphylla* ($2n$=ca. 256, Harriman 1965). However, the counts have to be considered a priori as approximate and inaccurate due to the clumping of numerous very small chromosomes (Harriman 1965). This is illustrated by more than a three-fold variation in chromosome number ($2n = 74$–256) between counts in different root tips in *C. diphylla* (Harriman 1965). Future analyses of DAPI-stained meiotic chromosomes coupled with GISH and/or FISH localization of centromeric satellite repeats can elucidate chromosome numbers and the origin of the high polyploid *Cardamine* species.

Base chromosome numbers vary from $x = 4$–17 with more than one-third of the taxa having karyotypes based on $x = 8$ (Warwick and Al-Shehbaz 2006), implying that $x = 8$ is most likely an ancestral chromosome number of the whole family. Base chromosome numbers are practical in recognizing diploids ($2x$) from higher ploidy levels ($3x$, $4x$, etc.) within a given taxon. However, frequent auto- and allopolyploid events increasing the number of chromosome sets have been followed by species- and lineage-specific chromosome reshuffling. Chromosome fusions and fissions are causing an intra-generic numeric variation known as descending and ascending dysploidy. Therefore, several crucifer genera (e.g., *Brassica*, *Cochlearia*, *Diplotaxis*, *Erysimum*, and *Physaria*) are polybasic, i.e., characterized by multiple base chromosome numbers (Warwick and Al-Shehbaz 2006). This makes the base number concept impractical in some genera. Moreover, generic base numbers based on a lowest chromosome count available do not reflect the true nature of diploid-like genomes which were often influenced by paleo- and mesopolyploid events followed by subsequent diploidization (see below).

1.3.4 Hybridization and Polyploidy

Like in other plant families, polyploidy, hybridization and introgression have significantly shaped genome evolution in many crucifer species groups and genera. Two key review papers on polyploidy in Brassicaceae are available: Marhold and Lihová (2006) provided a comprehensive review on polyploid evolution in the family, and a chromosome number database was compiled by Warwick and Al-Shehbaz (2006). The importance of polyploid speciation is clearly demonstrated in some Brassiceae genera comprising exclusively polyploid species such as *Crambe* ($2n = 30$, 60, and 120), *Moricandia* ($2n = 22$, 28, 56, and 84), and *Vella* ($2n = 34$, 68, and 102) or in genera with polyploid taxa prevailing over diploids (e.g., *Aethionema*, *Braya*, *Draba*). The most detailed polyploidy pattern distribution has been elaborated for the genus *Draba* comprising more than 350 species and therefore representing one of the largest genera of Brassicaceae (Jordon-Thaden and Koch 2008). Here, polyploidization and speciation rates are exceeding by far those described for all other plants (Jordon-Thaden and Koch 2008). Numerous cases of inter-species hybridization and polyploid formation reported in Brassicaceae are evolutionary young events (ca. 10^2–10^5 years old) which should be distinguished from much older paleo- and mesopolyploid whole-genome duplications. In crucifers, new polyploid species are generated by autopolyploidy or through inter-species hybridization. Hybrid speciation includes either hybrids between genomes of different chromosome numbers, often undergoing subsequent duplication (allopolyploidy) or (perhaps rarer) hybrids between species with the same chromosome numbers (homoploid hybrid speciation or recombinational speciation) (see Mallet 2007, Soltis et al. 2007, Hegarty and Hiscock 2008 for reviews).

Drawing a distinct line between autopolyploidy and allopolyploidy is not always straightforward and in some cases not feasible due to a problematic definition of a species (cf. Rieseberg and Willis 2007, Mallet 2007). Numerous reports on even-numbered intraspecific cytotypes (cf. Warwick and Al-Shehbaz 2006; e.g., *Calepina irregularis*: $2n=2x$, $4x=16$, 32; *Isatis tinctoria*: $2n=2x$, $4x=14$, 28) are suggesting that autopolyploidy is common in Brassicaceae. This is consistent with the present view on autopolyploidy as a significant, though underestimated mode of polyploid speciation in plants (Soltis et al. 2007). Taxonomic treatment of autopolyploid derivates and their respective progenitors is problematic. Most frequently diploid and autopolyploid populations possess indistinguishable phenotypes and are treated as cytotypes without a taxonomic rank (e.g., diploid and autopolyploid cytotypes within four species of the *Cardamine digitata* aggregate, Jørgensen et al. 2008); less common is their circumscription as subspecies (autotetraploid *Biscutella laevigata* subsp. *laevigata*, $2n=4x=36$, Tremetsberger et al. 2002, autotetraploid *Cardamine amara* subsp. *austriaca*, $2n=4x=32$, Marhold 1999) or separate species (*Cochlearia officinalis*, $2n=4x=24$ and *C. anglica*, $2n=8x=42$; Koch et al. 1998a,

1999b; *Cardamine matthioli*, $2n=2x=16$ and *C. majovskii*, $2n=4x=32$; Lihová and Marhold 2003). The most cited example of autopolyploid speciation in Brassicaceae is that of the tetraploid *B. laevigata* described by Manton (1937). *B. laevigata* is the most common *Biscutella* species comprising diploid ($2n=2x=18$) and tetraploid ($2n=4x=36$) cytotypes. The diploid cytotype is treated as several subspecies, whereas the tetraploid populations confined to higher altitudes of the Alps were circumscribed as subsp. *laevigata* (Tremetsberger et al. 2002, Parisod and Besnard 2007). As the tetraploid populations show a tetrasomic inheritance and morphologically resemble different diploid subspecies, it was concluded that the tetraploid taxon originated via a polytopic autopolyploidy (Tremetsberger et al. 2002). The autopolyploid origin is also underlined by the occurrence of tetravalents and trivalents during meiosis I (Manton 1937). However, the multivalent formation should not be taken as a conclusive evidence of an autopolyploid origin as some autopolyploids can exhibit regular bivalent pairing and still display polysomic inheritance (Santos et al. 2003, Soltis et al. 2007). Analysis of meiotic pairing in established and newly generated *A. thaliana* autotetraploid ($2n=4x=20$) lines, revealed concomitant decrease and increase in multivalents and bivalents, respectively, suggesting a partial diploidization of meiosis (Santos et al. 2003). In presumably autotetraploid *Polyctenium fremontii* ($2n=4x=28$, Boechereae), although tetravalents are frequently observed in pachytene, regular bivalents prevail in metaphase I (T. Mandáková and M.A. Lysak, unpublished data). A regular bivalent pairing throughout the first meiotic division has been found in two apparently autotetraploid ($2n=4x=28$) species of the Calepineae, *C. irregularis*, and *Goldbachia laevigata* (Mandáková and Lysak 2008). As cross-species chromosome painting allows the identification of individual chromosomes during meiosis, pairing of homologous and homeologous chromosomes in autopolyploid plants can be analyzed with a high degree of precision.

Inter-species hybridization occurs frequently in Brassicaceae, particularly in some genera such as *Arabis*, *Boechera*, *Cardamine*, or *Rorippa*. Hybrids can be sterile, partially, or fully fertile and then having the capacity of backcrossing to the respective parental species. For instance, long-distance dispersal followed by hybridization has been suggested for the origin of the polyploid Australian and New Zealand *Lepidium* taxa. Phylogenetic analysis of noncoding chloroplast and nuclear (ITS) DNA regions showed that Australian/New Zealand taxa have most likely an allopolyploid and bicontinental origin, resulting from crosses between African and Californian *Lepidium* species (Mummenhoff et al. 2004). Other well-documented examples where interspecific hybrids have become newly established allopolyploid species include the composite genomes of *Diplotaxis muralis* ($2n=42$; parental species *Diplotaxis viminea*, $2n=20$ and *Diplotaxis tenuifolia*, $2n=22$; Eschmann-Grupe et al. 2004), three *Brassica* allotetraploids of the U's triangle (e.g., $2n=38$ in *B. napus*, parental subsp. *B. oleracea*, $2n=18$, and *Brassica rapa*, $2n=20$), *Arabidopsis kamchatica* ($2n=32$; parental subsp. *A. lyrata* and *A. halleri* subsp. *gemmifera*, both $2n=16$; Shimizu et al. 2005, Koch and Matschinger 2007, Schmickl et al. 2008), or *A. suecica* ($2n=26$; parental subsp. *A. arenosa*, $2n=16$, 32 and *A. thaliana*, $2n=10$; Mummenhoff and Hurka 1994). In some hybrids, the genome incompatibility has been overcome by whole-genome duplication (autoallopolyploidy), either directly or via a triploid bridge and unreduced gametes. To such allopolyploid crucifer species belong *Draba ladina* ($2n=32$; parental subsp. *Draba aizoides* and *Draba tomentosa*, both $2n=16$; Widmer and Baltisberger 1999) or *Cardamine schulzii* ($2n=6x=48$), an autopolyploid derivate of the interspecific hybrid (*C.* × *insueta*, $2n=3x=24$) between *C. amara* and *Cardamine rivularis* (both $2n=16$) (Urbanska et al. 1997). In the genus *Boechera*, the potential sterility of the hybridogenous *B.* × *divaricarpa* (*Boechera holboellii* × *Boechera stricta*) and its negative evolutionary consequences have been overcome by the apomictic reproduction (Dobes et al. 2004a, b, Schranz et al. 2005). In European *Microthlaspi perfoliatum* with $2n=14$, 28, 42 cytotypes the situation is less clear and one of the putative parental taxa of the hexaploid cytotype most likely went extinct (Koch et al. 1998b, Koch and Hurka 1999, Koch and Bernhardt 2004).

Genomic in situ hybridization (GISH) refers to the specific identification of parental genomes in interspecific hybrids and allopolyploid species by simultaneous or subsequent hybridization of fluorescently labeled genomic DNA (gDNA) of the presumed parents to chromosomes of the composite genome. In Brassicaceae, most GISH studies were initially concentrated on the natural and synthetic allotetraploid *Brassica* species (*Brassica carinata*, *Brassica juncea*, *B. napus*) and artificial intergeneric hybrids between *B. napus* and other crucifer species mainly from the Brassiceae (Chèvre et al. 2007, reviewed by Lysak and Lexer 2006, Snowdon 2007). GISH data corroborated the phylogenetic relationships among the three "diploid" donor *Brassica* species with the two lineages separating *Brassica nigra* (BB genome, $2n=16$) from *B. oleracea* (CC, $2n=18$) and *B. rapa* (AA, $2n=20$) (Warwick and Sauder 2005 and references therein). GISH discriminated between A and C genomes in *B. carinata* (BBCC, $2n=34$) and between A and B genomes in *B. juncea* (AABB, $2n=36$) (Snowdon et al. 1997, Maluszynska and Hasterok 2005), whereas A and C genomes showing a high level of sequence homeology have not been discerned in *B. napus* (AACC, $2n=38$) (Snowdon et al. 1997). This was not confirmed by Howell et al. (2008) who showed that the A and C genomes can be distinguished clearly using gDNA of *B. oleracea* as the probe and gDNA of *B. rapa* DNA as a block. Despite a very close phylogenetic relationship between *Arabidopsis thaliana* (TT, $2n=10$) and *Arabidopsis arenosa* (AA, $2n=16$; AAAA, $2n=32$), GISH with increased amounts of gDNAs was successfully applied to identify both parental genomes in the natural and synthetic *Arabidopsis suecica* (AATT, $2n=26$) as well as in an artificial hybrid between *Arabidopsis thaliana* and *A. suecica* (Ali et al. 2004, Lysak and Lexer 2006). Recently, GISH was used to elucidate the genome composition of sexual ($2n=14$) and apomictic ($2n=15$) genotypes from the *Boechera holboellii* complex (Kantama et al. 2007). This species complex, including *B. holboellii*, *B. stricta*, and their presumed hybrid *B.* × *divaricarpa*, exhibits extensive karyological variation due to recurrent hybridization, introgression, and apomixis. Two-color GISH analysis using *B. holboellii* and *B. stricta* gDNAs revealed that the analyzed apomicts represent interspecific hybrids with different contribution of *B. holboellii*- and *B. stricta*-derived chromosomes (Kantama et al. 2007). As in most studies carried out for Brassicaceae species so far, GISH in *Boechera* was principally based on genome-specific pericentromeric tandem repeats (Kantama et al. 2007, see also Ali et al. 2004, Lysak and Lexer 2006).

These examples of GISH studies illustrate an untapped capacity of the technique to reveal the origin of hybridogenous taxa within taxonomically complicated crucifer groups (e.g., *Aethionema*, *Cardamine*, *Draba*, *Heliophila*; also see Marhold and Lihová 2006).

1.3.5 Genome and Chromosome Collinearity

Based on DNA markers, molecular phylogenetics is steadily improving our understanding of taxon-to-taxon relationships within Brassicaceae. Nevertheless, inter-species and infrafamiliar relationships may also be inferred from genome-wide comparisons. Such comparisons can be carried out through comparative genetic mapping, comparative cytogenetic analysis, or by comparing whole-genome sequences. As comparative genetic mapping in Brassicaceae is extensively covered by other Chapters 5 and 6 and recent reviews (e.g., Koch and Kiefer 2005, Lysak and Lexer 2006, Schranz et al. 2006, Snowdon 2007), only a brief account of key findings directly linked to genome and karyotype evolution of Brassicaceae is outlined herein.

Arabidopsis sequence data along with the wealth of genetic markers have been crucial for *Arabidopsis* becoming a reference genome in comparative genetic mapping across Brassicaceae.

The first wave of crucifer comparative genetics was marked by analyzing the extent of cross-species genome collinearity between the diploid *Brassica* species and between *Arabidopsis* and *Brassica* species, respectively (Lagercrantz and Lydiate 1996, Lagercrantz 1998). However, the budding field of whole-genome comparisons in Brassicaceae has been fostered significantly by genetic maps comparing *Arabidopsis* with three n=8 Camelineae taxa, *A. lyrata* subsp. *petraea* (Kuittinen et al. 2004), *A. lyrata* subsp. *lyrata* (Yogeeswaran et al. 2005) and *C. rubella* (Boivin et al. 2004). Despite some discrepancies among the three Camelineae karyotypes, particularly in the number of inferred inversion events due to the different marker density, all three maps were largely congruent and showed a strikingly high extent of large-scale collinearity with the five *A. thaliana* linkage groups (Boivin et al. 2004, Kuittinen et al. 2004, Koch and Kiefer 2005, Yogeeswaran et al. 2005, Lysak and Lexer 2006).

1.3.6 Revealing Chromosome Homeology Through Comparative Chromosome Painting

Chromosome painting points to the identification of large chromosome regions or whole chromosomes using chromosome-specific DNA probes. Fluorescently

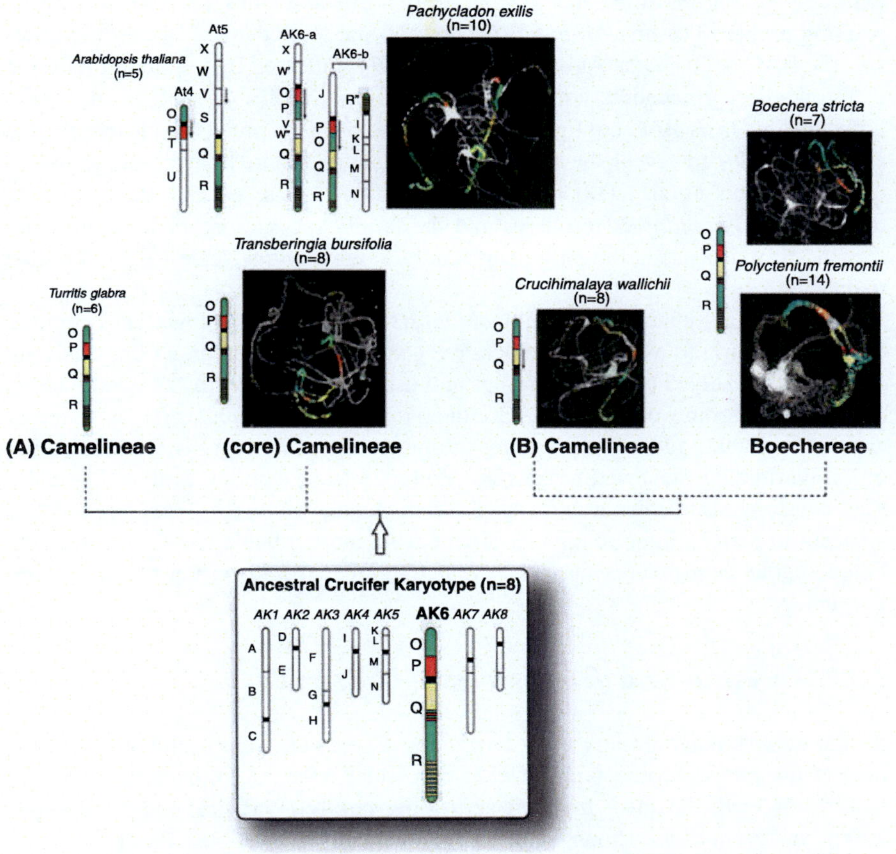

Fig. 1.5 Analysis of ancestral chromosome AK6 by CCP in tribes Boechereae and Camelineae *sensu lato*. Chromosome AK6 comprising genomic blocks O, P, Q, and R has been identified using fluorescently labeled *A. thaliana* BAC contigs (Alexa Fluor 488 – *green*, Cy3 – *yellow*, and Texas Red – *red*) hybridized to pachytene chromosomes of the analyzed species. Although the ancestral AK6 pattern remained conserved in several species, it was reshuffled in *A. thaliana*, *Crucihimalaya wallichii*, and *Pachycladon exilis* by species-specific chromosome rearrangements. Note three painted bivalents and one tetravalent in polyploid *P. exilis* and *Polyctenium fremontii*, respectively. Mandáková and Lysak (unpublished data), and modified according to Koch and Al-Shehbaz (2009), Lysak et al. (2006), and Schranz et al. (2006)

labeled painting probes are hybridized to chromosomes of the same species or chromosomes of related species to reveal shared (homeologous) chromosome regions. The latter approach was termed comparative chromosome painting (CCP). In Brassicaceae, several favorable conditions for establishing CCP have been met. Crucifer genomes possess only a small percentage of DNA repeats clustered mostly within pericentromeric regions as well as considerably well-preserved inter-species chromosome collinearity (see above). Moreover, *Arabidopsis* has been furnished with sets of chromosome-specific bacterial artificial clones (BACs) serving as painting probes. The first *Arabidopsis* chromosome to be painted by chromosome-specific BAC contigs was At4 (Lysak et al. 2001), followed by multicolor painting of the whole chromosome complement (Lysak et al. 2003, Pecinka et al. 2004). Later, *Arabidopsis* BAC contigs were applied as painting probes in closely related crucifer species to reveal the extent of evolutionarily shared chromosome homeology (Jackson et al. 2000, Comai et al. 2003, Lysak et al. 2003). Due to the available *Arabidopsis* resources and the family-specific organization of DNA repeats, Brassicaceae is the only plant family in which large-scale CCP is feasible to date.

In CCP experiments (Fig. 1.5), *Arabidopsis* BAC contigs arranged into painting probes according to available comparative genetic maps or on an ad hoc basis are differentially labeled by fluorochromes and usually hybridized to extended meiotic (pachytene) chromosomes. As BAC clones representing *Arabidopsis* centromeres are not available, and those covering the pericentromeric regions and containing cross-hybridizing dispersed repeats are excluded from painting probes, homeologous centromeres in other crucifer species remain unlabeled. Yet, *Arabidopsis* BACs adjacent to pericentromeric regions often enable more reliable identification of the homeologous centromeres than genetic markers dependent on the recombination frequency.

1.3.7 Ancestral Crucifer Karyotype (ACK, n=*8)*

As the eight linkage groups of *A. lyrata* and *C. rubella* were found almost identical at the gross chromosomal level (Koch and Kiefer 2005), base chromosome number of x=8, the most frequent across the family (Warwick and Al-Shehbaz 2006), and *A. thaliana* shown to be phylogenetically derived species (Koch et al. 1999a, 2000, 2001, Koch and Matschinger 2007), it was concluded that the five *A. thaliana* chromosomes (n=5) must have been derived from an ancestral n=8 karyotype resembling that of *A. lyrata*, *C. rubella* (Koch and Kiefer 2005), and that of *A. halleri*, as recently shown by Roosens et al. (2008). Since each comparison to the reduced *Arabidopsis* karyotype (n=5) was inevitably complicated by accounting

Fig. 1.6 Overview of karyotype evolution in Brassicaceae. Karyotypes have been reconstructed by comparative chromosome painting (CCP: Lysak et al. 2006, Mandáková and Lysak 2008, and unpublished data) and by comparative linkage mapping (Koch and Kiefer 2005, Parkin et al. 2005, Schranz et al. 2007). The tentative evolutionary scenario assumes the Ancestral Crucifer Karyotype (ACK, n=8) as being ancestral for Lineage I and II, respectively. ACK-like karyotypes of *Arabidopsis lyrata* and *Capsella rubella* from the core Camelineae are not shown as well as the karyotype of *Noccaea caerulescens* reshuffled by secondary pericentric inversions (cf. Mandáková and Lysak 2008). The 24 genomic blocks are indicated by *uppercase letters* (A–X) and colored according to their position on chromosomes AK1–AK8 of the ACK (Schranz et al. 2006). Downward-pointing *arrows* indicate the opposite orientation of genomic blocks compared with the position in the ACK

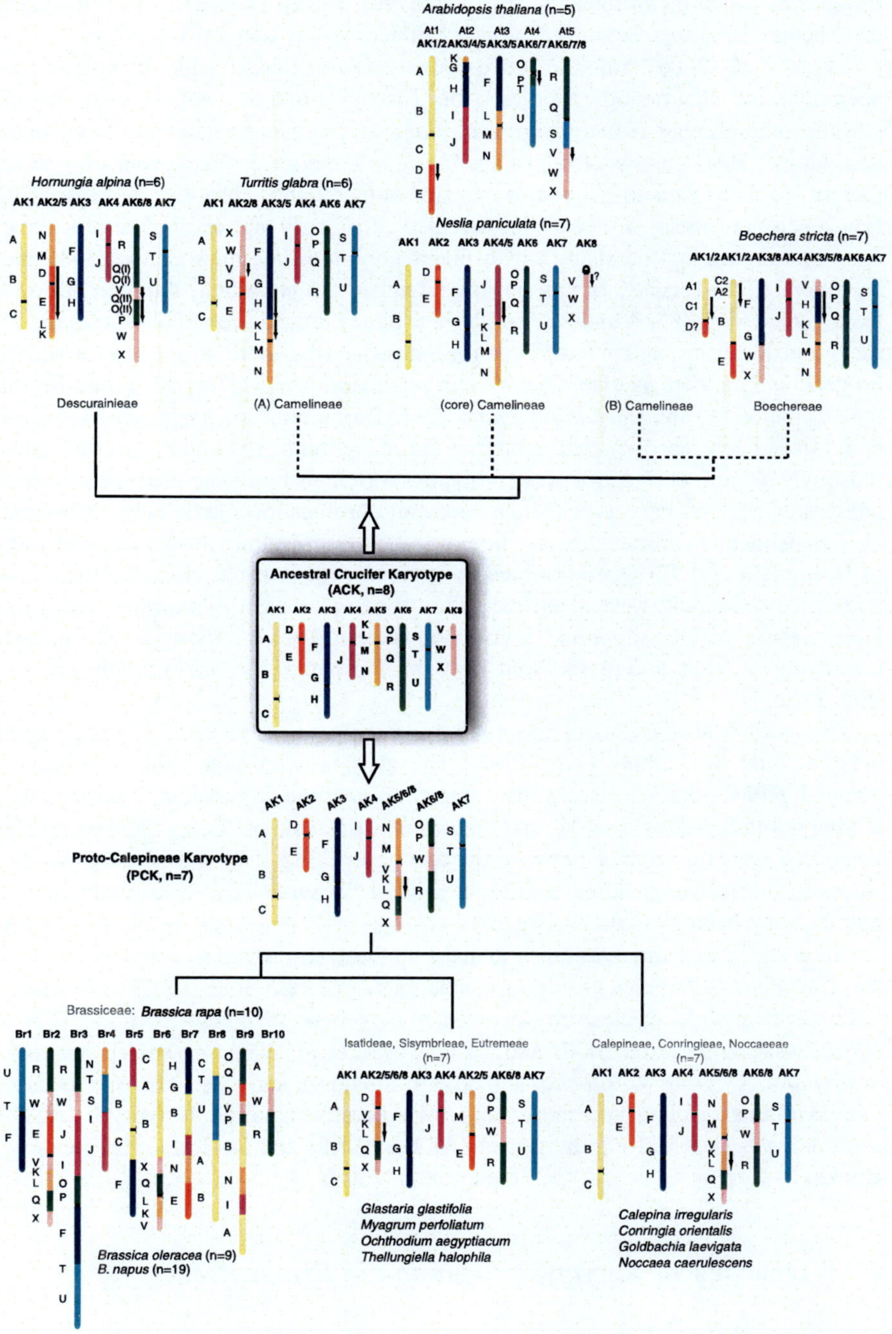

Fig. 1.6 (continued)

for species-specific chromosome rearrangements, shared ancestral $n=8$ karyotype had become an attractive working concept underlined by two crucial studies.

Lysak et al. (2006) aimed to prove if the *A. lyrata/Capsella*-like karyotype has been ancestral also for other species from Lineage I and to compare scenarios of chromosome number reduction in *A. thaliana* and its close relatives. To this end, *Arabidopsis* BAC contigs arranged according to the eight linkage groups (or parts thereof) of *A. lyrata* and *C. rubella* revealed homeologous chromosome regions in four species of Lineage I. The results revealed a strikingly high level of chromosome conservation among the species with diverse chromosome numbers ($n=5$–8) and corroborated the scenario of chromosome number reduction in *A. thaliana* inferred from genetic data (Koch and Kiefer 2005) as three chromosome fusions, two reciprocal translocations, and at least three inversions. Furthermore, in species for which no genetic data were available, i.e. *Neslia paniculata* ($n=7$), *Turritis glabra* ($n=6$, both Camelineae) and *Hornungia alpina* ($n=6$, Descurainieae), karyotype evolution could have been reconstructed assuming the reduction of the eight ancestral chromosomes toward $n=7$ and $n=6$ karyotypes. Although in all the analyzed species possessing derived karyotypes some ancestral chromosomes have been reshuffled by independent fusion events, six homologous chromosomes in *Neslia*, and four in *Hornungia* and *Turritis* resembled the structure of ancestral chromosomes, and large ancestral blocks were identified within chromosomes of *A. thaliana* (Fig. 1.6). Hence, these results supported a common ancestral $n=8$ karyotype of Lineage I, existing at least before the split between Descurainieae and Camelineae s.l. (Fig. 1.3).

The additional support for the ancestral karyotype came from a comparative genetic study by Parkin et al. (2005). The authors used over 1,000 genetically mapped RFLP loci to discern the extent of genome homeology between the allopolyploid *B. napus* ($2n=38$) and the *Arabidopsis* genome. Despite the relatively large phylogenetic distance between the two genera (Fig. 1.3) and species-specific chromosome rearrangements, a minimum of 21 conserved collinear chromosome segments has been identified between the two species. These shared genomic blocks could be duplicated and rearranged to make up the extant *B. napus* genome. Most of the *Arabidopsis–Brassica* chromosomal segments (Parkin et al. 2005) were found to be identical with collinear blocks revealed by genetic mapping (Koch and Kiefer 2005, Lysak and Lexer 2006) and CCP (Lysak et al. 2006) as shared between *A. thaliana*, *A. lyrata*, *C. rubella*, and other Camelineae and Descurainieae species. This synthesis resulted into the concept of Ancestral Crucifer Karyotype (ACK) comprising 8 ancestral chromosomes (AK1 to AK8) and 24 conserved genomic blocks (A–X) (Figs. 1.5 and 1.6; Schranz et al. 2006).

1.3.8 Overview of Karyotype Evolution in Brassicaceae

Considering that the Brassicaceae comprises some 3,700 species, we have information on the karyotype structure of only a handful of species with a bias toward the Camelineae. Despite this limitation some preliminary conclusions on karyotype evolution within Lineages I and II can be drawn from the available genetic, cytogenetic, and phylogenetic data (Fig. 1.6).

While still far from totally encompassing conclusions, an increasing body of evidence suggests the ACK as an ancestral karyotype for crucifer Lineage I as previously indicated by Lysak et al. (2006). Recently, this was corroborated by comparative genetic mapping in *B. stricta* (*n*=7, Schranz et al. 2007) as well as by CCP analysis in this species and the autotetraploid *P. fremontii* (*n*=14; T. Mandáková and M.A. Lysak, unpublished data), showing that karyotypes of the two Boechereae species can be reconstructed using ancestral chromosomes and genomic blocks of the ACK (Figs. 1.5 and 1.6). Furthermore, CCP in *Cardamine hirsuta* (*n*=8) indicated that ACK is likely an ancestral genome of the Cardamineae (T. Mandáková and M.A. Lysak, unpublished data). Hence, the ACK has been confirmed as an ancestral karyotype for Boechereae, Camelineae, Cardamineae, and Descurainieae and together these findings suggest that the whole Lineage I probably descended from the ACK (Fig. 1.6). However, taking the Camelineae as an example, the complexity of karyotype evolution in Brassicaceae can be seen in its full extent. Figure 1.5 shows the CCP analysis of ancestral chromosome AK6 in tribes Camelineae (*sensu lato*) and Boechereae. Whereas AK6 remained conserved in Boechereae and most Camelineae species (e.g. *Transberingia bursifolia*, see also Lysak et al. 2003, 2006), in other Camelineae species its structure has been altered by secondary intra-chromosomal (*Crucihimalaya wallichii*) as well as inter-chromosomal (*A. thaliana*, *Pachycladon exilis*) rearrangements. It remains to be seen if the cytogenetic data will provide an insight into the evolution of the polyphyletic Camelineae (Bailey et al. 2006, Koch and Al-Shehbaz 2009).

Recently, scenarios of karyotype evolution have been reconstructed in eight species from six *x*=7 tribes belonging to Lineage II or being closely affiliated with this lineage (Mandáková and Lysak 2008). An ancestral Proto-Calepineae Karyotype (PCK; *n*=7) was inferred as shared among all *x*=7 species analyzed. Hence, the PCK represents an ancestral karyotype of the Lineage II s.l. As the PCK shares five chromosomes and conserved associations of genomic blocks with the ACK, it was proposed that either both karyotypes descended from a common ACK-like genome or the PCK has been directly derived from the ACK (Fig. 1.6). More data are needed to resolve this controversy.

As there are no genetic and cytogenetic data available for the tribes comprising Lineage III, it remains elusive whether the ACK is likely to be an ancestral karyotype of all three crucifer lineages and clades not yet assigned to tribes. Also it remains to be tested if CCP using *Arabidopsis* BAC contigs is applicable to Aethionemeae, a sister group to the remaining Brassicaceae tribes (see Fig. 1.3).

Karyotype evolution has been reconstructed only in species with haploid or base chromosome number equal or lower than eight. Such scenarios imply chromosome number reduction from ACK or ACK-like karyotypes comprising eight chromosomes (*n*=8). As signs for centromere retention were found neither by sequence analysis in *Arabidopsis* (AGI 2000) nor by cytogenetics in *Arabidopsis* and other crucifer taxa (Lysak et al. 2006, Schranz et al. 2006, Schubert 2007, Mandáková and Lysak 2008), a loss of one to three centromeres has been assumed. In Lineages I and II, centromere elimination has been explained by a universal mechanism. The first step is a reciprocal translocation between a (sub)metacentric and acrocentric (telocentric) chromosome with translocation breakpoints located close to the chromosome ends. The translocation generates a "fusion" chromosome and a minichromosome comprising mainly of centromere of the acrocentric and two telomeres. The latter translocation product is supposed to be meiotically unstable and eliminated. Ectopic (non-allelic) intra- and inter-chromosomal recombination within the repeat arrays is the most likely molecular mechanism of the described rearrangements (Lysak et al. 2006, Schranz et al. 2006, Schubert 2007, Mandáková and Lysak 2008).

Sequence-based phylogenetic trees play an invaluable role in the navigation of cytogenetic analyses, and cytogenomic data (cytogenetic signatures) overlaid onto phylogenetic trees may provide novel markers corroborating the existing phylogenies or resolving conflicting phylogenetic signals. Chromosome rearrangements show only a low tendency toward convergent evolution and are generally considered as rare genomic changes (RGCs; Rokas and Holland 2000), although rare reversals and convergence due to the reuse of the same breakpoints (Dobigny and Yang 2008) have not been investigated in plants and cannot be ruled out. The parsimony assumption implies that associations of conserved blocks can be disrupted by rearrangements occurring independently in different lineages, whereas the same combination of chromosomal blocks (syntenic association) in two independent lineages is unlikely (Faraut 2008).

In Brassicaceae, even the scant data on karyotype evolution has provided some cytogenetic signatures supporting and re-defining the proposed phylogenetic topologies. In tribes Brassiceae, Isatideae, and Sisymbrieae of Lineage II, and closely related Calepineae, Conringieae, Eutremeae, and Noccaeeae, CCP analysis revealed the syntenic association of blocks O/P/W/R (chromosome AK6/8) and V/K/L/Q/X. These block associations may be a unique signature shared by Lineage II and several tribes within the currently unresolved polytomy suggesting their common ancestry (Fig. 1.6; Mandáková and Lysak 2008). Future comparative cytomolecular studies in additional crucifer species and groups not yet analyzed should uncover other phylogenetically informative cytogenetic signatures.

Acknowledgment We are thankful to T. Mandáková for sharing unpublished cytogenetic data. This work was supported by research grants from the Grant Agency of the Czech Academy of Science (KJB601630606 and IAA601630902) and the Czech Ministry of Education (MSM0021622415) to MAL and various grants over the last 10 years from the German Research Foundation (DFG) and the Austrian Science Fund (FWF) to MAK.

References

Arabidopsis Genome Initiative (2000) Analysis of the genome sequence of the flowering plant *Arabidopsis thaliana*. Nature 408:796–815

Al-Shehbaz IA (1984) The tribes of Cruciferae (Brassicaceae) in the southeastern United States. J Arnold Arbor 65:343–373

Al-Shehbaz IA (1985) The genera of Brassiceae (Cruciferae; Brassicaceae) in the southeastern United States. J Arnold Arbor 66:279–351

Al-Shehbaz IA (1988a) The genera of Arabideae (Cruciferae; Brassicaceae), the southeastern United States. J Arnold Arbor 69:85–166

Al-Shehbaz IA (1988b) The genera of Anchonieae (Hesperideae) (Cruciferae; Brassicaceae), the southeastern United States. J Arnold Arbor 69:193–212

Al-Shehbaz IA (1988c) The genera of Sisymbrieae (Cruciferae; Brassicaceae), the southeastern United States. J Arnold Arbor 69:213–237

Al-Shehbaz IA, Beilstein MA, Kellogg EA (2006) Systematics and phylogeny of the Brassicaceae (Cruciferae): an overview. Plant Syst Evol 259:89–120

Al-Shehbaz IA, Mummenhoff K (2005) Transfer of the South African genera *Brachycarpaea, Cycloptychis, Schlechteria, Silicularia,* and *Thlaspeocarpa* to *Heliophila* (Brassicaceae). Novon 15:385–389

Al-Shehbaz IA, Warwick SI (2007) Two new tribes (Donstostemoneae and Malcolmieae) in the Brassicaceae (Cruciferae). Harv Pap Bot 12(2):429–433

Ali HBM, Lysak MA, Schubert I (2004) Genomic in situ hybridization in plants with small genomes is feasible and elucidates the chromosomal parentage in interspecific *Arabidopsis* hybrids. Genome 47:954–960

Allan HH (1961) Cruciferae. Fl N Z 1:174–189

Angiosperm Phylogeny Group II (2003) An update of the Angiosperm Phylogeny Group classification for the orders and families of flowering plants: APG II. Bot J Linn Soc 141:399–436

Appel O, Al-Shehbaz IA (2002) Cruciferae. In: Kubitzki K, Bayer C (eds) The families and genera of vascular plants, pp 75–174. Springer, Berlin

Bailey CD, Koch MA, Mayer M, et al (2006) Toward a global phylogeny of the Brassicaceae. Mol Biol Evol 23:2142–2160

Beilstein MA, Al-Shehbaz IA, Kellogg EA (2006) Brassicaceae phylogeny and trichome evolution. Am J Bot 93:607–619

Beilstein MA, Al-Shehbaz IA, Mathews S, et al (2008) Brassicaceae phylogeny inferred from phytochrome A and *ndhF* sequence data: tribes and trichomes revisited. Am J Bot 95: 1307–1327

Blanc G, Hokamp K, Wolfe KH (2003) A recent polyploidy superimposed on older large-scale duplications in the *Arabidopsis* genome. Genome Res 13:137–144

Boivin K, Acarkan A, Mbulu RS, et al (2004) The *Arabidopsis* genome sequence as a tool for genome analysis in Brassicaceae. A comparison of the *Arabidopsis* and *Capsella rubella* genomes. Plant Physiol 135:735–744

Bowers JE, Chapman BA, Rong JK, Paterson AH (2003) Unravelling angiosperm genome evolution by phylogenetic analysis of chromosomal duplication events. Nature 422:433–438

Cheng BF, Heneen WK, Chen BY (1995) Mitotic karyotypes of *Brassica campestris* and *Brassica alboglabra* and identification of the *B. alboglabra* chromosome in an addition line. Genome 38:313–319

Chèvre AM, Adamczyk K, Eber F, et al (2007) Modelling gene flow between oilseed rape and wild radish. I. Evolution of chromosome structure. Theor Appl Genet 114:209–221

Comai L, Tyagi AP, Lysak MA (2003) FISH analysis of meiosis in *Arabidopsis* allopolyploids. Chromosome Res 11:217–226

Couvreur TLP, Franzke A, Al-Shehbaz IA, Bakker FT, Koch MA, Mummmenhoff K (2009) Molecular phylogenetics, temporal diversification and principles of evolution in the mustard family (Brassicaceae). Mol Biol Evol 27:55–71

De Bodt S, Maere S, Van de Peer Y (2005) Genome duplication and the origin of angiosperms. Trends Ecol Evol 20:591–597

De Candolle AP (1821) Mémoire sur la famille des Crucifères. Mém Mus Hist Nat 7(1):169–252

De Craene LPR, Haston E (2006) The systematic relationships of glucosinolate-producing plants and related families: a cladistic investigation based on morphological and molecular characters. Bot J Linn Soc 151:453–494

Devos KM, Brown JKM, Bennetzen JL (2002) Genome size reduction through illegitimate recombination counteracts genome expansion in *Arabidopsis*. Genome Res 12:1075–1079

Dobeš C, Mitchell-Olds T, Koch MA (2004a) Extensive chloroplast haplotype variation indicates Pleistocene hybridization and radiation of North American *Arabis drummondii*, *A.* × *divaricarpa*, and *A. holboellii* (Brassicaceae). Mol Ecol 13:349–370

Dobeš C, Mitchell-Olds T, Koch MA (2004b) Intraspecific diversification in North American *Arabis drummondii*, *A.* ×*divaricarpa*, and *A. holboellii* (Brassicaceae) inferred from nuclear and chloroplast molecular markers – an integrative approach. Am J Bot 91:2087–2101

Dobigny G, Yang F (2008) Comparative cytogenetics in the genomics era: cytogenomics comes of age. Chromosome Res 16:1–4

Easterly NW (1963) Chromosome numbers of some northwestern Ohio *Cruciferae*. Castanea 28:39–42

Eschmann-Grupe G, Hurka H, Neuffer B (2004) Extent and structure of genetic variation in two *Diplotaxis* species (Brassicaceae) with contrasting breeding systems. Plant Syst Evol 244: 31–43

Faraut T (2008) Addressing chromosome evolution in the whole-genome sequence era. Chromosome Res 16:5–16

Fransz P, de Jong JH, Lysak M, et al (2002) Interphase chromosomes in *Arabidopsis* are organized as well defined chromocenters from which euchromatin loops emanate. Proc Natl Acad Sci USA 99:14584–14589

Franzke A, German D, Al-Shehbaz IA, et al (2009) *Arabidopsis* family ties: molecular phylogeny and age estimates in the Brassicaceae. Taxon 58:425–437

Gaeta RT, Pires JC, Iniguez-Luy F, et al (2007) Genomic changes in resynthesized *Brassica napus* and their effect on gene expression and phenotype. Plant Cell 19:3403–3417

Galloway GL, Malmberg RL, Price RA (1998) Phylogenetic utility of the nuclear gene arginine decarboxylase: an example from Brassicaceae. Mol Biol Evol 15:1312–1320

Gandolfo MA, Nixon KC, Crepet WL (1998) A new fossil flower from the Turonian of New Jersey: *Dressiantha bicarpellata* gen. et sp. nov. (Capparales). Am J Bot 85:964–974

German DA, Al-Shehbaz IA (2008) Five additional tribes (Aphragmeae, Biscutelleae, Calepineae, Conringieae, and Erysimeae) in the Brassicaceae. Harv Pap Bot 13:165–170

Hall JC, Iltis HH, Sytsma KJ (2004) Molecular phylogenetics of core brassicales, placement of orphan genera *Emblingia*, *Forchhammeria*, *Tirania*, and character evolution. Syst Bot 29: 654–669

Hall JC, Sytsma KJ, Iltis HH (2002) Phylogeny of Capparaceae and Brassicaceae based on chloroplast sequence data. Am J Bot 89:1826–1842

Harriman NA (1965) The genus *Dentaria* L. (Cruciferae) in eastern North America. PhD. Thesis Vanderbilt University

Hayek A (1911) Entwurf eines Cruciferensystems auf phylogenetischer Grundlage. Beih Bot Centralbl 27:127–335

Hedge IC (1976) A systematic and geographical survey of the Old World Cruciferae. In: Vaughn JG, Macleod AJ, Jones BMG (eds) The biology and chemistry of the Cruciferae, pp 1–45. Academic Press, London

Hegarty MJ, Hiscock SJ (2008) Genomic clues to the evolutionary success of review polyploid plants. Curr Biol 18:435–444

Henry Y, Bedhomme M, Blanc G (2006) History, protohistory and prehistory of the *Arabidopsis thaliana* chromosome complement. Trends Plant Sci 11:267–273

Hewson HJ (1982) Brassicaceae. In: Briggs BG (ed) Flora of Australia, vol 8, pp 231–357. CSIRO Publishing, Canberra

Howell EC, Kearsey MJ, Jones GH, et al (2008) A and C genome distinction and chromosome identification in *Brassica napus* by sequential FISH and GISH. Genetics 180: 1849–1857

Jackson SA, Cheng Z, Wang ML, et al (2000) Comparative fluorescence in situ hybridization mapping of a 431-kb *Arabidopsis thaliana* bacterial artificial chromosome contig reveals the role of chromosomal duplications in the expansion of the *Brassica rapa* genome. Genetics 156:833–838

Jaillon O, Aury JM, Noel B, et al (2007) The grapevine genome sequence suggests ancestral hexaploidization in major angiosperm phyla. Nature 449:463–467

Janchen E (1942) Das System der Cruciferen. Öster Bot Zeitschr 91:1–28

Jordon-Thaden I, Koch MA (2008) Diversity patterns in the genus *Draba*: a first global perspective. Plant Ecol Divers 1(2):255–263

Jørgensen MH, Carlsen T, Skrede I, Elven R (2008) Microsatellites resolve the taxonomy of the polyploid *Cardamine digitata* aggregate (Brassicaceae). Taxon 57:882–892

Judd WS, Sanders RW, Donoghue MJ (1994) Angiosperm family pairs: preliminary phylogenetic analyses. Harv Pap Bot 5:1–51

Kantama L, Sharbel TF, Schranz ME, et al (2007) Diploid apomicts of the *Boechera holboellii* complex display large-scale chromosome substitutions and aberrant chromosomes. Proc Natl Acad Sci USA 104:14026–14031

Koch M (2003) Molecular phylogenetics, evolution and population biology in Brassicaceae. In: Sharma AK, Sharma A (eds) Plant genome: biodiversity and evolution, vol 1a (phanerogams), pp 1–35. Science Publishers, Enfield, NH

Koch MA, Al Shehbaz IA (2009) Molecluar systematics and evolution of "wild" crucifers (Brassicaceae or Cruciferae). In: Gupta S (ed) Biology and breeding of Crucifers. Taylor and Francis Group, London

Koch M, Al-Shehbaz IA, Mummenhoff K (2003) Molecular systematics, evolution, and population biology in the mustard family (Brassicaceae). Ann Missouri Bot Gard 90: 151–171

Koch M, Bernhardt KG (2004) Comparative biogeography of the cytotypes of annual *Microthlaspi perfoliatum* (Brassicaceae) in Europe using isozymes and cpDNA data: refugia, diversity centers, and postglacial colonization. Am J Bot 91:114–124

Koch M, Bishop J, Mitchell-Olds T (1999a) Molecular systematics of *Arabidopsis* and *Arabis*. Plant Biol 1:529–537

Koch MA, Dobeš C, Kiefer C, et al (2007) Supernetwork identifies multiple events of plastid *trn*F (GAA) pseudogene evolution in the Brassicaceae. Mol Biol Evol 24:63–73

Koch M, Haubold B, Mitchell-Olds T (2000) Comparative evolutionary analysis of chalcone synthase and alcohol dehydrogenase loci in *Arabidopsis*, *Arabis*, and related genera (Brassicaceae). Mol Biol Evol 17:1483–1498

Koch M, Haubold B, Mitchell-Olds T (2001) Molecular systematics of the Brassicaceae: evidence from coding plastidic *matK* and nuclear *Chs* sequences. Am J Bot 88:534–544

Koch M, Hurka H (1999) Isozyme analysis in the polyploid complex *Microthlaspi perfoliatum* (L.) F. K. Meyer: morphology, biogeography and evolutionary history. Flora 194:33–48

Koch M, Huthmann M, Hurka H (1998a) Molecular biogeography and evolution of the *Microthlaspi perfoliatum* s.l. polyploid complex (Brassicaceae): chloroplast DNA and nuclear ribosomal DNA restriction site variation. Can J Bot 76:382–396

Koch M, Huthmann M, Hurka H (1998b) Isozymes, speciation and evolution in the polyploid *Cochlearia* L. (Brassicaceae). Bot Acta 111:411–425

Koch M, Kiefer M (2005) Genome evolution among cruciferous plants – a lecture from the comparison of the genetic maps of three diploid species: *Capsella rubella*, *Arabidopsis lyrata* ssp. *petraea* and *Arabidopsis thaliana*. Am J Bot 92(4):761–767

Koch M, Kiefer C (2006) Molecules and migration: biogeographical studies in cruciferous plants. Plant Syst Evol 259(2–4):121–142

Koch M, Kiefer C, Vogel J, et al (2006) Three times out of Asia Minor – the phylogeography of *Arabis alpina* L. (Brassiaceae). Mol Ecol 15:825–839

Koch MA, Matschinger M (2007) Evolution and genetic differentiation among relatives of *Arabidopsis thaliana*. PNAS 104(15):6272–6277

Koch M, Mummenhoff K (2006) Evolution and phylogeny of the Brassicaceae. Plant Syst Evol 259:81–258

Koch M, Mummenhoff K, Hurka H (1999b) Molecular phylogenetics of *Cochlearia* (Brassicaceae) and allied genera based on nuclear ribosomal ITS DNA sequence analysis contradict traditional concepts of their evolutionary relationship. Plant Syst Evol 216:207–230

Koornneef M, Fransz P, de Jong H (2003) Cytogenetic tools for *Arabidopsis thaliana*. Chromosome Res 11:183–194

Kuittinen H, de Haan AA, Vogl C, et al (2004) Comparing the linkage maps of the close relatives *Arabidopsis lyrata* and *A. thaliana*. Genetics 168:1575–1584

Lagercrantz U (1998) Comparative mapping between *Arabidopsis thaliana* and *Brassica nigra* indicates that *Brassica* genomes have evolved through extensive genome replication accompanied by chromosome fusions and frequent rearrangements. Genetics 150:1217–1228

Lagercrantz U, Lydiate D (1996) Comparative genome mapping in *Brassica*. Genetics 144: 1903–1910

Leitch IJ, Chase MW, Bennett MD (1998) Phylogenetic analysis of DNA C-values provides evidence for a small ancestral genome size in flowering plants. Ann Bot 82(Suppl. A):85–94

Lihova J, Marhold K (2003) Taxonomy and distribution of the *Cardamine pratensis* group (Brassicaceae) in Slovenia. Phyton 43:241–261

Lysak M, Berr A, Pecinka A, et al (2006) Mechanisms of chromosome number reduction in *Arabidopsis thaliana* and related Brassicaceae species. Proc Natl Acad Sci USA 103: 5224–5229

Lysak MA, Cheung K, Kitschke M, et al (2007) Ancestral chromosomal blocks are triplicated in Brassiceae species with varying chromosome number and genome size. Plant Physiol 145: 402–410

Lysak MA, Fransz PF, Ali HBM, Schubert I (2001) Chromosome painting in *Arabidopsis thaliana*. Plant J 28:689–697

Lysak MA, Koch MA, Leitch IJ, et al (2009) The dynamic ups and downs of genome size evolution in Brassicaceae. Mol Biol Evol 26:85–98

Lysak ML, Koch MA, Pecinka A, et al (2005) Chromosome triplication found across the tribe Brassiceae. Genome Res 15:516–525

Lysak MA, Lexer C (2006) Towards the era of comparative evolutionary genomics in *Brassicaceae*. Plant Syst Evol 259:175–198

Lysak MA, Pecinka A, Schubert I (2003) Recent progress in chromosome painting of *Arabidopsis* and related species. Chromosome Res 11:195–204

Maere S, De Bodt S, Raes J, et al (2005) Modeling gene and genome duplications in eukaryotes. Proc Natl Acad Sci USA 102:5454–5459

Magallon S, Crane PR, Herendeen PS (1999) Phylogenetic pattern, diversity, and diversification of eudicots. Ann Miss Bot Gard 86:297–372

Magallon S, Sanderson MJ (2001) Absolute diversification rates in angiosperm clades. Evolution 55:1762–1780

Mallet J (2007) Hybrid speciation. Nature 446:279–283

Maluszynska J, Hasterok R (2005) Identification of individual chromosomes and parental genomes in *Brassica juncea* using GISH and FISH. Cytogenet Genome Res 109:310–314

Mandáková T, Joly S, Krzywinski M, Mummenhoff K, Lysak MA (2010) Fast diploidization in close mesopolyploid relatives of *Arabidopsis*. Plant Cell 22:2277–2290

Mandáková T, Lysak MA (2008) Chromosomal phylogeny and karyotype evolution in x=7 crucifer species (Brassicaceae). Plant Cell 20:2559–2570

Manton I (1932) Introduction to the general cytology of the Cruciferae. Ann Bot 46: 509–556

Manton I (1937) The problem of *Biscutella laevigata* L. II. The evidence from meisois. Ann Bot 1:439–462

Marais W (1970) Cruciferae. In: Codd LE, De Winter B, Killick DJB et al (eds) Flora of southern Africa, Vol. 13, pp 1–118. National Botanic Gardens, Kirstenbosch

Marhold K (1999) Taxonomic evaluation of the tetraploid populations of *Cardamine amara* (Brassicaceae) from the Eastern Alps and adjacent areas. Bot Helv 109:67–84

Marhold K, Lihová J (2006) Polyploidy, hybridization and reticulate evolution: lessons from the Brassicaceae. Plant Syst Evol 259:143–174

Ming R et al (2008) The draft genome of the transgenic tropical fruit tree papaya (*Carica papaya* Linnaeus). Nature 452:991–996

Mitchell-Olds T, Al-Shehbaz IA, Koch MA, et al (2005) Crucifer evolution in the post-genomic era. In: Henry RJ (ed) Plant diversity and evolution, pp 119–137. CAB International, Oxfordshire, UK

Montgomery FH (1955) Preliminary studies in the genus *Dentaria* in eastern North America. Rhodora 57:161–173

Mummenhoff K, Hurka H (1994) Subunit polypeptide composition of Rubisco and the origin of allopolyploid *Arabidopsis suecica* (Brassicaceae). Biochem Syst Ecol 22:807–812

Mummenhoff K, Linder P, Friesen N, et al (2004) Molecular evidence for bicontinental hybridogenous genomic constitution in *Lepidium* sensu stricto (Brassicaceae) species from Australia and New Zealand. Am J Bot 91:254–261

Oyama RK, Clauss MJ, Formanova N, et al (2008) The shrunken genome of *Arabidopsis thaliana*. Plant Syst Evol 273:257–271

Parisod C, Besnard G (2007) Glacial in situ survival in the Western Alps and polytopic autopolyploidy in *Biscutella laevigata* L. (Brassicaceae). Mol Ecol 16:2755–2767

Parkin IAP, Gulden SM, Sharpe AG, et al (2005) Segmental structure of the *Brassica napus* genome based on comparative analysis with *Arabidopsis thaliana*. Genetics 171: 765–781

Pecinka A, Schubert V, Meister A, et al (2004) Chromosome territory arrangement and homologous pairing in nuclei of *Arabidopsis thaliana* are predominantly random except for NOR-bearing chromosomes. Chromosoma 113:258–269

Prantl K (1891) Cruciferae. In: Engler A, Prantl K (eds) Die natürlichen Pflanzenfamilien, vol III, 2, pp 145–206. Verlag von Wilhelm Engelmann, Leipzig

Rana D, van den Boogaart T, O'Neill CM, Hynes L, Bent E, Macpherson L, Park JY, Lim YP, Bancroft I (2004) Conservation of the microstructure of genome segments in *Brassica napus* and its diploid relatives. Plant J 40:725–733

Rieseberg LH, Willis JH (2007) Plant speciation. Science 317:910–914

Rokas A, Holland WH (2000) Rare genomic changes as a tool for phylogenetics. Trends Ecol Evol 15:454–459

Rollins RC (1993) The Cruciferae of Continental North America. Stanford University Press, Stanford

Roosens NHCJ, Willems G, Gode C, et al (2008) The use of comparative genome analysis and syntenic relationships allows extrapolating the position of Zn tolerance QTL regions from *Arabidopsis halleri* into *Arabidopsis thaliana*. Plant Soil 306:105–116

Santos JL, Alfaro D, Sanchez-Moran E, et al (2003) Partial diploidization of meiosis in autotetraploid *Arabidopsis thaliana*. Genetics 165:1533–1540

Schmickl R, Jorgensen MH, Brysting A (2008) Phylogeographic implications for North American arctic *Arabidopsis lyrata*. Plant Ecol Divers 1(2):245–254

Schranz E, Dobes C, Koch MA, et al (2005) Sexual reproduction, hybridization, apomixis and polyploidization in the genus *Boechera* (Brassicaceae). Am J Bot 92(11):1797–1810

Schranz ME, Lysak MA, Mitchell-Olds T (2006) The ABC's of comparative genomics in the Brassicaceae: building blocks of crucifer genomes. Trends Plant Sci 11:535–542

Schranz ME, Mitchell-Olds T (2006) Independent ancient polyploidy events in the sister families Brassicaceae and Cleomaceae. Plant Cell 18:1152–1165

Schranz ME, Windsor AJ, Song B-H, Lawton-Rauh A, Mitchell-Olds T (2007) Comparative genetic mapping in *Boechera stricta*, a close relative of Arabidopsis. Plant Physiol 144: 286–298

Schubert I (2007) Chromosome evolution. Curr Opin Plant Biol 10:1–7

Schulz OE (1936) Cruciferae. In: Engler A, Harms H (ed) Die natürlichen Pflanzenfamilien, vol. 17B, pp 227–658. Verlag von Wilhelm Engelmann, Leipzig

Shimizu KK, Fujii S, Marhold K, et al (2005) *Arabidopsis kamchatica* (Fisch. ex DC.) K. Shimizu & Kudoh and *B. kamchatica* subsp. *kawasakiana* (Makino) K. Shimizu & Kudoh, new combinations. Acta Phytotax Geobot 56:163–172

Simillion C, Vandepoele K, Van Montagu MCE, et al (2002) The hidden duplication past of *Arabidopsis thaliana*. Proc Natl Acad Sci USA 99:7719–7723

Snowdon RJ (2007) Cytogenetics and genome analysis in *Brassica* crops. Chromosome Res 15: 85–95

Snowdon RJ, Köhler W, Friedt W, et al (1997) Genomic in situ hybridization in *Brassica* amphidiploids and interspecific hybrids. Theor Appl Genet 95:1320–1324

Soltis DE, Soltis PS, Schemske DW, et al (2007) Autopolyploidy in angiosperms: have we grossly underestimated the number of species? Taxon 56:13–30

Stevens PF (2001) onwards. Angiosperm Phylogeny Website.[and more or less continuously updated since]. http://www.mobot.org/MOBOT/research/APweb/. Accessed Version 9, June 2008

Tang HB, Bowers JE, Wang XY, et al (2008) Perspective – Synteny and collinearity in plant genomes. Science 320:486–488

Tremetsberger K, König C, Samuel R, et al (2002) Infraspecific genetic variation in *Biscutella laevigata* (Brassicaceae): new focus on Irene Manton's hypothesis. Plant Syst Evol 233: 163–181

Tsunoda S, Hinata K, Gómez-Campo C (1980) *Brassica* crops and wild allies, pp 1–354. Japan Scientific Societies Press, Tokyo

Urbanska KM, Hurka H, Landolt E, et al (1997) Hybridization and evolution in *Cardamine* (Brassicaceae) at Urnerboden, Central Switzerland: biosystematic and molecular evidence. Plant Syst Evol 204:233–256

Vaughan JG, Macleod AJ, Jones BMG (1976) The biology and chemistry of the cruciferae, pp 1–355. Academic Press, London

Warwick SI, Al-Shehbaz IA (2006) Brassicaceae: chromosome number index and database on CD-Rom. Plant Syst Evol 259:237–248

Warwick SI, Al-Shehbaz IA, Sauder CA (2006a) Phylogenetic position of *Arabis arenicola* and generic limits of *Eutrema* and *Aphragmus* (Brassicaceae) based on sequences of nuclear ribosomal DNA. Can J Bot 84:269–281

Warwick SI, Francis A, Al-Shehbaz IA (2006b) Brassicaceae: species checklist and database on CD-Rom. Plant Syst Evol 259:249–258

Warwick SI, Mummenhoff K, Sauder C, Koch MA, Al-Shehbaz IA (2010) Closing the gaps: Phylogenetic relationships in the Brassicaceae based on DNA sequence data of nuclear ribosomal ITS region. Plant Syst Evol 285(3–4):209–232

Warwick SI, Sauder C (2005) Phylogeny of tribe Brassiceae (Brassicaceae) based on chloroplast restriction site polymorphisms and nuclear ribosomal internal transcribed spacer and chloroplast *trn*L intron sequences. Can J Bot 83:467–483

Warwick SI, Sauder CA, Al-Shehbaz IA, et al (2007) Phylogenetic relationships in the tribes Anchonieae, Chorisporeae, Euclidieae, and Hesperideae (Brassicaceae) based on nuclear ribosomal ITS DNA sequences. Ann Miss Bot Gard 94:56–78

Warwick SI, Sauder CA, Al-Shehbaz IA (2008) Phylogenetic relationships in the tribe Alysseae (Brassicaceae) based on nuclear ribosomal ITS DNA sequences. Can J Bot 86:315–336

Widmer A, Baltisberger M (1999) Molecular evidence for allopolyploid speciation and a single origin of the narrow endemic *Draba ladina* (Brassicaceae). Am J Bot 86:1282–1289

Wikström N, Savolainen V, Chase MW (2001) Evolution of the angiosperms: calibrating the family tree. Proc R Soc Lond Ser B 268:2211–2220

Yang YW, Lai KN, Tai PY, et al (1999) Rates of nucleotide substitution in angiosperm mitochondrial DNA sequences and dates of divergence between *Brassica* and the other angiosperm lineages. J Mol Evol 48:597–604

Yogeeswaran K, Frary A, York TL, et al (2005) Comparative genome analyses of *Arabidopsis* spp.: inferring chromosomal rearrangement events in the evolutionary history of *A. thaliana*. Genome Res 15:505–515

Ziolkowski PA, Kaczmarek M, Babula D, et al (2006) Genome evolution in *Arabidopsis/Brassica*: conservation and divergence of ancient rearranged segments and their breakpoints. Plant J 47:63–74

Zunk K, Mummenhoff K, Hurka H (1999) Phylogenetic relationships in tribe Lepidieae (Brassicaceae) based on chloroplast DNA restriction site variation. Can J Bot 77:1504–1512

Chapter 2
Brassicaceae in Agriculture

Suzanne I. Warwick

Abstract This chapter reviews the agricultural role of the Brassicaceae (Cruciferae) or mustard family. The family includes many economically important edible and industrial oilseed, vegetable, condiment, and fodder crop, such as. It also includes the molecular plant model, such as *Arabidopsis thaliana*. Current crops are reviewed and new and underutilized crucifer crop species discussed. Proposed new uses for these crops, such as biofuel platforms or green manure covers or biofumigants, are also briefly reviewed. The family also contains a rich source of agronomic and economic traits in its highly diverse wild germplasm. Traits discussed in this chapter include morphological and chemical traits; physiological traits such as C_3–C_4 photosynthesis, cytoplasmic male sterility, apomixis, and regeneration or transformation ability; and tolerances to various stresses such as salt, heavy metals, cold, drought, herbicides, diseases, insect, and nematode pests. These traits are of potential value in crop improvement programs and many wild crucifers now serve as model species in their study.

Keywords Brassicaceae · Cruciferae · Wild germplasm · Agronomic and economic trait sources

Contents

2.1 Introduction 34
2.2 Taxonomy and Genetic Relationships of *Brassica* Crop Species 35
2.2.1 *B. oleracea* 36
2.2.2 *B. rapa* 37
2.2.3 *B. nigra* 38
2.2.4 *B. napus* 38
2.2.5 *B. carinata* 38

S.I. Warwick (✉)
Eastern Cereal and Oilseed Research Centre, Agriculture and Agri-Food Canada, Central Experimental Farm, Ottawa, ON, Canada K1A OC6
e-mail: warwicks@agr.gc.ca

R. Schmidt, I. Bancroft (eds.), *Genetics and Genomics of the Brassicaceae*, Plant Genetics and Genomics: Crops and Models 9, DOI 10.1007/978-1-4419-7118-0_2,

2.2.6 *B. juncea* 39
2.3 Other Crucifer Crops 39
2.3.1 *Camelina* 39
2.3.2 *Crambe* 39
2.3.3 *Eruca* 40
2.3.4 *Raphanus* 40
2.3.5 *Sinapis* 41
2.4 Underutilized Crucifer Crops 41
2.5 Brassicaceae as Sources of Agronomic and Economic Traits 42
2.5.1 Morphological Traits 42
2.5.2 Chemical Traits 43
2.5.3 C3–C4 Photosynthesis 44
2.5.4 Cytoplasmic Male Sterility 45
2.5.5 Breeding Systems and Apomixis 45
2.5.6 Plant Regeneration and Transformation 45
2.5.7 Salt and Heavy Metal Tolerances 46
2.5.8 Cold Tolerance 46
2.5.9 Drought Tolerance 47
2.5.10 Herbicide Resistance 47
2.5.11 Disease Resistance 48
2.5.12 Insect and Nematode Resistance 49
2.6 Conclusion 50
References 50

2.1 Introduction

The Brassicaceae (Cruciferae) or mustard family includes many economically important edible and industrial oilseed, vegetable, condiment, and fodder crop species. The most important edible oil crop is canola or oilseed rape (*Brassica napus*); while mustard condiment crops include: *Brassica juncea* – Indian mustard and *Sinapis alba* – white mustard. Many Brassica species are also important vegetable crops, e.g., cole crops (*Brassica oleracea*). Several species, e.g., *Brassica carinata*, *Camelina sativa*, *Crambe abyssinica*, *Eruca vesicaria*, have potential as new edible oil/protein crops, biodiesel fuel crops, or platforms for bioproducts or molecular farming (Gugel and Falk 2006, Warwick and Gugel 2003, Warwick et al. 2006b, 2007a). The family is also known for its more than 120 weedy species, several of which are important cosmopolitan agricultural weeds (e.g., wild mustard (*Sinapis arvensis*)), stinkweed (*Thlaspi arvense*) while others form crop-weed complexes (e.g., *Raphanus sativus–Raphanus raphanistrum*). Several of these related weeds are able to exchange genes, including transgenes, with crops under natural field conditions (reviewed in Warwick et al. 2003, 2008b), potentially increasing weediness. Several representatives of the family have achieved the well-accepted status of "model organisms" for genomic studies, including *Arabidopsis thaliana* and *Brassica* spp. (Parkin et al. 2005), and other model species have been recently proposed, e.g., *Capsella* and *Arabis* (Koch and Kiefer 2005, Schranz et al. 2006b).

Related wild or lesser known crop species in the family represent virtually untapped and unlimited genetic sources of agronomic and economic traits (Warwick et al. 2009) and several Brassicaceae biodiversity databases are available to guide genomic researchers, genetic resource managers and plant breeders in the selection of appropriate taxa for phylogenetic, genomic and agronomic screening studies. These include the following: a species checklist of 338 genera and 3,709 species currently recognized in the family (Warwick et al. 2006a); a chromosome number index (Warwick and Al-Shehbaz 2006) reporting on over 9,000 chromosome counts from 232 genera and 1,558 species; a summary of all trait genetic studies (Séguin-Swartz et al. 1997); and a guide to the cytodeme status, chromosome number, hybridization potential, life cycle, growth form, ecology, and geographical distribution of all species in the tribe Brassiceae (Warwick et al. 2009).

A comprehensive molecular phylogenetic overview of the Brassicaceae is now emerging (Bailey et al. 2006, Warwick et al. 2007b, 2008c). These studies support the division of the family into 34 tribes (Al-Shehbaz et al. 2006, Al-Shehbaz and Warwick 2007, German and Al-Shehbaz 2008). A phylogenetic framework is essential for comparative genomic studies within the Brassicaceae and for the identification of closely related species that could serve as agronomic and economic trait sources. Knowledge of natural phylogenetic relationships allows estimates of derived versus ancestral states for various morphological, cytological, and biochemical characters and estimates of evolutionary distances and divergence times between groups (Lysak and Lexer 2006, Schranz et al. 2006b). For example, after the major split from the basal tribe Aethionemeae, there would appear to be a rapid radiation event in the family at ca. 24 Myr leading to the various tribal lineages including the split between two of the best studied genera *Arabidopsis* (tribe Arabideae) and *Brassica* (tribe Brassiceae).

2.2 Taxonomy and Genetic Relationships of *Brassica* Crop Species

The genus *Brassica* belongs to the Brassiceae. A natural tribe of ca. 50 genera, it is distinguished by its two-segmented fruit and unique conduplicate arrangement of the cotyledons or first leaves in the seed (reviewed in Gómez-Campo 1980, 1999, Warwick and Sauder 2005). Molecular and hybridization data now indicate that close relatives of the *Brassica* crop species also include species currently placed in separate genera in three different subtribes – the Brassicinae, Raphaninae, and Moricandiinae. This group of closely related species corresponds closely to the *Brassica* coenospecies, defined by Harberd (1972) as the "group of wild species sufficiently related to the six cultivated species of *Brassica* to be potentially capable of experimental hybridization with them." The most closely related genera to the *Brassica* crop species include: *Ceratocnemum, Coincya, Cordylocarpus, Diplotaxis, Enarthrocarpus, Eruca, Erucastrum, Guiraoa, Hemicrambe, Hirschfeldia, Kremeriella, Moricandia, Morisia, Muricaria, Orychophragmus, Otocarpus, Raffenaldia, Raphanus, Rapistrum, Rytidocarpus, Sinapidendron, Sinapis*, and *Trachystoma*.

An old world genus, *Brassica* includes about 35 species of mostly annual herbs, with some perennial herbs and small shrubs. Cultivated brassicas are represented by six interrelated species, three of which are diploids – *Brassica nigra* ($2n = 16$, genome BB), *B. oleracea* ($2n = 18$, genome CC), and *Brassica rapa* ($2n = 20$, AA) and three amphidiploid derivatives – *B. carinata* ($2n = 34$, BBCC), *B. juncea* ($2n = 36$, AABB), and *B. napus* ($2n = 38$, AACC). The latter three are derived by hybridization and polyploidization of two of the diploid taxa. The genomic relationships of the six *Brassica* crop species, known as the triangle of U (1935) and shown in Fig. 2.1, has been confirmed by chromosome pairing and artificial synthesis of the amphidiploids, nuclear DNA content and sequence analysis, and the use of genome-specific markers. Both nuclear DNA sequence data and chloroplast restriction site data, suggest separate evolutionary pathways, with *B. rapa* and *B. oleracea* (including wild CC genome species) assigned to one group with *Diplotaxis erucoides* ($n = 7$) or a close relative as the primary progenitor, and *B. nigra* assigned to a second group with *S. arvensis* ($n = 9$) or a close relative as the primary progenitor species (Warwick and Black 1993, Warwick and Sauder 2005).

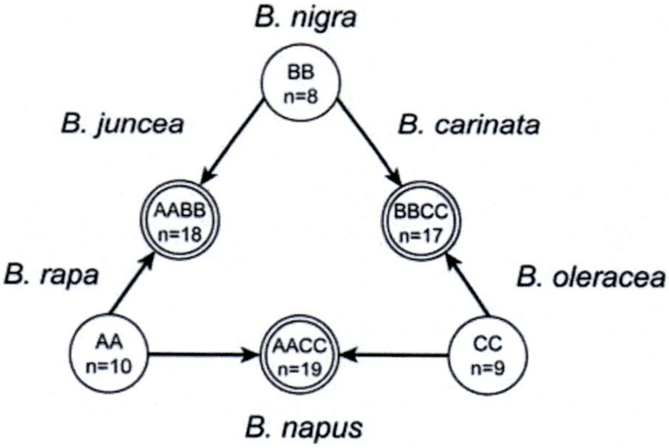

Fig. 2.1 Triangle of U showing the genetic relationships among the six cultivated species of *Brassica*. Adapted from U (1935)

Brassica species are believed to have originated in the Mediterranean-Middle Eastern area with a secondary center of origin and differentiation of *B. rapa* and *B. juncea* in China. Tremendous morphological variability is observed in the many subspecies, botanical varieties, and cultivar groups of *B. oleracea*, *B. rapa*, and *B. juncea*. Numerous parallel vegetable forms have been selected, and the three species have differentiated historically along similar lines (Prakash and Hinata 1980). The crop brassicas demonstrate great genetic and morphological diversity and plants yield edible roots, stems, leaves, buds, flowers, and seeds. In addition, some of the types are used as forage, sources of oil, or even ornamentals.

2.2.1 B. oleracea

Designated as cole crops, this species has a great diversity of morphotypes. *B. oleracea* is a member of the CC genome complex, which includes a number of interfertile Mediterranean species (*Brassica cretica*, *Brassica hilarionis*, *Brassica incana*, *Brassica insularis*, *Brassica macrocarpa*, *Brassica montana*, *Brassica rupestris*, and *Brassica villosa*), wild *B. oleracea* from coastal areas of western Europe, and *Brassica bourgeaui* from the Canary Islands.

The most important *B. oleracea* crops (taxonomic varieties in parentheses) are kales (var. *viridis*, var. *costata*, var. *medullosa*, var. *sabellica*) which develop a strong main stem and are used for their edible foliage; branching bush kales (var. *ramosa*), formerly much cultivated for their edible foliage; cabbages (var. *capitata* and var. *sabauda*) characterized by the formation of heads formed from tightly packed leaves; brussel sprouts (var. *gemmifera*) in which axillary buds form edible heads of tightly packed leaves; kohlrabi (var. *gongyloides*) cultivated for its above ground thickened stem; cauliflower and broccoli (var. *botrytis* and var. *italica*) cultivated for their thickened edible inflorescences; and chinese kale (var. *alboglabra*), a cultivated white-flowering crop grown in China, generally assumed to be an ancient import from the Mediterranean region and often treated as a separate species *Brassica alboglabra*.

Various origins of the different cultivated types have been proposed, including a single origin of all types from wild *B. oleracea* from western Europe, triple and even multiple origins involving related wild species of the CC complex. Molecular studies (Song et al. 1990) supported a monophyletic origin for the cultivated morphotypes of *B. oleracea* from a progenitor that was similar to wild *B. oleracea* consistent with morphological evidence that the earliest cultivated *B. oleracea* was probably a leafy kale from which the other cultivated types originated. Molecular evidence further suggested that selective introgression from other wild CC cytodeme members, *B. insularis* and *B. incana*, may also have contributed to the variability of cultivated *B. oleracea*.

2.2.2 B. rapa

B. rapa is highly polymorphic and contains many crops that have been domesticated over a long period in Europe as well as in Asia. Little is known about its true existence in the wild, plants found under natural conditions seem to be escapes from cultivation (spp. *sylvestris*). *B. rapa* is most closely related to *B. oleracea* and both have arisen from ancestral members of the C genome cytodeme.

The most important crops in *B. rapa* (subspecies in parentheses) are vegetable turnip (spp. *rapa*); fodder turnip (spp. *rapa*) which forms a leaf rosette and/or a turnip; turnip rape (spp. *oleifera*) and toria (spp. *dichotoma*), black seeded with annual spring and biennial winter types, used for oil extraction; yellow sarson (spp. *trilocularis*), annual, yellow seeded, used for oil extraction; Chinese cabbage (spp. *pekinensis*), Asiatic heading vegetable, with petioles winged; Pak choi (subsp. *chinensis*), a Chinese non-heading leaf vegetable, with petioles fleshy but not winged; Mizuma, mibuna, komatsuna, or leaf turnip (spp. *nipposinica*), Asiatic non-heading leafy vegetable, with many tillers and either pinnate (mizuma) or entire leaves (mibuna); broad-beak mustard or Chinese savoy (spp. *narinosa*), an Asiatic non-heading leafy vegetable, flat rosette of many small leaves; and broccoletto (formerly treated as *Brassica ruvo*, assigned to spp. *oleifera*), European vegetable with an enlarged, compact inflorescence. Most of the above-mentioned crops were described as separate species, but they readily intercross and hence belong to the $n = 10$ *B. rapa* cytodeme.

Various data (morphology, geographical distribution, isozymes, nuclear RFLPs, and AFLPs) have indicated a division of *B. rapa* into two main groups, perhaps corresponding to two independent centers of origin (Song et al. 1990, Warwick et al. 2008a). The primary center is Europe and includes turnip and turnip rape from which Asian sarson and toria types were derived. The second center is in China and contains the various Asian vegetables indicated above.

2.2.3 B. nigra

Black mustard, once widely grown as a condiment mustard, has largely been replaced by *B. juncea*. It is still grown as a condiment crop in parts of Asia. Although little information is available, the occurrence of land races in Europe, the Mediterranean, and the Ethiopian plateau indicates that *B. nigra* probably originated in central and southern Europe. It is presumed to have been introduced into India relatively recently. *B. nigra*, has evolved separately from the other two diploid *Brassica* species, and numerous data sets (cytological, isozyme, nuclear and chloroplast DNA restriction site, and sequence data) have suggested a closer genetic relationship to the genus *Sinapis*, particularly the weed species *S. arvensis* ($n = 9$), than to *B. rapa* and *B. oleracea*.

2.2.4 B. napus

Oilseed rape or canola is of comparatively recent origin and wild populations have not been found (Prakash and Hinata 1980). It is generally accepted that *B. napus* originated in southern Europe or the Mediterranean region, where the ranges of the two parental taxa *B. rapa* and *B. oleracea* overlap. Molecular data have provided evidence for multiple polyploid origins of *B. napus*, including crosses of *B. rapa* with *B. oleracea* and *B. rapa* with one of the wild C genome relatives, *B. montana* (Song and Osborn 1992). Two subspecies are generally recognized: the biennial vegetable rutabaga or swede (spp. *rapifera*) and an annual oilseed or fodder crop (spp. *napus* or spp. *oleifera*).

2.2.5 B. carinata

Abyssinian mustard is both an oilseed and a vegetable crop in Ethiopia with little differentiation into various crop types. Although wild types have not been located, it is believed to have originated in the Ethiopian plateau of northeast Africa, as a cross between wild-growing *B. nigra* and cultivated kale-like forms of *B. oleracea*.

The agronomic potential and germplasm diversity of *B. carinata* was reviewed in Warwick et al. (2006b).

2.2.6 B. juncea

Indian or brown mustard is grown in North America and Europe for condiment use, on the Indian subcontinent for seed oil and the Far East as a vegetable. Due to ecogeographical variation and human selection, a number of morphologically distinct forms are available, including oleiferous, semi-oleiferous, rapiferous, and leafy types.

There is some uncertainty as to the probable center of origin of *B. juncea*. It most likely originated in the Middle East or west Asian region, based on geographical sympatry of the parental taxa, *B. nigra* and *B. rapa*, and the presence of wild-growing *B. juncea* in this area. Other hypotheses suggest Asiatic origins with the center of major diversity in China. It seems likely that *B. juncea* may have arisen more than once as a result of hybridization, similar to that revealed for *B. napus*. Indeed, recent molecular studies have suggested more than one origin for each of the three varieties examined (vars. *multiseps, rapifera*, and *tsa-tsai*).

2.3 Other Crucifer Crops

2.3.1 Camelina

Camelina is an old world genus (tribe Camelineae) of 11 species. *C. sativa*, a native of southeastern Europe and southwestern Asia, had been gathered or cultivated as an oilseed for many centuries in Europe and the oil apparently used for culinary purposes and in lamps (Facciola 1990). The species has attracted renewed interest as an oil crop (Plessers et al. 1962, Putnam et al. 1993, Hebard 1998, Leonard 1998), with an adaptation to varied climate conditions and nutritionally poor soils and minimal input needs and with relatively high resistance to disease and pests. In Europe, where it is now widely grown, it has shown considerable potential in the food, animal feed, nutraceutical, paint, dye, cosmetic, and biofuel industries. In North America, it is being grown on a trial basis mainly for its potential as a biofuel in Saskatchewan, the Maritime Provinces, and the northern US Great Plains Regions (reviewed in Zubr 1997, Gugel and Falk 2006). Genetic mapping of agronomic traits has already been initiated in this species (Gehringer et al. 2006).

2.3.2 Crambe

Crambe, an old world genus of the tribe Brassiceae, is composed of approximately 30 species. *C. abyssinica* ($n = 45$) is an industrial oilseed crop, belonging to *Crambe* Section *Leptocrambe*. The seed oil is of considerable economic importance in industrial applications including use as erucamide (anti-block and slip agent in plastic films), coatings, lubricants (such as metal cutting oils, automatic transmission fluid supplement, and hydraulic fluid), and nylon. The use of *Crambe* meal as a protein source for feeds has also been investigated (Carlson and Tookey 1983). *C. abyssinica* is endemic to the Abyssinian highlands, and the name has been used not only for the wild Ethiopian population but also for the forms cultivated as an oilseed crop. It is derived from *Crambe hispanica* L. ($n = 30$), a widespread endemic of the Mediterranean region. Cultivation of *C. abyssinica* was apparently initiated in the USSR and has been investigated in many areas of the world, including the midwestern United States, the Netherlands, and Canada (e.g., Erickson and Bassin 1990). The agronomic potential and germplasm diversity of *Crambe* species is reviewed in Warwick and Gugel (2003).

2.3.3 Eruca

Eruca, an old world genus of the tribe Brassiceae, is composed of four species that are native to the Mediterranean region. One taxon is cultivated, *E. vesicaria* spp. *sativa* (frequently referred to as *E. sativa*). Subspecies *sativa* ($n = 11$, E genome), is an annual herb that has been cultivated since ancient times as a leafy vegetable (rocket or arugula), either for salad (Mediterranean, North America) or as a cooked green (Italy). It is also grown as a cold weather oilseed crop to produce jamba oil in Asia, mainly in India but also in Pakistan and Afghanistan, and is being considered as a protein meal supplement (Yaniv et al. 1998, Fagbenro 2004). The seed oil is used as an illuminant, lubricant, hair oil, vesicant, and for massage and pickling. Subspecies *vesicaria* occurs in the Mediterranean, whereas spp. *sativa* has been introduced and naturalized in many areas of the world. In some regions, such as in Mexico, naturalized populations are abundant and serious weeds. The agronomic potential and germplasm diversity of *Eruca* species is reviewed in Warwick et al. (2007a).

2.3.4 Raphanus

The genus *Raphanus*, an old world genus of tribe Brassiceae, is composed of two species: radish, *R. sativus* ($n = 9$, R genome), and wild radish, *R. raphanistrum* ($n = 9$). Radish has been cultivated for thousands of years and was grown extensively in ancient Egypt. *R. sativus* is not known in the wild, except for escapes forming weedy naturalized populations. There is some controversy as to the probable center of origin of *R. sativus*. It most likely originated in the Middle East or west Asian region, possibly from *R. raphanistrum*, although other suggestions indicate Asiatic origins with a center of major diversity in China. Important *R. sativus* crop varieties include small radish (var. *sativus* or *radicula*) grown for its edible root; black or large radish (var. *niger* or *longipinnatus*) grown for its roots, leaves, and young seed pods (believed to be the oldest type); mougri, rat-tailed, or aerial radish (var. *mougri* or *caudatus*) grown primarily for its edible young seed pods; and fodder or oilseed radish (var. *oleifera*) grown for animal fodder or green manure.

2.3.5 Sinapis

S. alba or white mustard, a Mediterranean species of tribe Brassiceae, is cultivated in many countries in Europe and North America. Seeds of this species are the main ingredient for mustard production (along with *B. juncea*) and for commercial mucilage production. In the last few years it has been increasingly cultivated as a fresh forage and green manuring plant in some countries in Europe and America.

2.4 Underutilized Crucifer Crops

Underutilized crucifer crops include *Diplotaxis* spp., rocket; *Lepidium sativum*, cress (e.g. Italy oilseed crop; Angelini et al. 1997); *Nasturtium officinale*, water cress; *Orychophragmus violaceus* (China oil crop; Li et al.1995, Huang et al. 1999, Luo et al. 1994, 1998a, 1998b) and root crops such as horseradish *Armoracia rusticana*, wasabi *Eutrema wasabi*, and maca *Lepidium meyenii*. Many crucifers are grown as ornamentals, and two such species are being developed as speciality industrial oil crops. *Lunaria annua*, for example, has 30–40% oil and high 44% erucic acid levels, as well as high (23%) concentrations of nervonic acid, which is used medically to treat multiple sclerosis (Marvin et al. 2000, Mastebroek and Marvin 2000, Walker et al. 2003), while *Matthiola incana* is rich in omega-3 linolenic acid (Ecker et al. 1992, Yaniv et al. 1997). *Lesquerella fendleri* is also another potential new speciality oil crop for arid lands in North America (Dierig et al. 2004, Salywon et al. 2005). *L. sativum* on the other hand is being considered for medicinal and functional food health properties (Mathews et al. 1993, Gokavi et al. 2004). Other wild Brassicaceae species, e.g., *Brassica fruticulosa*, are also being considered for vegetable diversification in Mediterranean regions (Branca 1995, Branca and Iapichino 1997).

Past folk medicinal use of other wild crucifer species (Specht and Diederichsen 2001) also suggests new unexplored crop opportunities. These include *Capsella bursa-pastoris* (Europe and Asia), *Cochlearia arctica* and *Cochlearia officinalis* (Europe); *Conringia orientalis* (Europe); *Descurainia sophia* (cultivated in Afghanistan, China; used for digestive troubles); *Erysimum cheiri* and *Erysium diffusum* (N India, Iraq, Russia); *Hesperis matronalis* (dame's rocket) (Europe); *Lepidium meyenii* (maca or Peruvian ginsing, South America); *Lobularia maritima* (India); and *Rorippa indica* (China, Vietnam – asthma remedy). Indeed two of these species, *D. sophia* and *R. indica*, have been recently used in *B. napus* germplasm enhancement in China (Guan et al. 2007a, 2007b). Diversification of crucifer crop use as fodder, green manure or cover crops are also of increasing agricultural interest (Lange et al. 1989, Mitchell et al. 1999, Bellostas et al. 2007, Larkin and Griffin 2007).

2.5 Brassicaceae as Sources of Agronomic and Economic Traits

2.5.1 Morphological Traits

Several morphological characters in the family are of agricultural interest or potential utility. Most obvious is resistance to pod shattering, which has been reported for *B. juncea* (Prakash and Chopra 1988a), *B. macrocarpa* and *B. hilarionis* (Mithen and Herron 1991), *Brassica tournefortii*, *C. orientalis* and *Hirschfeldia incana* (Salisbury 1989), and *Raphanus* spp. (Agnihotri et al. 1991). Trichomes (hairs) on the cotyledons and juvenile leaves are also traits of interest, given their role in hampering insect herbivory, and are found for, e.g., on the wild C genome species *Brassica incana* and *B. villosa*. Similarly, increased leaf thickness/waxiness provides drought and insect tolerance in *B. oleracea* and close relatives (Gómez-Campo et al. 1999, Stoner 1990).

Variation in growth form is limited and the family is primarily herbaceous with only 5% of the species typically woody (Al-Shehbaz 1984). These include the vine/woody climbers, e.g., *Heliophila scandens* (South Africa), *Lepidium scandens* (Australia), *Cremolobus peruvianus* (Peru); subshrubs (*Vella* spp.); large shrubs, e.g., *Foleyola* (N. Africa), *Parolinia* (Canary Islands); and small trees, e.g., *Farsetia somalensis* (NE Africa). Fruits are borne above ground, but geocarpy where the fruit is buried underground in a peanut-like manner has evolved independently in a handful of species: e.g., *Morisia monanthos* (Corsica and Sardinia), *Cardamine chenopodiifolia* (South America), and *Geococcus pusillus* (Australia).

The floral structure in the family is also highly conserved, i.e., four yellow petals arranged in a cross-shape, but exceptions in color (white, pink, purple) and shape (e.g., bilaterally symmetrical flowers of *Streptanthus*) are known. Of agronomic interest are species with small petals or with no petals (e.g., *C. orientalis*). Stamen number is usually six, with interesting exceptions of 2–4 stamens in *Lepidium* or 8–24 stamens in *Megacarpaea polyandra*. Nectary types vary with some species having lateral, median, or both, some glucose-, others sucrose dominant (Davis et al. 1998). In contrast, fruit type is highly variable in the family and is typically dehiscent bivalvate capsule (silique or silicule), but can be indehiscent, becoming lomentaceous or achenelike and only rarely nutlet, samara, schizocarp, or even a drupe. In plants with dehiscent siliques, dispersal is generally close to the parental plant, but transport by sea is enhanced by the corky fruit of *Cakile*, *Crambe*, *Raphanus* spp., tumbling action of the weed, *Anastatica hierochuntica* (Rose of Jericho) in the Saharo-Sindian region, explosive dehiscence as in various *Cardamine* spp.; while hooked hairs or spines on the fruits can aid animal dispersal. Seed size varies greatly from the smallest in *Mancoa mexicana* (Mexico) and Saharan species of *Diplotaxis* (at 0.02 and 0.05 mg, respectively) to the largest *Megacarpaea gigantea* (central Asia) weighing 90 mg; and measuring 1.8×1.5 cm (Al-Shehbaz 1984, 1986). A few crop members of the family have an incredible capacity for vegetative propagation, e.g., American water cress: *Neobeckia lacustris* and horseradish *A. rusticana*. It was this large underground biomass production that led to the suggested use of *A. rusticana* in phytoremediation efforts (Palmer et al. 2001).

2.5.2 Chemical Traits

Many genera of the Brassicaeae have been studied for their chemical constitution, especially for variation in oil content and seed fatty acid and glucosinolates composition. Kumar and Tsunoda (1980) reported on oil content and the fatty acid composition of 172 crucifer species representing 70 genera. Other surveys of wild species include wild *Brassica* spp. (Vioque et al. 1990, Yaniv et al. 1991, Ahuja et al. 1998, Velasco et al. 1998). High erucic acids levels (>45–50%) have been reported for *B. cretica*, *B. incana*, *B. rupestris*, and *B. villosa* (Yaniv et al. 1991, Velasco et al. 1998); *C. abyssinica* and *C. hispanica* (Yaniv et al. 1991, Mulder and Mastebroek 1996, Prakash and Bhat 2007); *E. vesicaria* (*E. sativa*) (Yaniv et al. 1991); *Erucastrum cardaminoides* and *Sinapidendron angustifolia* (Prakash and Bhat 2007); *S. alba* (Yaniv et al. 1994) and *S. arvensis* (Daun et al. 2003). High linoleic and/or linolenic acids have been reported in: *Brassica elongata* (Velasco et al. 1998); *C. sativa* (Budin et al. 1995, Shukla et al. 2002, Matthäus and Zubr 2000, Zubr and Matthäus 2002); *D. sophia* (Luo et al. 1999); *M. incana* (Ecker et al. 1992); *O. violaceus* (Wang et al. 1999); *Alyssum*, *Barbarea*, *Cardamine*, *Conringia*, and *L. sativum* (Prakash and Bhat 2007). High hydroxy fatty acids have been found in several *Lesquerella* and *Physaria* spp. (Salywon et al. 2005).

Glucosinolates (mustard oil glucosides) and their glucosinolate hydrolysis products provide the characteristic odors and flavors of crucifers. Their pharmacological role in the prevention of disease and in chemical defense against pathogens, herbivores, and weeds is attracting increasing attention (e.g. Angelini et al. 1998, Clauss et al. 2006). More than 96 glucosinolates have been reported in the Brassicaceae family, and many of these are unique to certain species and genera (Fahey et al. 2001). Many surveys of glucosinolates in wild crucifers have been conducted, including 51 crucifer spp. (Al-Shehbaz and Al-Shammary 1987); 259 crucifer spp. (Daxenbichler et al. 1991); 85 crucifer spp. (Bennett et al. 2004); 25 *Brassica* spp. (Horn and Vaughan 1983, Cole 1997, Velasco and Becker 2000); *B. oleracea* and 9 wild *Brassica* C genome spp. (Mithen et al. 1987a); 13 *Cakile* spp. (Rodman 1974, 1976); *C. sativa* (Schuster and Friedt 1998); 9 crucifer spp.: *C. bursa-pastoris*, *E. vesicaria* spp. *sativa*, *Erysimum allionii*, *E. cheiri*, *H. matronalis*, *L. fendleri*, *L. maritima*, *Matthiola longipetala* (Vaughn and Berhow 2005); *Diplotaxis tenuifolia* and *E. vesicaria* spp. *sativa* (Bennett et al. 2007); *Lepidium peruvianum* (Li et al. 2001); and 3 *Zilla* spp. (El-Menshawi et al. 1980).

Many species have potential for and could be grown for their value-added traits or production of pharmaceuticals. For example, in a survey of 91 crucifer spp. (Goffman et al. 1999), tocopherols, sources of vitamin E levels ranged from 68 mg/kg oil in *Diplotaxis viminea* to 2,479 mg/kg oil in *Schivereckia doerfleri*. The weed species *Lepidium draba*, extensively investigated for potential use in fighting disease (reviewed in Francis and Warwick 2008), ranked third in a survey of 700 plants for presence of compounds that could halt/delay the growth of cancer cells. Glucoraphanin, an alkenyl glucosinolate, which hydrolyzes to form the enzyme inducer sulforaphane, is purified from this species for its use as a dietary additive for cancer and high blood pressure treatments. Sulforaphane is also effective against pathogens such as bacteria, yeasts, fungi, mycoplasm, protozoans, nematodes, and viruses.

Our knowledge of other secondary metabolites in the family is limited, but deserves further attention. High concentrations of alkaloids (*L. annua*), cardenolides (*E. cheiri*), cucurbitacins (*Iberis amara*) are known (Al-Shehbaz 1984), and high concentration of cinnamoyl esters which is used in sunscreen reported in *L. fendleri* (Compton et al. 2004). Unusual floral pigments, acylated cyanidin glucosides were reported for *E. cheiri*, *L. maritima* and *L. annua* (Tatsuzawa et al. 2006). *A. rusticana*, for example, has served as a commercial source of peroxidases (Kushad et al. 1999). Flavonoid chemistry is less well known in the family, e.g., *Crambe* spp. (Aguinagalde and Gómez-Campo 1984); *Diplotaxis* spp. (Hussiney et al. 1998, Sánchez-Yélamo 1994); *Erucastrum* spp. (Sánchez-Yélamo 2001, 2004); *C. sativa*, *Crambe* spp., and *T. arvense* (Onyilagha et al. 2003). Sinapine levels have been screened in 23 crucifer species (Bouchereau et al. 1991) and in *D. tenuifolia*, *E. vesicaria* spp. sativa, and *L. sativum* (Özeker and Esiyok 1999) and in *Sinapis* and related species (Agerbirk et al. 2008).

Mucilage production in seeds is characteristic of many species in this family, particularly in those occupying droughty areas. A family wide survey would be worthwhile. Mucilage from the seeds of *S. alba* is one of the main commercial mucilage sources (Cui et al. 1993), but other crops have been investigated for this product, e.g., *L. sativum* (Mathews et al. 1993).

Crucifer species with particularly high glucosinolate levels have been investigated for their potential use as a biofumigant or fungicide/nematocide particularly in the invasive *Alliaria petiolata* (McCarthy and Hanson 1998, Roberts and Anderson 2001, Aminidehaghi et al. 2006, Cipollini and Gruner 2007); *A. rusticana* (Kotova et al. 1999); *Barbarea verna* (Curto et al. 2005); *Brassica.* (Turk and Tawaha 2003); *C. abyssinica* (Mohiuddin et al. 1990, Peterson et al. 2000); *D. sophia* (Yang and Mu 2006); *D. erucoides* (anti-microbial compounds) (Peláez et al. 1998); *E. vesicaria* spp. *sativa* (Curto et al. 2005); *Lepidium perfoliatum* (Aminidehaghi et al. 2006); *N. officinale* (anti-nemotode) (Kotova et al. 1999); *Rapistrum rugosum* (Curto et al. 2005); *S. arvensis* (anti-mosquito) (Bowers et al. 1997); *Zilla spinosa* (anti-weed and -rhizosphere fungi) (El-Khatib and Abd-Elaah 1998).

2.5.3 C3–C4 Photosynthesis

Most members of the Brassicaceae have typical C_3 photosynthesis (Uprety et al. 1995). The C_3–C_4 intermediate species *Moricandia arvensis*, *Moricandia nitens*, *Moricandia sinaica*, *Moricandia spinosa*, and *Moricandia suffruticosa* (Bauwe 1983, Razmjoo et al. 1996, Apel et al. 1997, Rylott et al. 1998) and *D. tenuifolia* (Apel et al. 1996, 1997, Peisker et al. 1998, Bang et al. 2003, Ueno et al. 2003) have been the subjects of considerable investigation as C_4 photosynthesis is believed to be a more efficient system particularly under drought conditions.

2.5.4 Cytoplasmic Male Sterility

Members of the family have provided valuable sources of novel cytoplasmic male sterility genes vital to the production of crop hybrid systems. These include *B. juncea* 126-1 (Sodhi et al. 2006); *B. juncea* hau (Wan et al. 2007); *B. napus* nap and pol (Brown 1999); *Brassica oxyrrhina* oxyrrhina (Prakash and Chopra 1988b, Kanada and Kato 1997); *B. tournefortii* (Pradhan et al. 1991, Pahwa et al. 2004); *Diplotaxis berthautii* (Malik et al. 1999); *Diplotaxis catholica* (Mohapatra et al. 1998, Pathania et al. 2003); *Diplotaxis erucoides* (Malik et al. 1999); *Diplotaxis harra* (Klimaszewska and Keller 1988); *Diplotaxis muralis* (Hinata and Konno 1979, Riungu and McVetty 2000, 2003a, b); *Diplotaxis siifolia* siifolia (Rao et al. 1994, Rao and Shivanna 1996); *Enarthrocarpus lyratus* (Banga et al. 2003b, Deol et al. 1999, 2003, Janeja et al. 2003); *E. vesicaria* subsp. *sativa* (Matsuzawa et al. 1999); *Erucastrum canariense* (Prakash et al. 2001, Banga et al. 2003a); *H. incana* (Horovitz and Galil 1972); *L. fendleri* (Dierig et al. 2001); *M. arvensis* moricandia (Kirti et al. 1998, Prakash et al. 1998, Bhat et al. 2005); *R. sativus* ogura (Ogura 1968, Yamagishi 1998, Murayama et al. 2004); and *Trachystoma ballii* trachystoma (Kirti et al. 1997).

2.5.5 Breeding Systems and Apomixis

Most members of the family are outcrossing and insect pollinated (wind pollination rare), and indeed the conserved architecture of the flower in the family is believed to be very closely linked to pollination by insects (Al-Shehbaz 1984). Autogamy or selfing is common in many of the weedy species, e.g., *Erucastrum gallicum*, while cleistogamy, i.e, self-fertilization without flower opening, always occurs in some species such as the submersed aquatic *Subularia aquatica.* The molecular basis of the cleistogamous trait has been recently reported for *Cardamine kokaiensis*, a close relative of *Arabidopsis* (Morinaga et al. 2008). With few exceptions, the flowers of the Brassicaeae are always perfect, but dioecism is reported in some *Lepidium* spp. from New Zealand and monoecism (male and female flowers on same plant) in the central Asian species *Megacarpaea megalocarpa* (Al-Shehbaz 1984). Of primary interest to crucifer breeding is the trait for apomixis which was first described in *Arabis* spp. (Roy and Rieseberg 1989, Roy 1995); additional taxa have been added – *Arabis gunnisoniana* (Taskin et al. 2004); *Arabis holboellii* (Naumova et al. 2001); and *Boechera* spp. (Schranz et al. 2005, 2006a).

2.5.6 Plant Regeneration and Transformation

Many crucifers have served as model species for improving plant regeneration and transformation systems, including *Alyssum borzaeanum* (Paunescu 2008), *Alyssum murale* (Vinterhalter et al. 2008), *Arabidopsis halleri* (Dal-Corso et al. 2005), *A. thaliana* (Gaj 2004), *B. carinata* (Verma et al. 2008), *B. juncea* (Dhawan et al 2000, Eapen 2007, Dutta et al. 2008, Prem et al. 2008, Wang et al. 2008a), *B. oleracea* (Munshi et al. 2007, Zhang et al. 2008), *B. napus* (Ali et al. 2007, Ben-Ghnaya et al. 2008, Haddadi et al. 2008, Munir et al. 2008, Verma et al. 2008), *B. rapa* (Gao et al. 2008), *C. sativa* (Tattersall and Millam 1999, Lu and Kang 2008); *C. abyssinica* and *C. hispanica* (Sonntag and Rudloff 2001, Sonntag and Gramenz 2004); *D. muralis* (Sikdar et al. 1990); *E. vesicaria* spp. *sativa* (Sikdar et al. 1987, Zhang et al. 2005); *Isatis indigotica* (Hu et al. 1999, Zhang et al. 2003, Xu et al. 2004); *L. fendleri* (Skarjinskaia et al. 2003, Wang et al. 2008b); *M. incana* (Mensuali-Sodi et al. 1994, Siemens et al. 1995); *M. arvensis* (Rashid et al. 1996, Craig et al. 1997); *M. nitens* (Tian and Meng 1998); *O. violaceus* (Hu et al. 1999); *R. indica* (Mandal and Sikdar 2003); *R. nasturtium–aquaticum* (Jin et al. 1999); *Thlaspi caerulescens* (Guan et al. 2008); and *Thellungiella halophila* (Li et al. 2007). The transformation of members of the Brassicaceae is described in Chapter 18 by Sparrow et al., this volume.

2.5.7 Salt and Heavy Metal Tolerances

Several species in the family exhibit distinct salt tolerance, including those found in coastal strand habitats, e.g. *Cakile* spp. (Boyd and Barbour 1986, Megdiche et al. 2007), *Crambe maritima* and *R. raphanistrum* ssp. *maritimus*. Salt tolerant taxa include desert plants *L. fendleri* (Dierig et al. 2004) and *E. vesicaria* subsp. *sativa* (Ashraf and Noor 1993, Ashraf 1994). *Thellungiella salsuginea* (=*T. halophila*) occupies saline flat habitats and now serves as a model system for studying this trait (Inan et al. 2004).

Many species of the family are tolerant to heavy metals and have the capacity to hyperaccumulate Cd, Ni, Pb, Se, Sr, and/or Zn (Boyd et al. 1994, Kruckeberg and Reeves 1995, Boyd and Martens 1998, Palmer et al. 2001, Prasad and Freitas 2003, Ghaderian et al. 2007, Przedpelska and Wierzbicka 2007). To date, 90 species from 11 genera have been reported, including *Alyssum* (48 spp.), *Thlaspi* (28 spp.), *Bornmuellera* (4 spp.), *Arabidopsis* (3 spp.), *Arabis* (1 sp.), *Cardamine* (1 sp.), *Cochlearia* (1 sp.), *Peltaria* (1 sp.), *Pseudosempervivum* (1 sp.), *Stanleya* (1 sp.), *Streptanthus* (1 sp.). *T. caerulescens* serves as the model plant species for heavy metal tolerance studies (Assunção et al. 2003). *B. juncea* is one of the most tolerant species (Belimov et al. 2007) and a proposed crop species for use in phytoremediation efforts of sites contaminated with heavy metals.

2.5.8 Cold Tolerance

Cold tolerance is an important agronomic trait especially for northern temperate climates. Laroche et al. (1992) reported on cold tolerance in the weed species *Barbarea vulgaris*, *D. sophia*, and *T. arvense*, while recent studies have focused on a "Yukon ecotype" of *T. salsuginea* (Wong et al. 2005). Cold tolerance genes have been detected in microarray analyses of *T. salsuginea* (Taji et al. 2004, Gong et al. 2005, Griffith et al. 2007) and *T. arvense* (Sharma et al. 2007). Other arctic or alpine species in the family are also potential sources of cold tolerance traits. For example, *Arabis*, *Crucihimalaya*, and *Draba* species are adapted to alpine areas up to 6,000 m in the Himalayas, Alps, Rockies, Andes, while *Romanschulzia* (Mexico and Central America) and *Oreophyton* (East Africa) are adapted to high mountains of the tropics (Al-Shehbaz 1984). *Draba*, *Eutrema*, and *Parrya* species occupy arctic habitats, while *Pringlea antiscorbutica*, also the subject of cold tolerance studies (Hennion et al. 2006), occupies sub-antarctic regions. Only a few members of the Brassiceae occupy alpine habitats (elevations >2,000–2,500 m above snow line), *Brassica nivalis* (Mt. Olympus, Greece); *Brassica jordanoffii* (Mt Pirin Planina, Bulgaria); *Coincya richeri* (Alps in France and Italy) (Leadlay and Heywood 1990), and *Erucastrum abyssinicum* and *Erucastrum pachypodium* (Ethiopian Highlands) (Al-Shehbaz 1985).

2.5.9 Drought Tolerance

T. salsuginea is also a model species for studying drought tolerance (Wong et al. 2005). Several species in the tribe Brassiceae show remarkable drought tolerance, including *B. carinata* and *B. juncea* (Mishra et al. 1999); *B. tournefortii* (Salisbury 1989, Prakash and Bhat 2007); *Carrichtera annua* (Boaz et al. 1990); *Diplotaxis acris* and *D. harra* (Boaz et al. 1990, Prakash and Bhat 2007); *Enarthrocarpus strangulatus*, *Erucaria boveana*, *Erucaria microcarpa* and *Erucaria uncata*, *Pseuderucaria clavata*, *Savignya parviflora*, *Schouwia purpurea* (Boaz et al. 1990); *E. vesicaria* spp. *sativa* (Sun et al. 1991, 1999, Prakash and Bhat 2007). Many species occur in the Saharan Desert, e.g., *Foleyola billotii*, *Fortuynia* spp., *Physorhynchus* spp., and *Z. spinosa* (Warwick et al. 2009), while *Lesquerella* spp. occur in the deserts of North America (Ravetta and Soriano 1998, Ploschuk et al. 2001, Prakash and Bhat 2007). *Moricandia* species also show drought tolerance and their higher water use efficiency has been associated with the C3–C4 intermediate photosyntheic pathway (McVetty et al. 1989).

2.5.10 Herbicide Resistance

Several weedy crucifer species have developed herbicide resistance (reviewed in Heap 2009, Warwick et al. 2005). A wild biotype of *B. rapa* (Maltais and Bouchard 1978) served as the trait source for development of triazine-resistant lines of *B. napus* in the 1980s. Triazine-resistant biotypes have also been reported in *C. bursa-pastoris*, *R. raphanistrum*, and *S. arvensis*. Acetolactate synthase (ALS) inhibitor-resistant biotypes have been reported in 12 species from seven different countries: *B. tournefortii*, *Camelina microcarpa*, *D. sophia*, *D. tenuifolia*, *Neslia paniculata*, *R. raphanistrum*, *R. sativus*, *R. rugosum*, *S. arvensis*, *Sisymbrium orientale*, S. *thellungii*, and *Thlaspi arvensis*. The latter are generally due to target site mutations of the ALS gene (Hanson et al. 2004, Warwick et al. 2005, Christoffers et al. 2006), but a metabolism-based resistant *S. arvensis* biotype has also been detected (Veldhuis et al. 2000). Paraquat-resistant biotypes have only been reported in *Lepidium virginicum* (Canada: Smisek et al. 1998) and auxin-resistant biotypes in *S. arvensis* (Canada: Heap and Morrison 1992) and *S. orientale* (Australia). There is much interest in incorporating the *S. arvensis* auxin-resistance trait in *B. napus*. The physiological, biochemical, molecular, and genetic basis of the trait is known (Zheng and Hall 2001, Jugulam et al. 2005); and auxinic-resistant microspore-derived doubled haploid *S. arvensis* plants produced (Mithila and Hall 2007).

2.5.11 Disease Resistance

Wild species in the family have provided an invaluable source of disease resistance traits for crop improvement. White rust (*Albugo candida*) resistance has been reported as common in *B. carinata* and more limited in *B. rapa*, *B. juncea*, *B. nigra* (Liu and Rimmer 1991, Gulati et al. 1991, Kolte et al. 1991); *Brassica maurorum* (Chrungu et al. 1999); *E. vesicaria* spp. *sativa* (Bansal et al. 1997); and *R. sativus* (Williams and Pound 1963, Kolte et al. 1991). Resistance to black leaf spot (*Alternaria* spp.) has also been widely reported in the family (Sharma et al. 2002): *A. petiolata*, *B. vulgaris* (Westman and Dickson 1998); *B. elongata* and *B. fruticulosa* (Siemens 2002); *B. maurorum* (Chrungu et al. 1999); *B. nigra* (Westman and Dickson 1998, Westman et al. 1999); *Brassica souliei* (Siemens 2002); *Brassica spinescens* (Agnihotri et al. 1991); *C. sativa* and *C. bursa–pastoris* (Conn et al. 1988, reviewed in Tewari 1991, Westman and Dickson 1998, Westman et al. 1999, Siemens 2002, Pedras et al. 2003b); *Coincya* spp. and *D. catholica* (Prakash and Bhat 2007); *D. erucoides* and *D. tenuifolia* (Siemens 2002, Klewer et al. 2003); *E. vesicaria* subsp. *sativa* (Conn and Tewari 1986, Tewari 1991); *Hemicrambe fruticulosa*, *H. matronalis*, *N. paniculata*, and *R. sativus* (Siemens 2002); *S. alba* (Brun et al. 1987, Sharma and Singh 1992, Siemens 2002, Pedras et al. 2003b); and *S. arvensis* (Siemens 2002). Blackleg (*Leptosphaeria maculans*) resistance has been reported in *A. thaliana* (Brun and Tribodet 1995, Chen and Séguin-Swartz 1997, 1999); *B. carinata*, *B. juncea*, and *B. nigra* (Rimmer and van den Berg 1992); *B. elongata* and *B. fruticulosa* (Siemens 2002); *B. insularis*, *Brassica atlantica*, and *B. macrocarpa* (Mithen et al. 1987b, Mithen and Herron 1991, Mithen and Magrath 1992); *C. sativa* (Siemens 2002, Li et al. 2005); *Coincya monensis* (Winter et al. 1999, 2002, 2003); *E. vesicaria* and *Eruca pinnatifida* (Tewari et al. 1996, Siemens 2002); *Carrichtera*, *Diplotaxis*, *Hirschfeldia*, *Raphanus*, *Rapistrum*, and *Sinapis* weedy species in Australia (Salisbury 1987); *C. monensis* (Siemens 2002); *D. muralis*, and *D. tenuifolia* (Chen and Séguin-Swartz 1997, 1999); *H. incana* (Siemens 2002); *R. raphanistrum* (Chen and Séguin-Swartz 1999); *R. sativus* (Siemens 2002); *S. alba* (Gugel and Séguin-Swartz 1997); *S. avensis* (Siemens 2002, Winter et al. 1999, 2002, 2003); *Sisymbrium loeselii* (Chen and Séguin-Swartz 1997, 1999); and *T. arvense* (Pedras et al. 2003a). Studies in *T. arvense* have indicated for, e.g., that two antifungal phytoalexins wasalexinA and arvelalexin are responsible for the resistance. Downy mildew (*Peronospora parasitica*) tolerance has been reported in *B. oleracea* wild accessions (Greenhalgh and Mitchell

1976) and *E. vesicaria* (Singh and Kolte 1999). Clubroot (*Plasmodiophora brassicae*) tolerance has been reported in *A. thaliana* (Rehn et al. 2004); *A. rusticana* (Prakash and Bhat 2007); *C. bursa–pastoris* (Siemens 2002), and *Raphanus* spp. (Crute et al. 1980, Long et al. 1992). Sclerotinia stem rot (*Sclerotinia sclerotiorum*) tolerance has been reported in *C. bursa–pastoris* (Chen et al. 2007), *E. vesicaria* subsp. *sativa* (Guan et al. 2004) and *E. gallicum* (Lefol et al. 1997, Gugel et al. 1997). Turnip mosaic virus resistance has been reported in *A. petiolata* and *H. matronalis* (Stobbs and Stirling 1990) and *S. alba* and *S. arvensis* (Mamula et al. 1997). Wilt disease (*Verticillium dahliae*) resistance has been reported in *A. rusticana* (Atibalentja and Eastburn 1998). Black rot (*Xanthomonas campestris*) resistance has been reported in *B. nigra* (Marthe et al. 2004); *B. rapa* (Ignatov et al. 1999); *A. petiolata, B. vulgaris, B. juncea, B. nigra, Erysimum hieraciifolium*, and *M. incana* (Westman and Dickson 1998). Resistance to peppery leaf spot (bacteria *Pseudomonas syringae*) was recently reported in two *B. juncea* germplasm accessions (Wechter et al. 2007) indicating the often unexplored wealth of diversity in crop germplasm collections.

2.5.12 Insect and Nematode Resistance

Resistance to flea beetles (*Phyllotreta cruciferae* and *Phyllotreta striolata*), the most important insect pests of oilseed crucifer crops in North America, has been reported in *A. thaliana* (Prakash and Bhat 2007); *B. incana* and *B. villosa* (Bodnaryk, personal communication, Warwick et al. 2009); *B. juncea* (Bodnaryk 1997); *C. sativa* (Pachagounder et al. 1998, Soroka et al. 2003, Henderson et al. 2004); *C. bursa–pastoris* (Prakash and Bhat 2007); *C. abyssinica* (Anderson et al. 1992, Soroka et al. 2003, Henderson et al. 2004); *C. hispanica* and *C. glabrata* (Soroka et al. 2003); *S. alba* (Lamb 1980, Bodnaryk and Lamb 1991, Bodnaryk 1997, Gavloski et al. 2000, Henderson et al. 2004); and *T. arvense* (Gavloski et al. 2000); whereas *B. vulgaris* was shown to be resistant to the European flea beetle *Phyllotreta nemorum* (Renwick 2002). Resistance to the lepidopteran pests diamond-back moth (*Plutella xylostella*) has been described in *B. vulgaris* (Renwick 2002, Lu et al. 2004, Badenes-Perez et al. 2005); *B. juncea* (Ruwandi and Gillott 1998); *B. oleracea* (Stoner 1990, Ramachandran et al. 1998); *B. napus* (Ramachandran et al. 1998); *C. abyssinica* (Kmec et al. 1998); and *R. raphanistrum* (Lehtila and Strauss 1999); and to the cabbage butterfly (*Pieris* spp.) in *Erysimum cheiranthoides* and *I. amara* (Renwick 2002) and *P. napi oleracea* in *A. petiolata* and *B. vulgaris* (Renwick 2002). Resistance to the cabbage aphid (*Brevicoryne brassicae*) was reported for *B. fruticulosa* and *B. spinescens* (Cole 1994, Singh et al. 1994, Ellis and Farrell 1995); *E. vesicaria* subsp. *sativa* (Singh et al. 1994); and *S. alba* (Thompson 1963) and to the mustard aphid (*Lipaphis erysimi*) in *B. carinata, B. nigra, B. juncea*, and *Eruca sativa* (Rana et al. 1995, Lal et al. 1997, Chander and Bakhetia 1998). Cabbage white fly (*Aleyrodes proletella*) resistance was described for *B. cretica, B. fruticulosa, B. incana, B. insularis, B. spinosa, B. villosa* (Ramsey and Ellis 1994). Cabbage root fly (*Delia radicum*) resistance was reported for *B. fruticulosa* and C genome species *B. incana*, *B. macrocarpa*, *B. spinescens*, *B. villosa* (Ellis et al. 1999); *C. sativa* (Soroka et al. 2003) and *S. alba* (Jyoti et al. 2001). Cabbage seed pod weevil (*Ceutorhynchus obstrictus*) resistance was described for *B. juncea*, *B. nigra*, *B. tournefortii*, *C. sativa*, and *S. alba* (Ulmer and Dosdall 2006, Carcamo et al. 2007). Mustard Sawfly (*Athalia proxima*) resistance was shown for *C. sativa* (Singh and Sachan 1997).

Resistance has been reported in *R. sativus* and *S. alba* to the beet cyst nematode, *Heterodera schachtii* (Thierfelder et al. 1991, Lelivelt and Krens 1992, Lelivelt et al. 1993) and root-knot nematode (*Meloidogyne* spp.) (Buente et al. 1997, Pattison et al. 2006).

2.6 Conclusion

Wild Brassicaceae germplasm offers numerous future prospects for agronomic and economic traits and new crops. Characterization of the wild germplasm, however, is still incomplete and available information fragmentary. A systematic investigation of such resources would improve efficicacy of their utilization in crucifer crop breeding programs (Prakash and Bhat 2007). Efforts should be made to maintain and expand wild germplasm collections before these species become extinct due to habitat loss or where access is restricted due to global strife. More collections are needed from centers of variability and centers of origin, and an emphasis should also be placed on non-native areas of the species range both for weedy species and neglected landraces (e.g., Farnham et al. 2008). Future introgression of desirable genes/traits will require greater knowledge of the genetic basis or inheritance of these traits, as well as information on precise chromosomal locations. The development of genetic maps for key relatives and or new crucifer crops (see chapters to follow) will be of increasing importance. Some traits can be introduced through artificial hybridization (Warwick et al. 2009), while traits arising from more distant sources will need to be isolated and introduced through transgenic means. Production of addition lines in various crop genome backgrounds will be needed and in situ hybridization techniques will be helpful in identifying such additions with more precision. The ultimate goal will be the pyramiding of traits in crop cultivars. New crop opportunities exist among the many species currently cultivated for ornamental or vegetable/salad use and diversification of crucifer crop use whether as fodder, green manure or cover crops, biofumigants, etc., also offer exciting future agricultural possibilities.

References

Agerbirk N, Warwick SI, Hansen PR et al (2008) *Sinapis* phylogeny and evolution of glucosinolates and specific nitrile degrading enzymes. Phytochemistry 69:2937–2949

Agnihotri A, Shivanni KR, Lakshmikumaran MS et al (1991) Micropropagation and DNA analysis of wide hybrids of cultivated *Brassica*. Proc GCIRC 8th Rapeseed Congr Abst.: p 151

Aguinagalde I, Gómez-Campo C (1984) The phylogenetic significance of flavonoids in *Crambe* (Cruciferae). Bot J Linn Soc 89:277–288

Ahuja I, Malik CP, Raheja RK et al (1998) Physiological and biochemical changes in fruit development of *Brassica oxyrrhina* and *Brassica tournefortii*. Phytomorphology 48: 399–404

Ali H, Ali Z, Ali H, Mehmood S, Ali W (2007) In vitro regeneration of *Brassica napus* L. cultivars (Star, Cyclone and Westar) from hypocotyls and cotyledonary leaves. Pak J Bot 39:1251–1256

Al-Shehbaz IA (1984) The tribes of Cruciferae (Brassicaceae) in the southeastern United States. J Arnold Arbor 65:343–373

Al-Shehbaz IA (1985) The genera of Brassiceae (Cruciferae: Brassicaceae) in the southeastern United States. J Arnold Arbor 66:279–351

Al-Shehbaz IA (1986) The genera of Lepideae (Cruciferae: Brassicaceae) in the southeastern United States. J Arnold Arbor 67:265–311

Al-Shehbaz IA, Al-Shammary KI (1987) Distribution and chemotaxonomic significance of glucosinolates in certain middle-eastern Cruciferae. Biochem Syst Ecol 15:559–569

Al-Shehbaz IA, Warwick SI (2007) Two new tribes (Dontostemoneae and Malcolmieae) in the Brassicaceae (Cruciferae). Harv Pap Bot 12:429–433

Al-Shehbaz IA, Beilstein MA, Kellogg EA (2006) Systematics and phylogeny of the Brassicaceae (Cruciferae): an overview. Plant Syst Evol 259:89–120

Aminidehaghi M, Rezaeinodehi A, Khangholi S (2006) Allelopathic potential of *Alliaria petiolata* and *Lepidium perfoliatum*, two weeds of the Cruciferae family. J Plant Dis Prot Special Issue 20:455–462

Anderson MD, Peng C, Weiss MJ (1992) Crambe, *Crambe abyssinica* Hochst, as a flea beetle resistant crop (Coleoptera: Chrysomelidae). J Econ Entomol 85:594–600

Angelini LG, Moscheni E, Colonna G et al (1997) Variation in agronomic characteristics and seed oil composition of new oilseed crops in central Italy. Ind Crops Prod 6:313–323

Angelini L, Lazzeri L, Galletti S et al (1998) Antigerminative activity of three glucosinolate-derived products generated by myrosinase hydrolysis. Seed Sci Technol 26:771–780

Apel P, Hillmer S, Pfeffer M et al (1996) Carbon metabolism type of *Diplotaxis tenuifolia* (L.) DC. (Brassicaceae). Photosynthetica 32:237–243

Apel P, Horstmann C, Pfeffer M (1997) The *Moricandia* syndrome in species of the Brassicaceae – evolutionary aspects. Photosynthetica 33:205–215

Ashraf M (1994) Organic substances responsible for salt tolerance in *Eruca sativa*. Biol Plant 36:255–259

Ashraf M, Noor R (1993) Growth and pattern of ion uptake in *Eruca sativa* Mill. under salt stress. Ange Bot 67:17–21

Assunção AGL, Schat H, Aarts MGM (2003) *Thlaspi caerulescens*, an attractive model species to study heavy metal hyperaccumulation in plants. New Phytol 159:351–360

Atibalentja N, Eastburn DM (1998) *Verticillium dahliae* resistance in horseradish germplasm from the university of illinois collection. Plant Dis 82:176–180

Badenes-Perez FR, Shelton AM, Nault BA (2005) Using yellow rocket as a trap crop for diamondback moth (Lepidoptera: Plutellidae). J Econ Entomol 98:884–890

Bailey CD, Koch MA, Mayer M et al (2006) Toward a global phylogeny of the Brassicaceae. Mol Biol Evol 23:2142–2160

Bang SW, Mizuno Y, Kaneko Y et al (2003) Production of intergeneric hybrids between the C_3–C_4 intermediate species *Diplotaxis tenuifolia* (L.) DC. and *Raphanus sativus* L. Breed Sci 53: 231–236

Banga SS, Banga SK, Bhaskar PB et al (2003a) Alloplasmic line of *Brassica napus* L. with *Erucastrum canariense* cytoplasm is male sterile. Proc GCIRC 11th Int Rapeseed Congr 1:pp 324–325

Banga SS, Deol JS, Banga SK (2003b) Alloplasmic male-sterile *Brassica juncea* with *Enarthrocarpus lyratus* cytoplasm and the introgression of gene(s) for fertility restoration from cytoplasm donor species. Theor Appl Genet 106:1390–1395

Bansal VK, Tewari JP, Tewari I et al (1997) Genus *Eruca*: a potential source of white rust resistance in cultivated brassicas. Plant Genet Resour Newsl 109:25–26

Bauwe H (1983) Comparative phylogenetic age of C_3–C_4 intermediate species of *Moricandia* determined by isoelectric focusing and amino acid composition of small subunit of ribulose 1,5-bisphosphate carboxylase/oxygenase. Photosynthetica 17:442–449

Belimov AA, Safronova VI, Demchinskaya SV et al (2007) Intraspecific variability of cadmium tolerance in hydroponically grown Indian mustard (*Brassica juncea* (L.) Czern.) seedlings. Acta Physiol Plant 29:473–478

Bellostas N, Sorensen JC, Sorensen H (2007) Profiling glucosinolates in vegetative and reproductive tissues of four *Brassica* species of the U-Triangle for their fumigation potential. J Sci Food Agric 87:1586–1594

Ben-Ghnaya A, Charles G, Branchard M (2008) Rapid shoot regeneration from thin cell layer explants excised from petioles and hypocotyls in four cultivars of *Brassica napus* L. Plant Cell Tissue Organ Cult 92:25–30

Bennett RN, Mellon FA, Kroon PA (2004) Screening crucifer seeds as sources of specific intact glucosinolates using ion-pair high-performance liquid chromatography negative ion electrospray mass spectrometry. J Agric Food Chem 52:428–438

Bennett RN, Carvalho R, Mellon FA et al (2007) Identification and quantification of glucosinolates in sprouts derived from seeds of wild *Eruca sativa* L. (salad rocket) and *Diplotaxis tenuifolia* L. (wild rocket) from diverse geographical locations. J Agric Food Chem 55:67–74

Bhat SR, Prakash S, Kirti PB et al (2005) A unique introgression from *Moricandia arvensis* confers male fertility upon two different cytoplasmic male-sterile lines of *Brassica juncea*. Plant Breed 124:117–120

Boaz M, Plitmann U, Heyn CC (1990) The ecogeographic distribution of breeding systems in the Cruciferae (Brassicaceae) of Israel. Isr J Bot 39:31–42

Bodnaryk RP (1997) Will low-glucosinolate cultivars of the mustards *Brassica juncea* and *Sinapis alba* be vulnerable to insect pests? Can J Plant Sci 77:283–287

Bodnaryk RP, Lamb RJ (1991) Mechanisms of resistance to the flea beetle, *Phyllotreta cruciferae* (Goeze), in yellow mustard seedlings, *Sinapis alba* L. Can J Plant Sci 71:13–20

Bouchereau A, Hamelin J, Lamour I et al (1991) Distribution of sinapine and related compounds in seeds of *Brassica* and allied genera. Phytochemistry 30:1873–1881

Bowers WS, Sener B, Evans PH et al (1997) Activity of Turkish medicinal plants against mosquitoes *Aedes aegypti* and *Anopheles gambiae*. Insect Sci Appl 16:339–342

Boyd RS, Barbour MG (1986) Relative salt tolerance of *Cakile edentula* (Brassicaceae) from lacustrine and marine beaches. Am J Bot 73:236–241

Boyd RS, Martens SN (1998) Nickel hyperaccumulation by *Thlaspi montanum* var. *montanum* (Brassicaceae): a constitutive trait. Am J Bot 85:259–265

Boyd RS, Shaw JJ, Martens SN (1994) Nickel hyperaccumulation defends *Streptanthus polygaloides* (Brassicaceae) against pathogens. Am J Bot 81:294–300

Branca F (1995) Studies on some wild Brassicaceae species utilizable as vegetables in the Mediterranean areas. Plant Genet Resour Newsl 104:6–9

Branca F, Iapichino G (1997) Some wild and cultivated Brassicaceae exploited in Sicily as vegetables. Plant Genet Resour Newsl 110:22–28

Brown GG (1999) Unique aspects of cytoplasmic male sterility and fertility restoration in *Brassica napus*. J Hered 90:351–356

Brun H, Tribodet M (1995) Pathogenicity of *Leptosphaeria maculans* isolates on one ecotype of *Arabidopsis thaliana*. Cruciferae Newsl Eucarpia 17:74–75

Brun H, Pleiss J, Renard M (1987) Resistance of some crucifers to *Alternaria brassicae* (Berk.) Sacc. Proc GCIRC 7th Int Rapeseed Congr:pp 1222–1227

Budin JT, Breene WM, Putnam DH (1995) Some compositional properties of camelina (*Camelina sativa* L. Crantz) seeds and oils. J Am Oil Chem Soc 72:309–315

Buente R, Mueller J, Friedt W (1997) Genetic variation and response to selection for resistance to root-knot nematodes in oil radish (*Raphanus sativus* ssp. *oleiferus*). Plant Breed 116:263–266

Carcamo H, Olfert O, Dosdall L et al (2007) Resistance to cabbage seedpod weevil among selected Brassicaceae germplasm. Can Entomol 139:658–669

Carlson KD, Tookey HL (1983) Crambe meal as a protein source for feeds. J Am Oil Chem Soc 60:1979–1985

Chander H, Bakhetia DRC (1998) Evaluation of some cruciferous genotypes at seedling stage for resistance to mustard aphid, *Lipaphis erysimi* (Kalt.) under screen house and field conditions. J Insect Sci 11:19–25

Chen CY, Séguin-Swartz G (1997) A comparative study of the response of wild crucifers to the blackleg fungus, *Phoma lingam*. Can J Plant Pathol 19:107

Chen CY, Séguin-Swartz G (1999) Reaction of wild crucifers to *Leptosphaeria maculans*, the causal agent of blackleg of crucifers. Can J Plant Pathol 21:361–367

Chen H, Wang H, Li Z (2007) Intertribal crosses between *Brassica* species and *Capsella bursa-pastoris* for the improvement of oil quality and resistance to *Sclerotinia sclerotiorum* of *Brassica* crops. Proc GCIRC 12th Int Rapeseed Congr 1:pp 411–413

Christoffers MJ, Nandula VK, Howatt KA et al (2006) Target-site resistance to acetolactate synthase inhibitors in wild mustard (*Sinapis arvensis*). Weed Sci 54:191–197

Chrungu B, Verma N, Mohanty A et al (1999) Production and characterization of interspecific hybrids between *Brassica maurorum* and crop brassicas. Theor Appl Genet 98:608–613

Cipollini D, Gruner B (2007) Cyanide in the chemical arsenal of garlic mustard, *Alliaria petiolata*. J Chem Ecol 33:85–94

Clauss MJ, Dietel S, Schubert G et al (2006) Glucosinolate and trichome defenses in a natural *Arabidopsis lyrata* population. J Chem Ecol 32:2351–2373

Cole RA (1994) Isolation of a chitin-binding lectin with insecticidal activity in chemically-defined synthetic diets from two wild brassica species with resistance to cabbage aphid *Brevicoryne brassicae*. Entomol Exp Appl 72:181–187

Cole RA (1997) The relative importance of glucosinolates and amino acids to the development of two aphid pests *Brevicoryne brassicae* and *Myzus persicae* on wild and cultivated brassica species. Entomol Exp Appl 85:121–133

Compton DL, Laszlo JA, Isbell TA (2004) Cinnamoyl esters of lesquerella and castor oil: novel sunscreen active ingredients. J Am Oil Chem Soc 81:945–951

Conn KL, Tewari JP (1986) Hypersensitive reaction induced by *Alternaria brassicae* in *Eruca sativa*, an oil yielding crucifer. Can J Plant Pathol 8:348

Conn KL, Tewari JP, Dahiya JS (1988) Resistance to *Alternaria brassicae* and phytoalexin-elicitation in rapeseed and other crucifers. Plant Sci 56:21–25

Craig W, Wiegand A, O'Neill CM et al (1997) Somatic embryogenesis and plant regeneration from stem explants of *Moricandia arvensis*. Plant Cell Rep 17:27–31

Crute IA, Gray AR, Crisp P et al (1980) Variation in *Plasmodiophora brassicae* and resistance to clubroot disease in *Brassicas* and allied crops. Plant Breed 50:91–104

Cui W, Eskin NAM, Biliaderis CG (1993) Chemical and physical properties of yellow mustard (*Sinapis alba* L.) mucilage. Food Chem 46:169–176

Curto G, Dallavalle E, Lazzeri L (2005) Life cycle duration of *Meloidogyne incognita* and host status of Brassicaceae and Capparaceae selected for glucosinolate content. Nematology 7: 203–212

Dal-Corso G, Borgato L, Furini A (2005) In vitro plant regeneration of the heavy metal tolerant and hyperaccumulator *Arabidopsis halleri* (Brassicaceae). Plant Cell Tissue Organ Cult 82: 267–270

Daun J, Barthet V, Scarth R (2003) Erucic acid levels in *Sinapis arvensis* L from different parts of the world. Proc GCIRC 11th Int Rapeseed Congr 1:pp 290–292

Davis AR, Pylatuik JD, Paradis JC et al (1998) Nectar-carbohydrate production and composition vary in relation to nectary anatomy and location within individual flowers of several species of Brassicaceae. Planta 205:305–318

Daxenbichler ME, Spencer GF, Carlson DG et al (1991) Glucosinolate composition of seeds from 297 species of wild plants. Phytochemistry 30:2623–2638

Deol JS, Ahuja I, Banga SS (1999) Physiological investigations of male sterility caused by *Enarthrocarpus lyratus* cytoplasm in Brassicas. Crop Improv 26:156–162

Deol JS, Shivanna KR, Prakash S et al (2003) *Enarthrocarpus lyratus*-based cytoplasmic male sterility and fertility restorer system in *Brassica rapa*. Plant Breed 122:438–440

Dhawan AK, Jain A, Inderjeet SJ (2000) An efficient plant regeneration protocol from seedling explants of *Brassica juncea* RH-781, a freeze tolerant cultivar. Cruciferae Newsl Eucarpia 22:21–22

Dierig DA, Tomasi PM, Ray DT (2001) Inheritance of male sterility in *Lesquerella fendleri*. J Am Soc Hortic Sci 126:738–743

Dierig DA, Tomasi PM, Salywon AM et al (2004) Improvement in hydroxy fatty acid seed oil content and other traits from interspecific hybrids of three *Lesquerella* species: *Lesquerella fendleri*, *L. pallida*, and *L. lindheimeri*. Euphytica 139:199–206

Dutta I, Saha P, Das S (2008) Efficient *Agrobacterium*-mediated genetic transformation of oilseed mustard [*Brassica juncea* (L.) Czern.] using leaf piece explants. In Vitro Cell Dev Biol Plant 44:401–411

Eapen S (2007) Genetic transformation in *Brassica juncea* (L.) Czern & Coss using wild types and shooter mutants of *Agrobacterium tumefaciens* and the regeneration response. Plant Cell Biotechnol Mol Biol 8:205–208

Ecker R, Yaniv Z, Zur M et al (1992) Embryonic heterosis in the linolenic acid content of *Matthiola incana* seed oil. Euphytica 59:93–96

El-Khatib AA, Abd-Elaah GA (1998) Allelopathic potential of *Zilla spinosa* on growth of associate flowering plants and some rhizosphere fungi. Biol Plant 41:461–467

El-Menshawi B, Karawya M, Wassel G et al (1980) Glucosinolates in the genus *Zilla* (Brassiceae). J Nat Prod 43:534–536

Ellis PR, Farrell JA (1995) Resistance to cabbage aphid (*Brevicoryne brassicae*) in six brassica accessions in New Zealand. N Z J Crop Hort Sci 23:25–29

Ellis PR, Pink DAC, Barber NE et al (1999) Identification of high levels of resistance to cabbage root fly, *Delia radicum*, in wild *Brassica* species. Euphytica 110:207–214

Erickson DB, Bassin P (1990) Rapeseed and Crambe: alternative crops with potential industrial uses. Bull Kans Agric Exp Stn 656:1–33

Facciola S (1990) Cornucopia – a source book of edible plants, p 677. Kampong Publications, Vista, CA

Fagbenro OA (2004) Soybean meal replacement by roquette (*Eruca sativa* Miller) seed meal as protein feedstuff in diets for African catfish, *Clarias gariepinus* (Burchell 1822), fingerlings. Aquac Res 35:917–923

Fahey JW, Zalcmann AT, Talalay P (2001) The chemical diversity and distribution of glucosinolates among plants. Phytochemistry 56:5–51

Farnham M, Davis E, Morgan J et al (2008) Neglected landraces of collard (*Brassica oleracea* L. var. viridis) from the Carolinas (USA). Genet Resour Crop Evol 55:797–801

Francis A, Warwick SI (2008) The biology of Canadian weeds. 3. *Lepidium draba* L, *L. chalepense* L, *L. appelianum* Al-Shehbaz (updated). Can J Plant Sci 88:379–401

Gaj MD (2004) Factors influencing somatic embryogenesis induction and plant regeneration with particular reference to *Arabidopsis thaliana* (L.) Heynh. Plant Growth Regul 43:27–47

Gao HL, Li Y, Song YP, Gao SY, Wang JJ (2008) In vitro culture and regeneration system of non-heading Chinese cabbage (*Brassica campestris* ssp. *chinensis* Makino). Xibei Zhiwu Xuebao 28:963–968

Gavloski JE, Ekuere U, Keddie A et al (2000) Identification and evaluation of flea beetle (*Phyllotreta cruciferae*) resistance within Brassicaceae. Can J Plant Sci 80:881–887

Gehringer A, Friedt W, Lühs W et al (2006) Genetic mapping of agronomic traits in false flax (*Camelina sativa* subsp. *sativa*). Genome 49:1555–1563

Ghaderian SM, Mohtadi A, Rahiminejad R et al (2007) Hyperaccumulation of nickel by two *Alyssum* species from the serpentine soils of Iran. Plant Soil 293:91–97

German DA, Al-Shehbaz IA (2008) Five additional tribes (Aphragmeae, Biscutelleae, Calepineae, Conringieae, and Erysimeae) in the Brassicaceae (Cruciferae). Harv Pap Bot 13:165–170

Goffman FD, Thies W, Velasco L (1999) Chemotaxonomic value of tocopherols in Brassicaceae. Phytochemistry 50:793–798

Gokavi SS, Malleshi NG, Guo M (2004) Chemical composition of garden cress (*Lepidium sativum*) seeds and its fractions and use of bran as a functional ingredient. Plant Foods Hum Nutr 59: 105–111

Gómez-Campo C (1980) Morphology and morphotaxonomy of the tribe Brassiceae. In: Tsunoda S, Hinata K, Gómez-Campo C (eds) Brassica crops and wild allies, pp 3–31. Japan Science Societies Press, Tokyo

Gómez-Campo C (1999) Taxonomy. In: Gómez-Campo C (ed) The biology of *Brassica* coenospecies, pp 3–32. Elsevier Science BV, Amsterdam

Gómez-Campo C, Tortosa ME, Tewari I et al (1999) Epicuticular wax columns in cultivated *Brassica* species and in their close wild relatives. Ann Bot 83:515–519

Gong Q, Li P, Ma S et al (2005) Salinity stress adaptation competence in the extremophile *Thellungiella halophila* in comparison with its relative *Arabidopsis thaliana*. Plant J 44: 826–839

Greenhalgh JG, Mitchell ND (1976) The involvement of flavour volatiles in the resistance of downy mildew of wild and cultivated forms of *Brassica oleracea*. New Phytol 77:391–398

Griffith M, Timonin M, Wong ACE et al (2007) *Thellungiella*: an *Arabidopsis*-related model plant adapted to cold temperatures. Plant Cell Environ 30:529–538

Guan CY, Li FQ, Li X et al (2004) Resistance of rocketsalad (*Eruca sativa* Mill.) to stem rot (*Sclerotinia sclerotiorum*). Sci Agric Sinica 37:1138–1143

Guan R, Jiang S, Xin R et al (2007a) Studies on rapeseed germplasm enhancement by use of cruciferous weed *Descurainia sophia*. Proc GCIRC 12th Int Rapeseed Congr 1: pp 261–265

Guan R, Jiang S, Xin R et al (2007b) Studies on rapeseed germplasm enhancement by use of cruciferous weed *Rorippa indica*. Proc GCIRC 12th Int Rapeseed Congr 1: pp 329–332

Guan ZQ, Chai TY, Zhang YX, Xu J, Wei W, Han L, Cong L (2008) Gene manipulation of a heavy metal hyperaccumulator species *Thlaspi caerulescens* L. via *Agrobacterium*-mediated transformation. Mol Biotechnol 40:77–86

Gugel RK, Falk KC (2006) Agronomic and seed quality evaluation of *Camelina sativa* in western Canada. Can J Plant Sci 86:1047–1058

Gugel RK, Séguin-Swartz G (1997) Introgression of blackleg resistance from *Sinapis alba* into *Brassica napus*. Brassica 97, Int Soc Hortic Sci Symp Brassicas/10th Crucifer Genetics Workshop, 23–27 Sept 1997, Rennes, France, Abst.: p 222

Gugel RK, Séguin-Swartz G, Warwick SI (1997) Transfer of blackleg resistance from *Erucastrum gallicum* to *Brassica rapa*. Can J Plant Pathol 19:109

Gulati SC, Varma NS, Mani N et al (1991) Resistance to white rust (*Albugo candida*) in Indian mustard. Proc GCIRC 8th Int Rapeseed Congr:pp 256–261

Haddadi P, Moieni A, Gh K, Abdollahi MR (2008) Effects of gibberellin, abscisic acid and embryo desiccation on normal plantlet regeneration, secondary embryogenesis and callogenesis in microspore culture of *Brassica napus* L. cv. PF704. Int J Plant Prod 2:153–162

Hanson BD, Park KW, Mallory-Smith CA et al (2004) Resistance of *Camelina microcarpa* to acetolactate synthase inhibiting herbicides. Weed Res 44:187–194

Harberd DJ (1972) A contribution to the cytotaxonomy of *Brassica* (Cruciferae) and its allies. Bot J Linn Soc 65:1–23

Heap IM, Morrison IN (1992) Resistance to auxin-type herbicides in wild mustard (*Sinapis arvensis* L.) populations in western Canada. Weed Sci Soc Am Abst 32:55

Heap IM (2009) International survey of herbicide-resistant weeds. (http://www.weedscience.com)

Hebard A (1998) *Camelina sativa* – a pleasurable experience or another false hope? Lipid Technol 10:81–83

Henderson AE, Hallett RH, Soroka J (2004) Prefeeding behavior of the crucifer flea beetle, *Phyllotreta cruciferae*, on host and nonhost crucifers. J Insect Behav 17:17–39

Hennion F, Frenot Y, Martin-Tanguy J (2006) High flexibility in growth and polyamine composition of the crucifer *Pringlea antiscorbutica* in relation to environmental conditions. Physiol Plant 127:212–224

Hinata K, Konno N (1979) Studies on a male sterile strain having the *Brassica campestris* nucleus and the *Diplotaxis muralis* cytoplasm I. On the breeding procedure and some characteristics of the male sterile strain. Jpn J Breed 29:305–311

Horn PJ, Vaughn JG (1983) Seed glucosinolates of fourteen wild *Brassica* species. Phytochemistry 22:465–470

Horovitz A, Galil J (1972) Gynodioecism in east Mediterranean *Hirschfeldia incana*, Cruciferae. Bot Gaz 133:127–131

Hu Q, Andersen SB, Hansen LN (1999) Plant regeneration capacity of mesophyll protoplasts from *Brassica napus* and related species. Plant Cell Tissue Organ Cult 59:189–196

Huang BQ, Luo P, Li YW (1999) Domestication and cultivation of *Orychophragmus violaceus* as a new oil crop in China. Cruciferae Newsl Eucarpia 21:13–14

Hussiney HA, El-Missiry MM, Ismail SI (1998) Flavonoids of *Diplotaxis harra* (Forssk.) Boiss. and *Diplotaxis acris* (Forssk.) Boiss. Egypt J Pharmacol Sci 38:131–136

Ignatov A, Kuginuki Y, Hida K (1999) Disease reaction to *Xanthomonas campestris* pv. *campestris* races 1, 4 and 5 in weedy and cultivated *Brassica rapa* L. Cruciferae Newsl Eucarpia 21: 123–124

Inan G, Zhang Q, Li P et al (2004) Salt cress. A halophyte and cryophyte *Arabidopsis* relative model system and its applicability to molecular genetic analyses of growth and development of extremophiles. Plant Physiol 135:1718–1737

Janeja HS, Banga SK, Bhaskar PB et al (2003) Alloplasmic male sterile *Brassica napus* with *Enarthrocarpus lyratus* cytoplasm: introgression and molecular mapping of an *E. lyratus* chromosome segment carrying a fertility restoring gene. Genome 46:792–797

Jin RG, Liu YB, Tabashnik BE et al (1999) Tissue culture and Agrobacterium-mediated transformation of watercress. Plant Cell Tissue Organ Cult 58:171–176

Jugulam M, McLean MD, Hall JC (2005) Inheritance of picloram and 2, 4-D resistance in wild mustard (*Brassica kaber*). Weed Sci 53:417–423

Jyoti JL, Shelton AM, Earle ED (2001) Identifying sources and mechanisms of resistance in crucifers for control of cabbage maggot (Diptera: Anthomyiidae). J Econ Entomol 94: 942–949

Kanada I, Kato M (1997) Effect of *Brassica oxyrrhina* cytoplasm on *Raphanus sativus*. Breed Sci 47:57–65

Kirti PB, Baldev A, Gaikwad K et al (1997) Introgression of a gene restoring fertility to CMS (*Trachystoma*) *Brassica juncea* and the genetics of restoration. Plant Breed 116:259–262

Kirti PB, Prakash S, Gaikwad K et al (1998) Chloroplast substitution overcomes leaf chlorosis in a *Moricandia arvensis* – based cytoplasmic male sterile *Brassica juncea*. Theor Appl Genet 97:1179–1182

Klewer A, Scheunemann R, Sacristán MD (2003) Incorporation of blackspot resistance from different origins into oilseed rape. Proc GCIRC 11th Int Rapeseed Congr 1:pp 65–67.

Klimaszewska K, Keller WA (1988) Regeneration and characterization of somatic hybrids between *Brassica napus* and *Diplotaxis harra*. Plant Sci 58:211–222

Kmec P, Weiss MJ, Milbrath LR et al (1998) Growth analysis of crambe. Crop Sci 38:108–112

Koch MA, Kiefer M (2005) Genome evolution among cruciferous plants: a lecture from the comparison of the genetic maps of three diploid species – *Capsella rubella*, *Arabidopsis lyrata* subsp. *petraea*, and *A. thaliana*. Am J Bot 92:761–767

Kolte SJ, Bordoloi DK, Awasthi RP (1991) The search for resistance to major diseases of rapeseed and mustard in India. Proc GCIRC 8th Int Rapeseed Congr:pp 219–225

Kotova VV, Cozari EG, Kononkov PF et al (1999) Use of antagonistic plants against plant-parasite nematodes. Sel'skokhozyaistvennaya Biol 0(3):24–32

Kruckeberg AR, Reeves RD (1995) Nickel accumulation by serpentine species of *Streptanthus* (Brassicaceae): field and greenhouse studies. Madroño 42:458–469

Kumar PR, Tsunoda S (1980) Variation in oil content of fatty acid composition among seeds from the Cruciferae. In: Tsunoda S, Hinata K, Gómez-Campo C (eds) *Brassica* crops and wild allies, pp 235–252. Japan Scientific Societies Press, Tokyo

Kushad MM, Guidera M, Bratsch AD (1999) Distribution of horseradish peroxidase activity in horseradish plants. HortScience 34:127–129

Lal MN, Singh SS, Singh VP (1997) Screening of Brassica germplasms for resistance against mustard aphid, *Lipaphis erysimi* (Kalt.). J Entomol Res 21:371–375

Lamb RJ (1980) Hairs protect pods of mustard (*Brassica hirta* "Gisilba") from flea beetle feeding damage. Can J Plant Sci 60:1439–1440

Lange W, Toxopeus H, Lubberts JH et al (1989) The development of raparadish *Brassicoraphanus* $2n = 38$, a new crop in agriculture. Euphytica 40:1–14

Larkin RP, Griffin TS (2007) Control of soilborne potato diseases using *Brassica* green manures. Crop Prot 26:1067–1077

Laroche A, Geng XM, Singh J (1992) Differentiation of freezing tolerance and vernalization responses in Cruciferae exposed to a low temperature. Plant Cell Environ 15:439–445

Leadlay EA, Heywood VH (1990) The biology and systematics of the genus *Coincya* Porta & Rigo ex Rouy (Cruciferae). Bot J Linn Soc 102:313–398

Lefol C, Séguin-Swartz G, Morrall RAA (1997) Effect of petal age on infection by *Sclerotinia sclerotiorum* in *Erucastrum gallicum*. Brassica 97, Int Soc Hortic Sci Symp Brassicas/10th Crucifer Genetics Workshop, 23–27 Sept. 1997, Rennes, France, Abst.:p 188

Lehtila K, Strauss SY (1999) Effects of foliar herbivory on male and female reproductive traits of wild radish, *Raphanus raphanistrum*. Ecology 80:116–124

Lelivelt CL, Krens FA (1992) Transfer of resistance to the beet cyst nematode *Heterodera schachtii* Schm. into the *Brassica napus* L. gene pool through intergeneric somatic hybridization with *Raphanus sativus* L. Theor Appl Genet 83:887–894

Lelivelt CLC, Leunissen EHM, Frederiks HJ et al (1993) Transfer of resistance of the beet cyst nematode (*Heterodera schachtii* Schm.) from *Sinapis alba* L. (white mustard) to the *Brassica napus* L. gene pool by means of sexual and somatic hybridization. Theor Appl Genet 85: 688–696

Leonard EC (1998) Camelina oil: α-linolenic source. Inform 9:830–838

Li G, Ammermann U, Quiros CF (2001) Glucosinolate contents in maca (*Lepidium peruvianum* Chacon) seeds, sprouts, mature plants and several derived commercial products. Econ Bot 55:255–262

Li H, Barbetti MJ, Sivasithamparam K (2005) Hazard from reliance on cruciferous hosts as sources of major gene-based resistance for managing blackleg (*Leptosphaeria maculans*) disease. Field Crops Res 91:185–198

Li HQ, Xu J, Chen L, Li MR (2007) Establishment of an efficient *Agrobacterium tumefaciens*-mediated leaf disc transformation of *Thellungiella halophila*. Plant Cell Rep 26:1785–1789

Li XF, Yang Y, Wang YP (1995) Preliminary assessment of two new cruciferous oil plant germplasms distributed in China. Chin J Bot 7:156–163

Liu Q, Rimmer SR (1991) Inheritance of resistance in *Brassica napus* to an Ethiopian isolate of *Albugo candida* from *Brassica carinata*. Can J Plant Pathol 13:197–201

Long ML, Xing GM, Okubo H et al (1992) Cross compatibility between *Brassicoraphanus* (*Brassica oleracea* X *Raphanus sativus*) and cruciferous crops, and rescuing the hybrid embryos through ovary and embryo culture. J Fac Agric Kyushu Univ 37:29–39

Lu C, Kang J (2008) Generation of transgenic plants of a potential oilseed crop *Camelina sativa* by *Agrobacterium*-mediated transformation. Plant Cell Rep 27:273–278

Lu J, Liu S, Shelton AM (2004) Laboratory evaluations of a wild crucifer *Barbarea vulgaris* as a management tool for the diamondback moth *Plutella xylostella* (Lepidoptera: Plutellidae). Bull Entomol Res 94:509–516

Luo P, Lan ZQ, Li ZY (1994) *Orychophragmus violaceus*, a potential edible-oil crop. Plant Breed 113:83–85

Luo P, Huang BQ, Lan ZQ et al (1998a) A study on the vegetable resource *Orychophragmus violaceus*. Sichuan Daxue Xuebao Ziran Kexueban 35:638–641

Luo P, Huang BQ, Yin JM et al (1998b) A new forage genetic resource *Orychophragmus violaceus* (L.) O.E.Schulz. Genet Res Crop Evol 45:491–494

Luo P, Lan ZQ, Gao HB et al (1999) *Descurainia sophia* – a neglected cruciferous plant resource. Cruciferae Newsl Eucarpia 21:15–16

Lysak MA, Lexer C (2006) Towards the era of comparative evolutionary genomics in Brassicaceae. Plant Syst Evol 259:175–198

Malik M, Vyas PO, Rangaswamy NS et al (1999) Development of two new cytoplasmic male-sterile lines in *Brassica juncea* through wide hybridization. Plant Breed 118:75–78

Maltais B, Bouchard CJ (1978) Une moutarde des oiseaux (*Brassica rapa* L.) resistante a l'atrazine. Phytoprotection 59:117–119

Mamula D, Juretic N, Horvath J (1997) Susceptibility of host plants to belladonna mottle and turnip yellow mosaic tymoviruses: multiplication and distribution. Acta Phytopathol Entomol Hungarica 32:289–298

Mandal P, Sikdar SR (2003) Plant regeneration from mesophyll protoplasts of *Rorippa indica* (L.) Hiern, a wild crucifer. Curr Sci 85:1451–1454

Marthe F, Richter K, Schrader O et al (2004) Cabbage (*Brassica oleracea*) with new resistance to black rot (*Xanthomonas campestris* pv. *campestris*) from black mustard (*Brassica nigra*). Brassica 2004, 4th ISHS Symposium Brassicas/Proc 14th Crucifer Genet Workshop, p 175.

Marvin HJP, Mastebroek HD, Becu DMS et al (2000) Investigation into the prospects of five novel oilseed crops within Europe. Outlook Agric 29:47–53

Mastebroek HD, Marvin HJP (2000) Breeding prospects of *Lunaria annua* L. Ind Crops Prod 11:139–143

Mathews S, Singhal RS, Kulkarni PR (1993) Some physicochemical characteristics of *Lepidium sativum* (haliv) seeds. Nahrung 37:69–71

Matsuzawa Y, Mekiyanon S, Kaneko Y et al (1999) Male sterility in alloplasmic *Brassica rapa* L. carrying *Eruca sativa* cytoplasm. Plant Breed 118:82–84

Matthäus B, Zubr J (2000) Variability of specific components in *Camelina sativa* oilseed cakes. Ind Crops Prod 12:9–18

McCarthy BC, Hanson SL (1998) An assessment of the allelopathic potential of the invasive weed *Alliaria petiolata* (Brassicaceae). Castanea 63:68–73

McVetty PBE, Austin RB, Morgan CL (1989) A comparison of the growth, photosynthesis, stomatal conductance and water use efficiency of *Moricandia* and *Brassica* species. Ann Bot 64:87–94

Megdiche W, Ben Amor N, Bebez A et al (2007) Salt tolerance of the annual halophyte *Cakile maritima* as affected by the provenance and the developmental stage. Acta Physiol Plant 29:375–384

Mensuali-Sodi A, Brea M, Panizza M et al (1994) In vitro-regenration of shoots in *Matthiola incana* L. from seedling explants of different age. Gartenbauwissenschaft 59:77–80

Mishra RS, Abdin MZ, Uprety DC (1999) Interactive effects of elevated CO_2 and moisture stress on the photosynthesis, water relation and growth of *Brassica* species. J Agron Crop Sci 182:223–229

Mitchell JP, Thomsen CD, Graves WL et al (1999) Cover crops for saline soils. J Agron Crop Sci 183:167–178

Mithen RF, Herron C (1991) Transfer of disease resistance to oilseed rape from wild *Brassica* species. Proc GCIRC 8th Int Rapeseed Congr:pp 244–249

Mithen RF, Magrath R (1992) Glucosinolates and resistance to *Leptosphaeria maculans* in wild and cultivated *Brassica* spp. Plant Breed 108:60–68

Mithen RF, Lewis BG, Heaney RK et al (1987a) Glucosinolates of wild and cultivated *Brassica* species. Phytochemistry 26:1969–1973

Mithen RF, Lewis BG, Heaney RK et al (1987b) Resistance of leaves of *Brassica* species to *Leptosphaeria maculans*. Trans Br Mycol Soc 88:525–531

Mithila J, Hall JC (2007) Production of an auxinic herbicide-resistant micro spore-derived doubled haploid wild mustard (*Sinapis arvensis* L.) plant. Crop Prot 26:357–362

Mohapatra T, Kirti PB, Kumar VD et al (1998) Random chloroplast segregation and mitochondrial genome recombination in somatic hybrid plants of *Diplotaxis catholica* + *Brassica juncea*. Plant Cell Rep 17:814–818

Mohiuddin S, Qureshi RA, Qureshi SA et al (1990) Studies on the repellent activity of some indigenous plant oils against *Tribolium castaneum* (Herbst.). Pak J Sci Ind Res 33: 326–328

Morinaga S, Nagano AJ, Miyazaki S et al (2008) Ecogenomics of cleistogamous and chasmogamous flowering: genome wide gene expression patterns from cross-species microarray analysis in *Cardamine kokaiensis* (Brassicaceae). J Ecol 96:1086–1097

Mulder JH, Mastebroek HD (1996) Variation for agronomic characteristics in *Crambe hispanica*, a wild relative of *Crambe abyssinica*. Euphytica 89:267–278

Munir M, Rashid H, Rauf M, Chaudhry Z, Bukhari MS (2008) Callus formation and plantlets regeneration from hypocotyl of *Brassica napus* by using different media combinations. Pakistan J Bot 40:309–315

Munshi MK, Roy PK, Kabir MH, Ahmed G (2007) In vitro regeneration of cabbage (*Brassica oleracea* L. var. *capitata*) through hypocotyl and cotyledon culture. Plant Tissue Cult Biotechnol 17:131–136

Murayama K, Yahara T, Terachi T (2004) Variation of female frequency and cytoplasmic male-sterility gene frequency among natural gynodioecious populations of wild radish (*Raphanus sativus* L.). Mol Ecol 13:2459–2464

Naumova TN, van der Laak J, Osadtchiy J et al (2001) Reproductive development in apomictic populations of *Arabis holboellii* (Brassicaceae). Sex Plant Reprod 14:195–200

Ogura H (1968) Studies on the new male-sterility in Japanese radish, with special reference to the utilization of this sterility towards the practical raising of hybrid seeds. Mem Fac Agric Kagoshina Univ 6:39–78

Onyilagha J, Bala A, Hallett R et al (2003) Leaf flavonoids of the cruciferous species, *Camelina sativa*, *Crambe* spp., *Thlaspi arvense* and several other genera of the family Brassicaceae. Biochem Syst Ecol 31:1309–1322

Özeker E, Esiyok D (1999) Identification of phenolic compounds in seeds of different rocket species (*Eruca sativa* and *Diplotaxis tenuifolia*) and land cress (*Lepidium sativum*). Cruciferae Newsl Eucarpia 21:21–22

Pachagounder P, Lamb RJ, Bodnaryk RP (1998) Resistance to the flea beetle *Phyllotreta cruciferae* (Coleoptera: Chrysomelidae) in false flax, *Camelina sativa* (Brassicaceae). Can Entomol 130:235–240

Pahwa RS, Banga SK, Gogna KPS et al (2004) Tournefortii male sterility system in *Brassica napus*. Identification, expression and genetic characterization of male fertility restorers. Plant Breed 123:444–448

Palmer CE, Warwick SI, Keller W (2001) Brassicaceae (Cruciferae) family, plant biotechnology, and phytoremediation. Int J Phytoremediation 3:245–287

Parkin IAP, Gulden SM, Sharpe AG et al (2005) Segmental structure of the *Brassica napus* genome based on comparative analysis with *Arabidopsis thaliana*. Genetics 171:765–781

Paunescu A (2008) Histological investigation of the secondary somatic embryogenesis of *Alyssum borzaeanum* (Brassicaceae). Phytol Balc 14:111–117

Pathania A, Bhat SR, Kumar VD et al (2003) Cytoplasmic male sterility in alloplasmic *Brassica juncea* carrying *Diplotaxis catholica* cytoplasm: molecular characterization and genetics of fertility restoration. Theor Appl Genet 107:455–461

Pattison AB, Versteeg C, Akiew S et al (2006) Resistance of Brassicaceae plants to root-knot nematode (*Meloidogyne* spp.) in northern Australia. Int J Pest Manag 52:53–62

Pedras MSC, Chumala PB, Suchy M (2003a) Phytoalexins from *Thlaspi arvense*, a wild crucifer resistant to virulent *Leptosphaeria maculans*: structures, syntheses and antifungal activity. Phytochemistry 64:949–956

Pedras MSC, Montaut S, Zaharia IL et al (2003b) Transformation of the host-selective toxin destruxin B by wild crucifers: probing a detoxification pathway. Phytochemistry 64:957–963

Peisker M, Heinemann I, Pfeffer M (1998) A study on the relationship between leaf conductance, CO_2 concentration and carboxylation rate in various species. Photosynth Res 56:35–43

Peláez F, Collado J, Arenal F et al (1998) Endophytic fungi from plants living on gypsum soils as a source of secondary metabolites with antimicrobial activity. Mycol Res 102:755–761

Peterson CJ, Cosse A, Coats JR (2000) Insecticidal components in the meal of *Crambe abyssinica*. J Agric Urban Entomol 17:27–36

Plessers AG, McGregor WG, Carson RB et al (1962) Species trials with oilseed plants II. *Camelina*. Can J Plant Sci 42:452–459

Ploschuk EL, Windauer L, Ravetta DA (2001) Potential value of traits associated with perennial habit in the development of new oil-seed crops for arid lands. A comparison of *Lesquerella fendleri* and *L. mendocina* subjected to water stress. J Arid Environ 47:373–386

Pradhan AK, Mukopadhyay A, Pental D (1991) Identification of the putative cytoplasmic donor of a cms system in *Brassica juncea*. Plant Breed 106:204–208

Prakash S, Bhat SR (2007) Contribution of wild crucifers in *Brassica* improvement: past accomplishment and future perspectives. Proc GCIRC 12th Int Rapeseed Congr 1:pp 213–215

Prakash S, Chopra VL (1988a) Introgression of resistance to pod shatter in *Brassica napus* from *Brassica juncea* through non-homologous recombination. Plant Breed 101:167–168

Prakash S, Chopra VL (1988b) Synthesis of alloplasmic *Brassica campestris* and induction of cytoplasmic male sterility. Plant Breed 101:253–255

Prakash S, Hinata K (1980) Taxonomy, cytogenetics and origin of crop Brassicas, a review. Opera Bot 55:1–57

Prakash S, Kirti PB, Bhat SR et al (1998) A *Moricandia arvensis*-based cytoplasmic male sterility and fertility restoration system in *Brassica juncea*. Theor Appl Genet 97:488–492

Prakash S, Ahuja I, Upreti HC et al (2001) Expression of male sterility in alloplasmic *Brassica juncea* with *Erucastrum canariense* cytoplasm and the development of a fertility restoration system. Plant Breed 120:479–482

Prasad MNV, Freitas HMO (2003) Metal hyperaccumulation in plants – Biodiversity prospecting for phytoremediation technology. Electron J Biotechnol http://www/ejbiotechnology.info/content/vol6/issue 3/index.html

Prem D, Gupta K, Sarkar G, Agnihotri A (2008) Activated charcoal induced high frequency microspore embryogenesis and efficient doubled haploid production in *Brassica juncea*. Plant Cell Tissue Organ Cult 93:269–282

Przedpelska E, Wierzbicka M (2007) *Arabidopsis arenosa* (Brassicaceae) from lead-zinc waste heap in southern Poland – a plant with high tolerance to heavy metals. Plant Soil 299: 43–53

Putnam DH, Budin JT, Field LA et al (1993) *Camelina*: a promising low-input oilseed. In: Janick J, Simon JE (eds) New crops: exploration, research, and commercialization, pp 314–322. Wiley, New York, NY

Ramachandran S, Buntin GD, All JN et al (1998) Diamondback moth (Lepidoptera: Plutellidae) resistance of *Brassica napus* and *B. oleracea* lines with differing leaf characteristics. J Econ Entomol 91:987–992

Ramsey AD, Ellis PR (1994) Resistance in wild brassicas to the cabbage whitefly, *Aleyrodes proletella*. ISHS Symposium on Brassicas, 9th Crucifer Genetics Workshop, Lisbon, Portugal, Abst.: p 32

Rana JS, Khokhar KS, Singh H (1995) Relative susceptibility of *Brassica* species to mustard aphid, *Lipaphis erysimi* (Kalt.). J Insect Sci 8:96–97

Rao GU, Shivanna KR (1996) Development of a new alloplasmic CMS *Brassica napus* in the cytoplasmic background of *Diplotaxis siifolia*. Cruciferae Newsl Eucarpia 18:68–69

Rao GU, Batra-Sarup V, Prakash S et al (1994) Development of new cytoplasmic male-sterility system in *Brassica juncea* through wide hybridization. Plant Breed 112:171–174

Rashid H, Toriyama K, Hinata K (1996) Transgenic plant production from leaf disc of *Moricandia arvensis* using *Agrobacterium tumefaciens*. Plant Cell Rep 15:799–803

Ravetta DA, Soriano A (1998) Alternatives for the development of new industrial crops for Patagonia. Ecol Austral 8:297–307

Razmjoo K, Toriyama K, Ishii R et al (1996) Photosynthetic properties of hybrids between *Diplotaxis muralis* DC, a C_3 species, and *Moricandia arvensis* (L.) DC, a C_3–C_4 intermediate species in Brassicaceae. Genes Genet Syst 71:189–192

Rehn F, Arbeiter A, Siemens J (2004) The gene RPB1 confers resistance of *Arabidopsis thaliana* to the obligate biotrophic parasite *Plasmodiophora brassicae*. Brassica 2004, 4th ISHS Symposium Brassicas/14th Crucifer Genet Workshop, Abst.: p 137

Renwick JAA (2002) The chemical world of crucivores: Lures, treats and traps. Entomol Exp Appl 104:35–42

Rimmer SR, van den Berg CJG (1992) Resistance of oilseed *Brassica* spp. to blackleg caused by *Leptosphaeria maculans*. Can J Plant Pathol 14:56–66

Riungu TC, McVetty PBE (2000) *Diplotaxis muralis* (mur) cytoplasmic male sterility system maintainer occurrence and frequency in summer rape. Can J Plant Sci 80:587–589

Riungu TC, McVetty PBE (2003a) Development and evaluation of *Diplotaxis muralis* (mur) cytoplasmic male sterility system in summer rape. Can J Plant Sci 83:261–269

Riungu TC, McVetty PBE (2003b) Inheritance of maintenance and restoration of the *Diplotaxis muralis* (mur) cytoplasmic male sterility system in summer rape. Can J Plant Sci 83: 515–518

Roberts KJ, Anderson RC (2001) Effect of garlic mustard (*Alliaria petiolata* (Beib.) Cavara and Grande)) extracts on plants and arbuscular mycorrhizal (AM) fungi. Am Midl Nat 146: 146–152

Rodman JE (1974) Systematics and evolution of the genus *Cakile* (Cruciferae). Contrib Gray Herb 205:3–146

Rodman JE (1976) Differentiation and migration of *Cakile* (Cruciferae): seed glucosinolate evidence. Syst Bot 1:137–148

Roy BA (1995) The breeding systems of six species of *Arabis* (Brassicaceae). Am J Bot 82: 869–877

Roy BA, Rieseberg LH (1989) Evidence for apomixis in *Arabis*. J Hered 80:506–508

Ruwandi A, Gillott C (1998) Resistance of *Brassica*, especially *B. juncea* (L.) Czern, genotypes to the diamondback moth, *Plutella xylostella* (L.). Crop Prot 17:85–94

Rylott EL, Metzlaff K, Rawsthorne S (1998) Developmental and environmental effects on the expression of the C_3–C_4 intermediate phenotype in *Moricandia arvensis*. Plant Physiol 118:1277–1284

Salisbury PA (1987) Blackleg resistance in weedy crucifers. Cruciferae Newsl Eucarpia 12:90

Salisbury PA (1989) Potential utilization of wild crucifer germplasm in oilseed *Brassica* breeding. Proc. 7th Austral Rapeseed Agron Breeders Workshop, pp 51–53

Salywon AM, Dierig DA, Rebman JP et al (2005) Evaluation of new *Lesquerella* and *Physaria* (Brassicaceae) oilseed germplasm. Am J Bot 92:53–62

Sánchez-Yélamo MD (1994) A chemosystematic survey of flavonoids in the Brassicinae: *Diplotaxis*. Bot J Linn Soc 115:9–18

Sánchez-Yélamo MD (2001) Study of flavonoid patterns in some species of *Erucastrum* (Brassicinae). Cruciferae Newsl Eucarpia 23:5–6

Sánchez-Yélamo MD (2004) Taxonomic relationships among *Erucastrum* and *Brassica* species based on flavonoid compounds. Cruciferae Newsl Eucarpia 25:13–14

Schranz ME, Dobes C, Koch MA et al (2005) Sexual reproduction, hybridization, apomixis, and polyploidization in the genus *Boechera* (Brassicaceae). Am J Bot 92:1797–1810

Schranz ME, Kantama L, de-Jong H et al (2006a) Asexual reproduction in a close relative of *Arabidopsis*: a genetic investigation of apomixis in *Boechera* (Brassicaceae). New Phytol 171:425–438

Schranz ME, Lysak MA, Mitchell-Olds T (2006b) The ABC's of comparative genomics in the *Brassicaceae*: building blocks of crucifer genomes. Trends Plant Sci 11:535–542

Schuster A, Friedt W (1998) Glucosinolate content and composition as parameters of quality of *Camelina* seed. Ind Crops Prod 7:297–302

Séguin-Swartz G, Warwick SI, Scarth R (1997) Cruciferae: compendium of trait genetics. Agric Agri-Food Res Branch Tech Bull No. 1997-3E and pdf file; http://www.brassica.info)

Sharma G, Kumar VD, Haque A et al (2002) *Brassica* coenospecies: a rich reservoir for genetic resistance to leaf spot caused by *Alternaria brassicae*. Euphytica 125:411–417

Sharma N, Cram D, Huebert T et al (2007) Exploiting the wild crucifer *Thlaspi arvense* to identify conserved and novel genes expressed during a plant's response to cold stress. Plant Mol Biol 63:171–184

Sharma TR, Singh BM (1992) Transfer of resistance to *Alternaria brassicae* in *Brassica juncea* through interspecific hybridization among *Brassica*. J Genet Breed 46:373–378

Shukla VKS, Dutta PC, Artz WE (2002) *Camelina* oil and its unusual cholesterol content. J Am Oil Chem Soc 79:965–969

Siemens J (2002) Interspecific hybridisation between wild relatives and *Brassica napus* to introduce new resistance traits into the oilseed rape gene pool. Czech J Genet Plant Breed 38:155–157

Siemens J, Köhn C, Sacristán MD (1995) Plant regeneration from root explants of *Matthiola incana* (L.) R.Br. and transformation attempts. Cruciferae Newsl Eucarpia 17:42–43

Sikdar SR, Chatterjee G, Das S et al (1987) Regeneration of plants from mesophyll protoplasts of the wild crucifer *Eruca sativa* Lam. Plant Cell Rep 8:722–725

Sikdar SR, Sengupta S, Das S et al (1990) Plant regeneration from mesophyll protoplasts of *Diplotaxis muralis*, a wild crucifer. Plant Cell Rep 6:486–489

Singh MP, Kolte SJ (1999) Differential reactions of various crucifer host species against isolates of *Peronospora parasitica*. J Mycol Plant Pathol 29:118–121

Singh R, Ellis PR, Pink DAC et al (1994) An investigation of the resistance to cabbage aphid in brassica species. Ann Appl Biol 125:457–465

Singh SP, Sachan GC (1997) Effect of different temperatures and host plants on the developmental behaviour of mustard sawfly, *Athalia proxima*. Indian J Ent 59:34–40

Skarjinskaia M, Svab Z, Maliga P (2003) Plastid transformation in *Lesquerella fendleri*, an oilseed Brassicaceae. Transgenic Res 12:115–122

Smisek A, Doucet C, Jones M et al (1998) Paraquat resistance in horseweed (*Conyza canadensis*) and Virginia pepperweed (*Lepidium virginicum*) from Essex County, Ontario. Weed Sci 46: 200–204

Sodhi YS, Chandra A, Verma JK et al (2006) A new cytoplasmic male sterility system for hybrid seed production in Indian oilseed mustard *Brassica juncea*. Theor Appl Genet 114:93–99

Song K, Osborn TC (1992) Polyphyletic origins of *Brassica napus*: new evidence based on organelle and nuclear RFLP analyses. Genome 35:992–1101

Song K, Osborn TC, Williams PH (1990) *Brassica* taxonomy based on nuclear restriction fragment length polymorphisms (RFLPs). 3. Genome relationships in *Brassica* and related genera and the origin of *B. oleracea* and *B. rapa* (syn. *campestris*). Theor Appl Genet 79:497–506

Sonntag K, Gramenz J (2004) In vitro regeneration system in *Crambe* via protoplast culture. Cruciferae Newsl Eucarpia 25:101–102

Sonntag K, Rudloff E (2001) Preliminary studies on genetic transformation of *Crambe abyssinica*. Cruciferae Newsl Eucarpia 23:33–34

Soroka J, Gugel R, Elliott R et al (2003) Resistance of crucifer species to insect pests. Proc GCIRC 11th Int Rapeseed Congr 3:pp 1031–1033

Specht CE, Diederichsen A (2001) Cruciferae. In: Hanelt P (ed) Mansfeld's encyclopedia of agricultural and horticultural crops vol 3, pp 1413–1481. Springer, Berlin

Stobbs LW, Stirling A (1990) Susceptibility of Ontario weed species to turnip mosaic virus. Can J Plant Pathol 12:255–262

Stoner K (1990) Glossy leaf wax and plant resistance to insects in *Brassica oleracea* under natural infestation. Environ Entomol 19:730–739

Sun WC, Pan QY, An XH et al (1991) *Brassica* and *Brassica*-related oilseed crops in Gansu, China. Proc GCIRC 8th Int Rapeseed Congr:pp 1130–1135

Sun WC, Yang Q, Zhang J et al (1999) Assessment on drought tolerance of *Eruca sativa* genotypes from northwestern China. Proc GCIRC 10th Int Rapeseed Congr, Contribution #628 on CD ROM; also available at http://www.regional.org.au/au/gcirc

Taji T, Seki M, Satou M et al (2004) Comparative genomics in salt tolerance between *Arabidopsis* and *Arabidopsis* -related halophyte salt cress using *Arabidopsis* microarray. Plant Physiol 135:1697–1709

Taskin KM, Turgut K, Scott RJ (2004) Apomictic development in *Arabis gunnisoniana*. Isr J Plant Sci 52:155–160

Tatsuzawa F, Saito N, Shinoda K et al (2006) Acylated cyanidin 3-sambubioside-5-glucosides in three garden plants of the Cruciferae. Phytochemistry 67:1287–1295

Tattersall A, Millam S (1999) Establishment and in vitro regeneration studies of the potential oil crop species *Camelina sativa*. Plant Cell Tissue Organ Cult 55:147–149

Tewari JP (1991) Current understanding of resistance to *Alternaria brassiceae* in Cruciferae. Proc GCIRC 8th Int Rapeseed Congr:pp 471–476

Tewari JP, Bansal VK, Tewari I et al (1996) Reactions of some wild and cultivated accessions of *Eruca* against *Leptosphaeria maculans*. Cruciferae Newsl Eucarpia 18:130–131

Thierfelder A, Hackenberg E, Nichterlein K et al (1991) Development of nematode-resistant rapeseed genotypes via interspecific hybridization. Proc GCIRC 8th Int Rapeseed Congr: pp 269–273

Thompson KF (1963) Resistance to the cabbage aphid (*Brevicoryne brassicae*) in brassica plants. Nature 198:209

Tian ZH, Meng JL (1998) Plant regeneration from cultured protoplasts of *Moricandia nitens*. Plant Cell Tissue Organ Cult 55:217–221

Turk MA, Tawaha AM (2003) Allelopathic effect of black mustard (*Brassica nigra* L.) on germination and growth of wild oat (*Avena fatua* L.). Crop Prot 22:673–677

U N (1935) Genome analysis in *Brassica* with special reference to the experimental formation of *B. napus* and peculiar mode of fertilization. Jpn J Bot 7:389–452

Ueno O, Bang SW, Wada Y et al (2003) Structural and biochemical dissection of photorespiration in hybrids differing in genome constitution between *Diplotaxis tenuifolia* (C3–C4) and radish (C3). Plant Physiol 132:1550–1559

Ulmer BJ, Dosdall LM (2006) Glucosinolate profile and oviposition behavior in relation to the susceptibilities of Brassicaceae to the cabbage seedpod weevil. Entomol Exp Appl 12: 203–213

Uprety DC, Shyam-Prakash P, Abrol YP et al (1995) Variability for photosynthesis in *Brassica* and allied genera. Indian J Plant Physiol 38:207–213

Vaughn SF, Berhow MA (2005) Glucosinolate hydrololysis products from various plant sources: pH effects, isolation and purification. Ind Crops Prod 21:193–202

Velasco L, Becker HC (2000) Variability for seed glucosinolates in a germplasm collection of the genus *Brassica*. Genet Resour Crop Evol 47:231–238

Velasco L, Goffman F, Becker HC (1998) Variability for the fatty acid composition of the seed oil in a germplam collection of the genus *Brassica*. Genet Resour Crop Evol 45:371–382

Veldhuis LJ, Hall LM, O'Donovan JT et al (2000) Metabolism-based resistance of a wild mustard (*Sinapis arvensis* L.) biotype to ethametsulfuron-methyl. J Agric Food Chem 48:2986–2990

Verma SS, Chiinnusarny V, Bansal KC (2008) A simplified floral dip method for transformation of *Brassica napus* and *B. carinata*. J Plant Biochem Biotechnol 17:197–200

Vinterhalter B, Savic J, Platisa J, Raspor M, Ninkovic S, Mitic N, Vinterhalter D (2008) Nickel tolerance and hyperaccumulation in shoot cultures regenerated from hairy root cultures of *Alyssum murale* Waldst. & Kit. Plant Cell Tissue Organ Cult 94:299–303

Vioque J, Pastor J, Vioque E (1990) Analysis of the fatty acids and sterols of the oils of four *Brassica* species. Lagascalia 16:95–103

Walker RL, Walker KC, Booth EJ (2003) Adaptation potential of the novel oilseed crop, Honesty (*Lunaria annua* L.), to the Scottish climate. Ind Crops Prod 18:7–15

Wan Z, Fu T, Tu J et al (2007) Genetic classification of a newly identified cytoplasmic male sterility *hau* CMS system in *Brassica napus* L. Proc. Proc GCIRC 12th Int Rapeseed Congr 1: pp 18–24

Wang J, Li Y, Liang C (2008a) Recovery of transgenic plants by pollen-mediated transformation in *Brassica juncea*. Transgenic Res 17:417–424

Wang W, Wang C, Huang BL, Huang B (2008b) *Agrobacterium tumefaciens*-mediated transformation of *Lesquerella fendleri* L., a potential new oil crop with rich lesquerolic acid. Plant Cell Tissue Organ Cult 92:165–171

Wang YP, Lan LF, Li XF et al (1999) A preliminary assessment of some wild cruciferous oil plant(s) in the western Sichuan of China and their utilization. Cruciferae Newsl Eucarpia 21: 29–30

Warwick SI, Al-Shehbaz IA (2006) Brassicaceae: chromosome number index and database on CD-Rom. Pl Syst Evol 259:237–248

Warwick SI, Black LD (1993) Molecular relationships in subtribe Brassicinae (Cruciferae, tribe Brassiceae). Can J Bot 71:906–918

Warwick SI, Gugel R (2003) Genetic variation in the *Crambe abyssinica* – *C. hispanica* – *C. glabrata* complex. Genet Resour Crop Evol 50:291–305

Warwick SI, Sauder C (2005) Phylogeny of tribe Brassiceae based on chloroplast restriction site polymorphisms and nuclear ribosomal internal transcribed spacer (ITS) and chloroplast *trn*L intron sequences. Can J Bot 83:467–483

Warwick SI, Francis A, Gugel RK (2009) Guide to wild germplasm of Brassica and allied crops (tribe Brassiceae, Brassicaceae. 3rd Ed. Agric Agri-food Res Branch Publ, Ottawa,ON, Canada. Contribution No. 991475 and pdf files at http://www.brassica.info

Warwick SI, Simard MJ, Légère A et al (2003) Hybridization between transgenic *Brassica napus* L. and its wild relatives: *Brassica rapa* L, *Raphanus raphanistrum* L, *Sinapis arvensis* L., and *Erucastrum gallicum* (Willd.) O. E. Schulz. Theor Appl Genet 107:528–539

Warwick SI, Sauder C, Beckie HJ (2005) Resistance in Canadian biotypes of wild mustard (*Sinapis arvensis* L.) to acetolactate synthase (ALS)-inhibiting herbicides. Weed Sci 53:631–639

Warwick SI, Francis A, Al-Shehbaz IA (2006a) Brassicaceae: species checklist and database on CD-Rom. Plant Syst Evol 259:249–258

Warwick SI, Gugel R, McDonald T et al (2006b) Genetic variation and agronomic potential of Ethiopian mustard (*Brassica carinata*) in western Canada. Genet Resour Crop Evol 53:297–312

Warwick SI, Gugel RK, Gómez-Campo C et al (2007a) Genetic variation in the *Eruca vesicaria* (L.) Cav. Plant Genet Resour Charact Util 5:142–153

Warwick SI, Sauder C, Al-Shehbaz IA et al (2007b) Phylogenetic relationships in the Brassicaceae tribes Anchonieae, Chorisporeae, Euclidieae, and Hesperideae based on nuclear ribosomal ITS DNA sequences. Ann Mo Bot Gard 94:56–78

Warwick SI, James T, Falk KC (2008a) AFLP-based molecular characterization of *Brassica rapa* and diversity in Canadian spring turnip rape cultivars. Plant Genet Resour Charact Util 6: 11–21

Warwick SI, Légère A, Simard MJ et al (2008b) Do escaped transgenes persist in nature? The case of an herbicide resistance transgene in a weedy *Brassica rapa* population. Mol Ecol 17: 1387–1395

Warwick SI, Sauder CA, Al-Shehbaz IA (2008c) Phylogenetic relationships in the tribe Alysseae (Brassicaceae) based on nuclear ribosomal ITS DNA sequences. Can J Bot 86:315–336

Wechter WP, Farnham MW, Smith JP et al (2007) Identification of resistance to peppery leaf spot among *Brassica juncea* and *Brassica rapa* plant introductions. HortScience 42: 1140–1143

Westman AL, Dickson MH (1998) Disease reaction to *Alternaria brassicicola* and *Xanthomonas campestris* pv. *campestris* in *Brassica nigra* and other weedy crucifers. Cruciferae Newsl Eucarpia 20:87–88

Westman AL, Kresovich S, Dickson MH (1999) Regional variation in *Brassica nigra* and other weedy crucifers for disease reaction to *Alternaria brassicicola* and *Xanthomonas campestris* pv. *campestris*. Euphytica 106:253–259

Williams PH, Pound GS (1963) Nature and inheritance of resistance to *Albugo candida* in radish. Phytopathology 53:1150–1154

Winter H, Gaertig S, Diestel A et al (1999) Blackleg resistance of different origin transferred into *Brassica napus*. Proc GCIRC 10th Int Rapeseed Congr., Contribution #593 on CD ROM; also available at http://www.regional.org.au/au/gcirc

Winter H, Snowdon RJ, Bellin U et al (2002) Blackleg resistance gene transfer into *Brassica napus* from related species. 13th Crucifer Genetics Workshop, Davis, California, USA. Abst.:p 15.

Winter H, Diestel A, Gärtig S et al (2003) Transfer of new blackleg resistances into oilseed rape. Proc GCIRC 11th Int Rapeseed Congr 1:19–21.

Wong CE, Li Y, Whitty BR et al (2005) Expressed sequence tags from the Yukon ecotype of *Thellungiella* reveal that gene expression in response to cold, drought and salinity shows little overlap. Plant Mol Biol 58:561–574

Xu TF, Zhang L, Sun XF et al (2004) Production and analysis of organic acids in hairy-root cultures of *Isatis indigotica* Fort. Biotechnol Appl Biochem 39:123

Yamagishi H (1998) Distribution and allelism of restorer genes for Ogura cytoplasmic male sterility in wild and cultivated radishes. Genes Genet Syst 73:79–83

Yang C, Mu X (2006) Allelopathic effects of companion weed *Descurainia sophia* on wheat. Yingyong Shengtai Xuebao 17:2389–2393

Yaniv Z, Elber Y, Zur M et al (1991) Differences in fatty acid composition of oils of wild cruciferae seed. Phytochemistry 30:841–843

Yaniv Z, Schafferman D, Amar Z (1998) Tradition, uses and biodiversity of rocket (*Eruca sativa*, Brassicaceae in Israel). Econ Bot 52:394–400

Yaniv Z, Schafferman D, Elber Y et al (1994) Evaluation of *Sinapis alba*, native to Israel, as a rich source of erucic acid in seed oil. J Industr Crops 2:137–142

Yaniv Z, Schafferman D, Zur M et al (1997) Evaluation of *Matthiola incana* as a source of omega-3-linolenic acid. Ind Crops Prod 6:285–289

Zhang L, Xu TF, Sun XF et al (2003) Factors influencing shoot regeneration from cotyledons of tetraploid *Isatis indigotica* Fort. In Vitro Cell Dev Biol Plant 39:459–462

Zhang T, Cao ZY, Wang XY (2005) Induction of somatic embryogenesis and plant regeneration from cotyledon and hypocotyl explants of *Eruca sativa* Mill. In Vitro Cell Dev Biol Plant 41:655–657

Zhang W, Fu Q, Dai X, Bao M (2008) The culture of isolated microspores of ornamental kale (*Brassica oleracea* var. *acephala*) and the importance of genotype to embryo regeneration. Sci Hort 117:69–72

Zheng HG, Hall JC (2001) Understanding auxinic herbicide resistance in wild mustard: physiological, biochemical, and molecular genetic approaches. Weed Sci 49:276–281

Zubr J (1997) Oil-seed crop: *Camelina sativa*. Ind Crops Prod 6:113–119

Zubr J, Matthäus B (2002) Effects of growth conditions on fatty acids and tocopherols in *Camelina sativa* oil. Ind Crops Prod 15:155–162

Chapter 3
The Non-coding Landscape of the Genome of *Arabidopsis thaliana*

Georg Haberer, Yu Wang, and Klaus F.X. Mayer

Abstract Sequencing and analysis of the genome of *Arabidopsis thaliana* has marked a milestone for plant genomics. Since analytical approaches to elucidate the gene function of the 27,000 genes and genome-wide functional studies, such as transcriptional characteristics, have revolutionized our functional and molecular understanding of a plants lifestyle. Classical target of such studies is the genic- and protein-coding space of the genome. However, the non-genic genomic space contains a multitude of important sequence elements that are essential for proper and regulated functioning of the genome. The analysis of cis elements as important information-containing sequence elements has been complicated by the experimental and algorithmic difficulties to detect them. For small RNAs and microRNAs hardly anything was known until turn of the century but since insights into their molecular nature and their important role in regulation of genome dynamics and biological processes has truly revolutionized not only plant biology. In this chapter we aim to give an overview about our current knowledge and analytical approaches to analyze these types of non-coding sequence elements as well as knowledge about their biological roles and molecular functioning.

Keywords Cis-regulatory element · Promoter · Non-coding RNA · *Arabidopsis thaliana*

Contents

3.1 Introduction 68
3.1.1 An Introduction to Cis Elements 69
3.1.2 The Core Promoter 71

K.F.X. Mayer (✉)
MIPS, Institute for Bioinformatics and Systems Biology, Helmholtz Center Munich, 85764 Neuherberg, Germany
e-mail: k.mayer@helmholtz-muenchen.de

Georg Haberer and Yu Wang contributed equally to the manuscript

R. Schmidt, I. Bancroft (eds.), *Genetics and Genomics of the Brassicaceae*, Plant Genetics and Genomics: Crops and Models 9, DOI 10.1007/978-1-4419-7118-0_3,

3.1.3 The Proximal and Distal Promoter 76
3.1.4 Detection of Cis-regulatory Elements 76
3.1.5 Cis Elements: Conclusion and Outlook 87
3.1.6 The Arabidopsis Non-coding RNA Landscape 88
3.1.7 Long ncRNAs 90
3.1.8 Small RNAs 91
3.1.9 Non-coding RNA: Conclusions 110
References 111

3.1 Introduction

Arabidopsis thaliana (Arabidopsis) is a genome reference model not only for Brassicaceae and dicotyledonous plants but also for the whole plant kingdom. Important aspects for genome references include a finished genome with high sequence quality standards, a vital and broad research community, and most importantly a highly curated and knowledge-enriched genome annotation and data availability. All these aspects hold for *Arabidopsis*. Structural genome annotation continues to be a critical aspect and for the time being is not a fully resolved problem. However, with the advent of next generation sequencing technologies and technologies such as RNA-Seq, algorithmic attacks based on intrinsic characteristics are now massively supported by extrinsic and experimental data that now can be generated economically. This now facilitates the calling of protein-coding genes and quality has improved significantly.

The classic focus of any genome analysis – not only in plants – is the structural detection and description of protein-coding genes and, for more complex genomes, the analysis and description of repeat content and structure. Intense maintenance and inclusion of new experimental data and experimental findings into the initial *Arabidopsis* genome annotation has developed *Arabidopsis* to a reference genome of the highest quality, with roughly 27,000 annotated protein-coding genes (http://arabidopsis.org). However, thus far only a fraction of the genome content has been functionally assigned. Even in the small and information-dense *Arabidopsis* genome only 33% of the genome is protein coding and 13% is annotated as repetitive elements of different classes (http://arabidopsis.org). Thus more than half of the genome constitutes sequences that have not been annotated and do not have assigned functional roles. Thus, among others, long known and described sequence elements like promoters and promoter elements thus far are not standard intrinsic parts of genome analysis and genome-annotation resources. In addition, over the last decade an exhaustive and Nobel Prize-winning tribe of new important genome outputs has been found and described. The exhaustive classes of non-coding transcripts include non-coding RNAs, microRNAs, and siRNAs (Lee et al. 1993, Hamilton and Baulcombe 1999, Wightman et al. 1993, Ruvkun et al. 2004). Many of them have been shown to play important and critical regulatory roles in a range of different biological processes underpinning their importance.

Genomic elements, promoter elements and, at least in part, non-coding RNAs, share small size and complications for their ab initio detection and description. Nevertheless these element types are important components for the functioning and regulation of the genome and thus are priority elements to analyze, describe, and include into next-generation genome data resources. In this chapter we aim to give an overview about characteristics, analytical and experimental approaches, and associated data resources for these two types of elements.

3.1.1 An Introduction to Cis Elements

In higher organisms, establishment of a master body plan, cell differentiation, growth, and proper interaction with a changing environment require tight regulation of gene and protein activities. Multiple-independent regulation can be achieved at any of the steps that lead from an inactive, chromatin-suppressed gene to its active protein product. Mechanisms include, among others, post-transcriptional controls like mRNA degradation and stability via miRNAs, or post-translational modifications of protein activity by, for example, membrane anchoring or phosphorylation. One of the first – and pivotal – mechanisms, however, is control of transcription: the mRNA production from a gene. Transcriptional control itself is accomplished by the interplay of a particular gene region, the promoter, and transcription factors binding to this region. Promoters are regions close to the transcription initiation site (TIS) and hence are often termed cis-active, while transcription factors (TF) are typically encoded by unlinked loci and thus are acting in trans.

In higher plants like *Arabidopsis*, more than 5% of the genes encode for transcription factors, underlining the relevance of a tight and manifold transcriptional control (Riechmann et al. 2000). TFs are composed of one or more DNA-binding domains (DBD), a trans-activating domain (TAD), and optionally, signal-sensing domains (SSD). The latter domains provide sites for activation or inactivation of TFs including ligand binding, phosphorylation, and protein–protein interactions sites. The trans-activating domain makes contact to the basal transcriptional machinery, either directly or by co-factor bridges (Fig. 3.1). DNA-binding domains attach the TF to specific DNA sequences of a promoter thereby enabling the TF to act in cis. Binding sites of TFs are called response (RE) or cis-regulatory elements (CRE). Promoters are typically composed of many CREs but not all sites have to be bound by a TF for an active promoter. In fact, diverse states of a promoter can be realized by varying sets of TFs. On the other hand, a single TF can therefore trigger different cellular responses in diverse promoter architectures dependent on the type of additionally recruited TFs. This combinatorial control of transcription enables manifold responses by a limited set of TFs and CREs. Promoters define the expression potential of a gene and are an integral part of the genomic blueprint. TFs in turn are interpreters of the regulatory code and realize a particular cellular state.

Promoters are in fact composite or hierarchical meta-structures divided into a core, proximal, and an optional distal promoter part. The core promoter surrounds the transcription start site (TSS) and has a small size of approximately 100 bp or

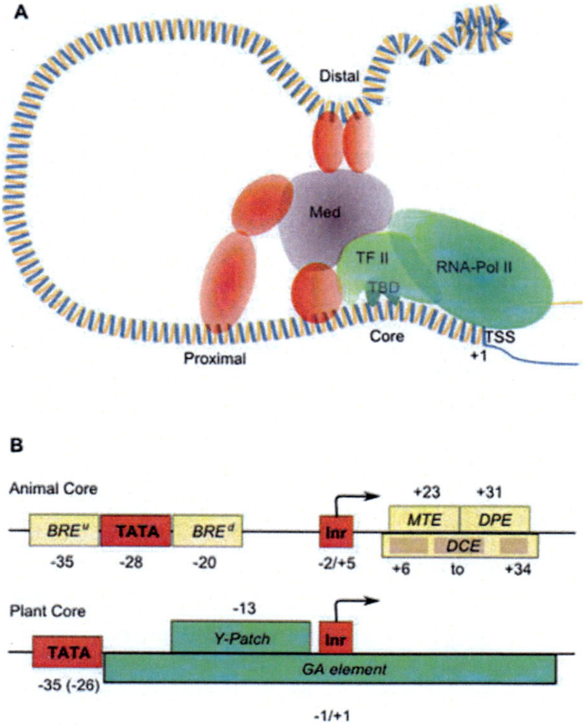

Fig. 3.1 Transcription initiation and core promoter architecture in eukaryotes. (**a**) The cartoon depicts a highly simplified view of transcription initiation in eukaryotes. The basal transcriptional machinery, including the general transcription factors (TF II; *green*) and including the TATA-box-binding protein (TBD), bind to specific sub-elements of the core promoters. Recruitment of RNA-Pol II (*green*) forms the pre-initiation complex (Kapranov et al. 2007) and enables basal transcription levels. Additional transcription factors (Thibaud-Nissen et al. 2006) bound in a sequence-specific manner cis-regulatory elements of the proximal and distal promoter. They modulate either by direct contacts or via a bridging complex, the Mediator (Med; *purple*), transcriptional output of the PIC. In (**b**), metazoan and plant core promoters are compared. Common sub-functions like the TATA-box and the initiator (Inr) are colored in *red*. Animal and plant-specific sub-elements of core promoters are shown in *yellow* and *green*, respectively. For further details, see text

less. The proximal promoter has a well-defined localization 5′-adjacent to the core promoter. In contrast, position of the distal promoter is less defined. It can extend to up to several thousand base pairs upstream of the TIS and even include functional elements in introns and downstream of the respective gene. Plant promoters generally comprise an upstream sequence of a few hundred up to several thousand base pairs in which a number of cis-regulatory modules (CRMs) are interspersed. CRMs themselves are clusters of combined multiple cis-active sites, the cis-regulatory, or response elements. CREs are the elementary building blocks and are short DNA motifs of a size of 5–20 bp that are bound by TFs in a sequence-specific manner. In most cases, binding tolerates a certain degree of sequence variability. The degree of variability itself may in turn differ between individual sites or nucleotides of the motif and can be deduced from several known binding sites. Conservation between binding sites, the motif specificity, can be quantitatively expressed as information content or relative entropy which is directly proportional to the TF-binding affinity for a motif (Berg and von Hippel 1987, Stormo and Fields 1998).

Assembly of the pre-initiation complex (PIC) at the core promoter, including general transcription factors (GTFs), co-factors, and a DNA-dependent RNA polymerase, provides a low-level basal transcription rate. Though required, it is generally not sufficient for the gene to be fully active. Instead, modulation of the basal transcription rate and specific spatio-temporal expression patterns are realized by binding of sequence-specific transcription factors to sites located in the distal and proximal promoter (Fig. 3.1). Other proteins lacking DNA-binding domains – namely methylases, chromatin re-modelers, and kinases – can provide additional input to this large regulatory complex. As a consequence, final transcriptional output results from a large, highly complex signal processor containing many dozens up to a few hundred components.

This part of the review will focus on our knowledge about cis-active elements in *Arabidopsis*. We will concentrate on RNA polymerase II-dependent transcription, as the majority of coding genes are driven by Pol II promoters. The second players, the trans-activators, will be described only if required to understand aspects of the cis-logic. Many other processes including silencing or chromatin remodeling significantly influence transcript levels. Though we are fully aware of this fact, even a superficial description of these mechanisms is out of scope of this chapter. In the next paragraphs, we will try and pull together the large number of reports of motif discoveries in this model plant. In the last decade, experimental as well as computational surveys have considerably enhanced our understanding of cis-regulation in *Arabidopsis* and both approaches will be discussed.

3.1.2 The Core Promoter

The core promoter serves as an assembly platform for the PIC including the general transcription factors TFIIA, TFIIB, TFIID, TFIIE, TFIIF, TFIIH, and the RNA polymerase II (Butler and Kadonaga 2002, Thomas and Chiang 2006). It determines both position and direction, i.e., start site and codogenic strand of transcription. Most of our knowledge of how core promoters interact with the general transcriptional machinery is derived from studies in yeast, fly, and human. Seven sub-sequences have been identified in animal core promoters (Fig. 3.1b). These elements are well delineated and are typically described by a consensus sequence. They have optimal distances and strand orientations to the transcription start site (TSS). Some elements like the downstream promoter (DPE) and the downstream core element (DCE) are mutually exclusive. However, in general, they operate in diverse compositions resulting in core promoter subtypes. Assembly of the PIC usually begins with binding of TFIID to the TATA-box, the Initiator Inr, and/or the DCE. TFIID itself consists of several components that recognize different core promoter elements: the TATA-box-binding protein (TBP) and a group of co-factors, the TBP-associated factors (TAFs) of which a subset binds the initiator, DPE and DCE sub-functions of the core promoter. After TFIID binding, the residual GTFs and RNA-Pol II form the final PIC via either the pre-assembled RNA-Pol II holoenzyme (two-component pathway) or a sequential assembly pathway.

As far as we know from similarity and domain searches, higher plants essentially possess the same protein complex for the basal transcriptional machinery as animals and yeast. Experimental studies about basal transcription initiation and its regulation in plants, however, lag far behind those available for fungal and animal systems. This is especially distressing as bioinformatic analysis revealed, besides conserved sub-elements, important differences in core promoter structures between plants and animals (Fig. 3.1b). Thanks to several large-scale studies, however, we have today detailed experimental knowledge of in vivo TSSs for the majority of *Arabidopsis* genes, including alternative initiation sites (Seki et al. 2002, Yamamoto et al. 2009, Tanaka et al. 2009). Sequence as well as distance and orientation of the TATA-box to the TSS seems to be well conserved between both kingdoms but plants apparently lack most other functional sub-elements including MTE, DPE, DCE, and the up- and downstream TFIIB-recognition elements (BREs) (Molina and Grotewold 2005, Yamamoto et al. 2007a). Similar to fungal and animal initiators, the plant Inr surrounds the TSS but differs in size and its consensus sequence. The general eukaryotic initiator has a pyrimidine-rich sequence, YYA(+1)nWYY, where nucleotide A represents the transcription start (+1). In a study aligning known higher plant promoters, two distinct initiators were initially proposed for dicots and monocots (Shahmuradov et al. 2003). The dicot initiator with a consensus sequence WnTCA(+1)W matches the 5′-part of the general consensus. In contrast, a consensus Ann(+1)CA has been reported for the monocot Inr that does not seem to be related to previously reported initiator sequences. However, the underlying data basis of 70 unrelated monocotyledonous promoters may have been too small to unveil and specify such weak signals like initiators. A recent study investigated local distribution of short sequences (LDSS) profiles in more than 10,000 core promoters of *A. thaliana* and *Oryza sativa* (Yamamoto et al. 2007b). The authors identified a common dimeric motif surrounding the TSS. The motif was named YR-rule according to the most prominent nucleotides at the –1,+1 positions of the TSS and can be interpreted as a less specific variant of the pyrimidine-rich initiator. Diverse variants or specifications of the YR-initiator are positively correlated with average gene expression levels (Yamamoto et al. 2009). Two hexameric positive LDSS profiles, the TATA-box and a $(CT)_n$/pyrimidine-rich hexamer, were found to be strikingly enriched at specific positions in regions referring to core promoters. Interestingly, only a subset of possible permutations of $(CT)_6$ hexamers exhibited a localized enrichment at a peak position at –13 bp distant to the TSS. The term Y-patch was coined for this group of pyrimidine-rich sequences. The Y-patch seems to be plant specific as mammalian (mouse and human) promoters were not associated with pyrimidine-rich sequences in a LDSS analysis (Yamamoto et al. 2007a). Position and sequence composition resembles a previously reported core element in filamentous fungi, the CT-block, that determines start and efficacy of transcription (Punt et al. 1990). Another TC-rich sequence, located in the core promoter of the tudor gene in *Drosophila*, serves as a nucleation site for a PIC which is specific for neuronal and germ cells (Holmes and Tjian 2000). Interestingly, the complex involves a *Drosophila*-specific variant of a TBP, the TATA-box-binding protein-related factor 1 (TRF1), suggesting that the Y-patch in plants may also be accessible for isoforms of TPB. In rice and *Arabidopsis*, the Y-patch is found in a range from –10 to –60 bp of the TSS, with a peak location at –13 bp between the TATA-box and the Inr element. Reverse complements of the TATA-box and the Y-patch showed no positional enrichment indicating that TATA-box and Y-patch are strongly strand sensitive. Approximately one-third and one-half of the *Arabidopsis* promoters contained a TATA-box and/or a Y-patch around their peak positions, respectively. Another study estimated about 70% of the *Arabidopsis* promoters lacking a TATA-box (Molina and Grotewold 2005). This large number is in agreement with studies in fly and human, for which 57% and 68–78% TATA-less promoters have been reported. In a comprehensive survey investigating alternative TSS in *Arabidopsis* and rice, different abundances for TATA-boxes between up- and downstream TSSs have been reported in both plants (Tanaka et al. 2009). In contrast to upstream start sites, downstream TSS did not show enriched relative entropies for this core promoter motif. Compared to upstream TSSs, general core promoter structures of alternative downstream TSS were highly distinct in their nucleotide compositions and in their GC- and AT-skews around the TSS (Tanaka et al. 2009). The eminent differences between the two types of alternative TSS suggest specialized mechanisms for the transcription from downstream TSSs. It is currently unknown what elements operate in and drive transcription from downstream core promoters. A third plant-specific core promoter element, the GA element, has been identified by LDSS analysis of more than 150,000 TSS tags that have been generated from cap-trapper transcript isolation followed by massively parallel signature sequencing (CT-MPSS) (Yamamoto et al. 2009). In contrast to TATA-box- and Y-patch-containing core promoters, GA elements are broad-type TSS clusters extending up to a few dozen base pairs up- and downstream from the Inr. Their location overlaps with preferred positions of a TATA-box and Y-patch and hence they are mutually exclusive to these elements. In *Arabidopsis*, GA elements constitute approximately 21% of all genic promoters. Besides these elements, LDSS analysis as well as word searches applying expectation-maximization and Gibbs-sampling identified a handful of oligonucleotide sequences between –50 and +50 bp of the TSS that showed sequence-specific positional enrichments in *Arabidopsis* (Molina and Grotewold 2005, Yamamoto et al. 2007b). Positional constraints and their widespread occurrence suggest that these elements may either be part of the core promoter or represent binding sites for widely used TFs in proximal promoters.

The existence of distinct core promoter elements and diverse promoter compositions by these elements raises questions about a ubiquitous and uniform basal transcription complex and the whole purpose of promoter subtypes in plants. In metazoans, several promoter core elements as well as multiple paralogues for the TFIID complex exist. The factors and assembly pathways forming a transcriptionally competent PIC are dependent on the promoter. Distinct combination of core promoter elements and their interaction with or utilization of distinct isoforms of the general transcriptional machinery have been shown to play a central role in gene-, tissue-, and lineage-specific expression in metazoans (Butler and Kadonaga 2002, Thomas and Chiang 2006, Holmes and Tjian 2000, Hochheimer et al. 2002). In plants, we have only limited knowledge about promoter subtypes and potential selective interactions with GTF variants. However, a few findings indicate that distinct compositions of PICs as well as functionally diverse core promoter subtypes exist in higher plants. In *Arabidopsis*, genes responsive to abiotic stresses generally have TATA-box-containing core promoters. In contrast, GA elements seem not enriched in regulated genes (Yamamoto et al. 2009). Similar to animals, the genomes of maize and *Arabidopsis* contain at least two TBPs (Gasch et al. 1990). In maize, quantitative expression differences in tissues have been observed for two TBPs indicating possible non-overlapping functions (Vogel et al. 1993). Gene duplications with potential sub-functionalization have also been observed for other components of TFIID (Lago et al. 2004). Similarity searches against yeast, fly, and human orthologs identified 15 TAFs in *Arabidopsis* of which seven (TAF 1,4,6,11,12,14,15) are present as gene duplications in the genome (Lago et al. 2004, Lawit et al. 2007). Yeast two-hybrid interaction mappings imply functionality for many of these bona fide TAFs and revealed highly conserved contact sites within the TFIID complex between plants and other eukaryotes (Lawit et al. 2007). A couple of mutants indicate that individual TAFs in *Arabidopsis* play a critical role to regulate specific biological processes. The stg1 (salt tolerance during germination 1) mutant encodes TAF10 and is more sensitive to salt for a hypomorphic mutant while it confers higher salt tolerance in neo- and hypermorphic alleles (Gao et al. 2006). Loss-of-function mutations in the TAF1 gene reduce histone H3 acetylation in light-responsive promoters, indicating that TAF1 is a co-activator specific for light-inducible promoters (Bertrand et al. 2005). An amorphic T-DNA insertion in TAF6 is gametophytic lethal (Lago et al. 2005). It specifically affects pollen tube growth but does not distort development of pollen or female gametophyte. The mutant phenotypes of the three TAFs suggest that control of a specific subset of genes is impaired by these basal transcription factors. Both TAF6 and TAF1 have duplicated copies in the genome, and the mutant phenotypes of TAF6, TAF1 and 1b strongly support that these copies are non-redundant. In metazoans, several TAFs make direct contact with core promoter elements including the initiator, DPE, and DCE. At present, we have a very poor understanding how GTFs interact with plant-specific components of core promoters. The observation of genetically distinct TAFs in *Arabidopsis*, however, supports the existence of specialized GTFs that target specific gene sets.

Variability of the PIC may be further extended by multiple forms of general co-activators. The mediator complex bridges regulatory factors and the basal transcriptional machinery thereby rendering it responsive to activators (Fig. 3.1a) (Sikorski and Buratowski 2009). The mediator is a large protein complex consisting of at least 24 subunits and it has been actually assigned to the GTFs by some authors. In *Arabidopsis*, this complex has been co-purified with RNA-Pol II (Backstrom et al. 2007). The *Arabidopsis* mediator contained both plant-specific and well-known conserved components of which some are duplicated in the genome. Two mutant phenotypes, *STRUWWELPETER (SWP)* and *PHYTOCHROME AND FLOWERING TIME 1 (PFT1)*, suggest roles in specific biological processes for at least some mediator components. SWP encodes the mediator subunit med14 and shows reduced cell numbers and increased cell sizes in leaves and shoots (Autran et al. 2002). However, no obvious phenotype has been observed in mutant roots. Genetic analysis of PFT1, the *Arabidopsis* mediator component Med25, identified this basal regulatory factor as negative regulator of phyB and as positive regulator of jasmonate signaling (Kidd et al. 2009).

In mammals, an additional type of basic promoter, the CpG islands, has been linked to transcriptionally active DNA. Promoter delineation and gene predictions have been successfully based on this general feature. CpG islands frequently encompass a region of several hundred base pairs and thus are larger than the core promoter described above. It is assumed that they act as several weak promoters from which transcription is initiated at several positions. In plants, computational searches have detected no CpG islands applying mammalian parameters. However, Rombauts et al. have denoted that the landscape of CpG/CpNpG islands differs substantially between promoters and other genetic elements (introns and coding exons) in *Arabidopsis* (Rombauts et al. 2003). The authors propose that the failure to detect CpG islands in plants may simply result from improper thresholds. Alternatively, it has been proposed that the previously described GA element may function as mammalian CpG counterparts in higher plants (Yamamoto et al. 2009). Its high frequency in *Arabidopsis* core promoters and its relatively broad extent compared to TATA-boxes and Y-patches argue for this interpretation. On the other hand, important differences to CpG islands exist. GA elements are not associated with housekeeping genes and do not seem to be targets for methylation. Thus, GA elements may be rather plant-specific realizations of broad-type core promoters. In summary, there is no experimental evidence for CpG islands in plants and this topic remains an open question.

Many animal promoters contain a common motif, the CAAT-box, which is bound by the transcriptional activator nuclear factor Y (NF-Y). The motif actually is located in the proximal promoter. Its widespread occurrence in approximately 25–30% of eukaryotic promoters and its position immediate upstream to the core promoter justify a discussion in this paragraph. CAAT-boxes in yeast and animals comprise a sequence of 13 bp with a consensus sequence CYYCCAATSRGMG. The pentamer core CCAAT has been shown to be essential for binding. Alignments detected a considerably smaller consensus CAAT in approximately 30% of angiosperm promoters (Shahmuradov et al. 2003). The shorter consensus suggests that the CAAT-box in plants has significantly evolved from NF-Y-binding sites in fungi and animals. NF-Y binds the motif as a trimer, and proximity to the core promoter facilitates direct activation of GTFs. In contrast to many other eukaryotes possessing only one or two copies for each of the three NF-Y subunits, *Arabidopsis* contains 10 NF-YA, 13 NF-YB, and 13 NF-YC homologs (Edwards et al. 1998, Siefers et al. 2009). Promoter:β-glucuronidase fusions exhibited marked tissue-specific expression patterns for most paralogous subunits (Siefers et al. 2009).

The distinct expression indicates functional diversification and specialization of NF-Y in *Arabidopsis*. Consistent with diversified roles of NF-Ys in *Arabidopsis*, individual subunits have functions in flowering time, embryo maturation, and meristem development.

In summary, biochemical and mutant analysis suggest that assembly and interaction of the basal transcriptional machinery at and with core promoters is far more than a simple on–off switch in plants. Comparable to metazoans, distinct core promoter subtypes and multiple isoforms of the general transcription apparatus may have evolved in plants to control transcriptional competency for distinct gene subsets. On this note, core promoters and promoter-selective assembly pathways of GTFs could play a role not only to toggle transcription but also to direct and specify expression patterns.

3.1.3 The Proximal and Distal Promoter

Proximal and distal promoters comprise regulatory sites that interact with specific TFs to modulate levels and specify patterns of transcriptional gene activity. Though proximal and distal promoters differ in their location to the core promoter, there is no clear-cut border between a proximal promoter and 5′-upstream sequences of a distal promoter. Thus, description of these two promoter parts is combined in the following paragraphs and special characteristics of each will be mentioned where appropriate.

Functional elements and modules are generally assumed to be scarcely distributed in proximal and – even more pronounced – in distal promoters and separated by long stretches of non-functional sequence. The sizes of these intermediate sequences and the sequence variability of binding sites render motif discovery a search for the needle in the haystack. It is not surprising that researchers came up with many ideas how to sort the sheep from the goat. In the following paragraphs, we will summarize and focus on results and solutions rather than on methodological or mathematical discussions. Readers further interested in the latter topics will, however, find sufficient suggestions in the cited literature to delve into these important aspects of motif discovery. Nevertheless, at the beginning of the section about "Detection of Unknown Sites" we will briefly introduce some concepts for motif models and detection approaches as they are used in computational biology.

3.1.4 Detection of Cis-regulatory Elements

3.1.4.1 Experimental Approaches

A large number of experimental methods have been designed to detect CREs in promoters. Since decades, promoter deletion studies, site-directed mutagenesis, gel-shift assays, and DNAseI footprints have been successfully deployed to experimentally uncover CREs in gene promoters of interest. Though these methods excel in resolution and accuracy, cost and complexity restrict their applicability to

Table 3.1 Web-based databases of plant CREs and promoters

Resource	Web address
AGRIS	http://arabidopsis.med.ohio-state.edu/
ATCOECIS	http://bioinformatics.psb.ugent.be/ATCOECIS/
AthaMap	http://www.athamap.de/
Athena	http://www.bioinformatics2.wsu.edu/cgi-bin/Athena/cgi/home.pl
PLACE	http://www.dna.affrc.go.jp/
PlantCare	http://bioinformatics.psb.ugent.be/webtools/plantcare/html/
PlantPAN	http://plantpan.mbc.nctu.edu.tw/
EPD	http://www.epd.isb-sib.ch/
ppDB	http://ppdb.gene.nagoya-u.ac.jp/cgi-bin/index.cgi
Transfac	http://www.gene-regulation.com/pub/databases.html

single case studies. Nevertheless, many dozens of experimentally characterized and verified CREs have been identified by various studies in plants. There are far too many studies on individual *Arabidopsis* promoters to be detailed in this chapter. Excellent overviews are provided by a couple of resources that collect and publish experimental TF-binding sites (Table 3.1). JASPER and the commercial resource TRANSFAC contain CREs from many species including plants (Vlieghe et al. 2006, Wingender et al. 2000). Both resources enable advanced motif representations like position-specific scoring matrices (see below). Several databases have specialized in plant CREs. PLACE and PlantCARE are collections of sites found both in mono- and dicots, while AthaMap, ATCOECIS, and AGRIS exclusively focus on *Arabidopsis* motifs (Higo et al. 1999, Lescot et al. 2002, Bulow et al. 2009, Davuluri et al. 2003, Vandepoele et al. 2009). AthaMap extends matches of known DNA motifs to all promoters regardless of their functional verification while PLACE and AGRIS report only literature verified sites. Users should be aware that some sites are based on observations of single instances and many motifs are redundant between collections and some even within a database. Nonetheless, these databases represent a highly valuable resource for wet-lab and computational biologists. In addition, researchers have access to a large number of bona fide plant promoters including experimental TSSs through the European Promoter Database EPD and the Plant Promoter Database ppDB (Cavin Perier et al. 1998, Yamamoto and Obokata 2008).

Given the difficulties of experimental promoter studies, many studies explored alternative methods to overcome their limitations. The ChIP-chip technology and its variants allow genome-wide identification of potential binding sites of a TF (Ren et al. 2000). Today, co-immunoprecipitation of bound DNA fragments and their computational analysis with alignment tools or motif finders allocate a comprehensive binding site collection for a particular TF of interest. Importantly, the retrieved fragments estimate motif variability and enrich for functional sites, i.e., sequences that are bound in vivo by a particular TF. A large-scale ChIP-on-chip experiment detected in vivo binding sites for most of the 141 known TFs in yeast (Lee et al. 2002). The number of TFs in the *Arabidopsis* genome is about an order of magnitude higher and transcriptional states are far more complex in multicellular organisms. Nevertheless, several surveys revealed highly promising results of this technology in *Arabidopsis*. The TGA family of transcriptional activators typically binds to the core consensus TGACG and comprises 10 members in *Arabidopsis*. Besides other TGA paralogues, TGA2 is required for the systemic acquired resistance (SAR) in *Arabidopsis*. A ChIP-chip experiment identified 51 targets of TGA (Thibaud-Nissen et al. 2006). The majorities of these regions contained an extended core TGACGTCA and were adjacent to SA inducible genes suggesting that this detected motif is a native TGA2-binding site. Several binding sites were located outside the presumed promoter and hence would have been missed in more traditional approaches. This highlights the importance of whole genome-binding studies. In vivo binding studies using a high-density oligomer microarray revealed more than 3,000 putative binding sites for Hy5, a central regulator of light responses in *Arabidopsis* (Lee et al. 2007). Targets were expectedly enriched for early light-responsive genes. Consistent with previous reports, motif searches identified a G-box (CACGTG) and its CG- and CA-hybrids (GACGTG and GACGTA, respectively) as the probable DNA motif that is bound by Hy5. For PHYTOCHROME INTERACTING FACTOR 3-LIKE5 (PIL5), binding analysis of a ChIP-chip experiment also revealed the G-box as main target in *Arabidopsis* promoters (Oh et al. 2009). Though TGA2 and PIL5 target the same CRE, the two transcriptional activators regulate different biological processes and targets. Additional factors and the genomic context likely contribute to specificity of G-boxes.

A recent study identified genome-wide binding sites of the MADS-box transcription factor SEPALLATA3 (SEP3) using chromatin immunoprecipitation followed by ultrahigh throughput sequencing (ChIP-SEQ) (Kaufmann et al. 2009). The MADS-box family regulates a large number of developmental processes in higher plants and the SEPALLATA subfamily is essential for floral organ identity. ChIP-SEQ identified more than 4,000 SEP3-binding sites and >3,000 SEP3 targets in the genome. Consistent with its role as transcriptional activator, most SEP3 binding sites were located in the proximity of genes. Sites were preferentially located in the proximal promoter. However, positional enrichment was also detected in regions directly downstream of the 3′-UTR and in intronic sequences. MADS-domain proteins bind to a motif consisting of two half sites, the CArG-box. Two consensus sequences are associated with the CArG-box, the SRF-type $CC[W]_6GG$, and the related MEF2-type $CTA[W]_4TAG$. Many plant MADS factors bind both types and intermediate motif forms [45,46]. Interestingly, strong SEP3-binding sites showed a preference for the SRF-type CArG-box indicating that particular MADS-box proteins may bind in vivo specific variants of the CArG motif (Kaufmann et al. 2009). Similar specificities have been reported for AGL15, an embryo-specific MADS-box TF in *Arabidopsis* that preferentially binds the MEF2 variant, and AGL1 and 2 (Huang et al. 1996, Tang and Perry 2003). Though many plant MADS-box proteins can bind to various forms of the CArG-box, binding preferences reported in vivo might explain – at least in part – how specificity is maintained in higher plants for this expanded and diversified gene family.

ChIP-chip and variations of it determine in vivo binding sites for a single known TF on genome scale. An inverse problem is the identification of unknown direct regulators binding to a given promoter. Recently, yeast one-hybrid system was successfully used to screen a collection of *Arabidopsis* transactivators that bind to regulatory regions of CCA1, a major component of the circadian clock (Pruneda-Paz et al. 2009). Screens of tiled promoter fragments isolated a TCP-like TF, the CCA1 HIKING EXPEDITION (CHE) protein. Subsequently it was shown that CHE binds to a site in the CCA1 promoter that matches the class-I TCP-binding sites with the consensus GGNNCCCAC. Yeast one-hybrid and ChIP-chip offer exciting opportunities to experimentally study interfaces of cis- and trans-regulatory elements of the *Arabidopsis* transcriptome at an unprecedented detail and scale.

3.1.4.2 Computational Approaches

Despite the progress in the genome-wide experimental detection of TF-binding sites, computational motif discovery is currently by far the most common and widely applied method to uncover CREs in *Arabidopsis* on a large scale. An overwhelming number of approaches and methods have been developed, and a complete listing of all protocols and their variations is out of scope of this chapter. Instead, this section will focus on applied CRE discovery in *Arabidopsis* and will only briefly survey fundamental methodological differences as far as they are relevant to the application part.

Detection of Known Sites

Computational motif discovery can be divided into two principal approaches: searches for instances of known TF-binding sites and de novo discovery of previously unknown motifs. Detection of known sites is straightforward and many motif databases or resources offer this feature. However, most motifs have low information content and searches will frequently detect non-functional sites by chance. To minimize false positives, many researchers back up genome-wide computational discoveries of known motifs by additional supporting evidences.

One possible way to corroborate results uses the observation that many motifs act in concert. Many responses to the phytohormone abscisic acid (ABA) are communicated by at least two distinct CREs, the ABA response element (ABRE) with a ACGT core, and the CE3 coupling element (CE) with a GMCGCGTGKC consensus. The two single CREs were combined in a genome-wide scan for ABA and stress-responsive promoters in *Arabidopsis* (Zhang et al. 2005). Application of an ABRE-CE module provided higher sensitivities and selectivities compared to single instances. Starting from several module variants, an accurate as well as degenerate ABRE-CE module was determined. In addition, the identified ABA- and stress-inducible genes were supported by RT-PCR experiments and/or previously published microarray results. Another study demonstrated that stepwise refinements by in silico and experimental evidences can reveal cell-type-specific promoters and genes (Won et al. 2009). Genome-wide searches for a previously defined root-hair specific element (RHE) with a consensus of wwmnTGnn(n)yGCACGw were subsequently filtered by analysis of root-hair defective mutants and promoter assays and identified a total of 19 root-hair specific genes in *Arabidopsis* (Kim et al. 2006). Interestingly, phylogenetic footprinting revealed that RHE is a conserved regulatory module in angiosperms despite major differences in the root architecture between mono- and dicotyledonous plants (Kim et al. 2006). Mapping of known motifs and integration to experimental expression data, local enrichment and/or co-occurrence analysis have been integrated to various extents by a couple of *Arabidopsis* tools for CRE detection, including AthaMAP, the *A. thaliana* regulatory element analyzer AtREA, Athena and PlantPAN, and the commercial product TRANSFAC (Wingender et al. 2000, Bulow et al. 2009, Choudhury and Lahiri 2008, O'Connor et al. 2005, Chang et al. 2008).

Detection of Unknown Sites

In comparison to mapping known motif sites, de novo motif discovery is a challenging problem and an important topic of active research. De novo detections can be classified by their implementation how motifs are modeled and by the sequence information on which motif evidence is based on (Fig. 3.2b). Biologists typically use two different representations to characterize cis-regulatory elements, either word-based models, including exact k-mers and consensus sequences, or position-specific scoring or weight matrices (PSSM;PWM) (Fig. 3.2a). Apparently, exact words are the most simple motif representation and cannot capture any motif variability. Consensus sequences display nucleotide variability within a motif in a qualitative manner using the full IUPAC alphabet while PSSMs report background adjusted probabilities for each nucleotide and position in a motif. The latter probabilistic sequence models are based on experimental observations that (most) positions in a binding site contribute independently to the binding energy (Berg and von Hippel

Fig. 3.2 Motif representations and classification of motif finders. **a** Shows two popular data models for motifs for a higher plant TATA-box, taken from (Molina and Grotewold 2005). The matrix denotes for each position the frequency of the nucleotides A,C,G,T. Position-specific scoring matrices (PSSMs) are background normalized motif models of the frequency matrix. The sequence logo in the *top* panel, here shown for a uniform background of equal nucleotide probabilities, graphically display PSSM contents. In the *lower line* of the matrix, the consensus sequence of the respective frequency matrix is shown. Clearly, consensus sequences represent motif variation with less precision. Exact words like TATAAATA cannot measure any variability. **b** Popular motif finders can be classified according to the type of evidence feature detected motifs are based on (*vertical separator*) and the type of motif models is searched for (*horizontal separator*). Alignments of evolutionary conserved regions implicitly create frequency matrices while phylogenetic footprints of FootPrinter and network-level conservation are words or consensus sequences. An analogous bipartition is also given for several tools discovering over-represented motifs in functionally related, co-expressed data sets. Seeder is a mixture, starting with exact seeds which are subsequently extended to PSSMs. Three programs, PhyloCon, PhyME, and PhyloGibbs, tightly and algorithmically integrate two sources of evidences. RSAT is a toolbox of several separate methods that can in part communicate with each other

A

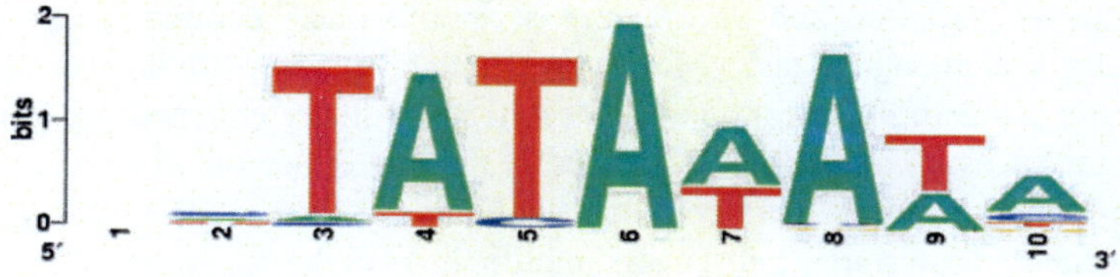

	-2	-1	1	2	3	4	5	6	7	8
A	0.248	0.243	0.058	0.917	0.000	0.998	0.493	0.943	0.417	0.655
C	0.242	0.434	0.030	0.000	0.049	0.001	0.000	0.001	0.020	0.093
G	0.126	0.111	0.000	0.001	0.000	0.000	0.000	0.038	0.000	0.100
T	0.384	0.213	0.911	0.083	0.951	0.001	0.507	0.018	0.563	0.153
	t	c	T	A	T	A	T/A	A	T/A	A

B

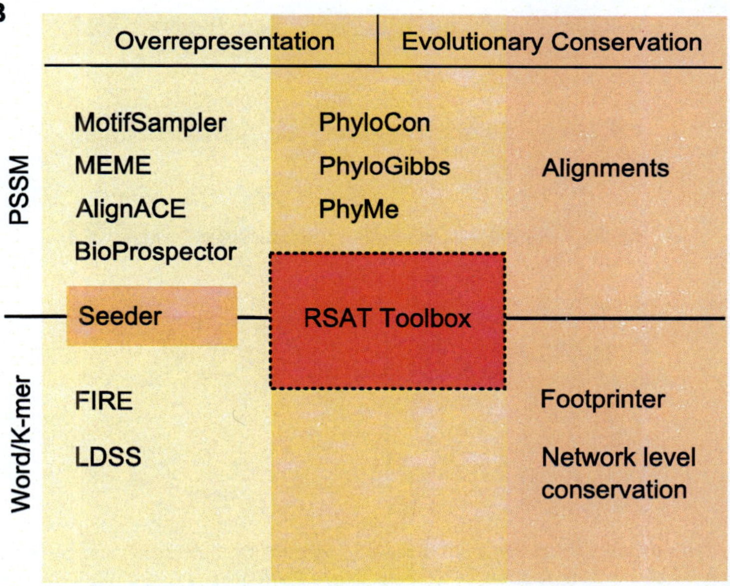

Fig. 3.2 (contd.)

1987, Stormo and Fields 1998). There are advantages and disadvantages to any of the three representations. PSSMs are embedded in a statistical framework and are the most precise and advanced representations of a DNA motif. In the frequent case of only a few known binding sites, however, they may easily be too specific or may represent a rare binding site variant, thus lacking sensitivity. In addition, the larger number of free parameters for a PSSM can impede de novo motif discovery. In contrast, word and consensus searches have no statistical power per se and may fail to report functional variants of a motif. On the other hand, these representations have a well-defined search space for which very fast and exact look-up algorithms exist and global optimality is guaranteed.

Sequence variation and small sizes of motifs make computational detection of CREs a challenging task. Most motif finders therefore reduce the complexity of the problem by selecting sequences that are likely enriched for (functional) motif instances. Grouping of sequences is thereby either based on a shared feature or function or by their common evolutionary history. The latter approach, phylogenetic footprinting, discovers motifs in orthologous promoter sequences (Fig. 3.3a). Orthologues in closely related species are assumed to participate in evolutionary conserved transcriptional networks. Functional sequences responsible for expression patterns are conserved due to selective constraints while non-functional promoter parts have diverged in their sequence. Hence, CREs can be

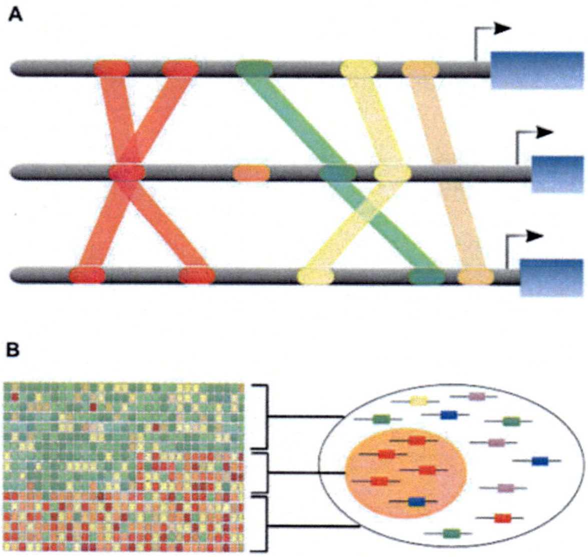

Fig. 3.3 Information sources for motif discovery. Two major information sources are shown to illustrate principles of computational CRE detection. **a** Phylogenetic footprinting compares evolutionary conserved promoter or upstream regions between closely related species. Non-functional sequences in gray diverge while functional sequences like CREs (*colored boxes*) are under selective constraints and can be uncovered as conserved sub-elements. The figure also displays some limitations of this approach. Alignments may miss motifs due to partially duplicated/deleted copies (Thibaud-Nissen et al. 2006), small scale inversions (*green* versus *yellow*) or underscore motifs that are completely deleted in one or more species (*light brown*). Phylogenetic footprinting will also not detect any sites which have been evolved de novo in one species (*orange*). **b** In the second approach, motifs are detected in functionally related gene sets, typically a cluster of co-expressed genes. Motifs responsible for a specific expression pattern (*red boxes*) can be discovered by their over-representation. Inappropriate experimental setups, technical noise in microarrays, and post-transcriptional regulation of mRNA levels can result in inhomogeneous clusters. Depending on the impact of such effects, co-expressed clusters may not even come close to approximate co-regulated genes

detected by multiple alignments or searches for words common to promoters of all or most species. The second type of sequence information, groups of genes selected by shared features or functions, is typically collected from microarray experiments. In this approach, co-expressed genes are assumed to be co-regulated. CREs are identified in promoter sequences of co-expressed genes by their statistical over-representation.

Both types of sequence information have limitations. Phylogenetic footprinting may miss motifs due to sequence inversions, repeats or multiple motif copies, and diversification of orthologs, for example by species–specific gene duplications. Microarrays measure steady-state mRNA levels that are affected by cis-active transcriptional control as well as by post-transcriptional mechanisms regulating mRNA stability. Some experimental setups may violate the assumption of a simple regulatory circuitry causing the observed cluster structures. Mixed cell and tissue samples, long-term or pleiotropic effects triggering transcription factor cascades, may considerably complicate the interpretation of expression clusters. Thus, co-expressed genes may not always be transcriptionally co-regulated. A general problem for all motif detection approaches is false or missing annotations of promoter and gene boundaries that may hide and even corrupt motif discoveries. In *Arabidopsis*, however, long-lasting manual curations, the availability of whole genome tiling arrays, and detailed large transcript mappings have provided researchers with a high-quality gene annotation (Seki et al. 2002, Yamamoto et al. 2009, Tanaka et al. 2009, Matsui et al. 2008, Poole 2007).

Current algorithms for motif discovery differ in the type of employed sequence information and motif models they search for. In fact, all four possible combinations of either orthologous or co-expressed sequence information with word-based or sequence-probabilistic motif models have been realized (Fig. 3.2b). Popular and widely applied tools are MEME, AlignACE, Seeder, MotifSampler, and FootPrinter (Bailey et al. 2009, Hughes et al. 2000, Thijs et al. 2001, Blanchette and Tompa 2003, Fauteux et al. 2008). Several other algorithms, like PhyloCon, PhyloGibbs, and PhyME, integrate two sources of information, co-expression, and conservation (Wang and Stormo 2003, Siddharthan et al. 2005, Sinha et al. 2004). They report equal and higher performances but require more comprehensive input data. These approaches started and reflect recent popular tendencies to integrate several evidences for motif predictions to enhance sensitivity and specificity. Additional support for motifs is thereby derived from cis-regulatory modules, positional constraints, evolutionary conservation, and/or co-expression. Other tools like regulatory sequence analysis tool (RSAT) bundle independent methods to investigate and to integrate multiple evidences (Thomas-Chollier et al. 2008, Defrance et al. 2008).

Assessments of prediction accuracy for the diverse motif finders revealed that there is no single best method for motif discovery (Prakash and Tompa 2005, Tompa et al. 2005). Each program has its own strength and weakness for the wide range of existing diverse biological questions and problems. To address a particular problem, experimental setup or expected result type, researchers frequently develop custom-made solutions and adjustments. For example, in a microarray survey of transcriptional responses to acquired systemic response (SAR) in *Arabidopsis*, promoters containing a W-box were shown to play a central role (Maleck et al. 2000). The W-box with a consensus TGACY is bound by the large family of WRKY transcription factors. The degenerated pentamer with an invariant TGAC core is expected to occur by chance approximately every few hundred base pairs. Thus, almost every *Arabidopsis* promoter contains a W-box motif in its proximal promoter. The low information content raises the question as to the specificity of this motif. However, modeling motif copy number per promoter under a binomial distribution revealed a co-expression group for which induced SAR tightly correlated with statistically increased copy numbers in their promoters. Serial wiring of multiple short binding sites is one way to render promoters specific. In *Arabidopsis*, a specific transcriptional program responsive to SAR is linked to increased occurrences of WRKY-binding sites in the proximal promoter. A simple count of present and absent boxes would have missed the underlying motif.

The SAR experiment above was the beginning of a series of microarray-driven CRE discoveries in *Arabidopsis*. In *Arabidopsis* thousands of individual large-scale expression measurements cover a wide range of biological conditions and today provide the most comprehensive "expression atlas" for a higher plant (Craigon et al. 2004, Schmid et al. 2005). From the beginning, CRE discovery has gone along with identification of co-expressed groups and transcriptional changes. Two exemplary studies illustrate how novel cis-active motifs are extracted from such analysis. Transcriptional responses of *Arabidopsis* plants that have been exposed to mechanical wounding was monitored by whole genome microarrays (Walley et al. 2007). Importantly, the study focused on very rapid responses within 5 min for CRE detection. This experimental setup thereby avoids complications from superimposed later effects, for example, transcription factors that have been activated or silenced by immediate early response. A rapid stress-response element (RSRE) with the consensus CGCGTT has been identified in promoters early responsive to wounding. Interestingly, promoter constructs consisting of several RSRE copies conferred in vivo responses to multiple stress conditions, including a wide range of biological elicitors and cold stress. The motif likely has a key role in the coordination of responses to a broad range of environmental stresses. Several-large-scale microarray surveys that investigated reprogramming of the *Arabidopsis* transcriptome by diurnal rhythms have identified central circadian clock regulatory elements. A nonameric oligonucleotide AAAATATCT strictly correlated with a set of co-expressed genes that showed a periodic peak expression at the subjective end of day (Harmer et al. 2000). Transgenic reporter lines and site-directed mutagenesis demonstrated that this evening element (EE) was sufficient to confer clock regulation. Remarkably, sequence of EE was highly invariant in the evening-phased co-expression group. EE is bound by the central clock component, CCA1, which also binds to the closely related CCA1-binding site (CBS; AAAAAATCT). Interestingly, an integrative survey of three circadian clock microarray studies revealed over-representation of EE in evening-phased genes but no correlation of CBS to any of the analyzed clock phases. Though CBS does not seem phase specific, it is generally enriched in promoters of clock-regulated genes. These findings nicely illustrate that even highly similar cis-active sequences may sharply differentiate transcriptional programs. However, artificial promoters with multiple CBS sites were able to drive reporter gene activity in an evening-phase manner. Though these findings may represent artifacts, further studies – especially on in vivo concentration and binding affinities of CCA1 – are required to understand the suggested functional difference of CBS and EE. To date, multiple/diverse CREs that confer transcript accumulations in specific phases of the circadian clock have been uncovered in *Arabidopsis* by bioinformatic analysis of microarray data. A network discovery pipeline revealed *Arabidopsis* transcriptome changes which were induced by cycled light and temperature conditions. Data serialization and word-based pattern discovery detected in time-shifted data sets a couple of sequences which were enriched in co-expressed clusters (Michael et al. 2008). A G-box-like motif and the newly discovered morning element (ME; AACCACGAAAAT) correlated with elevated expression at dawn and/or early day, while a GATA element was enriched in afternoon and evening phases (Michael et al. 2008, Covington et al. 2008). The GATA element is proposed to act synergistic to the previously described EE (Harmer and Kay 2005). Other elements – a Telobox-like motif with the sequence AAACCC, a hexameric sequence AGCCC, and the protein box element (PBX; ATGGGCC) – were highly abundant in promoters of midnight-phased genes (Michael et al. 2008, Covington et al. 2008).

The identification of clock responsive CREs exemplifies the successful alliance of large-scale expression data analysis and computational motif discovery. Complementary to these approaches, large-scale or genome-wide motif searches have detected numerous candidate motifs in *Arabidopsis* by their evolutionary conservation in orthologous promoters. Following a general trend in CRE discovery, many studies support detected phylogenetic footprints by additional evidences. A large-scale analysis in *Brassica oleracea* and *A. thaliana* exploited alignments conserved in orthologous upstream sequences and co-expression information to derive numerous candidate TF-binding sites (Haberer et al. 2006). Motifs were subsequently obtained from clustering phylogenetic footprints. Approximately 50% of known motifs have been retrieved suggesting a large coverage and sensitivity of the approach. In addition, occurrence of many motifs correlated with particular biological processes. Maybe the most noteworthy feature of the *Brassica–Arabidopsis* comparison was the use of unordered *Brassica* genomic survey sequences with a low genome coverage of about 0.25-fold. Next generation sequencing will likely produce numerous such views of low coverage on plant genomes. The study provides a preview of how large numbers of incomplete genomes in combination with a few high-quality genomes can be utilized for motif discovery (Haberer et al. 2006).

The previous trial applied PhyloCon that algorithmically interlaces co-expression and sequence conservation. There are also large-scale surveys that successfully employed evolutionary conservation either as selector or classifier. A motif search comparing finished genomes of *Arabidopsis* and Poplar combined searches for over-represented PSSMs in co-expressed genes and a subsequent evolutionary filter that was based on network-level conservation (Vandepoele et al. 2006). Contrary to phylogenetic footprinting that detects conserved sites on gene-by-gene basis, the latter approach operates at a system level and computes significant motifs by their increased global co-conservation between two genomes (Elemento et al. 2007). The co-expression analysis of a large set of microarray experiments in *Arabidopsis* followed by a network-level filter based on intersection of *Arabidopsis* and poplar upstream regions identified 80 motifs and 139 regulatory modules consisting of two or three single elements. Interestingly, constrained positional relations were observed for some modules indicating compact and cooperative architectures of these modules in *Arabidopsis*. Most of these modules were novel and significantly associated to specific biological processes. In a follow-up study, it has been shown that co-expression alone may perform poorly to infer gene functions and that integration of CRE information significantly enhances the functional transfer (Vandepoele et al. 2009). In the previously mentioned LDSS analysis, positional constrained CRE candidates were further evaluated by their conservation in rice (Yamamoto et al. 2007). Approximately 30% of *Arabidopsis* LDSS octamers overlapped with rice LDSS motifs by sequence similarity and positional enrichment. Localized distribution of many motifs that were enriched in the first 100 bp upstream of the TSS elegantly demonstrated the concept of proximal promoters in *Arabidopsis* and higher plants.

Many large (and small)-scale motif discoveries are carried out on promoter or upstream sequence comprising a sequence size between 1 or 2 kb. In *Arabidopsis*, this restriction is reasonable in the light of a mean gene size of about 2 kb and a compact genome size of approximately 125 Mb. It should be noted, however, that distal promoters may easily extend such spans and that some parts of a gene promoter may also be localized downstream or in intronic sequences. Exclusion of these regions may miss important regulatory units. Correct spatio-temporal expression of AGAMOUS (AG), a MADS-box gene involved on flower development, requires intragenic control sequences (Sieburth and Meyerowitz 1997). Fusions of the 5′-upstream sequence including either the first or third coding exon with a GUS reporter localized these control elements to a 3 kb long intron. Two transcription factors, LEAFY and WUSCHEL, have been shown to bind to sites within this second AG intron (Busch et al. 1999, Lohmann et al. 2001). Phylogenetic shadowing, a variant of phylogenetic footprinting that employs a large array of closely related sequences to delineate motifs (Boffelli et al. 2003), revealed three regions of low divergence in this large intron in 29 species of the Brassicaceae family (Hong et al. 2003). These regions were conserved in other dicot species and harbor several CREs, including two CArG-boxes, three WUSCHEL-binding sites, and two CAAT boxes. Mutations in these three regions and in the identified CRE candidates showed functionality and diverse effects on expression levels and spatio-temporal patterns of AG transcipts (Hong et al. 2003). Intronic CREs may be a general characteristic of MADS-box genes in *Arabidopsis* as, for example, AGL6 and AGL13 also harbor cis-regulatory motifs in their large intron (Schauer et al. 2009). The four motifs identified by phylogenetic shadowing are conserved at least in the Brassicaceae and are specific for AGL6 and AGL13. Searches in size-restricted upstream regions may also overlook motifs in distal promoters. Phylogenetic footprints of the CRABS CLAW (CRC) promoter region of three Brassicaceae have identified five distinct conserved regulatory modules that span a region of more than 3 kb upstream to the TSS (Lee et al. 2005). Staggered promoter deletion constructs with a GUS reporter demonstrated specific roles of all of the five modules for CRC expression in valves, sepals, pedicels, and nectaries. In mammals and flies, long-range enhancers can function more than 100 kb distant to a TSS. In *Arabidopsis*, distal promoters are generally assumed to be more compact and long-range enhancers have not yet been described. However, activation tagging has demonstrated that cis elements can in principle convey gene activation in vivo through long distances in *Arabidopsis* (Ren et al. 2004). Ectopic overexpression of the C2H2 zinc finger TELOMERASE ACTIVATOR 1 (TAC1) in the tac1-1D allele resulted from the insertion of 35S promoter more than 78 kb downstream of the *TAC1* genomic location. The inserted enhancer thereby activated specifically *TAC1* transcription but not any of the intermediate genes. A variant of phylogenetic footprinting, intragenomic footprinting, discovers conserved non-coding sequences (CNS) by their retention in promoters of genes that have been duplicated in the most recent alpha genome duplication of *Arabidopsis* (Haberer et al. 2004). As many higher plant genomes have undergone multiple segmental or whole genome duplications, intragenomic footprinting has been proposed as an alternative to classic phylogenetic footprinting that is applied to orthologous sequences. A large-scale study applying intragenomic footprinting to alpha-duplicated pairs identified almost 15,000 *Arabidopsis* CNS of which many represent regulatory motifs (Thomas et al. 2007). A total of 254 genes, so-called Bigfoot genes, were characterized by notably CNS-rich upstream regions (Thomas et al. 2007, Freeling et al. 2007). The median distance of CNS to the start codon was more than 3 kb, indicating large distal promoters for these genes. In an enumerative survey of heptamer words, variants of the core G-box were by far the most enriched motifs associated with "Bigfoot" promoters. Though functionality of these motifs remains to be shown, motifs and enhancers operating on long distances may have been overseen in *Arabidopsis* since experimental design thus far was not directed toward such motifs. The previous study, however, also suggests that only a small portion of genes may be under control of long-range enhancers, consistent with compact sizes of intergenic regions and experimental promoter studies in *Arabidopsis*.

3.1.5 Cis Elements: Conclusion and Outlook

Arabidopsis has immensely broadened our understanding of virtually any aspect of plant biology and the impact of *Arabidopsis* research cannot be overstated. This is equally true for the study of transcriptional control. Extensive resources have been generated, including a microarray compendium comprising hundreds of conditions and databases and tools which allow researchers detailed computer-aided studies of their gene-of-interest. To date, a vast number of single case studies as well as large-scale experimental and computational approaches have discovered several hundred CREs. Many new developments may lead to a quantum leap for motif discovery. The rapidly increasing number of finished plant genomes and the stunning progress in sequencing provides a rich and large-scale source to uncover evolutionary conserved binding sites by phylogenetic footprinting and shadowing. In addition, routine application of the ChIP-chip and ChIP-seq technology, one-hybrid assays and advances in computational methods hold out the promise of a once complete assortment of cis-active motifs in *Arabidopsis*.

Yet, important questions still remain unanswered. Our knowledge about core promoters in *Arabidopsis* is still in its infancy. So far we do not understand the function of different core promoter subtypes nor do we understand the function of plant-specific core elements. Striking differences between up- and downstream TSS and their core promoters are only understood in a descriptive manner. And even with the accomplishment of a complete set of cis-motifs, we will not be able to understand transcriptional control of biological processes. TF-binding sites are the smallest components of the cis-regulatory code in the genome. Motifs are combined to modules, modules to promoters, and promoters make up regulatory networks that correspond to biological processes. Moreover, integration of promoter models with chromatin remodeling which has a large impact on transcriptional competency and with mechanisms regulating RNA stability will be required to understand and correctly interpret observed steady-state transcript levels. Deciphering of the regulatory code has just begun.

3.1.6 The Arabidopsis Non-coding RNA Landscape

Non-coding RNAs (ncRNA) recently emerged as an important landmark on the genomic landscape of eukaryotes. The widespread involvement of ncRNAs in various cellular activities has not been appreciated until recently (reviewed in Prasanth and Spector 2007, Mercer et al. 2009). It is the frenetic pace of technology advancement that enables us to probe the cellular universe in an unmatched resolution. Many new discoveries have been made in the last years about ncRNAs in the model organism *A. thaliana*, which are the subjects of this section.

In contrast to messenger RNA (mRNA), non-coding RNA (ncRNA) refers to an RNA molecule in a cell that does not encode for protein. Roughly, ncRNAs can be functionally classified into two categories (Fig. 3.4). The first category consists of so-called housekeeping ncRNAs (Szymanski and Barciszewski 2002) or infrastructural ncRNAs (Mattick and Makunin 2006). These are ubiquitously expressed at stable levels in different cellular contexts and are involved in the viability of the cell. The representatives of these ncRNAs are rRNA, tRNA, snoRNA, vault RNA (vRNA, Kickhoefer et al. 2003), etc. (the upper half of Fig. 3.4). The functions of infrastructural RNAs have been well studied (reviewed in Storz 2002, Mattick and Makunin 2006) and therefore not discussed further here. Many of these ncRNA families can be found in Rfam, a large collection of multiple sequence alignments and covariance models for ncRNAs (Gardner et al. 2009).

The second category is regulatory ncRNA, including small RNAs, macro-ncRNAs, antisense transcripts, and many another ncRNAs that are regulators at transcriptional or post-transcriptional level (the second half of Fig. 3.4). Small

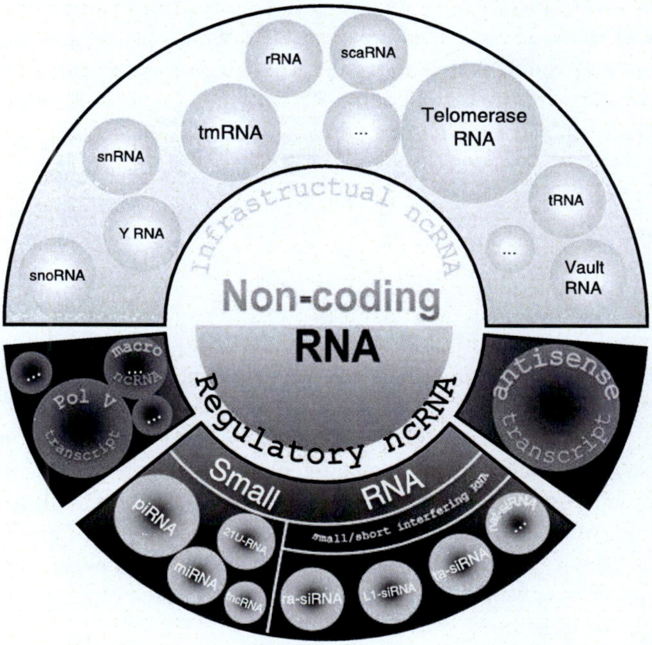

Fig. 3.4 Classification of non-coding RNA families. Non-coding RNA families can be roughly classified into the following functional categories: infrastructural RNAs (represented by the *upper half* of the figure) and regulatory RNAs (represented by the *lower half* of the figure). This section focuses on recently discovered non-coding regulatory RNAs, which include small RNAs, antisense transcripts, macro-ncRNA, Pol V transcripts, and other (Hammell et al. 2000) ncRNAs. Small RNAs consist of piwi-associated RNA, microRNA, short interfering RNA, 21U RNA, and tiny non-coding RNA. Abbreviations: L1-siRNA, long interspersed element-1-specific siRNAs; macro-ncRNA, macro-non-coding RNA; miRNA, microRNA; nat-siRNA, nature antisense transcripts derived short interfering RNA; piRNA, piwi-associated RNA; rRNA, ribosomal RNA; ra-siRNA, repeat-associated short interfering RNA; scaRNA, small Cajal body-specific RNA; siRNA, short interfering RNA; snRNAs, small nuclear RNAs; snoRNAs, small nucleolar RNAs; tRNA, transfer RNA; ta-siRNA, trans-acting short interfering RNA; tmRNA, transfer-messenger RNA; tncRNA, tiny non-coding RNA;Y RNA, Y chromosome RNA

regulatory RNAs have been identified in various organisms, where they generally function by base-pairing with complementary sequences in other RNAs or DNA. In eukaryotes, small RNAs are less than ~35 nucleotides (nt) long, which can be subdivided into several classes according to origins. Small/short interfering RNA (siRNA) is derived from a double-strand RNA, either from inverted repeats, transposons, or natural antisense transcripts, which is then named as repeat-associated siRNA (ra-siRNA), long interspersed element-1 specific siRNAs (L1-siRNA), and natural antisense transcript siRNA (nat-siRNA), respectively. Other small RNAs include microRNA (miRNA), piwi-associate RNA (piRNA), 21U RNA, and tiny ncRNA (tncRNA) (reviewed in Ghildiyal and Zamore 2009). We will focus upon these ncRNAs and their functions.

In the following sections, we are going to take readers to a genomic landscape of non-coding RNAs in the model organism *A. thaliana*, to see how much we have learned about non-coding RNAs so far and what kind of biological process they are involved. We start with long non-coding RNAs which include antisense transcripts and Pol V transcripts in *Arabidopsis*.

3.1.7 Long ncRNAs

3.1.7.1 Natural Antisense Transcripts

In plants, whole genome tiling array experiments not only identified many un-annotated protein-coding genes, but also discovered that about 7,600 (Yamada et al. 2003) or 12,090 (Stolc et al. 2005) genes in *A. thaliana* and about 24% genes in *O. sativa* (Li et al. 2007) had antisense transcripts. These results suggest that anti-sense expression is widely spread in plant genomes. Many *A. thaliana* genes reside on the genome in an overlapping fashion, forming cis-NATs pairs. Based on TAIR 6.0 annotation, Jin et al. (2008) tested expression patterns of 1,057 cis-NAT pairs in *Arabidopsis*. They found a subset of cis-NAT pairs displayed negatively correlated expression profiles as well as inverse differential expression changes under at least one experimental condition.

3.1.7.2 Transcripts Generated by RNA Polymerase V

Arabidopsis harbors two RNA polymerases, Pol VI and Pol V (Wierzbicki et al. 2008), which are implicated in small RNA-mediated chromatin-based gene silencing. In *Arabidopsis*, transcription of intergenic non-coding regions by Pol V promotes heterochromatin formation and silencing of nearby genes, (Wierzbicki et al. 2008). Pol V transcripts were detected in six intergenic non-coding regions by RT-PCR by comparing low abundance transcripts of wild-type plants with those of Pol V mutants. Pol V transcripts are at least 200 nt long and do not contain poly A tails. They physically interact with AGO4 and possibly guide AGO4 to target loci through base-paring with associated siRNAs (Wierzbicki et al. 2009). Consequently, repressive epigenetic modifications of target regions are induced and in addition nearby genes and retroelements are transcriptionally silenced as well. Although Pol V is specific to flowering plants, it is tempting to speculate that the pervasive intergenic transcripts detected in eukaryotic genomes (Kapranov et al. 2007) may play a similar role in modifying chromatin structure.

In general, the functions of the majority of long ncRNAs are not elucidated very well. There might be many new discoveries remaining to be made on the functional role of long ncRNAs. In contrast, in the last decade, small RNAs, especially miRNAs, has been shown to be crucial players for plant development, defense against viral and bacterial infection, and stress responses. They have been exhaustively studied in *Arabidopsis* and in following we aim to highlight and summarize knowledge accumulated about the molecular function of this class of non-coding genetic elements.

3.1.8 Small RNAs

Small RNAs (sRNAs) are 17–35 nt long regulatory non-coding RNAs that include microRNAs (miRNAs), small/short interfering RNAs (siRNAs), and Piwi-interacting RNAs (piRNAs). These small molecules regulate gene expression, control transposon activities, maintain genome integrity, and help to protect plant cells from invading viruses. Although sRNAs have been already extensively reviewed, recent progresses in the deep sequencing technology (review in Meyers et al. 2006) have revealed many new species of sRNAs. We will summarize the latest finding about sRNAs and, particularly, present miRNA and several other small regulatory RNAs such as salt-induced natural antisense transcript siRNA (nat-siRNAs) and trans-acting siRNA (ta-siRNA).

3.1.8.1 MicroRNAs

MicroRNAs (miRNAs) are about 21 nt long endogenous non-coding RNAs that regulate a large number of protein-coding genes at the post-transcriptional level. miRNA was first discovered in *Caenorhabditis elegans* in 1992 (Lee et al. 1993) while the first plant miRNA was reported about 10 years later (Llave et al. 2002a, b, Reinhart et al. 2002). Many miRNAs are deeply conserved in each kingdom, for instance, let-7 is conserved in worms, insects, and mammals while miR156 is conserved in land plants, from moss to flowering plants. Currently the microRNA database – miRBase database (release 14) – contains 10,883 miRNA entries. The total number of human miRNA is at least ~800, which makes it one of the largest gene families in human genome. Numerous miRNA prediction methods have been developed based on the secondary structure prediction, phylogenetic conservation, and thermodynamic stability of stem loops (Fig. 3.5, a representative of conserved miRNAs: ath-miR164c). While these computational methods are indispensable in whole genome miRNA gene scanning, the predicted miRNA candidates have to pass "Expression criteria," which requires the detection of the predicted miRNA sequences by experimental techniques, either Northern blotting, cDNA cloning, microarray analysis, or deep sequencing (Ambros et al. 2003a, Meyers et al. 2008).

miRNA Biogenesis

The majority of animal miRNAs are located in the introns of protein-coding genes (Kim and Kim 2007). These miRNAs are most likely co-transcribed with their "host gene" and the processing of intronic miRNAs does not interfere with splicing events (Kim and Kim 2007). Many intergenic animal miRNAs form polycistronic clusters (Altuvia et al. 2005) facilitating coordinated expression. The long primary miRNA transcripts (pri-miRNAs) are transcribed by RNA polymerase II (Lee et al. 2004) or by RNA polymerase III (in human, Borchert et al. 2006). The transcription of miRNA genes is likely controlled by cis-regulatory elements in the promoter regions of miRNA genes.

Animal pri-miRNAs typically consist of a stem loop with 5′- and 3′-flanking segments. Pri-miRNAs are processed by RNase III like enzymes Drosha and its double-strand RNA-binding partner DGCR8 (in vertebrates). DGCR8 recognizes

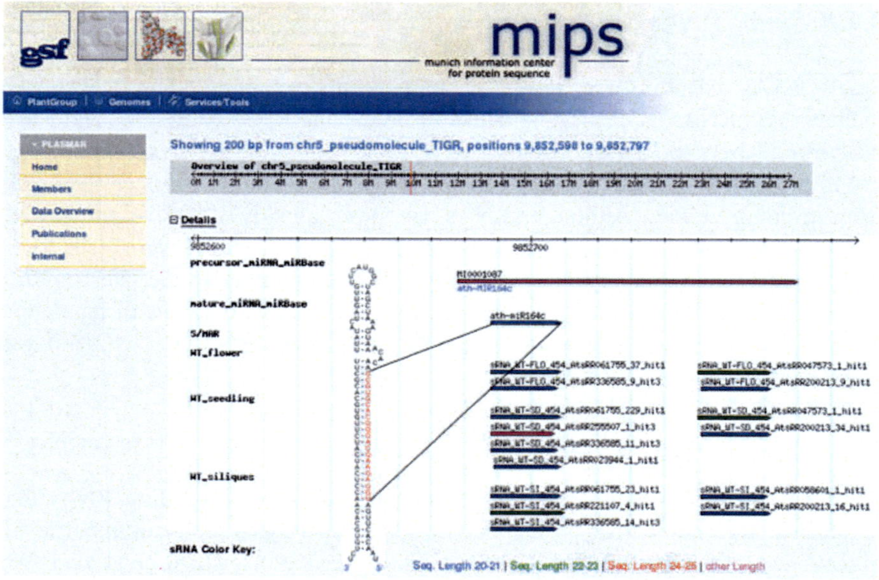

Fig. 3.5 A genome browser view of miR164c. The small RNA genome browser is hosted at MIPS: http://mips.helmholtz-muenchen.de/cgi-bin/proj/plant/gbrowse/gbrowse/siRNA/. The precursor, ath-MIR164d, and the mature miRNA, ath-miR164c, are shown in the figure, together with deep sequencing of sRNA profile in flowers, siliques, and seedlings of *A. thaliana* (accession: Columbia, Rajagopalan et al. 2006). Note that not only mature miRNA sequences but also miRNA* sequences were detected in different organ/tissues. The miRNA precursor has the typical hairpin structure with the mature miRNA sequence on one arm of the hairpin. Sequences are colored according to different lengths

the junction between the flanking segments and the stem loop. The Drosha-DGCR8 complex is therefore anchored precisely at the junction, placing Drosha on the stem-loop, ~11 bp away from the junction, where Drosha cleaves the stem loop to liberate a precursor of miRNA (pre-miRNA) (Han et al. 2006). Interestingly, some of miRNA-containing introns are so short that they encode only precursor of miRNA. This type of miRNA is termed "miRtrons" since their biogenesis does not require Drosha cleavage (Berezikov et al. 2007, Okamura et al. 2007, Ruby et al. 2007). Animal pre-miRNAs are typically ~70 nt long with a distinctive stem loop structure, whose one stem contains miRNA and the other contains miRNA*. Pre-miRNAs are exported to cytoplasm by the nuclear transport factor, exportin-5. In the cytoplasm, Dicer, another RNase III type protein, processes pre-miRNAs to generate ~22 nt miRNA:miRNA* duplexes with 2 nt 3′ overhangs. Mature miRNAs are then selected by an unknown mechanism and incorporated into the RNA-induced silencing complex (RISC). RISC is the central element of all RNA silencing pathways. It consists of at least one Argonaute protein and a small non-coding RNA. Depending on the incorporated small RNA, RISC can cleave mRNA, block protein synthesis, and introduce transcriptional gene silencing. miRNA*s were generally believed to be degraded quickly. However, recent reports find that in many tissues miRNA*s are constantly detected, suggesting miRNA*s may be functionally active in certain circumstances. In flies, miRNA*s are sorted into RNA interference pathway by being loaded into AGO2 (Ghildiyal et al. 2010, Okamura et al. 2009).

Many plant miRNA genes (reviewed in Voinnet 2009) are found in intergenic regions and are believed to represent independent transcriptional units. Recently, 63 TSSs of *Arabidopsis* miRNAs have been identified by 5′ RACE (Xie et al. 2005). A follow-up computational study predicted many transcription factor binding sites, some of which are over-represented in the plant miRNA promoter regions as compared to promoters of protein-coding genes (Megraw et al. 2006). Just like normal protein-coding genes, plant miRNA genes can have multiple transcription start sites (Song et al. 2007), exon–intron structure, and alternative 3′ ends. It has been shown that some of ath-miR164 primary transcripts have two exons (Nikovics et al. 2006).

Similar to animal miRNA biogenesis, plant pri-miRNA is processed by DICER-Like 1 enzyme (DCL1) and other proteins including a double-stranded RNA (dsRNA) binding-domain protein, Hyponastic Leaves1 (HYL1) and a zinc-finger-domain protein Serrate (SE). In *Arabidopsis,* DCL1 and HYL1 colocalize in discrete nuclear bodies, called D-bodies. Fang and Spector (2007) showed that an introduced pri-miRNA was recruited to D-bodies, indicating that D-bodies are involved in pri-miRNA processing in the cell nucleus. HYL1 may assist DCL1 to precisely process pri-miRNAs into pre-miRNAs. Some other unidentified factors may also participate in pri-miRNA processing in vivo. Comparing to ~70 nt animal pre-miRNAs, plant pre-miRNAs are in general larger. The average lengths of *A. thaliana* and *O. sativa* pre-miRNAs are 170 and 145 nt, respectively. It is believed that in *Arabidopsis*, DCL1 is mainly responsible for processing pre-miRNAs into miRNA:miRNA* duplexes. There are two exceptions (ath-miR822 and ath-miR839) where DCL4 takes the dicing role in miRNA maturation, implying a second biogenesis pathway of miRNAs. After miRNA:miRNA* duplexes are diced from pre-miRNAs, the duplexes are methylated on the ribose of the last nucleotide by the miRNA methyltransferase HEN1 and transferred to the cytoplasm by the HASTY (HST) transporter. In the cytoplasm, miRNA is mainly incorporated into the RISC along with ARGONAUTE1 (AGO1) or ARGONAUTE type proteins.

miRNA and Target Interaction

miRNAs recognize their targets by simple nucleotide base pairing. In animals, miRNAs mostly guide RISCs to the 3′-UTRs of target mRNAs and probably repress translation or destabilize mRNAs. In plants, miRNAs usually have near-perfect complementarities to their targets and guide RISCs to cleave the corresponding transcripts (reviewed by Brodersen and Voinnet 2009, Voinnet 2009). Recent studies have shown that translational repression is a general mode of plant miRNA action (Brodersen et al. 2008).

Target prediction has been fairly straightforward for plant miRNAs due to the near-perfect complementarities to their targets (Zhang 2005). Conversely, target prediction for animal miRNAs is intriguingly complicated because of the limited complementarity between animal miRNAs and their targets. Based on limited experimental data, the first generation miRNA target prediction programs used evolutionary conservation as a powerful tool to assist miRNA-binding sites identification. Although limited in numbers most experimentally verified binding sites show a strong complementarity bias to the 5′ ends of animal miRNAs. The perfect complementarity to the region covering seven nucleotides starting from either the first or the second nucleotide at the 5′ end of a miRNA has been argued as sufficient to identify a target site. However, a recent experiment seriously challenged the so-called "perfect seed matching rule." Didiano and Hobert (2006) showed that G:U base pairing is tolerated in the "seed" region of the *lsy-6* miRNA interaction with its in vivo target *cog-1*, and that 6- to 8-base pair perfect seed pairing is not a generally reliable predictor for an interaction of *lsy-6* with a 3′-UTR. Furthermore, they demonstrated that the predicted target sites of 13 *lsy-6* target genes are not functional in their sensor system. MicroRNA *lsy-6* can only interact with its target site in specific 3′-UTR contexts. Obviously, mRNA has its own secondary structure which may interfere with the binding of miRNAs to their target sites. Two non-sequence-specific contextual features beyond miRNA target sites were later proved to be critical determinants of miRNA-mediated 3′-UTR regulation (Didiano and Hobert 2008). The emerging picture of the limited complementarities between plant miRNAs and functional target sites (Brodersen et al. 2008, Dugas and Bartel 2008) argues toward the need to further develop prediction tools for plant miRNA targets.

Thus far no high-throughput experimental methods for miRNA target identification are available, which has motivated the continuous development of computational development of target prediction programs (reviewed in Alexiou et al. 2009). Current efforts of target verification have been focused on bringing in the cellular context of miRNAs, targets, and effector proteins such as Argonautes (Easow et al. 2007, Hammell et al. 2008, Chi et al. 2009, Hong et al. 2009).

Arabidopsis miRNA Families

There are 190 miRNA genes in *A. thaliana*, consisting of 91 families. Many *Arabidopsis* miRNA families are deeply conserved in land plants (Table 3.2, Axtell and Bartel 2005, Sunkar and Jagadeeswaran 2008). Some miRNA families expanded by gene duplication (Allen et al. 2004, Maher et al. 2006) with the largest one, ath-miR169, consisting of 14 family members. One scenario for de novo genesis of microRNA genes is by inverted duplication of target gene sequences. Several only recently evolved miRNA genes (only found in *Arabidopsis*) and other small RNA-generating loci were shown to contain extended similarities at the arms on each side of their respective foldback precursors to some protein-coding sequences, suggesting that these miRNAs and small RNA loci evolved from inverted gene duplication (Allen et al. 2004).

A large fraction of *Arabidopsis* conserved miRNA targets are transcription factors involved in the regulation of plant development. Therefore, conserved miRNAs participate in a range of plant developmental processes (Table 3.2). Some

Table 3.2 miRNA families in *A. thaliana* and their conservation in the plant kingdom

miRNA family	No. loci in *Arabidopsis*	Target family	Conservation							Function
			Eudicots	Monocots	Magnoliids	Gymnosperms	Ferns	Lycopods	Mosses	
miR156/ miR157	12	SPL	*	*	*	*	*	*	*	Retain the juvenile status of young plants
miR158	2	PPR								
miR159	6	MYB	*	*	*	*		*		Desensitize hormone signaling during germination
miR160	3	ARF	*	*	*	*	*	*	*	Regulate auxin signal transduction during plant development
miR161	1	PPR								
miR162	2	DCL	*	*						Maintain DCL1 at a funcationally sufficient level
miR163	1	SAMT								
miR164	3	NAC	*	*	*	*				Regulate plant development and aging-induced cell death
miR165/ miR166	9	HD-ZIPIII	*	*	*	*	*	*	*	Regulate leaf development, vascular patterning, and SAM function

Table 3.2 (continued)

miRNA family	No. loci in *Arabidopsis*	Target family	Conservation							Function
			Eudicots	Monocots	Magnoliids	Gymnosperms	Ferns	Lycopods	Mosses	
miR167	4	ARF	*	*	*	*			*	Regulate adventitious rooting, ovule, and anther development
miR168	2	AGO	*	*	*	*				Maintain AGO1 homeostasis
miR169	14	NF-Y	*	*	*	*				Involve in nutrient, drought response
miR170/ miR171	4	SCL	*	*	*	*	*	*	*	Involve in cell differentiation
miR172	5	AP2	*	*	*	*	*			Regulate flowering time and floral organ identity
miR173	1	TAS1,TAS2								Generate ta-siRNAs which target PPR proteins
miR319		TCP	*	*	*			*	*	Coordinate leaf development and physiology
miR390/ miR391	3	TAS3	*	*	*		*		*	Generate ta-siRNAs which target ARF3,4, involve in developmental timing and patterning
miR393	2	TIR1/AFB, bHLH	*	*		*				Involve in antibacterial resistance

Table 3.2 (continued)

miRNA family	No. loci in *Arabidopsis*	Target family	Conservation							Function
			Eudicots	Monocots	Magnoliids	Gymnosperms	Ferns	Lycopods	Mosses	
miR394	2	F-Box	*	*	*	*				
miR395	6	APS, AST	*	*						Regulate sulfur assimilation
miR396	2	GRF	*	*	*	*	*	*		Regulate leaf growth and development
miR397	2	LAC	*	*						Regulate nonessential copper proteins
miR398	3	CSD, CytC oxidase	*	*	*	*				Regulate copper homeostasis
miR399	6	E2-UBC	*	*						Regulate phosphate homeostasis
miR400	1	PPR								
miR402	1	ROS1-Like								
miR403	1	AGO	*							
miR408	1	LAC, PLC	*	*		*		*	*	Regulate nonessential copper proteins
miR447	3	2-PGK								
miR472	1	CC-NBS-LRR								

Table 3.2 (continued)

miRNA family	No. loci in *Arabidopsis*	Target family	Conservation							Function
			Eudicots	Monocots	Magnoliids	Gymnosperms	Ferns	Lycopods	Mosses	
miR771	1	eIF-2								
miR773	1	MET2								
miR774	1	F-box								
miR775	1	GT								
miR776	1	PK								
miR777	1	CIP4.1-like								
miR778	1	SUVH								
miR780	1	CHX								
miR823	1	CMT3								
miR824	1	MADS-Box								
miR827	1	SPX								
miR828	1	MYB,TAS4								
miR842	1	JR/MBP								
miR844	1	PK								
miR846	1	JR/MBP								

Table 3.2 (continued)

miRNA family	No. loci in *Arabidopsis*	Target family	Conservation							Function
			Eudicots	Monocots	Magnoliids	Gymnosperms	Ferns	Lycopods	Mosses	
miR856	1	CHX								
miR857	1	LAC								Regulate nonessential copper proteins
miR858	1	MYB								
miR859	1	F-Box								

Only miRNAs with known targets are listed in the table. For each family, the conservation is shown for Eudicots, Monocots, Magnoliids, Gymnosperms, Ferns, Lycopods, and Mosses. Note that the indicated conservation is rather conservative since the miRNA profiling might not be carried out in many plant species. One interesting observation is that conserved miRNA families tend to have more family members while non-conserved miRNAs tend to be singletons miRNAs are also implicated in stress response and antiviral or antibacterial defense. In the following key findings for individual miRNA families will be discussed and summarized.

miR156/miR157

Twelve loci in *A. thaliana* encode members of the miR156/miR157 family. All miR156 miRNAs are 20 nt while three miRNA 157 loci are 21 nt and miR157d is 20 nt long. Nevertheless, all mature miR156/miR157 sequences are almost identical, with only one nucleotide difference. Since they most likely regulate the same set of targets, SPL (SQUAMOSA promoter-binding-like) transcription factors, miR156 and miR157 are grouped into one family.

miR156 is highly expressed early in shoot development and the expression level of miRNA 156 decreases over time. High level of miR156 in young plants is associated with low SPL levels. Consequently, when miR156 abundance declines, SPL levels increase. miR156 targets 10 different SPL genes, including SPL3/4/5, SPL9/15, SPL2/10/11, and SPL6/SPL13a/b. SPL genes function in distinct pathways to promote different adult phase vegetative traits and flowering. By simultaneously repressing these SPL involved pathways, miR156 acts as a master regulator to retain the juvenile status of the young plants (Wang et al. 2009, Wu et al. 2009).

miR159

miR159 is induced by abscisic acid and drought treatment during germination in an ABI3 (abscisic acid-insensitive 3)-dependent manner (Reyes and Chua 2007). All three members of the miR159 family are expressed during germination. miR159 targets several MYB transcription factors. During seedling stress response, ABA-induced miR159 cleaves MYB33 and MYB101 transcripts to desensitize hormone signaling. In miR159a and miR159b double mutants, MYB33 and MYB65 expression levels are elevated, but the expression of five other GAMYB-like family members is not altered in mir159ab plants. Phenotypic analysis demonstrated miR159a and miR159b specifically target MYB33 and MYB65 (Allen et al. 2007).

miR160

miR160 targets three ARF (auxin-response transcription factor) genes, ARF10, ARF16, and ARF17. ARF genes are involved in the regulation of auxin signal transduction during plant development. Transgenic plants expressing a miRNA-resistant version of ARF10 and ARF17 demonstrate dramatic developmental defects, such as embryo and emerging leaf symmetry anomalies, leaf shape defects, premature inflorescence development, altered phyllotaxy along the stem, contorted flowers, twisted siliques, sterility, and root growth defects (Mallory et al. 2005, Liu et al. 2007). Derepression of ARF10 increases the sensitivity of seeds and seedlings to exogenous ABA during germination and post-germinative shoot formation, indicating that repression of ARF10 by miR160 is essential for normal plant development. In root, miR160 regulates ARF10 and ARF16, together with auxin, generating a pattern consistent with root cap development (Wang et al. 2005). In adventitious root, miR160, targeting ARF17 and miR167, targeting ARF6 and ARF8s forms a complex regulatory network to control adventitious root initiation (Gutierrez et al. 2009).

miR161/miR163

miR161 and miR163 are non-conserved microRNAs in *Arabidopsis*. The foldback arms of MIR163 have significant similarity to a segment of three SAMT-like genes, which are miR163 targets. There is also segmental similarity between MIR161 foldback arms and PPR target genes. Together, these data indicate that MIR161 and MIR163 genes evolved relatively recently by inverted duplication events associated with active expansion of target gene families. The resulting hairpin structures were then adapted to the miRNA biogenesis pathway (Allen et al. 2004). The duplication event may have included promoter sequences of the original genes since significant similarities between the promoters of MIR161/163 genes and their corresponding target genes were also detected (Wang et al. 2006).

miR162

miR162 is expressed at a relatively low level in *Arabidopsis* seedlings, leaves, and inflorescences. miR162 has two loci on the genome, MIR162a and MIR162b. miR162 is probably processed by DCL1 from the MIR162a primary transcript. miR162b has a very low abundance in wild-type plants, may compensate for the loss of function of MIR162a (Hirsch et al. 2006). Remarkably, miR162 targets DCL1 and establishes a negative feedback loop to regulate DCL1 abundance.

miR164

The miR164 family consists of three family members, miR164a, miR164b, and miR164c, which target NAC (no apical meristem (NAM), *Arabidopsis* transcription factor like family (ATAF), and CUP-SHAPED COTYLEDON (CUC)) genes. Transgenic plants with expression of miR164-resistant CUC1, CUC2 mRNA show alternation in embryonic, vegetative, and floral development. Conversely, constitutive overexpression of miR164 not only phenocopies cuc1 cuc2 double mutants but also leads to new phenotypes such as leaf and stem fusions (Laufs et al. 2004, Mallory et al. 2004a, Peaucelle et al. 2007, Raman et al. 2008). miR164abc triple mutants show a more severe disruption of shoot development than any single miR164 mutant (Sieber et al. 2007). miR164 single mutants have different phenotypes. miR164a and miR164b mutant plants express less miR164 and more NAC1. These plants have more lateral roots as compared to wild-type plants (Guo et al. 2005). miR164a mutant plants also have a higher level of leaf serration, which can be abolished by CUC2 inactivation, implying that the balance between co-expressed CUC2 and miR164a determines the extent of serration (Nikovics et al. 2006). miR164c mutants have extra petals in early-arising flowers (Baker et al. 2005). These observations suggest that the redundancy among miR164 genes is not complete and each miR164 gene has also specialized functions. In *Arabidopsis* leaves, miR164 expression gradually decreases with aging through negative regulation by ethylene-insensitive 2 (EIN2). Consequently, a miR164 target, Oresara1

(ORE1), is upregulated and contributes to aging-related cell death (Kim et al. 2009). Together, the results demonstrate that miR164 is an important regulatory component in plant development and aging-induced cell death.

miR165/miR166

In *Arabidopsis* there are two miR165 genes and seven miR166 genes. Their mature miRNA sequences differ from one another by a single nucleotide. The individual miR165/miR166 genes exhibits distinct temporal and spatial expression patterns in different plant organs (Table 3.3; Jung and Park 2007). In leaves, histone deacetylases (HDACs), HDT1/HD2A and HDT2/HD2B, ASYMMETRIC LEAVES, AS1 and AS2 act independently to control levels and/or patterns of miR165/166 distribution for the development of adaxial–abaxial leaf polarity (Ueno et al. 2007). AS1/AS2 may negatively regulate the expression of miR165/miR166 expressions (Fu et al. 2007). In shoot apical meristem (SAM), miR165/miR166 is negatively regulated in parallel by AGO10 and tasiR-ARF pathway (Liu et al. 2008b).

miR165/miR166 targets the class III homeodomain leucine-zipper (HD-ZIPIII) transcription factors, including PHB (PHABULOSA) and PHV (PHAVOLUTA). Mutations in the miR165/166 complementary sites of PHB and PHV genes cause severe developmental defects in leaves (Mallory et al. 2004b). Overexpression of miR165/miR166 causes dramatic reduction of the transcripts of five HD-ZIPIII transcription factors in *Arabidopsis* (Zhou et al. 2007).

miR167

miR167 controls expression patterns of both ARF6 and ARF8. MIR167a is expressed in ovules and anthers where wild-type ARF6 and ARF8 transcripts are normally not detected. However, transgenic plants with modified ARF6 (mARF6) and ARF8 (mARF8) target sites accumulate both mARF6 and mARF8 transcripts in ovules and anthers. These plants have sterile flowers due to arrested ovule development and anther indehiscence, indicating that miR167 regulates both female and male floral organ development (Wu et al. 2006). miR167 is also involved in adventitious rooting. miR167 and miR160 form a complex regulatory network with their corresponding targets ARF6, ARF8, and ARF17. These ARF transcription factors not only regulate each other's expression transcriptionally but also modulate miR167 and miR160 abundance (Gutierrez et al. 2009).

miR168

miR168 has two loci on the genome, miR168a, which is highly expressed with a predominantly 21 nt miR168 species, and miR168b, which is lowly expressed with an equal amount of 21 and 22 nt miR168 species (Vaucheret 2009). miR168 co-expresses with AGO1 and cleaves AGO1 mRNA. At post-transcriptional level, miR168 is stabilized by AGO (Vaucheret et al. 2006). This retro-regulation of miR168 and AGO1 is necessary to maintain AGO1 homeostasis, which is crucial for the proper functioning of other miRNAs and plant development (Vaucheret et al. 2004).

Table 3.3 Temporal and spatial expression of the miR165 and miR166 family

	Temporal expression							Spatial expression										
	Embryos				Late stages								Floral organs					
Name	Early globular	Late globular	Heart (abaxial)	Torpedo (abaxial)	SAM	Root-hypocotyl junctions	Cotyledon vasculatures (6-day old seedlings)	Leaf veins	Root tissues	Root tips	Rossette leaf tricomes	Immature siliques	Stamens	Ovules	Stigma	Receptacles	Sepals	Petals
miR165a																		
miR165b																		
miR166a																		
miR166b																		
miR166c																		
miR166d																		
miR166e																		
miR166f																		
miR166g	Unknown	Unknown	Unknown	Unknown														

- High-level expression
- Moderate-level expression
- Low-level expression
- No expression

In *Arabidopsis* there are two miR165 genes and seven miR166 genes. Their mature miRNA sequences differ from one another by a single nucleotide. The individual miR165/miR166 genes exhibit distinct temporal and spatial expression patterns in different plant organs. Expression characteristics are indicated by different gray shading of the respective table cells *miR169*

Using a deep sequencing approach miR169 miRNA was shown to be repressed upon phosphate limitation (Hsieh et al. 2009), while several miR169 genes were found to be repressed upon nitrogen limitation by a quantitative real-time polymerase chain reaction platform (Pant et al. 2009). miR169 targets Nuclear factor Y (NF-Y) transcription factor NFYA5 in *Arabidopsis*. NFYA5 is strongly induced by drought in vascular tissues and guard cells. Conversely, miR169a and miR169c were substantially downregulated by drought. Co-expression experiments suggest that miR169a was more efficient than other miR169 genes at repressing the NFYA5 mRNA level (Li et al. 2008).

miR170/miR171

miR171 targets three Scarecrow-like (SCL) transcription factors for cleavage (Llave et al. 2002). The miR171 gene is transcribed in a highly cell-type-specific manner and its transcription coincides with its site of endonucleolytic activity, suggesting miR171 is involved in cell differentiation (Parizotto et al. 2004).

miR172

miR172 promotes flowering and adult patterns of epidermal differentiation in leaves. miR172 downregulates a subfamily of APETALA2 (AP2) transcription factor genes mostly through a translational repression mechanism and mRNA cleavage (Aukerman and Sakai 2003, Chen 2004, Schwab et al. 2005). There are five loci of miR172 genes in *Arabidopsis*. Transcription of three miR172 genes, MIR172a, MIR172b, and MIR172c, are elevated steadily throughout the plant growth stages while the transcript levels of MIR172d and MIR172e are very low and do not exhibit any temporal regulation (Jung et al. 2007). SPL9 and SPL10 directly activate the transcription of MIR172b (Wu et al. 2009). The overall miR172 abundance is regulated by photoperiod via GIGANTEA(GI)-mediated miRNA processing (Jung et al. 2007). Together with miR156, miR172 is a part of regulatory feedback loops that integrate several distinct genetic pathways to guard the stable transformation from the juvenile to the adult phase (Wang et al. 2009, Wu et al. 2009).

miR173

miR173, miR390, and miR828 are unique from other miRNAs in their ability to trigger trans-acting small interfering RNAs (ta-siRNA) production from TAS1 and TAS2 (miR173), TAS3 (miR390), TAS4 (miR828). miR173 cleaves non-coding transcripts TAS1a, 1b, 1c and TAS2. The cleaved 3′ transcripts are stabilized, most likely by SUPPRESSOR OF GENE SILENCING 3 (SGS3), and converted to double-strand RNAs (dsRNAs) by RNA-DEPENDENT RNA POLYMERASE 6 (RDR6). Sequentially, dsRNAs are cleaved by DCL4 to produce phased 21 nt ta-siRNAs, which target different mRNAs, including pentatricopeptide repeat (PPR) proteins (Allen et al. 2005, Yoshikawa et al. 2005). miR173 can route any non-TAS transcripts into the TAS pathway to yield phased siRNAs, as long as these non-TAS transcripts contain a miR173 target site. This unique feature distinguishes miR173 from other conventional miRNAs, such as miR159, miR167, miR169, and miR171, whose target sites are also introduced into the very same non-TAS transcripts, although the transcripts are cleaved on these sites but non-phased siRNAs are detected (Montgomery et al. 2008b, Felippes and Weigel 2009). It is currently unknown how miR173 and its protein partners convey transcripts to TAS pathway.

miR319

Although miR319 and miR159 belong to one miRNA gene family by sharing 81% sequence identity, miR319 predominantly targets TCP (TEOSINTE BRANCHED/CYCLOIDEA/PCF) transcription factor genes while miR159 regulates MYB transcription factors. These specializations on targeting are due to both expression and sequence (although limiting) differences (Palatnik et al. 2007).

A miR319a mutant, jaw-D, which overexpresses miR319a, has highly crinkled leaves, indicating that miR319 controls leaf morphogenesis (Palatnik et al. 2003). In jaw-D plants, five TCP genes are downregulated as compared to wild-type plants. The miRNA-resistant forms of the TCP2 gene, when constitutively expressed in jaw-D plants, were able to rescue the leaf shape and curvature defects of jaw-D, suggesting that TCP mRNA degradation causes jaw-D phenotypes. miR319-regulated TCP transcription factors have also been shown to control biosynthesis of the hormone jasmonic acid and senescence (Schommer et al. 2008). Taken together, miR319 and its TCP targets regulate two sequential processes in leaf development: growth and senescence.

miR390/miR391

miR390 and miR391 are related miRNAs. The sequence difference between these two miRNAs is 5 nt. Therefore, miR390 and miR391 are grouped into one miRNA family. However, most functional analyses of this family is focused on miR390, which binds to AGO7 (Montgomery et al. 2008a) and targets TAS3 non-coding transcripts to generate ta-siRNAs to regulate several AUXIN RESPONSE FACTORs (including ARF3/ETTIN and ARF4) (Allen et al. 2005, Axtell et al. 2006, Fahlgren et al. 2006). Unlike miR173, which produces phased ta-siRNAs from 3′ cleaved transcripts, miR390 has two target sites on TAS3. Phased ta-siRNAs are produced between these two target sites. Interestingly, the 5′ target site has a central mismatch which prevents a 5′ cleavage of TAS3. This 5′ functional non-cleavage target site is essential for the TAS3 pathway as conversion to a cleavable miR390 target demolishes the production of ta-siRNAs from TAS3a locus (Axtell et al. 2006).

miR393

miR393 is involved in antibacterial resistance. miR393 targets transport inhibitor response 1 (TIR1) and auxin-signaling F-box proteins 2, 3 (AFB2, AFB3), which are receptors for auxin. Upon perception of pathogen-associated molecular patterns (PAMPs), for example, a flagellin-derived peptide, *Arabidopsis* plants upregulates miR393, which triggers downregulation of TIR1, AFB2, and AFB3. Repression of auxin signaling restricts bacteria growth, implying miRNA-mediated suppression of auxin signaling in plant resistance (Navarro et al. 2006). Conversely, some Pto DC3000 effectors counter plant defense by repressing the induction of miR393. In virulent Pto DC3000-treated plants, induction of the PAMP-responsive primary transcripts of miR393a and miR393b is suppressed (Navarro et al. 2008).

miR395

miR395 targets genes of two different families, a low-affinity sulfate transporter (SULTR2;1) and three ATP sulfurylases (APS1, APS3, and APS4) (Jones-Rhoades and Bartel 2004, Allen et al. 2005, Kawashima et al. 2009). Both of the gene families are parts of the sulfur assimilation pathway. Upon sulfur starvation (Franco-Zorrilla et al. 2007), five of six miR395 loci are induced in leaves and roots, except miR395f, whose expression level is beyond the limit of detection. All six miR395 genes are located on the chromosome 1, where miR395a,b,c are clustered together within ~5 kb and miR395d,e,f cluster together within ~4 kb. However, despite the close vicinity of the miR395 genes the genes do not show the same expression pattern, suggesting that these genes might be regulated differently at the transcriptional level (Kawashima et al. 2009). Some of the miR395 genes are strongly induced by –S in root tips and in both phloem companion cells of shoots and roots from young seedling. The induction of miR395 by –S is compromised in SULFUR LIMITATION1 (SLIM1, a key transcription factor in the sulfur assimilation pathway) mutant plants, suggesting that SLIM1 directly or indirectly regulates miR395.

miR396

miR396 targets growth-regulating factor (GRF) genes which are putative transcription factors with roles in leaf growth. miR396 genes are predominantly expressed in leaf and seedling. Transgenic plants that constitutively overexpress miR396 have narrow leaves due to reduction in cell number and have lower densities of stomata (Liu et al. 2009). miR396 can be induced either by *Psevdomonas syringae* pv. tomato (DC3000hrcC), which triggers a robust basal defense response in *Arabidopsis*, or by environmental stresses such as high salinity, drought, and cold (Liu et al. 2008a), indicating that induction of miR396 and consequently the downregulation of GRF genes are inherent reactions of *Arabidopsis* plants when challenged by environmental factors.

miR397

The miR397 family consists of two genes, miR397a and miR397b, both of which are localized on chromosome IV. miR397 is ubiquitously expressed in mature plants and in seedlings (Abdel-Ghany and Pilon 2008). miR397 targets copper-containing proteins such as laccases LAC2, LAC4, and LAC17 (Abdel-Ghany and Pilon 2008).

In response to copper deficiency, SQUAMOSA promoter-binding protein-like 7 (SPL7) activates several miRNAs including miR397 (Yamasaki et al. 2009). It has been proposed that the downregulation of copper-containing proteins by miRNAs allows plants to save copper for the most essential functions to deal with copper deficiency.

miR398

miR398 targets two Cu/Zn superoxide dismutases (cytosolic CSD1 and chloroplastic CSD2) (Sunkar et al. 2006, Yamasaki et al. 2007) and a subunit of the mitochondrial cytochrome *c* oxidase (COX5b-1) (Yamasaki et al. 2007). The miR398 level is negatively correlated with the CSD1 and CSD2 mRNA level in many adult plant tissues as well as in young, 2-week old seedlings, indicating that miR398 regulates the spatial and temporal expression pattern of CSD1 and CSD2 mRNAs (Sunkar et al. 2006). On one hand, miR398 expression is downregulated transcriptionally by oxidative stresses (Sunkar et al. 2006), ozone fumigation, and biotic stress (*P. syringae*) (Jagadeeswaran et al. 2009). On the other hand, miR398 can be induced by sucrose, resulting in decreased CSD1 and CSD2 mRNA and protein accumulation (Brodersen et al. 2008, Dugas and Bartel 2008). Intriguingly, plants expressing CSD1 and CSD2 mRNAs with engineered miR398 sites of more mismatches display increased mRNA accumulation, whereas CSD1 and CSD2 protein accumulation remains sensitive to miR398 levels, suggesting that miR398 can act as a translational repressor when target site complementarity is reduced (Dugas and Bartel 2008).

miR399

miR399 targets the mRNA of a putative ubiquitin-conjugating enzyme (UBC). Remarkably, the UBC mRNA has five miR399 target sites in the 5′-UTR. All six miR399 genes are induced in the vascular tissues of roots and leaves by low-phosphate stress to repress the UBC expression level (Fujii et al. 2005, Aung et al. 2006, Bari et al. 2006, Chiou et al. 2006). The primary transcripts of miR399 are also strongly induced, ranging from 1,000 to 10,000 fold changes, by low inorganic phosphate (Pi) and rapidly repressed after addition of Pi (Bari et al. 2006). In grafted *Arabidopsis* plants, where shoots constitutively overexpressing miR399 are grafted with wild-type roots, miR399 accumulates to a substantial level in roots while miR399 primary transcripts or precursors are not detectable (Lin et al. 2008, Pant et al. 2008). Taken together this suggests that miR399 can be translocated from shoots to roots.

3.1.8.2 Small/Short Interfering RNAs

Small/short interfering RNAs (siRNAs) are generated by dicer-mediated processive cleavage of double-stranded RNA (dsRNA). Like miRNAs, siRNAs are 6 along with Argonaute proteins. siRNAs are implicated in a variety of processes, including defense against viruses, establishment of heterochromatin, silencing of transposons and transgenes, and post-transcriptional regulation of mRNAs.

It was first reported in 1998 that a few molecules of a dsRNA can repress the expression of a gene homologous to the dsRNA in *C. elegans* (Fire et al. 1998). This phenomenon was termed RNA interference (RNAi). Later a study in plants suggested that the mechanism of dsRNA silencing might occur through the action of small ~25 nt double-stranded RNAs (Hamilton and Baulcombe 1999). The dsRNAs were processed into ~21 nt long siRNAs to induce the cleavage of a homologous transcript. Thereafter, siRNAs were widely used tools for gene silencing in eukaryotic cells. These siRNAs are normally referred to as exogenous siRNAs, either supplied by experimental modifications or more naturally contributed by invading viruses.

In *Arabidopsis*, recent works discovered several different types of endogenous siRNAs, including repeat-associated siRNAs (ra-siRNAs, including PolIV-dependent (p4)-siRNAs), natural antisense siRNAs (nat-siRNAs), and trans-acting siRNAs (tas-siRNAs).

3.1.8.3 Repeat-Associated siRNAs

Endogenous siRNAs have been found by large scale cDNA cloning efforts in *C. elegans* (Ambros et al. 2003b), *Drosophila* (Aravin et al. 2003) and *Arabidopsis* (Llave et al. 2002). Different types of endogenous siRNAs were identified. Repeat-associated siRNAs (ra-siRNA), also occasionally referred to as heterochromatin siRNAs (hc-siRNA) (Vaucheret 2006), are ~24 nt long, derived from repeat sequences related to transposons and commonly found in worm, fly, and plant, also in the unicellular eukaryotes such as *Trypanosoma brucei* and *Schizosaccharomyces pombe* (reviewed by Aravin and Tuschl 2005). Ra-siRNAs are involved in establishing and maintaining heterochromatin structure and in controlling transposons.

With the advent of deep sequencing technology, scientists were able to identify millions of small RNAs in *Arabidopsis*, most of which are ra-siRNAs (Lu et al. 2005, Rajagopalan et al. 2006, Kasschau et al. 2007, Mosher et al. 2009). The biogenesis of ra-siRNAs most likely requires a plant-specific polymerase, Pol IV, although a limited number of ra-siRNAs are still presented in Pol IV mutants, which may be created from other polymerases (Mosher et al. 2009). An SNF2 domain-containing protein (CLASSY1), together with Pol IV, may be responsible for the generation of single-stranded RNA transcripts, which in turn are subject to RNA-DEPENDENT RNA POLYMERASE 2 (RDR2)-dependent creation of double-strand RNAs (dsRNAs). The dsRNAs are then cleaved by DCL3 into 24 nt ra-siRNAs (Xie et al. 2004). Given their repetitive nature, ra-siRNAs can usually be mapped to multiple loci on the genome, which in turn makes the identification of their true origins difficult. Thus on a whole genome scale it is currently unknown which specific repeat, transposon and other intergenic loci give rise to the large amount of small RNAs (Fig. 3.6).

The surprising fact that millions of small RNAs are presented in a cell argues for the importance of small RNA function. However, in contrast to the well-studied

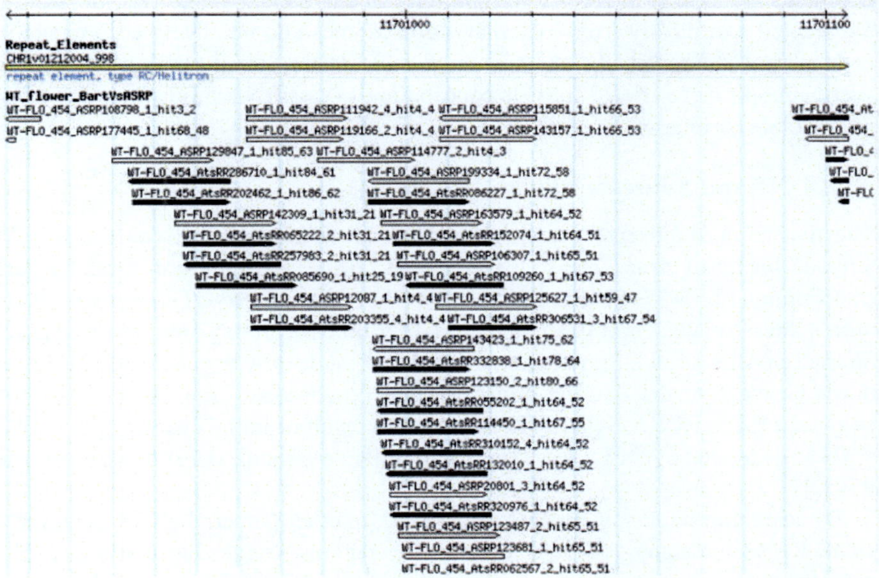

Fig. 3.6 A genome browser view of ra-siRNAs. A cluster of ra-siRNAs map to a repeat element of helitron on chromosome1. Here two set of ra-siRNAs, sequenced by two groups (Rajagopalan et al. 2006, Kasschau et al. 2007), are shown in two different colors, *white* and *black*. Note that these ra-siRNAs can be mapped to the genome multiple times (from 4 to 64 times). It is therefore impossible to assign their genomic origins unambiguously

miRNAs which account for only 5% of sRNA mass and less than 0.1% of the sequence complexity (Mosher et al. 2009), the majority of ra-siRNAs remains to be assigned to clear biological roles. Below, we summarize several representative cases of ra-siRNAs that are involved in maintaining genome integrity and regulating gene expressions.

24 nt ra-siRNAs with 5′ terminal adenosine are preferentially incorporated into RISC complexes along with AGO4 proteins (Mi et al. 2008). These ra-siRNAs are responsible for silencing transposons by guiding DNA methyltransferases together with AGO4. Additionally, AGO4 has the Asp–Asp–His catalytic motif which enables AGO4, guided by ra-siRNAs, to cleave target RNA transcripts. From some loci secondary siRNAs are created, which form a positive feedback loop to reinforce the silencing effects (Qi et al. 2006).

Most 24 nt ra-siRNAs are lost in the vegetative nucleus (VN) of *Arabidopsis* pollen while 21 nt ra-siRNAs from Athila retrotransposons are generated in VN and accumulate both in VN and sperm. Many transposable elements are reactivated in the VN which may due to the downregulation of the heterochromatin remodeler decrease in DNA methylation 1 (DDM1) and components of ra-siRNAs biogenesis machinery, such as Pol IV, RDR2, and DCL3 (Slotkin et al. 2009). It has been suggested that epigenetic reprogramming in VN may reveal intact TEs in the genome and 21 nt Athila siRNAs are created in VN in order to regulate Athila activity in sperms (Slotkin et al. 2009). Intriguingly, during seed development, ra-siRNAs are created specifically from maternal chromosomes and massively accumulate in endosperm (Mosher et al. 2009). It is possible that paternal chromosomes from sperms might be restricted in their contribution to ra-siRNA production due to an unfavorable chromatin structure.

3.1.8.4 Natural Antisense siRNAs

The first natural antisense transcript siRNA (nat-siRNA) was discovered in salt-stressed plants of *Arabidopsis* (Borsani et al. 2005). Salt-induced stress triggers transcription of SRO5 (SIMILAR TO RCD ONE 5), whose 3′-UTR overlaps 760 nt with a stress-related gene P5CDH (delta1-pyrroline-5-carboxylate dehydrogenase) on the opposite strand. Subsequentially a 24 nt nat-siRNA is generated that introduces cleavage of the P5CDH transcript. The biogenesis of the 24 nt nat-siRNA involves DCL2, RDR6, SGS3, and Pol IV. As a result of initial cleavage, 21 nt nat-siRNAs generated by DCL1 were formed to further cleave the constitutive transcript P5CDH.

By searching public small RNA databases, Jin et al. (Jin et al. 2008) found 828 small RNAs matching 165 NAT overlapping regions, suggesting that nat-siRNA regulation of antisense genes might be an important gene regulatory mechanism in *Arabidopsis*.

3.1.8.5 Trans-acting siRNAs

Trans-acting siRNAs (ta-siRNAs) are derived from TAS (Trans-acting siRNA) genes in *Arabidopsis*. TAS genes are long ncRNAs that are cleaved by miR173 (TAS1 and TAS2), miR390 (TAS3), and miR828 (TAS4) (Adenot et al. 2006, Fahlgren et al. 2006, Garcia et al. 2006). The ends of the cleaved RNA transcripts are stabilized by SUPPRESSOR OF GENE SILENCING3 (SGS3). The stabilized cleavage products are then converted to double-strand RNAs by RNA-DEPENDENT RNA POLYMERASE 6 (RDR6). The dsRNAs are processed by DCL4 into in-phased 21 nt ta-siRNAs.

In general, the 5′ half of ta-siRNAs shows a high level of complementarity with their target mRNAs. Like plant miRNAs, ta-siRNAs regulate the expression of their target mRNAs by guiding mRNA cleavage. TAS3-derived ta-siRNAs target AUXIN RESPONSE FACTORs (including ARF3 and ARF4; Adenot et al. 2006, Fahlgren et al. 2006, Garcia et al. 2006) to regulate developmental timing and patterning.

3.1.9 Non-coding RNA: Conclusions

In higher eukaryotes, protein-coding genes occupy only a mere fraction of a genome. The quantitative majority output of a genome is RNA transcripts that have no coding potential. These RNA transcripts include long non-coding RNAs and small RNAs, as presented in this chapter. Most importantly, what has been functionally characterized so far might well be only the tip of the iceberg. It is commonly accepted now that a large amount of transcripts of unknown functions (TUFs) exists in eukaryotic cells (see the RNA review issue of Cell, February 20, 2009 Volume 136, Issue 4). TUFs are usually located in unannotated regions of genomes or in antisense orientation to known protein-coding genes. They can be polyadenylated, nonpolyadenylated, or bimorphic, many of which have little coding potential. This new picture of eukaryotic transcriptome justifies that more research efforts are need to elucidate the function of these non-coding RNA transcripts.

References

Abdel-Ghany SE, Pilon M (2008) MicroRNA-mediated systemic down-regulation of copper protein expression in response to low copper availability in Arabidopsis. J Biol Chem 283:15932–15945

Adenot X, Elmayan T, Lauressergues D, Boutet S, Bouche N, Gasciolli V, Vaucheret H (2006) DRB4-dependent TAS3 trans-acting siRNAs control leaf morphology through AGO7. Curr Biol 16:927–932

Alexiou P, Maragkakis M, Papadopoulos GL, Reczko M, Hatzigeorgiou AG (2009) Lost in translation: an assessment and perspective for computational microRNA target identification. Bioinformatics 25:3049–3055

Allen RS, Li J, Stahle MI, Dubroue A, Gubler F, Millar AA (2007) Genetic analysis reveals functional redundancy and the major target genes of the Arabidopsis miR159 family. Proc Natl Acad Sci USA 104:16371–16376

Allen E, Xie Z, Gustafson AM, Carrington JC (2005) microRNA-directed phasing during trans-acting siRNA biogenesis in plants. Cell 121:207–221

Allen E, Xie Z, Gustafson AM, Sung GH, Spatafora JW, Carrington JC (2004) Evolution of microRNA genes by inverted duplication of target gene sequences in Arabidopsis thaliana. Nat Genet 36:1282–1290

Altuvia Y, Landgraf P, Lithwick G, Elefant N, Pfeffer S, Aravin A, Brownstein MJ, Tuschl T, Margalit H (2005) Clustering and conservation patterns of human microRNAs. Nucleic Acids Res 33:2697–2706

Ambros V, Bartel B, Bartel DP, Burge CB, Carrington JC, Chen X, Dreyfuss G, Eddy SR, Griffiths-Jones S, Marshall M, Matzke M, Ruvkun G, Tuschl T (2003a) A uniform system for microRNA annotation. RNA 9:277–279

Ambros V, Lee RC, Lavanway A, Williams PT, Jewell D (2003b) MicroRNAs and other tiny endogenous RNAs in C. elegans. Curr Biol 13:807–818

Aravin AA, Lagos-Quintana M, Yalcin A, Zavolan M, Marks D, Snyder B, Gaasterland T, Meyer J, Tuschl T (2003) The small RNA profile during Drosophila melanogaster development. Dev Cell 5:337–350

Aravin A, Tuschl T (2005) Identification and characterization of small RNAs involved in RNA silencing. FEBS Lett 579:5830–5840

Aukerman MJ, Sakai H (2003) Regulation of flowering time and floral organ identity by a MicroRNA and its APETALA2-like target genes. Plant Cell 15:2730–2741

Aung K, Lin SI, Wu CC, Huang YT, Su CL, Chiou TJ (2006) pho2, a phosphate overaccumulator, is caused by a nonsense mutation in a microRNA399 target gene. Plant Physiol 141: 1000–1011

Autran D, Jonak C, Belcram K, Beemster GT, Kronenberger J, Grandjean O, Inze D, Traas J (2002) Cell numbers and leaf development in Arabidopsis: a functional analysis of the STRUWWELPETER gene. EMBO J 21:6036–6049

Axtell MJ, Bartel DP (2005) Antiquity of microRNAs and their targets in land plants. Plant Cell 17:1658–1673

Axtell MJ, Jan C, Rajagopalan R, Bartel DP (2006) A two-hit trigger for siRNA biogenesis in plants. Cell 127:565–577

Backstrom S, Elfving N, Nilsson R, Wingsle G, Bjorklund S (2007) Purification of a plant mediator from Arabidopsis thaliana identifies PFT1 as the Med25 subunit. Mol Cell 26:717–729

Bailey TL, Boden M, Buske FA, Frith M, Grant CE, Clementi L, Ren J, Li WW, Noble WS (2009) MEME SUITE: tools for motif discovery and searching. Nucleic Acids Res 37:W202–W208

Baker CC, Sieber P, Wellmer F, Meyerowitz EM (2005) The early extra petals1 mutant uncovers a role for microRNA miR164c in regulating petal number in Arabidopsis. Curr Biol 15:303–315

Bari R, Datt Pant B, Stitt M, Scheible WR (2006) PHO2, microRNA399, and PHR1 define a phosphate-signaling pathway in plants. Plant Physiol 141:988–999

Berezikov E, Chung WJ, Willis J, Cuppen E, Lai EC (2007) Mammalian mirtron genes. Mol Cell 28:328–336

Berg OG, von Hippel PH (1987) Selection of DNA binding sites by regulatory proteins. Statistical-mechanical theory and application to operators and promoters. J Mol Biol 193:723–750

Bertrand C, Benhamed M, Li YF, Ayadi M, Lemonnier G, Renou JP, Delarue M, Zhou DX (2005) Arabidopsis HAF2 gene encoding TATA-binding protein (TBP)-associated factor TAF1, is required to integrate light signals to regulate gene expression and growth. J Biol Chem 280:1465–1473

Blanchette M, Tompa M (2003) Footprinter: a program designed for phylogenetic footprinting. Nucleic Acids Res 31:3840–3842

Boffelli D, McAuliffe J, Ovcharenko D, Lewis KD, Ovcharenko I, Pachter L, Rubin EM (2003) Phylogenetic shadowing of primate sequences to find functional regions of the human genome. Science 299:1391–1394

Borchert GM, Lanier W, Davidson BL (2006) RNA polymerase III transcribes human microRNAs. Nat Struct Mol Biol 13:1097–1101

Borsani O, Zhu J, Verslues PE, Sunkar R, Zhu JK (2005) Endogenous siRNAs derived from a pair of natural cis-antisense transcripts regulate salt tolerance in Arabidopsis. Cell 123: 1279–1291

Brodersen P, Sakvarelidze-Achard L, Bruun-Rasmussen M, Dunoyer P, Yamamoto YY, Sieburth L, Voinnet O (2008) Widespread translational inhibition by plant miRNAs and siRNAs. Science 320:1185–1190

Brodersen P, Voinnet O (2009) Revisiting the principles of microRNA target recognition and mode of action. Nat Rev Mol Cell Biol 10:141–148

Bulow L, Engelmann S, Schindler M, Hehl R (2009) AthaMap, integrating transcriptional and post-transcriptional data. Nucleic Acids Res 37:D983–D986

Busch MA, Bomblies K, Weigel D (1999) Activation of a floral homeotic gene in Arabidopsis. Science 285:585–587

Butler JE, Kadonaga JT (2002) The RNA polymerase II core promoter: a key component in the regulation of gene expression. Genes Dev 16:2583–2592

Cavin Perier R, Junier T, Bucher P (1998) The Eukaryotic Promoter Database EPD. Nucleic Acids Res 26:353–357

Chang WC, Lee TY, Huang HD, Huang HY, Pan RL (2008) PlantPAN: plant promoter analysis navigator, for identifying combinatorial cis-regulatory elements with distance constraint in plant gene groups. BMC Genomics 9:561

Chen X (2004) A microRNA as a translational repressor of APETALA2 in Arabidopsis flower development. Science 303:2022–2025

Chi SW, Zang JB, Mele A, Darnell RB (2009) Argonaute HITS-CLIP decodes microRNA-mRNA interaction maps. Nature 460:479–486

Chiou TJ, Aung K, Lin SI, Wu CC, Chiang SF, Su CL (2006) Regulation of phosphate homeostasis by MicroRNA in Arabidopsis. Plant Cell 18:412–421

Choudhury A, Lahiri A (2008) Arabidopsis thaliana regulatory element analyzer. Bioinformatics 24:2263–2264

Covington MF, Maloof JN, Straume M, Kay SA, Harmer SL (2008) Global transcriptome analysis reveals circadian regulation of key pathways in plant growth and development. Genome Biol 9:R130

Craigon DJ, James N, Okyere J, Higgins J, Jotham J, May S (2004) NASCArrays: a repository for microarray data generated by NASC's transcriptomics service. Nucleic Acids Res 32: D575–D577

Davuluri RV, Sun H, Palaniswamy SK, Matthews N, Molina C, Kurtz M, Grotewold E (2003) AGRIS: Arabidopsis gene regulatory information server, an information resource of Arabidopsis cis-regulatory elements and transcription factors. BMC Bioinformatics 4:25

Defrance M, Janky R, Sand O, van Helden J (2008) Using RSAT oligo-analysis and dyad-analysis tools to discover regulatory signals in nucleic sequences. Nat Protoc 3:1589–1603

Didiano D, Hobert O (2006) Perfect seed pairing is not a generally reliable predictor for miRNA-target interactions. Nat Struct Mol Biol 13:849–851

Didiano D, Hobert O (2008) Molecular architecture of a miRNA-regulated 3' UTR. RNA 14: 1297–1317

Dugas DV, Bartel B (2008) Sucrose induction of Arabidopsis miR398 represses two Cu/Zn superoxide dismutases. Plant Mol Biol 67:403–417

Easow G, Teleman AA, Cohen SM (2007) Isolation of microRNA targets by miRNP immunopurification. RNA 13:1198–1204

Edwards D, Murray JA, Smith AG (1998) Multiple genes encoding the conserved CCAAT-box transcription factor complex are expressed in Arabidopsis. Plant Physiol 117:1015–1022

Elemento O, Slonim N, Tavazoie S (2007) A universal framework for regulatory element discovery across all genomes and data types. Mol Cell 28:337–350

Fahlgren N, Montgomery TA, Howell MD, Allen E, Dvorak SK, Alexander AL, Carrington JC (2006) Regulation of AUXIN RESPONSE FACTOR3 by TAS3 ta-siRNA affects developmental timing and patterning in Arabidopsis. Curr Biol 16:939–944

Fang Y, Spector DL (2007) Identification of nuclear dicing bodies containing proteins for microRNA biogenesis in living Arabidopsis plants. Curr Biol 17:818–823

Fauteux F, Blanchette M, Stromvik MV (2008) Seeder: discriminative seeding DNA motif discovery. Bioinformatics 24:2303–2307

Felippes FF, Weigel D (2009) Triggering the formation of ta-siRNAs in Arabidopsis thaliana: the role of microRNA miR173. EMBO Rep 10:264–270

Fire A, Xu S, Montgomery MK, Kostas SA, Driver SE, Mello CC (1998) Potent and specific genetic interference by double-stranded RNA in Caenorhabditis elegans. Nature 391:806–811

Franco-Zorrilla JM, Valli A, Todesco M, Mateos I, Puga MI, Rubio-Somoza I, Leyva A, Weigel D, Garcia JA, Paz-Ares J (2007) Target mimicry provides a new mechanism for regulation of microRNA activity. Nat Genet 39:1033–1037

Freeling M, Rapaka L, Lyons E, Pedersen B, Thomas BC (2007) G-boxes, bigfoot genes, and environmental response: characterization of intragenomic conserved noncoding sequences in Arabidopsis. Plant Cell 19:1441–1457

Fu Y, Xu L, Xu B, Yang L, Ling Q, Wang H, Huang H (2007) Genetic interactions between leaf polarity-controlling genes and ASYMMETRIC LEAVES1 and 2 in Arabidopsis leaf patterning. Plant Cell Physiol 48:724–735

Fujii H, Chiou TJ, Lin SI, Aung K, Zhu JK (2005) A miRNA involved in phosphate-starvation response in Arabidopsis. Curr Biol 15:2038–2043

Gao X, Ren F, Lu YT (2006) The Arabidopsis mutant stg1 identifies a function for TBP-associated factor 10 in plant osmotic stress adaptation. Plant Cell Physiol 47:1285–1294

Garcia D, Collier SA, Byrne ME, Martienssen RA (2006) Specification of leaf polarity in Arabidopsis via the trans-acting siRNA pathway. Curr Biol 16:933–938

Gardner PP, Daub J, Tate JG, Nawrocki EP, Kolbe DL, Lindgreen S, Wilkinson AC, Finn RD, Griffiths-Jones S, Eddy SR, Bateman A (2009) Rfam: updates to the RNA families database. Nucleic Acids Res 37:D136–D140

Gasch A, Hoffmann A, Horikoshi M, Roeder RG, Chua NH (1990) Arabidopsis thaliana contains two genes for TFIID. Nature 346:390–394

Ghildiyal M, Xu J, Seitz H, Weng Z, Zamore PD (2010) Sorting of Drosophila small silencing RNAs partitions microRNA* strands into the RNA interference pathway. RNA 16:43–56

Ghildiyal M, Zamore PD (2009) Small silencing RNAs: an expanding universe. Nat Rev Genet 10:94–108

Guo HS, Xie Q, Fei JF, Chua NH (2005) MicroRNA directs mRNA cleavage of the transcription factor NAC1 to downregulate auxin signals for arabidopsis lateral root development. Plant Cell 17:1376–1386

Gutierrez L, Bussell JD, Pacurar DI, Schwambach J, Pacurar M, Bellini C (2009) Phenotypic plasticity of adventitious rooting in Arabidopsis is controlled by complex regulation of AUXIN RESPONSE FACTOR transcripts and microRNA abundance. Plant Cell 21:3119–3132

Haberer G, Hindemitt T, Meyers BC, Mayer KF (2004) Transcriptional similarities, dissimilarities, and conservation of cis-elements in duplicated genes of Arabidopsis. Plant Physiol 136: 3009–3022

Haberer G, Mader MT, Kosarev P, Spannagl M, Yang L, Mayer KF (2006) Large-scale cis-element detection by analysis of correlated expression and sequence conservation between Arabidopsis and Brassica oleracea. Plant Physiol 142:1589–1602

Hamilton AJ, Baulcombe DC (1999) A species of small antisense RNA in posttranscriptional gene silencing in plants. Science 286:950–952

Hammell M, Long D, Zhang L, Lee A, Carmack CS, Han M, Ding Y, Ambros V (2008) mirWIP: microRNA target prediction based on microRNA-containing ribonucleoprotein-enriched transcripts. Nat Methods 5:813–819

Han J, Lee Y, Yeom KH, Nam JW, Heo I, Rhee JK, Sohn SY, Cho Y, Zhang BT, Kim VN (2006) Molecular basis for the recognition of primary microRNAs by the Drosha-DGCR8 complex. Cell 125:887–901

Harmer SL, Hogenesch JB, Straume M, Chang HS, Han B, Zhu T, Wang X, Kreps JA, Kay SA (2000) Orchestrated transcription of key pathways in Arabidopsis by the circadian clock. Science 290:2110–2113

Harmer SL, Kay SA (2005) Positive and negative factors confer phase-specific circadian regulation of transcription in Arabidopsis. Plant Cell 17:1926–1940

Higo K, Ugawa Y, Iwamoto M, Korenaga T (1999) Plant cis-acting regulatory DNA elements (PLACE) database: 1999. Nucleic Acids Res 27:297–300

Hirsch J, Lefort V, Vankersschaver M, Boualem A, Lucas A, Thermes C, d'Aubenton-Carafa Y, Crespi M (2006) Characterization of 43 non-protein-coding mRNA genes in Arabidopsis, including the MIR162a-derived transcripts. Plant Physiol 140:1192–1204

Hochheimer A, Zhou S, Zheng S, Holmes MC, Tjian R (2002) TRF2 associates with DREF and directs promoter-selective gene expression in Drosophila. Nature 420:439–445

Holmes MC, Tjian R (2000) Promoter-selective properties of the TBP-related factor TRF1. Science 288:867–870

Hong RL, Hamaguchi L, Busch MA, Weigel D (2003) Regulatory elements of the floral homeotic gene AGAMOUS identified by phylogenetic footprinting and shadowing. Plant Cell 15: 1296–1309

Hong X, Hammell M, Ambros V, Cohen SM (2009) Immunopurification of Ago1 miRNPs selects for a distinct class of microRNA targets. Proc Natl Acad Sci USA 106:15085–15090

Hsieh LC, Lin SI, Shih AC, Chen JW, Lin WY, Tseng CY, Li WH, Chiou TJ (2009) Uncovering small RNA-mediated responses to phosphate deficiency in Arabidopsis by deep sequencing. Plant Physiol 151:2120–2132

Huang H, Tudor M, Su T, Zhang Y, Hu Y, Ma H (1996) DNA binding properties of two Arabidopsis MADS domain proteins: binding consensus and dimer formation. Plant Cell 8:81–94

Hughes JD, Estep PW, Tavazoie S, Church GM (2000) Computational identification of cis-regulatory elements associated with groups of functionally related genes in Saccharomyces cerevisiae. J Mol Biol 296:1205–1214

Jagadeeswaran G, Saini A, Sunkar R (2009) Biotic and abiotic stress down-regulate miR398 expression in Arabidopsis. Planta 229:1009–1014

Jin H, Vacic V, Girke T, Lonardi S, Zhu JK (2008) Small RNAs and the regulation of cis-natural antisense transcripts in Arabidopsis. BMC Mol Biol 9:6

Jones-Rhoades MW, Bartel DP (2004) Computational identification of plant microRNAs and their targets, including a stress-induced miRNA. Mol Cell 14:787–799

Jung JH, Park CM (2007) MIR166/165 genes exhibit dynamic expression patterns in regulating shoot apical meristem and floral development in Arabidopsis. Planta 225:1327–1338

Jung JH, Seo YH, Seo PJ, Reyes JL, Yun J, Chua NH, Park CM (2007) The GIGANTEA-regulated microRNA172 mediates photoperiodic flowering independent of CONSTANS in Arabidopsis. Plant Cell 19:2736–2748

Kapranov P, Cheng J, Dike S, Nix DA, Duttagupta R, Willingham AT, Stadler PF, Hertel J, Hackermuller J, Hofacker IL, Bell I, Cheung E, Drenkow J, Dumais E, Patel S, Helt G, Ganesh M, Ghosh S, Piccolboni A, Sementchenko V, Tammana H, Gingeras TR (2007) RNA maps reveal new RNA classes and a possible function for pervasive transcription. Science 316:1484–1488

Kasschau KD, Fahlgren N, Chapman EJ, Sullivan CM, Cumbie JS, Givan SA, Carrington JC (2007) Genome-wide profiling and analysis of Arabidopsis siRNAs. PLoS Biol 5:e57

Kaufmann K, Muino JM, Jauregui R, Airoldi CA, Smaczniak C, Krajewski P, Angenent GC (2009) Target genes of the MADS transcription factor SEPALLATA3: integration of developmental and hormonal pathways in the Arabidopsis flower. PLoS Biol 7:e1000090

Kawashima CG, Yoshimoto N, Maruyama-Nakashita A, Tsuchiya YN, Saito K, Takahashi H, Dalmay T (2009) Sulphur starvation induces the expression of microRNA-395 and one of its target genes but in different cell types. Plant J 57:313–321

Kickhoefer VA, Emre N, Stephen AG, Poderycki MJ, Rome LH (2003) Identification of conserved vault RNA expression elements and a non-expressed mouse vault RNA gene. Gene 309:65–70

Kidd BN, Edgar CI, Kumar KK, Aitken EA, Schenk PM, Manners JM, Kazan K (2009) The mediator complex subunit PFT1 is a key regulator of jasmonate-dependent defense in Arabidopsis. Plant Cell 21:2237–2252

Kim YK, Kim VN (2007) Processing of intronic microRNAs. EMBO J 26:775–783

Kim DW, Lee SH, Choi SB, Won SK, Heo YK, Cho M, Park YI, Cho HT (2006) Functional conservation of a root hair cell-specific cis-element in angiosperms with different root hair distribution patterns. Plant Cell 18:2958–2970

Kim JH, Woo HR, Kim J, Lim PO, Lee IC, Choi SH, Hwang D, Nam HG (2009) Trifurcate feed-forward regulation of age-dependent cell death involving miR164 in Arabidopsis. Science 323:1053–1057

Lago C, Clerici E, Dreni L, Horlow C, Caporali E, Colombo L, Kater MM (2005) The Arabidopsis TFIID factor AtTAF6 controls pollen tube growth. Dev Biol 285:91–100

Lago C, Clerici E, Mizzi L, Colombo L, Kater MM (2004) TBP-associated factors in Arabidopsis. Gene 342:231–241

Laufs P, Peaucelle A, Morin H, Traas J (2004) MicroRNA regulation of the CUC genes is required for boundary size control in Arabidopsis meristems. Development 131:4311–4322

Lawit SJ, O'Grady K, Gurley WB, Czarnecka-Verner E (2007) Yeast two-hybrid map of Arabidopsis TFIID. Plant Mol Biol 64:73–87

Lee JY, Baum SF, Alvarez J, Patel A, Chitwood DH, Bowman JL (2005) Activation of CRABS CLAW in the nectaries and carpels of Arabidopsis. Plant Cell 17:25–36

Lee RC, Feinbaum RL, Ambros V (1993) The C. elegans heterochronic gene lin-4 encodes small RNAs with antisense complementarity to lin-14. Cell 75:843–854

Lee J, He K, Stolc V, Lee H, Figueroa P, Gao Y, Tongprasit W, Zhao H, Lee I, Deng XW (2007) Analysis of transcription factor HY5 genomic binding sites revealed its hierarchical role in light regulation of development. Plant Cell 19:731–749

Lee Y, Kim M, Han J, Yeom KH, Lee S, Baek SH, Kim VN (2004) MicroRNA genes are transcribed by RNA polymerase II. EMBO J 23:4051–4060

Lee TI, Rinaldi NJ, Robert F, Odom DT, Bar-Joseph Z, Gerber GK, Hannett NM, Harbison CT, Thompson CM, Simon I, Zeitlinger J, Jennings EG, Murray HL, Gordon DB, Ren B, Wyrick

JJ, Tagne JB, Volkert TL, Fraenkel E, Gifford DK, Young RA (2002) Transcriptional regulatory networks in Saccharomyces cerevisiae. Science 298:799–804

Lescot M, Dehais P, Thijs G, Marchal K, Moreau Y, Van de Peer Y, Rouze P, Rombauts S (2002) PlantCARE, a database of plant cis-acting regulatory elements and a portal to tools for in silico analysis of promoter sequences. Nucleic Acids Res 30:325–327

Li WX, Oono Y, Zhu J, He XJ, Wu JM, Iida K, Lu XY, Cui X, Jin H, Zhu JK (2008) The Arabidopsis NFYA5 transcription factor is regulated transcriptionally and posttranscriptionally to promote drought resistance. Plant Cell 20:2238–2251

Li L, Wang X, Sasidharan R, Stolc V, Deng W, He H, Korbel J, Chen X, Tongprasit W, Ronald P, Chen R, Gerstein M, Wang Deng X (2007) Global identification and characterization of transcriptionally active regions in the rice genome. PLoS One 2:e294

Lin SI, Chiang SF, Lin WY, Chen JW, Tseng CY, Wu PC, Chiou TJ (2008) Regulatory network of microRNA399 and PHO2 by systemic signaling. Plant Physiol 147:732–746

Liu PP, Montgomery TA, Fahlgren N, Kasschau KD, Nonogaki H, Carrington JC (2007) Repression of AUXIN RESPONSE FACTOR10 by microRNA160 is critical for seed germination and post-germination stages. Plant J 52:133–146

Liu D, Song Y, Chen Z, Yu D (2009) Ectopic expression of miR396 suppresses GRF target gene expression and alters leaf growth in Arabidopsis. Physiol Plant 136:223–236

Liu HH, Tian X, Li YJ, Wu CA, Zheng CC (2008a) Microarray-based analysis of stress-regulated microRNAs in Arabidopsis thaliana. RNA 14:836–843

Liu Q, Yao X, Pi L, Wang H, Cui X, Huang H (2008b) The ARGONAUTE10 gene modulates shoot apical meristem maintenance and leaf polarity establishment by repressing miR165/166 in Arabidopsis. Plant J 40:1365–1369

Llave C, Kasschau KD, Rector MA, Carrington JC (2002a) Endogenous and silencing-associated small RNAs in plants. Plant Cell 14:1605–1619

Llave C, Xie Z, Kasschau KD, Carrington JC (2002b) Cleavage of Scarecrow-like mRNA targets directed by a class of Arabidopsis miRNA. Science 297:2053–2056

Lohmann JU, Hong RL, Hobe M, Busch MA, Parcy F, Simon R, Weigel D (2001) A molecular link between stem cell regulation and floral patterning in Arabidopsis. Cell 105:793–803

Lu C, Tej SS, Luo S, Haudenschild CD, Meyers BC, Green PJ (2005) Elucidation of the small RNA component of the transcriptome. Science 309:1567–1569

Maher C, Stein L, Ware D (2006) Evolution of Arabidopsis microRNA families through duplication events. Genome Res 16:510–519

Maleck K, Levine A, Eulgem T, Morgan A, Schmid J, Lawton KA, Dangl JL, Dietrich RA (2000) The transcriptome of Arabidopsis thaliana during systemic acquired resistance. Nat Genet 26:403–410

Mallory AC, Bartel DP, Bartel B (2005) MicroRNA-directed regulation of Arabidopsis AUXIN RESPONSE FACTOR17 is essential for proper development and modulates expression of early auxin response genes. Plant Cell 17:1360–1375

Mallory AC, Dugas DV, Bartel DP, Bartel B (2004a) MicroRNA regulation of NAC-domain targets is required for proper formation and separation of adjacent embryonic, vegetative, and floral organs. Curr Biol 14:1035–1046

Mallory AC, Reinhart BJ, Jones-Rhoades MW, Tang G, Zamore PD, Barton MK, Bartel DP (2004b) MicroRNA control of PHABULOSA in leaf development: importance of pairing to the microRNA 5' region. EMBO J 23:3356–3364

Matsui A, Ishida J, Morosawa T, Mochizuki Y, Kaminuma E, Endo TA, Okamoto M, Nambara E, Nakajima M, Kawashima M, Satou M, Kim JM, Kobayashi N, Toyoda T, Shinozaki K, Seki M (2008) Arabidopsis transcriptome analysis under drought, cold, high-salinity and ABA treatment conditions using a tiling array. Plant Cell Physiol 49:1135–1149

Mattick JS, Makunin IV (2006) Non-coding RNA. Hum Mol Genet 15(Suppl 1):R17–R29

Megraw M, Baev V, Rusinov V, Jensen ST, Kalantidis K, Hatzigeorgiou AG (2006) MicroRNA promoter element discovery in Arabidopsis. RNA 12:1612–1619

Mercer TR, Dinger ME, Mattick JS (2009) Long non-coding RNAs: insights into functions. Nat Rev Genet 10:155–159

Meyers BC, Axtell MJ, Bartel B, Bartel DP, Baulcombe D, Bowman JL, Cao X, Carrington JC, Chen X, Green PJ, Griffiths-Jones S, Jacobsen SE, Mallory AC, Martienssen RA, Poethig RS, Qi Y, Vaucheret H, Voinnet O, Watanabe Y, Weigel D, Zhu JK (2008) Criteria for annotation of plant MicroRNAs. Plant Cell 20:3186–3190

Meyers BC, Souret FF, Lu C, Green PJ (2006) Sweating the small stuff: microRNA discovery in plants. Curr Opin Biotechnol 17:139–146

Mi S, Cai T, Hu Y, Chen Y, Hodges E, Ni F, Wu L, Li S, Zhou H, Long C, Chen S, Hannon GJ, Qi Y (2008) Sorting of small RNAs into Arabidopsis argonaute complexes is directed by the 5' terminal nucleotide. Cell 133:116–127

Michael TP, Mockler TC, Breton G, McEntee C, Byer A, Trout JD, Hazen SP, Shen R, Priest HD, Sullivan CM, Givan SA, Yanovsky M, Hong F, Kay SA, Chory J (2008) Network discovery pipeline elucidates conserved time-of-day-specific cis-regulatory modules. PLoS Genet 4:e14

Molina C, Grotewold E (2005) Genome wide analysis of Arabidopsis core promoters. BMC Genomics 6:25

Montgomery TA, Howell MD, Cuperus JT, Li D, Hansen JE, Alexander AL, Chapman EJ, Fahlgren N, Allen E, Carrington JC (2008a) Specificity of ARGONAUTE7-miR390 interaction and dual functionality in TAS3 trans-acting siRNA formation. Cell 133:128–141

Montgomery TA, Yoo SJ, Fahlgren N, Gilbert SD, Howell MD, Sullivan CM, Alexander A, Nguyen G, Allen E, Ahn JH, Carrington JC (2008b) AGO1-miR173 complex initiates phased siRNA formation in plants. Proc Natl Acad Sci USA 105:20055–20062

Mosher RA, Melnyk CW, Kelly KA, Dunn RM, Studholme DJ, Baulcombe DC (2009) Uniparental expression of PolIV-dependent siRNAs in developing endosperm of Arabidopsis. Nature 460:283–286

Navarro L, Dunoyer P, Jay F, Arnold B, Dharmasiri N, Estelle M, Voinnet O, Jones JD (2006) A plant miRNA contributes to antibacterial resistance by repressing auxin signaling. Science 312:436–439

Navarro L, Jay F, Nomura K, He SY, Voinnet O (2008) Suppression of the microRNA pathway by bacterial effector proteins. Science 321:964–967

Nikovics K, Blein T, Peaucelle A, Ishida T, Morin H, Aida M, Laufs P (2006) The balance between the MIR164A and CUC2 genes controls leaf margin serration in Arabidopsis. Plant Cell 18:2929–2945

O'Connor TR, Dyreson C, Wyrick JJ (2005) Athena: a resource for rapid visualization and systematic analysis of Arabidopsis promoter sequences. Bioinformatics 21:4411–4413

Oh E, Kang H, Yamaguchi S, Park J, Lee D, Kamiya Y, Choi G (2009) Genome-wide analysis of genes targeted by PHYTOCHROME INTERACTING FACTOR 3-LIKE5 during seed germination in Arabidopsis. Plant Cell 21:403–419

Okamura K, Hagen JW, Duan H, Tyler DM, Lai EC (2007) The mirtron pathway generates microRNA-class regulatory RNAs in Drosophila. Cell 130:89–100

Okamura K, Liu N, Lai EC (2009) Distinct mechanisms for microRNA strand selection by Drosophila Argonautes. Mol Cell 36:431–444

Palatnik JF, Allen E, Wu X, Schommer C, Schwab R, Carrington JC, Weigel D (2003) Control of leaf morphogenesis by microRNAs. Nature 425:257–263

Palatnik JF, Wollmann H, Schommer C, Schwab R, Boisbouvier J, Rodriguez R, Warthmann N, Allen E, Dezulian T, Huson D, Carrington JC, Weigel D (2007) Sequence and expression differences underlie functional specialization of Arabidopsis microRNAs miR159 and miR319. Dev Cell 13:115–125

Pant BD, Buhtz A, Kehr J, Scheible WR (2008) MicroRNA399 is a long-distance signal for the regulation of plant phosphate homeostasis. Plant J 53:731–738

Pant BD, Musialak-Lange M, Nuc P, May P, Buhtz A, Kehr J, Walther D, Scheible WR (2009) Identification of nutrient-responsive Arabidopsis and rapeseed microRNAs by comprehensive real-time polymerase chain reaction profiling and small RNA sequencing. Plant Physiol 150:1541–1555

Parizotto EA, Dunoyer P, Rahm N, Himber C, Voinnet O (2004) In vivo investigation of the transcription, processing, endonucleolytic activity, and functional relevance of the spatial distribution of a plant miRNA. Genes Dev 18:2237–2242

Peaucelle A, Morin H, Traas J, Laufs P (2007) Plants expressing a miR164-resistant CUC2 gene reveal the importance of post-meristematic maintenance of phyllotaxy in Arabidopsis. Development 134:1045–1050

Poole RL (2007) The TAIR database. Methods Mol Biol 406:179–212

Prakash A, Tompa M (2005) Discovery of regulatory elements in vertebrates through comparative genomics. Nat Biotechnol 23:1249–1256

Prasanth KV, Spector DL (2007) Eukaryotic regulatory RNAs: an answer to the "genome complexity" conundrum. Genes Dev 21:11–42

Pruneda-Paz JL, Breton G, Para A, Kay SA (2009) A functional genomics approach reveals CHE as a component of the Arabidopsis circadian clock. Science 323:1481–1485

Punt PJ, Dingemanse MA, Kuyvenhoven A, Soede RD, Pouwels PH, van den Hondel CA (1990) Functional elements in the promoter region of the Aspergillus nidulans gpdA gene encoding glyceraldehyde-3-phosphate dehydrogenase. Gene 93:101–109

Qi Y, He X, Wang XJ, Kohany O, Jurka J, Hannon GJ (2006) Distinct catalytic and non-catalytic roles of ARGONAUTE4 in RNA-directed DNA methylation. Nature 443:1008–1012

Rajagopalan R, Vaucheret H, Trejo J, Bartel DP (2006) A diverse and evolutionarily fluid set of microRNAs in Arabidopsis thaliana. Genes Dev 20:3407–3425

Raman S, Greb T, Peaucelle A, Blein T, Laufs P, Theres K (2008) Interplay of miR164, CUP-SHAPED COTYLEDON genes and LATERAL SUPPRESSOR controls axillary meristem formation in Arabidopsis thaliana. Plant J 55:65–76

Reinhart BJ, Weinstein EG, Rhoades MW, Bartel B, Bartel DP (2002) MicroRNAs in plants. Genes Dev 16:1616–1626

Ren S, Johnston JS, Shippen DE, McKnight TD (2004) TELOMERASE ACTIVATOR1 induces telomerase activity and potentiates responses to auxin in Arabidopsis. Plant Cell 16: 2910–2922

Ren B, Robert F, Wyrick JJ, Aparicio O, Jennings EG, Simon I, Zeitlinger J, Schreiber J, Hannett N, Kanin E, Volkert TL, Wilson CJ, Bell SP, Young RA (2000) Genome-wide location and function of DNA binding proteins. Science 290:2306–2309

Reyes JL, Chua NH (2007) ABA induction of miR159 controls transcript levels of two MYB factors during Arabidopsis seed germination. Plant J 49:592–606

Riechmann JL, Heard J, Martin G, Reuber L, Jiang C, Keddie J, Adam L, Pineda O, Ratcliffe OJ, Samaha RR, Creelman R, Pilgrim M, Broun P, Zhang JZ, Ghandehari D, Sherman BK, Yu G (2000) Arabidopsis transcription factors: genome-wide comparative analysis among eukaryotes. Science 290:2105–2110

Rombauts S, Florquin K, Lescot M, Marchal K, Rouze P, van de Peer Y (2003) Computational approaches to identify promoters and cis-regulatory elements in plant genomes. Plant Physiol 132:1162–1176

Ruby JG, Jan CH, Bartel DP (2007) Intronic microRNA precursors that bypass Drosha processing. Nature 448:83–86

Ruvkun G, Wightman B, Ha I (2004) The 20 years it took to recognize the importance of tiny RNAs. Cell 116:S93–S96, 92 p following S96

Schauer SE, Schluter PM, Baskar R, Gheyselinck J, Bolanos A, Curtis MD, Grossniklaus U (2009) Intronic regulatory elements determine the divergent expression patterns of AGAMOUS-LIKE6 subfamily members in Arabidopsis. Plant J 59:987–1000

Schmid M, Davison TS, Henz SR, Pape UJ, Demar M, Vingron M, Scholkopf B, Weigel D, Lohmann JU (2005) A gene expression map of Arabidopsis thaliana development. Nat Genet 37:501–506

Schommer C, Palatnik JF, Aggarwal P, Chetelat A, Cubas P, Farmer EE, Nath U, Weigel D (2008) Control of jasmonate biosynthesis and senescence by miR319 targets. PLoS Biol 6:e230

Schwab R, Palatnik JF, Riester M, Schommer C, Schmid M, Weigel D (2005) Specific effects of microRNAs on the plant transcriptome. Dev Cell 8:517–527

Seki M, Narusaka M, Kamiya A, Ishida J, Satou M, Sakurai T, Nakajima M, Enju A, Akiyama K, Oono Y, Muramatsu M, Hayashizaki Y, Kawai J, Carninci P, Itoh M, Ishii Y, Arakawa T, Shibata

K, Shinagawa A, Shinozaki K (2002) Functional annotation of a full-length Arabidopsis cDNA collection. Science 296:141–145

Shahmuradov IA, Gammerman AJ, Hancock JM, Bramley PM, Solovyev VV (2003) PlantProm: a database of plant promoter sequences. Nucleic Acids Res 31:114–117

Siddharthan R, Siggia ED, van Nimwegen E (2005) PhyloGibbs: a Gibbs sampling motif finder that incorporates phylogeny. PLoS Comput Biol 1:e67

Sieber P, Wellmer F, Gheyselinck J, Riechmann JL, Meyerowitz EM (2007) Redundancy and specialization among plant microRNAs: role of the MIR164 family in developmental robustness. Development 134:1051–1060

Sieburth LE, Meyerowitz EM (1997) Molecular dissection of the AGAMOUS control region shows that cis elements for spatial regulation are located intragenically. Plant Cell 9:355–365

Siefers N, Dang KK, Kumimoto RW, Bynum WE 4th, Tayrose G, Holt BF 3rd (2009) Tissue-specific expression patterns of Arabidopsis NF-Y transcription factors suggest potential for extensive combinatorial complexity. Plant Physiol 149:625–641

Sikorski TW, Buratowski S (2009) The basal initiation machinery: beyond the general transcription factors. Curr Opin Cell Biol 21:344–351

Sinha S, Blanchette M, Tompa M (2004) PhyME: a probabilistic algorithm for finding motifs in sets of orthologous sequences. BMC Bioinformatics 5:170

Slotkin RK, Vaughn M, Borges F, Tanurdzic M, Becker JD, Feijo JA, Martienssen RA (2009) Epigenetic reprogramming and small RNA silencing of transposable elements in pollen. Cell 136:461–472

Song L, Han MH, Lesicka J, Fedoroff N (2007) Arabidopsis primary microRNA processing proteins HYL1 and DCL1 define a nuclear body distinct from the Cajal body. Proc Natl Acad Sci USA 104:5437–5442

Stolc V, Samanta MP, Tongprasit W, Sethi H, Liang S, Nelson DC, Hegeman A, Nelson C, Rancour D, Bednarek S, Ulrich EL, Zhao Q, Wrobel RL, Newman CS, Fox BG, Phillips GN Jr., Markley JL, Sussman MR (2005) Identification of transcribed sequences in Arabidopsis thaliana by using high-resolution genome tiling arrays. Proc Natl Acad Sci USA 102:4453–4458

Stormo GD, Fields DS (1998) Specificity, free energy and information content in protein-DNA interactions. Trends Biochem Sci 23:109–113

Storz G (2002) An expanding universe of noncoding RNAs. Science 296:1260–1263

Sunkar R, Jagadeeswaran G (2008) In silico identification of conserved microRNAs in large number of diverse plant species. BMC Plant Biol 8:37

Sunkar R, Kapoor A, Zhu JK (2006) Posttranscriptional induction of two Cu/Zn superoxide dismutase genes in Arabidopsis is mediated by downregulation of miR398 and important for oxidative stress tolerance. Plant Cell 18:2051–2065

Szymanski M, Barciszewski J (2002) Beyond the proteome: non-coding regulatory RNAs. Genome Biol 3: reviews0005

Tanaka T, Koyanagi KO, Itoh T (2009) Highly diversified molecular evolution of downstream transcription start sites in rice and Arabidopsis. Plant Physiol 149:1316–1324

Tang W, Perry SE (2003) Binding site selection for the plant MADS domain protein AGL15: an in vitro and in vivo study. J Biol Chem 278:28154–28159

Thibaud-Nissen F, Wu H, Richmond T, Redman JC, Johnson C, Green R, Arias J, Town CD (2006) Development of Arabidopsis whole-genome microarrays and their application to the discovery of binding sites for the TGA2 transcription factor in salicylic acid-treated plants. Plant J 47:152–162

Thijs G, Lescot M, Marchal K, Rombauts S, De Moor B, Rouze P, Moreau Y (2001) A higher-order background model improves the detection of promoter regulatory elements by Gibbs sampling. Bioinformatics 17:1113–1122

Thomas MC, Chiang CM (2006) The general transcription machinery and general cofactors. Crit Rev Biochem Mol Biol 41:105–178

Thomas BC, Rapaka L, Lyons E, Pedersen B, Freeling M (2007) Arabidopsis intragenomic conserved noncoding sequence. Proc Natl Acad Sci USA 104:3348–3353

Thomas-Chollier M, Sand O, Turatsinze JV, Janky R, Defrance M, Vervisch E, Brohee S, van Helden J (2008) RSAT: regulatory sequence analysis tools. Nucleic Acids Res 36:W119–W127

Tompa M, Li N, Bailey TL, Church GM, De Moor B, Eskin E, Favorov AV, Frith MC, Fu Y, Kent WJ, Makeev VJ, Mironov AA, Noble WS, Pavesi G, Pesole G, Regnier M, Simonis N, Sinha S, Thijs G, van Helden J, Vandenbogaert M, Weng Z, Workman C, Ye C, Zhu Z (2005) Assessing computational tools for the discovery of transcription factor binding sites. Nat Biotechnol 23:137–144

Ueno Y, Ishikawa T, Watanabe K, Terakura S, Iwakawa H, Okada K, Machida C, Machida Y (2007) Histone deacetylases and ASYMMETRIC LEAVES2 are involved in the establishment of polarity in leaves of Arabidopsis. Plant Cell 19:445–457

Vandepoele K, Casneuf T, Van de Peer Y (2006) Identification of novel regulatory modules in dicotyledonous plants using expression data and comparative genomics. Genome Biol 7:R103

Vandepoele K, Quimbaya M, Casneuf T, De Veylder L, Van de Peer Y (2009) Unraveling transcriptional control in Arabidopsis using cis-regulatory elements and coexpression networks. Plant Physiol 150:535–546

Vaucheret H (2006) Post-transcriptional small RNA pathways in plants: mechanisms and regulations. Genes Dev 20:759–771

Vaucheret H (2009) AGO1 homeostasis involves differential production of 21-nt and 22-nt miR168 species by MIR168a and MIR168b. PLoS One 4:e6442

Vaucheret H, Mallory AC, Bartel DP (2006) AGO1 homeostasis entails coexpression of MIR168 and AGO1 and preferential stabilization of miR168 by AGO1. Mol Cell 22:129–136

Vaucheret H, Vazquez F, Crete P, Bartel DP (2004) The action of ARGONAUTE1 in the miRNA pathway and its regulation by the miRNA pathway are crucial for plant development. Genes Dev 18:1187–1197

Vlieghe D, Sandelin A, De Bleser PJ, Vleminckx K, Wasserman WW, van Roy F, Lenhard B (2006) A new generation of JASPAR, the open-access repository for transcription factor binding site profiles. Nucleic Acids Res 34:D95–D97

Vogel JM, Roth B, Cigan M, Freeling M (1993) Expression of the two maize TATA binding protein genes and function of the encoded TBP proteins by complementation in yeast. Plant Cell 5:1627–1638

Voinnet O (2009) Origin, biogenesis, and activity of plant microRNAs. Cell 136:669–687

Walley JW, Coughlan S, Hudson ME, Covington MF, Kaspi R, Banu G, Harmer SL, Dehesh K (2007) Mechanical stress induces biotic and abiotic stress responses via a novel cis-element. PLoS Genet 3:1800–1812

Wang JW, Czech B, Weigel D (2009) miR156-regulated SPL transcription factors define an endogenous flowering pathway in Arabidopsis thaliana. Cell 138:738–749

Wang Y, Hindemitt T, Mayer KF (2006) Significant sequence similarities in promoters and precursors of Arabidopsis thaliana non-conserved microRNAs. Bioinformatics 22: 2585–2589

Wang T, Stormo GD (2003) Combining phylogenetic data with co-regulated genes to identify regulatory motifs. Bioinformatics 19:2369–2380

Wang JW, Wang LJ, Mao YB, Cai WJ, Xue HW, Chen XY (2005) Control of root cap formation by MicroRNA-targeted auxin response factors in Arabidopsis. Plant Cell 17:2204–2216

Wierzbicki AT, Haag JR, Pikaard CS (2008) Noncoding transcription by RNA polymerase Pol IVb/Pol V mediates transcriptional silencing of overlapping and adjacent genes. Cell 135: 635–648

Wierzbicki AT, Ream TS, Haag JR, Pikaard CS (2009) RNA polymerase V transcription guides ARGONAUTE4 to chromatin. Nat Genet 41:630–634

Wightman B, Ha I, Ruvkun G (1993) Posttranscriptional regulation of the heterochronic gene lin-14 by lin-4 mediates temporal pattern formation in C. elegans. Cell 75:855–862

Wingender E, Chen X, Hehl R, Karas H, Liebich I, Matys V, Meinhardt T, Pruss M, Reuter I, Schacherer F (2000) TRANSFAC: an integrated system for gene expression regulation. Nucleic Acids Res 28:316–319

Won SK, Lee YJ, Lee HY, Heo YK, Cho M, Cho HT (2009) Cis-element- and transcriptome-based screening of root hair-specific genes and their functional characterization in Arabidopsis. Plant Physiol 150:1459–1473

Wu G, Park MY, Conway SR, Wang JW, Weigel D, Poethig RS (2009) The sequential action of miR156 and miR172 regulates developmental timing in Arabidopsis. Cell 138:750–759

Wu MF, Tian Q, Reed JW (2006) Arabidopsis microRNA167 controls patterns of ARF6 and ARF8 expression, and regulates both female and male reproduction. Development 133:4211–4218

Xie Z, Allen E, Fahlgren N, Calamar A, Givan SA, Carrington JC (2005) Expression of Arabidopsis MIRNA genes. Plant Physiol 138:2145–2154

Xie Z, Johansen LK, Gustafson AM, Kasschau KD, Lellis AD, Zilberman D, Jacobsen SE, Carrington JC (2004) Genetic and functional diversification of small RNA pathways in plants. PLoS Biol 2:E104

Yamada K, Lim J, Dale JM, Chen H, Shinn P, Palm CJ, Southwick AM, Wu HC, Kim C, Nguyen M et al (2003) Empirical analysis of transcriptional activity in the Arabidopsis genome. Science 302:842–846

Yamamoto YY, Ichida H, Abe T, Suzuki Y, Sugano S, Obokata J (2007a) Differentiation of core promoter architecture between plants and mammals revealed by LDSS analysis. Nucleic Acids Res 35:6219–6226

Yamamoto YY, Ichida H, Matsui M, Obokata J, Sakurai T, Satou M, Seki M, Shinozaki K, Abe T (2007b) Identification of plant promoter constituents by analysis of local distribution of short sequences. BMC Genomics 8:67

Yamamoto YY, Obokata J (2008) ppdb: a plant promoter database. Nucleic Acids Res 36: D977–D981

Yamamoto YY, Yoshitsugu T, Sakurai T, Seki M, Shinozaki K, Obokata J (2009) Heterogeneity of Arabidopsis core promoters revealed by high-density TSS analysis. Plant J 60:350–362

Yamasaki H, Abdel-Ghany SE, Cohu CM, Kobayashi Y, Shikanai T, Pilon M (2007) Regulation of copper homeostasis by micro-RNA in Arabidopsis. J Biol Chem 282:16369–16378

Yamasaki H, Hayashi M, Fukazawa M, Kobayashi Y, Shikanai T (2009) SQUAMOSA promoter binding protein-Like7 Is a central regulator for copper homeostasis in Arabidopsis. Plant Cell 21:347–361

Yoshikawa M, Peragine A, Park MY, Poethig RS (2005) A pathway for the biogenesis of trans-acting siRNAs in Arabidopsis. Genes Dev 19:2164–2175

Zhang Y (2005) miRU: an automated plant miRNA target prediction server. Nucleic Acids Res 33:W701–W704

Zhang W, Ruan J, Ho TH, You Y, Yu T, Quatrano RS (2005) Cis-regulatory element based targeted gene finding: genome-wide identification of abscisic acid- and abiotic stress-responsive genes in Arabidopsis thaliana. Bioinformatics 21:3074–3081

Zhou GK, Kubo M, Zhong R, Demura T, Ye ZH (2007) Overexpression of miR165 affects apical meristem formation, organ polarity establishment and vascular development in Arabidopsis. Plant Cell Physiol 48:391–404

Chapter 4
Natural Variation in *Arabidopsis thaliana*

Maarten Koornneef, Matthieu Reymond, and Carlos Alonso-Blanco

Abstract *Arabidopsis thaliana* is a wild species widely distributed in diverse environments and current resources allow efficient quantitative analyses aimed to identify the genetic and molecular bases of adaptation. The study of natural genetic variation in this model plant has rapidly developed in the past 10 years, leading to the identification of hundreds of loci that are responsible for the variation of a plethora of traits and more than 30 of the underlying genes. This knowledge can be used for the identification of genes also relevant for crop breeding. Particularly, related species of *A. thaliana* such as *Brassica* sp. may benefit from this information because current genomic information is providing detailed knowledge of genetic synteny among these species. In this chapter, we summarize the approaches that are followed to dissect *A. thaliana* intraspecific variation. In addition, the main results obtained up to now are described considering current possibilities to transfer them to *Brassica* crops.

Keywords Arabidopsis · Brassica · Natural variation · Quantitative trait locus (QTL) · Polymorphism

Abbreviations

AIL Advanced Intercross line
AM Association Mapping
BIL Backcross Inbred Line
IL Introgression Line
LD Linkage Disequilibrium
QTL Quantitative Trait Locus
RIL Recombinant Inbred Line

M. Koornneef (✉)
Department of Plant Breeding and Genetics, Max Planck Institute for Plant Breeding Research, 50829 Cologne, Germany
e-mail: koornnee@mpipz.mpg.de

R. Schmidt, I. Bancroft (eds.), *Genetics and Genomics of the Brassicaceae*, Plant Genetics and Genomics: Crops and Models 9, DOI 10.1007/978-1-4419-7118-0_4,

Contents

4.1 Introduction 124
4.2 Geographical Distribution and Demographical History of *A. thaliana* 126
4.3 Genetic and Molecular Analysis of *A. thaliana* Natural Variation 127
4.4 Genetic Bases of Adaptation: QTL Underlying *A. thaliana* Natural Variation 130
4.5 Molecular Bases of Adaptation: Genes Underlying *A. thaliana* Natural Variation . . 137
4.6 The Use of *A. thaliana* Genetic Information in *Brassica* 139
References 141

4.1 Introduction

Genetic variation present among wild genotypes (accessions) of *Arabidopsis thaliana* collected from nature already attracted Prof. Laibach, the founder of *A. thaliana* research (Laibach 1943). In the past decade, increasing interest has been dedicated to this natural genetic variation mainly due to the efficiency to perform quantitative genetics in this model species (reviewed in Alonso-Blanco and Koornneef 2000). In addition, the possibility to do reverse genetics, which has already provided mutants in nearly every gene (Alonso et al. 2003), together with the availability of genomic tools such as nearly complete genomic sequences of several genotypes and whole genome microarrays, is facilitating the analyses of this variation down to gene and nucleotide levels.

A. thaliana is a wild species with a broad distribution in the northern hemisphere (Fig. 4.1a) (Al-Shebaz and ÓKane 2002, Hoffmann 2002). *A. thaliana* populations grow in very different world climates and habitats, including anthropogenic environments such as agricultural fields, stone walls, and roadsides, as well as naturally disturbed sites like rocky slopes and Mediterranean forests (Fig. 4.1b) (Hoffmann 2005, Picó et al. 2008). The large climatic and environmental variation among the locations where *A. thaliana* populations occur suggests that they are locally adapted, and part of the intraspecific variation observed for many traits is presumed to reflect adaptive variation that is now amenable to genetic and molecular analysis. Hence, *A. thaliana* natural variation provides an important resource to uncover the molecular bases of plant adaptation to different environments. Gene functions involved in plant survival and adaptation can be identified by artificially induced mutant analyses of different wild genotypes, where mutants with reduced fitness are easily selected. However, current mutant collections have been obtained using a limited number of *A. thaliana* laboratory strains such as Col (Columbia), L*er* (Landsberg *erecta*), Ws (Wassilevkajia), and C24, which only harbor a small portion of *A. thaliana* natural variation. Interestingly, it has been shown that around 9.4% of *A. thaliana* protein-coding genes are naturally absent or “knockout” mutated in some accessions (Clark et al. 2007), limiting the mutant spectra that can be found from each accession. Therefore, *A. thaliana* natural variation provides a relevant resource to find out such gene functions as well as those allelic variants that interact with the genetic background and/or the environment, or alleles showing

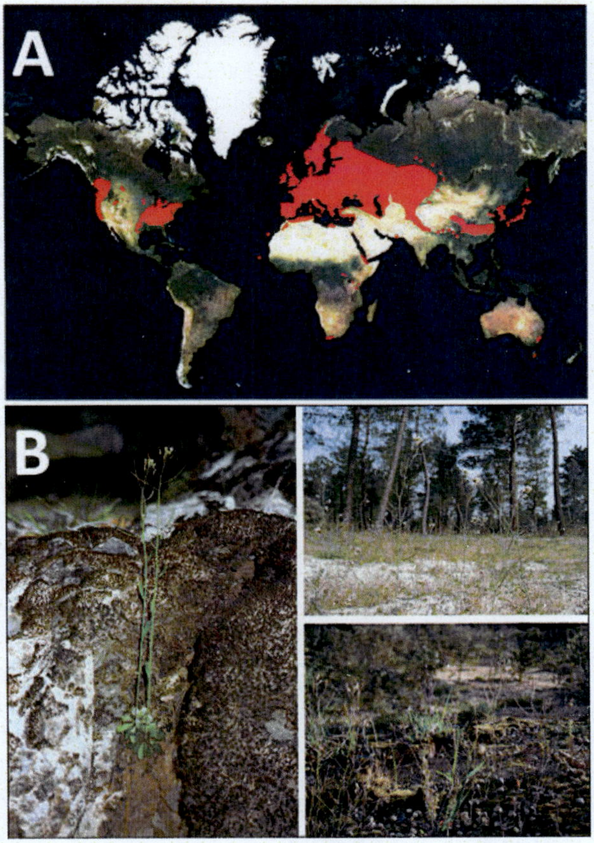

Fig. 4.1 Natural variation of *Arabidopsis thaliana*. **a** Geographic distribution range. **b** *A. thaliana* populations living on a rocky site (*left*), a pine forest (*top right*) and a holm oak forest, in the Iberian Peninsula

small effects on phenotype, particularly for traits related to plant adaptation (Benfey and Mitchell-Olds 2008).

A. thaliana natural variation is comparable to the variation exploited for plant breeding within crop species of the Brassicaceae or other taxonomic families. Most genetic variation present in cultivated species has also arisen spontaneously and has been maintained by artificial selection during the domestication and breeding processes. However, in contrast to the effects of natural selection on *A. thaliana* variation, artificial selection has mainly operated on certain traits related to plant growth or yield of specific organs (roots, leaves, fruits, and seeds) to increase fitness in agricultural environments. Despite differences in the selection processes, many traits of interest in plant breeding such as tolerance to biotic and abiotic environmental factors, biochemical (e.g., metabolite) compositions, flowering time, and seed dormancy show variation both in wild and cultivated plants and are relevant to the adaptation of both types of species to their particular environments. In addition, there is no major methodological difference in the strategies used to analyze natural variation in crops and wild species, because in both situations quantitative trait variations are due to the combined effects of multiple genes and the environment, which affect the expression of genes and traits. The genetic bases of natural variation are determined by quantitative genetic approaches, which in practice consist of Quantitative Trait Locus (QTL) analyses, where phenotypic variation is associated with allelic variation of molecular markers. As described in the chapters dealing with the genetics of *Arabidopsis* relatives such as *Arabidopsis lyrata*, *Capsella*, and *Brassica* crop species, minor methodological differences between species depend on specific biological requirements related to the induction of germination, flowering, and reproductive properties such as incompatibility, which condition the types of mapping populations, mutant collections, and transformation technologies.

In the context of comparison with Brassicaceae species, the analysis of natural variation of *A. thaliana* is particularly relevant in the following aspects: (i) *A. thaliana* is a wild species and therefore will enable the understanding of ecological and evolutionary processes that generate and maintain genetic variation related to plant adaptation of wild plant species with wide distribution and with complex population dynamics largely influenced by human activities (Mitchell-Olds and Schmitt 2006). (ii) In addition, the current position of *A. thaliana* as a model species facilitates the analysis of natural variation of general traits related to adaptation in other wild and cultivated Brassicaceae species of which many are outcrossers in contrast to *A. thaliana*, which is a selfer. Genes identified in *A. thaliana* can be used as candidate sequences to underlie variation in other plants, which will become particularly relevant when map positions of trait QTL are known in *A. thaliana* and crop species and syntenic genomic regions have been established. In the past 5 years there have been more than 300 articles that have been published on the analysis of natural variation of *A. thaliana*, including more than 30 reviews dealing with different aspects of this source of variation (Alonso-Blanco et al. 2009). In this revision we describe the specific resources available for the analysis of natural variation of this species and summarize current knowledge on the genetic and nucleotide polymorphisms underlying *A. thaliana* natural variation for some traits related with plant adaptation, of general interest for other Brassicaceae species.

4.2 Geographical Distribution and Demographical History of *A. thaliana*

A. thaliana is a native species of the Eurasian continent, its distribution range spanning Europe from the Mediterranean Basin to the Artic circle, North Africa, Central Asia, and Southeast Asia (Fig. 4.1). Human activity has led to its introduction to other regions, including North America, Japan, and East Australia (Al-Shebaz and ÓKane 2002, Hoffmann 2002). The center of origin of *A. thaliana* remains unknown, but it is speculated to be in Europe/North Africa or central Asia/Caucasus regions (Hoffmann 2005). Accessions from central Europe have long been available through the stock centers (http://www.arabidopsis.org, Chapter 16). In addition, new collections from other regions have recently become available, including accessions from central Asia, China, the Mediterranean Peninsulas, and the Caucasus (Jorgensen and Mauricio 2004, Stenoien et al. 2005, Schmid et al. 2006, Beck et al. 2008, He et al. 2007, Picó et al. 2008).

Current *A. thaliana* geographical distribution is the result of multiple migration/colonization and extinction/spatial contraction events since its divergence from the rest of the genus at the beginning of Pliocene, about 5.8 million years ago (Koch et al. 2001). *A. thaliana* demographical history is beginning to be inferred from geographical structure patterns of the presumably neutral genetic variation present in populations and/or accessions from different world regions. Population structure studies of worldwide collections have shown substantial geographical patterns at global scale (Sharbel et al. 2000, Nordborg et al. 2005, Schmid et al. 2006, François et al. 2008). These analyses have identified differentiated genetic groups mainly distributed in central Asia, northern and western Europe and southern Europe, suggesting that Asia and the Mediterranean regions contained *A. thaliana* glacial refugia. Frequency distribution of polymorphisms show two longitudinal gradients that support east-to-west and west-to-east migration waves, suggesting the colonization of central Europe from at least an Eastern refugium and an Iberian Peninsula refugium since the last glaciation 10,000 years ago (Schmid et al. 2006, François et al. 2008, Picó et al. 2008). In agreement, regional analyses of the variation present in the Mediterranean Peninsulas and the Caucasus have shown larger genetic diversity within populations of these regions than of other world regions (Beck et al. 2008, Picó et al. 2008). Regional analyses of nuclear and chloroplast genetic variation have also identified several differentiated groups within particular regions, which open various hypotheses related with the origin and expansion of *A. thaliana*. First, the presence of inferred ancestral haplotypes in the Caucasus region suggested that this might be *A. thaliana* center of origin (Beck et al. 2008). Second, genetic variation is geographically structured at regional scale in the Iberian Peninsula suggesting that there were several glacial refugia in this region with different contribution to the colonization of Europe (Picó et al. 2008). It has been proposed that only northern Iberian genetic groups might have contributed to the colonization of western Europe, while southern Iberian lineages are more related to accessions from the Mediterranean Basin reflecting Mediterranean expansion before or after the last glaciations (Picó et al. 2008). Third, several world regions that have been presumably recently colonized such as northern Europe and Japan show considerable genetic variation, suggesting their colonization from several sources (Stenoien et al. 2005, Jakobsson et al. 2007, Ostrowski et al. 2006, Picó et al. 2008). It is expected that genomic analyses of currently available accessions and new collections from underrepresented regions will further decipher *A. thaliana* demographical history and will help to understand the ecological and evolutionary relevance of traits presumably involved in adaptation to different environments.

4.3 Genetic and Molecular Analysis of *A. thaliana* Natural Variation

Genetic analysis of natural variation aims to determine the genetic architecture of quantitative traits, including estimation of the number of QTL, their position, their relative and absolute effects, their genetic interactions (QTL $\times$ QTL), and their interaction with environment (QTL × E) (Alonso-Blanco et al. 2006). This analysis is referred to as QTL mapping and it first requires experimental populations where genetic variation is segregating among lines. F_1 hybrid plants generated by controlled crosses between two parental accessions can be selfed to derive the corresponding F_2 population or crossed with a (recurrent) parent to obtain a backcross population. These kinds of populations have been used for QTL analysis of some traits (Kowalski et al. 1994, Kuittinen et al. 1997, Werner et al. 2005a) but they need to be genotyped in every study due to their high heterozygosity. In addition, they can be further selfed by single seed descent or backcrossed during several generations, to obtain populations of recombinant inbred lines (RILs) or introgression lines (ILs; also called backcrossed inbred lines, -BILs-) respectively. Both types of populations belong to the so-called permanent or immortal populations because they have very low heterozygosity and consequently, they can be propagated indefinitely and analyzed in replicates and/or in different environments for any trait. These are the most common mapping populations used in *A. thaliana* because they are easily generated due to its short generation time and its selfing nature. RIL populations are composed of genotypes that, on average, carry half of the genome from each parental accession. In contrast, IL populations consist of genotypes that carry a small number of chromosome segments corresponding to an average of less than 25% from one parental accession introgressed into the genetic background of another accession. Compared to RILs, IL populations show higher power to map small effect QTL, lower localization resolution, and they do not allow direct detection of genetic interactions among QTL (Keurentjes et al. 2007a). Currently there are 23 *A. thaliana* RIL populations publicly available, derived from 24 different accessions (Lister and Dean 1993, Alonso-Blanco et al. 1998, Deslandes et al. 1998, Wilson et al. 2001, Loudet et al. 2002, Clerkx et al. 2004, Mouchel et al. 2004, Magliano et al. 2005, Törjék et al. 2006, El-Lithy et al. 2006, Pfalz et al. 2007, Simon et al. 2008, ÓNeill et al. 2008). In addition, there are three genome-wide IL populations involving the most common laboratory strains (Keurentjes et al. 2007a, Törjék et al. 2008) and several other RIL populations are under development (http://www.inra.fr/internet/Produits/vast/RILs.htm). Most RIL populations have small sizes of less than 200 lines, but some of them contain more than 300 lines (Loudet et al. 2002, Törjék et al. 2006, Simon et al. 2008) enabling higher mapping accuracy and the detection of small effect QTL. Furthermore, permanent populations of higher complexity are beginning to be developed by randomly intercrossing two or more parental accessions, which are referred to as advanced intercross line (AIL) populations (Scarcelli et al. 2007). It is expected that such populations will allow the simultaneous analysis of QTL accounting for the variation present among multiple accessions.

QTL mapping requires the genotyping of experimental populations with markers every 5–10 cM, to obtain the corresponding genetic maps. Currently, a nearly unlimited number of single nucleotide polymorphism (SNP) markers can be obtained from the resequencing of 876 fragments in 96 accessions (Nordborg et al. 2005) or from the nearly full genome resequencing of 20 accessions by high-density arrays (Clark et al. 2007, polymorph.weigelworld.org/) and many others with new generation short read sequencing (Ossowski et al. 2008, Weigel and Mott 2009). The application of new deep sequencing technologies will further provide direct information on candidate polymorphisms from many accessions. These sequences can be surveyed to search for nucleotide polymorphisms using software programs specifically developed for that purpose, such as MSQT (Warthmann et al. 2007), and new parental accessions are then tested for polymorphism. In addition, large collections of microsatellite markers (Bell and Ecker 1994, Clauss et al. 2002, Loudet et al. 2002), InDels (Jander et al. 2002), and single feature polymorphisms (SFPs; Borevitz et al. 2007) are available (http://arabidopsis.org/), which can be genotyped with single marker techniques or high-throughput methods (www.1001genomes.org) such as microarray hybridization (Zhang et al. 2008). *A. thaliana* linkage maps show a total length of 412 ± 53 cM depending on the cross and marker coverage, but since its genome sequence is available, linkage maps constructed using sequenced markers can be precisely aligned. As expected from the small genome size and chromosome number, *A. thaliana* maps are the smallest among Brassicaceae species, which varies from 515 cM in *A. lyrata*; (Kuittinen et al. 2004) to 1,605 cM in *Brassica napus* (Sun et al. 2007). Finally, QTL mapping is carried out by statistical comparisons of phenotypic and genotypic data of experimental populations (Alonso-Blanco et al. 2006), which usually identify between 1 and 15 QTL per trait. QTL show very variable additive effects ranging from larger than 20% (e.g., biochemical traits such as seed soluble oligosaccharide content, Bentsink et al. 2000) to smaller than 5% (e.g., complex yield traits such as seed size; Alonso-Blanco et al. 1999). However, most QTL analyses have detected a larger number of moderate and minor effect QTL (<20% of phenotypic variance) than large effect QTL. Currently, the overall genome collinearity or synteny between the genetic maps of *A. thaliana* and several Brassicaceae such as *A. lyrata* or *Brassica* sp. is well established by using markers of known sequence (Kuittinen et al. 2004, Choi et al. 2007), which allows further comparison of QTL map positions among different species.

A. thaliana QTL correspond, on average, to genomic regions of 1.2–12 Mb containing 240–2,400 open reading frames. Molecular identification of genes and nucleotide polymorphisms underlying QTL involves thorough QTL fine mapping using new specific recombinant populations within the QTL region which can be derived from particular ILs (El-Assal et al. 2001) or heterogeneous inbred families (Loudet et al. 2008). Thereafter, a combination of different functional strategies allows the identification and test of candidate genes, including in silico analysis of genome sequences and predicted gene functions; phenotypic analysis and allelism tests with mutants available in reverse genetics platforms of T-DNA and transposon insertions (Alonso et al. 2003) or TILLING, where induced point mutations in candidate genes can be directly screened for among collections of plants derived from chemical mutagenesis experiments (Greene et al. 2003); generation and analysis of new mutants in specific accessions (e.g., *FLC, DOG1, and LPR1*, Michaels and Amasino 1999, Bentsink et al. 2006, Svistoonoff et al. 2007); sequencing and association analysis of nucleotide polymorphisms in candidate genes (e.g., *PhyC*, Balasubramanian et al. 2006); and expression analyses (e.g., *CBF2*,

Alonso-Blanco et al. 2005). In most cases, genetic complementation in transgenic plants carrying different natural alleles is allowing to conclusively demonstrate the genes and nucleotide polymorphisms causing *A. thaliana* trait variation.

Furthermore, a new type of permanent mapping population consisting of collections of wild accessions is being developed for so-called genome-wide linkage disequilibrium (LD) or association mapping (AM). It is expected that comparisons of phenotypic and genotypic data of these collections will enable detection of marker-trait associations due to linkage between markers and causal genes. This is based on the large number of historical recombination events that are segregating in such collections, determining that mainly closely linked nucleotide polymorphisms will show certain LD. Genome-wide sequencing analyses of 20 and 96 accessions indicate that *A. thaliana* LD decays within 10–25 kb (Nordborg et al. 2005, Kim et al. 2007). Thus, AM might locate the loci causing trait variation with an unprecedented accuracy within those LD intervals. However, this requires an extremely high density of markers of at least of 140,000 polymorphisms, which can now be achieved by microarray hybridization or high-throughput genome sequencing of accessions (Kim et al. 2007). In addition, the statistical power to detect significant associations depends on the number, frequency, and effect of the alleles at each causal gene. Given the large history and genetic variation of *A. thaliana* this power might be rather limited in worldwide collections of accessions. Moreover, AM shows a high proportion of spurious associations due to unknown genetic relationships among accessions, i.e., the population genetic structure (Aranzana et al. 2005). Currently, large efforts are devoted to develop new statistical methods and population designs that take into account genetic structure, which might reduce the rate of false-positive associations (Zhao et al. 2007). So far, no previously unknown gene responsible for quantitative trait variation has been identified in *A. thaliana* using genome-wide AM. However, this method appears as a very promising tool to find candidate genes within QTL confidence intervals identified in standard mapping populations.

4.4 Genetic Bases of Adaptation: QTL Underlying *A. thaliana* Natural Variation

There is a plethora of traits directly or indirectly related to plant adaptation that show natural variation in *A. thaliana*. Many of these traits have been dissected by QTL analyses and are briefly described below, including those listed in Table 4.1 for which the underlying genes have been identified. A relevant life-history trait extensively studied is flowering time, which has been analyzed in nearly all *A. thaliana* mapping populations. Since the genetic bases of this trait are well described, flowering time has been often used as a proof of concept to test new mapping methods (e.g. genome-wide association mapping, Aranzana et al. 2005, Zhao et al. 2007) or for comparisons of mapping power between RIL and NIL populations (Keurentjes et al. 2007a). Thus, it has been established that most late flowering accessions carry active alleles at the *FRI* and *FLC* loci, which genetically interact to delay flowering

Table 4.1 Genes involved in natural varation of *Arabidopsis thaliana*

Gene	Atg number	Traits affected	Molecular function	Functional polymorphism	References
Development and growth					
FRI	At4g00650	Flowering time	Unknown	Ins/del	Johansson et al. (2000)
FRL1	At5g16320	Flowering time	FRI paralogue	Nons subst	Schläppi (2006)
FRL2	At1g31814	Flowering time	FRI paralogue	Miss subst	Schläppi (2006)
FLC	At5g10140	Flowering time	MADS TF	TE insertion in intron	Michaels and Amasino (1999)
FLM	At1g77080	Flowering time	MADS TF	del (6.8 kbp)	Werner et al. (2005b)
HUA2/ART	At5g23150	Flowering time /shoot morphology	unknown	Miss subst	Doyle et al. (2005); Wang et al. (2007)
DOG1	At5g45830	Seed dormancy	Unknown	Promoter variation ?	Bentsink et al. (2006)
CBF2/DREB2	At4g25470	Frost tolerance	AP2 TF	Promoter del	Alonso-Blanco et al. (2005)
CRY2	At1g04400	Flowering time, seedling growth	Photoreceptor	Miss subst protein stability	El-Assal et al. (2001), Botto et al. (2003)
PHYA	At1g09570	Seedling growth	Photoreceptor	Miss subst, altered protein	Maloof et al. (2001)
PHYB	At2g18790	Seedling growth	Photoreceptor	Miss subst, altered protein	Filiault et al. (2008)
PHYC	At5g35840	Flowering time, seedling growth	Photoreceptor	Nonsense subst	Balasubramanian et al. (2006)
PHYD	At4g16250	Seedling growth, flowering time	Photoreceptor	del (14 bp)	Aukermann et al. (1997)
TZP	At5g43630	Blue light associated seedling growth	Zinc knuckle protein	8 bp insertion	Loudet et al. (2008)
GL1	At3g27920	Trichome formation	Myb TF	del (sbp)	Hauser et al. (2001)
CAL	At1g26310	Flower morphology	MADS TF	Miss subs	Kempin et al. (1995)
MUM2	At5g63800	Seed mucilage	Enzyme	del (44 bp) in exon	Macquet et al. (2007)

Table 4.1 (continued)

Gene	Atg number	Traits affected	Molecular function	Functional polymorphism	References
ER	At2g26330	Growth habit, stomatal patterning, disease tol.	LRR receptor	Unknown	Shpak et al. (2005); Masle et al. (2005)
LPR1	At1g23010	Root architecture under low P	Enzyme	del (16 bp) in promoter	Svistoonoff et al. (2007)
Plant growth					
BRX	At1g31880	Root growth	Novel TF	Nonsense subst	Mouchel et al. (2004)
INV	At1g12240	Sugar composition, root growth	Enzyme	Unknown	Sergeeva et al. (2006)
small effect QTL	At5g23170	Biomass accumulation	Protein kinase	Unknown	Kroyman and Mitchell-Olds (2005)
Hybrid necrosis and intraspecifc incompatibility					
DM1	At5g41740/50	Hybrid necrosis, disease resistance	NB-LRR protein	Stop codon and several deletions	Bomblies et al. (2007)
Mineral accumulation					
MOT1	At2g25680	Mo accumulation	Transporter	del (53 bp) in promoter	Baxter et al. (2008)
HKT1	At4g10310	Na accumulation	Na transporter	del (−725 and 687) in promoter	Rus et al. (2006)
APR2	At1g62180	S accumulation	Enzyme	Miss subst	Loudet et al. (2007)

Table 4.1 (continued)

Gene	Atg number	Traits affected	Molecular function	Functional polymorphism	References
Primary and secondary metabolism					
PAP1/MYB75	At1g56650	Anthocyanin synthesis under stress	Myb TF	Miss subst	Teng et al. (2005)
MAM1/2	At5g23010/20	Glucosinolates, insect tolerance	Enzyme	Gene deletion	Kroymann et al. (2003)
AOP1	At4G03070	Hydroxyalkyl glucosinolate synthesis	Enzyme	Unknown	Kliebenstein et al. (2001)
AOP2	At4G03060	Hydroxyalkyl glucosinolate synthesis	Enzyme	Unknown	Kliebenstein et al. (2001)
AOP3	At4G03050	Hydroxyalkyl glucosinolate synthesis	Enzyme	Unknown	Kliebenstein et al. (2001)
ESM	At3g14210	Glucosiniolate levels, insect tolerance	MyAP myrosinase-binding protein	Unknown	Zhang et al. (2006)
ESP	At1g54040	Glucosinolate type and insect tolerance	Epithiospecifier protein	del (124) and unknown miss subst	Lambrix et al. (2001)
Disease-resistance					
RTM1	At1g05560	Virus (TEV) movement	Jacalin lectin fam protein	Miss subst	Chrisholm et al. (2000)
RPM1/RPS3	At3g07040	Resist. to *Pseudomonas syringae*	CC-NB-LRR R protein	Gene deletion	Grant et al. (1995)
RPS4	At5g45250	Resist. to *P. syringae*	TIR-NB-LRR protein	Miss subst	Gassmann et al. (1999) and Bergelson et al. (2001)
RPS5	At1g12220	Resist. to *P. syringae*	CC-NB-LRR protein	Gene deletion	Warren et al. (1998) and Bergelson et al. (2001)

Table 4.1 (continued)

Gene	Atg number	Traits affected	Molecular function	Functional polymorphism	References
RPS2	At4g26090	Resist. to *P. syringae*	CC-NB-LRR protein	Indel/miss subst	Mindrinos et al. (1994)
RRS1	At5g45260	Resist. to *Ralstonia solanacearum*	TIR-NB-LRR WRY protein	Various miss subst	Deslandes et al. (2002)
RPP1/10/14	At3g44630	Resist. *Hyaloperonospora parasitica*	TIR-NB- LRR protein	Complex deletion	Botella et al. (1998)
RPP5	At4g16950	Resist. *H. parasitica*	TIR-NB- LRR protein	Complex deletion	Parker et al. (1997)
RPP8/HRT	At5g43470	Resist. *H. parasitica* and turnip crickle virus	CC-NB-LRR	Copy number	McDowell et al.(1998), Cooley et al. (2000)
RPP13	At3g46530	Resist. *H. parasitica*	CC-NB-LRR	Miss subst	Rose et al. (2004)
RPP27	At1g54470	Resist. *H. parasitica*	Receptor-like protein (RLP)	Miss subst and deletion	Tör et al. (2004)
RLM1	At1g64070	Resist. to *Leptosphaeria maculans*	TIR-NB-LRR R gene	Gene del/ and insertion	Staal et al. (2006)
RPW8	At3g50450/60/70	Resist. to powdery mildew	Atypical R gene	Gene copy variation and miss subst	Xiao et al. (2001, 2004)

TE: transposon element; miss: missense; nons: nonsense; subst: substitution; del: deletion; ins: insertion; ?: not demonstrated.

time (Koornneef et al. 2004, Shindo et al. 2006). In addition to *FRI* and *FLC* loci, more than 25 flowering time QTL have been detected in different populations (reviewed in Koornneef et al. 2004, Simon et al. 2008, ÓNeill et al. 2008) indicating that allelic variation at many other loci contribute to this variation. Currently the genes underlying seven other *A. thaliana* QTL have been already isolated and most of them are characterized by unique or very low frequency alleles (Table 4.1). Flowering time is regulated by environmental factors (photoperiod, light intensity, temperature, plant density, mineral nutrition, etc.) and significant QTL × E interactions have been detected under different controlled "laboratory" environments (Botto and Coluchio 2007, Zhao et al. 2007, Werner et al. 2005a, b). Furthermore, flowering time QTL have been also analyzed in mapping populations grown in outdoor field conditions, under fluctuating environments, and in contrasting seasons (Weinig et al. 2002, 2003, Korves et al. 2007) as well as in growth chambers that mimic environments closer to nature with respect to light and temperature factors (Li et al. 2006). However, QTL analyses under "natural" conditions require specific experimental developments to reduce human-mediated interference such as those proposed by Frenkel et al. (2008). Interestingly, although the major QTL for flowering time and other *A. thaliana* fitness-related traits found in laboratory environments are also detected under field conditions, a large number of environment specific QTL indicate that QTL × E is very common (Malmberg et al. 2005). Another life-history trait of general interest is seed dormancy, but this has been analyzed in a small number of *A. thaliana* populations, where more than 10 QTL have been detected (Alonso Blanco et al. 2003, Clerkx et al. 2004, Laserna et al. 2008).

At the morphological level, several traits related to development and growth have been also studied and QTL have been detected for root growth and architecture (Mouchel et al. 2004, Loudet et al. 2005, Fitz Gerald et al. 2006), vegetative growth (El-Lithy et al. 2004, Meyer et al. 2007), leaf architecture (Perez-Perez et al. 2002), leaf trichome density (Symonds et al. 2005), and floral morphology (Juenger et al. 2000). Comparative QTL mapping of organ and cell size have been also performed (Reymond et al. 2006, Tisné et al. 2008). Detailed QTL analyses have been carried out to dissect *A. thaliana* heterosis for biomass-related traits using triple testcross designs with RILs and ILs (Kusterer et al. 2007a, Melchinger et al. 2007). It has been shown that *A. thaliana* is also a very useful model to understand negative heterosis or hybrid necrosis, a particular intraspecific incompatibility, for which several QTL have been identified (Bomblies et al. 2007, Alcázar et al. 2009). Interestingly, several of these studies have shown that epistasis is a substantial genetic component of heterosis, hybrid necrosis, and other growth-related traits (Alcázar et al. 2009, Bomblies et al. 2007, Kusterer et al. 2007b, Reif et al. 2008, Reymond et al. 2006, Tisné et al. 2008).

QTL involved in the response of growth-related traits to biotic and abiotic factors have been also extensively studied in *A. thaliana*. QTL have been reported for growth responses to abiotic factors such as light (Magliano et al. 2005, Maloof et al. 2001, Botto et al. 2003, Botto and Coluccio 2007) or nutrients like nitrogen (Rauh et al. 2002, Loudet et al. 2003) or phosphate (Reymond et al. 2006). QTL analyses have been performed for tolerance to water stress (Hausmann et al. 2005, Juenger et al. 2005), salt stress (Quesada et al. 2002), freezing temperature (Alonso-Blanco et al. 2005), boron (Zeng et al. 2008), selenate (Zhang et al. 2006a), and ozone (Tamaoki et al. 2003). Comparison of *A. thaliana* accessions has shown significant natural genetic variation for plastic responses of growth traits across a CO_2 gradient (Tonsor and Scheiner 2007), but analysis of a RIL population did not reveal genetic variation for this plastic responses (Lau et al. 2007). A large number of studies have identified QTL accounting for variation in tolerance or resistance to fungal, bacterial, or viral pathogens (Denby et al. 2004, Kover et al. 2005, Kover and Cheverud, 2007, Perchepied et al. 2006, Göllner et al. 2008, Jubault et al. 2008, Rowe and Kliebenstein 2008, Sicard et al. 2008) or to herbivores and its relationship with glucosinolate content (Kliebenstein et al. 2002, Kroyman et al. 2003, Pfalz et al. 2007). Moreover, several studies have identified QTL involved in growth responses to different hormones (Borevitz et al. 2002, Lall et al. 2004, Millenaar et al. 2005).

QTL analyses have been also carried out for variation in chemical composition of different organs, including primary and secondary metabolites. In case of seed compounds, QTL have been characterized for oil content (Hobbs et al. 2004), phytate levels (Bentsink et al. 2003), and for soluble oligosaccharides (Bentsink et al. 2000). It has been shown that QTL accounting for primary metabolite content often co-locate with QTL for enzymatic activities (Sergeeva et al. 2006, Keurentjes et al. 2008). Among secondary metabolites, *A. thaliana* natural variation has been instrumentally important for resolving the biosynthetic pathway of glucosinolates (reviewed in Yan and Chen 2007). Currently, large efforts are also focused to identify QTL related to mineral nutrients and trace element composition, which together constitute the so-called ionome (Rus et al. 2006, Baxter et al. 2008, Harada et al. 2004, Loudet et al. 2008, Salt et al. 2008, Vreugdenhil et al. 2004).

A current challenge in the genetic dissection of natural variation is to link QTL involved in complex traits with QTL affecting the abundance of gene expression (transcript), metabolites, or proteins. These loci are referred to as eQTL, mQTL, and pQTL respectively, and allow the genetic integration of the different levels of molecular regulation with organism phenotypic variation. Genomic analyses of gene expression have been recently applied to mapping populations, showing enormous natural variation in global gene expression (DeCook et al. 2006, van Leeuwen et al. 2007, Vuylsteke et al. 2005, Tan et al. 2007). The first genome-wide eQTL mapping analyses in *A. thaliana* have shown that the expression variation of about 30–40% of the segregating genes is regulated by eQTL located on the gene encoding the mRNA (cis-eQTL). In addition, many of the eQTL located in other genomic regions (trans-eQTL) appear concentrated in a small number of regulatory hotspots that are integrated in regulatory networks (Kliebenstein et al. 2006, Keurentjes et al. 2007b, West et al. 2007). Similar genetic analyses have been carried out for the abundance of different kind of untargeted metabolites analyzed in metabolomic platforms (Keurentjes et al. 2006, Meyer et al. 2007, Rowe and Kliebenstein 2008). Thus, large numbers of mQTL have been mapped, showing that epistatic interactions control large portion of the variation (Rowe and Kliebenstein 2008). Further interactions between metabolism and growth have been shown when comparing mQTL and growth traits, since a specific combination of metabolites was associated with biomass but this was not due to co-location of biomass QTL and mQTL (Meyer et al. 2007). Comparative QTL mapping has also established a direct relationship between invertase activity and root growth, which was confirmed by mutant analysis (Sergeeva et al. 2006). Moreover, simultaneous analyses of eQTL and mQTL involved in carbohydrate metabolism have revealed new regulatory components that could not be detected by single trait QTL analyses (Keurentjes et al. 2008). It is expected that in the near future, similar high-throughput analyses will be carried out with proteins to map pQTL. Integration of the genetic variation for the various sources of "omics" data and the multiple organism phenotypic levels will enable the exploitation of *A. thaliana* natural variation for system biology studies (Benfey and Mitchell-Olds 2008). However, such integrative and comparative analyses await further developments of new bioinformatics genomic tools to collect, handle, and analyze the unprecedented amount of data that are being generated.

4.5 Molecular Bases of Adaptation: Genes Underlying *A. thaliana* Natural Variation

Around 30 genes underlying natural variation related to development and plant growth and more than a dozen genes involved in variation for plant pathogen resistance have been already identified in *A. thaliana* (Table 4.1). Most genes correspond to QTL with large effect on trait variation. However, one remarkable exception has approached the dissection of two closely linked QTL of small effect on growth rate, suggesting that some traits might have a much more complex polygenic architecture than that determined by QTL analysis (Kroymann and Mitchell-Olds 2005). The genes accounting for *A. thaliana* natural variation belong to all types of ontological classes, including genes-encoding transcription factors, signal transduction components, enzymes of primary and secondary metabolism, metal transporters, as well as genes with unknown functions (e.g., *FRI* and *DOG1*) (Table 4.1). Interestingly, the function of several of the genes involved in natural variation had not been previously found using mutant screens. The best known examples are the flowering time genes *FLC* and *FRI* that, as described above, interact genetically in a manner that active alleles are required in both genes to produce a strong late flowering phenotype (Michaels and Amasino 1999, Sheldon et al. 2000, Johanson et al. 2000). Since common laboratory strains carry natural loss of function alleles at one or both of these genes, they could not be identified in mutant analyses. Similarly, the *BRX* and *DOG1* loci identified unknown genes involved in the regulation of root growth and seed dormancy, respectively (Mouchel et al. 2004, Bentsink et al. 2006). In addition, several gain or change of function natural alleles of known genes have been identified, which had not been previously found in mutant analyses, such as those of photoreceptors *CRY2* (El-Assal et al. 2001), *PHYA* (Maloof et al. 2001), *PHYB* (Filiault et al. 2008), *PHYC* (Balasubramanian et al. 2006), and *PHYD* (Aukerman et al. 1997). In the case of *CRY2*, a specific gain of function allele was found to be caused by a single amino acid substitution with respect to the predominant wild-type allele (El-Assal et al. 2001). The nature of the various natural variants identified in photoreceptor genes and their phenotypic effects have been summarized in Filiault et al. (2008), showing that amino acid changes underlie natural variation, except for *PHYD* where a deletion led to a loss of function allele. Furthermore, several change of function alleles produced by mutations in regulatory regions affecting gene expression have been found for *FLC* (Gazzani et al. 2003, Michaels et al. 2003), *DOG1* (Bentsink et al. 2006) or *LPR1* (Svistoonoff et al. 2007). Some of the genes (*LPR1, MOT1, HKT1* and *CBF2*) have deletions in their promoter regions, thereby affecting steady-state transcript levels (Table 4.1). This high frequency of cis-acting regulatory polymorphisms is in agreement with the large number of eQTL that map in the vicinity of the genes encoding the mRNAs (cis-eQTL) (Keurentjes et al. 2007b, West et al. 2007). However, loss of function mutations appear as the commonest allelic variants accounting for many of the large effect QTL cloned until date (Table 4.1). Examples are the many independently arisen loss of function alleles of *FRI* (Hagenblad et al. 2004, Shindo et al. 2005), the seed mucilage gene *MUM2* (Macquet et al. 2007), the trichome formation gene *GL1* (Hauser et al. 2001), or the regulator of flavonoid biosynthesis *PAP1* (Teng et al. 2005). A particular class of these alleles that is also found corresponds to complete gene deletions, such as in the glucosinolate biosynthesis genes *MAM* (Kroymann et al. 2003), and *AOP* (Kliebenstein et al. 2001), the flowering time gene *FLM* (Werner et al. 2005b) and the *R* genes that are present in tandem repeats of variable number of copies (Alcázar et al. 2009). Although the phenotypic variation accounted by all *A. thaliana* QTL described until now is caused by variation at the level of nucleotide sequence, it is hypothesized that natural epigenetic variation might also be responsible for some heritable trait variation. This is supported by the considerable amount of natural variation found for DNA methylation (Riddle and Richards 2005, Zhang et al. 2006b, Zilberman et al. 2007, Vaughn et al. 2007) that is partly mediated by siRNAs, which also have been shown to differ among accessions (Zhai et al. 2008). Thus, it is postulated that methylation and heterochromatinization mediated by different kinds of siRNAs might cis- and trans-regulate gene expression and function in a more quantitative and reversible manner than sequence variation (Zhai et al. 2008).

Once the genes responsible of certain *A. thaliana* phenotypic variation have been isolated, a major current challenge is to demonstrate that their nucleotide polymorphisms are indeed involved in adaptation to particular environments. Given the large effect and the presence in a unique accession of several of the alleles identified, these might have originated from spontaneous mutations that occurred during laboratory multiplication. This has been shown for a loss of function allele of *HUA2* found only in subset of lines derived from the frequently used L*er* accession (Doyle et al. 2005). Similarly, the *LIGHT5* variant containing a small insertion was not found in sister lines derived from the same parental accession (Loudet et al. 2008). In addition, the *BRX* loss of function allele has not been found in genotypes collected in the same site where this variant was originally sampled many years before (Shindo et al. 2008). Thus, some of the alleles that are identified could also be deleterious alleles that are still segregating in natural populations. Analogous interpretations might be applied to natural loss of function alleles of the *ERECTA* and *GL1* genes, but the existence of several independent alleles suggests that they might have some advantage under particular environments (Hauser et al. 2001)

Demonstrating that the natural variation of these genes is involved in adaptation is currently addressed by combining population and evolutionary genetic analyses with detailed phenotypic studies of fitness-related traits. To this end, nucleotide diversity has been analyzed in collections of accessions for many genes such as *CRY2* (Olsen et al. 2004), *FRI* (Le Corre et al. 2002, Stinchcombe et al. 2004), *FLC* (Caicedo et al. 2004), *MAM2* (Kroymann et al. 2003) or several pathogen resistance genes like *RPM1* (Stahl et al. 1999), *RPS5* (Tian et al. 2003) and *RPS2* (Mauricio et al. 2003). The amount and pattern of nucleotide variation found in these genes have shown potential signatures of natural selection suggesting different evolutionary mechanisms to maintain such variation. However, demographical influences on the nucleotide pattern cannot be discarded and phenotypic analyses are necessary to test the fitness effect of such alleles. Thus, comparison of transgenic genotypes carrying the two alleles of *RPM1* in field conditions showed that the resistance allele reduces fitness in the absence of pathogen, which indicates that a cost of the resistance conferred by this gene contributes to maintain the *R* gene polymorphism (Tian et al. 2003). However, more complex trade-offs among different components of fitness might obscure the detection of fitness effects, as suggested for the *MAM2* gene and the differential effect of glucosinolates on specialist and generalist herbivores (Kroymann et al. 2003). Complementary to these approaches, it is expected that association between nucleotide variation in the causal genes and environmental factors will assist the detection of environments driving natural variation. Hence, it has been shown that variation at *FRI* and *FLC* contributes to a latitudinal cline of flowering time (Caicedo et al. 2004, Lempe et al. 2005, Stinchcombe et al. 2004). It is also expected that the new thorough collections of genotypes that are presently developed from several world regions (Jorgensen and Mauricio 2004, Stenoien et al. 2005, Schmid et al. 2006, Beck et al. 2008, He et al. 2007, Picó et al. 2008) will facilitate estimations of allele frequencies and geographical distributions and their association with environmental factors. Furthermore, phenotypic analyses aiming to test the involvement of natural variation in adaptation are currently carried out by several laboratories in common garden experiments, where different genotypes are grown under various natural conditions and assayed for fitness components. Thus, it has been shown that the shorter root variant caused by the loss of function allele of *BRX* has a slight fitness disadvantage in mixed plantings (Shindo et al. 2008). Finally, signatures of selection are also beginning to be tested at the level of local populations by comparing differentiation among populations for quantitative traits (Q_{ST}) and for nucleotide polymorphisms (F_{ST}) (Stenoien et al. 2005, Le Corre 2005). Hence, local adaptation might be evaluated while considering the genetic variation existing within populations, which is currently neglected.

4.6 The Use of *A. thaliana* Genetic Information in *Brassica*

A. thaliana information is assisting the genetic analysis of Brassicaceae, particularly that of *Brassica* crop species, mainly by providing candidate genes, based not only on gene sequence homology but also on similar QTL map positions, which can be efficiently identified when synteny among species is well established. This is effective because many genetic pathways affecting different common traits are rather conserved among species, and even natural variation has been shown to be caused by homologue genes in different species. This contribution is illustrated with the well-documented biochemical analyses of oil and glucosinolate composition, and with the study of flowering time, in different Brassicaceae species. QTL studies in *B. rapa* and *B. napus* have identified the fatty acid desaturase *FAD2* and *FAD3* genes that are orthologues of *A. thaliana* genes, as major determinants of the variation for oleic acid (Hu et al. 2006). In addition, the fatty acid elongase gene *FAE1* seems to contain allelic variation contributing to the content of seed erucic acid (Fourmann et al. 1998) and related compounds such as phytosterol and sinapate ester (Amar et al. 2008). The functionality of the *FAE1* loci in *B. napus* was recently confirmed by TILLING mutants of these genes (Wang et al. 2008). Similarly, *A. thaliana* has provided the candidate genes to identify the *BoGSL-ALK* gene controlling alkenyl glucosinolates in *Brassica oleracea* (Li and Quiros 2003), where a BAC contained the *AOP2* gene previously described in *A. thaliana* (Mithen et al. 1995). Analysis of extensive glucosinolate variation in *B. rapa* also identified several candidates based on comparison of QTL map positions and locations of *A. thaliana* homologue genes, including *AOP* genes (Lou et al. 2008) and a Myb transcription factor involved in regulation of glucosinolate biosynthesis (Hirai et al. 2007).

An important developmental trait that has been extensively analyzed in several Brassicaceae species is flowering time, for which QTL collocate with *FLC* homologues in *B. napus* (Long et al. 2007), *B. oleracea* (Okazaki et al. 2007), and *B. rapa* (Schranz et al. 2002, Lou et al. 2007), suggesting that *FLC*-like genes contribute to the natural variation observed in annual and biennial Brassicaceae species. This is supported by transgenic experiments showing that *B. napus FLC* genes show similar function as in *A. thaliana* (Tadege et al. 2001). In addition, it has been shown that the *FRI* gene of the wild species *A. lyrata* shows two allelic variants differing in several amino acid substitutions, which affect flowering in *A. thaliana* transgenic lines and are associated with *A. lyrata* flowering time variation (Kuittinen et al. 2008).

Comparative genetics among *A. thaliana* and other Brassicaceae species has been also carried out for several other traits where the underlying genes are known in *A. thaliana*, such as seed color and disease resistance. Yellow seeds are attractive for rapeseed because of their higher oil and meal protein content and their lower pigment contamination due to thinner seed coats. Comparative mapping has shown *TT10* as a candidate gene for this trait (Fu et al. 2007). In addition, it has been also suggested that the homologue of *TT12* gene could play a role in this variation present in several in *Brassica* species (Chai et al. 2008). Leucine-rich repeat (LRR)-like genes are obvious candidates for monogenic disease resistances as they are in almost any plant species (Table 4.1). This has been shown in *B. rapa* where comparative mapping identified a cluster of such genes on chromosome 4 of *A. thaliana* as candidates for clubroot resistance genes (Suwabe et al. 2006). Interestingly, despite *A. thaliana* accessions are commonly resistant to some diseases affecting *Brassica* crops such as black leg disease caused by *Leptosphaeria maculens*, it might also provide candidate genes for resistance against such diseases because it has been shown that *A. thaliana* resistance is acquired in different accessions by different *RLM* genes (Staal et al. 2006).

These examples illustrate the success of comparative QTL and gene mapping to provide candidate genes from *A. thaliana* to other Brassicaceae species for relative simple traits. It still remains unknown how useful comparative genetics will be for more complex quantitative traits such as growth, yield, or stress tolerances, where similar phenotypic variation might be generated by variants at many loci within each species. It might be expected that comparable traits are often controlled by related gene networks and the identification of genetic components of these networks in the model species will still remain an important resource of candidate genes for other species. In this respect, an important tool that will facilitate the selection of candidate genes for QTL involved in such complex traits is the detailed definition of the synteny among Brassicaceae species (Schranz et al. 2006). This has been established for most *Brassica, Capsella*, and *Arabidopsis* species described in this book, by comparing genetic maps based on DNA markers, which allow comparison of orthologue sequences (Kuittinen et al. 2004, Parkin et al. 2005, Choi et al. 2007, Suwabe et al. 2006, see Chapter 5 for details). However, genomic comparisons are not often straightforward due to the triplicated genomes and considerable gene loss occurred in *Brassica* species (Town et al. 2006). The availability in the very near future of full genome sequences of the crop species *B. rapa* and the wild species *A. lyrata* and *Capsella rubella* will sharply define the syntenic regions where homologue QTL and candidate genes among different species can be more accurately identified. This information will not only assist in the selection of genes candidate to underlie natural variation for complex traits, but enable the full expansion of comparative genomics and the development of new tools for understanding the molecular bases of adaptation in wild and crop Brassicaceae species.

Acknowledgments We thank colleagues and coworkers who provided unpublished information and Dr Jane Parker for updating us on the R genes. We apologize to those authors whose work could not be discussed due to space limitations. Research in our laboratories was supported by the ERA-PG program grants 034B ARABRAS to MK and MR and GEN2006-27786-E/VEG to CAB.

References

Al-Shehbaz IA, ÓKane SL (2002) Taxonomy and phylogeny of Arabidopsis (*Brassicaceae*). The Arabidopsis book. doi:1199/tab.001

Alcázar R, García AV, Parker JE et al (2009) Incremental steps toward incompatibility revealed by Arabidopsis epistatic interactions modulating salicylic acid pathway activation. Proc Natl Acad Sci USA 106:334–339

Alonso JM, Stepanova AN, Leisse TJ et al (2003) Genome-wide insertional mutagenesis of *Arabidopsis thaliana*. Science 301:653–657

Alonso-Blanco C, Aarts MG, Bentsink L et al (2009) What has natural variation taught us about plant development, physiology, and adaptation? Plant Cell 21:1877–1896

Alonso-Blanco C, Bentsink L, Hanhart CJ et al (2003) Analysis of natural allelic variation at seed dormancy loci of *Arabidopsis thaliana*. Genetics 164:711–729

Alonso-Blanco C, Blankesteijn-de Vries H, Hanhart C et al (1999) Natural allelic variation at seed size loci in relation to other life history traits of *Arabidopsis thaliana*. Proc Natl Acad Sci USA 96:4710–4717

Alonso-Blanco C, Gomez-Mena C, Llorente F et al (2005) Genetic and molecular analyses of natural variation indicate CBF2 as a candidate gene for underlying a freezing tolerance quantitative trait locus in Arabidopsis. Plant Physiol 139:1304–1312

Alonso-Blanco C, Koornneef M (2000) Naturally occurring variation in Arabidopsis: an underexploited resource for plant genetics. Trends Plant Sci 5:22–29

Alonso-Blanco C, Koornneef M, van Ooijen J (2006) QTL analysis. In: Salinas J, Sanchez-Serrano JJ (eds) Arabidopsis protocols. Methods in molecular biology, vol 323, pp 79–99. Humana Press, Inc., Totowa, NJ

Alonso-Blanco C, Peeters AJM, Koornneef M, Lister C, Dean C, van den Bosch N, Pot J, Kuiper MTR (1998) Development of an AFLP based linkage map of Ler, Col and Cvi *Arabidopsis thaliana* ecotypes and construction of a L*er*/Cvi recombinant inbred line population. Plant J 14:259–271

Amar S, Ecke W, Becker H et al (2008) QTL for phytosterol and sinapate ester content in *Brassica napus* L. collocate with the two erucic acid genes. Theor Appl Genet 116:1051–1061

Aranzana MJ, Kim S, Zhao K et al (2005) Genome-wide association mapping in Arabidopsis identifies previously known flowering time and pathogen resistance genes. PLoS Genet 1: 531–539

Aukerman MJ, Hirschfeld M, Wester L et al (1997) A deletion in the *PHYD* Gene of the Arabidopsis Wassilewskija ecotype defines a role for phytochrome D in red/far-red light sensing. Plant Cell 9:1317–1326

Balasubramanian S, Sureshkumar S, Agrawal M et al (2006) The *PHYTOCHROME C* photoreceptor gene mediates natural variation in flowering and growth responses of *Arabidopsis thaliana.* Nat Genet 38:711–715

Baxter I, Muthukumar B, Park HC et al (2008) Variation in molybdenum content across broadly distributed populations of *Arabidopsis thaliana* is controlled by a mitochondrial molybdenum transporter (*MOT1*). PLoS Genet 4:e1000004

Beck JB, Schmuths H, Schaal BA (2008) Native range genetic variation in *Arabidopsis thaliana* is strongly geographically structured and reflects pleistocene glacial dynamics. Mol Ecol 17: 902–915

Bell CJ, Ecker JR (1994) Assignment of 30 microsatellite loci to the linkage map of Arabidopsis. Genomics 19:137–144

Benfey PN, Mitchell-Olds T (2008) From genotype to phenotype: systems biology meets natural variation. Science 320:495–497

Bentsink L, Alonso-Blanco C, Vreugdenhil D et al (2000) Genetic analysis of seed-soluble oligosaccharides in relation to seed storability of Arabidopsis. Plant Physiol. 124:1595–1604

Bentsink L, Jowett J, Hanhart CJ et al (2006) Cloning of *DOG1*, a quantitative trait locus controlling seed dormancy in Arabidopsis. Proc Natl Acad Sci USA 103:17042–17047

Bentsink L, Yuan K, Koornneef M et al (2003) The genetics of phytate and phosphate accumulation in seeds and leaves of *Arabidopsis thaliana*, using natural variation. Theor Appl Genet 106:1234–1243

Bergelson J, Kreitman M, Stahl EA et al (2001) Evolutionary dynamics of plant R.genes. Science 292:2281–2285

Bomblies K, Lempe J, Epple P et al (2007) Autoimmune response as a mechanism for a Dobzhansky-Muller-type incompatibility syndrome in plants. PLoS Biol 5:e236

Borevitz JO, Hazen SP, Michael TP et al (2007) Genome-wide patterns of single-feature polymorphism in *Arabidopsis thaliana.* Proc Natl Acad Sci USA 104:12057–12062

Borevitz JO, Maloof JN, Lutes J et al (2002) Quantitative trait loci controlling light and hormone response in two accessions of *Arabidopsis thaliana.* Genetics 160:683–696

Botella MA, Parker JE, Frost LN et al (1998) Three genes of the Arabidopsis *RPP1* complex resistance locus recognize distinct *Peronospora parasitica* avirulence determinants. Plant Cell 10:1847–1860

Botto JF, Alonso-Blanco C, Garzaron I et al (2003) The Cape Verde Islands allele of *cryptochrome 2* enhances cotyledon unfolding in the absence of blue light in Arabidopsis. Plant Physiol 133:1547–1556

Botto JF, Coluccio MP (2007) Seasonal and plant-density dependency for quantitative trait loci affecting flowering time in multiple populations of *Arabidopsis thaliana*. Plant Cell Environ 30:1465–1479

Caicedo AL, Stinchcombe JR, Olsen KM et al (2004) Epistatic interaction between Arabidopsis *FRI* and *FLC* flowering time genes generates a latitudinal cline in a life history trait. Proc Natl Acad Sci USA 101:15670–15675

Chai Y-R, Lei B, Huang H-L et al (2008) *TRANSPARENT TESTA 12* genes from *Brassica napus* and parental species: cloning, evolution, and differential involvement in yellow seed trait. Mol Genet Genomics doi:10.1007/s00438-008-0399-1

Chisholm ST, Mahajan SK, Whitham SA et al (2000) Cloning of the Arabidopsis RTM1 gene, which controls restriction of long-distance movement of tobacco etch virus. Proc Natl Acad Sci USA 97:489–494

Choi S, Teakle G, Plaha P et al (2007) The reference genetic linkage map for the multinational *Brassica rapa* genome sequencing project. Theor App Genet 115:777–792

Clark RM, Schweikert G, Toomajian C et al (2007) Common sequence polymorphisms shaping genetic diversity in *Arabidopsis thaliana*. Science 317:338–342

Clauss MJ, Cobban H, Mitchell-Olds T (2002) Cross-species microsatellite markers for elucidating population genetic structure in Arabidopsis and Arabis (*Brassicaeae*). Mol Ecol 11: 591–601

Clerkx EJM, El-Lithy ME, Vierling E et al (2004) Analysis of natural allelic variation of Arabidopsis seed germination and seed longevity traits between the accessions Landsberg *erecta* and Shakdara, using a new recombinant inbred line population. Plant Physiol 135: 432–443

Cooley MB, Pathirana S, Wu HJ et al. (2000) Members of the Arabidopsis *HRT/RPP8* family of resistance genes confer resistance to both viral and oomycete pathogens. Plant Cell 12:663–676

DeCook R, Lall S, Nettleton D et al (2006) Genetic regulation of gene expression during shoot development in Arabidopsis. Genetics 172:1155–1164

Denby KJ, Kumar P, Kliebenstein DJ (2004) Identification of *Botrytis cinerea* susceptibility loci in *Arabidopsis thaliana*. Plant J 38:473–486

Deslandes L, Olivier J, Theulieres F et al (2002) Resistance to *Ralstonia solanacearum* in *Arabidopsis thaliana* is conferred by the recessive *RRS1-R* gene, a member of a novel family of resistance genes. Proc Natl Acad Sci USA 99:2404–2409

Deslandes L, Pileur F, Liaubet L et al (1998) Genetic characterization of *RRS1*, a recessive locus in *Arabidopsis thaliana* that confers resistance to the bacterial soilborne pathogen *Ralstonia solanacearum*. Mol Plant Microbe Interact 11:659–667

Doyle MR, Bizzell CM, Keller MR et al (2005) *HUA2* is required for the expression of floral repressors in *Arabidopsis thaliana*. Plant J 41:376–385

El-Assal SE-D, Alonso-Blanco C, Peeters AJM et al (2001) A QTL for flowering time in Arabidopsis reveals a novel allele of *CRY2*. Nat Genet 29:435–440

El-Lithy ME, Bentsink L, Hanhart CJ et al (2006) New arabidopsis recombinant inbred line populations genotyped using SNPWave and their use for mapping flowering-time quantitative trait loci. Genetics 172:1867–1876

El-Lithy ME, Clerkx EJ, Ruys GJ et al (2004) Quantitative trait locus analysis of growth-related traits in a new Arabidopsis recombinant inbred population. Plant Physiol 135:444–458

Filiault DL, Wessinger CA, Dinneny JR et al (2008) Amino acid polymorphisms in Arabidopsis *phytochrome B* cause differential responses to light. Proc Natl Acad Sci USA 105: 3157–3162

Fitz Gerald JN, Lehti-Shiu MD, Ingram PA et al (2006) Identification of quantitative trait loci that regulate Arabidopsis root system size and plasticity. Genetics 172:485–498

Fourmann M, Barret P, Renard M et al (1998) The two genes homologous to Arabidopsis *FAE1* co-segregate with the two loci governing erucic acid content in *Brassica napus*. Theor Appl Genet 96:852–858

François O, Blum MGB, Jakobsson M et al (2008) Demographic history of European populations of *Arabidopsis thaliana*. PLoS Genet 4(5):Public Library of Science, p e1000075

Frenkel M, Johansson Jänkänpää H, Moen J et al (2008) An illustrated gardener's guide to transgenic Arabidopsis field experiments. New Phytol 180:545–555

Fu F-Y, Liu L-Z, Chai Y-R et al (2007) Localization of QTLs for seed color using recombinant inbred lines of *Brassica napus* in different environments. Genome 50:840–854

Gassmann W, Hinsch ME, Staskawicz BJ (1999) The *Arabidopsis RPS4* bacterial-resistance gene is a member of the TIR-NBS-LRR family of disease-resistance genes. Plant J 20: 265–277

Gazzani S, Gendall AR, Lister C et al (2003) Analysis of the molecular basis of flowering time variation in Arabidopsis accessions. Plant Physiol 132:1107–1114

Grant MR, Godiard L, Straube E et al (1995) Structure of the Arabidopsis RPM1 gene enabling dual specificity disease resistance. Science 269:843–846

Greene EA, Codomo CA, Taylor NE et al (2003) Spectrum of chemically induced mutations from a large-scale reverse-genetic screen in Arabidopsis. Genetics 164:731–740

Göllner K, Schweizer P, Bai Y et al (2008) Natural genetic resources of *Arabidopsis thaliana* reveal a high prevalence and unexpected phenotypic plasticity of *RPW8*-mediated powdery mildew resistance. New Phytol 177:725–742

Hagenblad J, Tang C, Molitor J et al (2004) Haplotype structure and phenotypic associations in the chromosomal regions surrounding two *Arabidopsis thaliana* flowering time loci. Genetics 168:1627–1638

Harada H, Kuromori T, Hirayama T et al (2004) Quantitative trait loci analysis of nitrate storage in Arabidopsis leading to an investigation of the contribution of the anion channel gene, *AtCLC-c*, to variation in nitrate levels. J Exp Bot 55:2005–2014

Hauser M-T, Harr B, Schlotterer C (2001) Trichome distribution in *Arabidopsis thaliana* and its close relative *Arabidopsis lyrata*: molecular analysis of the candidate gene *GLABROUS1*. Mol Biol Evol 18:1754–1763

Hausmann NJ, Juenger TE, Sen S et al (2005) Quantitative trait loci affecting delta13C and response to differential water availability in *Arabidopsis thaliana*. Evolution 59:81–96

He F, Kang D, Ren Y et al (2007) Genetic diversity of the natural populations of *Arabidopsis thaliana* in China. Heredity 99:423–431

Hirai MY, Sugiyama K, Sawada Y et al (2007) Omics-based identification of Arabidopsis Myb transcription factors regulating aliphatic glucosinolate biosynthesis. Proc Natl Acad Sci USA 104:6478–6483

Hobbs DH, Flintham JE, Hills MJ (2004) Genetic control of storage oil synthesis in seeds of Arabidopsis. Plant Physiol 136:3341–3349

Hoffmann MH (2002) Biogeography of *Arabidopsis thaliana* (L.) Heynh. (*Brassicaceae*). J Biogeography 29:125–134

Hoffmann MH (2005) Evolution of the realized climatic niche in the genus: Arabidopsis (*Brassicaceae*). Evolution 59:1425–1436

Hu X, Sullivan-Gilbert M, Gupta M et al (2006) Mapping of the loci controlling oleic and linolenic acid contents and development of fad2 and fad3 allele-specific markers in canola (*Brassica napus L.*). Theor Appl Genet 113:497–507

Jakobsson M, Säll T, Lind-Halldén C et al (2007) The evolutionary history of the common chloroplast genome of *Arabidopsis thaliana* and *A. suecica*. J Evol Biol 20:104–121

Jander G, Norris SR, Rounsley SD et al (2002) Arabidopsis map-based cloning in the post-genome era. Plant Physiol 129:440–450

Johanson U, West J, Lister C et al (2000) Molecular analysis of *FRIGIDA*, a major determinant of natural variation in Arabidopsis flowering time. Science 290:344–347

Jorgensen S, Mauricio R (2004) Neutral genetic variation among wild North American populations of the weedy plant *Arabidopsis thaliana* is not geographically structured. Mol Ecol 13: 3403–3413

Jubault M, Lariagon C, Simon M et al (2008) Identification of quantitative trait loci controlling partial clubroot resistance in new mapping populations of *Arabidopsis thaliana*. Theor Appl Genet 117:191–202

Juenger TE, McKay JK, Hausmann N et al (2005) Identification and characterization of QTL underlying whole-plant physiology in *Arabidopsis thaliana*: delta13C, stomatal conductance and transpiration efficiency. Plant Cell Environ 28:697–708

Juenger T, Purugganan M, Mackay TFC (2000) Quantitative trait loci for floral morphology in *Arabidopsis thaliana*. Genetics 156:1379–1392

Kempin SA, Savidge B, Yanofsky MF (1995) Molecular basis of the cauliflower phenotype in Arabidopsis. Science 267:522–525

Keurentjes JJB, Bentsink L, Alonso-Blanco C et al (2007a) Development of a near-isogenic line population of *Arabidopsis thaliana* and comparison of mapping power with a recombinant inbred line population. Genetics 175:891–905

Keurentjes JJB, Fu J, de Vos CH et al (2006) The genetics of plant metabolism. Nat Genet 38: 842–849

Keurentjes JJB, Fu J, Terpstra IR et al (2007b) Regulatory network construction in Arabidopsis by using genome-wide gene expression quantitative trait loci. Proc Natl Acad Sci USA 104: 1708–1713

Keurentjes JJB, Koornneef M, Vreugdenhil D (2008) Quantitative genetics in the age of omics. Curr Opin Plant Biol 11:123–128

Kim S, Plagnol V, Hu TT et al (2007) Recombination and linkage disequilibrium in *Arabidopsis thaliana*. Nat Genet 39:1151–1155

Kliebenstein DJ, Lambrix VM, Reichelt M et al (2001) Gene duplication in the diversification of secondary metabolism: Tandem 2-oxoglutarate-dependent dioxygenases control glucosinolate biosynthesis in Arabidopsis. Plant Cell 13:681–693

Kliebenstein D, Pedersen D, Barker B et al (2002) Comparative analysis of quantitative trait loci controlling glucosinolates, myrosinase and insect resistance in *Arabidopsis thaliana*. Genetics 161:325–332

Kliebenstein D, West M, van Leeuwen H et al (2006) Identification of QTLs controlling gene expression networks defined a priori. BMC Bioinformatics 7:308

Koch M, Haubold B, Mitchell-Olds T (2001) Molecular systematics of the *Brassicaceae*: evidence from coding plastidic matK and nuclear Chs sequences. Am J Bot 88:534–544

Koornneef M, Alonso-Blanco C, Vreugdenhil D (2004) Naturally occurring genetic variation in *Arabidopsis thaliana*. Annu Rev Plant Biol 55:141–172

Korves TM, Schmid KJ, Caicedo AL et al (2007) Fitness effects associated with the major flowering time gene *FRIGIDA* in *Arabidopsis thaliana* in the field. Am Nat 169: 141–157

Kover PX, Cheverud J (2007) The genetic basis of quantitative variation in susceptibility of *Arabidopsis thaliana* to *Pseudomonas syringae* (*Pst* DC3000): evidence for a new genetic factor of large effect. New Phytol 174:172–181

Kover PX, Wolf JB, Kunkel BN et al (2005) Genetic architecture of *Arabidopsis thaliana* response to infection by *Pseudomonas syringae*. Heredity 94:507–517

Kowalski SP, Lan T-H, Feldmann KA et al (1994) QTL mapping of naturally-occurring variation in flowering time of *Arabidopsis thaliana*. Mol Gen Genet 245:548–555

Kroymann J, Donnerhacke S, Schnabelrauch D et al (2003) Evolutionary dynamics of an Arabidopsis insect resistance quantitative trait locus. Proc Natl Acad Sci USA 100: 14587–14592

Kroymann J, Mitchell-Olds T (2005) Epistasis and balanced polymorphism influencing complex trait variation. Nature 435:95–98

Kuittinen H, de Haan AA, Vogl C et al (2004) Comparing the linkage maps of the close relatives *Arabidopsis lyrata* and *A. thaliana*. Genetics 168:1575–1584

Kuittinen H, Mattila A, Savolainen O (1997) Genetic variation at marker loci and in quantitative traits in natural populations of *Arabidopsis Thaliana*. Heredity 79:144–152

Kuittinen H, Niittyvuopio A, Rinne P et al (2008) Natural variation in *Arabidopsis lyrata* vernalization requirement conferred by a *FRIGIDA* indel polymorphism. Mol Biol Evol 25:319–329

Kusterer B, Muminovic J, Utz HF et al (2007b) Analysis of a triple testcross design with recombinant inbred lines reveals a significant role of epistasis in heterosis for biomass-related traits in Arabidopsis. Genetics 175:2009–2017

Kusterer B, Piepho H-P, Utz HF et al (2007a) Heterosis for biomass-related traits in Arabidopsis investigated by quantitative trait loci analysis of the triple testcross design with recombinant inbred lines. Genetics 177:1839–1850

Laibach F (1943) *Arabidopsis thaliana* (L.) Heynh. als Objekt für genetische und entwicklungsphysiologische Untersuchungen. Bot Arch 44:439–455

Lall S, Nettleton D, DeCook R et al (2004) Quantitative trait loci associated with adventitious shoot formation in tissue culture and the program of shoot development in Arabidopsis. Genetics 167:1883–1892

Lambrix V, Reichelt M, Mitchell-Olds T et al (2001) The Arabidopsis epithiospecifier protein promotes the hydrolysis of glucosinolates to nitriles and influences *Trichoplusia ni* herbivory. Plant Cell 13:2793–2807

Laserna MP, Sanchez RA, Botto JF (2008) Light-related loci controlling seed germination in Ler × Cvi and Bay-0 × Sha recombinant inbred-line populations of *Arabidopsis thaliana*. Ann Bot 102:631–642

Lau JA, Shaw RG, Reich PB et al (2007) Strong ecological but weak evolutionary effects of elevated CO_2 on a recombinant inbred population of *Arabidopsis thaliana*. New Phytol 175:351–362

Le Corre V (2005) Variation at two flowering time genes within and among populations of *Arabidopsis thaliana*: comparison with markers and traits. Mol Ecol 14:4181–4192

Le Corre V, Roux F, Reboud X (2002) DNA polymorphism at the *FRIGIDA* gene in *Arabidopsis thaliana*: extensive nonsynonymous variation is consistent with local selection for lowering time. Mol Biol Evol 19:1261–1271

Lempe J, Balasubramanian S, Sureshkumar S et al (2005) Diversity of flowering responses in wild *Arabidopsis thaliana* strains. PLoS Genet 1:109–118

Li G, Quiros CF (2003) In planta side-chain glucosinolate modification in Arabidopsis by introduction of dioxygenase Brassica homolog *BoGSL-ALK*. Theor Appl Genet 106: 1116–1121

Li Y, Roycewicz P, Smith E et al (2006) Genetics of local adaptation in the laboratory: flowering time quantitative trait loci under geographic and seasonal conditions in Arabidopsis. PLoS One 1:1–8

Lister C, Dean C (1993) Recombinant inbred lines for mapping RFLP and phenotypic markers in *Arabidopsis thaliana*. Plant J 4:745–750

Long Y, Shi J, Qiu D et al (2007) Flowering time quantitative trait loci analysis of oilseed Brassica in multiple environments and genome-wide alignment with Arabidopsis. Genetics 177: 2433–2444

Lou P, Zhao J, He H et al (2008) Quantitative trait loci for glucosinolate accumulation in *Brassica rapa* leaves. New Phytol 179:1017–1032

Lou P, Zhao J, Kim JS et al (2007) Quantitative trait loci for flowering time and morphological traits in multiple populations of *Brassica rapa*. J Exp Bot 58:4005–4016

Loudet O, Chaillou S, Camilleri C et al (2002) Bay-0 × Shahdara recombinant inbred line population: a powerful tool for the genetic dissection of complex traits in Arabidopsis. Theor Appl Genet 104:1173–1184

Loudet O, Chaillou S, Merigout P et al (2003) Quantitative trait loci analysis of nitrogen use efficiency in Arabidopsis. Plant Physiol 131:345–358

Loudet O, Gaudon V, Trubuil A et al (2005) Quantitative trait loci controlling root growth and architecture in *Arabidopsis thaliana* confirmed by heterogeneous inbred family. Theor Appl Genet 110:742–753

Loudet O, Michael TP, Burger BT et al (2008) A zinc knuckle protein that negatively controls morning-specific growth in *Arabidopsis thaliana*. Proc Natl Acad Sci USA 105: 17193–17198

Loudet O, Saliba-Colombani V, Camilleri C, Calenge F, Gaudon V, Koprivova A, North KA, Kopriva S, Daniel-Vedele F (2007) Natural variation for sulfate content in *Arabidopsis thaliana* is highly controlled by *APR2*. Nat Genet 39:896–900

Macquet A, Ralet M-C, Loudet O et al (2007) A naturally occurring mutation in an Arabidopsis accession affects a β-D-Galactosidase that increases the hydrophilic potential of rhamnogalacturonan I in seed mucilage. Plant Cell 19:3990–4006

Magliano TMA, Botto JF, Godoy AV et al (2005) New arabidopsis recombinant inbred lines (Landsberg *erecta* × Nossen) reveal natural variation in phytochrome-mediated responses. Plant Physiol 138:1126–1135

Malmberg RL, Held S, Waits A et al (2005) Epistasis for fitness-related quantitative traits in *Arabidopsis thaliana* grown in the field and in the greenhouse. Genetics 171:2013–2027

Maloof JN, Borevitz JO, Dabi T et al (2001) Natural variation in light sensitivity of Arabidopsis. Nat Genet 29:441–446

Masle J, Gilmore SR, Farquhar GD (2005) The *ERECTA* gene regulates plant transpiration efficiency in Arabidopsis. Nature 436:866–870

Mauricio R, Stahl EA, Korves T et al (2003) Natural selection for polymorphism in the disease resistance gene *RPS2* of *Arabidopsis thaliana.* Genetics 163:735–746

McDowell JM, Dhandaydham M, Long TA et al (1998) Intragenic recombination and diversifying selection contribute to the evolution of downy mildew resistance at the *RPP8* locus of Arabidopsis. Plant Cell 10:1861–1874

Melchinger AE, Piepho H-P, Utz HF et al (2007) Genetic basis of heterosis for growth-related traits in Arabidopsis investigated by testcross progenies of near-isogenic lines reveals a significant role of epistasis. Genetics 177:1827–1837

Meyer RC, Steinfath M, Lisec J et al (2007) The metabolic signature related to high plant growth rate in *Arabidopsis thaliana.* Proc Natl Acad Sci USA 104:4759–4764

Michaels SD, Amasino RM (1999) *FLOWERING LOCUS C* encodes a novel MADS domain protein that acts as a repressor of flowering. Plant Cell 11:949–956

Michaels SD, He Y, Scortecci KC et al (2003) Attenuation of *FLOWERING LOCUS C* activity as a mechanism for the evolution of summer-annual flowering behavior in Arabidopsis. Proc Natl Acad Sci USA 100:10102–10107

Millenaar FF, Cox MCH, van Berkel YEM et al (2005) Ethylene-induced differential growth of petioles in Arabidopsis. Analyzing natural variation, response kinetics, and regulation. Plant Physiol 137:998–1008

Mindrinos M, Katagiri F, Yu G et al (1994) The *A. thaliana* disease resistance gene *RPS2* encodes a protein containing a nucleotide-binding site and leucine-rich repeats. Cell 78:1089–1099

Mitchell-Olds T, Schmitt J (2006) Genetic mechanisms and evolutionary significance of natural variation in Arabidopsis. Nature 441:947–952

Mithen R, Clarke J, Lister C et al (1995) Genetics of aliphatic glucosinolates. III. Side chain structure of aliphatic glucosinolates in *Arabidopsis thaliana.* Heredity 74:210–215

Mouchel ClF, Briggs GC, Hardtke CS (2004) Natural genetic variation in Arabidopsis identifies *BREVIS RADIX*, a novel regulator of cell proliferation and elongation in the root. Genes Dev 18:700–714

Nordborg M, Hu TT, Ishino Y et al (2005) The pattern of polymorphism in *Arabidopsis thaliana.* PLoS Biol 3:e196

Okazaki K, Sakamoto K, Kikuchi R et al (2007) Mapping and characterization of *FLC* homologs and QTL analysis of flowering time in *Brassica oleracea.* Theor Appl Genet 114: 595–608

Olsen KM, Halldorsdottir SS, Stinchcombe JR et al (2004) Linkage disequilibrium mapping of Arabidopsis *CRY2* flowering time alleles. Genetics 167:1361–1369

O'Neill C, Morgan C, Kirby J et al (2008) Six new recombinant inbred populations for the study of quantitative traits in *Arabidopsis thaliana.* Theor Appl Genet 116:623–634

Ossowski S, Schneeberger K, Clark RM, Lanz C, Warthmann N, Weigel D (2008) Sequencing of natural strains of *Arabidopsis thaliana* with short reads. Genome Res 18:2024–2033

Ostrowski MF, David J, Santoni S et al (2006) Evidence for a large-scale population structure among accessions of *Arabidopsis thaliana*: possible causes and consequences for the distribution of linkage disequilibrium. Mol Ecol 15:1507–1517

Parker JE, Coleman MJ, Szabò V et al (1997) The Arabidopsis downy mildew resistance gene *RPP5* shares similarity to the toll and interleukin-1 receptors with N and L6. Plant Cell 9: 879–894

Parkin IAP, Gulden SM, Sharpe AG et al (2005) Segmental structure of the *Brassica napus* genome based on comparative analysis with *Arabidopsis thaliana*. Genetics 171:765–781

Perchepied L, Kroj T, Tronchet M et al (2006) Natural variation in partial resistance to *Pseudomonas syringae* is controlled by two major QTLs in *Arabidopsis thaliana*. PLoS One 1:e123

Perez-Perez JM, Serrano-Cartagena J, Micol JL (2002) Genetic analysis of natural variations in the architecture of *Arabidopsis thaliana* vegetative leaves. Genetics 162:893–915

Pfalz M, Vogel H, Mitchell-Olds T et al (2007) Mapping of QTL for resistance against the crucifer specialist herbivore *Pieris brassicae* in a new Arabidopsis inbred line population, Da(1)-12xEi-2. PLoS One 2:e578

Picó FX, Mendez-Vigo B, Martinez-Zapater JM et al (2008) Natural genetic variation of *Arabidopsis thaliana* is geographically structured in the iberian peninsula. Genetics 180: 1009–1021

Quesada V, Garcia-Martinez S, Piqueras P et al (2002) Genetic architecture of NaCl tolerance in Arabidopsis. Plant Physiol 130:951–963

Rauh BL, Basten C, Buckler ES (2002) Quantitative trait loci analysis of growth response to varying nitrogen sources in *Arabidopsis thaliana*. Theor Appl Genet 104:743–750

Reif JC, Kusterer B, Piepho H-P et al (2008) Unravelling epistasis with triple testcross progenies of near isogenic lines. Genetics 181:247–257. doi:108.093047

Reymond M, Svistoonoff S, Loudet O et al (2006) Identification of QTL controlling root growth response to phosphate starvation in *Arabidopsis thaliana*. Plant Cell Environ 29:115–125

Riddle NC, Richards EJ (2005) Genetic variation in epigenetic inheritance of ribosomal RNA gene methylation in Arabidopsis. Plant J 41:524–532

Rose LE, Bittner-Edy PD, Langley CH et al (2004) The maintenance of extreme amino acid diversity at the disease resistance gene, *RPP13*, in *Arabidopsis thaliana*. Genetics 166:1517–1527

Rowe HC, Kliebenstein DJ (2008) Complex genetics control natural variation in *Arabidopsis thaliana* resistance to *Botrytis cinerea*. Genetics:doi:108.091439

Rus A, Baxter I, Muthukumar B et al (2006) Natural variants of *AtHKT1* enhance Na+ accumulation in two wild populations of Arabidopsis. PLoS Genet 2:e210

Salt DE, Baxter I, Lahner B (2008) Ionomics and the study of the plant ionome. Annu Rev Plant Biol 59:709–733

Scarcelli N, Cheverud JM, Schaal BA et al (2007) Antagonistic pleiotropic effects reduce the potential adaptive value of the *FRIGIDA* locus. Proc Natl Acad Sci USA 104:16986–16991

Schläppi MR (2006) *FRIGIDA LIKE 2* is a functional allele in Landsberg *erecta* and compensates for a nonsense allele of *FRIGIDA LIKE 1*. Plant Physiol 142:1728–1738

Schmid KJ, Torjek O, Meyer R et al (2006) Evidence for a large-scale population structure of *Arabidopsis thaliana* from genome-wide single nucleotide polymorphism markers. Theor Appl Genet 112:1104–1114

Schranz ME, Lysak MA, Mitchell-Olds T (2006) The ABC's of comparative genomics in the *Brassicaceae*: building blocks of crucifer genomes. Trends Plant Sci 11:535–542

Schranz ME, Quijada P, Sung S-B et al (2002) Characterization and effects of the replicated flowering time gene FLC in *Brassica rapa*. Genetics 162:1457–1468

Sergeeva LI, Keurentjes JJB, Bentsink L et al (2006) Vacuolar invertase regulates elongation of *Arabidopsis thaliana* roots as revealed by QTL and mutant analysis. Proc Natl Acad Sci USA 103:2994–2999

Sharbel TF, Haubold B, Mitchell-Olds T (2000) Genetic isolation by distance in *Arabidopsis thaliana*: biogeography and postglacial colonization of Europe. Mol Ecol 9:2109–2118

Sheldon CC, Rouse DT, Finnegan EJ et al (2000) The molecular basis of vernalization: the central role of *FLOWERING LOCUS C (FLC)*. Proc Natl Acad Sci USA 97:3753–3758

Shindo C, Aranzana MJ, Lister C et al (2005) Role of *FRIGIDA* and *FLOWERING LOCUS C* in determining variation in flowering time of Arabidopsis. Plant Physiol 138:1163–1173

Shindo C, Bernasconi G, Hardtke CS (2008) Intraspecific competition reveals conditional fitness effects of single gene polymorphism at the *Arabidopsis* root growth regulator *BRX*. New Phytol 180:71–80

Shindo C, Lister C, Crevillen P et al (2006) Variation in the epigenetic silencing of *FLC* contributes to natural variation in Arabidopsis vernalization response. Genes Dev 20:3079–3083

Shpak ED, McAbee JM, Pillitteri LJ, Torii KU (2005) Stomatal patterning and differentiation by synergistic interactions of receptor kinases. Science 309:290–293

Sicard O, Loudet O, Keurentjes JJB et al (2008) Identification of quantitative trait loci controlling symptom development during viral infection in *Arabidopsis thaliana*. Mol Plant Microbe Interact 21:198–207

Simon M, Loudet O, Durand S et al (2008) Quantitative trait loci mapping in five new large recombinant inbred line populations of *Arabidopsis thaliana* genotyped with consensus single-nucleotide polymorphism markers. Genetics 178:2253–2264

Staal J, Kaliff M, Bohman S et al (2006) Transgressive segregation reveals two Arabidopsis TIR-NB-LRR resistance genes effective against *Leptosphaeria maculans*, causal agent of blackleg disease. Plant J 46:218–230

Stahl EA, Dwyer G, Mauricio R et al (1999) Dynamics of disease resistance polymorphism at the *RPM1* locus of Arabidopsis. Nature 400:667–671

Stenoien HK, Fenster CB, Tonteri A et al (2005) Genetic variability in natural populations of *Arabidopsis thaliana* in northern Europe. Mol Ecol 14:137–148

Stinchcombe JR, Weinig C, Ungerer M et al (2004) A latitudinal cline in flowering time in *Arabidopsis thaliana* modulated by the flowering time gene *FRIGIDA*. Proc Natl Acad Sci USA 101:4712–4717

Sun Z, Wang Z, Tu J et al (2007) An ultradense genetic recombination map for *Brassica napus*, consisting of 13551 SRAP markers. Theor Appl Genet 114:1305–1317

Suwabe K, Tsukazaki H, Iketani H et al (2006) Simple sequence repeat-based comparative genomics between *Brassica rapa* and *Arabidopsis thaliana*: the genetic origin of clubroot resistance. Genetics 173:309–319

Svistoonoff S, Creff A, Reymond M et al (2007) Root tip contact with low-phosphate media reprograms plant root architecture. Nat Genet 39:792–796

Symonds VV, Godoy AV, Alconada T et al (2005) Mapping quantitative trait loci in multiple populations of *Arabidopsis thaliana* identifies natural allelic variation for trichome density. Genetics 169:1649–1658

Tadege M, Sheldon CC, Helliwell CA et al (2001) Control of flowering time by *FLC* orthologues in *Brassica napus*. Plant J 28:545–553

Tamaoki M, Matsuyama T, Kanna M et al (2003) Differential ozone sensitivity among Arabidopsis accessions and its relevance to ethylene synthesis. Planta 216:552–560

Tan X, Meyers B, Kozik A et al (2007) Global expression analysis of nucleotide binding site-leucine rich repeat-encoding and related genes in Arabidopsis. BMC Plant Biol 7:56

Teng S, Keurentjes J, Bentsink L et al (2005) Sucrose-specific induction of anthocyanin biosynthesis in Arabidopsis requires the *MYB75/PAP1* gene. Plant Physiol 139: 1840–1852

Tian D, Traw MB, Chen JQ et al (2003) Fitness costs of R-gene-mediated resistance in *Arabidopsis thaliana*. Nature 423:74–77

Tisné S, Reymond M, Vile D et al (2008) Combined genetic and modeling approaches reveal that epidermal cell area and number in leaves are controlled by leaf and plant developmental processes in Arabidopsis. Plant Physiol 148:1117–1127

Tonsor SJ, Scheiner SM (2007) Plastic trait integration across a CO_2 gradient in *Arabidopsis thaliana*. Am Nat 169:119–140

Town CD, Cheung F, Maiti R et al (2006) Comparative genomics of *Brassica oleracea* and *Arabidopsis thaliana* reveal gene loss, fragmentation, and dispersal after polyploidy. Plant Cell 18:1348–1359

Tör M, Brown D, Cooper A et al (2004) Arabidopsis downy mildew resistance gene *RPP27* encodes a receptor-like protein similar to *CLAVATA2* and tomato *Cf-9*. Plant Physiol 135:1–13

Törjék O, Meyer RC, Zehnsdorf M et al (2008) Construction and analysis of 2 reciprocal arabidopsis introgression line populations. J Hered 99:396–406

Törjék O, Witucka-Wall H, Meyer R et al (2006) Segregation distortion in Arabidopsis C24/Col-0 and Col-0/C24 recombinant inbred line populations is due to reduced fertility caused by epistatic interaction of two loci. Theor Appl Genet 113:1551–1561

Vaughn MW, Tanurdić M, Lippman Z et al (2007) Epigenetic natural variation in *Arabidopsis thaliana*. PLoS Biol 5:e174

Vreugdenhil D, Aarts MGM, Koornneef M et al (2004) Natural variation and QTL analysis for cationic mineral content in seeds of *Arabidopsis thaliana*. Plant Cell Environ 27:828–839

Vuylsteke M, van Eeuwijk F, Van Hummelen P et al (2005) Genetic analysis of variation in gene expression in *Arabidopsis thaliana*. Genetics 171:1267–1275

Wang Q, Sajja U, Rosloski S, Humphrey T, Kim MC, Bomblies K, Weigel D, Grbic V (2007) *HUA2* caused natural variation in shoot morphology of *A. thaliana*. Curr Biol 17: 1513–1519

Wang N, Wang Y, Tian F et al (2008) A functional genomics resource for *Brassica napus*: development of an EMS mutagenized population and discovery of *FAE1* point mutations by TILLING. New Phytol 180:751–765

Warren RF, Henk A, Mowery P et al (1998) A mutation within the leucine-rich repeat domain of the Arabidopsis disease resistance gene *RPS5* partially suppresses multiple bacterial and downy mildew resistance genes. Plant Cell 10:1439–1452

Warthmann N, Fitz J, Weigel D (2007) MSQT for choosing SNP assays from multiple DNA alignments. Bioinformatics 23:2784–2787

Weigel D, Mott R (2009) The 1001 genomes project for *Arabidopsis thaliana*. Genome Biol 10:107

Weinig C, Dorn LA, Kane NC et al (2003) Heterogeneous selection at specific loci in natural environments in *Arabidopsis thaliana*. Genetics 165:321–329

Weinig C, Ungerer MC, Dorn LA et al (2002) Novel loci control variation in reproductive timing in *Arabidopsis thaliana* in natural environments. Genetics 162:1875–1884

Werner JD, Borevitz JO, Uhlenhaut NH et al (2005a) *FRIGIDA*-independent variation in flowering time of natural *Arabidopsis thaliana* accessions. Genetics 170:1197–1207

Werner JD, Borevitz JO, Warthmann N et al (2005b) Quantitative trait locus mapping and DNA array hybridization identify an *FLM* deletion as a cause for natural flowering-time variation. Proc Natl Acad Sci USA 102:2460–2465

West MAL, Kim K, Kliebenstein DJ et al (2007) Global eQTL mapping reveals the complex genetic architecture of transcript-level variation in Arabidopsis. Genetics 175:1441–1450

Wilson IW, Schiff CL, Hughes DE et al (2001) Quantitative trait loci analysis of powdery mildew disease resistance in the *Arabidopsis thaliana* accession Kashmir-1. Genetics 158: 1301–1309

Xiao SY, Ellwood S, Calis E et al (2001) Broad-spectrum mildew resistance in *Arabidopsis thaliana* mediated by *RPW8*. Science 291:118–120

Xiao SY, Emerson B, Ratanasut K et al (2004) Origin and maintenance of a broad-spectrum diseases resistance locus in Arabidopsis. Mol Biol Evol 21:1661–1672

Yan X, Chen S (2007) Regulation of plant glucosinolate metabolism. Planta 226:1343–1352

Zeng C, Han Y, Shi L et al (2008) Genetic analysis of the physiological responses to low boron stress in *Arabidopsis thaliana*. Plant Cell Environ 31:112–122

Zhai J, Liu J, Liu B et al (2008) Small RNA-directed epigenetic natural variation in *Arabidopsis thaliana*. PLoS Genet 4:e1000056

Zhang L, Byrne PF, Pilon-Smits EAH (2006a) Mapping quantitative trait loci associated with selenate tolerance in *Arabidopsis thaliana*. New Phytol 170:33–42

Zhang Z, Ober JA, Kliebenstein DJ (2006) The gene controlling the quantitative trait locus *EPITHIOSPECIFIER MODIFIER1* alters glucosinolate hydrolysis and insect resistance in Arabidopsis. Plant Cell 18:1524–1536

Zhang X, Shiu S, Cal A et al (2008) Global analysis of genetic, epigenetic and transcriptional polymorphisms in *Arabidopsis thaliana* using whole genome tiling arrays. PLoS Genet 4:e1000032

Zhang X, Yazaki J, Sundaresan A et al (2006b) Genome-wide high-resolution mapping and functional analysis of DNA methylation in Arabidopsis. Cell 126:1189–1201

Zhao K, Aranzana MJ, Kim S et al (2007) An Arabidopsis example of association mapping in structured samples. PLoS Genet 3:e4

Zilberman D, Gehring M, Tran RK et al (2007) Genome-wide analysis of Arabidopsis thaliana DNA methylation uncovers an interdependence between methylation and transcription. Nat Genet 39:61–69

van Leeuwen H, Kliebenstein DJ, West MAL et al (2007) Natural variation among *Arabidopsis thaliana* accessions for transcriptome response to exogenous salicylic acid. Plant Cell 19: 2099–2110

Chapter 5
Chasing Ghosts: Comparative Mapping in the Brassicaceae

Isobel Parkin

Abstract The study of plant genome organization has benefited greatly from the application of comparative genetic mapping, which allows both the elucidation of chromosomal rearrangements resulting from speciation and the ability to transfer information and resources between species. A significant focus of comparative mapping in the Brassicaceae has been within the agronomically important species of the *Brassica* genera and between the *Brassica* crops and their well-characterized relative *Arabidopsis thaliana*. These studies have demonstrated the ghostly remnants of an hexaploid ancestor in the evolutionary past of the *Brassica* diploids that explain the observed levels of gene duplication within the genomes. Further, comparative mapping with *A. thaliana* has uncovered a segmental architecture of conserved ancestral blocks which can be replicated and rearranged to reflect the current genomes of all members of the Brassicaceae studied to date. The correspondence between the *A. thaliana* and *Brassica* genomic regions is being exploited to fine map, identify, and clone genes for economically valuable traits.

Keywords Homology · Collinearity · Conserved genome blocks · Polyploidy · Chromosomal rearrangements

Contents

5.1 Introduction 154
5.2 Common Terms Used in Comparative Mapping Studies 155
5.3 The Basics of Comparative Mapping 156
5.4 The Contribution of Polyploidy (Inter-specific Hybridization) to *Brassica* Genome Evolution 158
5.5 The Ghost of an Ancestral Hexaploid Genome 160
5.6 *A. thaliana*, a Model Genome for the Brassicaceae 160
 5.6.1 Across the A, B, and C Genomes 162

I. Parkin (✉)
Agriculture and Agri-Food Canada, Saskatoon Research Centre, Saskatoon, SK, Canada S7N 0X2
e-mail: isobel.parkin@agr.gc.ca

R. Schmidt, I. Bancroft (eds.), *Genetics and Genomics of the Brassicaceae*, Plant Genetics and Genomics: Crops and Models 9, DOI 10.1007/978-1-4419-7118-0_5,

5.6.2 Conserved Chromosome Landmarks 164
5.6.3 Rearrangement Hotspots 164
5.7 Exploiting Comparative Mapping for Trait Analysis 165
5.8 Extending the Comparisons to Related Species 166
5.9 The Promise of Sequenced Genomes 167
References 167

5.1 Introduction

Early efforts to understand plant genome organization and evolution relied upon coarse imaging of whole chromosome structures through microscopy. By interpreting the pairing structures formed during meiosis it was possible to infer how closely related chromosomes, and hence, genomes and species were. The advent of molecular markers and in particular in the 1980s the ability to visualize restriction fragment length polymorphisms (RFLP) as a tool in genetic linkage analysis enabled more comprehensive plant genome analysis and heralded the dawn of comparative mapping (reviewed in Gale and Devos (1998)).

The premise was simple: that common sets of markers could be used to identify related segments of DNA between not only species of the same tribe but also across wide taxonomic distances. Such analyses could even transcend the millions of years represented by the monocot–dicot species divide (Paterson et al. 2004, Tang et al. 2008). These data showed extensive conservation of gene content between species of the same tribe and it was demonstrated that taxonomically disparate species could in some instances be separated by a remarkably limited number of major or large effect chromosomal rearrangements. For example, in the Solanaceae, the tomato and potato genomes were differentiated by a mere five inversion events (Tanksley et al. 1992). The most comprehensive analysis of a single family has been completed for the Poaceae (reviewed in Devos and Gale (2000)), where the concept of ancestral blocks was first proposed (Moore et al. 1995).

With a relatively limited number of markers it was shown that the genome of the monocot model rice could be broken down into 19 conserved blocks of collinearity which upon rearrangement formed the genomes of a diversity of cereal species, despite a wide range in base chromosome number, genomic DNA content, and estimated divergence times of up to 60 million years. The addition of further markers and comparisons between additional species have more accurately defined the ancestral blocks, increasing the number to 30 and culminating in the "circle of cereals," an unified comparative map of the grasses, where the rice genetic linkage map is drawn at the origin of a set of concentric circles, each circle representing an additional cereal genome. This novel representation allows the relatively simple identification of related genome segments across species (Devos 2005).

In the Brassicaceae a number of genetic linkage maps have been generated over the past 20 years, largely focusing on the agronomically important members of U's triangle and utilizing different sets of markers that prevented cross-species comparison (Slocum et al. 1990, Landry et al. 1991, Kianian and Quiros 1992, Ferreira

Table 5.1 Description of selected published genetic linkage maps which have contributed to our understanding of genome evolution and organization within the *Brassica* genus of lineage II

Target species	Genomes under comparison[a]	References	Number of conserved loci (markers)[b]	Length of genetic map for target species (cM)
B. rapa (A)	A–A^{Bna}	Suwabe et al. (2008)	(44)	743
B. oleracea (C)	C–C^{Bna}	Bohuon et al. (1996)	129	875
B. oleracea (C)	C–At	Lan and Paterson (2000)	186	863.6
B. oleracea (C)	C–At	Lukens et al. (2003)	(131)	875[c]
B. nigra (B)	B–A^{Bna}–C^{Bna}	Lagercrantz and Lydiate (1996)	(158)	778
B. nigra (B)	B–At	Lagercrantz (1998)	284–160	751
B. napus (AC)	A^{Bna}–C^{Bna}	Parkin et al. (2003)	(162)	1,698.5
B. napus (AC)	A^{Bna}–C^{Bna}–At	Parkin et al. (2005)	1,232–550 (368)	1,968
B. juncea (AB)	A^{Bj}–B^{Bj}–A^{Bna}–B^{Bni}–C^{Bna}–At	Panjabi et al. (2008)	533	1,992.2

[a]The constituent genomes are indicated: A, B, or C for the diploids *B. rapa*, *B. oleracea*, and *B. nigra*, respectively; superscript *Bna*, *Bj*, or *Bni* indicate the respective diploid genome within the allopolyploid nucleus of *B. napus*, *B. juncea*, or the diploid *B. nigra*.
[b]The number of markers used to detect comparative loci is indicated in brackets. However, due to duplication within the genomes this will not necessarily reflect the number of conserved loci used in the comparison, which where available is preferentially indicated.
[c]The *B. oleracea* map used in Lukens et al. (2003) was taken from Bohuon et al. (1996).

et al. 1994, Uzunova et al. 1995, Piquemal et al. 2005). However, in the late 1990s the application of a common set of markers across the three diploid species and the amphidiploid genome derivatives allowed comparisons to be made within the *Brassica* genus of lineage II of the Brassicaceae (Table 5.1). More recently the sequencing of previously mapped RFLP probes and the use of sequence-based markers have identified related genomic regions of the dicot model *Arabidopsis thaliana* (lineage I) and defined a set of ancestral blocks which allow similar comparisons as those previously made in the Poaceae (Parkin et al. 2005, Panjabi et al. 2008). The knowledge gained from these comparative mapping studies in the Brassicaceae will be described, focusing largely on the *Brassica* genus, and more recent and future developments relevant to interpreting *Brassica* genome organization will be introduced.

5.2 Common Terms Used in Comparative Mapping Studies

The wide adoption of comparative mapping as a tool to study plant evolution has inevitably led to the creation of a new vocabulary to describe common themes. In some instances established genetics nomenclature has been appropriated and imbued with subtle differences of meaning, which can lead to confusion in the literature. The definition of some of the more commonly used terms is provided below.

Synteny: This describes the physical co-localization of genetic loci on the same chromosome within an individual or species and is often erroneously used in place of collinearity; syntenous loci although always physically linked are not necessarily genetically linked or arranged in a predictable pattern.

Collinearity: Strictly speaking this refers to multiple points found in a linear order; in comparative mapping "conserved linkage" or collinearity refers to regions of conserved marker content and order on two (or more) separate linkage groups (or chromosomes).

Homology: In the context of comparative mapping it is generally used to refer to chromosomal regions but sometimes individual genetic loci and indicates the shared ancestry of these homologous elements.

Homoeology: Refers to chromosomes, regions, or loci inherited from divergent but homologous genomes within an allopolyploid nucleus.

Polyploidy: Mode of evolution which involves the doubling of genome complements either through whole genome duplication (autopolyploidy) or through hybridization of two or more related but distinct genomes (allopolyploidy).

Orthologues and paralogues: The identification of orthologues, those sequences related by speciation, and paralogues, those sequences which originate through segmental or gene duplication, is contentious due to the dynamic nature of genome evolution. For example, the paralogue maybe favored and the "true" orthologue of a gene is lost over time due to adaptive pressures. To limit such confusion the more general term of homology will be largely used throughout the text.

5.3 The Basics of Comparative Mapping

Comparative mapping is a powerful tool which not only allows the study of genome evolution but also can be exploited to transfer knowledge and resources from model plant species to improve traits in related crops. Much of the comparative mapping work to date has utilized RFLP markers. Restriction fragment length polymorphism (RFLP) loci are revealed as differences in genomic fragment lengths resulting from sequence variation at enzyme restriction sites; RFLP probes are derived from labeled cDNA or genomic DNA and through hybridization identify homologous sequences within genomes, thus allowing the length polymorphisms to be visualized while simultaneously identifying conserved sequences across species. The relationship between genetic linkage maps is identified through the use of common sets of RFLP markers and the similarity of the species is assessed through the extent of conservation of marker content and order or collinearity (Fig. 5.1). The level of collinearity varies with the rates and modes of evolution which is unique to each plant lineage, ranging from almost complete alignment of linkage groups to the identification of many small blocks of similarity between genomes (Fig. 5.1). The breaks in the collinearity are indicative of chromosomal rearrangements, such as duplications, translocations, inversions, or transpositions, and detail the evolutionary history of each species. In any comparative mapping study, caution should

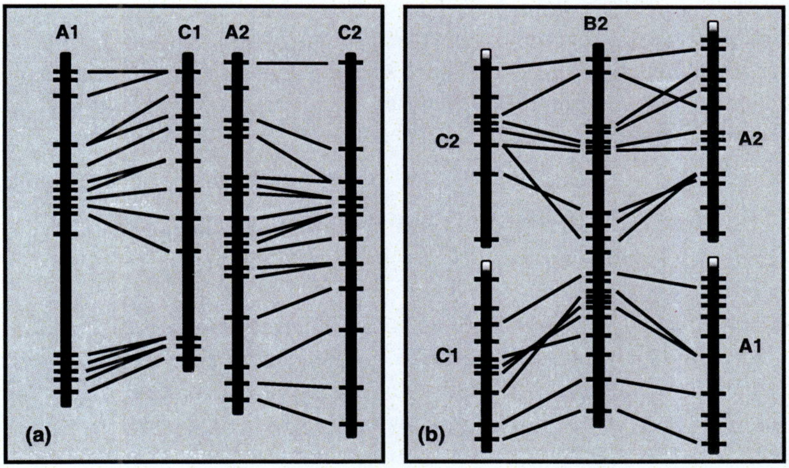

Fig. 5.1 Utilizing molecular markers to identify collinearity and uncover genome rearrangements. Comparing closely related species will identify limited rearrangements and greater stretches of conserved marker content and order, in (**a**) the *B. napus* A genome linkage groups A1 and A2 and their C genome homoeologues C1 and C2 are shown, these groups are collinear along their entire mapped lengths. However, when considering more distantly related species, for example, in (**b**) the B genome linkage group B2 (G5) is compared to the same *B. napus* linkage groups as in (**a**), the impacts of both genome duplication, B2 (G5) is collinear with segments of both A1 and C1, and chromosomal rearrangements such as translocations, chromosome fissions/fusions, and inversions are observed

be exercised due to the limitations imposed by genetic linkage analysis, in that only polymorphic regions of the genome can be observed and physical restriction of chromosome pairing can create clusters of coincident markers impeding the elucidation of marker order and hence collinearity.

Some of these limitations have been assuaged with the availability of fully sequenced genomes as points of reference in comparative mapping studies. For example, in the Brassicaceae, probes used for genetic mapping can be physically positioned on the *A. thaliana* genomic sequence based on conservation at the DNA sequence level. However, the question becomes what level of identity between two sequences indicates an evolutionary determined relationship and what is the impact of genome duplication, how similar are orthologues and paralogues found within the same genome to the ancestral homologue? By comparing the sequences of marker probes previously mapped in *B. oleracea* with genomic sequence from the crucifer model *A. thaliana,* Lukens et al. (2003) was able to define criteria by which to accept or reject the identification of orthologues (or the primary homologue) between the two species. These criteria were based on the observed distribution of sequence similarity scores (BLASTN) and the a priori knowledge of the expected nucleotide differences between paralogous (secondary homologues) sequences within the model genome itself (Lynch and Conery 2000), defining a significance of sequence similarity cut-off above which conserved sequences are expected to represent orthologues. This assumes that a query sequence from one genome is orthologous only to the other genome's sequence to which it is most similar and although not infallible, since strongly conserved motifs from replicated gene families can confound such analyses, such criteria provide a foundation for delineating conserved regions between genomes.

5.4 The Contribution of Polyploidy (Inter-specific Hybridization) to *Brassica* Genome Evolution

Polyploidy has played a significant role in the evolution of the Brassicaceae as it has done in many plant lineages. In the 1930s the eponymous U's triangle, as described in Chapter 1, was determined from cytological analyses of forced inter-specific hybrids (U 1935). Genetic linkage analysis has since confirmed the relationship between the six domesticated *Brassica* species, demonstrating at the molecular level that the three diploid species, *B. rapa* (A genome), *B. oleracea* (C), and *B. nigra* (B), had formed the three amphidiploid species, *B. napus* (AC), *B. juncea* (AB), and *B. carinata* (BC), through each possible pair-wise combination (Parkin et al. 1995, Bohuon et al. 1996, Panjabi et al. 2008, Suwabe et al. 2008).

The generation of new polyploids requires rapid diploidization within the nucleus to ensure stable pairing and inheritance; this can be achieved through significant chromosome rearrangement accentuated by sequence elimination or through genetic suppression of pairing between non-homologous chromosomes (or homoeologues) (Jenczewski and Alix 2004). Genetic linkage maps have been generated for *B. napus* and *B. juncea* as a result of normal diploid pairing between chromosomes of established allopolyploid lines and their modern day diploid relatives (Parkin et al. 1995, Axelsson et al. 2000). This has allowed the two diploid genomes to be identified within the allopolyploid nucleus and indicated that no major chromosomal rearrangements have occurred since the fusion of the *Brassica* A genome with either the B or the C genomes. This suggests that similar to wheat the *Brassica* allopolyploids thrived through the evolution of a heritable mechanism that suppresses illegitimate recombination events. However, although this is thought to be true for *B. napus*, where relatively high levels of homoeologous (or non-homologous) pairing have been observed in newly resynthesized lines, no such pairing was observed for resynthesized *B. juncea*. It appears that the divergence of the B genome from that of the A and C through both chromosomal rearrangements (see below) and genetic drift has been sufficient to limit pairing across these species, although it cannot be ruled out that the B genome retains a strong suppressor of illegitimate pairing. The linkage maps of the allopolyploids and those of the modern progenitor species represented the first comparative mapping data for the Brassicaceae.

To enable comparative mapping between species which have evolved through genome duplication ideally it is necessary to compare only homologous regions but with increased levels of duplication this becomes progressively more difficult. The first attempts to compare the genome of *B. oleracea* with *B. napus* led to the erroneous conclusion that there had been significant rearrangement of the C genome within the *B. napus* nucleus upon fusion with the A genome progenitor (Cheung et al. 1997). This confusion was the result of comparing the diploid genome with both the A and the C genomes within the allopolyploid species and thus identifying rearrangements which separated the two diploids. Such errors can be limited either through marker saturation or by inferring evolutionary relationships from evidence of pairing between similar chromosomes.

The homoeologous pairing events observed between the more closely related A and C genomes, identified through the utilization of a newly resynthesized *B. napus* line in mapping studies, have been exploited to identify the regions of primary homology between the two genomes (Parkin et al. 2003). These data were corroborated by the presence of extensive collinearity which determined that a minimum number of 16 chromosomal rearrangements were necessary to differentiate the A and C genomes (Fig. 5.2). The *Brassica* B genome has been shown to be phylogenetically distinct from the A and C diploids, which is reflected at both the sequence level (Sabhyata et al. 1996, Sharpe and Parkin, unpublished data) and in the absence of homoeologous pairing between the A and the B genomes (Axelsson et al. 2000). In contrast, although there remains some ambiguity in identifying the homologous regions across the three *Brassica* diploid genomes (A, B and C), comparative mapping has suggested a similar number of chromosomal rearrangements separate all three species (Fig. 5.2) (Lagercrantz and Lydiate 1996, Parkin et al. 2003, Panjabi et al. 2008). Also apparent from these initial studies of species from U's triangle was the high level of intra- and inter-chromosomal duplication observed within the *Brassica* diploid genomes suggesting the lineage had evolved from an ancient polyploid. This theory was cemented by subsequent comparisons to the genome of the

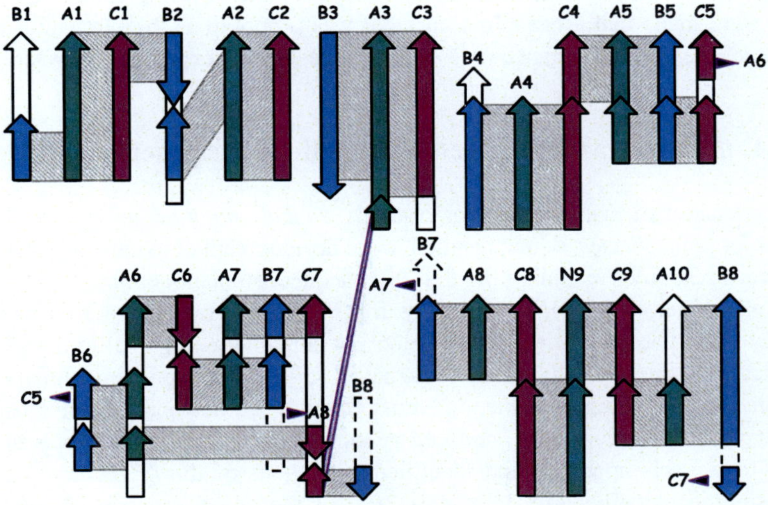

Fig. 5.2 Comparative mapping across the *Brassica* A (*green*), B (*blue*) and C (*purple*) diploid genomes. A schematic representation of the conserved genomic regions across the three genomes, the collinear segments are indicated by the *grey shaded regions* between linkage groups. The orientation of the linkage groups is indicated by the *arrows* (*pointing to top*) and is according to the published maps of Parkin et al. (2005) and Lagercrantz and Lydiate (1995). The B genome linkage groups are named according to the recent work of Panjabi et al. (2008)

plant model *A. thaliana*, a member of the Brassicaceae believed to have diverged from the *Brassica* species 14–24 million years ago (see Section 5.6 below) (Koch et al. 2000).

5.5 The Ghost of an Ancestral Hexaploid Genome

The use of RFLP markers to generate the first *Brassica* linkage maps was instrumental in uncovering evidence of whole genome duplication events within the diploid genomes. Up to 73% of RFLP probes have been shown to identify two or more loci within the diploid *Brassica* genomes (Lagercrantz and Lydiate 1996, Parkin et al. 2003). Further the duplicate loci were not randomly distributed across the genome but were found in collinear blocks of genetically linked loci. A recurring pattern was beginning to emerge with significant numbers of these conserved blocks being observed three times within the different genomes (Fig. 5.3). It should be noted that not all RFLP probes will reveal three loci in each diploid genome. Detection of all homologous regions is limited by not only the available polymorphism but also the on-going evolution of each conserved segment. Town et al. (2006) sequenced the three homologous regions from each of the triplicated copies of *B. oleracea* equivalent to two regions of the *A. thaliana* genome. This analysis provided the first comprehensive molecular study uncovering the impact of evolution upon the duplicated regions in *Brassica* genomes. It was demonstrated that each copy will be under local adaptive pressures that lead to sequence divergence and in some instances complete or partial loss of genomic sequences

Although the cumulative evidence from the *Brassica* diploids was highly suggestive, it was not until the first comparisons with the genome of the *A. thaliana* model plant were made that a hexaploid ancestor was proposed as the progenitor of the *Brassica* lineage (Lagercrantz 1998), an hypothesis which is now widely accepted.

5.6 *A. thaliana*, a Model Genome for the Brassicaceae

The benefactor of many decades of research, *A. thaliana*, emerged as the de facto model for plant species, which resulted in the development of extensive genetic and genomics resources including the first fully sequenced plant genome (Arabidopsis Genome Initiative 2000). Fortuitously for *Brassica* researchers, *A. thaliana* is closely related at the sequence level (~86% within coding regions) to modern day brassicas (Parkin et al. 2005); by sequencing RFLP probes previously mapped in *Brassica* species it was possible to identify through *in silico* analysis the most similar sequence(s) within the *A. thaliana* genome. Lukens et al. (2003) was the first study to use this approach and identified 34 regions of the *A. thaliana* genome with significant collinearity to almost 30% of the genetic map of *B. oleracea*. The increased density of markers, or possible points of comparison, developed for *B. napus* identified 21 blocks of genetically linked markers in the A and C genome of *B. napus* that were also found to be physically linked on the genome of *A. thaliana*

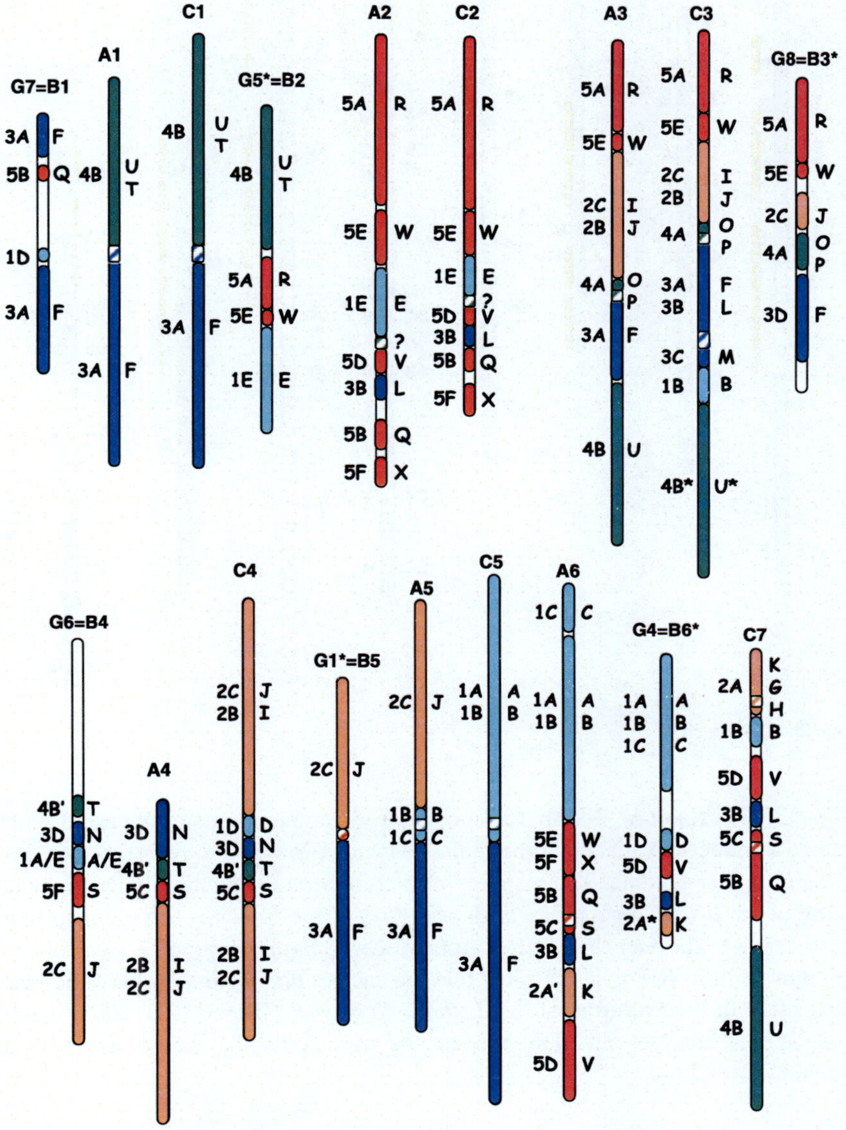

Fig. 5.3 How the ancestral blocks are arranged to construct the linkage groups of the A, B, and C genomes of *Brassica* species. The original designation of the blocks based on their collinearity with *A. thaliana* is shown to the *left* of the linkage groups (Parkin et al. 2005); the suggested block nomenclature according to collinearity with the lineage I ancestral karyotype is shown to the *right* (Schranz et al. 2006). The colors indicate collinearity with different *A. thaliana* chromosomes: *light blue* – chromosome 1; *orange* – chromosome 2; *dark blue* – chromosome 3; *green* – chromosome 4; and *red* – chromosome 5. The patterned blocks indicate the tentative positions of centromeres. The A and C linkage groups are based on the maps of Parkin et al. (2005) and the B genome is a derived composite based on information presented in Largercrantz and Lydiate (1995) and Panjabi et al. (2008), inverted linkage groups compared to these maps are indicated by an *asterisk*

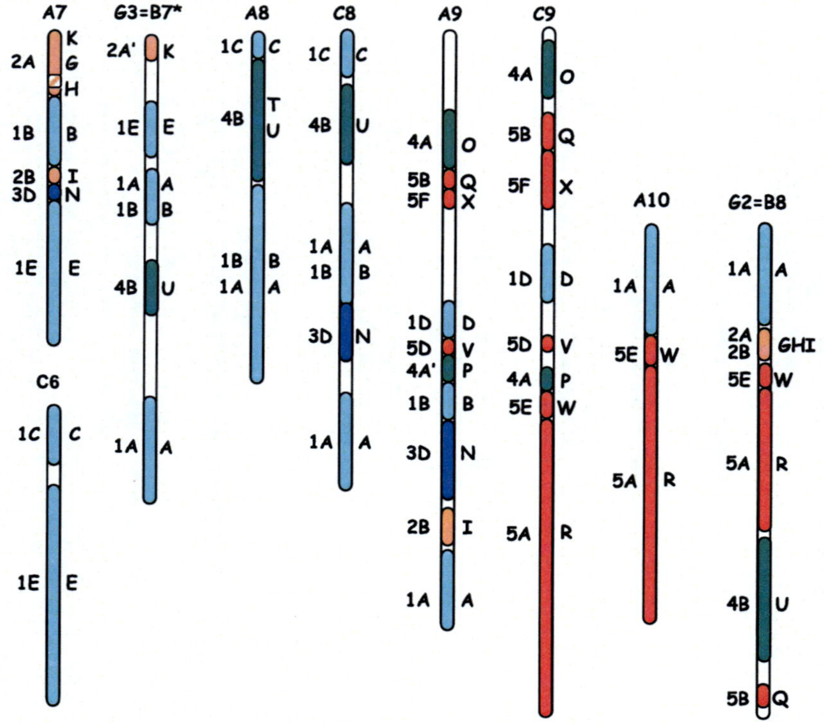

Fig. 5.3 (continued)

(Fig. 5.3) (Parkin et al. 2005). These conserved genomic units or ancestral blocks defined a framework that could be replicated and rearranged to represent over 80% of the *B. napus* genome. The mapping also corroborated the underlying triplicated nature of the *Brassica* genomes, with over 85% of the *B. napus* linkage map found in six copies. The fact that each region of the *A. thaliana* genome was found in six copies within *B. napus* indicated that the ancient duplication events which have taken place in the evolution of the *A. thaliana* genome (Blanc et al. 2003, Bowers et al. 2003, Henry et al. 2006) predate the triplication that has occurred in the *Brassica* lineage.

5.6.1 Across the A, B, and C Genomes

The comparison between the A and the C genomes has since been extended to the B genome to uncover a similar segmental arrangement of blocks (Panjabi et al. 2008). These data not only add to a previous study comparing *A. thaliana* with the B genome (Lagercrantz 1998) but also suggest the earlier work was flawed by the use of markers which later proved to be duplicated in the *A. thaliana* genome. There are still gaps in each of the diploid genomes where the relationship with *A. thaliana* is tenuous due to a dearth of markers; however, the use of the block-based architecture across all three genomes presents an opportunity to uncover conserved arrangements of blocks common to the *Brassica* lineage, which may provide insights into the organization of the *Brassica* progenitor. For example, the inversion of block 5E (W) to lie next to block 5A (R) (present on A2/C2, A3/C3, C9/A10, and B3/B8: Fig. 5.3) is common to all three genomes. Interestingly this arrangement is also conserved in a number of $x = 7$ taxa from lineage I but is not present in the $x = 8$ taxa, the chromosomal organization of which had been proposed as an ancestral karyotype for both lineage I and II species (Mandakova and Lysak 2008). By comparing the macrostructure across the three genomes the closer phylogenetic relationship of the A and C genomes is apparent with a number of linkage groups being homologous along almost their entire lengths (A1/C1, A2/C2, A3/C3) (Fig. 5.2). In contrast there is only one B genome linkage group which appears to be conserved with the A genome but not the C genome (B5/A5), although there is limiting data to suggest that B4 may be aligned completely with A4 and the lower half of C4. These comparisons have suggested direct mechanisms for the reduction of chromosome numbers between the genomes, for example, it appears that A7 and A8 have fused to form B7. When homoeologous regions between the A and the C had been defined previously there was one region of the A genome, the top of A10, which showed no homoeology with the C genome. However, it appears that the organization of A10 is completely conserved on linkage group B8 (Figs. 5.3 and 5.4); it remains ambiguous as to whether the region in question is missing from the C genome.

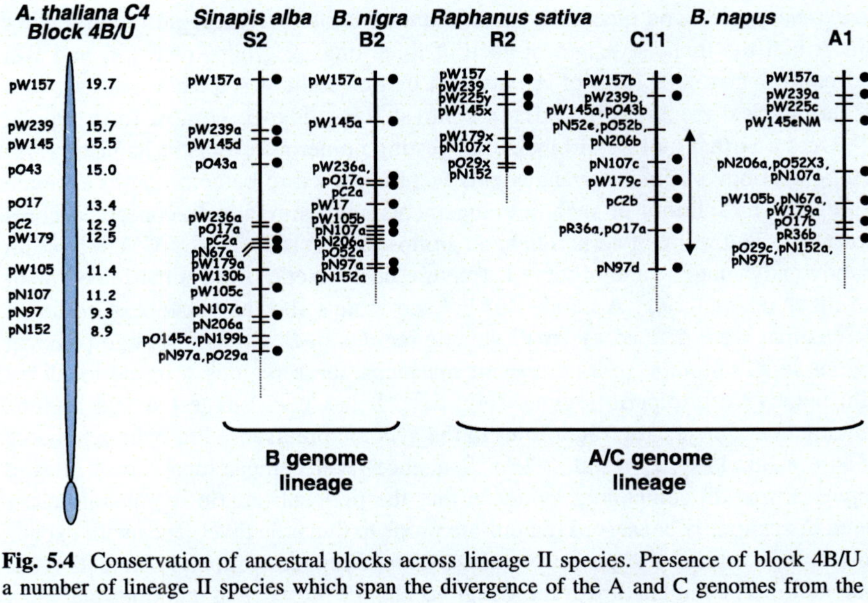

Fig. 5.4 Conservation of ancestral blocks across lineage II species. Presence of block 4B/U in a number of lineage II species which span the divergence of the A and C genomes from the B genome species

5.6.2 Conserved Chromosome Landmarks

Cytological studies in *Brassica* species have been notoriously difficult due to the small size of the chromosomes and it has only been with recent improvements in labeling and imaging technologies that the individual chromosomes can be differentiated in the nucleus (Howell et al. 2008). This has meant that alignment of the linkage maps with the chromosomal karyotype has remained elusive. Comparative mapping has allowed the tentative positions of centromeres to be inferred in the three diploid genomes based on the mapping of markers flanking the *A. thaliana* centromeric regions to conserved collinear regions of the *Brassica* genome (Fig. 5.3). In three instances these regions have been confirmed (C1, C2, and C4) by the genetic mapping of telocentric chromosomes in AAC triploid lines (Kelly and Lydiate, unpublished). These data imply that a significant number of the macro-rearrangements which differentiate the three diploid genomes are the result of chromosome fission and fusion at centromeric and telomeric sites.

5.6.3 Rearrangement Hotspots

Due to the extensive restructuring and the unknown architecture of the progenitor genome it is difficult to accurately determine the number of chromosomal rearrangements which led to the present organization of the *Brassica* diploids. However, careful study of the conserved blocks across the three genomes allows a number of observations. Some linkage groups appeared less susceptible to further rearrangement upon formation and certain block relationships were more stably inherited. The conserved blocks ranged dramatically in size both genetically (*Brassica* species) and physically (*A. thaliana*) with the largest region representing almost half the mapped length of each of three linkage groups (A1, C1, and B2) and equivalent to ~9 Mbp of *A. thaliana* chromosome 4 sequence (almost 50% of the physical length) (Fig. 5.3). In contrast some linkage groups, for example, A9, were a virtual mosaic of blocks, suggesting numerous events in its history. As mentioned above, some rearrangements were observed to be common to all three genomes. The stability of such rearrangements and the relatively conserved chromosome organization of certain linkage groups (e.g., A1/C1, A5/B5/C5) implies an adaptive advantage to the observed structure or a genetic or physical impediment to further restructuring. A number of linkage groups showed clusters of collinear regions that were defined by small genetic regions in *B. napus* but large physical regions in *A. thaliana,* in a number of instances, these regions were co-localized with putative centromeric regions (Fig. 5.3). It has been suggested that regions flanking centromeres are fragile sites in the genome predisposed to rearrangements (Moore et al. 1997, Qi et al. 2006). Sequence-level comparisons across lineage I species provide compelling evidence that the pericentromeric regions although found in apparently conserved regions are prone to dynamic divergence with expansion through insertion of genes, pseudogenes, and repetitive mobile elements (Hall et al. 2006). The fluid nature of these regions and the presence of repetitive elements could lead to a propensity for chromosomal rearrangements which would be consistent with the alignment of centromeric regions with the endpoints of the ancestral blocks (Schranz et al. 2006).

5.7 Exploiting Comparative Mapping for Trait Analysis

Before the absolute extent of genome conservation between *A. thaliana* and its *Brassica* relatives was determined it had been observed that these oilseeds shared common phenotypes. The simple genome, short generation time, and the availability of significant genetics and genomics resources for *A. thaliana* allowed genes controlling such phenotypes to be identified with relative ease. By identifying allelic variants or manipulating homologues of these genes in *B. napus* it was possible to affect similar traits in the crop as determined in the weed. An early example of this was the association of variant alleles for the *FATTY ACID ELONGASE* (*FAE1*) gene in *B. napus* with reduction in erucic acid in the seed, comparatively the *FAE1 A. thaliana* mutant had reduced levels of long chain fatty acids (Roscoe et al. 2001). The conservation of gene function between species suggested that *A. thaliana* could be a valuable source of candidate genes for traits of agronomic importance in *Brassica* crop species.

With the advent of the comparative mapping data it was possible to associate genomic regions in *Brassica* species underlying advantageous traits with conserved regions in *A. thaliana*. These comparisons allow exploitation of the *A. thaliana* genomic sequence for both the development of targeted markers and the identification of potential candidates controlling the expression of traits of interest (Qiu et al. 2006). Such analyses can be of particular value for the analyses of quantitative trait loci (QTL), where the control of the phenotype is complex being conferred by the presence of a number of loci of varying effect. As well as facilitating the identification of additional markers to saturate the QTL region for fine mapping, the comparative mapping data can indicate where the number of loci controlling a phenotype could be a reflection of the high level of duplication present within the *Brassica* genomes. For example, accumulation of aliphatic glucosinolates in the seed is controlled by at least three QTL loci in *B. napus* (Howell et al. 2003) which are localized to homologous regions of the genome (block C5E/W), suggesting that a duplicate gene family, rather than three unrelated genes, could be manipulated to impact a change.

Based on conserved map positions a number of candidate genes, previously characterized in *A. thaliana*, have now been correlated with QTL loci in *B. napus*, *B.oleracea*, and *B. juncea*, particularly for flowering time, inflorescence morphology, and seed glucosinolate biosynthesis (Osborn et al. 1997, Lan and Paterson 2000, Long et al. 2007, Bisht et al. 2009). The use of comparative mapping was also instrumental in cloning the gene responsible for a dwarf phenotype observed and genetically mapped in *B. rapa* (Muangprom and Osborn 2004). The phenotype was located in the conserved C2A/K block where a DELLA protein involved in gibberellic acid biosynthesis (RGA1) was identified. Furthermore, a mutant allele of the

B. rapa homologue was confirmed to reproduce the dwarf phenotype (Muangprom et al. 2005).

However, the use of comparative mapping data can be misleading in regions where the underlying α duplication, or the most recent duplication event in the history of the *A. thaliana* genome, and the subsequent triplication in the *Brassica* genomes causes difficulties in differentiating between conserved regions (Mayerhofer et al. 2005). In addition, as a consequence of random gene loss in the duplicate regions in both species, the generation of targeted markers can be unpredictable (Town et al. 2006).

5.8 Extending the Comparisons to Related Species

Comparative mapping between *A. thaliana* and its close relatives of lineage I has suggested that the widely studied model genome with its low chromosome number is actually an anomaly among its peers, evolved from a progenitor with a chromosome complement of seven or eight. Alignment of the *A. thaliana* genome with that of *Capsella rubella* (Boivin et al. 2004) and *A. lyrata* (Kuittinen et al. 2004) indicated the reduction in chromosome number was largely the result of chromosome fusions rather than elimination of genomic DNA. In addition, such alignments have shown that while *A. lyrata* and *C. rubella* demonstrate strong collinearity, at least seven major rearrangements, including inversion, chromosomal fusion, and translocation events, are specific to the *A. thaliana* genome (Yogeeswaran et al. 2005). The elucidation of the ancestral collinear blocks in *B. napus* (Parkin et al. 2005) and the painting of those blocks on $n = 7$ and $n = 8$ Brassicaceae species have contributed greatly to our understanding of evolutionary steps in the formation of important *Brassica* species (Schranz et al. 2006, Mandakova and Lysak 2008).

The *Brassica* genus is incredibly diverse in genome content, chromosome number, and morphological form and contains a relatively large number of agronomically important species. Despite this collected wealth of diversity, there is interest in capturing and exploiting traits and allelic variation found among related genera, one such example that has been validated for hybrid development in *B. napus* is the use of the ogura cytoplasmic male sterility system identified in *Raphanus sativa* (Primard-Brisset et al. 2005). To facilitate such applications genetic linkage studies using the core set of RFLP probes sequenced in Parkin et al. (2005) have been initiated in *R. sativa* (Bett and Lydiate 2003), *Sinapis alba* (Nelson and Lydiate 2006), and *Moricandia arvense* (Beschorner and Lydiate, unpublished). Perhaps not surprisingly, preliminary analyses of these data corroborate the existence of an ancestral hexaploid and indicate that the ancestral blocks are conserved across the genera (Fig. 5.4). These and additional studies in species of lineage II, from which the *Brassica* crops evolve, will be necessary to identify conserved block arrangements and elucidate the ancestral karyotype for this lineage (Mandakova and Lysak 2008).

5.9 The Promise of Sequenced Genomes

The genomic sequence of the first *Brassica* species, *B. rapa,* will be available shortly (Yang et al. 2005, Hong et al. 2008). This foundational resource will allow a comprehensive analysis of the relationship between a related crop and weed genome, uncovering at the micro-level the impacts of genome duplication and allowing precise identification of rearrangement endpoints, which could point to the evolutionary mechanisms driving such changes. With the advent of more efficient and cost-effective sequencing technologies it is possible to envision all members of U's triangle being scrutinized at the sequence level. Such analyses will empower the improvement of the constituent crops and cross-species comparisons will provide insights into the evolution of the different crop types.

References

Arabidopsis Genome Initiative (2000) Analysis of the genome sequence of the flowering plant *Arabidopsis thaliana*. Nature 408:796–815

Axelsson T, Bowman CM, Sharpe AG, Lydiate DJ, Lagercrantz U (2000) Amphidiploid *Brassica juncea* contains conserved progenitor genomes. Genome 43:679–688

Bett KE, Lydiate DJ (2003) Genetic analysis and genome mapping in Raphanus. Genome 46: 423–430

Bisht NC, Gupta V, Ramchiary N, Sodhi YS, Mukhopadhyay A, Arumugam N, Pental D, Pradhan AK (2009) Fine mapping of loci involved with glucosinolate biosynthesis in oilseed mustard (*Brassica juncea*) using genomic information from allied species. Theor Appl Genet 118: 413–421

Blanc G, Hokamp K, Wolfe KH (2003) A recent polyploidy superimposed on older large-scale duplications in the Arabidopsis genome. Genome Res 13:137–144

Bohuon EJR, Keith DJ, Parkin IAP, Sharpe AG, Lydiate DJ (1996) Alignment of the conserved C genomes of *Brassica oleracea* and *Brassica napus*. Theor Appl Genet 93:833–839

Boivin K, Acarkan A, Mbulu RS, Clarenz O, Schmidt R (2004) The Arabidopsis genome sequence as a tool for genome analysis in Brassicaceae. A comparison of the Arabidopsis and *Capsella rubella* genomes. Plant Physiol 135:735–744

Bowers JE, Chapman BA, Rong J, Paterson AH (2003) Unravelling angiosperm genome evolution by phylogenetic analysis of chromosomal duplication events. Nature 422:433–438

Cheung WY, Champagne G, Hubert N, Landry BS (1997) Comparison of the genetic maps of *Brassica napus* and *Brassica oleracea*. Theor Appl Genet 94:569–582

Devos KM (2005) Updating the 'crop circle'. Curr Opin Plant Biol 8:155–162

Devos KM, Gale MD (2000) Genome relationships: the grass model in current research. Plant Cell 12:637–646

Ferreira ME, Williams PH, Osborn TC (1994) RFLP mapping of *Brassica napus* using doubled haploid lines. Theor Appl Genet 89:615–621

Gale MD, Devos KM (1998) Comparative genetics in the grasses. Proc Natl Acad Sci USA 95:1971–1974

Hall AE, Kettler GC, Preuss D (2006) Dynamic evolution at pericentromeres. Genome Res 16: 355–364

Henry Y, Bedhomme M, Blanc G (2006) History, protohistory and prehistory of the *Arabidopsis thaliana* chromosome complement. Trends Plant Sci 11:267–273

Hong CP, Kwon SJ, Kim JS, Yang TJ, Park BS, Lim YP (2008) Progress in understanding and sequencing the genome of *Brassica rapa*. Int J Plant Genomics 2008:582837

Howell EC, Kearsey MJ, Jones GH, King GJ, Armstrong SJ (2008) A and C genome distinction and chromosome identification in *Brassica napus* by sequential fluorescence in situ hybridization and genomic in situ hybridization. Genetics 180:1849–1857

Howell PM, Sharpe AG, Lydiate DJ (2003) Homoeologous loci control the accumulation of seed glucosinolates in oilseed rape (*Brassica napus*). Genome 46:454–460

Jenczewski E, Alix K (2004) From Diploids to Allopolyploids: the emergence of efficient pairing control genes in plants. Crit Rev Plant Sci 23:21–45

Kianian SF, Quiros CF (1992) Generation of a *Brassica oleracea* composite RFLP map: linkage arrangements among various populations and evolutionary implications. Theor Appl Genet 84:544–554

Koch MA, Haubold B, Mitchell-Olds T (2000) Comparative evolutionary analysis of chalcone synthase and alcohol dehydrogenase loci in Arabidopsis, Arabis, and related genera (Brassicaceae). Mol Biol Evol 17:1483–1498

Kuittinen H, de Haan AA, Vogl C, Oikarinen S, Leppala J, Koch M, Mitchell-Olds T, Langley CH, Savolainen O (2004) Comparing the linkage maps of the close relatives *Arabidopsis lyrata* and *A. thaliana*. Genetics 168:1575–1584

Lagercrantz U (1998) Comparative mapping between *Arabidopsis thaliana* and *Brassica nigra* indicates that Brassica genomes have evolved through extensive genome replication accompanied by chromosome fusions and frequent rearrangements. Genetics 150:1217–1228

Lagercrantz U, Lydiate DJ (1995) RFLP mapping in *Brassica nigra* indicates differing recombination rates in male and female meioses. Genome 38:255–264.

Lagercrantz U, Lydiate DJ (1996) Comparative genome mapping in Brassica. Genetics 144: 1903–1910

Lan TH, Paterson AH (2000) Comparative mapping of quantitative trait loci sculpting the curd of *Brassica oleracea*. Genetics 155:1927–1954

Landry BS, Hubert N, Etoh T, Harada JJ, Lincoln SE (1991) A genetic map for *Brassica napus* based on restriction fragment length polymorphisms detected with expressed DNA sequences. Genome 34:543–552

Long Y, Shi J, Qiu D, Li R, Zhang C, Wang J, Hou J, Zhao J, Shi L, Park BS, Choi SR, Lim YP, Meng J (2007) Flowering time quantitative trait loci analysis of oilseed Brassica in multiple environments and genomewide alignment with Arabidopsis. Genetics 177: 2433–2444

Lukens L, Zou F, Lydiate D, Parkin I, Osborn T (2003) Comparison of a *Brassica oleracea* genetic map with the genome of Arabidopsis thaliana. Genetics 164:359–372

Lynch M, Conery JS (2000) The evolutionary fate and consequences of duplicate genes. Science 290:1151–1155

Mandakova T, Lysak MA (2008) Chromosomal phylogeny and karyotype evolution in x=7 crucifer species (Brassicaceae). Plant Cell 20:2559–2570

Mayerhofer R, Wilde K, Mayerhofer M, Lydiate D, Bansal VK, Good AG, Parkin IA (2005) Complexities of chromosome landing in a highly duplicated genome: toward map-based cloning of a gene controlling blackleg resistance in *Brassica napus*. Genetics 171: 1977–1988

Moore G, Foote T, Helentjaris T, Devos K, Kurata N, Gale M (1995) Was there a single ancestral cereal chromosome? Trends Genet 11:81–82

Moore G, Roberts M, Aragon-Alcaide L, Foote T (1997) Centromeric sites and cereal chromosome evolution. Chromosoma 105:321–323

Muangprom A, Osborn TC (2004) Characterization of a dwarf gene in *Brassica rapa*, including the identification of a candidate gene. Theor Appl Genet 108:1378–1384

Muangprom A, Thomas SG, Sun TP, Osborn TC (2005) A novel dwarfing mutation in a green revolution gene from *Brassica rapa*. Plant Physiol 137:931–938

Nelson MN, Lydiate DJ (2006) New evidence from *Sinapis alba* L. for ancestral triplication in a crucifer genome. Genome 49:230–238

Osborn TC, Kole C, Parkin IA, Sharpe AG, Kuiper M, Lydiate DJ, Trick M (1997) Comparison of flowering time genes in *Brassica rapa, B. napus* and *Arabidopsis thaliana*. Genetics 146: 1123–1129

Panjabi P, Jagannath A, Bisht NC, Padmaja KL, Sharma S, Gupta V, Pradhan AK, Pental D (2008) Comparative mapping of Brassica juncea and Arabidopsis thaliana using Intron Polymorphism (IP) markers: homoeologous relationships, diversification and evolution of the A, B and C Brassica genomes. BMC Genomics 9:113

Parkin IA, Gulden SM, Sharpe AG, Lukens L, Trick M, Osborn TC, Lydiate DJ (2005) Segmental structure of the Brassica napus genome based on comparative analysis with *Arabidopsis thaliana*. Genetics 171:765–781

Parkin IA, Sharpe AG, Keith DJ, Lydiate DJ (1995) Identification of the A and C genomes of amphidiploid *Brassica napus* (oilseed rape). Genome 38:1122–1131

Parkin IA, Sharpe AG, Lydiate DJ (2003) Patterns of genome duplication within the *Brassica napus* genome. Genome 46:291–303

Paterson AH, Bowers JE, Chapman BA, Peterson DG, Rong J, Wicker TM (2004) Comparative genome analysis of monocots and dicots, toward characterization of angiosperm diversity. Curr Opin Biotechnol 15:120–125

Piquemal J, Cinquin E, Couton F, Rondeau C, Seignoret E, Doucet I, Perret D, Villeger MJ, Vincourt P, Blanchard P (2005) Construction of an oilseed rape (*Brassica napus* L.) genetic map with SSR markers. Theor Appl Genet 111:1514–1523

Primard-Brisset C, Poupard JP, Horvais R, Eber F, Pelletier G, Renard M, Delourme R (2005) A new recombined double low restorer line for the Ogu-INRA cms in rapeseed (*Brassica napus* L.). Theor Appl Genet 111:736–746

Qi L, Friebe B, Gill BS (2006) Complex genome rearrangements reveal evolutionary dynamics of pericentromeric regions in the Triticeae. Genome 49:1628–1639

Qiu D, Morgan C, Shi J, Long Y, Liu J, Li R, Zhuang X, Wang Y, Tan X, Dietrich E et al (2006) A comparative linkage map of oilseed rape and its use for QTL analysis of seed oil and erucic acid content. Theor Appl Genet 114:67–80

Roscoe TJ, Lessire R, Puyaubert J, Renard M, Delseny M (2001) Mutations in the fatty acid elongation 1 gene are associated with a loss of beta-ketoacyl-CoA synthase activity in low erucic acid rapeseed. FEBS Lett 492:107–111

Sabhyata B, Madan Singh N, Malathi L (1996) Structural analysis of the rDNA intergenic spacer of *Brassica nigra*: evolutionary divergence of the spacers of the three diploid brassica species. J Mol Evol 43:460–468

Schranz ME, Lysak MA, Mitchell-Olds T (2006) The ABC's of comparative genomics in the Brassicaceae: building blocks of crucifer genomes. Trends Plant Sci 11:535–542

Slocum MK, Figdore SS, Kennard WC, Suzuki JY, Osborn TC (1990) Linkage arrangement of restriction fragment length polymorphism loci in *Brassica oleracea*. Theor Appl Genet 80: 57–64

Suwabe K, Morgan C, Bancroft I (2008) Integration of Brassica A genome genetic linkage map between *Brassica napus* and *B. rapa*. Genome 51:169–176

Tang H, Bowers JE, Wang X, Ming R, Alam M, Paterson AH (2008) Synteny and collinearity in plant genomes. Science 320:486–488

Tanksley SD, Ganal MW, Prince JP, de-Vicente MC, Bonierbale MW, Broun P, Fulton TM, Giovannoni JJ, Grandillo S, Martin GB, Messeguer R, Miller JC, Miller L, Paterson AH, Pineda O, Roder MS, Wing RA, Wu W, Young ND (1992) High density molecular linkage maps of the tomato and potato genomes. Genetics 132:1141–1160

Town CD, Cheung F, Maiti R, Crabtree J, Haas BJ, Wortman JR, Hine EE, Althoff R, Arbogast TS, Tallon LJ, Vigouroux M, Trick M, Bancroft I (2006) Comparative genomics of *Brassica oleracea* and *Arabidopsis thaliana* reveal gene loss, fragmentation, and dispersal after polyploidy. Plant Cell 18:1348–1359

U N (1935) Genomic analysis in Brassica with special reference to the experimental formation of *B. napus* and peculiar mode of fertilization. Jpn J Bot 7:389–452

Uzunova M, Ecke W, Weissleder K, Röbbelen G (1995) Mapping the genome of rapeseed (Brassica napus L.). I. Construction of an RFLP linkage map and localization of QTLs for seed glucosinolate content. Theor Appl Genet 90:194–204

Yang TJ, Kim JS, Lim KB, Kwon SJ, Kim JA, Jin M, Park JY, Lim MH, Kim HI, Kim SH, Lim YP, Park BS (2005) The Korea brassica genome project: a glimpse of the Brassica genome based on comparative genome analysis with Arabidopsis. Comp Funct Genomics 6:138–146

Yogeeswaran K, Frary A, York TL, Amenta A, Lesser AH, Nasrallah JB, Tanksley SD, Nasrallah ME (2005) Comparative genome analyses of Arabidopsis spp.: inferring chromosomal rearrangement events in the evolutionary history of *A. thaliana*. Genome Res 15:505–515

Chapter 6
Comparative Genome Analysis at the Sequence Level in the Brassicaceae

Chris Town, Renate Schmidt, and Ian Bancroft

Abstract In the world of plant genome sequencing, the cultivated *Brassica* species have been relatively under-resourced compared with other crop species largely due to their position in the economic hierarchy of perceived importance. Thus, with the completion of the *Arabidopsis thaliana* genome in the year 2000, the limited sequencing efforts undertaken in the *Brassica* crops and other species of the Brassicaceae have been largely restricted either to survey sequencing of various insert size clones or to finished sequences of small genomic regions, generally as bacterial artificial chromosome (BAC) clones. In this chapter, we review the sequencing efforts to date and how they have been used in comparative analysis with the Arabidopsis genome and with each other to begin to understand the genome organisation of members of the crucifer family, how they relate to one another, and how they may have evolved.

Keywords Collinearity · Comparative genome analysis · Duplication · Genome organisation · Polymorphism · Reference sequence · Sequence comparison · Survey sequencing

Abbreviations

AOP	2-oxoglutarate-dependent dioxygenase
BAC	Bacterial artificial chromosome
Col	Columbia
DH	Doubled haploid
EST	Expressed sequence tag
FISH	Fluorescence in situ hybridisation
GRP	Glycine-rich pollen surface protein
InDel	Insertion/deletion
kbp	Kilo base pairs
L*er*	Landsberg *erecta*

C. Town (✉)
J. Craig Venter Institute, 9704 Medical Center Drive, Rockville, MD 20850, USA
e-mail: cdtown@jcvi.org

R. Schmidt, I. Bancroft (eds.), *Genetics and Genomics of the Brassicaceae*, Plant Genetics and Genomics: Crops and Models 9, DOI 10.1007/978-1-4419-7118-0_6,

LRR	Leucine-rich repeat
Mbp	Mega base pairs
MITE	Miniature inverted repeat transposable elements
NB-LRR	Nucleotide-binding leucine-rich repeat
rDNA	Ribosomal DNA
R gene	Disease resistance gene
RACE	Rapid amplification of cDNA ends
RLK	Receptor-like kinase
SNP	Single nucleotide polymorphism
TAIR	The Arabidopsis Information Resource
TRIM	Terminal repeat retrotransposons in miniature

Contents

6.1 Introduction/Overview . . . 172
6.2 The *A. thaliana* Reference Genome . . . 173
6.3 Comparative Analysis of *A. thaliana* Accessions . . . 174
6.4 Sequence Comparisons Between *A. thaliana* and Near Relatives . . . 176
6.5 Sequence Comparisons Between *A. thaliana* and *Brassica* Species . . . 181
6.5.1 *B. rapa* . . . 182
6.5.2 *B. oleracea* . . . 184
6.5.3 *B. napus* . . . 185
6.6 Sequence Relationships Between *Brassica* Genomes . . . 188
6.6.1 *Brassica* A Genomes: *B. rapa* and *B. napus* . . . 188
6.6.2 *Brassica* C Genomes: *B. oleracea* and *B. napus* . . . 188
6.7 Comparative Analysis of *B. napus* Accessions . . . 189
6.8 Summary . . . 190
References . . . 191

6.1 Introduction/Overview

The Brassicaceae lend themselves well to study different aspects of genome plasticity as differences in genome size and chromosome number are readily observed. An approximately 16-fold variation of genome sizes has been established for this family (Lysak et al. 2009). The size of the *Arabidopsis thaliana* genome was estimated as 160 Mbp and even species of the same genus that diverged approximately 5 million years ago such as *Arabidopsis halleri* and *Arabidopsis lyrata* (Koch et al. 2000, 2001) have genomes that are about 50–60% larger. Moreover, they have $2n = 16$ chromosomes rather than $2n = 10$ as *A. thaliana* (Johnston et al. 2005). *Capsella rubella* and *Boechera stricta* diverged approximately 10–14 million years ago from *A. thaliana* and have $2n = 16$ and $2n = 14$ chromosomes, respectively (Koch et al. 2000, 2001). Their genomes are of similar size to those of *A. halleri* and *A. lyrata* (Oyama et al. 2008). Differences with respect to chromosome number and genome size have also been observed for the *Brassica* species that diverged from the *Arabidopsis* lineage approximately 20 million years ago (Yang et al. 1999,

Koch et al. 2001). *Brassica nigra*, *Brassica oleracea*, and *Brassica rapa* have, for example, $2n = 16$, 18, and 20 chromosomes, respectively. Their genome sizes range from 530 to 700 Mbp, whereas values up to 1,285 Mbp have been found for the amphidiploids *Brassica juncea*, *Brassica napus*, and *Brassica carinata* (Johnston et al. 2005).

Importantly, the influence of recent and ancient polyploidisation events on genome evolution can also be studied in the crucifer family. Different pairwise combinations of *B. rapa* (AA), *B. nigra* (BB), and *B. oleracea* (CC) yielded the amphidiploid species *B. juncea* (AABB), *B. napus* (AACC), and *B. carinata* (BBCC) (U 1935). The hybridisation of *B. oleracea* and *B. rapa* to form the *B. napus* genome represents a very recent event, most probably occuring less than 10,000 years ago. Results of detailed genetic, physical mapping and cytological studies of the extant *Brassica* diploid species lend support for a hexaploidisation event in the *Brassica* lineage that postdated the divergence from the *Arabidopsis* lineage (Lagercrantz and Lydiate 1996, Lagercrantz 1998, O'Neill and Bancroft 2000, Lysak et al. 2005, Rana et al. 2004, Parkin et al. 2005, Lysak et al. 2007). Three even more ancient polyploidisation events are shared by the *Brassica* and *Arabidopsis* lineages (Tang et al. 2008).

Comparative genetic mapping between species of this family readily revealed collinear blocks even though the species differed with respect to genome size, base chromosome number, and ploidy. The genome structures of *A. thaliana*, *A. lyrata*, *C. rubella*, and *B. rapa* can be derived from an ancestral karyotype of $n = 8$ by rearranging 24 collinear blocks (Schranz et al. 2006). Such experiments have been indispensable in describing conservation or differences of gross chromosomal structure in closely related species but much more detailed comparisons require sequence information of orthologous regions. The genome of *A. thaliana* was the first of any plant to be sequenced (Arabidopsis Genome Initiative 2000). In addition to paving the way for many fundamental discoveries in plant biology, it has also served as a surrogate reference genome for other members of the Brassicaceae. Sequence information in species of the Brassicaceae has been generated by sequencing selected medium or large inserts from clone libraries in their entirety, by establishing end sequence information for clones of genomic libraries or by whole-genome shotgun approaches. Comparisons of these sequences with the annotated genome of *A. thaliana* allowed an assessment of conserved sequences. Moreover, the gene and repetitive DNA content were analysed in the different species as well as the spacing of genes and their orientation relative to each other. It was also feasible to perform detailed analyses of exon/intron structures. These comparative analyses pointed to the mechanisms that bring about differences in gene arrangements and exon/intron structures of genes.

6.2 The *A. thaliana* Reference Genome

A. thaliana has one of the smallest known nuclear genomes in higher plants. By December 2000, the nucleotide sequence of the majority of the genome of the accession Col-0 (Columbia-0) had been determined by a clone by clone approach at accuracy rates between 99.99 and 99.999%. The established genome sequence amounted to 115.5 Mbp in total and spanned the 10 chromosome arms from either the telomeres or the ribosomal DNA repeats to the satellite sequences of the centromeric regions. The regions containing highly repetitive sequences organised in tandem such as the ribosomal DNA repeat regions and the centromeric satellites were not completely sequenced (Arabidopsis Genome Initiative 2000). The centromere cores consist of long arrays of 178 bp repeats interspersed with *Athila* retroelements and are flanked by regions that consist of numerous classes of repetitive sequences. However, expressed genes are also found within the genetically defined centromeres (Copenhaver et al. 1999, Hall et al. 2004).

The first annotation of gene structures took into account similarities to known proteins and expressed sequence tags (ESTs), but to achieve the genome-wide annotation of 25,498 genes, ab initio gene prediction had to be used in addition (Arabidopsis Genome Initiative 2000). Since then the Arabidopsis genome has been completely reannotated. Newly identified genes and pseudogenes were added, the annotation of regions with homology to transposons was improved and a large proportion of the original gene models were refined (Haas et al. 2005). As a follow-up of this work, several updates of the genome annotation have been released by The Arabidopsis Information Resource (TAIR, Swarbreck et al. 2008, http://www.arabidopsis.org/). Currently the annotation efforts are concentrated on the non-coding portions of the genome (Haberer et al., Chapter 3, this book). All five chromosomes are characterised by a high density of protein-coding genes, approximately one gene every 4.5 kbp, but a lower gene density is found in heterochromatic regions. A large proportion of Arabidopsis genes are present in more than one copy in the genome, only 35% of the predicted protein genes occur once in the genome. This high redundancy is mainly explained by two features, tandem arrays of duplicated gene sequences and segmental duplications (Arabidopsis Genome Initiative 2000). Evidence for three paleopolyploidy events has been found in the *Arabidopsis* lineage, the most ancient one represents a triplication whereas the remaining two are duplications (Tang et al. 2008). The most recent event, termed α (Bowers et al. 2003), took place in the lineage of the Brassicaceae, probably less than 10 million years after the divergence of the Brassicaceae and Cleomaceae (Schranz and Mitchell-Olds 2006). In the segmental duplications, a set of genes in common to each of the duplicated segments is found interspersed with genes that are unique to any of the duplications (Blanc et al. 2000, Arabidopsis Genome Initiative 2000). This pattern can be explained by dispersed loss of genes in duplicated chromosome segments (Ku et al. 2000).

6.3 Comparative Analysis of *A. thaliana* Accessions

Genome size measurements revealed a 1.1-fold variation with respect to mean C-values when 21 different *A. thaliana* accessions that had been collected throughout the entire Eurasian range (Schmuths et al. 2004) were analysed.

A comparison of the reference genome sequence of the accession Columbia with sequenced PCR products derived from other Arabidopsis accessions and/or populations can be readily used to determine polymorphisms. In a large-scale study of this kind, 876 short fragments, representing 0.48 Mbp in total, were analysed in 96 Arabidopsis individuals. More than 17,000 single nucleotide polymorphisms (SNPs) and insertion/deletion (InDel) polymorphisms were identified. A negative correlation between gene density and the local polymorphism level was observed, whereas a positive correlation was found in regions with segmental duplications (Nordborg et al. 2005). The same collection of accessions was used to assess variation in the leucine-rich repeat (LRR) region of 27 disease resistance (*R*) genes. The *R* genes as a whole exhibited higher nucleotide diversity and more recombination than the set of 876 loci that had been sampled to represent the entire genome (Bakker et al. 2006). An evaluation of a set of 27 defense response genes that may be activated downstream of *R*-gene mediated resistance revealed contrasting results. The defense response genes displayed overall lower levels of polymorphism than the reference set of the 876 loci (Bakker et al. 2008).

Alignment of sequence data generated via whole-genome shotgun sequencing approaches with the reference sequence of the Columbia accession enables genome-wide detection of polymorphisms. A whole-genome shotgun sequence of the Landsberg *erecta* (L*er*) accession representing approximately twofold genome coverage was used for such an approach. A total of 56,670 polymorphisms were identified. The Columbia and Landsberg *erecta* accessions differed by 37,344 SNPs and 18,579 small InDels (less than or equal to 100 bp). Larger InDels spanning more than 100 bp up to 38 kbp were also found. On average, one bioinformatically predicted SNP was found for every 3.3 kbp and one InDel for every 6.6 kbp (Jander et al. 2002). A detailed analysis of approximately 2,000 insertions greater than 100 bp in the Columbia genome relative to the Landsberg *erecta* sequence shed light on the genetic mechanisms leading to these changes. The differences between the two genomes were classified into three main categories. Approximately a quarter of the events each were explained by unequal recombination and transposable elements, respectively. Illegitimate recombination was implicated in almost 45% of the cases. A large subset of InDels, approximately 30%, affected the coding regions of genes. InDels in coding regions were mainly brought about by recombination events, only a small fraction of 5% was caused by transposable elements (Ziolkowski et al. 2009).

A set of 20 accessions that were selected to represent maximal genetic diversity were used for re-sequencing by hybridisation (Clark et al. 2007). Per accession an average of nearly 100,000 SNPs were detected. In all strains a total of one million non-redundant SNPs were identified by accepting moderate false discovery rates. Tracts of reduced hybridisation spanning at least 200 bp were also recorded in order to predict regions that were either deleted or highly dissimilar in sequence in a particular accession with respect to the reference genome. In total, 13,470 of such predicted polymorphic regions were detected with a median size of 589 bp and a maximum size of 41.2 kbp across the 20 accessions. Approximately 4% of the genome was found to be highly dissimilar or deleted relative to the Columbia genome sequence. SNPs that affected the exon/intron structure of a gene by introducing premature stop codons, generating nonfunctional splice donor or acceptor sites or altering initiation or stop codons were classified as large-effect SNPs. A large fraction of the *A. thaliana* protein-coding genes, 9.4%, were affected by such large-effect SNPs or predicted polymorphic regions. Such types of changes were overrepresented in tandemly duplicated genes. Interestingly, individual gene families were affected to a different extent by major-effect changes. Nucleotide-binding leucine-rich repeat (NB-LRR) genes were found particularly to be prone to these changes; 60% of the genes in this family were affected. A high frequency of genes with such changes was also found for the F-box and receptor-like kinase (RLK) gene families. Predicted polymorphic regions co-incided with a comparatively large proportion of the NB-LRR and RLK genes, thus indicating a skew towards a higher-than-average level of polymorphism in these two gene families which play a role in disease resistance (Clark et al. 2007). These results corroborated the results which Bakker et al. (2006) had established for a set of 27 NB-LRR genes. An improved detection method for polymorphic regions from data generated via re-sequencing by hybridisation revealed patterns of polymorphism at an even higher resolution (Zeller et al. 2008). For example, it was shown that the average polymorphism in intergenic regions is lowest immediately upstream of the transcription start site, whereas maximal levels are reached within 450 bp upstream of the transcribed sequences.

Very detailed comparative sequence information of Arabidopsis accessions was generated by applying Illumina sequencing-by-synthesis technology. Short reads accounting for 15- to 25-fold genome coverage were generated for three different Arabidopsis accessions, Col-0, Bur-0, and Tsu-1, and aligned to the reference genome sequence of the Columbia genotype. Over 800,000 unique SNPs and almost 80,000 non-redundant 1- to 3-bp InDels were detected; moreover, approximately 2,000 potential errors were pinpointed in the reference sequence. More than 3.4 Mbp of the Bur-1 and Tsu-1 genomes were found to be highly dissimilar, deleted or duplicated when compared to the Columbia sequence. InDels of up to 641 bp were identified when sequence assemblies of these regions were matched to the reference sequence. Almost half of the SNPs in coding sequences result in changes of amino acids, moreover Bur-0 or Tsu-1 differed from Col-0 by more than 1,800 potential frameshifts (Ossowski et al. 2008). This study sets the stage for a comprehensive assessment of sequence variation in Arabidopsis and a project has been initiated which aims at the analysis of 1,001 accessions (Weigel and Mott 2009).

6.4 Sequence Comparisons Between *A. thaliana* and Near Relatives

The annotated sequence of the Columbia reference genome has been exploited to study genome collinearity of small orthologous genomic regions in several genera closely related to *A. thaliana*, such as *Capsella*, *Olimarabidopsis*, and *Boechera*, that diverged from Arabidopsis approximately 10–14 million years ago (Koch et al. 2000, 2001). Moreover, other *Arabidopsis* species, like *Arabidopsis arenosa,*

A. lyrata, and *A. halleri*, have also been used for comparative analyses. Their divergence from *A. thaliana* has been estimated at approximately 5 million years ago (Koch et al. 2000). Comparative genetic mapping of *A. lyrata*, *B. stricta*, and *C. rubella* relative to the *Arabidopsis thaliana* genome revealed collinear chromosomal segments (Boivin et al. 2004, Kuittinen et al. 2004, Yogeeswaran et al. 2005, Schranz et al. 2007). These results together with those from phylogenetic analyses were used to infer the number and nature of chromosome rearrangements that underlie the chromosome number reduction from $n = 8$ to $n = 5$ that took place in the lineage leading to *A. thaliana* (Yogeeswaran et al. 2005, Lysak et al. 2006).

Comparisons of two small segments of the *C. rubella* genome to the corresponding segments in the *A. thaliana* genome located in the euchromatic regions of chromosomes 1 and 4 revealed almost complete microcollinearity (Fig. 6.1). Eleven protein-coding genes and one t-RNA gene were found in each species. The genes were present in the same order and orientation relative to each other. The number and position of intron sequences were conserved in both species. Despite the fact that intergenic regions and introns were differently sized in the two species, the regions analysed were of similar size (Acarkan et al. 2000, Rossberg et al. 2001). BLAST-2-sequences alignments of the *A. thaliana* region on chromosome 1 with its counterpart in *C. rubella* revealed that approximately three quarters of the *C. rubella* sequences were found in high scoring sequences pairs. More than 50% of the high scoring sequence pairs consisted of protein-coding exon sequences (Fig. 6.1). These findings were corroborated when a small set of random short *C. rubella* sequences was compared to the corresponding Arabidopsis sequences in the same manner. It was determined that protein-coding exon sequences were clearly over-represented among the aligned sequences when compared to intron and intergenic sequences. The sequence identity of aligned protein-coding exon sequences was, at 90%, approximately 10% higher than the value for aligned intron

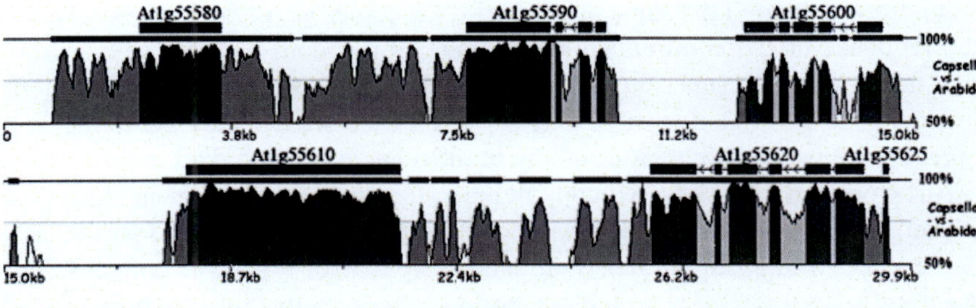

Fig. 6.1 Collinear arrangement of genes in orthologous regions of the *A. thaliana* and *C. rubella* genomes. A region on *A. thaliana* chromosome 1 containing the orthologue of the tomato *Lateral suppressor* gene (At1g55580) was compared to the corresponding region in the *C. rubella* genome (Rossberg et al. 2001) using zPicture with a 100-bp sliding window (http://zpicture.dcode.org/). The *y*-axis shows sequence identity between the two genomic regions in percentage ranging from 50 to 100%. The exon/intron structure of the *A. thaliana* genes is shown at the *top*. *Dark grey boxes* denote exons, whereas introns are indicated as *arrowheads*. *Dark*, *light*, and *medium grey shading* specify sequence conservation in exons, introns, and intergenic regions, respectively. Sequence identity is generally higher in exons compared to the values for introns and intergenic regions

or intergenic sequences, in contrast to a 10-fold reduced frequency of small InDels (1–21 bp) in exons (Boivin et al. 2004). A value of 90.3% sequence identity was determined for aligned *A. thaliana* and *B. stricta* coding sequences and evidence for sequence conservation beyond protein-coding exon sequences was also found for this species pair. Analysis of informative alignments upstream of coding sequences revealed an average identity of 71.4% in the regions 500 bp upstream of the translation initiation codons and the highest sequence identity values were found in the areas directly adjacent to the ATG codons (Windsor et al. 2006).

Comparing the *A. thaliana* region on chromosome 4 with its *C. rubella* counterpart disclosed one deviation from microcollinearity, one of the genes had undergone a tandem duplication in *C. rubella* since the two species diverged (Acarkan et al. 2000). A very recent change of gene copy number was also identified in *A. halleri*. The region around the *HMA4* gene of *A. thaliana* corresponds to a complex triplication in the *A. halleri* genome which is comprised of three copies of *HMA4*, fragments of two genes that are located downstream of *HMA4* in *A. thaliana* and several transposon and retrotransposon insertions (Hanikenne et al. 2008).

Differences in the number of tandemly repeated genes in different species were also noted when the regions harbouring a cluster of glycine-rich pollen surface proteins (GRPs or oleopollenins) were analysed in *A. arenosa*, *Boechera drummondii*, *C. rubella*, and *Olimarabidopsis pumila* (Fiebig et al. 2004, Schein et al. 2004). The analysed region of *C. rubella* harbours 18 genes in collinear arrangement compared to *A. thaliana*, but one of the genes from the *GRP* cluster was missing; moreover, one of the genes in the syntenic region was classified as a pseudogene (Fiebig et al. 2004). In *B. drummondii* several genes of the *GRP* cluster were not found, but stretches of weak sequence similarity to two Arabidopsis genes in collinear positions indicated that these genes had been deleted in the *Boechera* lineage (Schein et al. 2004). In contrast, the genes in the region around the *GRP* cluster and all oleopollenin gene copies were completely collinear in *A. thaliana* and *O. pumila*. In the tetraploid *A. arenosa* two homeologous regions were identified. Both regions harboured the same number of *GRP* copies as *A. thaliana*; however, a 2 kbp sequence insertion introducing premature stop codons had altered the structure of one of the *GRP* genes on one of the homeologues. In the other homeologous segment a gene was found inserted between two of the oleopollenin genes that was not present in the syntenic regions of any of the other species studied.

Apart from variations of *GRP* copy number, the analysis of the gene clusters in the different species revealed that the second exon of the *GRP* genes was particularly variable through duplication, deletion, and divergence of repetitive domains. The sequence divergence in these protein regions was significantly higher than that found for the flanking genes (Fiebig et al. 2004).

In addition to the small deviations from microcollinearity described above, another event was noted which distinguished the *A. thaliana* region around the *GRP* gene cluster from the syntenic regions in the other species analysed. The transposon that was located adjacent to the *GRP* gene cluster in *A. thaliana* was not found in any other species studied (Fiebig et al. 2004). Deviations from collinearity due to mobile elements are frequently reported in microsynteny studies (Bennetzen 2000,

Schmidt 2002), but only occasionally are the insertions of transposons or retrotransposons found in genes. For example, the orthologue of *A. thaliana* gene At4g16690 was interrupted by two large insertions in *C. rubella* which share homology with mobile elements (Schmidt et al. 2003).

Collinearity around the *MIR319a* locus of *A. thaliana* was assessed in *A. halleri* and *C. rubella*. In *A. halleri* a 131-kbp sequence contig was studied; all 18 genes were found in collinear arrangement with their counterparts in the corresponding *A. thaliana* region that spanned 76.3 kbp. The comparison of a 147-kbp segment of *C. rubella* revealed 30 genes, of which 5 were in common with the region studied in *A. halleri*. Small differences in gene repertoire were noted when the *C. rubella* sequence contig was compared to the orthologous region of *A. thaliana* that spanned 134 kbp. Three out of the 34 genes that were present in the Arabidopsis regions were not found in *C. rubella* and a fourth one was partially deleted. Moreover, the two regions differed by an inversion that encompassed more than 10 genes (Warthmann et al. 2008).

In summary, the comparative study of small orthologous euchromatic regions derived from several species that diverged at most 10–14 million years ago generally revealed conserved gene repertoire and order; moreover, the spacing of genes was similar. Nevertheless, evidence was found for small deviations from microcollinearity such as duplications or deletions of gene sequences, inversions and translocations of single genes or small groups of genes, pseudogene formation, and differences with respect to the presence of mobile elements. Similar results were also obtained when such analyses were carried out for species of the Brassicaceae with a more distant relationship to *A. thaliana*, such as *Arabis alpina* (Wang et al. 2009), *Raphanus sativus* (Desloire et al. 2003, Brown et al. 2003), *Sisymbirum irio* (Fiebig et al. 2004), *Thlaspi arvense* (Zhou et al. 2007), and *Thellungiella halophila* (Deng et al. 2009, Nah et al. 2009). These data are also consistent with the results of microcollinearity studies that were established for other plant families, most notably the grasses (Bennetzen 2000, Schmidt 2002). Thus, bearing in mind that small deviations from microcollinearity were readily found, although only a very limited number of chromosome segments have been studied, it is clear that numerous small differences will be observed across the genomes of different species. Whole-genome shotgun sequencing programs are under way for several crucifers including *A. halleri*, *A. lyrata*, *Boechera divericarpa*, *Boechera holboellii*, *B. stricta*, *B. rapa*, *C. rubella*, and *Thellungiella halophila* (http://www.jgi.doe.gov/). Thus, a much more detailed assessment of fine-grained changes in crucifer genomes can be carried out once this sequence information is available.

In contrast, the comparative analysis of pericentromeric regions revealed a rather different picture; pronounced changes in fine-scale structure were observed when sequence contigs of *A. arenosa*, *C. rubella*, and *O. pumila* corresponding to the pericentromeric region of *A. thaliana* chromosome 3 were assessed (Hall et al. 2006). Presence of centromeric satellite sequences and the results of fluorescence in situ hybridisations (FISH) provided evidence that the identified sequences adjoin centromere satellite arrays in all three species. Six expressed *A. thaliana* genes were found in the same order in *A. arenosa* and *C. rubella*; however, one of the *C. rubella* genes was found in an inverted orientation. The six genes were also conserved in *O. pumila* but in this region an additional gene was found that corresponded to an *A. thaliana* gene located on chromosome 1. For two other genes with expression evidence in *A. thaliana*, only remnants were found in the other species. The conserved gene order was in marked contrast to the differences that were observed with respect to the spacing of the conserved genes in the four species. The *A. thaliana* genes were spread across a 643-kbp region, whereas their homologues in the other species were located in segments that spanned between 38 and 70 kbp. This disparity can partially be accounted for by the different abundance of mobile elements. Mobile elements comprise approximately 60% of this region of the *A. thaliana* genome, compared to values between 11 and 27% in the other species. Furthermore, many species-specific insertions were observed that were classified as genes or pseudogenes, because they were predicted by ab initio gene-finding programs or because they share sequence similarity to *A. thaliana* genes or parts thereof distributed across the genome. The 643-kbp region of *A. thaliana* contains, for example, 27 hypothetical or predicted genes; in the other species such genes and gene fragments were also observed but less frequently than in Arabidopsis. This rather pronounced degree of structural changes in the centromeric regions is accompanied by a particularly rapid evolution of the centromere satellite sequences. *A. thaliana* and *C. rubella* diverged only approximately 10–14 million years ago (Koch et al. 2000, 2001), nonetheless their centromere satellite sequences have diverged to such an extent that significant alignments could not be produced. The satellite sequences of *A. thaliana* and *A. arenosa* share only 73.3% sequence identity even though these species diverged within the last 5 million years (Hall et al. 2005).

The comparative analysis of small orthologous segments derived from different species has been very informative. However, in order to allow more general conclusions it is preferable to collect information derived from many different genomic regions. This has been achieved by combining small- and medium-scale shotgun sequencing approaches of a species of interest with comparative analyses to the *A. thaliana* genome (Windsor et al. 2006, Oyama et al. 2008).

A medium-scale shotgun sequencing approach was chosen to assess microsynteny of the *B. stricta* and *A. thaliana* genomes. Windsor et al. (2006) end sequenced approximately 23,000 *Boechera* genomic clones with an average insert size of about 13 kbp, generating 33.5 Mbp of high-quality sequence. Clones with significant sequence similarity to annotated repetitive elements, rDNA (ribosomal DNA), and organellar DNAs were excluded from subsequent analyses. From the remaining sequences, they identified 6,334 non-redundant clones where both ends had high-quality matches (BLASTn E< = 1e-30) to the *A. thaliana* genome and another 6,129 clones in which only one end showed a significant match. Almost half of the *B. stricta* matches fall within *A. thaliana* coding sequences. Among the clones with paired end matches to the *A. thaliana* genome, three-quarters had matches less than 50 kbp apart and were considered to be examples of microsynteny. The mean separation of the *B. stricta* hits on the Arabidopsis genome was 11,949 bp, a number that is significantly different from the average insert size of the clones (13,187 bp; $P < 0.005$) indicating that in these regions the *A. thaliana* genome is approximately 10%

A. thaliana genome spanning between 30 and 500 kbp the *B. rapa* BAC clones were assigned to the corresponding region in the *A. thaliana* genome. In total, more than 4,000 *B. rapa* BAC clones were mapped relative to the *A. thaliana* genome in this manner covering more than 90 Mbp of the euchromatic regions in the Arabidopsis genome. These results suggest that a large proportion of the gene space in the *B. rapa* genome can be efficiently targeted for sequencing using this strategy that has been termed comparative tiling (Yang et al. 2005). A recent study underscores the efficacy of this approach. A set of 589 minimally overlapping comparatively tiled BAC clones were chosen for sequencing and 410 sequence contigs were generated that covered 65.8 Mbp in total. These contigs encompassed more than 90% of the euchromatic regions in *A. thaliana* and approximately 30% of the euchromatic areas in *B. rapa*. The *B. rapa* sequence contigs showed overall a twofold higher density of transposable elements and a slightly lower average gene density than the euchromatic *A. thaliana* counterparts. Gene length is shorter in *B. rapa* than in *A. thaliana;* likewise, the number of exons per gene was lower. A comparison of the *B. rapa* sequence contigs with the *A. thaliana* reference genome revealed extensive synteny and provided strong support for a whole genome triplication in the *Brassica* species after divergence from *A. thaliana*. Nonetheless, it was estimated that the *B. rapa* genome, with approximately 53,000 genes, contains only about twice as many protein-coding genes as the *A. thaliana* genome. Different factors were found to contribute to the extensive gene loss in *B. rapa* after the whole genome triplication. First, loss of some triplicated regions was observed. Second, differential gene loss was prevalent in triplicated segments throughout the entire gene space. Third, tandemly duplicated genes were not found as frequently in *B. rapa* as in *A. thaliana* (Mun et al. 2009).

BAC end mapping was also used to identify breakpoints in Arabidopsis–*Brassica* collinearity by identifying BAC clones in which the end sequences either showed similarity to regions on an Arabidopsis chromosome that are more than 500 kbp apart or in which the two ends matched to regions located on different chromosomes (Trick et al. 2009a). Such BAC clones were sequenced if at least two other BAC clones confirmed the assignment to two non-collinear blocks in the Arabidopsis genome. A detailed comparison of the arrangement of the gene sequences found in these BAC sequences revealed three groups of clones. Two groups consisted of clones in which large-scale collinearity discontinuities were not verified. In one of these groups the matches of end sequences to non-contiguous sequences in Arabidopsis had to be attributed to the presence of a single non-collinear gene or gene fragment at one of the BAC ends most probably indicative of a small-scale rearrangement. The other group consisted of clones in which multiple gene models showed similarity to two segments in the *A. thaliana* genome that were related due to duplications predating the divergences of the *Brassica* and *Arabidopsis* lineages. However, almost half of the clones analysed contained a minimum of two different runs of genes that mapped to non-contiguous regions in the Arabidopsis genome and thus encompassed regions belonging to different collinear blocks. Additionally, 20 BAC sequences or sequence contigs with collinearity discontinuities were identified during the ongoing *B. rapa* genome sequencing project (Mun and Park,

Chapter 15, this book). Of the 50 cases analysed in total, 19 and 31 of the discontinuities were classified as inter-chromosomal and intra-chromosomal rearrangements, respectively (Trick et al. 2009a). Interestingly, only few of the chromosomal breakpoints that were identified were in agreement with the results of comparative genetic mapping results (Parkin et al. 2005, Schranz et al. 2006). However, these discrepancies were mainly due to the fact that the comparative sequence studies identified many relatively small segments which would escape detection in most comparisons involving genetic linkage mapping.

Lagercrantz (1998) noted that the map positions of interstitial telomere repeats often coincided with collinearity breakpoints in *B. nigra*. However, detailed inspection of the sequences around the collinearity discontinuities in *B. rapa* did not reveal evidence for centromeric or telomeric repeat sequences. In only one of the regions studied were interstitial telomere repeats found, but at some distance from the breakpoint. Instead, gene fragments with homology to regions in the *A. thaliana* genome other than those linked together were often found at the discontinuities (Trick et al. 2009a).

6.5.2 B. oleracea

Bancroft and colleagues used a series of hybridisation probes from the *FCA* region of *A. thaliana* chromosome 4 to isolate BACs comprising corresponding regions of the *B. oleracea* (var. *alboglabra* A12) genome (O'Neill and Bancroft 2000). Because the region on *A. thaliana* chromosome 4 from which the probes were derived was segmentally duplicated on chromosome 5 (α duplication), the result was six regions. Sequencing these regions established the relationships both between coding sequences that arose from the genome triplication in the *Brassica* lineage and those related by the α duplication that was shared by the *Arabidopsis* and *Brassica* lineages (Town et al. 2006).

Sequence identity between coding sequences related by the α duplication, in both *A. thaliana* and *B. oleracea* is approximately 70%. In contrast, coding sequences of homeologous genes within each triplicated region show 80–86% nucleotide identity to one another, and also to the corresponding Arabidopsis genes, indicating that the genome triplication in this species, like in *B. rapa*, took place shortly after the divergence of the *Brassica* and *Arabidopsis* lineages. There is insufficient data to determine whether the triplication occurred as a single event or as two successive duplication events.

Analysis of segmentally duplicated genes in *A. thaliana* frequently revealed divergence of transcription profiles (Blanc and Wolfe 2004, Haberer et al. 2004). Recent work supports the notion that there has been diversification of expression patterns of paleologous genes in *B. oleracea* since the triplication event and that this diversification varies between different morphotypes (Iniguez-Luy et al. unpublished).

As in *B. rapa*, there is gene loss from the three homeologous regions after the triplication event such that out of 38 loci, only 7 are retained in three copies, 20 as duplicates, and 11 as single-copy genes. Given that in some cases only parts of genes were found in collinear position to their Arabidopsis counterparts in the *B. oleracea* regions it is likely that at least a fraction of the gene loss is due to deletion events. Four genes corresponded to genes located elsewhere in the Arabidopsis genome and most likely inserted into their current position in the *B. oleracea* genome subsequent to the triplication in the *Brassica* lineage. However, while there is conserved collinearity between genes in *A. thaliana* and *B. oleracea* as in *B. rapa*, some of the regions of the *B. oleracea* genome analysed in this study, in contrast to *B. rapa*, are greatly expanded when compared to *A. thaliana* due to the presence of transposable element-related sequences (mainly retrotransposons), predicted hypothetical genes, and gene fragments (remnants) that correspond to *A. thaliana* genes located elsewhere in the genome.

Whole genome shotgun (WGS) sequencing of the *B. oleracea* TO1000 genome was carried out as an adjunct to the annotation of the Arabidopsis genome (Ayele et al. 2005, Katari et al. 2005). Alignment of the sequence reads with the Arabidopsis genome provided strong support for many Arabidopsis gene models and resulted in the creation of 19 new genes with updates to another 17 gene models (Ayele et al. 2005). The *Brassica–Arabidopsis* sequence alignments were also used to train Twinscan that led to the prediction of many novel genes. Over 300 novel predictions in previously unannotated regions were subsequently validated by RACE (Moskal et al. 2007). A comparison of the *B. oleracea* sequence contigs with the annotated Arabidopsis genome revealed approximately 4,000 putative orthologous upstream sequences. This information was used in combination with expression profiling data from *A. thaliana* for large-scale discovery of *cis*-regulatory elements in upstream regions of genes (Haberer et al. 2006).

The *B. oleracea* WGS sequence was exploited to compare the transposable element content in *A. thaliana* and *B. oleracea* in a genome-wide fashion. Importantly, it was shown that both species share almost all lineages of transposable elements; however, nearly all types of transposable elements are more numerous in the *B. oleracea* genome (Zhang and Wessler 2004). Particularly studies of species belonging to the grasses unveiled the important contribution of transposable elements to genome size variation (Bennetzen 2000, Schmidt 2002). Hence, it was not surprising that the fivefold larger *B. oleracea* genome showed a higher abundance of transposable elements in comparison to *A. thaliana*. However, by comparing the density of particular families of transposable elements (copy number per Mbp) in both genomes several types of transposons were identified with copy numbers higher than expected for the proportional increase in genome size (Zhang and Wessler 2004). Consistent with the findings in the Poaceae (Bennetzen 2000, Schmidt 2002), retroelement amplification contributed to genome size expansion in Brassicaceae; however, *Pong*-like and CACTA elements also showed significantly higher densities in *B. oleracea* than in *A. thaliana*.

6.5.3 B. napus

The work described above on *B. rapa* (Yang et al. 2006) and *B. oleracea* (Town et al. 2006) focussed on two different regions of the genome and did not permit a direct comparison between the genomes of *B. rapa* (A genome) and *B. oleracea* (C genome). The analysis of O'Neill and Bancroft (2000) was extended to detect orthologous contigs in *B. rapa* and *B. napus* (Rana et al. 2004). Sequencing of BACs representing orthologous regions across these species (Cheung et al. 2009) revealed the relationships between the corresponding regions of the A and C genomes (Fig. 6.2) and permits comparison between orthologous regions of the A genome of *B. rapa* ssp. *trilocularis* RO18 and the C genome of *B. oleracea* ssp. *alboglabra* A12DH and between orthologous regions of the A and C genomes of the allotetraploid *B. napus* var. Tapidor.

Both comparisons lead to the same conclusions:

(a) There is essentially complete collinearity and content in the A and C genomes between the genes that encode proteins of known function. Sequence identity within the coding sequences is in the range 94–97% and intron number is conserved between orthologous genes. These results are consistent with the recent divergence of the progenitor species *B. rapa* and *B. oleracea*, which was estimated at 4 million years ago (Inaba and Nishio 2002).
(b) The main difference between the A and C genomes lies in the intergenic regions. Over four of the five regions sequenced across the four genomes, the C genome was on average 35% larger than the A genome, a number that agrees well with the reported size ratio of the two genomes (700 Mbp for *B. oleracea* versus 530 Mbp for *B. rapa*, Johnston et al. 2005). The fifth region (contig E) is an outlier with approximately 70 kbp of the A genome corresponding to about 350 kbp of the C genome.

In *B. rapa*, most of the intergenic regions are comprised of anonymous sequence with occasional regions (ca. 1.5 per 100 kbp) being identified as being related to transposable elements. By contrast, annotation of the intergenic regions in *B. oleracea* identifies both transposable elements and hypothetical genes (those identified only by computer algorithms without database support). The transposable element-related regions are related to retroelements of the Ty1/copia-like or Ty3/gypsy-like classes.

Stearoyl-ACP desaturase genes were used as probes to identify *B. napus* BAC clones that correspond to the *A. thaliana* chromosome 3 region which contains a tandem triplet of these genes. Two regions carrying a single copy of stearoyl-ACP desaturase genes each were found and were assigned to the A and C genome, respectively. Only 14 of the 25 genes in the *A. thaliana* genome segment are conserved in the two corresponding regions of the *B. napus* genome. The A and C genome regions showed collinear order of conserved genes but they were differentiated by multiple insertion–deletion events. For example, two copies each were found to correspond to At3g02590 in the *B. napus* regions. However, one of the copies in the A genome carried a retroelement insertion (Cho et al. 2010).

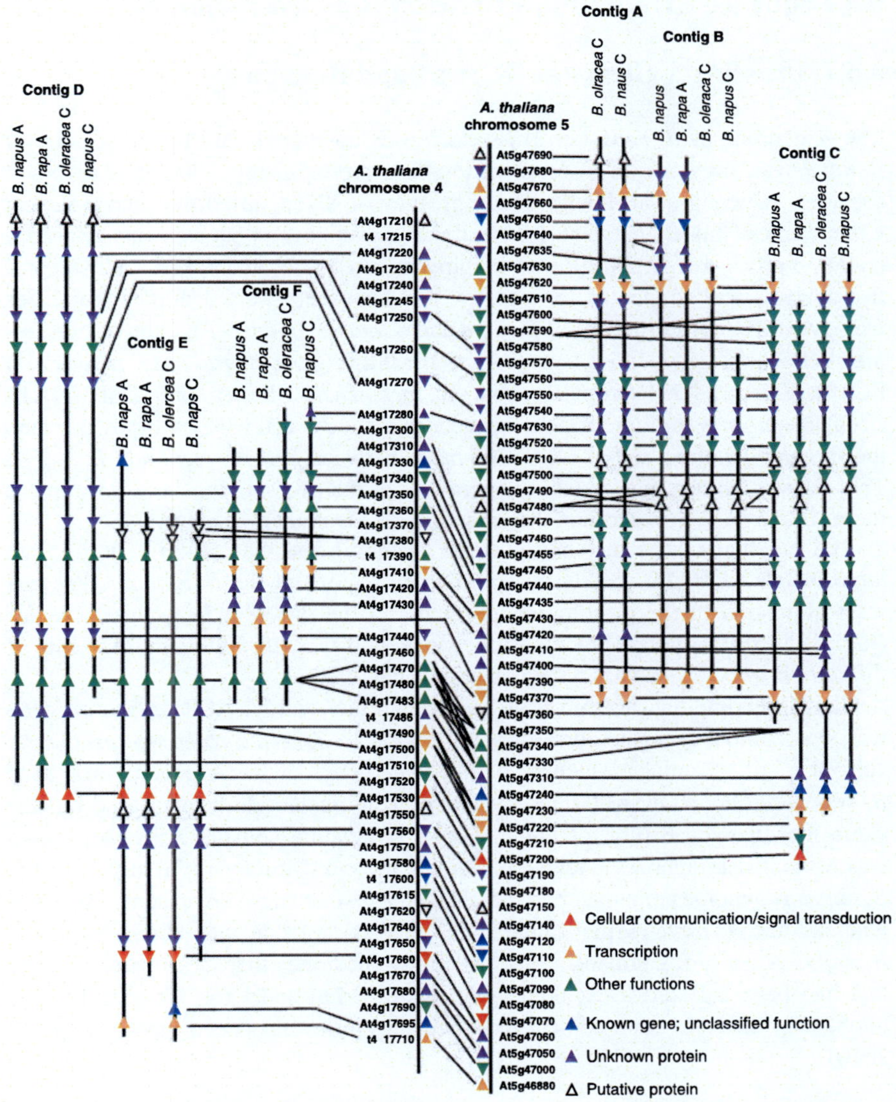

Fig. 6.2 Orthology relationships across the A and C genomes of *B. rapa*, *B. oleracea*, and *B. napus*, and their relationship to a segmentally duplicated region in *A. thaliana*. Sequence contigs are depicted as *vertical lines* and *horizontal lines* point to homologous gene sequences. Genes are represented as *triangles*. The coding strand of the genes can be inferred from the orientation of the triangles. Colour-coding of the triangles serves to indicate the functional classification of the predicted protein sequences. Details are given in the key at the *bottom* of the diagram. The figure is reproduced from Cheung et al. 2009, with permission

6.6 Sequence Relationships Between *Brassica* Genomes

6.6.1 Brassica *A Genomes:* B. rapa *and* B. napus

The A genome of *B. rapa* ssp. *trilocularis* was compared with the A genome of *B. napus* var. Tapidor over five regions totalling approximately 270 kbp of common sequence (Cheung et al. 2009). The frequency of SNPs varied between 0.82 and 1.98%, and of InDels between 1.72 and 4.39 per kbp. The SNP frequency within coding regions was slightly lower, ranging between 0.72 and 1.49%. A comparative analysis was conducted for a region in which there were available sequences from two different subspecies of *B. rapa* and *B. napus*. In the ~17.5 kbp of overlapping sequences between all three BACs: JBr38I20 (from *B. rapa* ssp. *trilocularis* RO18), KBrH138O03 (from *B. rapa* ssp. *pekinensis* Chiifu), and JBnB169A13 (from *B. napus* var. Tapidor), the sequences differed at 0.63% of bases between the *B. rapa* genomes and at ~1.7% of bases between each *B. rapa* and *B. napus*. This showed, in this region, that the genomes of the two subspecies of *B. rapa* are more similar to each other than they are to the A genome of *B. napus*.

The vast majority of InDels was small (spanning less than 50 bp), but multiple large InDels were also readily observed when regions of the *B. napus* A genome were compared to the orthologous segments in the different *B. rapa* subspecies (Fig. 6.3a, b). Such large InDels were even found when orthologous segments of different *B. rapa* subspecies were compared (Fig. 6.3c).

The A-genome region from *B. napus* around the stearoyl-ACP desaturase gene was also compared to the corresponding region in *Brassica rapa* ssp. *pekinensis* cultivar Chiifu. Coding sequences of *B. napus* genes of the A genome and their *B. rapa* orthologues were 97.5% identical at the nucleotide level. Based on this value the divergence time of the *B. napus* A genome and the *B. rapa* genome was estimated at approximately 2 million years ago. Moreover, the region of the *B. napus* A genome contained two mobile elements that were not present in *B. rapa*, whereas the *B. rapa* region contained a gene model that was absent from the *B. napus* A genome segment. Thus, this study illustrates that the *B. rapa* cultivar that had been selected for genome sequencing (Mun and Park, Chapter 15, this book) may differ in gene content from breeding lines of oilseed rape (Cho et al. 2010).

6.6.2 Brassica *C Genomes:* B. oleracea *and* B. napus

The C genome of *B. oleracea* ssp. *alboglabra* A12DH was compared with the C genome of *B. napus* var. Tapidor DH over six regions totalling approximately 800 kbp of common sequence (Cheung et al. 2009). The frequency of SNPs varied between 0.39 and 1.96%, and of InDels between 0.55 and 3.87 per kbp. The SNP frequency within coding regions was slightly lower, ranging between 0.16 and 1.57%. The comparison of regions of the *Brassica* C genome and the orthologous segments of the *B. oleracea* genome revealed numerous InDels that were larger than

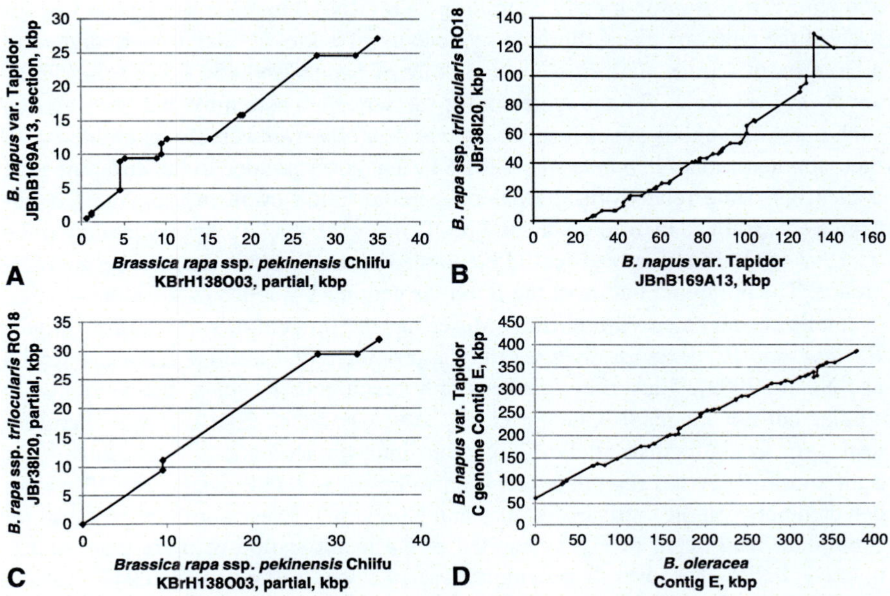

Fig. 6.3 Sequence comparisons between different *Brassica* genomes. MUMmer-based sequence alignments between orthologous *Brassica* regions (Town et al. 2006, Cheung et al. 2009). Points mark breaks in collinearity. *Vertical lines* represent sequence present in the BAC shown on the *y*-axis that is absent from the BAC shown on the *x*-axis while *horizontal lines* indicate the converse. Due to the scale of the plots and some data filtering, only InDels spanning a few hundred bp or more can be seen. (**a**) Comparison of *B. napus* var. Tapidor A genome BAC JBnB169A13 and BAC KBrH138O03 from *B. rapa* ssp. *pekinensis* Chiifu. (**b**) Comparison of BAC JBr38I20 from *B. rapa* ssp. *trilocularis* RO18 to *B. napus* var. Tapidor A genome BAC JBnB169A13. (**c**) Comparison of BAC JBr38I20 from *B. rapa* ssp. *trilocularis* RO18 and BAC KBrH138O03 from *B. rapa* ssp. *pekinensis* Chiifu. (**d**) Comparison of two C genome E contigs from *B. oleracea* ssp. *alboglabra* A12 and *B. napus* var. Tapidor

a few hundred base pair (Fig. 6.3d), similar to the results that were established for the comparisons of orthologous A genome segments (Fig. 6.3a–c).

One explanation for the variation in SNP and InDel rates over different sequenced contigs for both the A and the C genomes is that the *B. napus* var. Tapidor genome is a mosaic of A and C genome segments that have been introgressed during breeding from a variety of A and C genomes each of which differs to a different degree from the *B. rapa* ssp. *pekinensis* Chiifu and the *B. oleracea* ssp. *alboglabra* A12DH genomes (which are highly inbred or a doubled haploid, respectively) with which the comparisons are being made (Cheung et al. 2009).

6.7 Comparative Analysis of *B. napus* Accessions

The power of massively parallel sequencing was exploited for the large-scale discovery of SNPs between the two doubled haploid lines of *B. napus* cultivars Tapidor and Ningyou 7. Approximately 20 million 35 bp RNA-seq reads were generated for each of the cultivars using Illumina sequencing (Trick et al. 2009b) and aligned to a large set of sequence assemblies consisting of Sanger and 454 ESTs from different *Brassica* species. The parameters for the sequence assembly had been chosen such that orthologous genes from *B. rapa* and *B. oleracea* were co-assembled with their counterparts in *B. napus* whereas ESTs that corresponded to the different gene copies that were related via the paleohexaploidisation event in the *Brassica* lineage were arranged in different assemblies (Trick et al. 2009c). Approximately three quarters of the Illumina reads could be aligned to the *Brassica* unigene set as a reference. The polyploid nature of the *B. napus* genome complicates the identification of allelic sequence polymorphisms. Moreover, the progenitor species of *B. napus*, *B. rapa*, and *B. oleracea*, diverged only approximately 4 million years ago (Inaba and Nishio 2002). Thus, the sequences of homeologous *B. napus* genes are highly similar but not identical (Cheung et al. 2009, Cho et al. 2010). It was therefore necessary to distinguish allelic SNPs from inter-homeologue polymorphisms. At a minimum RNA-seq read depth of four, more than 40,000 candidate SNPs were detected between the cultivars Tapidor and Ningyou 7. Even more stringent criteria (minimum read depth of eight) resulted in the identification of more than 20,000 putative SNPs. The overall polymorphism rate was estimated at 0.047–0.084% (Trick et al. 2009b).

6.8 Summary

The sequence of *A. thaliana* has acted as a surrogate reference genome for other members of the Brassicaceae for the last decade. The value of this approach has been confirmed as sequence data for other members of the family have begun to emerge. As would be expected, the more closely related species show not only higher sequence identity in both coding and non-coding regions, but also a higher proportion of conserved collinear genes between any two species. Nevertheless, many small-scale deviations from collinearity distinguish genomes of different species and even accessions of a single species. Pericentromeric regions show much more pronounced changes in structure and sequence than areas in the euchromatin.

BAC sequencing in both *B. rapa* and *B. oleracea* provides strong evidence for an initially triplicated *Brassica* genome shortly after the separation of the *Arabidopsis* and *Brassica* lineages followed by interspersed gene loss, so that in either diploid species only a fraction of the initial gene complement are retained in three copies. Much of this gene loss occurred before the speciation of *B. rapa* and *B. oleracea* since there is a high degree of conservation of the content and order of genes that can be annotated with known function. The size difference between the *Brassica* A and C genomes appears to be largely due to insertion of transposable elements throughout the genome that occurred after speciation.

In the allotetraploid *B. napus*, the A and C genomes have clearly retained their identities and are very similar to the A and C genomes of the two diploids (*B. rapa* and *B. oleracea*) indicating that there has been little, if any exchange between homeologous chromosomes in the short domestication history of this crop.

Acknowledgments Work on Brassicaceae genomics in the authors' laboratories has been supported by grants from the United States National Science Foundation (DBI-9813586 and DBI-0638536), the German Federal Ministry of Education and Research (BMBF), and the UK Biotechnology and Biological Sciences Research Council.

References

Acarkan A, Rossberg M, Koch M, Schmidt R (2000) Comparative genome analysis reveals extensive conservation of genome organisation for *Arabidopsis thaliana* and *Capsella rubella*. Plant J 23:55–62

Arabidopsis Genome Initiative (2000) Analysis of the genome sequence of the flowering plant *Arabidopsis thaliana*. Nature 408:796–815

Ayele M, Haas BJ, Kumar N et al (2005) Whole genome shotgun sequencing of *Brassica oleracea* and its application to gene discovery and annotation in Arabidopsis. Genome Res 15:487–495

Bakker EG, Toomajian C, Kreitman M, Bergelson J (2006) A genome-wide survey of *R* gene polymorphisms in Arabidopsis. Plant Cell. 18:1803–1818

Bakker EG, Traw MB, Toomajian C et al (2008) Low levels of polymorphism in genes that control the activation of defense response in *Arabidopsis thaliana*. Genetics 178:2031–2043

Bennetzen JL (2000) Comparative sequence analysis of plant nuclear genomes: microcolinearity and its many exceptions. Plant Cell 12:1021–1029

Blanc G, Barakat A, Guyot R et al (2000) Extensive duplication and reshuffling in the Arabidopsis genome. Plant Cell 12:1093–1101

Blanc G, Wolfe KH (2004) Functional divergence of duplicated genes formed by polyploidy during Arabidopsis evolution. Plant Cell 16:1679–1691

Boivin K, Acarkan A, Mbulu RS et al (2004) The Arabidopsis genome sequence as a tool for genome analysis in Brassicaceae. A comparison of the Arabidopsis and *Capsella rubella* genomes. Plant Physiol 135:735–744

Bowers JE, Chapman BA, Rong J, Paterson AH (2003) Unravelling angiosperm genome evolution by phylogenetic analysis of chromosomal duplication events. Nature 422:433–438

Brown GG, Formanová N, Jin H et al (2003) The radish *Rfo* restorer gene of Ogura cytoplasmic male sterility encodes a protein with multiple pentatricopeptide repeats. Plant J 35:262–272

Cheung F, Trick M, Drou N et al (2009) Comparative analysis between homoeologous genome segments of *Brassica napus* and its progenitor species reveals extensive sequence-level divergence. Plant Cell 21:1912–1928

Cho K, O'Neill CM, Kwon SJ et al (2010) Sequence-level comparative analysis of the *Brassica napus* genome around two stearoyl-ACP desaturase loci. Plant J 61:591–599

Clark RM, Schweikert G, Toomajian C et al (2007) Common sequence polymorphisms shaping genetic diversity in *Arabidopsis thaliana*. Science 317:338–342

Copenhaver GP, Nickel K, Kuromori T et al (1999) Genetic definition and sequence analysis of Arabidopsis centromeres. Science 286:2468–2474

Deng Z, Li Y, Xia R et al (2009) Structural analysis of 83-kb genomic DNA from *Thellungiella halophila*: sequence features and microcolinearity between salt cress and *Arabidopsis thaliana*. Genomics 94:324–332

Desloire S, Gherbi H, Laloui W et al (2003) Identification of the fertility restoration locus, *Rfo*, in radish, as a member of the pentatricopeptide-repeat protein family. EMBO Rep 4:588–594

Fiebig A, Kimport R, Preuss D (2004) Comparisons of pollen coat genes across Brassicaceae species reveal rapid evolution by repeat expansion and diversification. Proc Natl Acad Sci USA 101:3286–3291

Gao M, Li G, McCombie WR, Quiros CF (2005) Comparative analysis of a transposon-rich *Brassica oleracea* BAC clone with its corresponding sequence in *A. thaliana*. Theor Appl Genet 111:949–955

Gao M, Li G, Potter D et al (2006) Comparative analysis of methylthioalkylmalate synthase (MAM) gene family and flanking DNA sequences in *Brassica oleracea* and *Arabidopsis thaliana*. Plant Cell Rep 25:592–598

Gao M, Li G, Yang B, McCombie WR, Quiros C (2004) Comparative analysis of a Brassica BAC clone containing several major aliphatic glucosinolate genes with its corresponding Arabidopsis sequence. Genome 47:666–679

Grant MR, McDowell JM, Sharpe AG et al (1998) Independent deletions of a pathogen-resistance gene in Brassica and Arabidopsis. Proc Natl Acad Sci USA 95:15843–15848

Haas BJ, Wortman JR, Ronning CM et al (2005) Complete reannotation of the Arabidopsis genome: methods, tools, protocols and the final release. BMC Biol 3:7

Haberer G, Hindemitt T, Meyers BC et al. (2004) Transcriptional similarities, dissimilarities, and conservation of cis-elements in duplicated genes of Arabidopsis. Plant Physiol 136:3009–3022

Haberer G, Mader MT, Kosarev P et al (2006) Large-scale cis-element detection by analysis of correlated expression and sequence conservation between Arabidopsis and *Brassica oleracea.* Plant Physiol 142:1589–1602

Hall AE, Keith KC, Hall SE et al (2004) The rapidly evolving field of plant centromeres. Curr Opin Plant Biol 7:108–114

Hall AE, Kettler GC, Preuss D (2006) Dynamic evolution at pericentromeres. Genome Res 16: 355–364

Hall SE, Luo S, Hall AE, Preuss D (2005) Differential rates of local and global homogenization in centromere satellites from Arabidopsis relatives. Genetics 170:1913–1927

Hanikenne M, Talke IN, Haydon MJ et al (2008) Evolution of metal hyperaccumulation required *cis*-regulatory changes and triplication of *HMA4*. Nature 453:391–395

Hong CP, Plaha P, Koo DH et al (2006) A Survey of the *Brassica rapa* genome by BAC-end sequence analysis and comparison with *Arabidopsis thaliana.* Mol Cells 22:300–307

Inaba R, Nishio T (2002) Phylogenetic analysis of Brassiceae based on the nucleotide sequences of the *S*-locus related gene, *SLR1*. Theor Appl Genet 105:1159–1165

Jander G, Norris SR, Rounsley SD et al (2002) Arabidopsis map-based cloning in the post-genome era. Plant Physiol 129:440–450

Johnston JS, Pepper AE, Hall AE et al (2005) Evolution of genome size in Brassicaceae. Ann Bot 95:229–235

Katari MS, Balija V, Wilson RK et al (2005) Comparing low coverage random shotgun sequence data from *Brassica oleracea* and *Oryza sativa* genome sequence for their ability to add to the annotation of *Arabidopsis thaliana*. Genome Res 15:496–504

Koch MA, Haubold M, Mitchell-Olds T (2000) Comparative evolutionary analysis of chalcone synthase and alcohol dehydrogenase loci in Arabidopsis, Arabis, and related genera (Brassicaceae). Mol Biol Evol 17:1483–1498

Koch M, Haubold B, Mitchell-Olds T (2001) Molecular systematics of the Brassicaceae: evidence from coding plastidic *matK* and nuclear *Chs* sequences. Am J Bot 88:534–544

Ku HM, Vision T, Liu J, Tanksley SD (2000) Comparing sequenced segments of the tomato and Arabidopsis genomes: large-scale duplication followed by selective gene loss creates a network of synteny. Proc Natl Acad Sci USA 97:9121–9126

Kuittinen H, de Haan AA, Vogl C et al (2004) Comparing the linkage maps of the close relatives *Arabidopsis lyrata* and *A. thaliana*. Genetics 168:1575–1584

Lagercrantz U (1998) Comparative mapping between *Arabidopsis thaliana* and *Brassica nigra* indicates that Brassica genomes have evolved through extensive genome replication accompanied by chromosome fusions and frequent rearrangements. Genetics 150: 1217–1228

Lagercrantz U, Lydiate DJ (1996) Comparative genome mapping in Brassica. Genetics 144: 1903–1910

Lysak MA, Berr A, Pecinka A et al (2006) Mechanisms of chromosome number reduction in *Arabidopsis thaliana* and related Brassicaceae species. Proc Natl Acad Sci USA 103: 5224–5229

Lysak MA, Cheung K, Kitschke M, Bures P (2007) Ancestral chromosomal blocks are triplicated in Brassiceae species with varying chromosome number and genome size. Plant Physiol 145:402–410

Lysak MA, Koch MA, Beaulieu JM et al (2009) The dynamic ups and downs of genome size evolution in Brassicaceae. Mol Biol Evol 26:85–98

Lysak MA, Koch MA, Pecinka A, Schubert I (2005) Chromosome triplication found across the tribe Brassiceae. Genome Res 15:516–525

Moskal WA Jr, Wu HC, Underwood BA et al (2007) Experimental validation of novel genes predicted in the un-annotated regions of the Arabidopsis genome. BMC Genomics 8:18

Mun JH, Kwon SJ, Yang TJ et al (2009) Genome-wide comparative analysis of the *Brassica rapa* gene space reveals genome shrinkage and differential loss of duplicated genes after whole genome triplication. Genome Biol 10:R111

Nah G, Pagliarulo CL, Mohr PG et al (2009) Comparative sequence analysis of the *SALT OVERLY SENSITIVE1* orthologous region in *Thellungiella halophila* and *Arabidopsis thaliana*. Genomics 94:196–203

Nordborg M, Hu TT, Ishino Y et al (2005) The pattern of polymorphism in *Arabidopsis thaliana*. PLoS Biol 3:e196

O'Neill CM, Bancroft I (2000) Comparative physical mapping of segments of the genome of *Brassica oleracea* var. *alboglabra* that are homoeologous to sequenced regions of chromosomes 4 and 5 of *Arabidopsis thaliana*. Plant J 23:233–243

Ossowski S, Schneeberger K, Clark RM et al (2008) Sequencing of natural strains of *Arabidopsis thaliana* with short reads. Genome Res 18:2024–2033

Oyama R, Clauss MJ, Formanová N et al (2008) The shrunken genome of *Arabidopsis thaliana*. Plant Syst Evol 273:257–271

Parkin IA, Gulden SM, Sharpe AG et al (2005) Segmental structure of the *Brassica napus* genome based on comparative analysis with *Arabidopsis thaliana*. Genetics 171:765–781

Quiros CF, Grellet F, Sadowski J et al (2001) Arabidopsis and Brassica comparative genomics: sequence, structure and gene content in the *ABI1-Rps2-Ck1* chromosomal segment and related regions. Genetics 157:1321–1330

Rana D, van den Boogaart T, O'Neill CM et al (2004) Conservation of the microstructure of genome segments in *Brassica napus* and its diploid relatives. Plant J 40:725–733

Rossberg M, Theres K, Acarkan A et al (2001) Comparative sequence analysis reveals extensive microcolinearity in the *lateral suppressor* regions of the tomato, Arabidopsis, and Capsella genomes. Plant Cell 13:979–988

Schein M, Yang Z, Mitchell-Olds T, Schmid K (2004) Rapid evolution of a pollen-specific oleosin-like gene family from *Arabidopsis thaliana* and closely related species. Mol Biol Evol 21: 659–669

Schmidt R (2002) Plant genome evolution: lessons from comparative genomics at the DNA level. Plant Mol Biol 48:21–37

Schmidt R, Acarkan A, Boivin K et al (2003) The sequence of the Arabidopsis genome as a tool for comparative structural genomics in Brassicaceae. In: Nagata T, Tabata S (eds) Biotechnology in agriculture and forestry, vol 52, Brassica and Legumes, pp 19–36. Springer, Berlin/Heidelberg

Schmuths H, Meister A, Horres R, Bachmann K (2004) Genome size variation among accessions of *Arabidopsis thaliana*. Ann Bot 93:317–321

Schranz ME, Lysak MA, Mitchell-Olds T (2006) The ABC's of comparative genomics in the Brassicaceae: building blocks of crucifer genomes. Trends Plant Sci 11:535–542

Schranz ME, Mitchell-Olds T (2006) Independent ancient polyploidy events in the sister families Brassicaceae and Cleomaceae. Plant Cell 18:1152–1165

Schranz ME, Windsor AJ, Song BH et al (2007) Comparative genetic mapping in *Boechera stricta*, a close relative of Arabidopsis. Plant Physiol 144:286–298 [Erratum in: Plant Physiol 144:1690]

Swarbreck D, Wilks C, Lamesch P et al (2008) The Arabidopsis Information Resource (TAIR): gene structure and function annotation. Nucleic Acids Res 36 (Database issue):D1009–D1014

Tang H, Bowers JE, Wang X et al (2008) Synteny and collinearity in plant genomes. Science 320:486–488

Town CD, Cheung F, Maiti R et al (2006) Comparative genomics of *Brassica oleracea* and *Arabidopsis thaliana* reveal gene loss, fragmentation, and dispersal after polyploidy. Plant Cell 18:1348–1359

Trick M, Cheung F, Drou N et al (2009c) A newly-developed community microarray resource for transcriptome profiling in Brassica species enables the confirmation of Brassica-specific expressed sequences. BMC Plant Biol 9:50

Trick M, Kwon SJ, Choi SR et al (2009a) Complexity of genome evolution by segmental rearrangement in *Brassica rapa* revealed by sequence-level analysis. BMC Genomics 10:539

Trick M, Long Y, Meng J et al (2009b) Single nucleotide polymorphism (SNP) discovery in the polyploid *Brassica napus* using Solexa transcriptome sequencing. Plant Biotechnol J 7: 334–346

U N (1935) Genomic analysis in Brassica with special reference to the experimental formation of *B. napus* and peculiar mode of fertilisation. Jpn J Bot 7:389–452

Wang R, Farrona S, Vincent C et al (2009) *PEP1* regulates perennial flowering in *Arabis alpina*. Nature 459:423–427

Warthmann N, Das S, Lanz C, Weigel D (2008) Comparative analysis of the *MIR319a* microRNA locus in Arabidopsis and related Brassicaceae. Mol Biol Evol 25:892–902

Weigel D, Mott R (2009) The 1001 genomes project for *Arabidopsis thaliana*. Genome Bio 10:107

Windsor AJ, Schranz ME, Formanová N et al (2006) Partial shotgun sequencing of the *Boechera stricta* genome reveals extensive microsynteny and promoter conservation with Arabidopsis. Plant Physiol 140:1169–1182

Yang TJ, Kim JS, Kwon SJ et al (2006) Sequence-level analysis of the diploidization process in the triplicated *FLOWERING LOCUS C* region of *Brassica rapa*. Plant Cell 18: 1339–1347

Yang TJ, Kim JS, Lim KB et al (2005) The Korea Brassica genome project: a glimpse of the Brassica genome based on comparative genome analysis with Arabidopsis. Comp Funct Genomics 6:138–146

Yang YW, Lai KN, Tai PY, Li WH (1999) Rates of nucleotide substitution in angiosperm mitochondrial DNA sequences and dates of divergence between Brassica and other angiosperm lineages. J Mol Evol 48:597–604

Yogeeswaran K, Frary A, York TL et al (2005) Comparative genome analyses of Arabidopsis spp.: inferring chromosomal rearrangement events in the evolutionary history of *A. thaliana*. Genome Res 15:505–515

Zeller G, Clark RM, Schneeberger K et al (2008) Detecting polymorphic regions in *Arabidopsis thaliana* with resequencing microarrays. Genome Res 18:918–929

Zhang X, Wessler SR (2004) Genome-wide comparative analysis of the transposable elements in the related species *Arabidopsis thaliana* and *Brassica oleracea*. Proc Natl Acad Sci USA 101:5589–5594

Zhou N, Robinson SJ, Huebert T et al (2007) Comparative genome organization reveals a single copy of *CBF* in the freezing tolerant crucifer *Thlaspi arvense*. Plant Mol Biol 65:693–705

Ziolkowski PA, Koczyk G, Galganski L et al (2009) Genome sequence comparison of Col and L*er* lines reveals the dynamic nature of Arabidopsis chromosomes. Nucleic Acids Res 37: 3189–3201

Chapter 7
Structural and Functional Evolution of Resynthesized Polyploids

J. Chris Pires and Robert T. Gaeta

Abstract Polyploidy is widespread among the flowering plants. While many extant plant species show evidence of polyploidy in their genomes, there is still much to be learned regarding the role it has played in phenotypic evolution and speciation. The *Brassica* genus has polyploidy at multiple levels: the genomes of diploid species show evidence of repeated rounds of ancient polyploidization, and the agronomically important diploid species (*Brassica rapa*, *Brassica oleracea*, and *Brassica nigra*) may hybridize to form allopolyploids (*Brassica napus*, *Brassica juncea*, and *Brassica carinata*). The phenotypic diversity among these six domesticated species is spectacular. Research has provided evidence that gene and genome redundancy contribute significantly to the variation observed among *Brassica* species. The relative ease by which *Brassica* allopolyploids can be resynthesized has allowed them to emerge as an efficient model for studying the consequences of polyploidization. Studies on resynthesized *B. napus* polyploids have reported on homoeologous genome rearrangements and epigenetic changes, as well as changes in gene expression, protein expression, and phenotypic variation in the early generations following hybridization and polyploidization. In contrast, studies in resynthesized *B. juncea* polyploids show little evidence for rapid changes, although a recent report indicated changes within the organelle genomes. The differences between these two resynthesized allopolyploids may be attributed to the differing degrees of similarity between their diploid progenitor genomes (those of *B. napus* being more similar to each other than those of *B. juncea*). It may also be due to differences in genetic variation for homoeologous pairing control in these two species. These studies demonstrate that polyploidization may have immediate genetic and phenotypic consequences, particularly in *B. napus*. The variation that results from a polyploidization event may be critical in the early establishment and evolution of new polyploids; however, selection must have played a critical role in the establishment of natural *Brassica* polyploids, as their genomes appear relatively stable. Future studies of resynthesized *Brassica* allopolyploids should include tests for the effects of progenitor genotypes,

---

J.C. Pires (✉)
Division of Biological Sciences, 371 Life Sciences Center, University of Missouri, Columbia, MO 65211, USA
e-mail: piresjc@missouri.edu

R. Schmidt, I. Bancroft (eds.), *Genetics and Genomics of the Brassicaceae*, Plant Genetics and Genomics: Crops and Models 9, DOI 10.1007/978-1-4419-7118-0_7,

selection for fertility under field-style conditions, and use genomic approaches that have up to now been limited to studies of polyploid *Arabidopsis suecica.*

Keywords Chromosome rearrangement · Epigenetic · Gene expression · Homoeolog · Polyploidy · Polyploidization

Abbreviation

HNRT Homoeologous non-reciprocal transposition

Contents

7.1 Polyploidy Is a Pervasive Phenomenon in Flowering Plants 196
7.2 Ancient Whole Genome Duplications in the Brassicaceae 197
7.3 Resynthesized *Brassica* and *Arabidopsis* Polyploids 199
7.3.1 Phenotypic Effects in Resynthesized Polyploids 200
7.3.2 Genetic and Epigenetic Changes in Resynthesized Polyploids 201
7.4 Conclusions and Future Research in Resynthesized Polyploids 205
References 206

7.1 Polyploidy Is a Pervasive Phenomenon in Flowering Plants

Polyploidy refers to the presence of three or more chromosome sets within a species, and the genomes of many eukaryotic organisms display evidence of polyploid ancestry (Comai 2005). Heritable polyploidy can result from balanced genome-wide duplications involving the same genome (autopolyploidy) or by the interspecific combination of two or more distinct genomes (allopolyploidy). Aneuploidy refers to chromosome numbers that are not exact multiples of the haploid number (e.g., the presence of additional copies of or absence of one or more chromosomes). Autopolyploids and allopolyploids may form by several mechanisms, including somatic genome doubling, the union of unreduced gametes, or triploid bridge (Ramsey and Schemske 1998, 2002, Soltis and Soltis 2003, Comai 2005). Non-heritable forms of polyploidy (i.e., involving endoreduplication) also abound in eukaryotes, and this phenomenon can be important for adjusting genome dosage within a specific cell type to meet the metabolic needs of different tissues within an otherwise diploid organism (Comai 2005).

While polyploidy is observed in animal and plant lineages, it is generally tolerated to a greater degree in plants, fish, and frogs compared to mammals and birds (LeComber and Smith 2004, Mable 2004, Comai 2005, Otto 2007). Estimates vary, but some reports suggest that up to 80% of extant flowering plant species may have undergone recent polyploid events in their ancestry (Masterson 1994, Ramsey and Schemske 1998, 2002, Gottlieb 2003, Otto and Whitton 2000, DeBodt et al. 2005, Meyers and Levin 2006, Rieseberg and Willis 2007, Wood et al. 2009). However, numerous diploid plant species are diploidized paleopolyploids (e.g., *Arabidopsis* and maize; Wolfe 2001), and with the recent discovery of numerous ancient polyploidy events, at some deep level all angiosperms may be polyploid (reviewed in Van de Peer et al. 2009, Soltis et al. 2009). These ancient whole genome duplications have been hypothesized to have contributed to the diversification of many developmental and regulatory genes in flowering plants (DeBodt et al. 2005, Otto 2007, Lynch and Wagner 2008, Edger and Pires 2009, Freeling 2009).

Polyploidy is an important mechanism for speciation in plants, with 15% of angiosperm and 31% of fern speciation events accompanied by ploidy increase (Wood et al. 2009). In addition, while there are a few examples of plant polyploids that appear to have a single origin (e.g., wheat and *Arabidopsis*), there is evidence that a significant proportion of polyploids have recurrent or multiple origins (e.g., *Tragopogon, Mimulus,* and *Brassica napus*) (reviewed in Soltis and Soltis 2003, Tate et al. 2009b). The role of polyploidy in the evolution and domestication of crops cannot be overstated, with many of our most important crops being known polyploids (e.g., wheat, oat, sugar cane, soybean, banana, potato, coffee, tobacco, and cotton) (Leitch and Bennett 1997, Matzke et al. 1999, Chen 2007, Dubcovsky and Dvorak 2007).

Polyploids often display novel morphological, ecological, cytological, and physiological variation relative to their progenitors (Levin 1983, Ramsey and Schemske 2002). Phenotypic variation for traits like flowering time, pest resistance, organ biomass, and drought tolerance might have contributed to niche exploitation and speciation during the evolution of many polyploids (Rieseberg and Willis 2007). Comai et al. (2000) reported that newly resynthesized *A. suecica* allopolyploids exhibited phenotypic instability, and studies in resynthesized *B. napus* allopolyploids have demonstrated the emergence of de novo phenotypic variation (Schranz and Osborn 2000, 2004, Pires et al. 2004, Gaeta et al. 2007). The mechanisms that lead to novel variation in a new polyploid are largely unknown; however, data from a growing number of studies suggest the involvement of genetic and epigenetic changes, altered gene expression, allele-dosage effects, and distorted regulatory interactions (Soltis and Soltis 1995, Guo et al. 1996, Osborn et al. 2003b, Soltis and Soltis 2003, Adams and Wendel 2005, Chen and Ni 2006, Hegarty and Hiscock 2008, Leitch and Leitch 2008, Doyle et al. 2008, Pignatta and Comai 2009, Rapp et al. 2009, Soltis and Soltis 2009). There remain many unanswered questions regarding the population dynamics of these events and how they contribute to changes in gene expression and novel phenotypic variation.

7.2 Ancient Whole Genome Duplications in the Brassicaceae

Studies of the *Arabidopsis* genome have led to great advances in our understanding of genome evolution in angiosperms. In particular, the small, tractable genome of *Arabidopsis* has provided valuable resources for genomic studies in *Brassica* species, for which fully annotated genome sequences are still unavailable. *Arabidopsis* is approximately 87% similar to *B. oleracea* at the sequence level in exons (Cavell et al. 1998, Parkin et al. 2005). The relatively small genome of *Arabidopsis* shows evidence for three genome-wide duplication events, referred to as γ (~220 MYA), β (~160 MYA), and α (~40 MYA) events (Vision et al.

2000, Simillion et al. 2002, Blanc et al. 2003, Bowers et al. 2003, DeBodt et al. 2005). The recent α duplication event is specific to members of the mustard family (Brassicaceae) (Schranz et al. 2006). Gene duplications that have arisen during these events are thought to be a major source of novel variation and gene functions (Moore and Purugganan 2003). The fate of duplicated genes varies, but may include loss of gene function (pseudogene formation) and gene loss, the partitioning of the ancestral function among the duplicates (subfunctionalization), new functions (neofunctionalization), and dosage compensation (Lynch and Force 2000, Lynch et al. 2001, Moore and Purugganan 2003, DeBodt et al. 2005, Otto 2007, Lynch and Wagner 2008, Edger and Pires 2009, Freeling 2009). The important plant model *A. thaliana* ($2n = 10$) is the maternal parent of a natural allotetraploid *A. suecica* ($2n = 26$) (Comai et al. 2000), which formed after hybridization with the autotetraploid *A. arenosa*. Thus, natural allopolyploids of *Arabidopsis* exhibit both ancient and recent levels of genome duplication.

The genus *Brassica* includes three cultivated diploid species; *B. rapa* (A genome, $n = 10$), *B. nigra* (B genome, $n = 8$), and *B. oleracea* (C genome, $n = 9$) (U 1935, Prakash and Hinata 1980). These species diverged from *Arabidopsis* approximately 15–20 MYA, and have since experienced an additional whole genome triplication (paleohexaploid event) (Lagercrantz 1998, Yang et al. 1999, Lysak et al. 2005, Parkin et al. 2005, Town et al. 2006, Cheung et al. 2009). Comparative genomic analyses between *Brassica* and *Arabidopsis* have identified large blocks of genetic synteny (up to 9 Mb in length), although these blocks are often replicated and rearranged in *Brassica* (Parkin et al. 2005, Town et al. 2006, Schranz et al. 2006, Cheung et al. 2009). For example, in *B. oleracea* and *B. napus,* numerous genes have been deleted from duplicated regions of the genome, and insertions or transpositions (including some of plastid origin) have been detected (Town et al. 2006, Cheung et al. 2009).

Many replicated genes have been lost following genome duplications in diploid *Brassica* species; however, some duplicates have been retained which have important effects on phenotypic variation. One of the best examples comes from analyses of flowering-time in *Arabidopsis* and *Brassica.* In *Arabidopsis* flowering is repressed in a dosage sensitive manner by expression of the transcription factor *FLC,* which is present as a single copy gene (Michaels and Amasino 1999, 2000). Four orthologous genes are detectable in *B. rapa* (*BrFLC1*, *BrFLC2*, *BrFLC3*, and *BrFLC5*), three of which are thought to have arisen following polyploidization (Schranz et al. 2002). Alleles of these loci have additive affects on flowering time in *B. rapa,* allowing for increased allelic combinations (Schranz et al. 2002). Cultivars of *Brassica* demonstrate a wide range of flowering time phenotypes, and it is possible that dosage affects among replicated flowering loci contribute significantly to the variation observed (Osborn 2004).

The three *Brassica* diploid species can hybridize with one another forming three cultivated allopolyploid species (*B. juncea*, AABB, $n = 36$; *B. carinata*, BBCC, $n = 34$; and *B. napus*, AACC, $n = 38$). U (1935) was the first to hypothesize on the relationships between the diploids and allopolyploid *Brassica* species, and many studies have since confirmed this based on phenotypic, isozyme, rDNA, and nuclear

RFLP data (Dass and Nybom 1967, Vaughan 1977, Coulhart and Denford 1982, Quiros et al. 1985, 1986, Takahata and Hinata 1986, Song et al. 1988). *B. napus* cultivars (i.e., oilseeds and rutabagas) are thought to have polyphyletic origins, suggesting that extant lineages may have derived from diverse ancestry and multiple hybridization events (Song and Osborn 1992, Gomez-Campo and Prakash 1999, Flannery et al. 2006, Graham King pers. comm.). Similarly, there are data suggesting that *B. juncea* and *B. carinata* may also have multiple origins (Song and Osborn 1992). Multiple origins have also been reported in allopolyploids of *Tragopogon*, *Mimulus*, *Rubus*, and many other species (reviewed by Soltis and Soltis 2003, Tate et al. 2009b).

As a result of ancient and recent polyploid events, there may be significant genetic variation among extant *Brassica* cultivars, as seen among *B. oleracea* morphotypes (Lukens et al. 2004). With respect to domesticated allopolyploid *B. napus,* research suggests that the genomes are essentially the sum of their diploid progenitors, demonstrating evidence for few genomic rearrangements (Parkin et al. 1995, Axelsson et al. 2000, Rana et al. 2004, Cheung et al. 2009). In contrast, studies in resynthesized *B. napus* show that the time period immediately subsequent to allopolyploidization can be tumultuous and dynamic, involving genetic and epigenetic changes, extensive chromosomal rearrangements, and changes in gene expression and phenotype (Song et al. 1995, Pires et al. 2004, Lukens et al. 2006, Gaeta et al. 2007, Cifuentes et al. 2010, Szadkowski et al. 2010, Gaeta and Pires 2010). Interestingly, of the few genomic locations for which genome rearrangements have been detected in natural *B. napus*, they are the same areas prone to genetic changes in resynthesized lines (see Gaeta et al. 2007). Similarly, in *A. suecica'* resynthesized allotetraploids display genome rearrangements, changes in gene expression, and changes in phenotypes, some of which appear to have occurred in natural lines (Comai et al. 2000, Pontes et al. 2004, Madlung et al. 2005, Madlung and Comai 2004, Wang et al. 2006, Ha et al. 2009).

7.3 Resynthesized *Brassica* and *Arabidopsis* Polyploids

The precise diploid progenitors of extant domesticated allopolyploid *Brassica* species are unknown, making it impossible to understand what happened during the early stages following polyploidization. Alternatively, *Brassica* allopolyploids can be resynthesized in the laboratory from known progenitors, providing an efficient model for studying the early evolution of new species (U 1935, Inomata 1978, Bajaj et al. 1986, Osborn et al. 2003b). Most data on polyploidization in *Brassica* has been obtained from studies in *B. napus*, although a couple of studies on *B. juncea* and *B. carinata* have been conducted. *B. napus* can be resynthesized by hybridization of *B. oleracea* and *B. rapa* followed by embryo rescue and genome doubling, and several studies in recent decades have used such polyploids as models for analyzing early polyploid evolution (e.g., genetic, phenotypic, and transcriptional changes) (Song et al. 1995, Pires et al. 2004, Schranz and Osborn 2000, 2004, Lukens et al. 2006, Albertin et al. 2006, Gaeta et al. 2007). *B. juncea* resynthesized from *B. rapa* and *B. nigra* have similarly been analyzed for phenotypes and genome evolution

(Prakash 1973, Song et al. 1995, Axelsson et al. 2000, Srivastava et al. 2004). *B. carinata* can be resynthesized in a similar manner from *B. nigra* and *B. oleracea*, and a few studies have reported on their synthesis and phenotypes (Sarla and Raut 1988, Jourdan and Salazar 1993). *Arabidopsis* allopolyploids can similarly be resynthesized by hybridization of *A. thaliana* and *A. arenosa* (see Comai et al. 2000, Wang et al. 2006).

7.3.1 Phenotypic Effects in Resynthesized Polyploids

Polyploid plants often have larger cells and may have larger stature, leaves, flowers, stomata, and fruit relative to their diploid progenitors (Stebbins 1950, 1971, Grant 1975, Levin 2002); however, the effects of polyploidy on overall biomass may vary among and within species (Tal 1980, Abel and Becker 2007). Phenotypes in polyploids often exceed the range of the parents, and polyploids may have increased resistance to environmental stresses relative to their parents (Levin 1983). Novel phenotypes in a new polyploid may be critical for niche exploitation and establishment (Leitch and Leitch 2008).

In many species, allotetraploidization has an obvious effect on phenotypes in both natural and resynthesized allopolyploids, probably due to increased heterozygosity obtained through intergenomic hybridization. This has been observed in, *B. napus*, *B. carinata*, and *B. juncea*. On the contrary, analyses of homozygous resynthesized *B. oleracea* and *B. rapa* autotetraploids found that total vegetative biomass in the tetraploids was equal to or lower than the diploids (Abel and Becker 2007). Similar observations of de novo phenotypic variation following allopolyploidization have been made in resynthesized allopolyploids of the related crucifer *Arabidopsis* and in other species (Comai et al. 2000, Ramsey and Schemske 2002). Comai et al. (2000) reported variation for morphology, fertility, and flowering time in F_2 allotetraploids and noted that many phenotypes were unstable over several generations.

In resynthesized *B. napus*, phenotypes can often be obtained that transgress mid-parent or high parent levels (Schranz and Osborn 2000, 2004, Pires et al. 2004, Gaeta et al. 2007). In addition, phenotypic variation increases among lineages within several generations of self-pollination, leading to novel variants in flowering time, flower size, plant height, leaf morphology, leaf size, and cuticle waxes (Schranz and Osborn 2000, 2004, Gaeta et al. 2007). This has been observed among lineages derived from a common set parents, or that were otherwise genetically identical at the time of their formation. Similar morphological variation and heterosis have been recorded in studies of *B. juncea* (Srivastava et al. 2004) and *B. carinata* (Jourdan and Salazar 1993, Sarla and Raut 1988). Many of the traits that show novel variation are of agronomic interest; however, it is unknown whether novel or transgressive phenotypes contributed to the establishment of domesticated *Brassica* polyploids. Although resynthesized allopolyploids may resemble their natural cousins, there remain many questions regarding the genetic or epigenetic mechanisms responsible for novel phenotypes in polyploids.

7.3.2 *Genetic and Epigenetic Changes in Resynthesized Polyploids*

7.3.2.1 Genetic Changes and Their Effects on Gene Expression and Phenotypes

Genetic changes in newly formed allopolyploid species may generate novel gene expression and phenotypic variation (Osborn et al. 2003b, Soltis and Soltis 2003, Adams and Wendel 2005, Chen and Ni 2006, Hegarty and Hiscock 2008, Leitch and Leitch 2008, Doyle et al. 2008, Pignatta and Comai 2009, Rapp et al. 2009, Soltis and Soltis 2009). Genetic changes have been reported in resynthesized *Brassica*, *Triticum*, *Arabidopsis*, Triticale, and *Nicotiana* allopolyploids, but have not been detected in *Spartina* or *Gossypium* (Kenton et al. 1993, Song et al. 1995, Feldman et al. 1997, Liu et al. 2001, Ozkan et al. 2001, Shaked et al. 2001, Adams et al. 2003, 2004, Pontes et al. 2004, Adams and Wendel 2005, Salmon et al. 2005, Lim et al. 2006, Gaeta et al. 2007, Bento et al. 2008, Eilam et al. 2008, Ma and Gustafson 2008, Anssour et al. 2009, Feldman and Levy 2009). Among these studies there are examples of both stochastic and directed genetic changes, including gene conversions, deletions, and homoeologous rearrangements.

In polyploids of the Brassicaceae genetic changes can arise rapidly following formation. In *A. suecica*, genome rearrangements involving rDNA loci have been reported, and it has been hypothesized that homoeologous recombination or transposon-mediated chromosome breakage may have been the cause (Pontes et al. 2004). Song et al. (1995) was the first to report rapid genetic changes in resynthesized *B. napus* allopolyploids (Soltis and Soltis 1995). More recently, an analysis of a population of 50 independently resynthesized *B. napus* allopolyploids found that on average ~14 (range 2–29) genetic changes occurred per line by the S5 generation, most of which resulted from homoeologous rearrangements (Gaeta et al. 2007). In *B. juncea,* extensive genetic changes were reported in one study (Song et al. 1995); however, a subsequent study found no evidence for genetic changes and suggested that disomic inheritance predominated (Axelsson et al. 2000). It should be noted that the experimental design and genotypes utilized in these two studies were quite different; consequently the nature of genome stability following polyploidization in *B. juncea* is still unclear and requires further study. Resynthesized *B. napus* allopolyploid genomes may be unstable than resynthesized *B. juncea* because of the relative similarity between the *B. napus* progenitor diploid genomes (see below). Genetic changes have not been reported in autopolyploid *Brassica*, and an analysis of *B. oleracea* synthetic tetraploids found very few changes in the proteome relative to diploids (Albertin et al. 2005).

Genetic changes in resynthesized *Brassica* allopolyploids appear to correspond to chromosome pairing. Cytogenetic analyses typically find more chromosome pairing between the A and C genomes relative to the B genome (Attia and Robbelen 1986a, b, Attia et al. 1987). This is consistent with the phylogenetic evidence that *B. rapa* and *B. oleracea* are monophyletic (i.e., *rapa/oleracea* lineage) and diverged from *B. nigra* (i.e., the *nigra* lineage) (Warwick and Black 1991, Lysak et al. 2005). However, the degree of intergenomic pairing among any two of the *Brassica* genomes may be somewhat dependent upon genotype (Attia et al. 1987, Sarla and Raut 1988). In *B. napus* homoeologous regions of the A (*B. rapa*) and C (*B. oleracea*) genomes can pair and recombine at meioses (Osborn et al. 2003a, Jenczewski et al. 2003, Leflon et al. 2006, Liu et al. 2006, Nicolas et al. 2007, 2008, 2009). A genetic locus controlling chromosome pairing (*PrBn*) has been identified in *B. napus* allohaploids (reviewed by Jenczewski and Alix 2004), perhaps analogous to the pairing control locus in wheat (*Ph-1*). In wheat, mutations in *Ph-1* result in extensive homoeologous recombination and chromosomal rearrangements (Okamoto 1957, Riley and Chapman 1958, Sears 1977, Griffiths et al. 2006). The details of how *PrBn* functions remain elusive; however, variation for pairing control may be an important factor determining the extent to which homoeologous exchanges occur in newly formed *Brassica* polyploids (Qi et al. 2007, Cifuentes et al. 2010, Szadkowski et al. 2010, Gaeta and Pires 2010).

The term *homoeologous non-reciprocal transposition* (HNRT) has been used to describe segregating intergenomic exchanges that are detectable by RFLP analysis in resynthesized *B. napus*. Although the initial events leading to these genetic changes are believed to be reciprocal exchanges, they are only detectable by molecular marker analysis when one form or the other becomes fixed; hence use of the term "non-reciprocal" may be somewhat misleading (Gaeta and Pires 2010). HNRTs represent the loss of one homoeologous locus and coincident duplication of the other, and may be interstitial or terminal on the rearranged chromosomes. HNRTs were first reported in natural mapping populations (Parkin et al. 1995, Sharpe et al. 1995, Udall et al. 2005) and have been observed at a much greater frequency in newly resynthesized *B. napus* (Pires et al. 2004, Gaeta et al. 2007). Some of the loci that undergo HNRTs in natural populations are prone to HNRT in resynthesized polyploids, suggesting there are hotspots for recombination between the progenitor genomes (Udall et al. 2005, Gaeta et al. 2007). Rearrangements have been detected on all chromosomes in *B. napus*; however, they are often observed among homoeologous linkage groups A1/C1, A2/C2, and A3/C3, which are homoeologous along their entire length and hence the least divergent between the progenitor genomes (Udall et al. 2005, Gaeta et al. 2007).

Genetic changes (e.g., insertions, deletions, duplications, translocations) have direct consequences on gene expression or phenotypes in polyploid species. Kashkush et al. (2002) reported that genetic changes correlated with cDNA amplified fragment length polymorphisms (cDNA-AFLPs) in resynthesized wheat allotetraploids. Similarly, analysis of gene expression in recently formed *Tragopogon* allotetraploids detected gene loss and differential expression of homoeologous genes caused by genetic changes (Tate et al. 2006, 2009a, b, Buggs et al. 2009). In *B. napus* homoeologous rearrangements cause changes in the contribution of homoeologous transcripts to gene expression (loss and gain of parental transcripts) (Gaeta et al. 2007). Nonstochastic, nonadditive changes in protein expression have also been observed (Albertin et al. 2006); however, it is unknown whether these changes result from genetic or epigenetic mechanisms. Homoeologous associations during meiosis may also lead to aneuploid gametes, as evidenced by decreases in fertility observed within several generations of self-pollination following polyploidization (Gaeta et al. 2007, Gaeta and Pires 2010). Chromosome number variation may also occur. For example, plants that are nullisomic–tetrasomic for specific chromosomes have been detected in mapping populations of *B. napus* (Udall et al. 2005). Similar examples of aneuploidy have been reported in recent studies of *Tragopogon* (Lim et al. 2008).

Specific chromosomal rearrangements have specific phenotypic effects in *B. napus*. HNRTs have been associated with yield QTL (Osborn et al. 2003a) and disease resistance QTL (Zhao et al. 2006) in mapping populations. In resynthesized *B. napus* HNRTs at duplicated *FLC* loci have been associated with flowering time variation. In one study, divergent selection for flowering time was conducted on the progenies of a newly formed *B. napus* allopolyploid (Pires et al. 2004). The resulting early- and late-flowering lines carried reciprocal forms of HNRTs at *FLC* loci on chromosomes A3/C3. Analysis of a population of nearly 50 independently resynthesized *B. napus* allopolyploids found that most genetic changes resulted from HNRTs and total genetic changes were correlated with overall phenotypic variation (Gaeta et al. 2007). This study also found that flowering time variation was significantly affected by HNRTs that occurred among *FLC* loci (28% of variation due to rearrangements at *B. oleracea FLC1* and *FLC3*). Collectively, there is significant evidence that homoeologous recombination can lead to novel allele combinations and phenotypic variation in newly formed allopolyploid *B. napus*.

Novel variation is hypothesized to contribute to the adaptability and establishment of newly formed polyploids (Osborn et al. 2003b, Soltis et al. 2009, Adams and Wendel 2005, Chen and Ni 2006, Leitch and Leitch 2008, Pignatta and Comai 2009, Rapp et al. 2009). However, many deleterious genetic rearrangements were probably lost during natural selection in wild species. In a study of nearly 50 resynthesized allopolyploids it was found that increased phenotypic variation came at the cost of decreased fertility (Gaeta et al. 2007). This is often observed in new polyploids, and in the case of *B. napus* may be the result of polysomic inheritance or some rearrangements may simply have negative effects on fitness. In view of the wide variety of *Brassica* morphotypes that have been selected by humans, it is likely that domestication of particular plants contributed to the genomic and phenotypic structure of natural *B. napus*. Comparative genomic studies between extant domesticated *B. napus* and its presumed progenitors suggests few genetic changes at the genome microscale (Rana et al. 2004, Cheung et al. 2009). These data may suggest that differences in homoeologous pairing control existed in the true progenitors of extant *B. napus* lineages relative to recently resynthesized lines. It is also possible, that during the formation of existing lineages genetic changes and genome instability were common, and what is observed now are the consequences of thousands of years of natural and artificial selection favoring stability and fertility. Resynthesized lines have never been selected for stability and fertility, and such experiments could shed light on some of these questions (Gaeta and Pires 2010).

7.3.2.2 Epigenetic Changes in Resynthesized Polyploids

Epigenetic modifications affect both gene expression and chromosome biology in many species (Henderson and Jacobsen 2007). Epigenetic changes are common in newly formed allopolyploid species and may involve DNA methylation changes, histone modifications, and RNA interference (RNAi) (Liu et al. 2001, Osborn et al. 2003a, Madlung and Comai 2004, Adams and Wendel 2005, Lukens et al. 2006, Gaeta et al. 2007). For example, Chen and Pikaard (1997) observed that the progenitor rRNA expression in newly formed *B. napus* was nonadditive and correlated with changes in DNA methylation and histone acetylation of progenitor loci. Cytosine methylation changes in wheat allopolyploids correlated with non-additive expression of several genes (Kashkush et al. 2002). In *A. suecica* allopolyploids, the silencing of transcription factors and the activation of silent transposons have been linked to changes in cytosine methylation (Lee and Chen 2001, Madlung et al. 2002, Hazzouri et al. 2008). In addition, it has been shown that genes can be silenced and later reactivated in the early generations following *A. suecica* synthesis, suggesting epigenetic mechanisms (Comai et al. 2000). The release of silent transposons might have secondary effects on genetic structure and in some cases may contribute to chromosomal breakage and rearrangement (McClintock 1984, Madlung et al. 2004). Adams et al. (2003) reported reciprocal silencing of homoeologous genes in different organs of allotetraploid cotton, suggesting that epigenetic changes following polyploidization may subfunctionalize the expression patterns of homoeologous genes. Collectively, these studies suggest that epigenetic changes following polyploidization are common in different species, and in some cases may directly affect the expression patterns of specific genes.

In a population of resynthesized *B. napus*, changes in cytosine methylation (particularly in a CpG context) were observed within a generation after polyploidization (Lukens et al. 2006, Gaeta et al. 2007). The CpG changes were measured by comparing *Hpa*II and *Msp*I restriction fragment patterns. By the S5 generation, most of the early methylation changes remained, but a few new changes were detected that had occurred during selfing. Total methylation changes were not correlated with overall qualitative changes in gene expression or phenotypic variability, as was the case for genomic rearrangements. However, this epigenetic analysis was limited to a survey of DNA methylation changes using a single pair of methylation-sensitive enzymes (*Hpa*II and *Msp*I), and therefore does not refute the possibility that epigenetic changes following polyploidization in *B. napus* have effects on gene expression and phenotypes. Systematic analyses of global chromatin and DNA modifications in resynthesized *B. napus* have not been conducted. Further studies are needed in *Brassica* allopolyploids to understand the types, distribution, and effects of epigenetic changes following polyploidization.

7.3.2.3 Proteome Changes in Resynthesized Polyploids

Genomic, transcriptional, and phenotypic analysis of resynthesized *B. napus* suggest that the union of the A and C genomes following hybridization and polyploidization is dynamic, yet variable among individual lineages (Gaeta et al. 2007). In a separate study on other resynthesized *B. napus* plant materials, proteomic analysis of stems and roots found that non-additive changes occur at the protein level in up to 35% of 1,600 polypeptides analyzed (Albertin et al. 2006, 2007). However, less than 1% of polypeptides completely disappeared from the proteome and changes did not occur more or less often in any particular functional category or metabolic pathway (Albertin et al. 2006, 2007). Interestingly, 90% of the non-additive protein expression resulted from hybridization and only 3% was associated with genome doubling. This suggests that some genetic and/or epigenetic change occur immediately following hybridization but before whole genome duplication. Some proteins demonstrated stochastic changes in expression, while others were organ specific (Albertin et al. 2006); thus, the overall structure of the *B. napus* proteome remained stable even with the observed nonadditive changes. Overall, the connection between the potential genetic, epigenetic, transcriptional, and proteome changes following polyploidization is yet to be elucidated.

7.4 Conclusions and Future Research in Resynthesized Polyploids

Resynthesized polyploids from a wide range of species have been important for studying genomic, epigenetic, proteomic, and phenotypic changes. Recent studies of resynthesized polyploids include *Nicotiana* (Lim et al. 2006, Anssour et al. 2009), *Solanum* (Stupar et al. 2007), *Tragopogon* (reviewed in Tate et al. 2009b), *Medicago* (Li et al. 2009), *Lilium* (Khan et al. 2009), *Aegilops* (Shcherban et al. 2008), *Eragrostis* (Ochogavía et al. 2009), *Panicum* (Kaushal et al. 2009), Triticale (Bento et al. 2008, Ma and Gustafson 2008), oats (Ueno and Morikawa 2007), and wheat (Ozkan et al. 2001, Shaked et al. 2001, Eilam et al. 2008, Pumphrey et al. 2009, reviewed by Feldman and Levy 2009, Liu et al. 2009). An enormous wealth of knowledge has been gathered on resynthesized allopolyploids in the family Brassicaceae such as *A. suecica* (Pontes et al. 2004, Madlung et al. 2002, 2004, Wang et al. 2006, Ha et al. 2009, Beaulieu et al. 2009, Yu et al. 2009, Wright et al. 2009), *B. napus* (Song et al. 1995, Chen and Pikaard 1997, Lukens et al. 2006, Albertin et al. 2006, 2007, Gaeta et al. 2007, Chen et al. 2008, Nicolas et al. 2008, 2009, Xu et al. 2009, Cifuentes et al. 2010, Szadkowski et al. 2010, Gaeta and Pires 2010), *B. carinata* (Song et al. 1995), *B. juncea* (Axelsson et al. 2000, Yadav et al. 2009), and *Brassica* tri-hybrids (Ge and Li 2007). Some investigations have also involved crosses of *Brassica* species with other genera such as *Capsella* (Chen et al. 2009), *Raphanus* (Chen and Wu 2008), and *Orychophragmus* (Li and Ge 2007, Ge et al. 2009).

A. suecica and *B. napus* have emerged as an important model for studying polyploidy. Like many natural polyploid species, resynthesized *A. suecica* and *B. napus* are phenotypically distinct from their progenitors. This novel variation can result from not only increased heterozygosity and gene dosage but also large-scale chromosome rearrangements and epigenetic remodeling. How these changes in newly formed polyploids relate to natural polyploids is still largely unknown. It is likely that during the evolution of many natural polyploids, genomic and epigenetic remodeling stabilizes during the processes of diploidization. In the case of *B. napus* in which large-scale rearrangements are frequently detected in resynthesized lines, polysomic inheritance must have been overcome

Future studies in *B. napus* could focus on crosses between synthetic polyploids and wild varieties for assessments of chromosome pairing and genome stability. Similarly, with plant cytogenetic advances in recent years, a thorough analysis of diverse natural varieties for chromosome pairing variation and homoeologous rearrangements should be possible using progenitor-specific probes (Pires and Xiong, unpublished data). This might be easier than using molecular markers, which have difficulty distinguishing rearrangements in heterozygous backgrounds. Perhaps it is time for studies of resynthesized *Brassica* polyploids to be tested in wild conditions or under specific selection schemes (similar to that employed by Pires et al. 2004). Resynthesized polyploid seed could be broadcast in fields or in multiple environments and allowed to open pollinate and "evolve" for several generations. Resynthesized lines could also be put under artificial selection for traits like drought tolerance, heat stress, and herbivore/pest resistance. Such selected lines could be analyzed for genomic content, disomic inheritance, and phenotypes. These studies would help us understand what happens in the early generations of resynthesized polyploids outside of the greenhouse or growth chamber.

Genetically modified *B. napus* is often cultivated adjacent to wild *B. rapa* varieties, and the two species can hybridize readily leading to movement of transgenic loci into the wild (Hansen et al. 2003). Indeed the role of homoeologous rearrangements in trait introgression needs to be explored further. Similarly, wide hybridization among other members of the Brassicaceae may be informative from the perspective of chromosomes pairing, recombination, and gene introgression (Li and Ge 2007, Ge et al. 2009, Chen and Wu 2008, Chen et al. 2009).

Studies in *B. napus* have been very thorough regarding analysis of many independently resynthesized polyploids, as well as genetic, transcriptional, phenotypic, and proteomic analyses. However, there are still many more questions to be answered in allopolyploid *B. juncea* and *B. carinata.* With regard to epigenetic changes following polyploidization, very little is known in all three species, except that DNA methylation changes occur across the genome in *B. napus.* With regard to transcriptional analysis, future studies should focus on the development and production of homoeolog-specific microarrays (Rapp et al. 2009) and deep sequencing (Ha et al. 2009).

References

Abel S, Becker HC (2007) The effect of autopolyploidy on biomass lines of *Brassica rapa* and *Brassica oleracea*. Plant Breed 126:642–643

Adams KL, Cronn R, Percifield R, Wendel JF (2003) Genes duplicated by polyploidy show unequal contributions to the transcriptome and organ-specific reciprocal silencing. Proc Natl Acad Sci USA 100:4649–4654

Adams KL, Percifield R, Wendel JF (2004) Organ-specific silencing of duplicated genes in a newly synthesized cotton allotetraploid. Genetics 168:2217–2226

Adams KL, Wendel JF (2005) Polyploidy and genome evolution in plants. Curr Opin Plant Biol 8:135–141

Albertin W, Alix K, Balliau T, Brabant P, Davanture M, Malosse C, Valot B, Thiellement H (2007) Differential regulation of gene products in newly resynthesized *Brassica napus* allotetraploids is not related to protein function nor subcellular localization. BMC Genomics 21:56

Albertin W, Balliau T, Brabant P, Chevre AM, Eber F, Malosse C, Thiellement H (2006) Numerous and rapid nonstochastic modifications of gene products in newly synthesized *Brassica napus* allotetraploids. Genetics 173:1101–1113

Albertin W, Brabant P, Catrice O, Eber F, Jenczewski E, Chevre AM, Thiellement H (2005) Autopolyploidy in cabbage (*Brassica oleracea* L) does not alter significantly the proteomes of green tissues. Proteomics 5:2131–2139

Anssour S, Krügel T, Sharbel TF, Saluz HP, Bonaventure G, Baldwin IT (2009) Phenotypic, genetic and genomic consequences of natural and synthetic polyploidization of *Nicotiana attenuata* and *Nicotiana obtusifolia*. Ann Bot 103:1207–1217

Attia T, Busso C, Robbelen G (1987) Digenomic triploids for an assessment of chromosome relationships in the cultivated diploid *Brassica* species. Genome 29:326–330

Attia T, Robbelen G (1986a) Meiotic pairing in haploids and amphidiploids of spontaneous versus synthetic origin in rape, *Brassica napus L.* Can J Genet Cytol 28:330–334

Attia T, Robbelen G (1986b) Cytogenetic relationship within the cultivated *Brassica* analyzed in amphidiploids from three diploid ancestors. Can J Genet Cytol 28:323–329

Axelsson T, Bowman CM, Sharpe AG, Lydiate DJ, Lagercrantz U (2000) Amphidiploids *Brassica juncea* contains conserved progenitor genomes. Genome 43:679–688

Bajaj YPS, Mahajan SK, Labana KS (1986) Interspecific hybridization of *Brassica napus* and *B. juncea* through ovary, ovule, and embryo culture. Euphytica 35:103–109

Beaulieu J, Jean M, Belzile F (2009) The allotetraploid *Arabidopsis thaliana-Arabidopsis lyrata* subsp. *petraea* as an alternative model system for the study of polyploidy in plants. Mol Gen Genet 281:421–435

Bento M, Pereira HS, Rocheta M, Gustafson P, Viegas W, Silva M (2008) Polyploidization as a retraction force in plant genome evolution: sequence rearrangements in Triticale. PLoS One 3(1):e1402

Blanc G, Hokamp K, Wolfe KH (2003) A recent polyploidy superimposed on older large-scale duplications in the *Arabidopsis* genome. Genome Res 13:137–144

Bowers JE, Chapman BA, Rong J, Paterson AH (2003) Unraveling angiosperm genome evolution by phylogenetic analysis of chromosomal duplication events. Nature 422:433–438

Buggs RJA, Doust AN, Tate JA, Koh J, Soltis K, Feltus FA, Paterson AH, Soltis PS, Soltis DE (2009) Gene loss and silencing in *Tragopogon miscellus* (Asteraceae): comparison of natural and synthetic allotetraploids. Heredity 103:73–81

Cavell AC, Lydiate D, Parkin IA, Dean C, Trick M (1998) Collinearity between a 30-centimorgan segment in *Arabidopsis thaliana* chromosome 4 and duplicated regions within the *Brassica napus* genome. Genome 41:62–69

Chen ZJ (2007) Genetic and epigenetic mechanisms for gene expression and phenotypic variation in plant polyploids. Annu Rev Plant Biol 58:377–406

Chen HF, Ge XH, Du XZ, Zhao ZG, Li ZY (2009) Genetic and histological characterization of a novel recessive genic male sterile line of *Brassica napus* derived from a cross with *Capsella bursa-pastoris*. Euphytica 167:31–37

Chen X, Li M, Shi J, Fu D, Qian W, Zou J, Zhang C, Meng J (2008) Gene expression profiles associated with intersubgenomic heterosis in *Brassica napus*. Theor Appl Genet 117: 1031–1040

Chen ZJ, Ni Z (2006) Mechanisms of genomic rearrangements and gene expression changes in plant polyploids. Bioessays 28:240–252

Chen ZJ, Pikaard CS (1997) Epigenetic silencing of RNA polymerase I transcription; a role for DNA methylation and histone modification in nucleolar dominance. Genes Dev 11: 2124–2136

Chen HG, Wu JS (2008) Characterization of fertile amphidiploid between *Raphanus sativus* and *Brassica alboglabra* and the crossability with *Brassica* species. Genet Resour Crop Evol 55:143–150

Cheung F, Trick M, Drou N, Lim Y-P, Park J-Y, Kwon S-J, Kim J-A, Scott R, Pires JC, Paterson AH, Town C, Bancroft I (2009) Comparative analysis between homoeologous genome segments of *Brassica napus* and its progenitor species reveals extensive sequence-level divergence. Plant Cell 21:1912–1928

Cifuentes M, Grandont L, Moore M, Chevre AM, Jenczewski E (2010) Genetic regulation of meiosis in polyploid species: new insights into an old question. New Phytol 186(1):29–36

Comai L (2005) The advantages and disadvantages of being polyploid. Nat Rev Genet 6: 836–846

Comai L, Tyagi AP, Winter K, Holmes-Davis R, Reynolds SH, Stevens Y, Byers B (2000) Phenotypic instability and rapid gene silencing in newly formed *Arabidopsis* allotetraploids. Plant Cell 12:1551–1567

Coulhart M, Denford KE (1982) Isozyme studies in *Brassica*. I. Electrophoretic techniques for leaf enzymes and comparison of *B. napus*, *B. campestris*, and *B. oleracea* using phosphoglucoisomerase. Can J Plant Sci 62:621–630

Dass H, Nybom H (1967) The relationships between *Brassica nigra*, *B. campestris*, and *B. oleracea*, and their amphidiploid hybrids studies by means of numerical chemotaxonomy. Can J Genet Cytol 9:880–890

DeBodt S, Maere S, Van de Peer Y (2005) Genome duplication and the origin of angiosperms. Trends Ecol Evol 20:592–597

Doyle JJ, Flagel LE, Paterson AH, Rapp RA, Soltis DE, Soltis PS, Wendel JF (2008) Evolutionary genetics of genome merger and doubling in plants. Annu Rev Genet 42:443–461

Dubcovsky J, Dvorak J (2007) Genome plasticity a key factor in the success of polyploid wheat under domestication. Science 316:1862–1866

Edger PP, Pires JC (2009) Gene and genome duplications: the impact of dosage sensitivity on the fate of nuclear genes. Chromosome Res 17:699–717

Eilam T, Anikster Y, Millet E, Manisterski J, Feldman M (2008) Nuclear DNA amount and genome downsizing in natural and synthetic allopolyploids of the genera *Aegilops* and *Triticum*. Genome 51:616–627

Feldman M, Levy AA (2009) Genome evolution in allopolyploid wheat-a revolutionary reprogramming followed by gradual changes. J Genet Genomics 36:511–518

Feldman M, Liu B, Segal G, Abbo S, Levy AA, Vega JM (1997) Rapid elimination of low-copy DNA sequences in polyploid wheat: a possible mechanism for differentiation of homoeologous chromosomes. Genetics 147:1381–1387

Flannery ML, Mitchell FJG, Coyne S, Kavanagh TA, Burke JI, Salamin N, Dowding P, Hodkinson TR (2006) Plastid genome characterization in *Brassica* and Brassicaceae using a new set of nine SSR. Theor Appl Genet 113:1221–1231

Freeling M (2009) Bias in plant gene content following different sorts of duplication: tandem, whole-genome, segmental, or by transposition. Annu Rev Plant Biol 60:433–453

Gaeta RT, Pires JC (2010) Homoeologous recombination in allopolyploids: the polyploid ratchet. New Phytol 186(1):18–28

Gaeta RT, Pires JC, Iniguez-Luy F, Leon E, Osborn TC (2007) Genomic changes in resynthesized *Brassica napus* and their effects on gene expression and phenotype. Plant Cell 19:1–15

Ge XH, Li ZY (2007) Intra- and intergenomic homology of B-genome chromosomes in trigenomic combinations of the cultivated *Brassica* species revealed by GISH analysis. Chromosome Res 15:849–861

Ge XH, Wang J, Li ZY (2009) Different genome-specific chromosome stabilities in synthetic *Brassica* allohexaploids revealed by wide crosses with *Orychophragmus*. Ann Bot 104:19–31

Gomez-Campo C, Prakash S (1999) The phylogeny of *Brassica* and allied genera. Domestication of cultivated Brassicas and allies. In: Gomez-Campo C (ed) Biology of *Brassica* Coenospecies, vol 4. Elsevier, Amsterdam

Gottlieb LD (2003) Plant polyploidy: gene expression and genetic redundancy. Heredity 91:91–92
Grant V (1975) Genetics of flowering plants. Columbia University Press, New York, NY
Griffiths S, Sharp R, Foote TN, Bertin I, Wanous M, Reader S, Colas I, Moore G (2006) Molecular characterization of *Ph1* as a major chromosome pairing locus in polyploid wheat. Nature 439:749–752
Guo M, Davis D, Birchler JA (1996) Dosage effects on gene expression in maize ploidy series. Genetics 142:1349–1355
Ha M, Lu J, Tian L, Ramachandran V, Kasschau KD, Chapman EJ, Carrington JC, Chen X, Wang XJ, Chen ZJ (2009) Small RNAs serve as a genetic buffer against genomic shock in *Arabidopsis* interspecific hybrids and allopolyploids. Proc Natl Acad Sci USA 106:17835–17840
Hansen LB, Siegismund HR, Jorgensen RB (2003) Progressive introgression between *Brassica napus* (oilseed rape) and *B. rapa*. Heredity 91:276–283
Hazzouri KM, Mohajer A, Dejak SI, Otto SP, Wright SI (2008) Contrasting patterns of transposable-element insertion polymorphism and nucleotide diversity in autotetraploid and allotetraploid *Arabidopsis* species. Genetics 179:581–592
Hegarty MJ, Hiscock SJ (2008) Genomic clues to the evolutionary success of polyploid plants. Curr Biol 18:R435–R444
Henderson IR, Jacobsen SE (2007) Epigenetic inheritance in plants. Nature 447:418–424
Inomata N (1978) Production of interspecific hybrids between *Brassica campestris* and *Brassica oleracea* by culture in vitro of excised ovaries II. Effects of coconut mil and casein hydrolysate on the development of excised ovaries. Jpn J Genet 53:1–11
Jenczewski E, Alix K (2004) From diploids to allopolyploids: the emergence of efficient pairing control genes in plants. CRC Crit Rev Plant Sci 23:21–45
Jenczewski E, Eber F, Grimaud A, Huet S, Lucas MO, Monod H, Chevre AM (2003) PrBn, a major gene controlling homoelogous pairing in oilseed rape (*Brassica napus*) haploids. Genetics 164:645–653
Jourdan PS, Salazar E (1993) *Brassica carinata* resynthesized by protoplast fusion. Theor Appl Genet 86:567–572
Kashkush K, Feldman M, Levy AA (2002) Gene loss, silencing and activation in newly synthesized wheat allotetraploids. Genetics 160:1651–1659
Kaushal P, Agrawal A, Malaviya DR, Siddiqui SA, Roy AK (2009) Ploidy manipulation in guinea grass (*Panicum maximum* Jacq., Poaceae) utilizing a hybridization-supplemented apomixis-components partitioning approach (HAPA). Plant Breed 128:295–303
Kenton A, Parokonny AS, Gleba YY, Bennet MD (1993) Characterization of the *Nicotiana tabacum* L. genome by molecular cytogenetics. Mol Genet Genomics 240:159–169
Khan N, Zhou S, Ramanna MS, Arens P, Herrera J, Visser RGF, Van Tuyl JM (2009) Potential for analytic breeding in allopolyploids: an illustration from Longiflorum × Asiatic hybrid lilies (*Lilium*). Euphytica 166:399–409
Lagercrantz U (1998) Comparative mapping between *Arabidopsis thaliana* and *Brassica nigra* indicates that *Brassica* genomes have evolved through extensive genome replication accompanied by chromosomes fusions and frequent rearrangements. Genetics 150: 1217–1228
LeComber SC, Smith C (2004) Polyploidy in fishes: patterns and processes. Biol J Linn Soc 82:431–442
Lee HS, Chen ZJ (2001) Protein-coding genes are epigenetically regulated in *Arabidopsis* polyploids. Proc Natl Acad Sci USA 98:6753–6758
Leflon M, Eber F, Letanneur JC, Chelysheva L, Coriton O, Huteau V, Ryder CD, Barker G, Jenczewski E, Chevre AM (2006) Pairing and recombination at meiosis of *Brassica rapa* (AA) × *Brassica napus* (AACC) hybrids. Theor Appl Genet 113:1467–1480
Leitch LJ, Bennett MD (1997) Polyploidy in angiosperms. Trends Plant Sci 2:470–476
Leitch AR, Leitch IJ (2008) Genomic plasticity and the diversity of polyploid plants. Science 320:481–483
Levin DA (1983) Polyploidy and novelty in flowering plants. Am Nat 122:1–25

Levin DA (2002) The role of chromosomal change in plant evolution. Oxford University Press, Oxford

Li ZY, Ge XH (2007) Unique chromosome behavior and genetic control in *Brassica* × *Orychophragmus* wide hybrids: a review. Plant Cell Rep 26:701–710

Li X, Wei Y, Nettleton D, Brummer C (2009) Comparative gene expression profiles between heterotic and non-heterotic hybrids of tetraploid *Medicago sativa*. BMC Plant Biol 9:107. doi:10.1186/1471–2229–9–107

Lim KY, Soltis DE, Soltis PS, Tate J, Matyasek R, Srubarova H, Kovarik A, Pires JC, Xiong Z, Leitch AR (2008) Rapid chromosome evolution in recently formed polyploids in *Tragopogon* (Asteraceae). PLoS One 3:e3353

Lim KY, Souckova-Skalicka K, Sarasan V, Clarkson JJ, Chase MW, Kovarik A, Leitch AR (2006) A genetic appraisal of a new synthetic *Nicotiana tabacum* (Solanaceae) and the Kostoff synthetic tobacco. Am J Bot 93:875–883

Liu ZQ, Adamczyk K, Manzanres-Sauleux M, Eber F, Lucas MO, Delmourme R, Chevre AM, Jenczewski E (2006) Mapping PrBn and other quantitative trait loci responsible for the control of homoeolgous chromosome pairing in oilseed rape (*Brassica napus* L.) haploids. Genetics 174:1583–1596

Liu B, Brubaker CL, Mergeai G, Cronn RC, Wendel JF (2001) Polyploid formation in cotton is not accompanied by rapid genomic changes. Genome 44:321–330

Liu B, Xu C, Zhao N, Qi B, Kimatu JN, Pang J, Han F (2009) Rapid genomic changes in polyploid wheat and related species: implications for genome evolution and genetic improvement. J Genet Genomics 36:519–528

Lukens LN, Pires JC, Leon E, Vogelzang R, Oslach L, Osborn TC (2006) Patterns of sequence loss and cytosine methylation within a population of newly resynthesized *Brassica napus* allopolyploids. Plant Physiol 140:336–348

Lukens LN, Quijada PA, Udall J, Pires JC, Schranz ME, Osborn TC (2004) Genome redundancy and plasticity within ancient and recent *Brassica* crop species. Biol J Linn Soc 82: 665–674

Lynch M, Force A (2000) The probability of duplicate gene preservation by subfunctionalization. Genetics 154:459–473

Lynch M, O'Hely M, Walsh B, Force A (2001) The probability of preservation of newly arisen gene duplicate. Genetics 159:1789–1804

Lynch VJ, Wagner GP (2008) Resurrecting the role of transcription factor change in developmental evolution. Evolution 62:2131–2154

Lysak MA, Koch MA, Pecinka A, Schubert I (2005) Chromosome triplication found across the tribe Brassiceae. Genome Res 15:516–525

Ma XF, Gustafson JP (2008) Allopolyploidization-accommodated genomic sequence changes in triticale. Ann Bot 101:825–832

Mable BK (2004) 'Why polyploidy is rarer in animals than in plants': Myths and mechanisms. Biol J Linn Soc 82:453–466

Madlung A, Comai L (2004) The effect of stress on genome regulation and structure. Ann Bot (Lond.) 94:481–495

Madlung A, Masuelli RW, Watson B, Reynolds SH, Davison J, Comai L (2002) Remodeling of DNA methylation and phenotypic and transcriptional changes in synthetic *Arabidopsis* allotetraploids. Plant Physiol 129:733–746

Madlung A, Tyagi AP, Watson B, Jiang H, Kagochi T, Doerge RW, Martienssen R, Comai L (2005) Genomic changes in synthetic *Arabidopsis* polyploids. Plant J 41:221–230

Masterson J (1994) Stomatal size in fossil plants: evidence of polyploidy in majority of angiosperms. Science 264:421–424

Matzke MA, Scheid OM, Matzke AJM (1999) Rapid structural and epigenetic changes in polyploid and aneuploid genomes. Bioessays 21:761–767

McClintock B (1984) The significance of responses of the genome to challenge. Science 226: 792–801

Meyers LA, Levin DA (2006) On the abundance of polyploids in flowering plants. Evolution 60:1198–1206

Michaels SD, Amasino RM (1999) *FLOWERING LOCUS C* encodes a novel MADS domain protein that acts as a repressor of flowering. Plant Cell 11:949–956

Michaels SD, Amasino RM (2000) Memoris of winter: vernalization and the competence to flower. Plant Cell Environ 23:1145–1153

Moore R, Puruggannan MD (2003) The early stages of duplicate gene evolution. Proc Natl Acad Sci USA 100:15682–15687

Nicolas SD, Le Mignon G, Eber F, Coriton O, Monod H, Clouet V, Huteau V, Lostanlen A, Delourme R, Chalhoub B, Ryder CD, Chevre AM, Jenczewski E (2007) Homoeologous recombination plays a major role in chromosome rearrangements that occur during meiosis of *Brassica napus* haploids. Genetics 175:487–503

Nicolas SD, Leflon M, Liu Z, Eber F, Chelysheva L, Coriton O, Chevre AM, Jenczewski E (2008) Chromosome 'speed dating' during meiosis of polyploid *Brassica* hybrids and haploids. Cytogenet Genome Res 120:331–338

Nicolas SD, Leflon M, Monod H, Eber F, Coriton O, Huteau V, Chevre AM, Jenczewski E (2009) Genetic regulation of meiotic cross-overs between related genomes in *Brassica napus* haploids and hybrids. Plant Cell 21:373–385

Ochogavía AC, Cervigni G, Selva JP, Echenique VC, Pessino SC (2009) Variation in cytosine methylation patterns during ploidy level conversions in *Eragrostis curvula*. Plant Mol Biol 70:17–29

Okamoto M (1957) Asynaptic effect of chromosome V. Wheat Inf Serv 5:6

Osborn TC (2004) The contribution of polyploidy to variation in *Brassica* species. Physiol Plant 121:531–536

Osborn TC, Butrulle DV, Sharpe AG, Pickering KJ, Parkin IAP, Parker JS, Lydiate DJ (2003a) Detection and effects of a homoeologous reciprocal transposition in *Brassica napus*. Genetics 165:1569–1577

Osborn TC, Pires JC, Birchler JA, Auger DL, Chen ZJ, Lee H, Comai L, Madlung A, Doerge RW, Colot V, Martienssen RA (2003b) Understanding mechanisms of gene expression in polyploids. Trends Genet 19:141–147

Otto SP (2007) The evolutionary consequences of polyploidy. Cell 131:452–462

Otto SP, Whitton J (2000) Polyploid incidence and evolution. Annu Rev Genet 34:401–437

Ozkan H, Levy AA, Feldman M (2001) Allopolyploidy-induced rapid genome evolution in the wheat (*Aegilops-Triticum*) group. Plant Cell 13:1735–1747

Parkin IAP, Gulden SM, Sharpe AG, Lukens L, Trick M, Osborn TC, Lydiate DJ (2005) Segmental structure of the *Brassica napus* genome based on comparative analysis with *Arabidopsis thaliana*. Genetics 171:765–781

Parkin IAP, Sharpe AG, Keith DJ, Lydiate DJ (1995) Identification of the A and C genomes of amphidiploid Brassica napus (oilseed rape). Genome 38:1122–1131

Pignatta D, Comai L (2009) Parental squabbles and genome expression: lessons from the polyploids. J Biol 8:43

Pires JC, Zhao J, Schranz ME, Leon EJ, Quijada PA, Lukens LN, Osborn TC (2004) Flowering time divergence and genomic rearrangements in resynthesized *Brassica* polyploids (Brassicaceae). Biol J Linn Soc 82:675–688

Pontes O, Neves N, Silva M, Lewis MS, Madlung A, Comai L, Viegas W, Pikaard CS (2004) Chromosomal locus rearrangements are a rapid response to formation of the allotetraploid *Arabidopsis suecica* genome. Proc Natl Acad Sci USA 101:18240–18245

Prakash S (1973) Artificial *Brassica juncea* Coss. Genetica 44:249–263

Prakash S, Hinata K (1980) Taxonomy, cytogenetics and origin of crop *Brassicas*, a review. Opera Bot 55:42–47

Pumphrey M, Bai J, Laudencia-Chingcuanco D, Anderson O, Gill BS (2009) Nonadditive expression of homoeologous genes is established upon polyploidization in hexaploid wheat. Genetics 181:1147–1157

Qi L, Friebe B, Zhang P, Gill BS (2007) Homoeologous recombination, chromosome engineering and crop improvement. Chromosome Res 15:3–19

Quiros CF, Kianian SF, Ochoa O, Douches D (1985) Genome evolution in *Brassica*: use of molecular markers and cytogenetic stocks. Cruciferae Newsl 10:21–23

Quiros CF, Kianian SF, Ochoa O, Douches D (1986) Evolutionary trends in *Brassica*; gathering evidence from chromosome addition lines. Cruciferae Newsl 11:22–23

Ramsey J, Schemske DW (1998) Pathways, mechanisms, and rates of polyploid formation in flowering plants. Annu Rev Ecol Syst 29:467–501

Ramsey J, Schemske DW (2002) Neopolyploidy in flowering plants. Annu Rev Ecol Syst 33: 589–639

Rana D, van den Boogaart T, O'Neill CM, Haynes L, Bent E, Macpherson L, Park JY, Lim YP, Bancroft I (2004) Conservation of the microstructure of genome segments in *Brassica napus* and its diploid relatives. Plant J 40:725–733

Rapp RA, Udall JA, Wendel JF (2009) Genomic expression dominance in allopolyploids. BMC Biol 7:18. doi:10.1186/1741-7007-7-18

Rieseberg LH, Willis JH (2007) Plant speciation. Science 317:911–914

Riley R, Chapman V (1958) Genetic control of the cytologically diploid behaviour of hexaploid wheat. Nature 182:713–715

Salmon A, Ainouche ML, Wendel JF (2005) Genetic and epigenetic consequences of recent hybridization and polyploidy in *Spartina (Poaceae)*. Mol Ecol 14:1163–1175

Sarla N, Raut RN (1988) Synthesis of *B. carinata* from *B. nigra* × *B. oleracea* hybrids obtained by ovary culture. Theor Appl Genet 76:846–849

Schranz ME, Lysak MA, Mitchell-Olds T (2006) The ABC's of comparative genomics in the Brassicaceae: building blocks of crucifer genomes. Trends Plant Sci 11:535–542

Schranz ME, Osborn TC (2000) Novel flowering time variation in the resynthesized polyploid *Brassica napus*. J Hered 91:242–246

Schranz ME, Osborn TC (2004) De novo variation in life-history traits and responses to growth conditions of resynthesized polyploid *Brassica napus* (Brassicaceae). Am J Bot 91:174–183

Schranz ME, Quijada P, Sung SB, Lukens L, Amasino R, Osborn TC (2002) Characterization and effects of the replicated flowering time locus gene *FLC* in *Brassica rapa*. Genetics 162: 1457–1468

Sears ER (1977) An induced mutant with homoeologous pairing in common wheat. Can J Genet Cytol 19:585–593

Shaked H, Kashkush K, Ozkan H, Feldman M, Levy AA (2001) Sequence elimination and cytosine methylation are rapid and reproducible responses of the genome to wide hybridization and allopolyploidy in wheat. Plant Cell 13:1749–1759

Sharpe AG, Parkin IAP, Keith DJ, Lydiate DJ (1995) Frequent nonreciprocal translocations in the amphidiploid genome of oilseed rape (*Brassica napus*). Genome 38:1112–1121

Shcherban AB, Badaeva ED, Amosova AV, Adonina IG, Salina EA (2008) Genetic and epigenetic changes of rDNA in a synthetic allotetraploid, *Aegilops sharonensis* × *A. umbellulata*. Genome 51:261–271

Simillion C, Vandepoele K, Van Montagu MCE, Zabeau M, Van de Peer Y (2002) The hidden duplication past of *Arabidopsis thaliana*. Proc Natl Acad Sci USA 99:13627–13632

Soltis DE, Albert VA, Leebens-Mack J, Bell CD, Paterson AH, Zheng C, Sankoff D, de Pamphilis CW, Wall PK, Soltis PS (2009) Polyploidy and angiosperm diversification. Am J Bot 96: 333–348

Soltis DE, Soltis PS (1995) The dynamic nature of polyploid genomes. Proc Natl Acad Sci USA 92:8089–8091

Soltis DE, Soltis PS (2003) Advances in the study of polyploidy since plant speciation. New Phytol 161:173–191

Soltis PS, Soltis DE (2009) The role of hybridization in plant speciation. Annu Rev Plant Biol 60:561–588

Soltis DE, Soltis PS, Tate JA (2003) Advances in the study of polyploidy since Plant Speciation. New Phytologist 161:173–191

Song K, Lu P, Tang K, Osborn TC (1995) Rapid genome change in synthetic polyploids of *Brassica* and its implications for polyploid evolution. Proc Natl Acad Sci USA 92:7719–7723

Song K, Osborn TC (1992) Polyphyletic origins of *Brassica napus*: new evidence based on organelle and nuclear RFLP analyses. Genome 35:992–1001

Song KM, Osborn TC, Williams PH (1988) *Brassica* taxonomy based on nuclear restriction fragment length polymorphisms (RFLPs). Theor Appl Genet 75:784–794

Srivastava PS, Narula A, Srivastava S (2004) Plant biotechnology and molecular markers. Kluwar, New Delhi

Stebbins GL (1950) Variation and evolution in plants. Columbia University Press, New York, NY

Stebbins GL (1971) Chromosomal evolution in higher plants. Edward Arnold, London

Stupar RM, Bhaskar PB, Yandell BS, Rensink WA, Hart AL, Ouyang S, Veilleux RE, Busse JS, Erhardt RJ, Buell CR, Jiang J (2007) Phenotypic and transcriptomic changes associated with potato autopolyploidization. Genetics 176:2055–2067

Szadkowski E, Eber F, Huteau V, Lodé M, Huneau C, Belcram H, Coriton O, Manzanares-Dauleux M, Delourme R, King G, Chalhoub B, Jenczewski E, Chèvre AM (2010) The first meiosis of resynthesized *Brassica napus*, a genome blender. New Phytol 186(1):102–112

Takahata Y, Hinata K (1986) A consideration of the species relationships in subtrive Brassicinae (Cruciferae) in view of cluster analysis of morphological characters. Plant Species Biol 1: 79–88

Tal M (1980) Physiology of polyploids. In: Lewis WH (ed) Polyploidy: biological relevance. Plenum Press, New York, NY

Tate JA, Joshi P, Soltis KA, Soltis PS, Soltis DE (2009a) On the road to diploidization? Homoeolog loss in independently formed populations of the allopolyploid *Tragopogon miscellus* (Asteraceae). BMC Plant Biol 9:80

Tate JA, Ni Z, Scheen A, Koh J, Gilbert CA, Lefkowitz D, Chen ZJ, Soltis PS, Soltis DE (2006) Evolution and expression of homeologous loci in *Tragopogon miscellus* (*Asteraceae*), a recent and reciprocally formed allopolyploid. Genetics 173:1599–1611

Tate JA, Symonds VV, Doust AN, Buggs RJA, Mavrodiev E, Majure LC, Soltis PS, Soltis DE (2009b) Synthetic polyploids of *Tragopogon miscellus* and *T. mirus* (Asteraceae): 60 years after Ownbey's discovery. Am J Bot 96:979–988

Town CD, Cheung F, Maiti R, Crabtree J, Haas BJ, Wortman JR, Hine EE, Althoff R, Arbogast TS, Tallon LJ, Vigouroux M, Trick M, Bancroft I (2006) Comparative genomics of *Brassica oleracea* and *Arabidopsis thaliana* reveal gene loss, fragmentation, and dispersal after polyploidy. Plant Cell 18:1348–1359

U N (1935) Genome analysis in *Brassica* with special reference to the experimental formation of *B. napus* and peculiar mode of fertilization. Jpn J Bot 7:389–452

Udall JA, Quijada PA, Osborn TC (2005) Detection of chromosomal rearrangements in four mapping populations of *Brassica napus L.* Genetics 169:967–979

Ueno M, Morikawa T (2007) Production of synthetic polyploid oats and detection of C genome rearrangements by GISH and FISH. Breed Sci 57:339–343

Van de Peer Y, Maere S, Meyer A (2009) The evolutionary significance of ancient genome duplications. Nat Rev Genet 10:725–732

Vaughan JG (1977) A multidisciplinary study of the taxonomy and origin of *Brassica* crops. Bioscience 27(1):35–40

Vision TJ, Brown DG, Tanksley SD (2000) The origins of genomic duplications in *Arabidopsis*. Science 290:2114–2117

Wang J, Tian L, Lee HS, Wei NE, Jiang H, Watson B, Madlung A, Osborn TC, Doerge RW, Comai L, Chen ZJ (2006) Genomewide nonadditive gene regulation in *Arabidopsis* allotetraploids. Genetics 172:507–517

Warwick SI, Black LD (1991) Molecular systematics of *Brassica* and allied genera (Subtribe Brassicinae, Brassiceae): chloroplast genome and cytodeme congruence. Theor Appl Genet 82:81–92

Wolfe KH (2001) Yesterday's polyploids and the mystery of diploidization. Nat Rev Genet 2: 333–341

Wood TE, Takebayashi N, Barker MS, Mayrose I, Greenspoon PB, Rieseberg LH (2009) Proc Natl Acad Sci USA 106:13875–13879
Wright KM, Pires JC, Madlung A (2009) Mitotic instability in resynthesized and natural polyploids of the genus *Arabidopsis* (Brassicaceae). Am J Bot 96:1656–1664
Xu Y, Zhong L, Wu X, Fang X, Wang J (2009) Rapid alterations of gene expression and cytosine methylation in newly synthesized *Brassica napus* allopolyploids. Planta 229:471–483
Yadav P, Bhat SR, Prakash S, Mishra LC, Chopra VL (2009) Resynthesized *Brassica juncea* lines with novel organellar genome constitution obtained through protoplast fusion. J Genet 88: 109–112
Yang YW, Lai KN, Tai PY, Li WH (1999) Rates of nucleotide divergence in angiosperm mitochondrial DNA sequences and dates of divergence between *Brassica* and other angiosperm lineages. J Mol Evol 48:597–604
Yu Z, Haage K, Streit VE, Gierl A, Torres Ruiz RA (2009) A large number of tetraploid *Arabidopsis thaliana* lines, generated by a rapid strategy, reveal high stability of neo-tetraploids during consecutive generations. Theor Appl Genet 118:1107–1119
Zhao J, Udall JA, Quijada PA, Grau CR, Meng J, Osborn TC (2006) Quantitative trait loci for resistance to Sclerotinia sclerotiorum and its association with a homoeologous non-reciprocal transposition in *Brassica napus* L. Theor Appl Genet 112:509–516

Chapter 8
Genetics of *Brassica rapa* L.

Nirala Ramchiary and Yong Pyo Lim

Abstract *Brassica rapa* ($2n = 20$, AA) is one among the six economically important cultivated *Brassica* species of U's triangle. The presence of wide genetic and morphological diversity in the form of several subspecies in different geographical regions allows growing of this species for producing leafy vegetables, vegetable oils, turnips roots, turnip greens, turnip tops, and fodder turnip. In addition, it is one of the diploid progenitor parental species which contributed A genome to the important oilseed crops, *Brassica juncea* ($2n = 36$, AABB) and *Brassica napus* ($2n = 38$, AACC). Due to the presence of wide genetic variability, genetic analysis for important traits and improvement in this crop are possible. While conventional genetic analyses and breeding could achieve success in improvement of many traits, but for the majority of the agronomically important traits recent advances in genetic marker techniques supplemented heavily for this crop. Molecular markers have been used to analyze genetic diversity in *B. rapa*, mapping of different traits in the genome in segregating populations generated for investigations of specific traits of interest. Several mapping studies have led to the identification and tagging of many traits for marker-assisted breeding, and in some cases map-based cloning of the responsible genes have been done, e.g. the flowering time gene *FLC*. However, traditional mapping of quantitative trait loci (QTL) is often not sufficient to develop effective markers for trait introgression or for identification of the genes responsible. The ongoing sequencing work of *B. rapa* genome will undoubtedly give great insight into the genetics underlying both simple and complex traits in this crop. Comparative mapping with the closest relative *Arabidopsis thaliana* and identification of candidate genes, use of EST-derived SSRs and SNPs from *Arabidopsis* and other *Brassica* species for high density mapping, isolation, and development of gene-based functional molecular markers would also be very much helpful for this crop. This chapter reviews the current use of available molecular marker technologies in *B. rapa* genetic analysis and breeding for important morphological,

Y.P. Lim (✉)
Department of Horticulture, Chungnam National University, Daejeon 305-764, South Korea
e-mail: yplim@cnu.ac.kr

R. Schmidt, I. Bancroft (eds.), *Genetics and Genomics of the Brassicaceae*, Plant Genetics and Genomics: Crops and Models 9, DOI 10.1007/978-1-4419-7118-0_8,

agronomic, quality, abiotic and disease resistance QTL/gene mapping and their utilization in marker-assisted breeding.

Keywords *Brassica rapa* · Molecular markers · Genetics · Quantitative trait locus/loci(QTL) · Comparative mapping · *Arabidopsis thaliana*

Abbreviations

AFLP	Amplified fragment length polymorphism
BAC	Bacterial artificial chromosome
B. rapa	*Brassica rapa*
CAPS	Cleaved amplified polymorphic sequence
DH	Doubled haploid
ESTP	Expressed sequences tag polymorphism
FISH	Fluorescence in situ hybridization
QTL	Quantitative trait locus/loci
RAPD	Random amplified polymorphic DNA
RFLP	Restriction fragment length polymorphism
RIL	Recombinant inbred lines
SCAR	Sequence characterized amplified regions
SNP	Single nucleotide polymorphism
SRAP	Sequenced-related amplified polymorphism
SSR	Simple sequence repeat
STS	Sequence-tagged-site

Contents

8.1 Introduction 216
8.2 *B. rapa* Breeding and Trait Genetics 217
8.3 Molecular Markers 218
8.3.1 Molecular Markers in Diversity Studies 219
8.3.2 Molecular Markers and Development of Genetic Linkage Maps in *B. rapa* 219
8.3.3 Molecular Markers and Trait Genetics 233
8.4 Comparative Mapping and Identification of Candidate Genes for Important Traits . 249
8.5 Conclusions and Perspectives 251
References 252

8.1 Introduction

Brassica rapa (AA, $2n = 20$) is one among 3,705 species complex belonging to 360 genera of Brassicaceae family. It is believed to be the first domesticated species among six economically important brassicas, i.e., *Brassica oleracea* (CC, $2n = 18$), *Brassica nigra* (AA, $2n = 16$), *Brassica napus* (AACC, $2n = 38$), *Brassica carinata* (BBCC, $2n = 34$), and *Brassica juncea* (AABB, $2n = 36$) (Prakash and Hinata

1980, Gomez Campo and Prakash 1999, Gomez campo 1999). *B. rapa* has the distinction of being grown for production of condiments, vegetables, vegetable oils, and fodder crops worldwide. In addition, it is the progenitor species of A genome to the major oilseed amphidiploid *Brassica* species, *B. napus* and, *B. juncea* (U 1935). *B. rapa* L. is believed to have originated in Mediterranean region and Central Asia where evidence of natural growing of this species is still found in these areas (Gomez Campo 1999, Prakash and Hinata 1980, Gomez Campo and Prakash 1999). Sinskaia (1928) also suggested the development of *B. rapa* into two directions forming European and the Asian types. While European oleiferous type originated in Mediterranean region, the Asian form originated in Afghanistan, Central Asia, and northwest India. Morphological, cytogenetics, and molecular marker analysis also gave conclusive evidence of Europe and eastern Asia to be two independent centers of origin of *B. rapa* (Denford and Vaughan 1977, Song et al. 1988b, Zhao et al. 2004b). Large number of morphological differences is observed in *B. rapa* that might have resulted due to the long history of breeding for different traits along with natural selection for adaptation in different geographical regions due to which a large number of subspecies are recognized. Oleiferous and turnip forms developed in Europe while in eastern Asia comprising China and west Asia evolved into leafy form and oleiferous form. Various types of leafy vegetables of *B. rapa* such as Chinese cabbage (*B. rapa* ssp. *pekinensis*), non-heading pak choi (*B. rapa* ssp. *chinensis*), and mizuna are found widely in East Asia particularly in China, Korea and in Japan that is why this region is known as secondary center for genetic diversity of *B. rapa*. In China and Korea, Chinese cabbage is the number one leafy vegetable crop. These leafy *Brassica* vegetables supply vitamins, minerals, dietary fibers to human health. In Korea, Chinese cabbage is the most important vegetable crop and regarded as a national food which is eaten in the form of fermented pickles "Kimchee." The oleiferous form of *B. rapa* (*B. rapa* ssp. *oleifera*) is grown for production of vegetable oils in China, Canada, India, and in northern Europe and is the third *Brassica* oilseed crop. Indian form developed into mainly oleiferous form such as yellow sarson, brown sarson and toria type (Singh 1958) that are grown in large areas in eastern and northeastern parts of India during winter season. Turnip rape is the commonly cultivated oilseed *B. rapa* in Europe. Recently, in China, oleiferous form of *B. rapa* has been used to diversify *B. napus* germplasm since *B. rapa* has advantage of having wide variability and greater source of genetic potential for yield and other traits. Turnip (*B. rapa* ssp. *rapifera*) is grown widely in countries having temperate climate such as in Canada, Europe, and in the USA for its roots, turnip greens, turnip tops, and fodder production, respectively.

8.2 *B. rapa* Breeding and Trait Genetics

Conventional *B. rapa* breeding had been given major emphasis on development of high-yielding cultivars including hybrids, improvement of fatty acid composition, oil content, and diseases resistance. Breeding of oilseed *B. rapa* in India resulted into development of improved cultivars of yellow sarson, brown sarson, and toria types (Gomez Campo and Prakash 1999). Chinese cabbage breeding in Japan and Korea could develop cultivars which could be grown year round in different seasons. Presently several Chinese cabbage hybrids have been developed in Japan and China. However, lack of effective pollination control mechanism limits the production of hybrid *B. rapa*. The development of *Brassica* cultivars lacking erucic acid (<2%) and containing low levels of glucosinolates (<20 μM/g of seed) was the main objective in 1970s. Genetic analysis in *B. rapa* showed a single gene inheritance of erucic acid trait (Kirk and Hurlstone 1983). The first low erucic acid *B. rapa* cultivar developed was "Span" (Downey et al. 1975) and breeding of low glucosinolates *B. rapa* cultivar resulted in the development of first cultivar "Candle" in Canada. Along with increasing yield, seed quality (low glucosinolates and zero erucic acid), seed oil content, breeding for yellow seed coat color (due to its low fiber and high oil content), tolerance to abiotic stresses such as heat, freezing tolerance, and disease resistance (white rust, clubroot disease resistance breeding) in *B. rapa* cultivars have been major aims. While conventional genetic analyses and breeding could achieve success in improvement of many traits, but for the majority of the agronomically important traits recent advances in genetic marker techniques supplemented heavily for this crop. The recent development in molecular genetics particularly the development of different types of molecular markers have been promising in genetic analysis, mapping of trait loci (QTL/genes) in the genome, and manipulation of desired traits in brassica crops (Snowdon and Friedt 2004).

8.3 Molecular Markers

The association of morphological and isozyme markers with important traits has been used in crop breeding, the earliest example being given by Sax (1923) who showed the association of seed coat color with seed size in beans. Similarly, morphological and isozyme markers have been used in *B. rapa* (Yarnell 1956, Williams and Hill 1986, Chevre et al. 1995). However, due to their low abundance in the genome and dependence on the environmental conditions for expression they are not used any longer. The developments in molecular genetics particularly the genetic marker techniques have revolutionized plant genome research and crop improvement in the last two decades. With the advent of molecular markers such as restriction Fragment Length Polymorphism (RFLP), Amplified Fragment Length Polymorphism (AFLP), Simple Sequence Repeats (SSRs), Single nucleotide Polymorphism (SNPs) etc. several molecular linkage maps have been developed for almost all the cultivated crops around the world. Molecular markers are very powerful tools to study the genetic architecture of crop plants, mapping of agronomically important qualitative and quantitative traits, precise transfer of important trait from QTL one genotype to another through marker-assisted breeding, map-based cloning of useful genes besides their use in diversity and evolutionary studies. With the help of dense genetic maps developed in many crops, numerous agronomically important gene(s)/QTL have been mapped/tagged and these genes have been transferred to elite crop varieties through marker-assisted selection (Snowdon and

Friedt 2004). The development and use of molecular markers in Brassicas started in late 1980s and since then different types of molecular markers have been developed and utilized for brassica breeding (Song et al. 1988a, b, Quiros et al. 1994, Kresovich et al. 1995, Demeke et al. 1992, Lowe et al. 2004, Choi et al. 2007, Kim et al. 2009).

8.3.1 Molecular Markers in Diversity Studies

Genetic diversity reflecting into phenotypic diversity is the most important in crop breeding for improving yield, quality, and biotic and abiotic stress tolerance. Genetic variation is the pre-requirement of crop-breeding program. Since its development, molecular markers have been used to study the genetic diversity and evolutionally relationships in brassicas. Restriction Fragment Length Polymorphism (RFLP), observed due to the variable length of DNA fragment, was the first DNA marker to be used for genetic diversity and evolutionary study in *B. rapa* and other *Brassica* species (Song et al. 1988a, b, 1990). Isozyme markers were used by Zhao and Becker (1998) to study the genetic diversity of *B. rapa* and *B. napus* belonging to European and Chinese origin. Cluster analysis using isozyme data showed two distinct groups of Chinese and European origin *B. rapa*, which further form subgroups classifying winter and spring type as seen in *B. napus*. Recently, the use of isozyme and RFLP markers is being replaced by other PCR-based markers such as random amplified polymorphic DNA (RAPD), amplified fragment length polymorphism (AFLP), and simple sequence repeats or microsatellites (SSR) markers. RAPD and AFLP are extensively used to study the genetic diversity in Brassica species including *B. rapa* (Demeke et al. 1992, Thormann et al. 1994, Das et al. 1999, Chen et al. 2000, He et al. 2002, Guo et al. 2002, Zhao et al. 2004a). He et al. (2002) used RAPD markers to study the genetic diversity in vegetable and oleiferous *B. rapa* originated from China. In their study, they observed distinct groups of vegetables and oil-yielding *B. rapa* which further formed subgroups (15 in oleiferous type) depending on the geographical origin of collection. They further reported that diversity level in spring type was higher than that of the winter type. Simple sequence repeats, being co-dominant, multi-allelic, robust, highly polymorphic, and high abundance, are used largely for diversity studies and for construction of molecular linkage maps in *B. rapa* (Choi et al. 2004, Suwabe et al. 2004a, Choi et al. 2007, Kim et al. 2009).

8.3.2 Molecular Markers and Development of Genetic Linkage Maps in B. rapa

Development of detailed genetic linkage maps in plants is a very useful tool for studying genome structure and evolution, mapping important traits, marker-assisted breeding, and finally map-based cloning of genes of interest (Beckmann and Soller 1986, Tanksley et al. 1989). Development of linkage maps in *Brassica* species is almost as old as development of molecular markers. To date several molecular linkage maps in *B. rapa* have been developed using molecular markers such as RFLP, RAPD, AFLP, and SSRs and mapping populations such as F_2, DH, RIL, and different backcross generation derived from inter- and intra-subspecies cross. The first *B. rapa* linkage map was developed by Song et al. (1991) using F_2 population derived from a cross between Chinese cabbage and spring broccoli. A total of 188 RFLP markers identifying 280 polymorphic loci were mapped in ten linkage groups (LGs) covering 1,850 cM. They observed extensive sequence duplication by the detection of two or more segregating loci (36.7% of the total) conserved on different linkage groups suggesting large chromosome fragments to be present in multiple copies. McGrath and Quiros (1991) developed a genetic linkage map using F_2 population from turnip × pak choi cross and 49 RFLP and isozyme markers covering a total of 262 cM. Chyi et al. (1992) used F_2 population derived from a cross between Sarson and canola for construction of linkage map with 360 RFLP loci. Their map covered a total of 1,876 cM in ten linkage groups (LGs) with an average marker interval of 5.2 cM. Teutonico and Osborn (1994) constructed *B. rapa* map using 139 RFLP and F_2 populations derived from turnip rape Per and the spring yellow Sarson R500. Ajisaka et al. (1995) developed a linkage map in Chinese cabbage F_2 population using 117 RAPD and isozyme makers. Nozaki et al. (1997) also developed *B. rapa* linkage map using F_2 population and 52 RAPD markers. Their map covered a total of 733 cM. Map based on F_2 populations derived from a cross between Chinese cabbage cultivar Aijiahuang and turnip cultivar Qisihai was developed by Lu et al. (2002a) using 131 AFLP and RAPD markers. The map covered a total genome length of 1,810.9 cM. Many other studies have used F_2 population for development of linkage map in *B. rapa* for different purpose (Schilling 1991, Matsumoto et al. 1998, Ajisaka et al. 1999, Zhang et al. 2000, Table 8.1.)

Of late, permanent/immortal mapping populations such as Recombinant Inbred lines (RILs) and doubled haploid (DH) populations derived from diverse parental crosses of *B. rapa* has been used for development of genetic linkage map and mapping and tagging of important traits. RIL and DH populations have the distinction of being able to retain the same genotypic constitution permanently and hence offer advantages for replicated trials. The first genetic map developed using RIL in *B. rapa* was by Kole et al. (1997, 1996) derived from the cross involving turnip rape Per and yellow sarson R500. This population was developed from the F_2 population used for mapping earlier by Teutonico and Osborn (1994). The map covered a total genome length of 890 cM with 144 RFLP markers giving an average marker density of 6 cM. A genetic map comprising 83 RFLP markers and covering 1,138.1 cM in total ten LGs were developed by Novakova et al. (1996). *B. rapa* genetic map comprising 265 AFLP and 87 RAPD was developed by Yu et al. (2003a) using RILs Chinese cabbage populations. The map covered a total of 2,665.7 cM distance with an average marker interval of 7.6 cM.

Doubled haploid populations have been used by different workers for development of genetic map. Choi et al. (2004) constructed a genetic map in DH population derived from a cross involving diverse Chinese cabbage parental Inbred lines, Chiifu, and Kenshin. A total of 644 markers comprising AFLP,

Table 8.1 Details of genetic linkage map and traits QTL/gene mapping studies in *B. rapa*

Parental information	Mapping population	Marker used	Total markers	Genome coverage (cM)	Objectives of study/traits investigated	Gene/QTL identified	References
Yorii Spring (Turnip) × Kwan Hoo Choi (Pakchoi)	F_2	RFLP Isozyme	49	262	Construction of linkage map, study of chromosome syntheny between *B. oleracea* and *B. rapa*	–	Mcgrath and Quiros (1991)
B. rapa ssp *oleifera* × Rapid cycling *B. rapa*	F_2	RFLP	58	700	Construction of linkage map		Schilling (1991)
Michiili (Chinese cabbage) × Spring broccoli	F_2	RFLP	280	1,850	Construction of linkage map, study of genome duplication, and conservation of duplicated chromosomal blocks	–	Song et al. (1991)
Sarson × Canola	F_2	RFLP	360	1,876	Construction of linkage map	–	Chyi et al. (1992)
Chinese cabbage × Chinese cabbage	F_2	RAPD Isozyme	117	860	Construction of linkage map	–	Ajisaka et al. (1995)
Turnip rape cv Per (biennial) × Yellow Sarson cv R500 (annual)	F_3	RFLP	139	1,785	Map construction, Comparison of linkage map between *B. rapa*, *B. oleracea, B. napus*, and *Arabidopsis thaliana*, mapping of yellow seed color, erucic acid, and pubescence traits	Mapped 1 QTL each on LG4, LG5 and LG1 for pubescence (*Pub*), yellow seed color (*YIs*), and erucic acid trait (*Eru*), respectively	Teutonico and Osborn (1994)

Table 8.1 (continued)

Parental information	Mapping population	Marker used	Total markers	Genome coverage (cM)	Objectives of study/traits investigated	Gene/QTL identified	References
Turnip rape cv Per × Yellow Sarson cv R500	F_3	RFLP	139	1,785	Mapping of vernalization trait, comparison of linkage groups containing QTL for vernalization and flowering time traits between *B. rapa* and *B. napus*	2 QTL, one each on LG2 and LG8 identified	Teutonico and Osborn (1995)
Michiili (Chinese cabbage) × Spring broccoli	F_2	RFLP	220	1,593	28 Morphological traits related to leaf, stem, and flowering	48 QTL identified in 9 linkage groups	Song et al. (1995)
Turnip rape cv Per × Yellow Sarson cv R500 *B. rapa* ssp. *rapifera* and *B. rapa* ssp. *oleifera*	$F_{2.3}$	RFLP	144	1,785	Mapping of white rust, (*Albugo candida*) resistance gene	One resistance locus *ACA1* on LG4 co-mapped with leaf Pubescence (*Pub1*) locus	Kole et al. (1996)
Turnip rape × Yellow Sarson	RILs	RFLP	144	890	Construction of linkage map	–	Kole et al. (1997)
Turnip rape × Turnip rape	F_2	RFLP RAPD	166	519	Construction of linkage map	–	Tanhuanpaa et al. (1996a)
Jo4002 × Sv3402	F_2	RAPD/ SCAR	104	–	Linkage analysis of RAPD markers with oleic acid traits using bulk segergant analysis	One locus *Fad2* identified on LG6	Tanhuanpaa et al. (1996b)

Table 8.1 (continued)

Parental information	Mapping population	Marker used	Total markers	Genome coverage (cM)	Objectives of study/traits investigated	Gene/QTL identified	References
K -151 (yellow sarson) × No4003 *B. campestris* ssp. *oleifera X B. alboglara*	BC_3	RAPD	42	–	Cosegreation analysis of RAPD marker with yellow seed coat color	Identified linked RAPD markers with seed coat color	Chen et al. (1997)
Turnip rape cv Per × Yellow Sarson cv R500	RILs	RFLP	144	890	Development of a genetic linkage map using RFLP markers in RI population of *B. rapa* and comparison with F_2 map	–	Kole et al. (1997)
Chinese cabbage × Mizuna *B. rapa* L.var *pekinensis* and *B. rapa* L var *japonica*	F_2	RAPD	52	733	Study of morphological traits such as leaf shape, leaf hairiness, period of bolting, self-incompatibility, and NS-glycoprotein	Detected two QTL for leaf shape, leaf hairiness one locus, two loci for bolting and one locus each for self-incompatibility and NS-glycoprotein	Nozaki et al. (1997)
Chinese cabbage × Chinese cabbage	DH	RFLP	63	735	Mapping of clubroot resistance locus and orange yellow pigmentation in head inner leaves and petals	Identified one locus each for clubroot resistance (*CRa*) on LG 3 and yellow - orange-pigmentation(*Oy*) on LG1	Matsumoto et al. (1998)
Turnip × Chinese Cabbage	F_2	RAPD	99	1,632	Construction of genetic linkage map	–	Zhang et al. (2000)

Table 8.1 (continued)

Parental information	Mapping population	Marker used	Total mark-ers	Genome coverage (cM)	Objectives of study/traits investigated	Gene/QTL identified	References
G309 (normal bolting) × DH27(extreme late bolting)	F_2	RAPD	527	–	Identification of RAPD markers linked to bolting trait using bulk segreganat analysis	Identified a QTL for bolting time using RAPD markers	Ajisaka et al. (2001)
Rc50a7 (early flowering) × (late flowering) *B. rapa* ssp. *chinensis*	F_2	RFLP cDNA *FLC*	26 4 1	-	Comparative mapping of flowering time QTL/genes in *B. rapa, B. oleracea, B. nigra, B. juncea*, and *A. thaliana*	Mapped *CO* and *FLC* homologues in four species and identified homologous relationships with *Arabidopsis* chromosomes 5	Axelsson et al. (2001)
Turnip rape cv Per (biennial) × Yellow Sarson cv R500 (annual)	BC_3S_1	RFLP *CO* *FLC*	12	–	Comparative mapping between *B. rapa* containing vernalizaton responsive gene *VRF2* region with *Arabidopsis thaliana* homologous Chromosome	Identified that *VRF2* in *B. rapa* is identical to *FLC* in *A. thaliana*	Kole et al. (2001)
Aijiaohuang × Qisihai *B rapa* L. ssp.*chinensis* × *B rapa* L ssp.*rapifera*	F_2	AFLP RAPD	131	1,811	Construction of genetic map, mapping of QTL for morphological traits	Identified 24 QTL on 8 different linkage groups	Lu et al. (2002a)

Table 8.1 (continued)

Parental information	Mapping population	Marker used	Total markers	Genome coverage (cM)	Objectives of study/traits investigated	Gene/QTL identified	References
Turnip rape cv Per × Yellow Sarson cv R500	BC_3S_1	RAPD, SSSR *FLC*	–	–	Functional validation and mapping of flowering time QTL with *FLC* homologues	Mapped, *FLC1, FLC2, FLC3,* and *FLC-5* in QTL region	Schranz et al. (2002)
93651–2 (low linolenic acid) × Sv3402(high)	F_2	RFLP(*Fad3*) RAPD	27	–	Mapping of loci involved in linolenic acid biosynthesis	Mapped three QTL and candidate gene *Fad2* in one QTL region	Tanhuanpaa and Schulman (2002)
Chinese cabbage × Chinese cabbage	RILs	AFLP RAPD	352	2,666	Mapping heat tolerance loci	Identified 5 QTL(ht1-ht5)	Yu et al. (2003a)
Chinese cabbage × Chinese cabbage	RILs	AFLP RAPD	352	2,666	QTL mapping of bolting, plant height and leaf-related traits	A total of 55 QTL detected in the whole genome	Yu et al. (2003b)
Chinese cabbage × Chinese cabbage	DH	AFLP	255	884	Construction of genetic linkage map	–	Wang et al. (2004)
CrGC1-21(dwarf) × CrGC1-33 (tall, wild type) (Rapid cycling *B. rapa*)	BC_1S_1	RFLP	92	–	Mapping of *Brassica* dwarfing gene, comparative mapping with *A. thaliana* and identification of candidate gene	Mapped major gene for dwarfing, *DWF2* on bottom of R06	Muangprom and Osborn (2004)
G004(resistant) × Hakusai Chukanbohon Nou 7'(A9709, susceptible) (Both *B. rapa* ssp. *pekinensis*)	F_2	SSR	94	–	Association analysis of SSR markers with clubroot resistance loci	Identified SSR markers linked to clubroot resistance locus *Crr1* and *Crr2*	Suwabe et al. (2003)

Table 8.1 (continued)

Parental information	Mapping population	Marker used	Total markers	Genome coverage (cM)	Objectives of study/traits investigated	Gene/QTL identified	References
Shinki(resistant) × 94SK (Susceptible) (*B. rapa* ssp. *pekinensis*)	F_2	AFLP	–	–	Identification of AFLP markers linked to clubroot resistance locus, conversion of AFLP markers into PCR-based SCAR marker	Mapped *CRb*, a gene confering resistance to clubroot disease along with SCAR and CAPS marker	Piao et al. (2004)
N-WMR-3 (resistant turnip) × A9709 (susceptible Chinese cabbage)	F_3	RAPD	–	–	Association analysis of RAPD markers with clubroot resistance locus through bulk segregant analysis	Identified a novel locus, *Crr3* associated with clubroot resistance	Hirai et al. (2004)
A9408 (late parent) × Homei P09 (early parent) *B. rapa* ssp. *pekinensis*	DH	AFLP	248	1,096.6	QTL mapping of bolting time	Mapped ten QTL (BT1-BT10) in 7 linkage groups	Nishioka et al. (2005)
N-WMR-3 (resistant turnip) × A9709(susceptible Chinese cabbage)	F_2/F_3	STS	23	–	Fine mapping of clubroot resistance locus *Crr3*	Mapped *Crr3* in LG2 and alignment with *Crr3* QTL region in *B. rapa* to the top arm of *Arabidopsis* chromosome 3	Saito et al. (2006)

Table 8.1 (continued)

Parental information	Mapping population	Marker used	Total markers	Genome coverage (cM)	Objectives of study/traits investigated	Gene/QTL identified	References
Chinese cabbage × Chinese cabbage	DH	RAPD RFLP SSR	262	1,005	Comparative mapping between *B. rapa* and *Arabidopsis* chromosomes, mapping and alignment of clubroot resistance loci containing region of *B. rapa* with *Arabidopsis* chromosome	Alignment of club root resistance loci *Crr1, Crr2*, and *Crr4* and found that first two loci were homologous to same region of *Arabidopsis* chromosome 4	Suwabe et al. (2006)
Chinese cabbage × Chinese cabbage	F_2/F_3	RFLP, SSR	545	1,287	Construction of genetic linkage map using EST-derived RFLP probe and mapping of 5 paralogous *FLC* genes, identification of genome duplication in *B. rapa* genome	Mapping of *5 FLC* genes – *FLC-1, FLC-2, 2 FLC-3*, and *FLC-5*	Kim et al. (2006)
R-o-18 × B162 (*B. rapa* ssp. *oleifera*)	F_2	AFLP SSR	246	664	QTL mapping for race 1 and race 4 *Xanthomonas campestris* pv *campestris* resistance	Six QTL conferring resistance to both the races (race 1 and race 4) were mapped	Soengas et al. (2007)
Chinese cabbage × Chinese cabbage	DH	AFLP RAPD SSR			Construction of reference genetic map for multinational *B. rapa* genome sequencing project	–	Choi et al. (2007)

Table 8.1 (continued)

Parental information	Mapping population	Marker used	Total mark-ers	Genome coverage (cM)	Objectives of study/traits investigated	Gene/QTL identified	References
Rapid cycling *B. rapa* × Rapid cycling *B. rapa*	F_2	SSR	22	–	Mapping of anthocyaninless biosynthesis locus and comparative mapping with *Arabidopsis* chromosomes	Mapped anthocyaninless(*anl*) in R09 and identified homologous *Arabidopsis* chromosome containing candidate genes	Burdzinski and Wendell (2007)
SPAN × BARI-6 *B. rapa* ssp. *oleifera* × *B. rapa* ssp. *trilocularis*	F_2, F_3, BC_1	SRAP SCAR SNP	48	–	Development of SRAP, SNP, and multiplexed SCAR markers for the seed coat color in *B. rapa* L.	Developed SNPs and multiplex SCAR markers for the major seed coat color gene *Br1*	Rahman et al. (2007)
K10 (resistant) × Q5 (susceptible) C9 (resistant) × 6R (susceptible) *B. rapa* ssp. *pekinensis*	$F_{2.1}$, $F_{2.2}$	AFLP RAPD SSR STS	318 187	640.3 485.9	Mapping of isolate specific clubroot resistance genes	Mapped two Clubroot resistances QTL *CRc* and *CRk* on R03 and R02, respectively. *CRc* locus is new locus and independent of previous QTL reported	Sakamoto et al. (2008)

Table 8.1 (continued)

Parental information	Mapping population	Marker used	Total mark-ers	Genome coverage (cM)	Objectives of study/traits investigated	Gene/QTL identified	References
YS-143 × PC-175 (DH38) YS-143 × VT-115 (DH30) *B. rapa* ssp. *oleifera* × *B. rapa* ssp. *chinensis*	DH 38 DH30	AFLP SSR	321 300	688 763	Map integration, mapping of QTL for glucosinolates accumulation in *B. rapa* leaves, and comparative mapping of homologous *Arabidopsis* chromosomes for candidate gene identification	Detected 16 QTL for total and individual aliphatic glucosinolates, 3 QTL for indolyl, and 3 QTL for aromatic glucosinolates, respectively, in 2 DH mapping populations	Lou et al. (2008)
Y177 × Y195 *B. rapa* ssp. *pekinensis*	DH	AFLP SSR SRAPs ESTPs	287	1,090	QTL mapping for mineral accumulation and shoot dry biomass under different Zn conditions	Mapped 10 QTL in *B. rapa* linkage map	Wu et al. (2008)

RAPD, ESTP, CAPS, and SSR gave total genome coverage of 1,131 cM with an average of one marker in 1.8 cM distance. In another study, Suwabe et al. (2004) constructed genetic map using Chinese cabbage DH population and 262 markers (SSR, RFLP, and RAPD) distributed in ten linkage groups giving a genome coverage of 1,005.5 cM and average marker distance of 3.7 cM. Wang et al. (2004) also developed a genetic map in Chinese cabbage using DH populations and 263 AFLP loci. Their map covered a total of 883.7 cM in ten LGs. A detailed genetic linkage map of *B. rapa* has been constructed containing 545 sequence-tagged loci covering 1,287 cM, with an average mapping interval of 2.4 cM (Kim et al. 2006). Till now several new *B. rapa* maps have been constructed (Suwabe et al. 2006, Soengas et al. 2007, Burdzinski and Wendell 2007, Rahman et al. 2007, Sakamoto et al. 2008, Lou et al. 2008, Wu et al. 2008) (Table 8.1).

However, even though many genetic maps have been constructed in *B. rapa* there is no detailed alignment/correlation between maps developed in different studies. So it was necessary to produce a common *B. rapa* reference genetic map using common set of molecular markers for a detailed study at the genome level and for mapping and tagging of traits of interest. With this objective, Choi et al. (2007) described the construction of a first reference genetic linkage map for the *Brassica* A genome using 78 doubled haploid lines derived from a cross involving two diverse Chinese cabbage (*B. rapa* ssp. *pekinensis*) inbred lines, "Chiifu-401- 42" (C) and "Kenshin-402-43" (K). Due to its small genome size with the extensive genetics and genomics resources available relative to the other *Brassica* species, *B. rapa* Chinese cabbage inbred line Chiifu-401-42 has been selected as the sequencing template in 2003 for the first whole *Brassica* A genome sequencing of Brassicas by Multinational *Brassica* Genome Project (MBGP). The map consists of a total of 556 markers comprising 278 AFLP, 235 SSR, 25 RAPD, and 18 ESTP, STS, and CAPS markers. The ten linkage groups were designated A1 to A10 based on the alignment with *B. napus* linkage map using common SSR markers. The total length of the linkage map was 1,182 cM with an average interval of 2.83 cM between adjacent loci (Fig. 8.1). The development of this map now helps to anchor sequence contigs for the international *B. rapa* Genome Sequencing Project (BrGSP) and formed the base for integration of genome sequence and genetic information which enables the international research community to share resources and data for the improvement of *B. rapa* and related research. Recently this version I reference genetic map was updated with the addition of 188 more BAC-anchored SSR markers, and this newly developed map is designated as version II reference genetic map of *B. rapa* A genome (Kim et al. 2009). The updated map consists a total of 722 markers comprising 269 AFLP, 411 SSR, 24 RAPD, 8 STS, 7 ESTP, and 3 CAPS markers loci in ten linkage groups (R01/A01-R10/A10) and spans a total of 1,123.4 cM with an average distance of 1.6 cM between two loci. The sequences of the anchored BACs enabled the identification of 30 blocks of conserved synteny, totaling 527.6 cM in length, between the genomes of *B. rapa* and *Arabidopsis*. The updated BAC anchored version II reference map will continue to serve as a basis to integrate the genetic, physical, and chromosome maps of *B. rapa* for the *BrGSP* besides being useful for evolutionary studies such as polyploidization, speciation, and genome duplication in the brassicas.

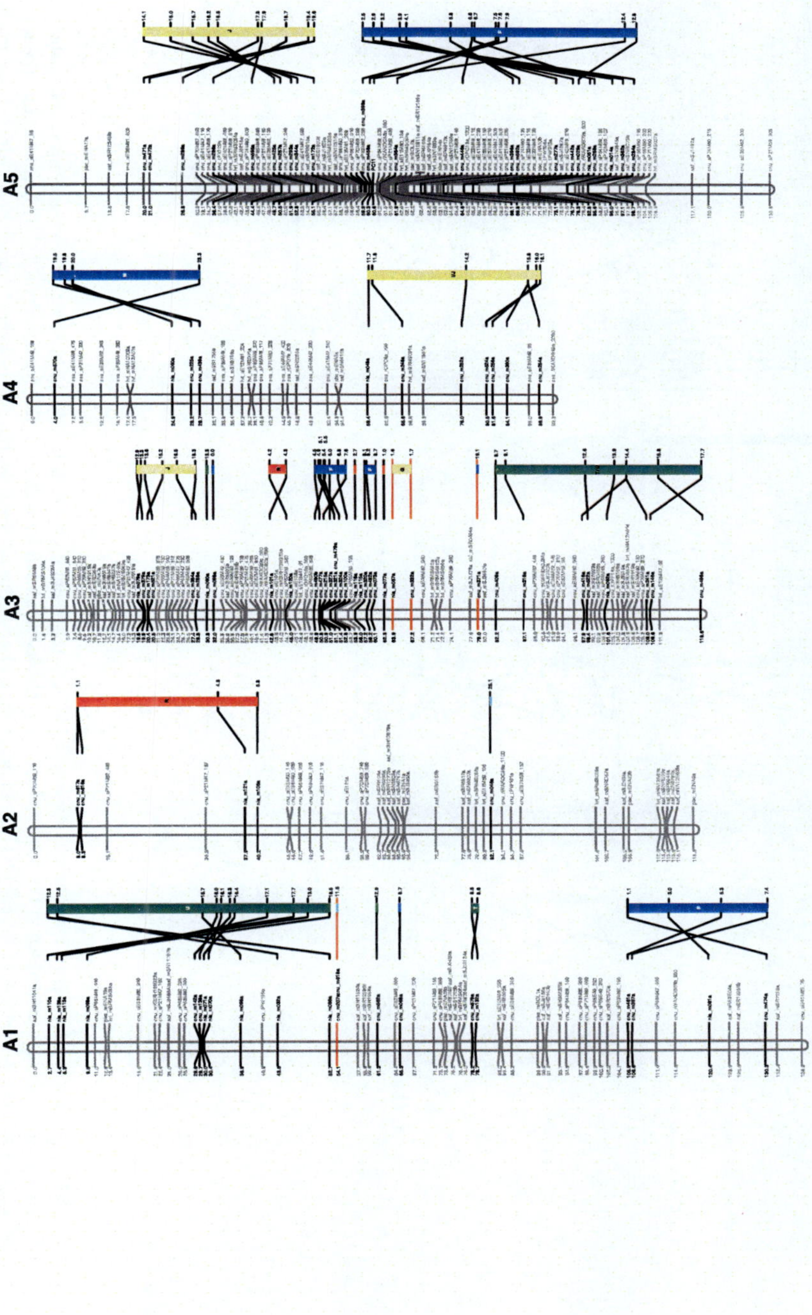

Fig. 8.1 The second generation reference genetic map of *Brassica rapa* developed by Kim et al. (2009). Cumulative recombination distances are shown to the left and marker loci to the right of the linkage groups. SSR markers developed from the BAC sequences are designated in bold strokes. The colored bars to the right of the linkage groups indicate *Arabidopsis* chromosomes (chromosome 1: light blue; chromosome 2: yellow; chromosome 3: dark blue;

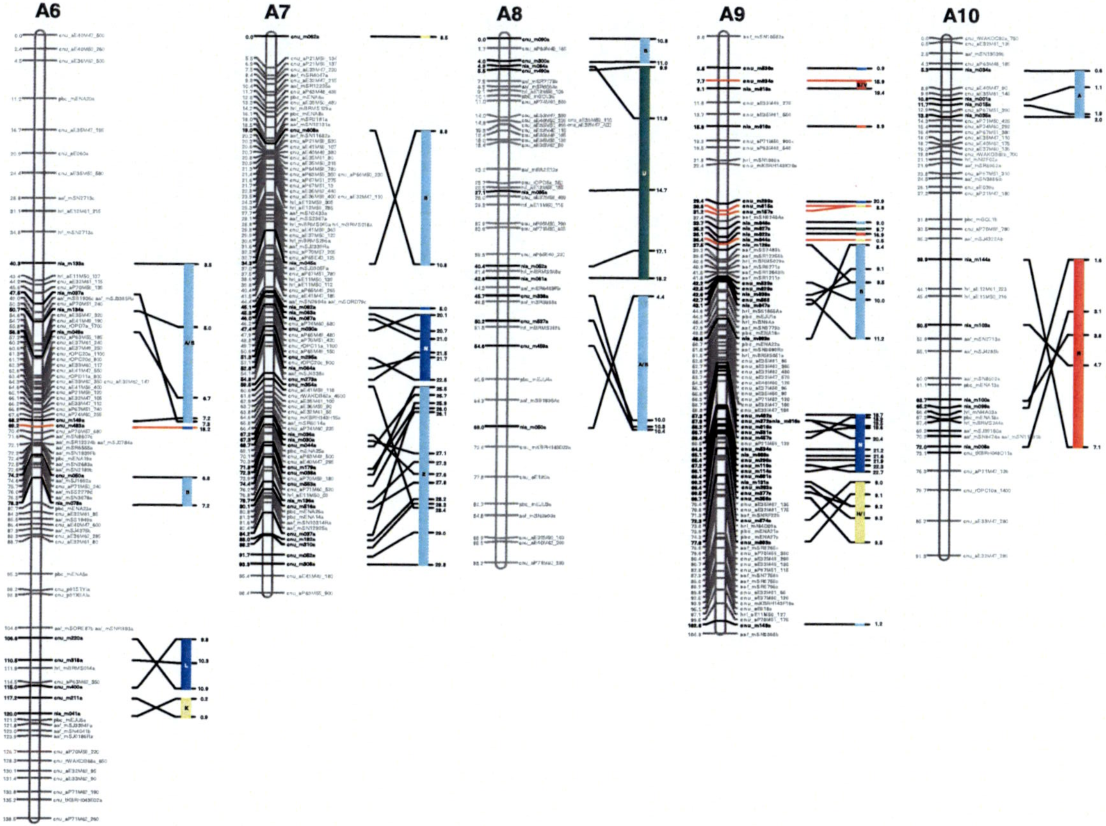

Fig. 8.1 (continued) chromosome 4: green; chromosome 5: red), representing the synteny blocks between the two genomes. The synteny blocks identified by Schranz et al. (2006) are embedded in the colored bars. New blocks identified by Kim et al. (2009), not identified by Schranz et al. (2006) are marked with red lines. The numbers to the right of the colored bars indicate aligned positions on *Arabidopsis* chromosomes in mega base pairs (Mb). Arabidopsis chromosomes are not to scale

8.3.3 Molecular Markers and Trait Genetics

Majority of the traits related to yield; quality, morphology, and biotic and abiotic stress tolerance are inherited quantitatively as they are governed by many genetic loci mapping throughout the genome. Due to many genes governing them in addition to being highly influenced by the environmental conditions, phenotypic selection through visual observation in breeding program is difficult. Recent advances in molecular markers and development of detailed molecular genetics linkage maps are very useful for identifying, tagging, and mapping of genomic regions containing loci for the traits of interest and precise transfer of these traits. This advantage is currently being used in many of the crop plants including *B. rapa* for mapping/tagging of loci for important agronomic and quality traits and marker-assisted transfer of useful genes. Various studies involving mapping of QTL/gene in *B. rapa* for flowering time, morphology, quality, heat resistance, and biotic and abiotic resistance are described in detail below (summary given in Table 8.1).

8.3.3.1 Mapping of Bolting, Flowering, and Vernalizaton Requirement

Flowering trait is important for crop plants for adaptation to different environmental conditions in different regions. Loci involved in flowering time and its related traits have been mapped in *B. rapa* genome by several workers (Song et al. 1995, Teutonico and Osborn 1995, Osborn et al. 1997, Axelsson et al. 2001, Schranz et al. 2002). Song et al. (1995) first mapped flowering-related traits such as days to bud (DB), days to flower (DF), and days from bud to flower (DBF) using 95 F_2 population-derived RFLP linkage map developed from Chinese cabbage cultivar "Michiili" and an accession of Spring broccoli. Five QTL for days to bud (on LG 3A/3R, 6A/6R, 7A/7R, and 8A/8R) and one QTL for days to flower on linkage group 3A (3R) were detected. On the same region containing QTL for DB and DF, one QTL for DBF was also observed. However, the major QTL for DB and DF are located between marker loci 113 and 102 and 200a and 29a, respectively, suggesting that different major genes controlling DB and DF traits are located in different region of the same chromosome.

Brassica cultivars are classified as biennial or annual based on their requirement for a period of cold treatment to induce flowering and the flowering traits are influenced both by genetical and environmental factors. Teutonico and Osborn (1995) using F_2 and F_3 populations developed from a cross between Per (biennial) and R500 (annual) oilseed *B. rapa* cultivars and RFLP map, identified two major QTL one each on LG2 and LG8 for days to flowering time (DTF) flanked by RFLP markers *COR6.6a-wg3h2a* on LG 2 and *ec5f3-ec5a7a* on LG 8, respectively. These QTL explained 44.6 and 21.7% of the phenotypic variation for DTF trait. Comparison of these QTL regions with common RFLP markers mapped for flowering time in *B. napus* (Ferreira et al. 1995) showed that *B. rapa* LG 2 was homologous to LG 9 of *B. napus*, harboring the largest effect flowering time QTL and vernalization requirement and LG 12 of *B. napus* was homologues with LG8 of *B. rapa*. Later, these two QTL were designated *VFR1* and *VFR2*, respectively (Osborn et al. 1997).

It was observed that *VFR1* QTL region was homologous to *Arabidopsis chromosome* 5 containing *FLC, FY,* and *Co* genes for flowering while *VRF2* corresponded to the *Arabidopsis* chromosome 4 region containing *FRI* locus (Osborn et al. 1997).

Kole et al. (2001) used BC_3S_1 population derived from the same parental cross to precisely determine the allelic effect and map position of *VFR2*, identified in the above two studies. Homozygous genotypic classes for *VFR2* locus differed by 43 and 95 days to flowering in the field and growth chamber, respectively; while the heterozygous for this locus was intermediate, suggesting additive gene action. Further analysis with RFLP markers and the *Arabidopsis FLC* cDNA as marker showed perfect co-segregation of *VFR2* with *FLC* and not with *Co*.

The first cloning of genes involved in *B. rapa* flowering time was done by Schranz et al. (2002). Four *B. rapa* homologs (*BrFLC*) of the MADS-box flowering-time regulator *FLC* were cloned and genetically mapped in *B. rapa*. *BrFLC1* was mapped to QTL region on R10 reported by Kole et al. (2001), while *BrFLC2* was mapped using 78 BC_3S_1 within confidence intervals of flowering-time QTL on R2. *BrFLC3* and *BrFLC5* were mapped onto the same linkage group R03, 100 BC_1S_1 population derived from original parental cross Per $\times$ R500. Study of main and interaction effect of two major flowering time loci *FR1* on R02 and *VFR2* on R10 using 326 F_2 lines derived from crossing two BC_3S_1 homozygous plants (*fr1/fr1,VFR2/VFR2* $\times$ *FR1/R1,vfr2/vfr2*) showed 87% phenotypic variation for flowering time. Ninety-eight percent of this genetic variation was due to the individual additive effects of *BrFLC1* (72.2%) and *BrFLC2* (25.4%), similar to the results for the populations with each gene segregating alone. Dominance at *BrFLC1* was significant in the F_2 population, as were some of the epistatic interactions.

Recently, Lou et al. (2008) mapped QTL for flowering time in four different populations developed from wide crosses between *B. rapa* accessions. The $F_{2/3}$ (RC-CC) population was produced from a cross between a rapid cycling line RC-144 and a vegetable-type Chinese cabbage line CC-156. Double haploid populations DH-38 and DH-30 were developed from crosses between the oil-type Yellow sarson YS-143, and the vegetable types Pak choi PC-175 and vegetable turnip VT-115, respectively. The backcross (BC_1) population of 136 plants [(VT- 115 XYS-143) XVT-115] was developed from a cross between one F_1 plant (VT-1153xYS-143), and one plant of parental accession VT- 115. 178 $F_{2/3}$ plants, 71 lines from population DH-38, 64 lines from population DH-30, and 136 BC_1 plants were used for flowering time and other traits evaluation. A total of eight QTL, three in $F_{2/3}$, three in DH-38, four in DH-30, and one in BC_1 were identified in four populations in seven linkage groups. However, out of these many, one major FLQTL-2 (17.7–59.3%) on R02 was detected in all the four populations. In general the phenotypic variation explained by QTL loci in $F_{2/3}$ population was lower (8.9–24.6%) than the phenotypic variation explained in DH populations (13.4–59.3%).

8.3.3.2 Mapping of Plant Height

Plant height is an important trait in oilseed brassicas. Breeding of reduced plant height is necessary in some oilseed brassicas such as in *B. juncea* and in *B. napus* because most of the cultivars are tall in nature which are prone to lodging and leading to yield losses and difficulty in harvesting. Song et al. (1995) mapped QTL for plant height at vegetative stage and first flower stage in *B. rapa* and detected two QTL on linkage groups 4A and 9A, and one QTL on linkage group 5A for former and later trait. Two and three QTL were also found for stem length (SL) and the ratio of length to diameter (SI), respectively.

Dwarf genes have been found to be associated with increased yields, higher fertility, early maturity, and high tillering capacity and this potential has been extensively used in rice (semi dwarf gene, *sd-1*) and in wheat (*Rht-B1b* and *Rht-D1b*) breeding leading to "Green Revolution" in late 1960s and early 1970s (Hedden 2003, Khush 2001). It has been shown that dwarf or reduced height is due to mutation in genes in the biosynthesis pathway or response pathway of gibberellic acid (GAs) which causes GA deficiency and dwarf phenotypes, and exogenous GA application can restore wild-type phenotype in these mutants. Several genes associated with GA biosynthesis and response pathway have been characterized in *A. thaliana.* Muangprom and Osborn (2004) evaluated phenotypic effect and genetic characteristics of dwarf gene *dwf2* in *B. rapa* and found that *dwf2* was insensitive to exogenous GA3 and controlled by semi-dominant single gene. Near-isogenic lines that were homozygous or heterozygous for *dwf2* had 47.4 or 30.0% reduction in plant height, respectively, and had the same or significantly higher numbers of primary branches than the wild-type line. They mapped *DWF2* gene to the bottom of linkage group R6 in a region having homology to the top of *A. thaliana* chromosome 2 containing *RGA* (repressor of *ga1*-3) gene. This novel dwarf mutant gene from *B. rapa* (*Brrga1*-d) caused by substitution of a conserved amino acid in the C-terminal domain of a DELLA protein gene was later cloned and characterized by Muangprom et al. (2005). *Brrga1-d*, retained its repressor function and accumulates to high levels but it did not show interaction with a protein component required for degradation, suggesting that the mutated amino acid causes dwarfism by preventing an interaction needed for its degradation. Transformation of *Arabidopsis* and oilseed rape (*B. napus*) with this novel gene confers non deleterious dwarf phenotypes, indicating its potential usefulness in other crop species. In another experiment Lou et al. (2008) identified three QTL affecting plant height on R02, R03, and R07 in the RC-CC $F_{2/3}$ population derived from a cross between a rapid cycling line RC-144 and a vegetable-type Chinese cabbage line CC-156, explaining 23.9, 15.7, and 8.9% of the phenotypic, variation, respectively. Information derived from above studies would be useful in isolation and mapping of dwarf gene from related *Brassica species* and dwarf brassica breeding.

8.3.3.3 Mapping of Root Traits

B. rapa is known for its wide diversity of vegetable and oleiferous forms including distinct and enlarged root of turnip (*B. rapa* ssp. *rapifera*) and non-enlarged root of Chinese cabbage (*B. rapa* L. ssp. *chinensis* (L.) Hanelt) and oleiferous forms. Turnip (*Brassica rapa* L. ssp. *rapifera*) is a major vegetable within the genus *Brassica*, commonly grown in temperate climates for its succulent storage root especially in Europe and East Asia (Chen 2001). Previous studies reported that the development of storage root is a complex trait and influenced by interaction of environmental, genetic, and physiological factors besides depending upon supply of sucrose and growth regulators from the shoots (Gupta et al. 2001, Rouhier and Usuda 2001). Genetic analysis further showed the complex and continuous inheritance pattern of root traits indicating the involvement of several genes (Beebe et al. 2006). Recently, genetic analysis and QTL mapping using DNA markers *B. rapa* ssp. *rapifera* root traits has been given attention. Lou et al. (2008) mapped QTL for turnip width, turnip weight, and turnip length in doubled haploid (DH-30) population (derived from a cross VT-115 XYS-143) and backcrossed population derived from a cross [(VT-115XYS-143) × VT-115]. A major QTL for each of the turnip traits was detected on top of the R02 which co-localized with the major flowering time QTL. The QTL TuQTL-1 explained about 24.0 and 36.7–40.0% of phenotypic variation in BC_1 and in DH-30 populations, respectively. In another experiment, Lu et al. (2008) used $F_{2/3}$ populations derived from a cross between Chinese cabbage (*B. rapa* ssp. *chinensis*) and turnip (*B. rapa* ssp. *rapifera*), which differed greatly in root characters. QTL mapping was done using the AFLP and RAPD markers map developed from F_2 population and by taking the trait value of F_3 families from each of the F_2 lines for taproot thickness, taproot length, and taproot weight. A total of 18 QTL were detected for the three tap root traits studied, of which seven QTL were for taproot thickness, five QTL were for taproot length, and six QTLs for taproot weight. The phenotypic variation explained by individual QTL ranged from 8.4 to 27.4%. The four major QTLs, *qTRT4b*for taproot thickness and *qTRW4* for taproot weight, explained 27.4 and 24.8% of the total phenotypic variance, respectively.

8.3.3.4 Mapping of Agronomic and Morphological Traits

B. rapa is well known for its wide variation for morphological traits which are grown as vegetables, oilseeds, and fodder. Genetic analysis and identification of chromosomal regions containing these morphological traits would be helpful in breeding for desired traits. Song et al. (1995) in an F_2 population derived from a cross between Chinese cabbage cv "Michihili" and Spring broccoli analyzed for variation in leaf, stem, and flowering characteristics. Genetic analysis and QTL mapping identified single dominant major gene for leaf pubescence and leaf lobes. Mapping of QTL related to leaf petiole morphology such as petiole length, petiole width (PW), and petiole thickness (PT) indicated that several minor and major genes governed these traits. Two QTL were detected for leaf length (LL) located on linkage groups 9R (9A) and 10R (10A). Leaf lamina length (LAL) and ratio of lamina length to width (LI) were found to be controlled mainly by a major gene at the *Lob* locus and a minor QTL located on the linkage group 7R(7A). QTL mapping for 12 morphological traits identified 51 QTL by Zhang (1999) while Yu et al. (2003b) identified 50 putative QTL for different traits. These includes, five QTL each for plant growth and plant diameter, seven QTL for leaf length, four QTL for leaf width, six QTL for leaf length/width ratio, seven QTL for petiole length, and four QTL for petiole width. Recently, Lou et al. (2007) used multiple segregating populations derived from parental crosses involving the three main groups of *B. rapa*, i.e., the oleiferous, leafy, and turnip types for genetic dissection of morphological traits. A total of 20 morphological traits including flowering time, six seed-related traits (seed pod length, seed pod width, beak length, seed color, seed weight), two growth-related traits (plant height, branches), six leaf-related traits (leaf edge shape, leaf trichomes, leaf number, leaf lobes, lamina length, lamina width, petiole length, leaf area, leaf index), and six turnip-related traits (turnip formation, turnip shoots, turnip length, turnip width, turnip weight) were studied. QTL mapping of these traits in three populations identified four QTL for seedpod traits and one QTL for seed coat color (SCQTL-1) in the two DH populations, and two QTL for seed weight in the RC-CC $F_{2/3}$ population. The proportion of total variation explained by each QTL ranged from 11.1 to 38.1% for seed pod traits and 10.2 – 17.6% for seed weight. One QTL flanked by two markers, E33M51-7CC and BRMS037 on R01, was detected for the number of leaf trichomes. Ten QTL distributing over seven linkage groups explaining phenotypic variation ranging from 6.5 to 26.4% were detected in the $F_{2/3}$ and DH populations. In these seven genomic regions two or more leaf traits were detected, e.g., lamina width co-segregated with other leaf traits. The QTL, QTL-1 on R02, LQTL-3 on R03, LQTL-4 on R05, LQTL-5 on R06, and LQTL-7 on R07 were detected in multiple populations, related to multiple traits, and appeared to be the major QTL affecting leaf size. LQTL-2 on R02 and LQTL-6 on R06 affected only leaf edge shape and represent loci for leaf serration.

8.3.3.5 Genetics and Mapping of Seed Coat Color

Along with yield improvement in terms of quantity, quality of brassica oilseeds is important. Most of the cultivated *B. rapa* is brown seeded while for commercial purpose, oilseed *B. rapa* with brown seeded is not preferred due to dark color in oil. Yellow seeds have a thinner seed coat than black seeds, thereby leading to lower hull proportion and higher oil and protein content in *Brassica* crops. Development of pure yellow-seeded oilseed *Brassica* cultivars is an important breeding objective, because yellow seeds have increased oil and protein content due to the thinner seed coat (Abraham and Bhatia 1986, Downey and Robbelen 1989). Genetic control of seed coat color is reported to be maternally inherited and controlled by one, two, and more than two genes, respectively, depending on the genotypes of the parents used in the genetic study (Ahmed and Zuberi 1971, Teutonico and Osborn 1994, Stringam 1980). The fist mapping of seed coat color gene in *B. rapa* was done by Teutonico and Osborn (1994) in an F_3 population derived from a cross between biennial Per with annual R500 cultivar. Genetic study using F_3 lines showed 3:1 segregation of brown and yellow color suggesting segregation of alleles at one locus. Linkage analysis with RFLP markers showed mapping of yellow seed coat color (YI) onto linkage group 5 between the markers *m456b* and *ec3c8b*. Chen et al. (1997) using *B. campestris-albograbra* additions lines linked RAPD markers with seed coat color. For seed coat color study they used, 20 lines each from BC_3 and BC_4 generations and found one RAPD marker out of 19 *albograbra* chromosome 1 specific markers to be linked with yellow seed coat color in addition lines. Further analysis could locate the locus on the terminal region of the alien chromosome.

Recently, Zhang et al. (2008) in a study using Chinese cabbage DH lines derived from a cross between a hairy, black-seeded DH line, "Y177-12", and a glabrous, yellow-seeded DH line, "Y195-93", reported cloning of a gene involved in glabrous and yellow seed color traits. Initial bulk segregant analysis using 1,190 sequence-related amplified polymorphism (SRAP) markers in four DNA bulks from 16 (4×4) glabrous and yellow-seeded DH lines and four others from 16 (4×4) hairy and black-seeded lines could identify only 48 to be polymorphic. These 48 SRAP markers were later used to amplify in 16 glabrous and yellow seeded and 16 hairy and black-seeded DH lines, of which only 13 SRAP could further be found polymorphic between these bulks. Finally, SRAP marker YR431 was found to be closely linked with both traits and converted to Sequenced Characterized Amplified Region (SCAR) marker. BAC clone containing this marker was sequenced and homology search identified a candidate *Brassica* ortholog of *TRANSPARENT TESTA GLABRA 1* (*TTG1*). Further sequence analysis of the alleles from hairless, yellow-seeded and hairy, black-seeded DH lines in *B. rapa* showed that a 94-base deletion was found in the hairless, yellow-seeded DH lines. Transformation of an *Arabidopsis ttg1* mutant with both the *TTG1* homologs from the black and yellow-seeded *B. rapa* lines showed that the *TTG1* homolog from the black-seeded *B. rapa* recovered the *Arabidopsis ttg1* mutant, while the yellow-seeded homolog did not, suggesting that the deletion in the *Brassica TTG1* homolog had led to the yellow-seeded natural mutant. This is the first report of cloning seed coat color gene in *B. rapa* which also simultaneously controls hairiness trait.

8.3.3.6 Mapping of Anthocyanin Pigmentation

Anthocyanins are known for protection of plants from photoinhibition, UVB light, and for modification of captured light quality and quantity (Barker et al. 1996, Klaper et al. 1996, Dodd et al. 1998). They are flavonoid pigments which give purple color to plant leaves and stems besides playing a role in metal accumulation. In *A. thaliana*, genes related to anthocyanin pigments have been identified and characterized. Among the genes identified are, *ANTHOCYANINLESS1* (*ANL1*), responsible for anthocyanin production; *ANTHOCYANINLESS2* (*ANL2*), gene affecting anthocyanin distribution; *ANTHOCYANIN11* (AT1G12910), contributing to anthocyanin production; *TRANSPARENT TESTA 9* (TT9), which is involved in flavonoid biosynthesis, and three pigmentation genes, i.e., AT1G56650, AT5G13930, and LAB (Kubo et al. 1999). Rapid cycling *B. rapa* anthocyaninless mutants don't produce purple color. Anthocyaninless trait in *B. rapa* is recessive and easily scorable since it is expressed at all stages of life cycle (Burdzinski and Wendell 2007). They used segregating population from three-generation pedigrees to assess linkage between the *anl* locus and the microsatellite markers derived from a cross between rapid cycling *B. rapa*, a true breeding anthocyaninless (*anl*) strain, and a true-breeding purple strain. Genetic analysis in F_2 population showed monogenic inheritance of this trait. Linkage analysis using 22 polymorphic microsatellite markers identified three markers namely *Bn9A, BRMS-024b*, and *Ra2-G05* to be associated with *anl* locus with more than five LOD value and forming a linkage group of 46.9 cM. Based on previous map position of *Bn9A* this *anl* containing linkage group is identified to be *B. rapa* linkage group R9 (A9). Based on comparative alignment between *B. rapa* chromosomes 1 and *Arabidopsis* chromosome 1, the presence of *Arabidopsis* ortholog of anthocyanin pigment gene *AN11* has been predicted on *B. rapa*.

8.3.3.7 Mapping of Self-Incompatibility

Most of the *B. rapa* cultivars are cross-pollinated because of sporophytically controlled self-incompatible (SI) nature which is caused by *S locus* with multiple alleles. Till date around 30 alleles have been identified in *B. rapa* (Nou et al. 1993). Two highly polymorphic *S* locus genes, *SLG* (*S* locus glycoprotein) and *SRK* (*S* receptor kinase), have been identified, both of which are expressed predominantly in the stigmatic papillar cell. *SRK* is the determinant of the *S* haplotype specificity of the stigma. Development of S locus-specific DNA marker is useful for practical *B. rapa* hybrid breeding through marker-assisted selection. Nishio et al. (1994) developed a simple method of detecting polymorphism in S locus glycoprotein gene, *SLG,* in Chinese cabbage and cabbage inbred lines using PCR RFLP for *S6* and *S2 SLG*. PCR-RFLP of *SLG* proved to be useful for identification of breeding lines as well as for S allele identification in cruciferous vegetables and in F_1 hybrid the PCVR-RFLP showed sum of both the parental bands there by increasing the identification of F_1 hybrid and purity of seeds. Nozaki et al. (1997) in a F_2 population derived from a cross between Chinese cabbage inbred line (*B. campestris* L. var. *pekinensis*, 2n = 20) and an inbred line of Mizu-na (*B. campestris* L. var. *japonica* 2n = 20) analyzed the association of RAPD and isozyme markers with the self-incompatibility (the S-glycoprotein) and the NS-glycoprotein locus. QTL mapping and linkage analysis detected self-incompatibility S-glycoprotein QTL on LG2 and a RAPD marker, F09-1040 was found to be linked with a recombination value of 20.2% with this locus, while NS-glycoprotein locus was located on linkage LG2 and linked with ACP-1 with a recombination value of 26.2%. Nishio et al. (1996) analyzed polymorphism of class I- and class II-specific *SLG* (the S-locus glycoprotein gene) gene in *B. campestris* by PCR-RFLP using *SLG-specific* primers and southern hybridization using a probe of S^5 *SLG* cDNA, a class-II *SLG*. The high *SLG* polymorphism detected by their investigation suggests the usefulness of the PCR-RFLP method for the identification of S alleles in breeding lines and for listing S alleles in *B. campestris.*

8.3.3.8 Mapping of Embryogenic Ability in Microspore Culture

Microspore culture technique for producing haploid or double haploid is a very important technique and integral part of plant breeding and genetic analysis in many crop species especially in cross-pollinated crops like *B. rapa*, for rapid fixation of genotype combination and production of homozygous inbred lines for hybrid seed production. Presently this technique is being used in producing doubled haploid plants for genetic analysis and mapping. Embryo production in microspore culture is reported to be genotype dependent (Chuong et al. 1987, Takahata 1997). Ajisaka et al. (1999) in F_2 population derived from "Homei 09" (highly responsive to microspore) × "Siloga S2" (low responsive) studied embryogenic ability and RAPD genotype of the each of the F_2 plants and found that the plants carrying "Homei 09" showed a higher ability of embryogenesis than those of "Siloga S2". Several parts of the linkage map showed a conspicuous distortion toward "Homei 09" alleles. Comparison of the distortion with those in the embryogenesis of the F_2 suggested that at least one part of the linkage map affected the embryogenic process in microspore culture. Genes in these areas of "Homei 09" may play a key role in triggering the embryogenesis or in embryo development at the early stage of microspore culture.

In another experiment, genetic analysis of embryogenic ability in Chinese cabbage and *B. napus* was investigated using F_2 lines, and QTL mapping in a microspore-derived (MD) and F_2 populations from the cross between "Ho Mei" (high-responsive parent in microspore embryogenesis) and "269" (low-responsive parent) in Chinese cabbage, and between "Lisandra" (high-responsive parent) and "Kamikita" (low-responsive parent) in oilseed rape. Analysis of segregation distortion showed 27% of the RAPD markers to be distorted in both the crops even though 71–75% of these distorted markers segregated in 3:1 Mendelian ratio in the F_2 population. Association analysis between distorted markers and microspore embryogenesis of the F_2 population showed that seven (OPB15-700, OPB20-1400, OPH06-1200, OPD10-550, OPA13-1200, RA1273C, OPE13-1600) and three (A2-1600, OPA10-1200, OPA13-1200) markers are associated with embryogenic ability in Chinese cabbage and oilseed rape, respectively. These markers showed additive effects on embryo yields and the plants having more alleles of the high-responsive parent produced higher embryo yields.

8.3.3.9 Mapping of Mineral Accumulation

Leaves of vegetable form of *B. rapa* such as Chinese cabbage and bok-choi are consumed as vegetables in China, Korea, and in Japan. *B. rapa* leaves are rich source of essential mineral nutrients and micronutrients besides being the supplement of vitamins (Ma et al. 2007). It would be beneficial to breed high-mineral content in *B. rapa* leaves if we identify the genetic factor underlying mineral accumulation in *B. rapa* vegetables. Wu et al. (2007) in a study involving genotypes of *B. rapa* vegetables found variation for Zn, Fe, and Mn to be 8-, 6-, and 2.5-fold, respectively, and 2-fold variation for shoot or root dry biomass production under Zn-deficient or toxic (excess Zn concentration) conditions. Similar kind of study in *B. oleracea* vegetables showed 2-fold genotypic variation for Ca, Mg, K, Fe, and Zn concentrations and for shoot dry biomass ratio-based Zn efficiency for *B. napus* and *B. juncea* (Grewal et al. 1997, Kopsell et al. 2004, Broadley et al. 2008). Identification of genetic factors controlling mineral accumulation was studied through QTL mapping approach in *A. thaliana* leaves for P, Cs, and K (Bentsink et al. 2003, Payne et al. 2004, Harada and Leigh 2006). In *B. oleracea* also QTL for Ca and Mg concentrations were mapped by Broadley et al. (2008). In a recent study, Wu et al.

(2008) using a DH mapping population of 183 plants derived from a cross involving Chinese cabbage parents Y177, originated from a winter-type Japanese cultivar "Jianchun" and Y195, derived from a summer type Chinese cultivar "Xiayang," studied the genetics of mineral accumulation in leaves and growth response to Zn. Analysis was done for the concentration of 11 elements in leaves, including four essential micro-elements (Fe, Zn, Mn, and Cu), five macro-elements (Na, Ca, K, P, and Mg), and also for a non-essential element (Sr) and a toxic element (Al). Shoot dry Biomass (SDB) under normal, deficient, and excessive Zn nutritional conditions were also investigated in hydroponic conditions. QTL analysis identified ten QTL, each explaining 11.1–17.1% for the Na, Mg, P, Al, Fe, Mn, Zn, and Sr concentration variance, by multiple-QTL model (MQM) mapping. One common QTL was found affecting SDB under normal, deficient, and excessive Zn nutritional conditions while an additional QTL for SDB was identified under Zn excess stress condition.

8.3.3.10 Mapping of Fatty Acid Composition

Brassica species namely *B. napus*, *B. juncea*, *B. rapa*, and *B. carinata* are grown for oil production. Fatty acid composition of seed determines not only the quality and quantity of oil but also physical and chemical properties of oil. The aim of many breeding strategies of spring turnip rape and other oilseed brassicas was the production of high-quality oil by reducing erucic acid (< 2%) and linolenic acid and increasing oleic acid and linoleic acid for longer keeping quality and more thermostable oils. The normal amount of oleic acid in the seed of spring turnip rape is reported to be around 60%, whereas oleic acid concentrations higher than 70% have already been reached in rapeseed (Tanhuanpaa et al. 1995). Most of the cultivars of oilseed *B. rapa* are having high erucic acid in seed oil (upto 50% or more). Content of erucic acid is important for some industrial uses. The inheritance of erucic acid in *B.rapa* has been reported to be governed by single gene as compared to the two genes in amphidiploid *B. juncea* and in *B. napus* (Kirk and Hurlstone 1983, Teutonico and Osborn 1994, Gupta et al. 2004).

Teutonico and Osborn (1994) using F_2 and F_3 populations mapped one locus governing this trait in linkage group 1. The F_2 and F_3 families were derived from parental lines Per (having negligible amount of erucic acid) and R500 (>50% erucic acid). Genetic analysis using the phenotype of F_3 families showed 1:2:1 segregation, suggesting single gene inheritance. Recently, Rahman et al. (2008) developed high-throughput genome-specific and gene-specific molecular markers for erucic acid genes in *B. napus* based on nucleotide differences between high and low erucic containing parents in *Bn-FAE1.1* gene in A genome and *Bn-FAE1.1* located C genome. High-throughput markers were validated in *B. napus* and *B.rapa* germplasm containing high and low erucic acid.

Tanhuanpaa et al. (1996b) in a study involving F_2 population derived from cross involving Jo4002 (low oleic acid) × Jo4072 (high oleic acid) of the *B. rapa* ssp. *oleifera* identified a genetic locus for oleic acid content trait. Bulk segregant analysis using 109 RAPD markers could differentiate the high and low oleic acid containing bulk with only eight RAPD markers. Linkage analysis in F_2 population showed mapping of six of these eight markers in one linkage group that harbors gene for oleic acid content. The co-dominant RAPD marker OPH-17 was found to be the closest marker linked to oleic acid, which was later converted to SCAR markers. This locus was further fine mapped using more markers in the linkage map. A *Brassica* ortholog of the *Arabidopsis FAD2* gene, which encodes an enzyme that desaturates oleic acid to linoleic acid, was cloned, and sequence analysis showed that wild-type and genotypes with high oleic acid differ with respect to only one nucleotide that changes leucine to proline (Tanhuanpaa et al. 1998). Allele-specific markers were designed and mapped to the QTL region of *B. rapa* which was consistent with the QTL corresponding to the *Arabidopsis fad2* gene. The allele-specific markers developed was very effective in selection for plants with high oleic acid content derived from Jo4072 because they were located exactly at the locus and can differentiate between homo- and heterozygotes.

Genetic study of linolenic acid content in 90 F_2 population of turnip rape (*B. rapa* ssp. *oleifera*) derived from low linolenic acid line 93651–2 and a high linolenic acid line Sv3402 revealed three loci for this trait (Tanhuanpaa and Schulman 2002). Bulk segregant analysis was done to identify RAPD markers linked to high and low linolenic acid content. Candidate gene approach was adapted to clone the *B. rapa* homolog of rapeseed *FAD3* gene. Sequence analysis found three nucleotide substitutions in the allele conferring low linolenic acid. A final 27 markers showing polymorphic band between low and high linolenic acid bulk and parental lines was used for linkage analysis and QTL mapping. These 27 markers comprising one *FAD3* RFLP marker, 5 codominant RAPDs, 5 RAPDs with the visible allele derived from 93651–2, and 16 RAPDs were mapped in the three linkage groups LG3 (6 markers), LG9 (7 markers), and LG10(16 markers), respectively. QTL analysis detected 3 QTL, one each on LG3, LG9, and LG10 explaining phenotypic variation of 32.8, 37.6, and 23.8%, respectively. Candidate gene *FAD3* was mapped within the QTL region in LG10. The allele-specific primers designed based on three nucleotide differences between high and low linolenic acid content can be used for marker-assisted selection for this trait in *B. rapa*.

8.3.3.11 Mapping of Glucosinolates Traits

Glucosinolates are sulfur-containing secondary metabolites predominantly found in plants of the order Brassicales/Capparales. Glucosinolates and their breakdown products have been recognized for their effects on plant defense against generalist herbivores, anti-nutritional properties in animals, and for its ability to give distinctive flavor and taste to cruciferous plant products (Mithen 2001, Kliebenstein et al. 2002). Due to its anti-nutritional properties in animals, "double zero" rapeseed, *B. rapa* and *B. juncea* cultivars having low erucic acid (<2%) and low glucosinolates (<20 μM/g of seed) contents have been bred in Canada (Rosa et al. 1997). However, recent studies conducted by several workers have indicated that several glucosinolates hydrolysis products, such as iso-thiocyanates, particularly breakdown products from sulforapahnine that are formed upon tissue damage, have been proved to be an anti-cancer effect when provided in the human diet (Fahey et al. 1998, Mithen et al. 2000, Talalay and Fahey 2001). Glucosinolates are divided into three types depending on the precursor amino acid from which they are derived, i.e., aliphatic derived from methionine, indolyl derived from tryptophan, and aromatic derived from tyrosine or phenylalanine. In brassicas, aliphatic glucosinolates are predominantly found (more than 85%). The biosynthesis of aliphatic glucosinolates in *A. thaliana* has been well characterized (Halkier and Gershenzon 2006). Cloning and characterization of *GSL-Elong* (involved in side chain elongation) and *GSL-ALK* (side chain modification) have been done in *B. oleracea* (Li and Quiros 2002, 2003). Detail genetic analysis of glucosinolates biosynthesis pathway including QTL mapping and validation of glucosinolates QTL with candidate genes have been done in *B. juncea* (Ramchiary et al. 2007, Bisht et al. 2009). However, little is known about the regulation of the biosynthesis and degradation in *B. rapa*.

Recently, Lou et al. (2008) reported mapping of QTL for glucosinolates accumulation in *B. rapa* leaves using two segregating populations, i.e., DH38, derived from a cross between yellow sarson R500 and pak choi variety HK Naibaicai; and DH30, from a cross between yellow sarson R500 and Kairyou Hakata, a Japanese vegetable turnip variety. The maps in two populations were developed using AFLP and SSR markers. QTL analysis identified 16 loci controlling aliphatic glucosinolates accumulation with major QTL on R03 (A3) and R10 (A10), and the sum of these QTL explained 40 and 56% of the phenotypic variation of total leaf aliphatic glucosinolates content in DH38, and 75 and 47% in DH30, respectively (in two seasons). Three loci each controlling total indolic glucosinolate concentration and aromatic glucosinolates concentrations were also identified. They further identified probable candidate genes involved in the glucosinolates biosynthesis pathway through comparative mapping based on *Arabidopsis–B. rapa* synteny and mapping of candidate orthologous genes in *B. rapa* which may account for the identified QTL.

8.3.3.12 Mapping of Abiotic Stress Tolerance

Brassica crops including *B. rapa* during their growing season encounter heat and cold stresses such as high and low temperatures in extreme climates. For example, in Indian subcontinents, oilseed *B. juncea* and *B. rapa* are grown as rainfed crops during winter season. These crops receive dry and hot temperature during later part of life cycle. Vegetable *B. rapa* ssp. Chinese cabbage also grows well in cool climate instead of hot temperature but freezing temperature also interferes with the normal growing of crop. Therefore breeding for heat resistant and freezing tolerance oilseed and vegetable *B. rapa* is very important. Genetic analysis and QTL mapping were done in Chinese cabbage RILs population for heat damage index traits in the seedling stage (Yu et al. 2003c). Five QTL were mapped on LG3, LG8, and LG9, of which phenotypic variation explained by ht-2 was the highest. QTL mapping of acclimatized and non-acclimatized freezing tolerance to different regions was also done and found that most freezing tolerance were due to over dominance gene action. Kole et al. (2002) mapped QTL for winter survival, non-acclimated and acclimated freezing tolerances, and flowering time in DH population of *B. rapa* and RIL population of *B. napus* derived from annual and biennial cross. Evaluation and QTL mapping for winter survival in different winter seasons detected 6 of 16 QTLs in more than one winter. Comparison between *B. rapa* and *B. napus* for winter survival and freezing tolerance QTL positions showed conservation of genetic loci between species in homologous regions. These findings may be useful for cross-species transfer of winter survival and freezing tolerance trait breeding in brassicas. Recently Lee et al. (2008) conducted microarray analysis of *B. rapa* plants treated at different abiotic stresses such as cold (4°C), salt (250 mM NaCl), and drought (air-dry) using KBGP-24 K, microarray chip containing sequence information of about 24,000 unigenes. Of these many unigenes on microarray chip, 417 (1.7%), 202 (0.8%), and 738 (3.1%) were identified as responsive which differently expressed 5-fold or more at least once during a 48-h treatment with cold, salt, and drought, respectively. Further RT-PCR analysis showed that 56 transcription factor genes and 60 signaling pathway genes involved in various transcriptional regulatory mechanisms and common signaling pathway were found expressing together, respectively, under the abiotic stresses in *B. rapa*.

8.3.3.13 Genetics and Mapping of Disease Resistance

Brassica plants are infected by various diseases among which clubroot caused by *Plasmodiophora brassicae*, white rust caused by *A. candida*, and black rot caused by *X. campestris* are more prominent and cause major damage to crucifers including oilseed and vegetables *B. rapa*.

Genetics and Mapping of Clubroot Resistance Genes

Clubroot disease, caused by the obligate plant pathogen *P. brassicae* Wor., is one of the most economically important diseases of *Brassica* crops in the world especially in China, Korea, and Japan. The breeding of CR cultivars of Chinese cabbage (*B. rapa* ssp. *pekenensis*) has been impeded because most cultivars appear to be highly susceptible to clubroot disease (Yoshikawa 1981, Cho et al. 2002). The identification of resistant sources in European fodder turnip (*B. rapa* ssp. *rapifera*) (Karling 1968, Buczacki et al. 1975; Crute et al. 1983, Crisp et al. 1989) allowed the transfer of CR genes from CR European fodder turnip to Chinese cabbage (Yoshikawa 1981). The introgression of CR genes from CR European fodder turnip including Gelria R, Siloga, Debra, and Milan White has provided and broadened the genetic diversity of Chinese cabbage. The European Clubroot Differential (ECD) hosts 01-04 represent the +typical resistance to different physiological pathotypes of *P. brassicae* (Toxopeus and Janssen 1975, Buczacki et al. 1975). Therefore they have been also used as resistant sources in breeding CR cultivars either in Chinese cabbage or in *B. napus*.

In the previous study, at least three independent dominant genes, which conferred differential (race-specific or vertical) resistance to particular pathotypes of *P. brassicae*, were suggested in turnip genotypes (Wit and Van de Weg 1964, Tjallingii 1965, Toxopeus and Janssen 1975, James et al. 1978, Crute et al. 1980). Yoshikawa (1993) demonstrated that clubroot resistance of European fodder turnips, including Siloga, was controlled mainly by a major gene and few minor genes. James et al. (1978) revealed three independent dominant genes conferring resistance in three different *B. rapa* genotypes to race 6 of *P. brassicae.* Crute et al. (1980) also demonstrated that three genes control the resistance in turnip. This suggests that clubroot resistance in *B. rapa* is controlled by several genes independently. Since CR cultivars of Chinese cabbage were released by introducing resistance genes from CR European fodder turnip, it is believed that CR cultivars of Chinese cabbage have the possibility of introducing one to several genes from turnip. The recent published data obtained by marker-trait analysis confirmed that at least eight CR loci are present in *B. rapa* (Suwabe et al. 2004 and 2006, Hirai et al. 2004, Piao et al. 2004, Sakamoto et al. 2008).

Genetic analysis and mapping of CR genes are well studied in *B. rapa*. All eight possible CR genes present in *B. rapa* have been identified through QTL mapping by different research groups using a range of different resistant sources and marker system. Kuginuki et al. (1997) identified three random amplified polymorphic DNA (RAPD) markers linked to a CR locus using Siloga as resistant source and 36 DH lines derived from five F_1 plants. These markers were converted into sequence tagged site (STS) markers (Kikuchi et al. 1999). In subsequent mapping study, this locus was designated *Crr1* (Suwabe et al. 2003). Matsumoto et al. (1998) also mapped the clubroot resistance gene *CRa* to an interval of about 34 cM between two restriction fragment length polymorphism (RFLP) markers on linkage group 3 using ECD02 as a resistant source. Suwabe et al. (2003) identified *Crr1* and *Crr2* two loci originated from Siloga using F_2 population and simple sequence repeat (SSR) marker system. They concluded that these two loci are complementary for clubroot resistance. Upon inoculation of plants with mild and virulent isolates, they found that the plants with homozygous resistance locus showed higher resistance to clubroot than the plants having heterozygous resistance locus. In addition to *Crr1* and *Crr2*, *Crr4* a weak QTL was detected using the same F_2 population (Suwabe et al. 2006). Hirai et al. (2004) identified and mapped a novel locus named *Crr3*, which is originated from CR turnip, Milan White. Another CR locus, *CRb* derived from Gelria R, was mapped between an interval of 3 cM using sequence characterized amplified region (SCAR) markers converted from amplified fragment length polymorphism (AFLP) markers (Piao et al. 2004). Homology analysis of SCAR marker sequences with that of *Arabidopsis* sequence database identified that the region containing *CRb* locus is homologous to the central part of *A. thaliana* chromosome 4 (Fig. 8.2, Piao et al. 2009) which is also corresponding to the region of the *B. rapa* chromosome containing *Crr1* and *Crr2* clubroot resistance QTL. *B. rapa* BAC clones containing these SCAR markers have been identified and location of these BACs on chromosome spreads was seen in clubroot resistance cultivar CR Shinki through Fluorescent in situ hybridization (FISH) (Fig. 8.3, Piao et al. 2009). Co-localization of these markers was seen in *B. rapa* chromosomes. Cloning of *CRb* gene is in progress. This locus is independent of *Crr1, Crr2*, *Crr3*, and *Crr4* (Saito et al. 2006). However, *Crr1*, *Crr2*, and *CRb* are distributed on three different chromosomes in *B. rapa*, R8, R1, and R3, respectively (Suwabe et al. 2006, Saito

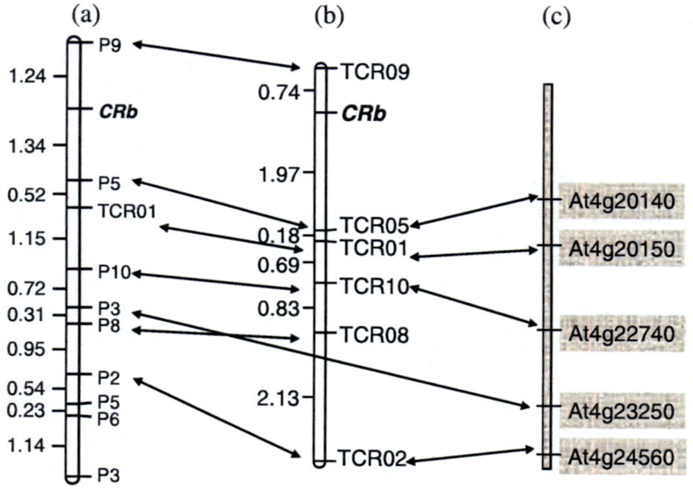

Fig. 8.2 Identification of homologous region of *B. rapa* linkage group containing *CRb* locus with that of *A. thaliana* chromosome 4 (Piao et al. 2004). (**a**) AFLP linkage group containing *CRb* locus, (**b**) linkage group converted to SCARs and CAPS markers, and (**c**) homologous *A. thaliana* chromosome 4. (Adapted from Piao et al. 2009)

et al. 2006). Based on this observation, evolution of CR genes has been suggested in either of the two possible ways: first, clubroot resistance was originally controlled by single major gene in ancestral genome, which later differentiated and diverged as functionally duplicate genes during the course of evolution in *Brassica* genome (Suwabe et al. 2006). Second, the resistance genes for clubroot were originally clustered in that region in ancestral genome which later distributed into different genomic regions following chromosomal rearrangement in *Brassica.* Current *Brassica* species, which diverged at 17–18 million years ago with *Arabidopsis*, are evolutionally believed to be derived from whole genome triplication and rearrangement of one ancestral genome (Lagercrantz 1998, O'Neill and Bancroft 2000, Yang et al. 2006). This hypothesis explains why these CR genes are dispersed and located on the different chromosome of *B. rapa*. Recently, Sakamoto et al. (2008) identified *CRk* and *CRc* loci derived from Debra using QTL analysis of two different F_2 populations and two different isolates. *CRk* is independent of the CR genes *Crr1*, *Crr2*, *CRa*, and *CRb*, but have the same QTL region with *Crr3* (Sakamoto et al. 2008). Totally, eight CR loci were mapped and distributed on the five different chromosomes of *B. rapa*. *Crr1 Crr2*, *Crr3*, *Crr4*, and *CRc* are mapped to R08, R01, R03, R06, R02 of the internationally agreed *B. rapa* reference genetic map, respectively. It is noteworthy that *CRa*, *CRb*, and *CRk* with *Crr3* are mapped on the same linkage group of R03, but they are not located on the same region, except for *CRk* and *Crr3*.

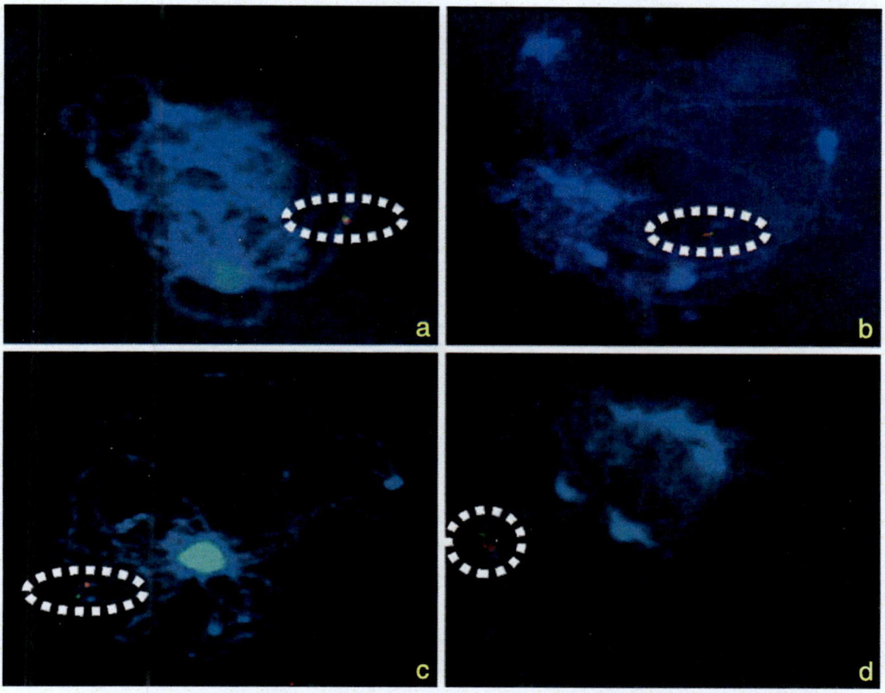

Fig. 8.3 FISH localization of BAC clones containing SCAR markers (SCAR05 and SCAR09) linked to clubroot resistance locus in *B. rapa* CR Shinki genome, (**a**) BAC clones KBrH060E03 (*green*) and KBrH097J16 (*red*) from same contig harboring SCAR05 marker showing localization in the same position in *B. rapa* CR Shinki chromosome 2, (**b**) BAC clones KBrH083J12 (*green*) and KBrH103M15 (*red*) from same contig harboring SCAR09 marker showing localization in same region, (**c**) BAC clones KBrH97J16 (*green*, harboring TCR05) and KBrH88B11 (*red*, harboring TCR09) from different contig linked to SCAR05 and SCAR09 markers showing localization in different region of the same chromosome, (**d**) BAC clones KBrH115F22 (*green*, harboring TCR05) and KBrH144K19 (*red*, harboring TCR09) from different contig linked to SCAR05 and SCAR09 markers showing localization in different region of the same chromosome. (Adapted from Piao et al. 2009)

Mapping of White Rust Resistance

White rust, caused by *A. candida* is economically important not only to *B. rapa* but also to other crucifers. Genetic analysis of *A. candida* resistance has been studied in different *Brassica* species and reported to be controlled by either single dominant major gene (Ebrahimi et al. 1976, Delwiche and Williams 1974, Tiwari et al. 1988, Kole et al. 1996), two or duplicate dominant genes (Verma and Bhowmik 1989, Santos et al. 2006) or minor genes (Edwards and Williams 1987, Kole et al. 1996, Edwards and Williams 1987). In *B. rapa* and other crucifers such as *B. juncea*, *B. napus*, and *B. carinata* race 2 (AC2) and race 7 (AC) are more common (Verma et al. 1975, Pound and Williams 1963, Petrie 1988, Rimmer et al. 2000, Ebrahimi et al. 1976, Delwiche and Williams 1974, 1981, Tiwari et al. 1988). Kole et al.

(1996) in a study using segregating F_2 and F_3 lines derived from a cross between Per (resistant) and R500 (susceptible) reported that resistance to *A. candida* race 2 (AC2) is governed by a single dominant major gene. Linkage analysis using 144 RFLP markers could map the resistance locus (*AC1*) on LG4 between the RFLP markers *ec2b3a* and *wg6cla*. In another experiment, genes for resistance to race (AC2) and race 7(AC7) of white rust (*A. candida*) in oilseed *B. rapa* were mapped using a recombinant inbred (RI) population and a genetic linkage map consisting of 144 restriction fragment length polymorphism (RFLP) markers and 3 phenotypic markers (Kole et al. 2002).

Phenotypic scoring as a measure of interaction phenotype (IP) on a 0–9 scale showed nearly identical IP for the two races and the population showed single major gene inheritance for this trait in young seedlings. QTL mapping detected the same major resistance locus for both races along with a second minor QTL effect for AC2 on linkage group 2 suggesting either a dominant allele at a single locus (*Aca1*) or two tightly linked loci controlling seedling resistance to both races of white rust in the biennial turnip rape cultivar Per. The map positions of white rust resistance genes in *B. rapa* were compared to the map positions in *B. napus* and finally to the physical map of the *Arabidopsis* genome to identify regions to target for comparative fine mapping using this model organism. Tanhuanpaa (2004) mapped a QTL explaining 18.4% of phenotypic variation for white rust resistance to race 7a and race 7v in F_2 population derived from a cross involving parents Bor4206 (susceptible) and Bor4109 (a Finnish resistant line). The white rust-resistant QTL was located in LG2 nearby the RAPD marker Z19a. Linkage analysis using resistance gene analogs for white rust resistance did not detect any association.

Mapping of Black Rot Resistance

Black rot of crucifers is a bacterial disease caused by *X. campestris* pv. *campestris* (Pammel) Dowson. The symptoms of black rot include marginal leaf chlorosis, necrosis, and darkening of leaf veins and vascular tissue within the stem. And finally wilting and necrosis in advanced stage of the disease development. Although disease has been reported to be more damaging in *B. oleracea*, turnip, turnip greens, and Chinese cabbage of *B. rapa* have been reported to be more seriously infected by this disease (Williams 1980, Schaad and Thaveeschai 1983, Ignatov et al. 2000). At least six races of *X. campestris* have been identified of which race1 and race 4 are reported to be more destructive in *B. oleracea* and in *B. rapa*. Relatively broad spectrum resistance plants compared to *B. oleracea* have been reported in *B. rapa* (Taylor et al. 2002). Soengas et al. (2007) studied the genetics of broad spectrum resistance in the *B. rapa* Chinese cabbage F_2 population derived from a cross involving broad spectrum resistant line B162 and a susceptible line R-o-18 to races 1 and 4 of the pathogen *X. campestris* pv. *campestris*. Phenotypic evaluation in 125 F_2 plants was done for the percentage of inoculation sites in which symptoms developed and the severity of symptoms per plant. The two measures of resistance were correlated and the QTL analysis showed detection of four QTL for two traits which were clustered on R06 (A06). The QTL detected on R06 (A06) were major (significant) QTL explaining about 24–64% phenotypic variation. QTL mapping using phenotypic data of F_2 plants detected two additional QTL for resistance to race 4 on R02 (A02) and R9 (A09) respectively.

Mapping of Turnip Mosaic Virus Resistance

A wide range of cultivated crops including cultivated *Brassica* species are infected by Turnip mosaic virus (TuMV; genus *Potyvirus*) and succumbed to major economic losses (Edwardson and Christie 1991, Shattuck 1992). Previous genetic analyses have identified dominant resistance genes for TuMV such as TuRB01, TuRB03, TuRB05 in *B. napus* (Walsh et al. 1999, Hughes et al. 2003), and TuRB01b in *B. rapa* (Rusholme et al. 2007). Further study by many workers revealed that avirulent determinant of resistance genes TuRB01, TuRB01b, and TuRB04 was due to cytoplasmic inclusion protein (Jenner et al. 2000, 2002, Walsh and Jenner 2002) while, for resistance genes TuRB03 and TuRB05, the viral avirulence determinant is due to P3 protein (Jenner et al. 2002, 2003). AFLP markers linked to TuMV resistance gene have been identified by Han et al. (2004) in a study using F_2 population derived from a cross between Brp0058 and Brp0181.

Genetic analysis and mapping of TuMV resistance genes were conducted using 120 first back-cross (BC_1) and their selfed progenies (BC_1S_1) derived from a cross involving *B. rapa* line RLR22 resistant to eight diverse turnip mosaic virus (TuMV) isolates and susceptible line R-o-18. The existence of two loci controlling resistance to TuMV isolate CDN 1 was established from contrasting patterns of segregation for resistance and susceptibility in the B1S1 families and identified two genes for resistance, i.e., recessive TuMV resistance 01 (*retr01*), had a recessive allele for resistance, was located on the upper portion of chromosome A4 while the second gene conditional TuMV resistance 01 (*ConTR01*), possessed a dominant allele for resistance and was located on the upper portion of chromosome A8. The first gene was epistatic to second gene and both the genes were found to control resistance to TuMV isolate CZE 1 suggesting the broad-spectrum resistance of RLR22. Further analysis confirmed that ConTR01, might be one {*eIF*(*iso*)4E} of the three eukaryotic initiation factor 4E (*eIF4E*) loci of *B. rapa* and recessive resistance gene retr01 might be one of the three loci of *eIF*(*iso*)4E in the A genome of *B. napus* derived from *B. rapa*. This study is the first to identify recessive gene (*retr01*) giving resistance to TuMV.

8.4 Comparative Mapping and Identification of Candidate Genes for Important Traits

Recent availability of cross-species transferable, robust molecular markers and the whole genome sequence information of *Arabidopsis* have been used for comparative genome analysis of *Brassica* species. Comparative mapping within and between Brassicas and *Arabidopsis* has revealed conservation of gene order in small chromosomal blocks despite inversions and large-scale deletions (Cavell et al. 1998, O'Neill and Bancroft 2000, Parkin et al. 2005, Panjabi et al. 2008). This might be due to the fact that Brassicas and *A. thaliana* belong to the same family, Brassicaceae, and around 85–90% sequence identity at the exonic region are found (Schmidt 2002). This information has been effectively used to align linkage groups containing important genes in *B. rapa* with that of *A. thaliana* for identification of candidate genes. The first report being the identification of orthologous *B. rapa* candidate gene *FLC* as the *VRF2* QTL by Kole et al. (1997, 2001) which was previously detected by Teutonico and Osborn (1995) and Osborn et al. (1997). Later, Schranz et al. (2002) isolated and validated by mapping three replicated *B. rapa FLC* genes (*BrFLC1, BrFLC2*, and *BrFLC5*) onto the three linkage groups, A2, A3, and A10, respectively, containing QTL for flowering time through candidate gene approach using *A. thaliana* sequence information. Recently, Bisht et al. (2009) through comparative mapping of aliphatic glucosinolates containing QTL regions of *B. juncea* with that of *A. thaliana* identified candidate genes such as *GSL-Elong, GSL-ALK, myb28*, and *myb29* genes that are involved in regulation of aliphatic glucosinolates biosynthesis pathway. Using sequence information of *A. thaliana*, orthologous genes from *B. juncea* was isolated and mapped to glucosinolates QTL regions. This validation experiment showing co-segregation with candidate genes and QTL region proved that comparative mapping and identification of *Arabidopsis* candidate genes is very much helpful in identification, isolation, and mapping of *Brassica* orthologous genes which in turn would be highly beneficial for genetic manipulation and breeding for desired traits in different *Brassica* species. Alignment of chromosomal region containing Clubroot resistance gene in *B. rapa* with that of *Arabidopsis* chromosomes, i.e., *Crr1*, *Crr2*, and *CRb* have been found to be syntenic to the central region of chromosome 4 of *A. thaliana* (Suwabe et al.2006 and Fig. 8.2). Recent study by Jubault et al. (2008) in *A. thaliana* has also identified one QTL loci for clubroot resistance in this region suggesting presence of functionally active candidate gene(s). This region of *Arabidopsis* chromosome has clusters of the disease-resistance genes containing leucine-rich repeats (LRRs) and nucleotide-binding sites (NBSs) motifs. Examples are *RPP* for resistance to *Peronospora parasitica* and *RPS* for a resistance to *Pseudomonas syringae* and *ACD*, acceleration of cell death in response to pathogen infection (Suwabe et al. 2006). It has been suggested that CR genes may be members of these resistance gene clusters. Saito et al. (2006) revealed that the genomic region around *Crr3* exhibits homology to the top of the long arm of *Arabidopsis* chromosome 3, possibly also *CRk* (Sakamoto et al. 2008). Fuchs and Sacristan (1996) mapped CR locus (*RPB1*) in *Arabidopsis* chromosome 1. QTL mapping for partial CR identified two QTL in chromosome 5 in F_2, and four QTL in RIL population, one each in chromosome 1 and 4, and two QTL in chromosome 5 of *A. thaliana* (Jubault et al. 2008). The CR QTL region in chromosome 5 co-localized with the region containing several resistance gene clusters. These resistance genes could be candidate for clubroot disease resistance. So fine mapping and detailed analysis of the expression level of these genes would help in identifying specific genes conferring resistance to clubroot disease.

8.5 Conclusions and Perspectives

Though several genetic analysis and mapping experiments have been done for important agronomic, quality and quantitative traits related to yield and quality in *B. rapa*, detailed genetic analysis and fine mapping to identify tightly linked molecular markers are necessary to make it more useful in breeding program. Out of several QTL mapped, only handful of QTL/gene(s) linked to molecular markers could be used in marker-assisted breeding program. The reason could be accounted for low-resolution mapping, linked markers are far away from the gene(s)/QTL of interest, inconsistency of detection of QTL, and use of non-reproducible marker system among many in addition to non-extensive genetic analysis and QTL mapping experiment in immortal RIL or DH populations in different environments and seasons because most of experiments are done in F_2 populations. The availability of reference genetic maps in *B. rapa* (Choi et al. 2007, Kim et al. 2009) and efficient cross-species transferable co-dominant molecular marker system (SSR) developed from BAC end sequences of *B. rapa* and other *Brassica* species, could be helpful in genetic analysis and mapping of important traits using common marker system. Alignment of linkage maps developed from different segregating population using common marker systems, identification and fine mapping of QTL region to identify candidate genes for traits of interest, and development of tightly linked molecular markers would be greatly helpful in understanding trait genetics and its manipulation in breeding program.

Continuous and rapid progress in development of public genetic markers, mapping populations, integrated genetic and physical maps, development of high-resolution map using EST-derived SSRs and SNP markers, and use of genome sequence information in addition to QTL/gene mapping information would be helpful in near future for breeding desired traits in *B. rapa*. Comparative mapping between *A. thaliana* and *Brassica* relatives could be used extensively for identification and isolation of orthologous candidate genes from *B. rapa* involved in morphology, quality, metabolism, and other important traits. This is being possible and highly advantageous since functions of most of the genes related to growth, morphology, quality, metabolism, and biotic and abiotic stresses are functionally characterized in model crucifer plant *A. thaliana* since its complete genome sequence became available in 2000. Further, near-complete sequence information derived from Multinational *Brassica rapa* Genome Sequencing project could be utilized for genomic studies and trait genetic analysis and manipulation. High-throughput marker genotyping technologies coupled with large-scale expression analysis using *Brassica* micro-array would also help in understanding the genes involved in traits of agronomic, morphology, quality, disease resistance, and abiotic and biotic stress tolerances and exploitation of allelic variation in *B. rapa* breeding.

References

Abraham V, Bhatia CR (1986) Development of strains with yellow seedcoat in Indian mustard *(Brassica juncea* Czern. & Coss.). Plant Breed 97:86–88

Ahmed SU, Zuberi MI (1971) Inheritance of seed coat color in *Brassica campestris*-variety Toria. Crop Sci 11:309–310

Ajisaka H, Kuginuki Y, Hida K, Enomoto S, Hirai M (1995) A linkage map of DNA markers in *Brassica campestris*. Breed Sci 45(Suppl.):195

Ajisaka H, Kuginuki Y, Shiratori M, Ishiguro K, Enomoto S, Hirai M (1999) Mapping of loci affecting the cultural efficiency of microspore culture of *Brassica rapa* L-syn. *campestris* L using DNA polymorphism. Breed Sci 49:187–192

Ajisaka H, Kuginuki Y, Yui S, Enomoto S, Hirai M (2001) Identification and mapping of a quantitative trait locus controlling extreme late bolting in Chinese cabbage (*Brassica rapa* L. ssp. *Pekinensis* syn. *campestris* L.) using bulked segregant analysis *A QTL controlling extreme late bolting in Chinese cabbage*. Euphytica 118:75–81

Axelsson T, Shavorskaya O, Lagercrantz U (2001) Multiple flowering time QTLs within several *Brassica* species could be the result of duplicated copies of one ancestral gene. Genome 44:856–864

Barker DH, Seaton GGR, Robinson SA (1996) Internal and external photoprotection in developing leaves of the CAM plant cotyledon orbiculata. Plant Cell Environ 20:617–620

Beckmann JS, Soller M (1986) Restriction fragment length polymorphisms and genetic improvement of agriculture species. Euphytica 35:111–124

Beebe SE, Rojas-Pierce M, Yan XL, Blair MW, Pedraza F, Munoz F et al (2006) Quantitative trait loci for root architecture traits correlated with phosphorus acquisition in common bean. Crop Sci 46:413–423

Bentsink L, Yuan K, Koornneef M, Vreugdenhil D (2003) The genetics of phytate and phosphate accumulation in seeds and leaves of *Arabidopsis thaliana* using natural variation. Theor Appl Genet 106:1234–1243

Bisht NC, Gupta V, Ramchiary N, Sodhi YS, Mukhopadhyay A, Arumugam N, Pental D, Pradhan AK (2009) Fine mapping of loci involved with glucosinolate biosynthesis in oilseed mustard (*Brassica juncea*) using genomic information from allied species. Theor Appl Genet. 118(3):413–421

Broadley MR, Hammond JP, King GJ, Astley D, Bowen HC, Meacham MC, Mead A, Pink DAC, Teakle GR, Hayden RM, Spracklen WP, White PJ (2008) Shoot calcium (Ca) and magnesium (Mg) concentration differ between subtaxa, are highly heritable, and associated with potentially pleitropic loci in *Brassica oleracea*. Plant Physiol 146:1707–1720

Buczacki ST, Toxopeus H, Mattusch P, Johnston TD, Dixon GR, Hobolth LA (1975) Study of physiological specialization in *Plasmodiophora brassicae*: proposals for attempted rationalization through an international approach. Trans Br Mycol Soc 65:295–303

Burdzinski C, Wendell DL (2007) Mapping the *Anthocyaninless* (*anl*) locus in rapid-cycling *Brassica rapa* (RBr) to linkage group R9. BMC Genet 8:64. doi:10.1186/1471–2156–8–64

Cavell A, Lydiate D, Parkin IA, Dean C, Trick M (1998) Collinearity between 30-centimorgan segment of *Arabidopsis thaliana* chromosome 4 and duplicated regions within the *Brassica napus* genome. Genome 41:62–69

Chen H (2001) Atlas of the traditional vegetables in China, pp 123–124. Zhejiang Science and Technology Publishing House, China

Chen YP, Cao JS, Miao Y, Ye WZ (2000) Analysis of genetic polymorphism in vegetable crops of *Brassica campestris* by RAPD markers. J Zhejiang Univ (Agric Life Sci) 26:131–136. in Chinese

Chen BY, Jizirgensen RB, Cheng BF, Heneen WK (1997) Identification and chromosomal assignment of RAPD markers linked with a gene for seed color in a *Brassica campestris-aLbogabra* addition line. Hereditas 126:133–138

Chevre AM, Delourme R, Eber F, Margale E, Quiros CF, Arus P (1995) Genetic analysis and nomenclature for seven isozyme systems in *Brassica nigra, B. oleracea* and *B. campestris*. Plant Breed 114:473–480

Cho KS, Han YH, Lee JT, Hur EJ, Yang TJ, Woo JG (2002) Pathogenic differentiation of *Plasmodiophora brassicae* and selection of Chinese cabbage cultivars resistant to clubroot disease in highland. Korean J Breed 34(3):168–173

Choi SR, Kim JH, Han TH, Piao ZY, Plaha P, Koo DH, Bang JW, King GJ, Teakle GR, Allender CJ, Lydiate DJ, Beynon E, Osborn TC, Lim YP (2004) Joint meeting of the 14th Crucifer Genetics Workshop and the 4th ISHS Symposium on *Brassicas* 2004, Chungnam National University Daejeon, Korea October 24–28, pp 151

Choi SR, Teakle GR, Plaha P, Kim JH, Allender CJ, Beynon E, Piao ZY, Soengas P, Han TH, King GJ, Barker GC, Hand P, Lydiate DJ, Batley J, Edwards D, Koo DH, Bang JW, Park BS, Lim YP (2007) The reference genetic linkage map for the multinational *Brassica rapa* genome sequencing project. Theor Appl Genet 115:777–792

Chuong PV, Deslauriers C, Kott LS, Beversdorf WD (1987) Effect of donor genotype and bud sampling on microspore culture of *Brassica napus*. Can J Bot 66:1676–1680

Chyi YS, Hoenecke ME, Sernyk JL (1992) A genetic map of restriction fragment length polymorphism loci for *Brassica rapa* (syn. *campestris*). Genome 25:746–757

Crisp P, Crute IR, Sutherland RA, Angell SM, Bloor K, Burgess H, Gordon PL (1989) The exploitation of genetic resources of Brassica oleracea in breeding for resistance to clubroot (*Plasmodiophora brassicae*). Euphytica 42:215–226

Crute IR, Gray AR, Crips P, Buczacki ST (1980) Variation in *Plasmodiophora brassicae* and resistance to clubroot disease in brassicas and allied crops -a critical review. Plant Breed Abstr 50:91–104

Crute IR, Phelps K, Barnes A, Buczacki ST, Crips P (1983) The relationship between genotypes of three Brassica species and collections of *Plasmodiophora brassicae*. Plant Pathol 32:405–420

Das S, Rajagopal J, Bhatia S, Srivastava PS, Lakshmikumaran M (1999) Assessment of genetic variation within *Brassica campestris* cultivars using amplified fragment length polymorphism and random amplification of polymorphic DNA markers. J Biosci 24:433–440

Delwiche PA, Williams PH (1974) Resistance to *Albugo candida* race 2 in *Brassica* species. Proc Am Phytopathol Soc 1:66

Demeke T, Adams RP, Chibbar R (1992) Potential taxonomic use of random amplified polymorphic DNA (RAPD): a case study in *Brassica*. Theor Appl Genet 84:990–994

Denford KE, Vaughan JG (1977) A comparative study of certain seed isoenzymes in the ten chromosome complex of *Brassica campestris* and its allies. Ann Bot 41:411–418

Dodd IC, Critchley C, Woodall GS, Stewart GR (1998) Photoinhibition in different colored juvenile leaves of Syzgium species. Exp Bot 49:1437–1445

Downey RK, Robbelen G (1989) *Brassica* species. In: Oil crops of the world, pp 339–362. McGraw-Hill, New York, NY

Downey RK, Stringam GR, McGregor DI, Stefansson BR (1975) Breeding rapeseed and mustard crops. In: Harapiak JT (ed) Oilseeds and pulse crops in Western Canada-A Symposium, Western Crop Fertizres, Calgary, Alberta, Canada, pp 157–183

Ebrahimi AG, Delwiche PA, Williams PH (1976) Resistance in *Brassica juncea* to *Peronospora parasitica* and *Albugo candida* race 2. Proc Am Phytopathol Soc 3:273

Edwards MD, Williams PH (1987) Selection for minor gene resistance to *Albugo candida* in a rapid cycling population of *Brassica campestris*. Phytopathology 77:527–532

Edwardson JR, Christie RG (1991). The Potyvirus Group. Volumes I–IV, Florida Agricultural Experiment Station Monograph 16

Fahey JW, Stephenson KK, Talalay P (1998) Glucosinolates, myrosinase,and isothiocyanates: three reasons for eating *Brassica* vegetables. ACS Symp Ser 701:16–22

Ferreira ME, Satagopan J, Yandell BS, Williams PH, Osborn TC (1995) Mapping loci controlling vernalization requirement and flowering time in *Brassica napus*. Theor Appl Genet 90:727–732

Fuchs H, Sacristán MD (1996) Identification of a gene in *Arabidopsis thaliana* controlling resistance to clubroot (*Plasmodiophora brassicae*) and characterization of the resistance response. Mol Plant Microbe Interact 9:91–97

Gomez Campo C (1999) Developments in plant genetics and breeding, vol 4. Biology of *Brassica* coenospecies. Elsevier, Amsterdam, Lausanne, New York, NY, Oxford, Shannon, Singapore, Tokyo

Gomez Campo C, Prakash S (1999) Origin and domestication. In: Gomez-Campo C(ed) Biology of *Brassica Coenospecies*. Developments in plant genetics and breeding, vol 4. Elsevier, Amsterdam

Grewal HS, Stangoulis JCR, Potter T, Graham RD (1997) Zinc efficiency of oilseed rape (*Brassica napus* and *B. juncea*) genotypes. Plant Soil 191:123–132

Guo JX, Zhou NY, Ma RC, Cao MQ (2002) Genetic diversity in *Brassica rapa* revealed by AFLP molecular markers. J Agric Biotechnol 10:138–143. in Chinese

Gupta V, Mukhopadhyay A, Arumugam N, Sodhi YS, Pental D, Pradhan AK (2004) Molecular tagging of erucic acid trait in oilseed mustard (*Brassica juncea*) by QTL mapping and single nucleotide polymorphisms in FAE1 gene. Theor Appl Genet 108:743–749

Gupta AK, Singh J, Kaur N (2001) Sink development, sucrose metabolising enzymes and carbohydrate status in turnip (*Brassica rapa* L.). Acta Physiol Plant 23:31–36

Halkier BA, Gershenzon J (2006) Biology and biochemistry of glucosinolates. Ann Rev Plant Biol 57:303–333

Han HP, Sun RF, Li F, Zhang SF, Niu XK (2004) AFLP markers linked to TuMV-resistance gene in Chinese cabbage (*Brassica rapa* L. ssp. *pekinensis*). Sci Agric Sin 37(4):539–544 in Chinese

Harada H, Leigh RA (2006) Genetic mapping of natural variation in potassium concentrations in shoots of *Arabidopsis thaliana.* J Exp Bot 57:953–960

He YT, Tu JX, Fu TD, Li DR, Chen BY (2002) Genetic diversity of germplasm resources of *Brassica campestris* L. in China by RAPD. Acta Agron Sin 28:697–703

Hedden P (2003) The genes of the green revolution. Trends Genet 19:5–9

Hirai M, Harada T, Kubo N, Tsukada M, Suwabe K, Matsumoto S (2004) A novel locus for clubroot resistance in *Brassica rapa* and its linkage markers. Theor Appl Genet 108:639–643

Hughes SL, Hunter PJ, Sharpe AG, Kearsey MJ, Lydiate DJ, Walsh JA (2003) Genetic mapping of a novel turnip mosaic virus resistance gene in *Brassica napus*. Theor Appl Genet 107:1169–1173

Ignatov AN, Kuginuki Y, Suprunova TP, Pozmogova GE, Seitova AM, Dorokhov DB, Hirai M (2000) RAPD markers linked to locus controlling resistance for race 4 of the black rot causative agent, *Xanthomonas campestris* pv. *Campestris* (Pamm.) Dow. in *Brassica rapa* L. Russ J Genet 36:281–283

James RV, Williams PH, Maxwell DP (1978) Inheritance and linkage studies related to resistance in *Brassica campestris* L. to *Plasmodiophora brassicae* race 6. Eucarpia Cruciferae Newsl 3:27

Jenner CE, Sanchez F, Nettleship SB, Foster GD, Ponz F, Walsh JA (2000) The cylindrical inclusion gene of turnip mosaic virus encodes a pathogenic determinant to the *Brassica* resistance gene TuRB01. Mol Plant Microbe Interact 13:1102–1108

Jenner CE, Tomimura K, Ohshima K, Hughes SL, Walsh JA (2002) Mutations in turnip mosaic virus P3 and cylindrical inclusion proteins are separately required to overcome two *Brassica napus* resistance genes. Virology 300:50–59

Jenner CE, Wang X, Tomimura K, Ohshima K, Ponz F, Walsh JA (2003) The dual role of the potyvirus P3 protein of turnip mosaic virus as a symptom and avirulence determinant in brassicas. Mol Plant Microbe Interact 16:777–784

Jubault M, Lariaagon C, Simon M, Delourme R, Manzanares-Dauleux M (2008) Identification of quantitative trait loci controlling partial clubroot resistance in new mapping populations of *Arabidopsis Thaliana*. Theor Appl Genet 117(2):191–202

Karling JS (1968) The Plasmodiophorales: including a complete host index, bibliography, and a description of diseases caused by species of this order, 2nd edn. Hafner, New York, NY

Khush GS (2001) Green revolution: the way forward. Nat Rev Genet 2:815–822

Kikuchi M, Ajisaka H, Kuginuki Y, Hirai M (1999) Conversion of RAPD markers for a clubroot resistance gene of *Brassica rapa* into sequence-tagged Sites (STSs). Breed Sci 49:83–88

Kim HR, Choi SR, Bae J, Hong CP, Lee SY, Hossain MD, Nguyen DV, Jin M, Park BS, Bang JW, Bancroft I, Lim YP (2009) Sequenced BAC anchored reference genetic map that reconciles the ten individual chromosomes of *Brassica rapa. BMC Genomics* 10:432

Kim JS, Chung TY, King GJ, Jin M, Yang TJ, Min YM, Kim H, Park BM (2006) A sequence-tagged linkage map of *Brassica rapa*. Genetics 174:29–39

Kirk JTO, Hurlstone CJ (1983) Variation and inheritance of erucic acid in *Brassica juncea*. Z Pflankzenzucht 90:331–338

Klaper R, Frankel S, Berenbaum MR (1996) Anthocyanin content and UVB sensitivity in *Brassica rapa*. Photochem Photobio 63:811–813

Kliebenstein DJ, Pedersen D, Barker B, Mitchell-Olds T (2002) Comparative analysis of quantitative trait loci controlling glucosinolates, myrosinase and insect resistance in *Arabidopsis thaliana*. Genetics 161:325–332

Kole C, Kole P, Vogelzang R, Osborn TC (1997) Genetic linkage map of a *Brassica rapa* recombinant inbred population. J Heredity 88:553–557

Kole C, Quijada P, Michaels SD, Amasino RM, Osborn TC (2001) Evidence for homology of flowering-time genes *VFR2* from *Brassica rapa* and *FLC* from *Arabidopsis thaliana*. Theor Appl Genet 102:425–430

Kole C, Teutonico R, Mengistu A, Williams PH, Osborn TC (1996) Molecular mapping of a locus controlling resistance to *Albugo candida* in *Brassica rapa*. Phytopath 86:367–369

Kole C, Williams PH, Rimmer SR, Osborn TC (2002) Linkage mapping of genes controlling resistance to white rust (Albugo candida) in *Brassica rapa* (syn. *campestris*) and comparative mapping to *Brassica napus* and *Arabidopsis thaliana*. Genome 45:22–27

Kopsell DE, Kopsell DA, LefsrudM G, Curran-Celentano J (2004) Variability in elemental accumulations among leafy *Brassica oleracea* cultivars and selections. J Plant Nutr 27:1813–1826

Kresovich S, Szewc-McFadden AK, Bilek SM, NcFerson JR (1995) Abundance and characterization of simple sequence repeats SSRs isolated from a size fractionated genomic library of *Brassica napus* L. rapeseed. Theor Appl Genet 91:206–211

Kubo H, Peeters AJM, Aarts MGM, Pereira A, Koornneef M (1999) ANTHOCYANINLESS2, a homeobox gene affecting anthocyanin distribution and root development in Arabidopsis. Plant Cell 11:1217–1226

Kuginuki Y, Ajisaka H, Yui M, Yoshikawa H, Hida K, Hirai M (1997) RAPD markers linked to a clubroot-resistance locus in *Brassica rapa L.* Euphytica 98:149–154

Lagercrantz U (1998) Comparative mapping between *Arabidopsis thaliana* and *Brassica nigra* indicates that *Brassica* genomes have evolved through extensive genome replication accompanied by chromosome fusions and frequent rearrangements. Genetics 150:1217–1228

Lee SC, Lim MH, Kim JA, Lee SI, Kim JS, Jin M, Kwon SJ, Mun JH, Kim YK, Kim HU, Hur Y, Park BS (2008) Transcriptome analysis in *Brassica rapa* under the abiotic stresses using *Brassica* 24 K Oligo Microarray. Mol Cells 26:595–605

Li JW (1981) The origins and evolution of vegetable crops in China. Sci Agric Sin 14:90–95

Li G, Quiros CF (2002) Genetic analysis, expression and molecular characterization of *BoGSL-ELONG*, a major gene involved in the aliphatic glucosinolate pathway of *Brassica* species. Genetics 162:1937–1943

Li G, Quiros CF (2003) In planta side-chain glucosinolate modification in Arabidopsis by introduction of dioxygenase brassica homolog BoGSL-ALK. Theor Appl Genet 106:1116–1121

Lou P, Zhao J, He H, Hanhart C, Pino Del Carpio D, Verkerk R, Custers J, Koornneef M, Bonnema G (2008) Quantitative trait loci for glucosinolate accumulation in *Brassica rapa* leaves. New Phytol 179:1017–1032

Lowe AJ, Moule C, Trick T, Edwards TJ (2004) Efficient large-scale development of microsatellites for marker and mapping applications in brassica crop species. Theor Appl Genet 108:1103–1112

Lu G, Cao JS, Chen H (2002a) Genetic linkage map of *Brassica campestris* L. using AFLP and RAPD markers. J Zhejiang Univ (Sci) 3:600–605

Lu G, Cao JS, Chen H, Xiang X (2002b) QTL mapping of some horticultural traits of Chinese cabbage. Sci Agric Sin 35:969–974. in Chinese

Lu G, Cao J, Yu X, Xiang X, Chen H (2008) Mapping QTLs for root morphological traits in *Brassica rapa* L. based on AFLP and RAPD markers. J Appl Genet 49:23–31

Ma G, Li Y, Jin Y, Du S, Kok FJ, Yang X (2007) Assessment of intake inadequacy and food source of zinc of people in China. Public Health Nutr 10:848–854

Matsumoto E, Yasui C, Ohi M, Tsukuda M (1998) Linkage analysis of RFLP markers fro clubroot resistance and pigmentation in Chinese cabbage (*Brassica rapa* ssp. *pekinensis*). Euphytica 104:79–86

McGrath JM, Quiros CF (1991) Inheritance of isozyme and RFLP markers in *Brassica campestris* and comparison with *B. oleracea*. Theo Appl Genet 82:668–673

Mithen R (2001) Glucosinolates – biochemistry, genetics and biological activity. Plant Growth Reg 34:91–103

Mithen RF, Dekker M, Verkerk R, Rabot S, Johnson IT (2000) The nutritional significance, biosynthesis and bioavailability of glucosinolates in human foods. J Sci Food Agric 80: 967–984

Mizushima U, Tsunoda S (1967) A plant exploration in *Brassica* and allied genera. Tohoku J Agr Res 17:247–277

Muangprom A, Osborn TC (2004) Characterization of a dwarf gene in *Brassica rapa*,including the identification of a candidate gene. Theor Appl Genet 108:1378–1384

Muangprom A, Thomas SG, Sun TP, Osborn TC (2005) A Novel Dwarfing Mutation in a Green Revolution Gene from *Brassica rapa*. Plant Physiol 137:931–938

Nishio T, Kusaba M, Watanabe M, Hinata K (1996) Registration of S-alleles in *Brassica campestris* L. by restriction fragment sizes of SLGs. Theor Appl Genet 92:388–394

Nishio T, Sakamoto K, Yamaguchi J (1994) PCR RFLP of S-Locus for identification of breeding lines in cruciferous vegetables. Plant Cell Rep 13:546–550

Nishioka M, Tamura K, Hayashi M, Fujimori Y, Ohkawa Y, Kuginuki Y, Harada K (2005) Mapping of QTLs for bolting time in Brassica rapa (syn. *Campestris*) under different environmental conditions. Breed Sci 55:127–133

Nou IS, Watanabe M, Isogai A, Hinata K (1993) Comparison of S- alleles and S-glycoproteins between two wild populations of *Brassica campestris* in Turkey and Japan. Sex Plant Reprod 6:79–86

Novakova B, Salava J, Lydiate D (1996) Construction of a genetic linkage map for *Brassica campestris* L. (syn. *Brassica rapa* L.). Genetika Slechteni 32:249–256

Nozaki T, Kumazaki A, Koba T, Ishikawa K, Ikehashi H (1997) Linkage analysis among loci for RAPDs isozyme and some agronomic traits in *Brassica campestris* L. Euphytica 95:115–123

Osborn TC, Kole C, Parkin IAP, Sharpe AG, Kuiper M, Lydiate DJ, Trick M (1997) Comparison of flowering time genes in *Brassica rapa*, *B. napus* and *Arabidopsis thaliana*. Genetics 146: 1123–1129

O'Neill CM, Bancroft I (2000) Comparative physical mapping of segments of the genome of *Brassica oleracea* var. *alboglabra* that are homoeologous to sequenced regions of chromosomes 4 and 5 of *Arabidopsis thaliana*. Plant J 23:233–243

Padilla G, Cartea ME, Rodrguez VM, Ordas A (2005) Genetic diversity in a germplasm collection of *Brassica rapa* subsp *rapa* L. from northwestern Spain. Euphytica 145:171–180

Panjabi P, Jagannath A, Bisht NC, Padmaja KL, Sharma S, Gupta V, Pradhan AK, Pental D (2008) Comparative mapping of *Brassica juncea* and *Arabidopsis thaliana* using Intron Polymorphism (IP) markers: homoeologous relationships, diversification and evolution of the A, B and C Brassica genomes. BMC Genomics 9:113

Parkin IAP, Gulden SM, Sharpe AG, Lukens L, Trick M, Osborn TC, Lydiate DJ (2005) Segmental structure of the *Brassica napus* genome based on comparative analysis with *Arabidopsis thaliana*. Genetics 171:765–781

Payne KA, Bowen HC, Hammond JP, Hampton CR, Lynn JR, Mead A, Swarup K, Bennett MJ, White PJ, Broadley MR (2004) Natural genetic variation in caesium (Cs) accumulation by *Arabidopsis thaliana*. New Phytol 162:535–548

Petrie GA (1988) Races of *Albugo candida* (white rust and stag head) on cultivated Cruciferae in Saskatchewan. Can J Plant Path 10:142–150

Piao ZY, Deng YQ, Choi SR, Park YJ, Lim YP (2004) SCAR and CAPS mapping of *CRb*, a gene conferring resistance to *Plasmodiophora* brassicae in Chinese cabbage (*Brassica rapa* ssp. *pekinensis*). Theor Appl Genet 108:1458–1465

Piao ZY, Ramchiary N, Lim YP (2009) Genetics of clubroot resistance in *Brassica* species. J Plant Growth Regul 28:252–264

Pound GS, Williams PH (1963) Biological races of *Albugo candida*. Phytopathology 53: 1146–1149

Prakash S, Hinata K (1980) Taxonomy, cytogenetics and origin of crop *Brassica*, a review. Opera Bot 55:1–57

Quiros CF, Hu J, Truco MJ (1994) DNA-based marker *Brassica* maps. In: Phillips RL, Vasil IK(eds) Advances in cellular and molecular biology of plants, Vol 1. DNA based markers in plants, pp 199–222. Kluwer Academic Publishers, Dordrecht

Rahman M, McVetty PBE, Li G (2007) Development of SRAP, SNP and multiplexed SCAR molecular markers for the major seed coat color gene in *Brassica rapa L.* Theor Appl Genet 115:1101–1107

Rahman M, Sun Z, McVetty PB, Li G (2008) High throughput genome-specific and gene-specific molecular markers for erucic acid genes in *Brassica napus* (L.) for marker-assisted selection in plant breeding. Theor Appl Genet 117:895–904

Ramchiary N, Bisht NC, Gupta V, Mukhopadhyay A, Arumugam N, Sodhi YS, Pental D, Pradhan AK (2007) QTL analysis reveals context-dependent loci for seed glucosinolate traits in oilseed *Brassica juncea:* importance of recurrent selection backcross (RSB) scheme for the identification of 'true' QTL. Theor Appl Genet 116:77–85

Rimmer SR, Mathur S, Wu CR (2000) Virulence of isolates of *Albugo candida* from western Canada to *Brassica* species.Can J. Plant Path 22:229–235

Rosa EAS, Hesney RK, Fenwick GR, Portas C (1997) Glucosinolates in crop plant. Hortic Rev 19:99–215

Rouhier H, Usuda H (2001) Spatial and temporal distribution of sucrose synthase in the radish hypocotyl in relation to thickening growth. Plant Cell Physiol 42:583–593

Rusholme RL, Higgins EE, Walsh JA, Lydiate DJ (2007) Genetic control of broad-spectrum resistance to turnip mosaic virus in *Brassica rapa* (Chinese cabbage). J Gen Virol 88:3177–3186

Saito M, Kubo N, Matsumoto S, Suwabe K, Tsukada M, Hirai M (2006) Fine mapping of the clubroot resistance gene, *Crr3*, in *Brassica rapa*. Theor Appl Genet 114:81–91

Sakamoto K, Saito A, Hayashida N, Taguchi G, Matsumoto E (2008) Mapping of isolate-specific QTL for clubroot resistance in Chinese cabbage (*Brassica rapa* L. ssp. *pekinensis*). Theor Appl Genet. doi:10.1007/s00122-008-0817–0

Santos MR, Dias JS, Silva MJ, Ferreira-Pinto MM (2006) Resistance to white rust in pak choi and Chinese cabbage at the cotyledon stage. Comm Agric Appl Biol Sci 71:963–971

Sax K (1923) The association of size differences with seed coat pattern and pigmentation in *Phaseolus vulgaris*. Genetics 8:552–560

Schaad NW, Thaveeschai N (1983) Black rot of crucifers in Thailand. Plant Dise 67:1231–1234

Schilling A (1991) Development of a molecular genetic linkage map in *Brassica rapa* (syn. *Campestris*) L. M.Sc. Thesis, University of Massachusetts, USA

Schmidt R (2002) Plant genome evolution: lessons from comparative genomics at the DNA level. Plant Mol Biol 48:21–37

Schranz ME, Quijada P, Sung SB, Lukens L, Amasino RM, Osborn TC (2002) Characterization and effects of the replicated time gene *FLC* in *Brassica rapa*. Genetics 162:1457–1468

Shattuck VI (1992) The biology, epidemiology and control of turnip mosaic virus. Plant Breed Rev 14:199–238

Singh D (1958) Rape and Mustard. Indian Central Oilseeds Committee. Examiner, Bombay, India, P105

Sinskaia EN (1928) The oleiferous plants and root crops of the family *Cruciferae*. Bull Appl Bot Genet Pl Breed 175:3–166

Snowdon RJ, Friedt W (2004) Molecular markers in *Brassica* oilseed breeding: current status and future possibilities. Plant Breed 123:1–8

Soengas P, Vicente JG, Pole JM, Pink DAC (2007) Identification of quantitative trait loci for resistance to *Xanthomonas campestris* pv. *campestris* in *Brassica rapa*. Theor Appl Genet 114:637–645

Song KM, Osborn TC, Williams PH (1988a) *Brassica* taxonomy based on nuclear restriction fragment length polymorphism (RFLPs). 1. Genome evolution of diploid and amphidiploids species. Theor Appl Genet 75:784–794

Song KM, Osborn TC, Williams PH (1988b) *Brassica* taxonomy based on nuclear restriction fragment length polymorphism (RFLPs). 2. Preliminary analysis of sub-species within *B.rapa* (syn. *campestris*) and *B. oleracea*. Theor Appl Genet 76:593–600

Song KM, Osborn TC, Williams PH (1990) *Brassica* taxonomy based on nuclear restriction fragment length polymorphisms (RFLPs) 3. Genome relationships in *Brassica* and related genera and the origin of *B. oleracea* and *B. rapa* (syn. *campestris*). Theor Appl Genet 76:497–506

Song K, Slocum MK, Osborn TC (1995) Molecular marker analysis of genes controlling morphological variation in *Brassica rapa* (syn. *campestris*). Theor Appl Genet 90:1–10

Song KM, Suzuki JY, Slocum MK, Williams PH, Osborn TC (1991) A linkage map of *Brassica rapa* base on restriction fragment length polymorphism loci. Theor Appl Genet 82:296–304

Stringam GR (1980) Inheritance in seed color in turnip rape. Can J Plant Sci 60:331–335

Suwabe K, Iketani H, Nunome T, Kage T, Hirai M (2002) Isolation and characterization of microsatellites in *Brassica rapa* L. Theor Appl Genet 104:1092–1098

Suwabe K, Iketani H, Nunome T, Ohyama A, Hirai M, Fukuoka H (2004a) Characteristics of microsatellites in *Brassica rapa* genome and their potentials of utilization for comparative genomics in cruciferae. Breed Sci 54:85–90

Suwabe K, Tsukazaki H, Iketani H, Hatakeyama K, Fujimura M, Konodo M, Nunome T, Fukuoka H, Hirai M, Matsumoto S (2004b) Joint meeting of the 14th Crucifer Genetics Workshop and the 4th ISHS Symposium on *Brassicas* 2004, Chungnam National University Daejeon, Korea October 24–28, pp 143

Suwabe K, Tsukazaki H, Iketani H, Hatakeyama K, Fujimura M, Nunome T, Fukuoka H, Matsumoto S, Hirai M (2003) Identification of two loci for resistance to clubroot (*Plasmodiophora brassicae* Woronin) in *Brassica rapa* L. Theor Appl Genet 107:997–1002

Suwabe K, Tsukazaki H, Iketani H, Hatakeyama K, Kondo M, Fujimura M, Nunome T, Fukuoka H, Hirai M, Matsumoto S (2006) Simple sequence repeat-based comparative genomics between *Brassica rapa* and *Arabidopsis thaliana*: the genetic origin of clubroot resistance. Genetics 173:309–319

Takahata Y (1997) Microspore culture. In: Kalia HR & Guputa SK (eds) Recent advances in oilseed *Brassicas*, pp 162–181. Kalyani Publisher, Ludhiana

Talalay P, Fahey JW (2001) Phytochemicals from cruciferous plants protect against cancer by modulating carcinogen metabolism. J Nutr 131:3027S–3033S

Tanhuanpaa PK (2004) Identification and mapping of resistance gene analogs and white rust resistance locus in *Brassica rapa* ssp. *oleifera*. Theor Appl Genet 108:1039–1046

Tanhuanpaa PK, Schulman A (2002) Mapping of genes affecting linolenic acid content in *Brassica rapa* ssp. *oleifera*. Mol Breed 10:51–62

Tanhuanpaa PK, Vilkke JP, Vikki HJ (1996b) Mapping a QTL for oleic acid concentration in spring turnip rape (*Brassica rapa* ssp. *oleifera*). Theor Appl Genet 92:952–956

Tanhuanpaa P, Vilkki J, Vihinen M (1998) Mapping and cloning of *FAD2* gene to develop allele-specific PCR for oleic acid in spring turnip rape (*Brassica rapa* ssp. *oleifera*). Mol Breed 4: 543–550

Tanhuanpaa PK, Vilkki JP, Vilkki HJ (1995) Association of RAPD marker with linolenic acid concentration in the seed oil of rapeseed (*B. napus* L). Genome 38:414–416

Tanhuanpaa PK, Vilkki JP, Vilkki HJ (1996a) A linkage map of spring turnip rape based on RFLP and RAPD markers. Agric Food Sci Finl 5:209–217

Tanksley SD, Young ND, Paterson AH, Bonierbale MW (1989) RFLP mapping in plant breeding. new tools for an old science. Bio/Technology 7:257–264

Taylor JD, Conway J, Roberts SJ, Astley D, Vicente JG (2002) Sources and origin of resistance to *Xanthomonas campestris* pv. *campestris* in *Brassica* genomes. Phytopathology 92:105–111

Teutonico RA, Osborn TC (1994) Mapping of RFLP and quantitative trait loci in *Brassica rapa* and comparison to the linkage maps of *B. napus, B. oleracea* and *Arabidopsis thaliana*. Theor Appl Genet 89:885–894

Teutonico RA, Osborn TC (1995) Mapping loci controlling vernalization requirement in *Brassica rapa*. Theor Appl Genet 91:1279–1283

Thormann CE, Ferreira ME, Carmago LEA, Tivang JG, Osborn TC (1994) Comparison of RFLP and RAPD markers to estimate genetic relationships within and among cruciferous species. Theor Appl Genet 88:973–980

Tiwari AS, Petrie GA, Downey RK (1988) Inheritance of resistance to *Albugo candida* race 2 in mustard (*Brassica juncea* (L.)Czern.). Can J Plant Sci 68:297–300

Tjallingii F (1965) Testing clubroot resistance in turnips in Netherlands and the physiologic specialization of *Plasmodiophora brassicae*. Euphytica 14:1–22

Toxopeus H, Janssen AMP (1975) Clubroot resistance in turnip. II. The slurry screening method and the clubroot races in the Netherlands. Euphytica 24:751–755

U N (1935) Genome analysis in *Brassica* with special reference to the experimental formation of *B. napus* and peculiar mode of fertilization. Jpn J Bot 7:389–452

Vavilov NI (1949) The origin, variation, immunity and breeding of cultivated plants. Chron Bot 13:1–364

Verma U, Bhowmik TP (1989) Inheritance of resistance to a *Brassica juncea* pathotype of Albugo candida in *B. napus*. Can J Plant Pathol 11:443–444

Verma PR, Harding H, Petrie GA, Williams PH (1975) Infection and temporal development of mycelium of *Albugo candida* in cotyledons of four *Brassica species*. Can J Bot 53: 1016–1020

Walsh JA, Jenner CE (2002) Turnip mosaic virus and the quest for durable resistance. Mol Plant Pathol 3:289–300

Walsh JA, Sharpe AG, Jenner CE, Lydiate DJ (1999) Characterization of resistance to turnip mosaic virus in oilseed rape (*Brassica napus*) and genetic mapping of TuRB01. Theor Appl Genet 99:1149–1154

Wang M, Zhang FL, Meng XD, Liu XC, Zhao XY, Fan ZC (2004) A linkage map construction for Chinese cabbage based on AFLP markers using DH population. Acta Agric Boreali-Sin 19:28–33

Williams PH (1980) Black rot: a continuing threat to world crucifers. Plant Dis 64:736–742

Williams PH, Hill CB (1986) Rapid cycling populations of *Brassica*. Science 232:1385–1389

Wit F (1964) Inheritance of reaction to clubroot in turnips. Hort Res 5:47–49

Wit F, Van de Weg M (1964) Clubroot resistance in turnips (*Brassica campestris* L.) I. Physiological races of the parasite and their identification in mixtures. Euphytica 13:9–18

Wu J, Schat H, Sun RF, Koornneef M, Wang XW, Aarts MGM (2007) Characterization of natural variation for zinc, iron and manganese accumulation and zinc exposure response in *Brassica rapa* L. Plant Soil 291:167–180

Wu J, Yuan YX, Zhang ZW, Zhao J, Song X, Li Y, Li X, Sun R, Koorneef M, Aaarts MGM, Wang XW (2008) Mapping QTLs for mineral accumulation and shoot dry biomass under different Zn nutritional conditions in Chinese cabbage (*Brassica rapa* L. ssp. *pekinensis*). Plant Soil 310:25–40

Yang TJ, Kim JS, Kwon SJ, Lim KB, Choi BS, Kim JA, Jin M, Park JY, Lim MH, Kim HI, Lim YP, Kang JJ, Hong JH, Kim CB, Bhak J, Bancroft I, Park BS (2006) Sequence-Level analysis of the diploidization process in the triplicated *FLOWERING LOCUS C* region of *Brassica rapa*. Plant Cell 18:1339–1347

Yarnell SH (1956) Cytogenetics of the vegetable crops. Crucifers Bot Rev 22:81–166

Yoshikawa H (1981) Breeding for clubroot resistance in Chinese cabbage. In: Talekar NS, Griggs TD (eds) Chinese cabbage. Proceedings of the 1st international symposium, Tsukuba, Japan, pp 405–413

Yoshikawa H (1993) Studies on breeding of clubroot resistance in cole crops. Bull Natl Res Inst Veg Ornam Plants Tea Jpn Ser A 7:1–165

Yu SC, Wang YJ, Zheng XY (2003a) A genetic linkage map of *Brassica campestris* L. ssp. *pekinensis* (syn. *B. rapa* L. ssp. *pekinensis*). Agric Sci China 2:49–55

Yu SC, Wang YJ, Zheng XY (2003b) Mapping and analysis QTL controlling heat tolerance in *Brassica campestris* L. ssp. *pekinensis*. Acta Hortic Sin 30:417–420

Yu SC, Wang YJ, Zheng XY (2003c) Mapping and analysis QTL controlling some morphological traits in Chinese cabbage (*Brassica campestris* L. ssp. *pekinensis*). Acta Genet Sin 30: 1153–1160

Zhang LG (1999) Construction of molecular genetic map and location of QTLs in *Brassica campestris* by F2 population of turnip × Chinese cabbage. Ph.D. Thesis. Northwest Agricultural University, Yangling, Shanxi Province, China

Zhang J, Lu Y, Yuan Y, Zhang X, Geng J, Chen Y, Cloutier S, McVetty PBE, Li G (2008) Map-based cloning and characterization of a gene controlling hairiness and seed coat color traits in *Brassica rapa*. Plant Mol Biol. doi:10.1007/s11103-008-9437–y

Zhang LG, Wang M, Chen H (2000) Construction of RAPDs molecular genetic map of Chinese cabbage. Acta Bot Sin 42(5):485–489

Zhao JY, Becker HC (1998) Genetic variation in Chinese and European oilseed rape (*B. napus*) and turnip rape (*B. campestris*) analysed with isozymes. Acta Agron Sin 24:213–220

Zhao JJ, Bonnema G, Wang XW, Xu ZY, Sun RF, Vreugdenhil D, Koorneef M(2004a) Genetic dissection of quality traits in *Brassica rapa* species for improved human health. In: joint meeting of the 14th Crucifer Genetics Workshop and 4th ISHS Symposium on Brassicas, 2004, Chungnam National University, Daejeon, Korea, 24–28 Oct, 2004, P112

Zhao J, Wang X, Deng B, Lou P, Wu J, Sun R, Xu Z, Vromans J, Koornneef M, Bonnema G (2005) Genetic relationships within *Brassica rapa* as inferred from AFLP fingerprints. Theor Appl Genet 110:1301–1314

Zhao JJ, Wang XW, Deng B, Lou P, Wu J, Sun RF, Xu ZY, Vroomans J, Koorneef M, Bonnema G (2004b) Phylogenetic relationships within *Brassica rapa* inferred from AFLP fingerprints. In: joint meeting of the 14th Crucifer Genetics Workshop and 4th ISHS Symposium on Brassicas, 2004, Chungnam National University, Daejeon, Korea, 24–28 Oct, 2004, P128

Chapter 9
The Genetics of *Brassica oleracea*

Carlos F. Quiros and Mark W. Farnham

Abstract *Brassica oleracea* is one of the most important species of the Brassicaceae family because the species includes some of the most economically important vegetables in the world. Common heading cabbage and cauliflower are the most widely grown crops of this species, but broccoli is also now emerging rapidly as a world vegetable. The wide center of origin for this species is the Mediterranean Basin, and primitive forebears of our modern *B. oleracea* crop forms have been cultivated and selected for several millennia. Undoubtedly, the diverse array of wild forms found in this species and other very closely related species played very important roles in stimulating the occurrence of morphological variation within and among the *B. oleracea* crops as they underwent development. In the years following the rediscovery of Mendel's work, many scientists studied the underlying genetic factors controlling the divergent morphologies within the species. This was of interest not only from a basic scientific standpoint, but also due to the practical necessity of understanding the complex sets of genes that combine and give rise to a specific crop form like heading cabbage or cauliflower. This knowledge is crucial in moving genes between crops in the process of breeding improved varieties. Secondary plant metabolites have emerged as key components of crops within this species because they appear to contribute added-value to the various crops by conferring intrinsic healthful effects on populations that consume these vegetables. Among the various components believed to confer a chemoprotective effect in *B. oleracea*, glucosinolates, and isothiocyanates have received the most attention in recent years and are considered in detail herein. The study of *B. oleracea* genetics has been greatly advanced during the modern era of gene study at the molecular level. Although the species has presented challenges, scientists focused on these crops are now mapping genes to specific chromosomes and the genome is well on its way to being sequenced. As knowledge advances at the molecular level, a fuller understanding of gene sequences and there relations to morphology, disease resistance, phytochemical make-up, and other important traits are being realized in *B. oleracea*.

C.F. Quiros (✉)
Department of Plant Sciences, University of California, Davis, CA 95616, USA
e-mail: cfquiros@ucdavis.edu

R. Schmidt, I. Bancroft (eds.), *Genetics and Genomics of the Brassicaceae*, Plant Genetics and Genomics: Crops and Models 9, DOI 10.1007/978-1-4419-7118-0_9,

Keywords Broccoli · Breeding · Cabbage · Cauliflower · Cole crops · Glucosinolates

Contents

9.1 Importance of *Brassica oleracea* Crops 262
9.2 Origin, Distribution, and Domestication 263
9.3 Taxonomy of *B. oleracea* Crops: Coenospecies and Cytodemes 264
9.4 Interspecific and Intergeneric Hybridizations 266
9.5 Genetics of Main Crop Morphotypes 266
9.5.1 Cabbage Traits 267
9.5.2 Kohlrabi Traits 268
9.5.3 Kale Traits 268
9.5.4 Brussel Sprouts Traits 269
9.5.5 Cauliflower and Broccoli Traits 269
9.6 Flower Color and Bolting 271
9.7 Secondary Metabolites: Glucosinolates (GSL) and Carotenoids 272
9.8 Disease and Insect Resistance 275
9.9 Chromosome Number Variation 277
9.9.1 Polyploidy 277
9.9.2 Aneuploidy 278
9.10 Monoploids and Anther/Microspore Culture 278
9.11 Genomic Tools: Markers, Genetic and Physical Maps 279
9.12 Map Development in *Brassica* 279
9.13 Synteny Maps 280
9.14 Genomics 281
9.15 Outlook 282
References 282

9.1 Importance of *Brassica oleracea* Crops

B. oleracea L. crops encompass a family of vegetables that are among the most important in the world. The three most commonly grown vegetables in this family are common cabbage, cauliflower, and broccoli. Cabbage is the most widely produced of the three (Table 9.1). Production of cauliflower is less than cabbage, and broccoli is the relative newcomer and emerging vegetable of this group. China and India are the biggest producers of cabbage and cauliflower (Table 9.1). Cabbage is very important in northern and eastern European countries while cauliflower is more important in southern Europe, the USA, and Mexico. World production statistics are not currently available for broccoli, but it is now the most important *B. oleracea* crop in the USA ranking among the top five vegetables in that country (USDA 2006). Its popularity in Europe and Asia has been steadily increasing as well.

Table 9.1 Ten countries with the greatest production (metric tons, MT) of cauliflower and cabbage. Estimates are based on most current UN-FAO statistics

	Cauliflower		Common cabbage	
Rank	Country	1,000 MT	Country	1,000 MT
1	China	7,739	China	34,101
2	India	4,800	India	6,000
3	Italy	513	Russian Fed.	3,985
4	Spain	438	Korea	3,300
5	France	395	Japan	2,200
6	USA	335	USA	2,156
7	Mexico	215	Ukraine	1,632
8	Pakistan	206	Poland	1,400
9	Poland	190	Indonesia	1,292
10	UK	160	Germany	975

Brussels' sprouts, kale, kohlrabi, and other *B. oleracea* crops are much less important from a worldwide perspective, but can be very important on a regional or country basis. Brussels' sprouts and kale are probably more important in much of Europe than broccoli but may not be grown at all in parts of Asia. Numerous other unique crops like Portuguese cabbage grown in Spain and Portugal or collard grown in the southeastern USA can be very important to regional economies. In southern Europe, where many primitive and wild forms of *B. oleracea* can still be found, relatively small-scale, unique crops can even be found in specific regions (Branca 2008). Some examples include landraces of purple and green cauliflower in Sicily as well as purple kohlrabi.

9.2 Origin, Distribution, and Domestication

Wild forms of *B. oleracea* are found in the Atlantic coasts of Europe, Northern France, and England, in particular, and related wild species are endemic of the Mediterranean basin. Because the cultivated *B. oleracea* crops were grown in close proximity to wild relatives, movement of genes from wild to cultivated types likely occurred by introgression. This factor might be as important as human selection in helping to create the wide morphological variation within *B. oleracea*. A great array of *B. oleracea* crops have been domesticated throughout Western Europe with the Greeks and Romans cultivating kales and nonheading cabbages since the first century AD if not before (Sauer 1993). As early as 2000 BC, local aborigines probably domesticated wild oleracea which were primitive kales or cabbages, now believed to be the first cultivated forms of *B. oleracea* (Chiang et al. 1993). These were later cultivated by the Celts whose name, "Bresic," for cabbage may be the likely progenitor of "Brassica." Heading cabbages and other leafy types probably had a common origin from the ancestral kales and nonheading cabbages (Herve 2003) and they were further developed in Portugal, Spain, and France. Conversely, Chinese kale (var. *alboglabra*) seems to have been domesticated in China from very early introductions from Europe (Herve 2003). The great diversity of morphological types that exists in *B. oleracea* and its relatives may be underlaid by diversity in karyotypic structure as well (Kianian and Quiros 1992).

Italy is the generally accepted center of origin for cauliflower and broccoli. The ancestral forms leading to these important modern *B. oleracea* crops might have been grown as early as III–IV BC in the west Mediterranean basin (Branca 2008). Broccoli might have originated from this lineage around 400–600 BC (Gomez-Campos and Hinata 1999) and cauliflower probably originated from broccoli (Gray 1982) based on the fact it is a very specialized form with reduced fitness, since the curd is formed in part by aborted floral meristems. Crisp and Tapsell (1992) postulate that ancestral broccoli was domesticated in the Mediterranean and then taken west to Sicily. Hybridization with wild oleracea might have aided in creation of biennial types. Following that, 500 years of selection to increase the size of the terminal head and mutation for curd formation was responsible for the development of cauliflowers (see below). Additional mutations have been responsible for cauliflower's adaptation to warmer regions resulting in the development of tropical types in India and Australia.

Brussels sprouts are perhaps the youngest *B. oleracea* crop and it probably originated in Belgium in the fourteenth century (Gomez-Campos and Hinata 1999) or later (Herve 2003). Little is known, about the origin of kohlrabi but it is evident in Central Europe since the sixteenth century (Diederichsen 2001). Purple landraces are cultivated in Sicily, so its domestication might have been associated to this region (La Malfa, G. personal com.).

9.3 Taxonomy of *B. oleracea* Crops: Coenospecies and Cytodemes

There is a lack of clarity about the taxonomic nomenclature of *Brassica* species and related genera in general, and this is reflected in confusion about *B. oleracea* in particular. Recent treatments (Al Shebaz et al. 2006, Gladis and Hammer 2001) of this topic attempt to shed light on this issue, but still many questions remain to be answered in this field. The most practical approach from the point of view of breeders and geneticists is to classify the group biologically, encompassing all the species and genera sufficiently related to the crop Brassicas as potentially capable of exchanging genes among them. These are referred in the broad sense as Brassica coenospecies (Gomez-Campos and Hinata 1999).

This discussion brings us to the concept of cytodeme, developed by Harberd (1972) to define the gene pool of a cultivated species. A cytodeme is a group of species sharing the same chromosome number that cross readily with each other, resulting in fertile hybrids. Therefore, the *oleracea* cytodeme includes the species *B. oleracea* and its nine wild relatives listed below. All these are found in the Mediterranean region (Snogerup 1980, Snogerup et al. 1990). These species with

The oleracea cytodeme
B. oleracea
B. rupestris
B. macrocarpa
B. insularis
B. montana
B. cretica
B. hilarionis
B. villosa
B. incana
B. bourgaei

n=9 chromosomes readily cross to produce fertile hybrids. All of them have the C genome but have modified karyotypes, typically due to a few chromosomal rearrangements (Kianian and Quiros 1992).

There is a great array of horticultural types in *B. oleracea*. Although the wild relatives of this species might have been involved in the development of the cultivated forms, there is no clear evidence that certain cultivated types have specific wild species as ancestors. For example, *B. cretica* has been postulated as the possible ancestor of cauliflower (Crisp and Tapsell 1992, Snogerup 1980). However, RFLP data indicate that this is probably not the case (Hosaka et al. 1990).

The cultivated forms of *B. oleracea* are classified in various taxonomic groups or varieties according to crop type as outlined by Diederichsen (2001) as follows:

The oleracea varieties
ramosa: thousand headed kale
viridis: borecole, collard
capitata: cabbage
gemmifera: Brussels sprouts
gongylodes: kohlrabi
botrytis: cauliflower, Romanesco
italica: sprouting broccoli
alboglabra: Chinese kale
sabellica: scotch kale
medullosa: marrow stem kale
sabauda: savoy cabbage
costata: Portuguese tronchuda types
palmifolia: palm kale

Since all these varieties belong to the same species, they all intercross readily. In some instances, a few minor chromosomal rearrangements between some of the crop types have been reported (Kianian and Quiros 1992)

9.4 Interspecific and Intergeneric Hybridizations

Reports of wide hybridization attempts in *Brassica* go back to the nineteenth century, when chromosome numbers of the species in the genus were unknown (Prakash and Chopra 1991). Karpechenko (1927) popularized wide hybridization experiments in *Brassica* by creating "raphanobrassica" after crossing radish and cabbage. His objective was to produce a double purpose crop, cabbage on top, and radish at the bottom, but unfortunately he obtained the opposite. However, later on this wide cross proved to be very useful for the generation of cytoplasmic male sterility (CMS) crops by transferring the radish Ogura cytoplasm to *B. oleracea* and *B. napus* (Pelletier et al. 1988).

On interspecific hybridizations, most of these have been done among species in the oleracea cytodeme but mostly for basic research studies (Kianian and Quiros 1992). In a couple of instances these crosses have been made to introgress traits of interest from the wild to the cultivated species. This has been done for glucosinolate and isothiocyanate content by crossing *B. oleracea* to *B. villosa* (Mithen et al. 2003) and *B. macrocarpa* (Quiros et al. unpublished). Therefore this area of research remains fairly unexploited, considering the great deal of variation present in the wild relatives of *B. oleracea* (Snogerup et al. 1990).

9.5 Genetics of Main Crop Morphotypes

B. oleracea is represented by important vegetables which in total provide nearly all parts of a plant that can be consumed: Leaves, terminal and axillary buds, stems and floral tissues. Breeders have been interested in learning the genetic basis of these morphological variations to assure fast recovery of the cultivated features typical of a given crop after crossing it to a different one. This is a common need for breeders working with this species because inter-crop crosses are commonly done to transfer traits of interest.

In general, crosses among varieties of different horticultural forms result in the breakdown of the horticultural characteristics of each variety. Typically, the different forms are conferred by multiple genes expressing different dominant relationships depending on the particular crosses. Much of the work related to this subject was done in the early 1900s. It was summarized in a general review by Yarnell (1956), who added additional information based on his own work. In many cases, the results were conflicting mainly due to a high heterozygotic nature of parents involved in the relevant crosses. More recently, with the advent of homozygous inbreds first, and then doubled haploids, it has been possible to gain a better understanding of the morphological variation of this species. Previous authors (Dickson and Wallace 1986, Chiang et al. 1993) have included lists of genes identified over time. We would argue that such lists must be considered carefully because they probably contain specific genes that are represented multiple times, genes are usually postulated based on one source or population that is never verified by others, and most of the cited factors have never been mapped as has been done in more heavily studied species (e.g., tomato or maize). For purposes of this chapter the following information tries to convey what is generally known and has been supported through actual experience.

9.5.1 Cabbage Traits

In an early examination of a cabbage × kale cross, Pease (1926) reported leaf folding as a faint indicator of heading and continuous variation for the two parental types in the F_2. Based on this work, two recessive factors, n_1 *and* n_2, were postulated for heading. Another early study (Allgayer 1928), which examined a kitchen kale × cabbage cross postulated four factors (one dominant and three recessive) necessary for heading. Pelofske and Baggett (1979) reported only 2% of F_2 plants forming heads in a broccoli × a cabbage inbred cross. They concluded that a variety of factors (e.g., clasping leaves) contribute to the overall expression of heading, and thus, the cabbage morphotype is multigenic in nature. Dickson and Wallace (1986) state that it is generally agreed (based on widespread experience) that heading is recessive to nonheading. Recent work of Farnham et al. (2005) in which a series of F_1 hybrids formed by crossing cabbage inbreds with nonheading collard inbreds were evaluated, confirm that heading is at least partially recessive.

Pease (1926) postulated several factors associated with cabbage heading including a short height factor (*t*) recessive to tall (*T*); a wide leaf (*W*) dominant over narrow leaf (*w*); sessile leaf (*pet*) recessive to a petiolate leaf (*Pet*); and an entire leaf (*En*) dominant to a lyrate leaf (*en*) In general, Pease classified cabbages as *ttWWpetpetEnEn*. Interestingly, Kennard et al. (1994) followed the segregation of a series of leaf, stem, flower, and flowering time traits and their associations with molecular markers in an F_2 population obtained by crossing a cabbage and broccoli and they confirmed most of the results obtained by Pease (1926) and Pelofske and Baggett (1979) regarding stem internode length and the linkage of head forming leaf overlap, petiole length, lamina length, as well as annual vs biennial habit, which mapped in linkage group 5C.

Yarnell (1956) classified head shape in three basic types: flat, round, and pointed. His conclusion that pointed is dominant over round and that many genetic factors are involved in determining shape was similar to that made later by Dickson and Wallace (1986).

Crinckled Savoy-type leaf opposed to smooth leaf is controlled by three or more genes (Dickson and Wallace 1986) and is at least partially dominant (Yarnell 1956). Red coloration due to increased anthocyanin content is also determined by at least two dominant factors (Yarnell 1956, Dickson and Wallace 1986). Sampson (1967) clarified earlier studies and cited two complementary dominant factors for color with several alleles at one locus. Gene *A*, with *D* produces red cabbage. A third gene, *B*, causes light red midrib, which in combination with *A* produces dark violet midrib. C is colorless, but together with *A,* results in dark violet midrib. E extends the dark violet color with *B*, and C has the same effect as *B* by itself. Therefore, red cabbage is *DABce*, while green cabbage with light red midrib is *daBCE*. Clearly, these and other studies indicate inheritance of color has quantitative attributes (Kristofersson 1924, Kwan 1934, Magruder and Myers 1933).

9.5.2 Kohlrabi Traits

The bulb or swollen stem of kohlrabi exhibits incomplete dominance in crosses with other crops with normal stem. Full expression of the swollen stem may be conditioned by at least three genes based on observed segregation of this trait in a large F_2 family and subsequent observations in the F_3 and backcross generations (Pease 1927). Cabbage head and kohlrabi bulb cannot be combined in the same plant (Yarnell 1956). The F_1 between alboglabra and kohlrabi shows shorter internodes and thicker stem at the top of the plant (G. King, personal com.). Bulb formation also shows partial dominance in synthetic allotetraploid hybrids of kohlrabi and turnip (Fig. 9.1).

Fig. 9.1 Allotetraploid kohlrabi-turnip showing partial enlargements of stem and root. Quiros, unpublished

9.5.3 Kale Traits

The curling of leaf margins as in scotch kale is considered polygenic in nature (Yarnell 1956). Pease (1926) relayed results of an earlier observer (Malinowski in 1916) who specifically postulated three recessive genes being responsible for curling. Results of more than one study showed that hybrids between cabbage and kale resulted in leaves more like those of cabbage; F_2 generations always display continuous variation (Yarnell 1956). The same color genes that operate in cabbage can also determine leaf color in various edible and ornamental kales. The thickened stem of marrow stem kale has been attributed to a single gene mutation that results in alteration of stem thickness (Yarnell 1956).

9.5.4 Brussel Sprouts Traits

In general, crosses of different *B. oleracea* crops to Brussels sprouts result in plants with intermediate levels of lateral bud formation that lack the formation of axillary heads characteristic of Brussels sprouts (Yarnell 1956). Thus, formation of lateral sprouts appears to be recessive. In addition, heading of axillary buds and the actual presence or absence of buds are probably inherited independently. The F_1 progeny produced by Sebastian et al. (2002) by crossing doubled haploid (DH) Brussels sprouts and DH cauliflower produced axillary buds like small broccolis (G. King, personal communication; Fig. 9.2)

Fig. 9.2 F_1 hybrid between DH cauliflower and DH Brussels sprout, courtesy of King and Okendon, unpublished

9.5.5 Cauliflower and Broccoli Traits

Two major genes homologous to *Arabidopsis Ap1-a* and *Cal-a* were found associated with curd formation in cauliflower by Smith and King (2002). They designated these genes *BoAp1-a (A)* and *BoCal-a (C)* and proposed a genetic model to explain the broccoli, cauliflower, and intermediate phenotypes: With their model cauliflower has a double recessive genotype for the two genes (*aacc*), broccoli is *A-C-*, whereas *AAC-* and *aaC-* are intermediate types. Labate et al. (2006) countered that the above model only explained 6% of observed phenotypic variation, concluding that other genes complementing or modifying *BoAp1-a* and *BoCal-a* must be present.

Fig. 9.3 Broccoli × cauliflower F_1

Their results are in agreement with Gao et al. (2007) who identified three QTL for curd formation, one of which corresponds to *BoAp1-a* . Gao's F_1 was branchy and intermediate, with small heads combining flower buds like the broccoli parent and arrested floral meristematic tissue as with the cauliflower parent (Fig. 9.3).

Smith (1999), personal communication by G. King, found a significant factor affecting fused inflorescence on C5, in alboglabra × broccoli and cauliflower × Brussel sprouts populations, with additional effects from other parts of the genome.

Lan and Paterson (2000) also followed the inheritance of various traits in populations from three crosses involving rapid cycling kale and different cauliflower parents. They detected 86 QTLs associated with curd-related traits. However, some of their results are problematic because none of their parental materials were homozygous and no F_1 phenotypes were reported. Duclos and Bjorkman (2008) followed the expression of a series of MADS-box genes from *A. thaliana* and two curd-specific genes. They provide a complex model of meristem arrest in cauliflower with temperature interactions, but were unable to pinpoint specific genes. *A. thaliana* is probably not a good model to explain floral arrest and development in *B. oleracea* wherein the cauliflower phenotype is much more complicated.

Cauliflower has four basic curd colors, white, green, purple, and orange. Crisp and Angell (1985) reported two recessive genes responsible for green curd that act to alter chlorophyll accumulation. However, more recent and extensive breeding work indicates that green curd is dominant to white and orange curd colors (N. Acciarri, pers. com.) The semidominant gene *Or* conditions orange color by altering carotenoid accumulation of curds (Dickson et al. 1988). Homozygous plants for the *Or* gene produce small curds and have stunted growth. Lu et al. (2006) cloned this gene concluding that it represents a gain-of- function-mutation conditioning enhanced β-carotene accumulation. Purple curds result from anthocyanin (cyanidine) accumulation, and this trait is controlled by a single partially dominant gene (Chiu et al. 2005). On the contrary, purple Sicilian cauliflowers, which are actually intermediate types between broccoli and cauliflower (Gray 1992), have purple that is recessive when crossed to white cauliflower (N. Acciarri, pers. com.). Three other genes called "bleaching genes" are postulated to control white curd under sun exposure (Crisp and Tapsell 1992). According to Watt (1966), the smooth curd of cauliflower is partially dominant to pyramidal curd of "Jesi" (white) and Romanesco (green), with the pyramidal curd phenotype being polygenic. However, in some crosses pyramidal curd is partially dominant to flat curd. (N. Acciarri, pers. com.)

Among the important *B. oleracea* crops, cauliflower and broccoli are probably the two most closely related ones, both originating in Italy. Kennard et al. (1994) found that flower clustering typical of green sprouting broccoli heads is recessive in the F_1 of cabbage and broccoli. This same observation has been made by Farnham numerous times (unpublished). A DH population of nearly 200 individual lines developed from a F_1 formed by crossing a simple rapid cycling *B. oleracea* with a broccoli (Iñiquez et al. 2009) contains about 5 – 10 lines that have a phenotype similar to the broccoli parent, reaffirming that the broccoli phenotype (e.g., relatively large reproductive head) is probably recessive and controlled by many genes. The F_1 was intermediate but more broccoli-like than anything else. An important attribute of modern broccoli cultivars is that they exhibit a relative lack of lateral shoot formation and only produce a large central head. Keyes and Honma (1986) concluded that a lack of lateral heads was dominant over lateral formation and postulated two genes controlling this trait.

9.6 Flower Color and Bolting

White flower, typical of alboglabra, is dominant over yellow and determined by gene *Wh*. With the more common crops, yellow flowers are more prevalent. An important attribute of *B. oleracea* crops is whether they are annuals or biennials that typically require vernalization. It is generally accepted that annual habit is dominant over biennial and also that the trait is polygenic with strong environmental interaction. Pelofske and Baggett (1979) followed the segregation of annual over biennial habit by crossing broccoli and cabbage inbreds. All the plants in the F_1 and BC_1 to broccoli and 92% of the F_2 were annuals. Fourty percent of BC_1 to cabbage were biennial. These results were also confirmed by Kennard et al. (1994).

Most students of *B. oleracea* agree that the Acephala group (e.g., var. viridis or medullosa) of crops includes some of the oldest cultivated forms of the species (Diederichsen 2001). The typical variants in this group are leafy greens that do not form vegetative or reproductive heads and that are consumed as harvested leaves. A modern example of such a crop might look something like American collards or a European kale. From these ancient types, the more popular modern forms like heading cabbage, broccoli, and cauliflower were selected. In reviewing the many studies focused on crop phenotype, many of which were conducted in the first half of the twentieth century, one generally sees that most of the modern *B. oleracea* crop types are typically controlled by recessive factors. This leads one to postulate that new crop forms likely came about first as mutants, and then as selected variants, arising from the more traditional and simple-phenotype crops. The presence of many wild types of *B. oleracea* in close proximity to the cultivated ones likely increased potential variation. After the primitive, but new variants were established, further selection by growers helped to enhance or modify the mutant type so that the modern phenotypes are now controlled by numerous genes.

9.7 Secondary Metabolites: Glucosinolates (GSL) and Carotenoids

A number of studies suggest that consumption of vegetables, including *B. oleracea* crops such as broccoli, cabbage, and kale, reduces the incidence of cancer in humans and other mammals (Block et al. 1992, Fahey et al. 1997, Steinmetz and Potter 1996). The likely constituents conferring chemoprotection by these vegetables are the glucosinolates (GSL) and their isothiocyanate (ITC) breakdown products. Although numerous GSL are identified in *B. oleracea crops* (Carlson et al. 1987, Kushad et al. 1999, Rosa et al. 1997), glucoraphanin is well recognized as a prominent aliphatic GSL abundant in broccoli heads (Farnham et al. 2000, Kushad et al. 1999) and seedling sprouts (Fahey et al. 1997) that has been studied in depth and likely instrumental in stimulating increased popularity of broccoli worldwide. The isothiocyanate sulforaphane, derived from the glucoraphanin by the action of the enzyme myrosinase, was found to confer protection against mammary tumor growth in rats after treatment with dimethyl benzanthracene, a carcinogenic agent (Zhang et al. 1992, 1994). A recent review of sulforaphane as an anticancer agent indicates that it likely acts against cancer through multiple mechanisms (Clarke et al. 2008).

Genetic studies in *A. thaliana* (Mithen et al. 1995, Mithen and Campos 1996) and *Brassica* sp. (Magrath et al. 1993, 1994, Chen 2003) support a proposed biosynthetic pathway for aliphatic GSL, as shown schematically in Fig. 9.4. The synthesis of these compounds is determined by a simple genetic system containing two distinct sets of genes, one determining side chain elongation and the second chemical modification of the side chains. Aliphatic GSL profiles vary among *A. thaliana* ecotypes and *Brassica* sp. These GSL are synthesized in the following sequence: methylsulfinylalkyl, alkenyl, and hydroxy types, which can be divided in *Brassica* species into three-carbon (3C), four-carbon (4C), and five-carbon (5C) groups based on their side-chain length. (*Arabidopsis* has additional long chain GSL: 6, 7, and 8C.)

In *A. thaliana,* several genes involved in the GSL pathway have been identified by genetic analysis, including *Gsl-elong, Gsl-alk, Gsl-ohp,* and *Gsl-oh* (Campos de Quiros et al. 2000, Magrath et al. 1994, Mithen et al. 1995, Mithen and Campos 1996). The *Gsl-elong* locus regulates side chain length, whereas *Gsl-alk* controls side chain desaturation. *Gsl-ohp* and *Gsl-oh* are responsible for side chain hydroxylation (Magrath et al. 1993, 1994). In *B. oleracea,* presence of homologs to *Gsl-pro, Gsl-oxid, Gsl-elong, Gsl-alk,* and *Gsl-oh* loci (Giamoustaris and Mithen, 1996,

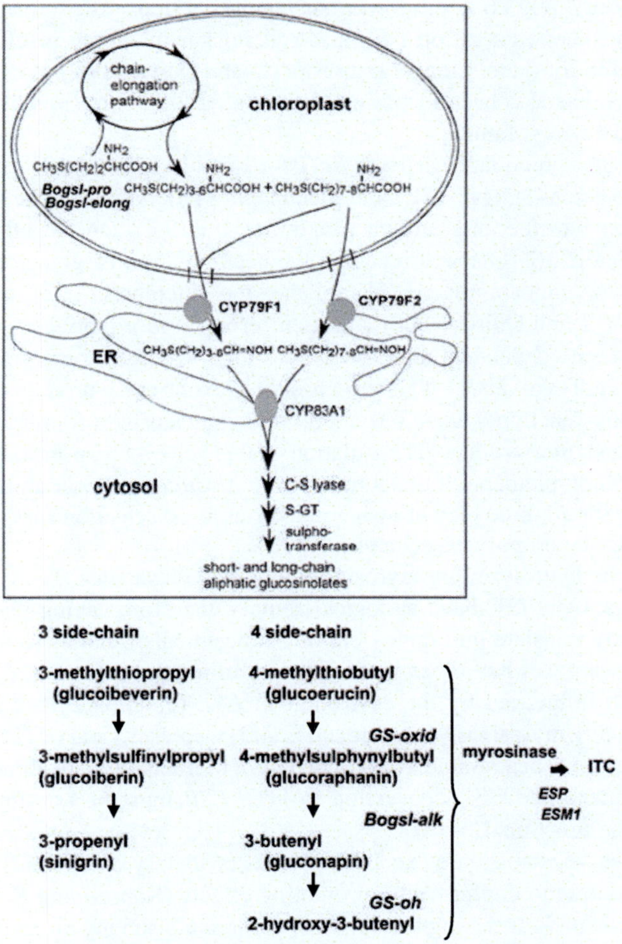

Fig. 9.4 GSL pathway based on *A. thaliana*, adapted from Chen et al. (2003)

Mithen et al. 1995) has been inferred from inspection of GSL profiles. In rapeseed (Magrath et al. 1993), a similar set of genes has also been proposed. Thus, genetic evidence suggests that biosynthesis of GSL is highly conserved in *Brassica* and *Arabidopsis*. However, many steps in side chain elongation, glycone formation, and aglycone modification remain to be characterized biochemically and genetically. With the present model, the 3C, 4C, and 5C GSL are closely related biosynthetically because all of them originate from the same precursor, methionine. Accordingly, it is expected that the presence of the dominant allele for the *Gsl-elong* gene will result in 4C GSL, whereas presence of the dominant allele for *Gsl-pro* will result in 3C GSL. A candidate gene for *BoGsl-elong* has been identified in broccoli, where the allele of this gene is functional. It is a member of the isopropyl malate synthase-like (*MAM-2*) gene family reported in *Arabidopsis*. The gene has 10 exons and a total length of about 4,000 kb. Cauliflower has a non-functional allele for *BoGsl-elong,* due to a 30 bp deletion in intron 1 and also splicing site mutations resulting in failure of splicing of intron 3 and a longer transcript (Li and Quiros 2002). A candidate gene for *BoGsl-pro* has also been identified (Gao et al. 2005), corresponding to another member of the MAM family.

Presence of a functional allele for *BoGsl-alk* results in synthesis of alkenyl glucosinolates, such as sinigrin (3C) and gluconapin (4C), which can be converted by hydroxylation into the anti-nutrient progoitrin (Fig. 9.4). On the other hand, the recessive allele *BoGsl-alk* will result in the accumulation of glucoraphanin (4C), a source for the anticarcinogen sulfuraphane, and glucoiberin (3C). *BoGsl-alk* has been cloned (Li and Quiros 2003) and it corresponds to a family of genes coding for 2-oxoglutarate-dependent dioxygenases (2-ODDs) located on chromosome 4 of *A. thaliana* (Hall et al. 2001). This gene has three exons and a product of 439 amino acids. The non-functional allele has a 2 bp deletion causing a frameshift mutation, resulting in a shorter product. The isolation of this gene by map-based cloning was confirmed by complementary transformation in *Arabidopsis*. Co-dominant markers are available for all three loci, allowing marker-assisted selection for *Brassica* lines of different GSL composition (Qiu et al. (2009).

GSL can be hydrolyzed by myrosinase into isothiocyanates (ITC), nitriles, and epithionitriles. Only ITC have biological activity, therefore having high GSL does not necessarily translate into high yield of nutraceuticals, such as sulforaphane. In *A. thaliana* two genes have been identified controlling the GSL hydrolysis process, which is also influenced by the environment: *ESP* (epithiospecifier protein) is a nitrile-enhancing myrosinase cofactor and *ESM1*, increases yield of ITC. Homologs to these genes have been found and mapped in *B. oleracea* (Qiu et al. unpublished). In order to maximize GSL conversion to ITC, *ESP* must be non-functional and *ESM1* must be functional.

B. oleracea vegetables are excellent sources of dietary carotenoids which have also been reported to confer health-promoting effects (Kopsell and Kopsell 2006). Kale has been cited for the highest levels of carotenoids among all leafy vegetables (Sommerburg et al. 1998). In general, lutein is the most abundant carotenoid in kale ranging from 4.8 to 13.4 mg/100 g fresh weight (FW) among different genotypes, while β-carotene is the next most abundant at levels from 3.8 to 10.0 mg/100 g FW (Kopsell et al. 2004). These authors observed significant genotype and environmental effects on lutein and β-carotene levels but no genotype by environment interaction. Recent work on broccoli (Farnham and Kopsell, unpublished) indicates that lutein is also the most prominent carotenoid in broccoli and that genotype has a highly significant effect on level of this carotenoid. It is likely that future efforts to select for lutein concentration are likely to be effective. The *Or*-gene that conditions orange cauliflower curd has led to extensive study of an altered gene and ultimately enzyme in the carotenoid biosynthetic pathway (Li et al. 2001, 2006); however, the orange mutants have not shed much light on our understanding of inter-varietal genetic variation for levels of specific carotenoids.

B. oleracea vegetables contain numerous other secondary plant metabolites that have been found associated with health-promoting effects. Examples include vitamin C and vitamin K (Kushad et al. 1999), polyphenolics (Vallejo et al. 2003), and seleno methionine complexes (Finley and Davis 2001). To date, little is known about genetic controls within the plant that affect different levels of these constituents.

9.8 Disease and Insect Resistance

Research on *B. oleracea* host plant resistance to disease has been conducted on a wide variety of diseases and crops (Koike et al. 2007). However, this section encompasses only four primary diseases for which more extensive research has been conducted and specific resistance genes have been identified.

Yellows of *B. oleracea* crops is an economically damaging disease caused by a specific form of Fusarium [*F. oxysporum* Schlechtend: Fr. F. sp. *Conglutinans* (Wollenweb)]. This soilborne fungus is persistent and only infects members of the Brassicaceae. Yellows is a warm-weather disease that does not develop well at temperatures below 16°C (Koike et al. 2007). Breeding for resistance to yellows in *B. oleracea* is one of the oldest examples in which genes for host plant resistance have been used effectively to control disease. "Wisconsin Hollander," an open-pollinated cabbage variety, was the first cultivar developed as resistant to yellows (Jones and Gilman 1915). This resistance was shown to be polygenic and was designated "Type B" (Armstrong 1933, Walker and Hooker 1945). A more important resistance, identified in "Wisconsin Ballhead" and ultimately designated "Type A" (Walker and Hooker 1945), was determined to be conditioned by a single dominant gene (Walker 1930). The dominant allele that conditions "Type A" resistance has been used very effectively in cabbage cultivars grown in North America for more than 70 years. This resistance has been characterized as a classic example of host plant resistance that is monogenic and also long lasting (Dixon 1981).

Clubroot disease, caused by the obligate parasite, *Plasmodiophora brassicae*, is one of the most destructive diseases of Brassica crops, occurring wherever they are grown (Koike et al. 2007). The causal agent of this disease possesses undefined variation that makes breeding for resistance problematic. Several *B. oleracea* sources of resistance to clubroot have been identified; however, few sources of resistance have been used effectively in cultivar development (Hirai 2006). Classical genetic studies that have examined inheritance of different *B. oleracea* sources of resistance typically find polygenic control of the respective traits (Laurens and Thomas 1993, Vriesenga and Honma 1971). In examining one source of club root resistance in cabbage, Chiang and Crete (1970) concluded the resistance was controlled by two genes that they designated *pb1* and *pb2*. Studying other sources, Voorrips and Kanne (1997) also identified two genes controlling resistance that were ultimately designated *pb3* and *pb4*. In separate mapping studies (Figdore et al. 1993, Landry et al. 1992, Voorips et al 1997; Granclement and Thomas 1996) aimed at marking clubroot resistance factors previously described, all groups identified at least two loci or QTL that were closely associated with their respective resistances. Chiang and Crete (1983) also transferred a single dominant gene for resistance to clubroot

(*P. brassicae* race 2) from *B. napus* to *B. oleracea*. Possible interrelationships between genes identified by different research groups have not been elucidated to date.

Downy mildew, caused by the oomycete *Hyaloperonospora* Constant. *parasitica* (Pers. :Fr) Fr., formerly *Peronospora parasitica* (Constantinescu and Fatehi 2002), is a destructive disease of *B. oleracea* that also occurs worldwide. This disease not only can be very destructive to young seedlings but can also damage crops at all stages through maturity (Koike et al. 2007). Natti et al. (1967) were the first to describe seedling resistance to downy mildew in *B. oleracea* that was conditioned by a single dominant gene. Other subsequent studies (Farnham et al. 2002, Jensen et al. 1999b, Mahajan et al. 1995) have also described seedling resistances controlled by similar gene action. In addition, Coelho and Monteiro (2003) identified a single dominant gene conditioning downy mildew resistance expressed at the mature plant stage. It is likely that these single factors for resistance represent race-specific genes (Natti et al. 1967, Jensen et al. 1999a); however, the exact role that pathogen race plays for this disease is uncertain due to a general and historical lack of homozygous differentials that could be useful in clarifying variation in *H. parasitica* virulence. Other sources of resistance are multigenic in nature (Jensen et al. 1999b, Wang et al. 2001), and Dickson and Petzholt (1993) hypothesized that modifying genes probably act in concert with major genes to confer variable levels of downy mildew resistance. Farinho et al. (2004) mapped the resistance locus identified by Coelho and Monteiro (2003) and characterized two markers linked in coupling to the resistance allele at distances of 3.1–3.6 cM. Giovannelli et al. (2002) developed Sequence Characterized Amplified Regions (SCAR) markers closely linked to the dominant resistance allele derived from “Everest” broccoli, and the SCARs were used to localize the resistance allele on a dense *B. oleracea* map and to show it is closely linked to the glucosinolate pathway gene *BoGsl-elong* (Gao et al. 2007).

Black rot is probably the most serious disease of crucifers worldwide, and it is a problem on all forms of *B. oleracea* (Koike et al. 2007). The causal agent of this disease is the aerobic, Gram-negative bacterium *Xanthomonas campestris* pv. *campestris*. The first report of host plant resistance to this disease was by Bain (1955), citing a single dominant gene(s) derived from the cabbage cultivars “Early Fuji” and “Huguenot.” Williams et al. (1972) confirmed the presence of a single major gene for black rot resistance in “Early Fuji” but they described it as recessive and designated it *f*. These authors explained variable segregation ratios among different populations involving “Early Fuji” due to the presence of different modifying genes, including one recessive factor and another dominant factor derived from other cabbage sources. Subsequent analysis of recombinant lines resulting from this work identified several QTL contributing to resistance with one having the largest effect possibly representing the *f* gene (Camargo et al. 1995). Dickson and Hunter (1987) analyzed resistance in cabbage PI436606 and described similar gene action (e.g., single recessive with modifiers) for this resistance as that by Williams et al. (1972) for “Early Fuji.” Resistance factors from the above sources have been used in the development of black rot resistant breeding materials; however, resistance is usually only partial and it is difficult to move into hybrid cultivars. Ignatov et al. (1999) described a vascular stem resistance gene to black rot in a kale (*SR1*) that appeared to segregate as a single dominant gene independent of the recessive factor from PI436606. Clearly, it is now better understood that inheritance of resistance will be influenced by *X. campestris* isolate or race (Ignatov et al. 1998). For example, when tested against race 3 (Vicente et al. 2002), resistance from PI 436606 was inherited as a single dominant gene, while it was previously described as recessive in nature (presumably against other races). Attempts have been made to move stronger resistance from other species like *B. carinata* (Tonguc et al. 2003) into *B. oleracea*, but to date these efforts have not resulted in stable resistance in the latter species.

Insect predation, especially by lepidopterous caterpillars, can be a serious problem in *B. oleracea* crop production. Compared to research on disease resistance, relatively little work has been done to identify or breed for insect resistance. In addition, specific genes for resistance are not generally recognized. Invariably, any resistances studied to date are considered quantitative in nature, often exhibiting relatively low heritability (Dickson and Eckenrode 1980). With several insect pests, relative insect resistance has been shown to be closely associated with a glossy leaf trait (Dickson and Eckenrode 1975, Stoner 1990, Farnham and Elsey 1995). Genes conferring a glossy leaf phenotype have not been characterized in depth, but they are typically controlled by a single gene, sometimes not only dominant, but also recessive (Stoner 1990).

9.9 Chromosome Number Variation

9.9.1 Polyploidy

Tetraploids have been induced in *B. oleracea* by chromosome doubling in several instances and in various crops such as kale (Chevre et al. 1989, Jenczewski et al. 2002), cabbage, sprouting broccoli, and Rapid Cycling var. alboglabra (Albertin et al. 2005) as well as in a few others (McNaughton 1973). Jenczewski et al. (2002) reported that autotetraploids of kales segregate in a tetrasomic fashion. However, chromosome pairing in these kales was diploid-like, but they had 10–20% reduction in pollen fertility (Chevre, personal communication). The authors suggested that this type of chromosome behavior is genotype-dependent considering that McNaughton (1973) observed quadrivalents in other tetraploid genotypes. Albertin et al. (2005) did not find changes in gene expression patterns in foliage tissue between diploid and tetraploids derived from the same genotype in their efforts to develop higher yielding *B. oleracea* fodder. Tetraploid fodder was obtained by crossing tetraploid cabbage × tetraploid kale induced by colchicine. These tetraploids were high yielding, but their low fertility was an obstacle for seed production (Olson and Ellestrom 1980).

Tetraploids are commonly observed among plants regenerated from anther and microspore cultures (Farnham 1998, Wang et al. 1999). Although many of these microspore-derived tetraploids are vigorous and appear to have viable pollen, fertility of these plants is invariably low, as has been observed in most other cases with tetraploids.

9.9.2 Aneuploidy

Monosomic addition lines (MAL) have been successfully produced by several laboratories (Quiros et al. 1987, McGrath and Quiros 1990, McGrath et al. 1990, Heneen and Jorgensen 2001, Heneen and Brismar 2001) allowing physical mapping of various genes, including the white flower gene *Wh*.

The basic scheme has been to cross natural or synthetic *B. napus* (AACC) to *B. rapa* (AA) to produce sequidiploids of genomic constitution AAC. These are backcrossed 4–6 times to the diploid parent selecting in each generation individuals carrying an extra C genome chromosome as an alien chromosome. The goal is to obtain the complete series of alien addition lines for each of the 9 C genome chromosomes on *B. rapa* diploid background: AA + C^1 to C^9.

These are maintained by selfing or backcrossing to the *B. rapa* parent, since obtention of disomic lines is very rare presumably due to poor transmission of the alien chromosome in pollen grains.

9.10 Monoploids and Anther/Microspore Culture

The first report of plant regeneration from cultures of immature pollen of an angiosperm described work using *B. oleracea* (Kameya and Hinata 1970). By the early 1980s, useful protocols for regenerating doubled haploids from anther cultures were established for this species (Keller 1984). Numerous studies helped to refine the application of anther culture and doubled haploid (DH) production for cabbage (Chiang et al. 1985, Roulund et al. 1990), broccoli (Arnison et al. 1990, Fabijanski et al. 1991), cauliflower (Boucalt et al. 1991, Yang et al. 1992), Brussels sprouts (Ockendon 1986, Ockendon and McClenaghan 1993), as well as other *B. oleracea* crops (Keller and Armstrong 1981). During this same period, successful regeneration of plants from isolated microspores was reported for *B. napus* (Lichter 1982). This Brassica species became a model for the direct culture of isolated microspores to regenerate DHs (reviewed by Cordewener et al. 1998). Ultimately, improved methods developed for *B. napus* were applied to *B. oleracea* (Duijs et al. 1992) and this technique is now used more often than anther culture by academic and industry-based researchers (personal communications). In total, the development of DHs of *B. oleracea* via anther and microspore cultures has become one of the most successful technologies that is now widely applied to the improvement of nearly all vegetables of this species (Schrijver 2002). Today, most *B. oleracea* breeding programs are as likely to develop inbreds from DH techniques as they are using conventional selection and inbreeding.

Although use of DH lines for crop improvement has become standard for *B. oleracea*, the use of doubled haploid populations for genetic studies has not progressed as rapidly. This is likely due to the fact that development of a relatively large (i.e., up to 200 lines) DH population is labor intensive and time consuming, requiring up to 18 months from culture initiation until DH line seed is obtained (Farnham 1998). Some of the first uses of DH populations for genetic studies were by Voorrips et al. (1997) who mapped two genes for resistance to clubroot and by Bohuon et al. (1996) who examined conserved C genome sequences between *B. oleracea* and *B. napus*. Wang et al. (2001) used a DH population to study inheritance of downy mildew resistance, and most recently, Iñiguez-Luy et al. (2009) developed a DH population from the F_1 of a cross between a DH broccoli and a DH rapid cycler to serve as an immortalized mapping population.

9.11 Genomic Tools: Markers, Genetic and Physical Maps

As in the majority of other crops, the evolution of genetic markers in *Brassica* and *B. oleracea* in particular has been from morphological followed by biochemical markers, mostly isozymes, and then DNA-based molecular markers, RFLP first and PCR-based markers next. The latest review on the subject was presented by Quiros and Paterson (2004). Since that time the effort has been focused mostly on the production of microsatellite or Single Sequence Repeat (SSR) markers (Tonguc and Griffiths 2004, Burgess et al. 2006, Louarn et al. 2007. Iñiguez-Luy et al. 2008). A microsatellite information exchange web page has been established to facilitate access to markers that are in the public domain (http://www.brassica.info/resource/markers/ssr-exchange.php).

9.12 Map Development in *Brassica*

The development of genetic maps in *Brassica* has a dual purpose: (1) understanding the origins of, and relationships among, the rapidly evolving genomes that are responsible for the morphological diversity and evolutionary success of the diploid cultivated *Brassica* species, and (2) utilization in applied genetics and breeding of the *Brassica* crops.

After the construction of the first substantial linkage map of *B. oleracea,* with isozyme loci, pioneered by Arus and Orton (1983), several others followed using a variety of molecular markers. Sebastian et al. (2002) established a consensus map for this species based on perpetuated individuals from two doubled haploid populations, (cauliflower × Brussels sprouts and broccoli × kale) constructed with 547 RFLP, AFLP, and SSR markers . The nine linkage groups from that map have been physically assigned to their respective chromosomes by Howell et al. (2002) using Fluorescent In Situ Hybridization FISH. With the availability of EST sequences from *Arabidopsis*, these have been used to construct several maps allowing partial comparison of the *A. thaliana* genome with the *B. oleracea* genome (Kowalski et al. 1994, Lan et al. 2000, Babula et al. 2003). This task was also accomplished by Li et al. (2003) using cDNA polymorphisms to construct a linkage map in *B. oleracea* followed by comparative physical mapping to *A. thaliana*.

Gao et al. (2007) reported the construction of a high-density genetic map based on the broccoli × cauliflower F_2 population used by Li et al. (2003), adding over 1,200 various types of PCR-based markers spanning 703 cM in nine linkage groups, designated LG1-LG9. It was developed in a F_2 segregating population of 143 individuals obtained by crossing doubled haploid plants of broccoli "Early-Big" and cauliflower "An-Nan Early." These markers are randomly distributed throughout the map, which includes a total of 1,062 genomic SRAP markers, 155 cDNA SRAP markers, 26 SSR markers, 3 broccoli BAC end sequences, and 11 known Brassica genes: *BoGsl-alk, BoGsl-elong, BoGsl-proa , BoGsl-prob, BoGslCS-lyase, BoGsl-oh, BoGslcyp79F1, BoGslS-GT* (glucosinolate pathway), *BoDM1* (resistance to downy mildew), *BoCALa, BoAP1a* (inflorescence architecture). *BoDM1*, and *BoGsl-elong* are linked on LG 2 at 0.8 cM, making it possible to use the glucosinolate gene as a marker for the disease resistance gene. By QTL analysis, it was found that three segments are involved in curd formation in cauliflower. The map was aligned to the C genome linkage groups and chromosomes of *B. oleracea* and *B. napus*, and anchored to the physical map of *A. thaliana*. This map adds over 1,000 new markers to *Brassica* molecular tools (Gao et al. 2007).

Iñiguez-Luy et al. (2009) recently constructed a map for *B. oleracea* using a DH population of approximately 150 lines developed from a broccoli × rapid cycling *B. oleracea* cross. A total of 120 RFLP probes and 146 SSR markers, as well as one phenotypic marker (flower color) have been mapped on this population to date. This map of 279 markers is distributed along nine linkage groups with a total distance of 891.4 cM and density of 3.2 cM/marker. This map represents a population that can be easily perpetuated due to a high level of self-compatibility for most of the individual DH lines.

Presently, efforts are being made to coordinate the mapping activities of many *Brassica* laboratories and to align the existing maps of *B. oleracea* and *B. napus,* assigning linkage groups to chromosomes and cross-referencing these to the *A. thaliana* physical map. Anchor probes based on *A. thaliana sequences*, called GST (genomic sequence tags) probes, are under development by the Brassica IGF (Investigating Gene Function) project (http://brassica.bbsrc.ac.uk/index.htm) in England. Additionally, efforts at Cold Spring Harbor and TIGR (http://www.tigr.org/tdb/e2k1/bog1/) in association with *Arabidopsis* sequence annotation yielded approximately 420,000 genomic shotgun sequences for *B. oleracea* that have been posted in GenBank and *Brassica* data bases (http://brassica.bbsrc.ac.uk/).

9.13 Synteny Maps

Synteny maps in the context of this chapter are those constructed by locating markers on chromosomes, without specifying order or distance, using specialized cytogenetic stocks. Physical mapping in *Brassica* has been partially accomplished by construction of synteny maps using alien addition lines for a few species (Quiros 2001, Quiros and Paterson 2004). These have been constructed for most of the

C-genome chromosomes by RFLP, isozymes, and RAPD markers. Genes for erucic acid content and a flower color gene have also been mapped.

Before the assignment of *B. oleracea* linkage groups to their respective chromosomes, it was possible to partially accomplish alignment to synteny maps based on the C-genome alien addition lines mentioned above (Quiros 2001, Heneen and Jorgensen 2001, Heneen and Brismar 2001). Although it was possible in most cases to physically assign linkage groups to chromosomes using such sets of lines, two major complications arising from this activity deserve comment. The first one involves the frequent lack of polymorphism of interspecific markers on the alien chromosomes in the intra-specific crosses used to develop the linkage maps. The second complication is the instability of the alien chromosomes that tend to suffer mostly terminal deletions.

Fluorescence *in situ* hybridization (FISH) to chromosome spreads, which has been mastered by a few laboratories, is proving quite useful for physical mapping. This technology makes it possible to identify all the chromosomes in the A, B, and C genomes based on specific probes and BAC clones (Jackson et al. 2000, Hasterok et al. 2001). Single gene mapping on chromosomes and extended DNA fibers has also been accomplished for the self-incompatibility related genes *SLG* and *SRL* (Fukui 2003).

On the whole, FISH using rDNA probes and DAPI stain has allowed the identification of virtually all mitotic chromosomes in the A and C genomes (Hasterok et al. 2001, Schrader et al. 2000, Snowdon et al. 2002). BAC clones as FISH probes are becoming increasingly useful for *Brassica* chromosome identification. These clones have been applied to meiotic (Ziolkowski and Sadowski 2002) and mitotic C genome chromosomes quite successfully. Howell et al. (2002) using mostly *B. oleracea* BAC clones as probes on mitotic chromosomes were able to assign and orient each linkage group of the consensus *B. oleracea* genetic map mentioned above to its respective chromosome.

9.14 Genomics

Qiu et al. (2009) sequenced five BAC clones of *B. oleracea* doubled haploid "Early Big" broccoli containing major genes in the aliphatic glucosinolate pathway, and comparatively analyzed them with analogous sequences in *A. thaliana* and *B. rapa*. Additionally, the analysis included published sequences from three other *B. oleracea* BAC clones and a contig of this species corresponding to segments in *A. thaliana* chromosomes IV and V. A total of 2,946 kb of *B. oleracea*, 1,069 kb of *B. rapa* sequence, and 2,607 kb of *A. thaliana* sequence were compared and analyzed. Conserved, collinearity for gene order and content was restricted to specific chromosomal segments, with frequent breaks in collinearity resulting in gene absence likely not due to gene loss but rearrangements. *B. oleracea* has the lowest gene density of the three species, followed by *B. rapa*. The genome expansion of the *Brassica* species, *B. oleracea*, in particular, is due to larger introns and gene spacers resulting from frequent insertion of DNA transposons and retrotransposons. These findings are discussed in relation to the possible origin and evolution of the Brassica genomes (see Chapter 6 by Town et al., this book).

9.15 Outlook

B. oleracea was one of the privileged species subjected to genetic research early on after the rediscovery of Mendel's laws of inheritance. This undoubtedly was due in part to the relative importance of the vegetables in this species. However, it was also due to the great morphological variation represented in the many different crops domesticated from this species, which sparked the curiosity of many breeders. It was also discovered early on that this species was a paleopolyploid, based mostly on cytogenetic research. Today, with approximately 50% of the genome sequenced based on shot gun reads , with the availability of highly saturated genetic maps and thousands of markers, and powerful statistical strategies, we still struggle to explain the inheritance of most traits intrinsic to each crop. Also, we still have a long way to go in understanding the structure and composition of the genome. With the current advances that we are seeing on genomics, we hope to be able to meet this challenge and to answer these questions in the years to come.

Acknowledgments We are indebted to Graham King for supplying Fig. 9.2., and to J. Erron Haggard for taking the picture used for Fig. 9.3.

References

Al-Shebaz IA, Beilstein MA, Kellogg EA (2006) Systematics and phylogeny of the Brassicaceae (Cruciferae): an overview. Pl Syst Evol 259:89–120

Albertin W, Brabant P, Catrice O, Eber F, Jenczewski E, Chèvre AM, Thiellement H (2005) Autopolyploidy in cabbage (Brassica oleracea L) does not alter significantly the proteomics of green tissue. Proteomics 5:2131–2139

Allgayer H (1928) Genetic investigations with garden cabbage by crossing trials by Richard Freedenberg. Z Indukt Abstamm Vererbungsl 47:191–260

Armstrong ME (1933) Fusarium resistance in Wisconsin Hollander cabbage. J Agric Res 47: 639–661

Arnison PG, Donaldson P, Jackson A, Semple C, Keller WA (1990) Genotype specific response of cultured broccoli (Brassica oleracea var. italica) anthers to cytokinins. Plant Cell Tissue Organ 20:217–222

Arus P, Orton T (1983) Inheritance and linkage relationships of isozyme loci in *Brassica oleracea.* J Hered 74:405–412

Babula B, Kaczmarck M, Barakat A, Delseny M, Quiros CF, Sadowski J (2003) Chromosomal mapping of B. oleracea based on Arabidopsis ESTs: complexity of comparative mapping. Mol Gen Genet 268:656–665

Bain DC (1955) Resistance of cabbage to black rot. Phytopathology 45:35–37

Block G, Patterson B, Subar A (1992) Fruit, vegetables and cancer prevention: a review of the epidemiological evidence. Nutr Cancer 18:1–29

Bohuon EJ, Keith DJ, Parkin IAP, Sharpe AG, Lydiate DJ (1996) Alignment of the conserved C genomes of Brassica oleracea and Brassica napus. Theor Appl Genet 93:833–839

Boucalt L, Chauvin JE, Margale E, Herve Y (1991) Etude de caracteres morphologiques et iso-enzymatiques sur des plantes issues de culture d' anthers chezle chou-fleur (*Brassica oleracea* var botrytis). Agronomie 11:727–736

Branca F (2008) Cauliflower and broccoli. In: J Prohens and F Nuez (eds.) Vegetables I, pp 147–182. Springer, New York, NY

Burgess B, Mountford H, Hopkins CJ, Love C, Ling AE, Spangenberg GC, Edwards D, Batley J (2006) Identification and characterization of single sequence repeat (SSR) markers derived *in silico* from Brassica oleracea genome shotgun sequences. Mol Ecol Notes 6: 1191–1194

Camargo LEA, Williams PH, Osborn TC (1995) Mapping of quantitative trait loci controlling resistance of Brassica oleracea to Xanthomonas campestris pv. campestris in the field and greenhouse. Phytopathology 85:1296–1300

Carlson DG, Daxenbichler M, VanEtten CH, Kwolek WF, Williams PH (1987) Glucosinolates in crucifer vegetables: broccoli, Brussels sprouts, cauliflower, collards, kale, mustard greens, and kohlrabi. J Amer Soc Hort Sci 112:173–178

Chen S, et al. (2003) CYP79F1 and CYP79F2 have distinct functions in the biosynthesis of aliphatic glucosinolates in Arabidopsis. Plant J 33:923–937

Chevre AM, Delourme R, Eber F, Thomas G, Baron F (1989) Etudes cytologiques de choux tetraploids (Brassica oleracea var. acephala) obtenues a partir de lignees diploides apres traitement a la cochicine. Agronomie 9:521–525

Chiang MS, Chong C, Landry RS, Crete R (1993) Cabbage Brassica oleracea subsp. capitata L. In: G Kalloo and BO Bergh (ed) Genetic improvement of vegetable crops, pp 113–155. Pergamon Press, Oxford

Chiang MS, Crete R (1970) Inheritance of clubroot resistance in cabbage (Brassica oleracea L. var. capitata L.). Can J Genet Cytol 12:253–256

Chiang MS, Crete R (1983) Transfer of resistance to race 2 of *Plasmodiophora brassicae* from *Brassica napus* to cabbage (*B. oleracea* ssp. *Capitata*). V. the inheritance of resistance. Euphytica 32:479–483

Chiang MS, Frechette S, Kuo CG, Chong C, Delafield SJ (1985) Embryogenesis and haploid plant production from anther culture of cabbage (Brassica oleracea var. capitata L.). Can J Plant Sci 65:1033–1037

Chiu L, Prior RL, Wu X, Li L 2005. Toward identification of the candidate gene controlling anthocyanin accumulation in purple cauliflower (brassica oleracea l. var. botrytis).American Society of Plant Biologists Annual Meeting. P 628

Clarke JD, Dashwood RH, Ho E (2008) Multi-targeted prevention of cancer by sulforaphane. Cancer Lett 269:291–304

Coelho PS, Monteiro AA (2003) Inheritance of downy mildew in mature broccoli plants. Euphytica 131:65–69

Constantinescu O, Fatehi J (2002) *Peronospora*-like fungi (Chromista, Peronosporales) parasitic on Brassicaceae and related hosts. Nova Hedwigia 74:291–338

Cordewener JHG, Custers JBM, Van Lookeren Campagne MM (1998) Microspore culture: a model for investigating the role of stress in the induction of embryogenesis. In: Chupeau Y, Caboche M and Henry Y (ed) Androgenesis and haploid plants, pp 54–68. Springer, Heidelberg

Crisp P, Angell. SM (1985) Genetic control of green colour in cauliflower. Ann Appl Biol 107:601

Crisp P, Tapsell CR (1992) Cauliflower. Brassica oleracea L. In: G Kalloo and BO Berg (ed) Genetic improvement of vegetable crops, pp 157–178. Pergamon Press, Oxford

de Quiros C, Magrath H,R, McCallum D, Kroymann J, Schnabelrauch D, Mitchell-Olds T, Mithen. R (2000) α-Keto acid elongation and glucosinolate biosynthesis in Arabidopsis thaliana. Theor Appl Genet 101:429–437

Dickson MH, Eckenrode CJ (1975) Variation in Brassica oleracea resistance to cabbage looper and imported cabbage worm in the greenhouse and field. J Econ Entomol 68:757–760

Dickson MH, Eckenrode CJ (1980) Breeding for resistance in cabbage and cauliflower to cabbage looper, imported cabbageworm, and diamondback moth. J Am Soc Hort Sci 105: 782–785

Dickson MD, Hunter JE (1987) Inheritance of resistance in cabbage seedlings to black rot. HortScience 22:108–109

Dickson MH, Lee CY, Blamble AE (1988) Orange-curd high carotene cauliflower inbreds, NY 156, NY 163, and NY 165. HortScience 23:778–779

Dickson MH, Petzholdt R (1993) Plant age and isolate source affect expression of downy mildew resistance in broccoli. HortScience 28:730–731

Dickson MH, Wallace DH (1986) Cabbage breeding. In: MJ Bassett (ed). Breeding vegetable crops, pp 395–432. AVI Publishing Co., Westport, CT

Diederichsen A (2001) Cruciferae: Brassica. In: Hanelt, P(ed) Institute of Plant Genetics and Crop Plant Research. Mansfeld's encyclopedia of agricultural and horticultural crops, pp 1435–1446. Springer, Berlin

Dixon GR (1981) Pathogens of crucifer crops. In: Vegetable crop diseases, pp 112–156. Macmillan Publishers LTD, London

Duclose DV, Bjorkman T (2008) Meristem identity gene expression during curd proliferation and flower initiation in Brassica oleracea. J Exp Bot 59:421–433

Duijs JG, Voorrips RE, Visser DL, Custers JBM (1992) Microspore culture is successful in most crop types of *Brassica oleracea* L. Eupytica 60:45–55

Fabijanski SF, Altosaar I, Arnison PG (1991) Heat shock response during anther culture of broccoli (*Brassica oleracea* var *italica).* Plant Cell Tiss Org Cult 26:203–212

Fahey JW, Zhang Y, Talalay P (1997) Broccoli sprouts: an exceptionally rich source of inducers of enzymes that protect against chemical carcinogens. Proc Natl Acad Sci 94: 10367–10372

Farinho M, Coelho P, Carlier J, Svetleva D, Monteiro A, Leitao J (2004) Mapping of a locus for adult plant resistance to downy mildew in broccoli (*Brassica oleracea* convar. *Italica).* Theor Appl Genet 109:1392–1398

Farnham MW (1998) Doubled-haploid broccoli production using anther culture: effect of anther source and seed set characteristics of derived lines. J Am Soc Hortic Sci 123:73–77

Farnham MW, Elsey KD (1995) Recognition of *Brassica oleracea* L. resistance against the silverleaf whitefly. HortScience 30:343–347

Farnham MW, Ruttencutter G, Smith JP, Keinath AP (2005) Hybridizing collard and cabbage may provide a means to develop collard cultivars. HortScience 40:1686–1689

Farnham MW, Stephenson KK, Fahey JW (2000) The capacity of broccoli to induce a mammalian chemoprotective enzyme varies among inbred lines. J Amer Soc Hort Sci 125:482–488

Farnham MW, Wang M, Thomas CE (2002) A single dominant gene for downy mildew resistance in broccoli. Euphytica 128:405–407

Figdore SS, Ferreira ME, Slocum MK, Williams PH (1993) Association of RFLP markers with trait loci affecting clubroot resistance and morphological characters in *Brassica oleracea* L. Euphytica 69:33–44

Finley JW, Davis CD (2001) Selenium (Se) from high-selenium broccoli is utilized differently than selenite, selenate and selenomethionine, but is more effective in inhibiting colon carcinogenesis. Biofactors 14:191–196

Fukui K (2003) *Brassica* and legume chromosomes. In: T Nagata, S Tabata (eds) Biotechnology in Agricultural and Forestry (BAF), 52, 114–141. Springer, Berlin

Gao M, Li G, Potter D, McCombie WR, Quiros CF (2005) Comparative analysis of methylthioalkylmalate synthase (MAM) gene family and flanking DNA sequences in *Brassica oleracea* and *A. thaliana*. Plant Cell Rep 25:592–598

Gao M, Li G, Yang B, Qiu D, Farnham M, Quiros C (2007) High-Density *Brassica oleracea* linkage map: identification of useful new linkages. Theor Appl Genet 115:277–287

Giamoustaris A, Mithen. R (1996) Genetics of aliphatic glucosinolates. IV. Side-chain modification in *Brassica oleracea*. Theor Appl Genet 93:1006–1010

Giovannellli JL, Farnham MW, Wang M, Strand AE (2002) Development of sequence characterized amplified region markers linked to downy mildew resistance in broccoli. J Amer Soc Hort Sci 127:597–601

Gladis Th, Hammer K (2001) Nomenclatural notes on the Brassica oleracea-group. Genet Res Crop Evol 48:7–11

Gomez-Campo C, Hinata H (1999) Biology of Brassica coenospecies. developments in plant genetics and breeding, 4. Elsevier, Amsterdam

Grandclement C, Thomas G (1996) Detection and analysis of QTLs based on RAPD markers for polygenic resistance to plasmodiophora brassicae Woron in *Brassica oleracea* L. Theor Appl Genet 93:86–90

Gray AR (1982) Taxonomy and evolution of broccoli (*Brassica oleracea* var. *italica)*. Econ Bot 36:397–410

Gray AR (1992) Broccoli. *Brassica oleracea* L. (Italica group). In: G Kalloo and BO Berg (ed) Genetic improvement of vegetable crops, pp 61–86. Pergamon Press, Oxford

Hall JC, Sytsma KJ, Iltis. HH (2001) Phylogeny of capparaceae and Brassicaceae based on chloroplast sequence data. Am J Bot 89:1826–1842

Harberd DJ (1972) A contribution to the cyto-taxonomy of *Brassica* (*Cruciferae*) and its allies. Bot J Linn Soc 65:1–23

Hasterok R, Jenkins G, Langdon T, Jones RN, Maluszynska J (2001) Ribosomal DNA is an effective marker of Brassica chromosomes. Theor Appl Genet 103:486–490

Heneen WK, Brismar K (2001) Maternal and embryonal control of seed colour by different *Brassica alboglabra* chromosomes. Plant Breed 120:325–329

Heneen WK, Jørgensen RB (2001) Cytology, RAPD, and seed colour of progeny plants from *Brassica rapa-alboglabra* aneuploids and development of monosomic addition lines. Genome 44:1007–1021

Herve Y (2003) Choux. In: Pitrat M and Foury C, (eds) History de legumes, des origins a l'oree du XXI siecle. INRA, Paris, France

Hirai M (2006) Genetic analysis of clubroot resistance in Brassica crops. Breed Sci 56: 223–229

Hosaka K, Kianian SF, McGrath JM, Quiros CF (1990) Development and chromosomal localization of genome specific DNA markers of *Brassica* and the evolution of amphidiploid and n=9 diploid species. Genome 33:131–142

Howell EC, Armstrong SJ, Barker GC, et al (2005) Physical organization of the major duplication on Brassica oleracea chromosome O6 revealed through fluorescence in situ hybridization with Arabidopsis and Brassica BAC probes. Genome 48:1093–1103

Howell BC, Barker GC, Jones GH, Kearsey MJ, King GJ, Kop EP, Ryder CD, Teakle GR, Vicente JG, Armstrong. SJ (2002) Integration of the cytogenetic and genetic linkage maps of *Brassica oleracea*. Genetics 161:1225–1234

Ignatov A, Kuginuki Y, Hida K (1998) Race-specific reaction to black rot in *Brassica oleracea.* Eur J Plant Path 104:8221–8827

Ignatov A, Kuginuki Y, Hida K (1999) Vascular stem resistance to black rot in *Brassica oleracea*. Can J Bot 77:442–446

Iñiguez-Luy F, Lukens L, Farnham MW, Amasino RM, Osborn TC (2009) Development of public immortal mapping populations, molecular markers and linkage maps for rapid cycling *Brassica rapa* and *Brassica oleracea*. Theor Appl Genet 120:31–43

Iñiguez-Luy F, Voort A, Osborn T (2008) Development of a set of public SSR markers derived from genomic sequence of rapid cycling *Brassica oleracea* L. genotype. Theor Appl Genet 117:977–985

Jackson SA, Cheng Z, Wang ML, Goodman HM, Jiang. J (2000) Comparative fluorescence in situ hybridization mapping of a 431-kb *Arabidopsis thaliana* bacterial artificial chromosome contig reveals the role of chromosomal duplications in the expansion of the *Brassica rapa* genome. Genetics 156:833–838

Jenczewski E, Eber F, Manzanares-Dauleux MJ, Chevre AM (2002) A strict diploid-like pairing regime is associated with tetrasomic heredity in induced autotetraploids of kale (*Brassica oleracea*). Plant Breed 121:177–179

Jensen BD, Hockenhull J, Munk L (1999a) Seedling and adult plant resistance to downy mildew (*Peronospora parasitica*) in cauliflower (*Brassica oleracea* convar. *botrytis* var. botrytis). Plant Pathol 48:604–612

Jensen BD, Vaerbak S, Munk L, Andersen SB (1999b) Characterization and inheritance of partial resistance to downy mildew, *Peronospora parasitica*, in breeding material of broccoli, *Brassica oleracea* convar. *botrytis* var. *italica*. Plant Breed 118:549–554

Jones LR, Gilman JC (1915) The control of cabbage yellows through disease resistance. Wis Agric Exp Stn Res Bull 38

Kameya T, Hinata K (1970) Induction of haploid plants from pollen grains of Brassica. Jpn J Breed 20:82–87

Karpechenko GD (1927) Polyploid hybrids of *Raphanus sativus* L. x *Brassica oleracea* L. Bul Appl Bot Genet Plant Breed 7:305–410

Keller WA (1984) Anther culture of Brassica. In: Cell culture and somatic cell genetics of plants, vol 1, pp 302–310. Academic Press, New York, NY

Keller WA, Armstrong KC (1981) Production of anther-derived dihaploid plants in autotetraploid marrowstem kale (*Brassica oleracea* var. *acephala*). Can J Genet Cytol 23:259–265

Kennard WC, Slocum MK, Figdore SS, Osborn TC (1994) Genetic analysis of morphological variation in *Brassica oleracea* using molecular markers. Theor Appl Genet 87:721–732

Keyes KA, Honma S (1986) Inheritance of lateral suppression and leaf numberin broccoli (*Brassica oleracea* L. *Italica* group). Eucarpia Cruciferae Newslett 11:43

Kianian SF, Quiros CF (1992) Trait inheritance, fertility and genomic relationships of some n=9 *Brassica* species. Genet Resour Crop Evol 39:165–175

Koike ST, Gladders T, Paulus AO (2007) Vegetable diseases: a colour handbook, p 448. Manson Publishing Ltd, London

Kopsell DA, Kopsell. DE (2006) Accumulation and bioavailability of dietary carotenoids in vegetable crops. Trends Plant Sci 11:499–507

Kopsell DA, Kopsell DE, Lefsrud MG, Curran-Celentano J, Dukach LE (2004) Variation in lutein, β-carotene, and chlorophyll concentrations among *Brassica oleracea* cultigens and seasons. HortScience 39:361–364

Kowalski SP, Lan T-H, Feldmann KA, Paterson AH (1994) Comparative mapping of *Arabidopsis thaliana* and *Brassica oleracea* chromosomes reveals islands of conserved organization. Genetics 138:499–510

Kristofferson KB (1924) Contributions to the genetics of *brassica oleracea*. Hereditas 5:297–364

Kushad MM, Brown AF, Kurlich AC, Juvik JA, Klein BP, Wallig MA, Jeffery EH (1999) Variation of glucosinolates in vegetable crops of *Brassica oleracea*. J Agr Food Chem 47:1541–1548

Kwan CC (1934) Inheritance of some plant characters in cabbage, Brassica oleracea var. capitata. J Agri Assoc China 126–127:81–127

Labate J, Robertson L, Baldo A, Bjorkman T (2006) Inflorescence identity genes alleles are poor predictors of inflorescence type in broccoli and cauliflower. J Amer Soc Hort Sci 1311:667–673

Lan TH, DelMonte TA, Reischmann KP, Hyman J, Kowalski SP, McFerson J, Kresovich S, Paterson AH (2000) An EST-enriched comparative map of *Brassica oleracea* and *Arabidopsis thaliana*. Genome Res 10:776–788

Lan TH, Paterson AH (2000) Comparative mapping of QTLs determining the plant size of *Brassica oleracea*. Theor Appl Genet 103:383–397

Landry BS, Hubert N, Crete R, Chiang MS, Lincoln SE, Etoh T (1992) A genetic map for *Brassica oleracea* based on RFLP markers detected with expressed DNA sequences and mapping of resistance genes to race 2 of *Plasmodiophora brassicae* (Woronin). Genome 35:409–420

Laurens F, Thomas G (1993) Inheritance of resistance to clubroot (*Plasmodiophora brassicae* Wor.) in kale (*Brassica oleracea* ssp. *acephala*). Hereditas 119:253–262

Li G, Gao M, Yang B, Quiros CF (2003) Gene to gene alignment between the *Brassica* and *Arabidopsis* genomes by transcriptional mapping. Theor App Genet 107:168–180

Li L, Lu S, Cosman KM, Earle ED, Garvin DF, O'Neill J (2006) β-carotene accumulation induced by the cauliflower *Or* gene is not due to an increased capacity of biosynthesis. Phytochemistry 67:1177–1184

Li L, Paolillo DJ, Parthasarathy MV, DiMuzio EM, Garvin DF (2001) A novel gene mutation that confers abnormal patterns of β-carotene accumulation in cauliflower (*Brassica oleracea* var. *botrytis*). Plant J 26:59–67

Li G, Quiros CF (2002) Genetic analysis, expression and molecular characterization of *BoGSL-ELONG*, a major gene involved in the aliphatic glucosinolate pathway of *Brassica* species. Genetics 162:1937–1943

Li G, Quiros CF (2003) *In planta* side-chain glucosinolate modification in Arabidopsis by introduction of dioxygenase Brassica homolog *BoGSL-ALK*. Theor Appl Genet 106:1116–1121

Lichter R (1982) Induction of haploid plants from isolated pollen of *Brassica napus*. Z Pflanzenphysiol 105:427–434

Louran S, Torp AM, Holme ID, Andersen SB, Jensen BD (2007) Database derived microsatellite markers (SSRs) for cultivar differentiation in *Brassica oleracea*. Genet Resour Crop Evol 54:1717–1725

Lu S, Van Eck J, Zhou X, Lopez AB, O'Halloran DM, Cosman KM, Conlin BJ, Paolillo DJ, Garvin DF, Vrebalov J, Kochian LV, Küpper H, Earle ED, Cao J, Li L (2006) The cauliflower *Or* Gene encodes a DnaJ Cysteine-rich domain-containing protein that mediates high-levels of β-Carotene accumulation. Plant Cell 18:3594–3605

Magrath R, Herron C, Giamoustaris A, Mithen R (1993) The inheritance of aliphatic glucosinolates in *Brassica napus*. Plant Breed 111:55–72

Magrath R, Morgen M, Bano F, Parkin I, Sharpe A, Lister C, Turner J, Dean C, Lydiate D, Mithen R (1994) Genetics of aliphatic glucosinolates. I. Side chain elongation in *Brassica napus* and *Arabidopsis thaliana*. Heredity 72:290–299

Magruder R, Myers CH (1933) The inheritance of some plant colors in cabbage. J Agr Res 47: 233–248

Mahajan V, Gill HS, More TA (1995) Inheritance of downy mildew resistance in Indian cauliflower (group III). Euphytica 86:1–3

McGrath JM, Quiros CF (1990) Generation of alien addition lines from synthetic *B. napus*: morphology, cytology, fertility, and chromosome transmission. Genome 33:374–383

McGrath JM, Quiros CF, Harada JJ, Landry BS (1990) Identification of *Brassica oleracea* monosomic alien chromosome addition lines with molecular markers reveals extensive gene duplication. Mol Gen Genet 223:198–204

McNaughton IH (1973) Synthesis and sterility of Raphanobrassica. Euphytica 22:70–88

Mithen R, Campos H (1996) Genetic variation of aliphatic glucosinolates in *Arabidopsis thaliana* and prospects for map based cloning. Entomol Exp Appl 80:202–205

Mithen RF, Clarke J, Lister. C, Dean C (1995) Genetics of aliphatic glucosinolates. III. Side chain structure of aliphatic glucosinolates in *Arabidopsis thaliana*. Heredity 74:210–215

Mithen R, Faulner K, Magrath R, Rose P, Williamson G, Marquez J (2003) Development of isothiocyanate-enriched broccoli, and its enhanced ability to induce phase 2 detoxification enzymes in mammalian cells. Theor Appl Genet 106:727–734

Natti JJ, Dickson MH, Atkin JD (1967) Resistance of *Brassica oleracea* varieties to downy mildew. Phytopathology 57:144–147

Ockendon DJ, McClenaghan R (1993) Effect of silver nitrate and 2,4 D on anther culture of Brussels sprouts (*Brassica oleracea* var. *gemmifera*). Plant Cell Tiss Org Cult 32:41–46

Okendon DJ (1986) Utilization of anther culture in breeding brussels sprouts. In: Genetic manipulation in plant breeding, pp 265–375. Walter de Gruyter and Co., Berlin

Olsson G, Ellestrom S (1980) Polyploid breeding in Europe. In: Hinata K and Gomez-Campo C (ed) Brassica crops and wild allies, pp 167–190. Japan Scientific Soc, Tokyo.

Pease MS (1926) Genetic studies in *Brassica oleracea*. J Genet 16:363–385

Pease MS (1927) Genetic studies in Brassica II. The Kohlrabil. J Genet 17:253–267

Pelletier G, et al (1988) Use of protoplasts in plant breeding: cytoplasmic aspects. Plant Cell Organ Cult 12:173–180

Pelofske PJ, Baggett JR (1979) Inheritance of internode length, plant form, and annual habit in a cross of cabbage and broccoli (*Brassica oleracea* var. *Capitata* L. and var. *Italica* Plenck). Euphytica 28:189–197

Prakash S, Chopra. VL (1991) Cytogenetics of crop *Brassicas* and their allies. In: Tsuchyia T and Gupta PK (ed) Chromosome engineering in plants:genetics, breeding, evolution, pp 161–180. Elsevier Science Publishers, The Netherland

Qiu D, Gao M, Li G, Quiros C (2009) Comparative sequence analysis for *Brassica oleracea* with similar sequences in *B. rapa* and *Arabidopsis thaliana*. Plant Cell Rept 28:649–658

Quiros CF (2001) DNA-based marker *Brassica* maps. In: Phillips RL and Vasil JK (ed) Advances in cellular and molecular biology of plants, vol 1. DNA based Markers in Plants, pp 201–238. Kluwer Academic Publ, Dordrecht, Boston, MA, London

Quiros CF, Ochoa O, Kianian SF, Douches D (1987) Analysis of the *Brassica oleracea* genome by the generation of *B. rapa-oleracea* chromosome addition lines: characterization by isozymes and rDNA genes. Theor Appl Genet 74:758–766

Quiros CF, Paterson AH (2004) Gene mapping and analysis. In: Pua EC, Douglas CJ (eds) Biotechnology in agriculture and forestry, Brassica, 54, 31–64. Springer, Berlin

Rosa EAS, Heaney RK, Fenwick GR, Portas CAM (1997) Glucosinolates in crop plants. Hort Rev 19:99–215

Roulund N, Hansted L, Andersen SB, Farestveit B (1990) Effect of genotype, environment and carbohydrate on anther culture response in head cabbage (*Brassica oleracea* L. convar. *capitata* (L.) Alef.). Euphytica 49:237–242

Sampson DR (1967) New light on the complexities of anthocyanin inheritance in *Brassica oleracea*. Can J Genet Cytol 9:352–358

Sauer JD (1993) Historical geography of crop plants: a selective roster. CRC Press, Boca Raton, FL

Schrader O, Budahn H, Ahne R (2000) Detection of 5S and 25S rRNA genes in *Sinapis alba, Raphanus sativus* and *Brassica napus* by double fluorescence in situ hybridization. Theor Appl Genet 100:665–669

Schrijver B (2002) Biotechnology applications in Brassica breeding. Proc of the 13th Crucifer Genetics Workshop. 23–26 March 2002. Davis, CA. p. 73.

Sebastian RL, Howell EC, King GJ, Marshall DF, Kearsey MJ (2002) An integratedAFLP and RFLP *Brassica oleracea* linkage map from two morphologically distinct doubled- haploid mapping populations. Theor Appl Genet 100:75–81

Smith LB, King GJ (2002) The distribution of *BoCAL*-a alleles in *Brassica oleracea* is consistent with a genetic model for curd development and domestication of cauliflower. Mol Breed 6: 603–613

Snogerup S (1980) The wild forms of the *Brassica oleracea* group (2n=18) and their possible relations to the cultivated ones. In: Tsonuda S, Hinata K, and Gomez-Campo C (ed) *Brassica* crops and wild allies: biology and breeding, pp 191–215. Japan Sci Soc Press, Tokyo

Snogerup S, Gustafsson M, von Bothmer R (1990) Brassica sect. Brassica (Brassicaceae) 1. Taxonomy and variation. Willdenowia 19:271–365

Snowdon RJ, Friedrich T, Friedt W, Köhler W (2002) Identifying the chromosomes of the A- and C-genome diploid *Brassica* species *B. rapa* (syn. *campestris*) and *B. oleracea* in their amphidiploid *B. napus*. Theor Appl Genet 104:533–538

Sommerburg O, Keune JEE, Bird AC, van Kuijk FJGM (1998) Fruits and vegetables that are sources for lutein and zeaxanthin: the macular pigment in human eyes. Br J Ophthalmol 82:907–910

Steinmetz KA, Potter JD (1996) Vegetables, fruit and cancer prevention: A review. J Amer Diet Assoc 96:1027–1039

Stoner KA (1990) Glossy leaf wax and plant resistance to insects in *B. oleracea* under natural infestation. Environ Entomol 19:730–739

Tonguc M, Earle ED, Griffiths PD (2003) Segregation distortion of Brassica carinata derived black rot resistance in *Brassica oleracea*. Euphytica 134:269–276

Tonguc M, Griffiths PD (2004) Genetic relationships of Brassica vegetables determined using database derived single sequence repeats. Euphytica 137:193–201

USDA (2006) Vegetables 2005 summary. National Agricultural Statistics Service. Vg 1–2 (06)

Vallejo F, Tomas-Barveran FA, Garcia-Viguera C (2003) Health-Promoting Compounds in Broccoli as Influenced by Refrigerated Transport and Retail Sale Period. J Sci Food Agric 83:1511–1516

Vicente JG, Taylor JD, Sharpe AG, Parkin IAP, Lydiate DJ, King GJ (2002) Inheritance of race-specific resistance to Xanthomonas campestri pv. campestris in Brassica genomes. Phytopathology 92:1134–1141

Voorrips RE, Jongerius MC, Kanne HJ (1997) Mapping of two genes for resistance to clubroot (*Plasmodiophora brassicae*) in a population of double haploid lines of *Brassica oleracea* by means of RFLP and AFLP markers. Theor Appl Genet 94:75–82

Voorrips RE, Kanne HJ (1997) Genetic analysis of resistance to clubroot (*Plasmodiophora brassicae*) in *Brassica oleracea*. I. analysis of symptom grades. Euphytica 93:31–39

Vriesenga JD, Honma S (1971) Inheritance of seedling resistance to clubroot in *Brassica oleracea* L. HortScience 6:395–396

Walker JC (1930) Inheritance of Fusarium resistance in cabbage. J Agr Res 40:721–745

Walker JC, Hooker WJ (1945) Plant nutrition in relation to disease development. I. cabbage yellows. Am J Bot 32:314–320

Wang M, Farnham MW, Nannes JSP (1999) Ploidy of broccoli regenerated from microspore culture versus anther culture. Plant Breed 118:249–252

Wang M, Farnham MW, Thomas CE (2001) Inheritance of true leaf stage downy mildew resistance in broccoli. J Amer Soc Hort Sci 126:727–729

Watt LE (1966) Conformation of cauliflower curds. Euphytica 15:111–115

Williams PH, Staub T, Sutton JC (1972) Inheritance of resistance in cabbage to black rot. Phytopathology 62:247–252

Yang Q, Chauvin JE, Herve Y (1992) A study of factors affecting anther culture of cauliflower (*Brassica oleracea* var. *botrytis*). Plant Cell Tiss Org Cult 28:289–296

Yarnell SH (1956) Cytogenetics of the vegetable crops II. Crucifers. Bot Rev 22:81–166

Zhang Y, Kensler TW, Cho CG, Posner GH, Talalay P (1994) Anticarcinogenic activities of sulforaphane and structurally related synthetic norbornyl isothiocyanates. Proc Natl Acad Sci USA 91:3147–3150

Zhang Y, Talalay P, Cho CG, Posner GH (1992) A major inducer of anticarcinogenic protective enzymes from broccoli: isolation and elucidation of structure. Proc Natl Acad Sci USA 89:2399–2403

Ziolkowski PA, Sadowski. J (2002) Fish-mapping of rDNA and *Arabidopsis* BACs on pachytene complements of selected brassicas. Genome 45:189–197

Chapter 10
The Genetics of *Brassica napus*

Federico L. Iniguez-Luy and Maria L. Federico

Abstract *Brassica napus* L. belongs to the Brassicaceae family of the Kingdom Plantae and is considered to be a newly formed species (5,000–10,000 mya) probably originating from independent and spontaneous inter-specific hybridizations between genotypes of turnip rape (*Brassica rapa*; AA, $2n = 20$) and cabbage/Kale (*Brassica oleracea*; CC, $2n = 18$). Genetically, *B. napus* is an allopolyploid (AACC, $2n = 38$) exhibiting disomic inheritance. Within the species, two botanical varieties have been defined: *B. napus* L. var *rapifera* (DC) Metzger ($2n = 4\times = 38$) and *B. napus* L. var *oleifera* Delile ($2n = 4\times = 38$). The latter has taken much of the attention and has become the second most cultivated oilseed crop (rapeseed) worldwide, after soybean. The appearance of annual and biannual rapeseed lines with low erucic acid (<2% in the oil) and low glucosinolates (<30 mg/g in the meals) has granted rapeseed CanOLA (Canadian Oil Low Acid) status as an excellent source for edible vegetable oil. The lipid profile of CanOLA oil is extremely well balanced (low in saturated fats, high in monosaturated fats, and rich in omega-3 fatty acids) making it the oil of preference by nutritionists worldwide. In this context, the commercial interest for rapeseed CanOLA has launched an impressive amount of genetics and genomics research which has made possible to make genetic gains in agronomical and quality traits through modern plant breeding. In fact, rapeseed ranks among the top crops for which molecular tools have been developed. To date, over 30 molecular linkage maps have been published using a range of different molecular marker types, population structures, and parental lines exhibiting different flowering time behaviors. These maps have proved extremely useful in order to dissect the genetic nature of the traits underlying the genetic variation found in rapeseed. This chapter will focus on the genetics and genomics aspects of rapeseed breeding describing the current knowledge on the origin of *B. napus*, genetics/genomic tools for the species, and specific target traits affecting *B. napus* oil production and quality.

F.L. Iniguez-Luy (✉)
Genomics and Bioinformatics Unit (UGB), Agri aquaculture Nutritional Genomic Center (CGNA), INIA-Carillanca, Km 10 Camino Cajon-Vilcun, PO Box 58D, Temuco, Chile
e-mail: federico.iniguez@cgna.cl

R. Schmidt, I. Bancroft (eds.), *Genetics and Genomics of the Brassicaceae*, Plant Genetics and Genomics: Crops and Models 9, DOI 10.1007/978-1-4419-7118-0_10,

Keywords *Brassica napus* · Origin · Oilseed crop · Genetic and genomic resources · Agronomical and nutritional traits

Abbreviations

AFLP	Amplified fragment length polymorphisms
ANF	Antinutritional factors
CanOlA	Canadian oil low acid
CO	Constans
DH	Doubled haploid
DNA	Deoxyribonucleic acid
ESTs	Expressed sequence tags
F1	First filial generation
F2	Second filial generation
FA	Fatty acid
FAD	Fatty acid desaturase
FAE	Fatty acid elongase
FISH	Fluorescence in situ hybridization
FLC	Flowering locus C
FRI	Frigida
GCA	General combining ability
GISH	Genomic in situ hybridization
Gly3P	Glycerol-3-phosphate
HEAR	High erucic acid rapeseed
HNRTs	Homoeologous non-reciprocal translocations
HO	High oleic
HRTs	Homoeologous reciprocal translocations
HS	High stearic
LEAR	Low erucic acid rapeseed
LL	Low linolenic
MAS	Marker-assisted selection
MGS	Microarray-based genomic selection
MMT	Million metric tons
mRNA	messenger Ribonucleic Acid
NGS	Next generation sequencing
NIRS	Near-infrared reflectance spectroscopy
NRTs	Non-reciprocal translocations
OP	Open pollinated
PCR	Polymerase chain reaction
QTL	Quantitative trait loci
RAPD	Randomly amplified polymorphic DNA
RFLP	Restriction fragment length polymorphisms
RIL	Recombinant inbred lines
RT	Reciprocal translocation

SAD	Stearoyl-acyl desaturase
SCA	Specific combining ability
SCAR	Sequenced characterized amplified regions
SNP	Single nucleotide polymorphisms
SRAP	Sequence-related amplified polymorphism
SSR	Simple sequence repeats

Contents

10.1 *Brassica napus* Origin and Domestication 293
10.2 *B. napus* and Its Importance as an Oilseed Crop 295
10.3 Status of the Genetics and Genomic Tools in Oil Rapeseed (*B. napus*) 295
  10.3.1 Genomic Tools I: Molecular Marker Technology in *B. napus* 295
  10.3.2 Genomic Tools II: Development of Genetic Linkage Maps 297
10.4 The Genetics of Specific Traits in Rapeseed *B. napus* 302
  10.4.1 Modified FA and Specialty Oil and Meal Profiles 302
  10.4.2 Oil Content 303
  10.4.3 Flowering Time Variation: Winter vs. Spring 304
  10.4.4 Hybrids, Population Development, and Seed Yield Improvement 306
  10.4.5 Other Important Oil Quality-Related Traits 309
10.5 Conclusions 311
References 312

10.1 *Brassica napus* Origin and Domestication

B. napus L. var *oleifera* Delile ($2n=4\times=38$), commonly known as rape, oilseed rape, Argentine rape, Swede rape, colza, raps, or rapeseed, and *B. napus* L. var *rapifera* (DC) Metzger ($2n=4\times=38$), commonly known as rutabaga, Swede, Swedish turnip belong to the Brassicaceae family of plants. *B. napus* is one of the six cultivated *Brassica* species that form the "U-triangle" (U 1935). Genetically, *B. napus* is an allopolyploid exhibiting disomic inheritance and is considered as a relatively new species (5,000–10,000 mya). Although its origins are not yet entirely clear, *B. napus* most likely originated from multiple spontaneous and independent inter-specific hybridizations between genotypes of turnip rape (*Brassica rapa*; AA, $2n=2\times=20$) and cabbage or kale (*Brassica oleracea*; CC, $2n=2\times=18$) growing side by side during medieval times (Prakash and Hinata 1980, Kimber and McGregor 1995, Gupta and Pratap 2007). However, it is also possible that a wild *Brassica* species ($n=9$) like *Brassica montana* Pourr., or an ancestral genome sharing similar characteristics, could have contributed to the maternally inherited C genome of *B. napus*. Evidence for this was gathered during the 1990s using chloroplast and mitochondrial DNA analysis (Song and Osborn 1992, Song et al. 1993,

1997). Moreover, in a recent study using microsatellite markers linked to the chloroplast genome, it was suggested that *B. montana* or a relatively close ancestor could have contributed more frequently as parental source representing the cytoplasm of the C genome (Allender et al. 2005). However, the same study did find, albeit at lower frequencies, similarities to the C genome of *B. oleracea*. This finding is consistent with *B. napus* having multiple origins. Unfortunately, *B. napus* is seldom found in the wild and a lack of wild relatives to relatively modern cultivars and landraces hinders an exhaustive phylogenetic analysis that could allow accurately determining the original parental diploid genomes of this species. Nonetheless, *B. napus* has received much attention as a model plant to understand the consequences of polyploidy in higher organisms, a subject that is dealt with elsewhere in this book (Shahidi 1990, Rana et al. 2004, Gaeta et al. 2007).

The center of origin for the parental diploid *Brassica* A and C genome species is located around the Mediterranean basin (mainly as *B. oleracea* forms), Middle East (mainly as *B. rapa* forms), and Asia Minor (*B. rapa*) (Gomez-Campo 1999, Song et al. 1988a, b, Warwick and Black 1993). Their introduction into Europe and the Far East could have taken place through various migrations out of Eastern Turkey, from 5000 BC through to perhaps AD 500. These multiple migrations granted not only diploid *Brassica* crops an important role as sources of food, medicine, forage, and ornamentals but also a place in history. Indeed, the introduction of cultivated forms of *B. rapa*, *B. oleracea*, and *B. napus* to eastern and western cultures might have been at the same time as the wheel, cultivated grains and domestic animals were being introduced (Damania et al. 1997). Thus, availability of parental species (*B. rapa*, *B. oleracea*, and/or *B. montana*) for the inter-specific hybridization that gave rise to *B. napus* situates as the most likely geographical center of origin the Mediterranean basin, northern or western Europe, and the Middle East. The evolution of *B. napus* from perhaps an outcast/rogue looking plant in turnip rape, cabbage, or kale fields to the important role it plays in agriculture can perhaps be attributed to the high levels of phenotypic and genotypic variation that has been used for selection during domestication. It is reasonable to imagine that such levels of variation have been recognized and widely exploited by past plant breeders. The domestication of the species from the early stages has been directed toward its uses as a source of edible and industrial oils and as a source of feed. Records from ancient civilizations in central Asia and the Middle East described the use of rapeseed oil as a fuel source not only for edible purposes but also for artificial illumination (Applequist and Ohlson 1972). Similarly, there is also evidence that rapeseed was cultivated in India as early as 3000 years ago (Hougen and Stefansson 1983, Damania et al. 1997). Massive cultivation in Europe dates from the thirteenth century, attaining an important role during the industrial revolution due to its excellent quality as lubricant oil. Gupta and Pratap (2007) described in detail the historical background of *B. napus* oil and forage types, from antiquity to modern times and from the early medicinal and feeding uses to the advanced agricultural practices of today. This chapter will focus on the genetics and genomics of rapeseed grown for oil. For information on Swede breeding and genetics refer to McNaughton and Thow (1972), Bradshaw and Griffiths (1990), and Ramsay et al. (2001).

10.2 *B. napus* and Its Importance as an Oilseed Crop

The appearance of "double low" rapeseed lines during the 1970s, i.e., with low erucic acid (<2% in the oil) and low glucosinolates (<30 mg/g in the meal), granted rapeseed as an excellent source for edible vegetable oil worldwide (Shahidi 1990, Buzza 1995). Canadian scientists and the Canadian vegetable oil industry marketed this oil under the denomination name of CanOLA (Canadian Oil Low Acid) and it was declared suitable for human and animal consumption by the United States Food and Drug Administration in 1985 (FDA 1995). The lipid profile of CanOLA oil is extremely well balanced for a vegetable oil (low in saturated fats, high in monosaturated fats, and rich in omega-3 fatty acids). This has made it the oil of preference by nutritionists across the globe (Stringam et al. 2003). In addition, the rapid rise of rapeseed as a source of vegetable oil has been accompanied by the development and introduction of hybrid rapeseed varieties first by the public sector and then massively by the private seed companies (see Section 10.5.4).

During the past 30 years, the cultivation and processing of rapeseed has grown rapidly becoming the second, after soybean, for the production of oil seeds and oil meals (protein source) and the third, after soybean and palm kernel, as vegetable source for edible oil (USDA 2009). In 2008, the worldwide production of rapeseed reached 58 million metric tons (MMT). The principal oilseed producers in 2009 were the European Union (EU-27 countries) with 21 MMT, followed by China (13 MMT), Canada (10 MMT), and India (7 MMT). The total oil and meal production during 2009 reached 22 MMT and 33 MMT, respectively (USDA 2009). In addition, the market prices associated to seed, oil, and meal from rapeseed canola increased substantially during the late 2000s. This was due to combination of high demands on food and energy, a reduction in international stocks, and a lower stock-to-use-ratio for rapeseed oil and meal (FAO 2008). During 2007–2008, rapeseed oil reached an average price of 1,426 USD per ton, while the prices for rapeseed reached 754 USD per ton (USDA 2009).

The cultivation of rapeseed is attractive to producers worldwide because (i) it gives a good capital return when compared to other annual crops, (ii) the prices are linked to different commercial variables than those for cereal production, (iii) it is possible to utilize the existing agricultural machinery used in cereal production, and (iv) it offers excellent yields when used in a rotation scheme with cereals. These facts coupled with the extraordinary oil quality have positioned rapeseed CanOLA as one of the top producing oilseed crops worldwide.

10.3 Status of the Genetics and Genomic Tools in Oil Rapeseed (*B. napus*)

10.3.1 Genomic Tools I: Molecular Marker Technology in B. napus

The use of molecular markers in plant breeding and genetics has grown immensely since its humble beginnings in the 1980s to the impressive high-throughput technologies of today. Various types of molecular markers have been used to construct linkage maps and associate genotypic differences with desirable phenotypic variation (Takeda and Matsuoka 2008). The utilization of different maker types has been based on the availability of resources and also varied in an objective-dependent fashion (e.g., germplasm evaluation vs. linkage mapping vs. selection of breeding lines). Regardless their usefulness, each of the marker types mentioned above exhibit a common theme: they rely entirely on differences in the DNA sequence of tested individuals (Vignal et al. 2002). Consequently, the best marker types are those that target directly at the variation encountered in the genomic DNA sequence of crops species. The *Brassica* research community has not been indifferent to the development of molecular markers for breeding and genetic studies (Snowdon and Friedt 2004). In fact, it has played an important role in testing and using molecular markers. Examples of the most recent mapping studies in *Brassica* crops containing the A and C genomes include Suwabe et al. (2006), Kim et al. (2006), Choi et al. (2007), and Iniguez-Luy et al. (2009) in *B. rapa;* Li et al. (2003), Babula et al. (2003), Gao et al. (2007), and Iniguez-Luy et al. (2009) in *B. oleracea*; and Parkin et al. (2005), Udall et al. (2005), and Piquemal et al. (2005) in *B. napus*. Additionally, molecular markers have been used to understand and organize the structure of the six cultivated *Brassica* genomes (Osborn and Lukens 2003, Parkin et al. 2005, Mayerhofer et al. 2005, Gaeta et al. 2007). Developed molecular marker types include restriction fragment length polymorphism (RFLP, Teutonico and Osborn 1994), amplified fragment length polymorphisms (AFLP, Sebastian et al. 2000), randomly amplified polymorphic DNA (RAPD, Foisset et al. 1996), sequence characterized amplified regions (SCAR, Iniguez-Luy et al. 2006), sequence-related amplified polymorphism (SRAP, Li and Quiros 2001), and simple sequence repeats (SSR, Iniguez-Luy et al. 2008).

The availability of sequence data for *Brassica* crops is now stirring researchers into developing newer and more efficient molecular markers. Such markers can be obtained by investigating the variation found in genomic sequences. Recently, SNPs (single nucleotide polymorphisms) have enabled to explore directly the variation in nucleotide base constitution hidden in genomic DNA sequences. Indeed, SNP markers have become the marker of choice in human and animal genetics revolutionizing the area of medical diagnosis (International HapMap Consortium et al. 2007). They are extremely helpful at defining the individual (e.g., varieties or parental lines in crops) by establishing a haplotype map (Rafalski 2002, McCarthy et al. 2008). In addition, SNPs can become useful to determine how individuals interact with drugs and biotic or abiotic stimuli when they are located directly at coding or regulatory regions of a gene (Collins et al. 1998). SNPs are very prolific and occur in unlimited numbers, hence, each SNP found as a single copy DNA can be utilized as a potential useful marker. SNP markers can be mapped to specific chromosomal regions in segregating populations or in turn be used in associative mapping studies of non bi-parental populations (International HapMap Consortium et al. 2007, Ganal et al. 2009).

SNP discovery in *B. napus* is still at its infancy compared to other crops like maize, rice, and barley, (Ganal et al. 2009) but growing rapidly. To date, only one publication has been reported on a genome-wide SNP discovery in *B napus*. Trick and coauthors (2009) elegantly described the use of a Solexa sequencing system to generate approximately 20 million ESTs from each of two rapeseed cultivars (Tapidor and Ningyou) and the identification of 23,330–41,593 putative SNPs between them. They developed computational tools to assess potential polymorphisms from the generated sequences and relied on a set reference sequence made of approximately 94,000 unigenes of different *Brassica* species. The majority of detected polymorphisms were indicative of transcription from homoeologous genes from the two parental genomes examined (hemi-SNPs) and could be used for mapping studies. It is worth mentioning here that SNP discovery in allopolyploid crops is challenging since it is required to discriminate between informative SNPs present within genomes from not informative SNPs that exist between genomes. For this reason, the method of choice to discover SNPs in allopolyploid crops has generally been amplicon resequencing (Westermeier et al. 2009).

New DNA sequencing technologies, also known as next-generation sequencing (NGS), offer a faster, more systematic, and cost-effective way of accessing genetic variation (Margulies et al. 2005, Wheeler et al. 2008). Their use on the transcriptome "is an efficient method for the identification of sequence variation between oilseed rape lines," as stated by Trick et al. (2009). However, a major challenge to broadly apply these NGS technologies continues to be the targeted DNA enrichment of large and complex eukaryotic genomes. If a researcher wants to target specific areas of the genome using amplicon resequencing, for example, sequencing a thousand genes would require designing and synthesizing several thousands of PCR primers and performing several thousands of PCR reactions, a costly and lengthy process. Recently, a microarray-based genomic selection (MGS) method has addressed this challenge allowing for the isolation of user-defined unique genomic sequences in one single enrichment step providing cost and time savings (Albert et al. 2007, Olson 2007, Okou et al. 2007). This method was developed by NimbleGen Systems Inc. and is also known as sequence capture microarray. A public effort using this technology plans to build a sequence capture microarray using sequence regions of the *B. napus* genome that have been previously linked to a phenotype of interest (QTL). The development of such an array and corresponding bioinformatic tools offers the possibility of the future development of a high-throughput SNP genotyping platform for this crop, as outlined schematically in Fig. 10.1.

10.3.2 Genomic Tools II: Development of Genetic Linkage Maps

Part of the development and implementation of modern molecular techniques in *B. napus* have been directed to understand and exploit the genetic control of morphological diversity within the species. Such efforts have mainly been devoted to improve the agronomical performance of rapeseed (e.g. seed yield) and the nutritional aspects of its products such as oil and meal. In *B. napus*, several independent genetic maps have been constructed to date deriving from more than 30 balanced/bi-parental populations. Table 10.1 lists 34 examples of public mapping studies in

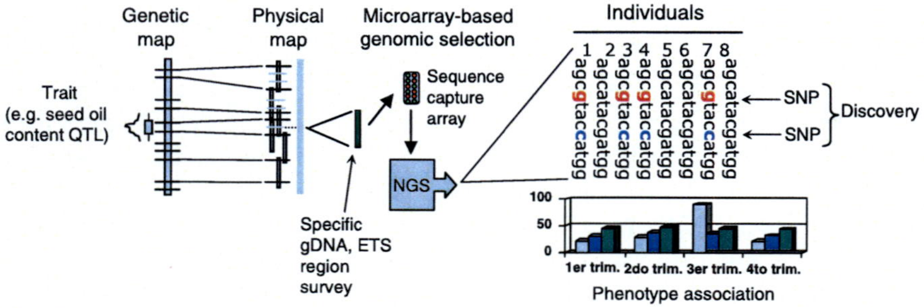

Fig. 10.1 Rationale behind the enrichment of specific regions associated to traits of interest for SNP discovery using sequence capture microarrays. Genomic regions associated to a phenotype of interest are assembled and tiled into custom microarrays chips. gDNA from different genotypes are hybridized and captured using these custom microarrays. NGS is used to discover SNP in captured DNA. *B. napus* haplotypes are established and associations between SNP and phenotypes are determined. QTL: Quantitative Trait Loci; gDNA: genomic Deoxyribonucleic Acid; EST: Expressed Sequence Tags; NGS: New Generation Sequencing; SNP: Single Nucleotide Polymorphisms

B. napus including types of populations developed and linkage maps constructed since 1991. These studies not only led to the identification of genomic regions that explained part of the observed genetic variation like seed and yield components (Butruille et al. 1999a, Quijada et al. 2006, Radoev et al. 2008, Shi et al. 2009), oil quality and quantity (Jourdren et al. 1996a, b, Hu et al. 1999, Schierholt et al. 2001, Howell et al. 2003, Barker et al. 2007, Jiang et al. 2009), genic male sterility (Delourme et al. 1994, Brown et al. 2003, He et al. 2008, Tang et al. 2009), disease resistance (Pilet et al. 1998a, b, Pilet et al. 2001, Zhao and Meng 2003, Mayerhofer et al. 2005, Zhao et al. 2005, Boys et al. 2007, Diederichsen et al. 2009), abiotic stress (Kole et al. 2002, Zhao et al. 2008), flowering time (Ferreira et al. 1994, Mei et al. 2009), and seed color (Somers et al. 2001, Rahman 2001) but also yielded valuable insights into the genomic structure and evolution of *Brassica* crops. Among other aspects of *B. napus* genetics, such results have highlighted the extensive duplicated nature, as well as frequent appearance of chromosomal rearrangements among the *Brassica* species (Cheung et al. 1997, Lagercrantz and Lydiate 1996, Lagercrantz et al. 1996, Lagercrantz 1998, Osborn et al. 1997, Osborn et al. 2003b, Parkin et al. 2005, Udall et al. 2005, Piquemal et al. 2005). This subject is dealt with elsewhere in this book (please refer to Chapter 5).

Highly dense genetic maps in *B. napus* have been achieved by constructing and integrating individual maps into consensus linkage maps. Efforts to integrate and develop consensus linkage maps go back to the mid-1990s. For example, Parkin et al. (1995) integrated two genetic maps, one based on a spring × winter double haploid (DH) population (Sharpe et al. 1995) and a resynthesized *B. napus* × winter DH population, using 211 common loci. Butruille et al. (1999b) created a composite linkage map by enriching inbred backcross linkage estimates with estimates previously mapped in a winter × spring cross (Ferreira et al. 1994). Lombard and Delourme (2001) combined data from three DH populations using 253 markers mapped in at least two individual DH populations. More recently, Udall and

Table 10.1 Mapping studies of *B. napus* L.

Study[a]	Mapping population[b]	Germplasm	No. of individuals	No. of markers	Marker type[c]	Map length (cM)	No. of LG
Landry et al. (1991)	F2	Spring × spring (Westar × Topas)	90	120	R	1,413	19
Ferreira et al. (1994)	DH	Winter × spring (Major × Stellar)	105	132	R	1,016	22+6p
Uzunova et al.(1995)	DH	Winter × spring (Mansholt × Samouri)	151	153	R	1,441	19
Sharpe et al. (1995)	DH	Spring × winter (N-o-1 × N-o-9)	92	277	R	1,741	19
Parkin et al. (1995)	DH	Synthetic × winter (Syn1 × N-o-9)	50	399	R	1,656	19
Cloutier et al. (1995)	2DH and F2	Spring × spring (Topas × Westar)	76DH+90F2	145	R	524–863	16
Foisset et al. (1996)	DH	Winter × spring (Darmor × Yudal)	152	254	R+I+Ra	1,765	19
Howell et al. (1996)	IBL	Winter × winter (Tapidor × Victor)	200	158	R	1,204	19
Jourdren et al. (1996a)	DH	Spring × spring (Stellar × Drakkar)	150	12	Ra	N.I.P.	2
Cheung et al. (1997)	DH	N.I.P. (90-DHW-1855-4 × 87-DHS-002)	95	342	R+Ra+St	1,954	19
Fray et al. (1997)	BC1	Spring × winter (N-o-11 × Tapidor)	83	76	R	N.I.P.	14
Kelly et al. (1997)	BC1	Spring × spring (N-o-1 × N-o-93)	180	243	R	N.I.P.	19
Pilet et al. (1998b)	DH	Winter × spring (Darmor × Yudal)	152	288	R+Ra	1,954	19
Butruille et al. (1999b)	IBL	Winter × spring (Ceres × Marnoo × Westar)	124+119	204	R	N.I.P.	20
Udall et al. (2005)	DH	Chinese × spring (RV289 × P1804)	162	132	R	1,460	19
Udall et al. (2005)	DH	Synthetic × spring (TO1147 xP1804)	162	142	R+Sr	1,668	19
Udall et al. (2005)	DH	Winter-intro. × spring (MF216 × P1804)	170	223	R	1,443	19
Udall et al. (2005)	DH	Winter-intro. × spring (RV128 × P1804)	164	206	R	1,473	19
Mayerhofer et al. (2005)	BC1	Synthetic × spring (DH12075 × PSA12)	90	N.I.P.	R+Sa	N.I.P.	19
Mayerhofer et al. (2005)	BC1	Synthetic × spring (Shiralee × PSA12)	100	N.I.P.	R+Sa	N.I.P.	19
Piquemal et al. (2005)	6 F2	Spring × winter (see table footnote)	574	292	Sr+Sc	2,619	19
Zhao et al. (2005)	DH	Winter × Chinese (Sollux × Gaoyou)	282	125	Sr	1,196	21
Li et al. (2006)	F2	Chinese × spring (SI-1300 × Eagle)	184	162	Ss+Sr	2,27.3	21
Qiu et al. (2006)	DH	Chinese × winter (Ningyou 7 × Tapidor)	188	277	A+R+St+Sr	1,685	19
Chen et al. (2007)	DH and IF2	Spring × Chinese (Quantun × No.2127-17)	258DH+129IF2	397	Sr+Sa	1,747.4	20

Table 10.1 (continued)

Study[a]	Mapping population[b]	Germplasm	No. of individuals	No. of markers	Marker type[c]	Map length (cM)	No. of LG
Fu et al. (2007)	RI	Chinese × Chinese (Zhongyou821 × GH 06)	N.I.P.	509	Sr+R+Sa	1,923	26
Rygulla et al. (2007)	DH	Winter × winter (307-406-1 × 307-230-2)	163	304	R+A+Sr	1,739	19
Sun et al. (2007)	DH	Spring × chinese (Westar × Zhongyou821)	58	1,634	Sa	1,604.8	19
Amar et al. (2008)	DH	Winter × winter (M. H. R. × Samourai)	148	185	R+A	1,739	20
Cai et al. (2008)	DH	Spring × spring (Hyola 401 × Q2)	45	248	Sr+A+Sa	1,634.7	19
Radoev et al. (2008)	DH	Winter × synthetic (Express617 × R539)	250	379	Sr+A	2,045	19
Feng et al. (2009)	F2	Chinese × Chinese (117AB × 7,605)	282	277	Sr+A	2,667.7	21
Mei et al. (2009)	F2	Chinese × Chinese (2,091 × 99CDAM)	145	241	Sr+A	2,094.6	20
Cheng et al. (2009)	DH	Chinese × Chinese (No. 2127 × ZY821)	88	153	Sr	1,821.3	19

N.I.P. = no information provided
[a]Reference to published maps in peer review journals
[b]Type of mapping population (filial generation 2 = F2; double haploid = DH; inbred backcross = IBL; backcross = BC)
[c]Marker types used in linkage map construction (R = RFLP; I = isozyme; Ra = RAPD; St = Sequence tagged site; Sr = SSR; Sa = SRAP; Sc = SCAR; A = AFLP)

coauthors (2005) integrated four linkage maps and found that on average, each individual map (derived from DH populations) contributed six loci per linkage group to the consensus map. Therefore, the number of RFLP loci detected increased from 1.65 to 2.24 loci/probe in the each individual maps to 3.02 loci/probe in the consensus map. This approach has been effective at stacking molecular marker loci onto linkage groups, but especial consideration needs to be taken when using consensus maps for QTL analysis as the real position of a QTL may be erroneously estimated since the positioning of molecular markers that have been mapped in only one reference map may not be correct in a consensus map. Therefore, researchers are aiming toward the development of highly dense single genetic linkage maps (Shi et al. 2009).

Interestingly, the development of genetics maps in *B. napus* using RFLP markers provided researches with the ability to track chromosomal translocations. This is important since the presence of genetic rearrangements in different germplasm sources could have an effect on quantitative and qualitative traits. Non-reciprocal translocations (NRTs) were documented using RFLPs to generate genetic maps of two segregating populations of DH lines (Sharpe et al. 1995, Parkin et al. 1995). These results suggested that novel homoeologous exchange events occurred during meiosis of F1 plants generating novel NRTs in a few individual DH lines. A reciprocal translocation (RT) has been identified in spring *B. napus* based on analysis of five mapping populations between spring and winter parents (Osborn et al. 2003). Monomorphic fragments between the two mapping parents segregated among the population genotypes, but they were not allelic. These fragments mapped to homoeologous regions of A7 and C6. Pachytene spreads from anther meiotic cells from a winter × spring F1 hybrid confirmed the presence of the RFLP-described RT (Osborn et al. 2003). More recently, Udall et al. (2005) reported the detection of novel homoeologous non-reciprocal transpositions (HNRTs), preexisting HNRTs, and homoeologous reciprocal transpositions (HRTs) in four mapping populations using RFLP genome analysis. Novel or de novo HNRTs were found to be the most abundant type of chromosomal rearrangements. QTL analysis for seed yield in the genetic backgrounds described by Udall et al. (2005) found that HNRT and HRT coincided with genomic regions that explain hybrid seed yield in spring lines introgressed with winter alleles (Quijada et al. 2006). Furthermore, Quijada et al. (2006) hypothesized that chromosomal rearrangements may account for allelic variation for some of the observed seed yield QTL (linkage groups A7, A10, C1, C3, C6, C9) and that this could be extended to provide a general explanation for a portion of the seed yield heterosis in rapeseed. For instance, higher seed yield of hybrids compared to their inbred parental lines could be due, in part, to increased intergenomic heterozygosity in regions containing HNRTs (see section 10.5.4). Such findings suggested that chromosomal rearrangements can have a major impact on genetic variation for seed yield and other complex traits in hybrid cultivars of rapeseed. Although, molecular markers can be very useful at identifying chromosomal rearrangements and their putative association with traits, it would be interesting to determine the break points and hot spots for such events using cytological means. New advances in FISH and GISH techniques using *B. napus* chromosomes will make possible to determine rearrangements that could be selected for desirable hybrid combinations. These experiments are underway (Christopher J. Pires, personal communication).

10.4 The Genetics of Specific Traits in Rapeseed *B. napus*

10.4.1 Modified FA and Specialty Oil and Meal Profiles

Depending on the fatty acid (FA) profile found in the seeds of particular cultivars, rapeseed oil is used for both edible and industrial purposes. Typically, the FA profile of edible rapeseed (modern "00" types or CanOLA) oil contains 60% of oleic acid (18:1n9, ω-9), 20% linoleic acid (18:2n6, ω-6), 10% α-linolenic acid (18:3n3, ω-3), 7% saturated fats and is low (1%) in erucic acid (22:1n9) (Wittkop et al. 2009). Industrial rapeseed oil ("++" types or traditional rapeseed), on the contrary, benefits from high levels of erucic acid (around 50%) and is used to manufacture lubricants, water repellents, waxes, etc. Traditional varieties are also known as HEAR (High Erucic Acid Rapeseed). Nowadays, however, the main industrial use of rapeseed oil is biodiesel production driven by a growing demand for renewable fuels (USDA, 2009). For this industrial end-use, oils low in erucic acid are also suitable.

Two homoeologous loci (E^A and E^C) are involved in the genetic control of erucic acid content in *B. napus* (Harvey and Downey 1964). These two loci have been mapped and together they explain 90% of the total variation (Jourdren et al. 1996a). Low Erucic Acid Rapeseed (LEAR, "0" type) varieties have been bred through the introgression of two recessive alleles (e^A and e^C) at these two loci. These alleles were first and solely found in a natural spring rapeseed mutant, Liho (Downey and Craig 1964). Molecular characterization has shown that E^A and E^C correspond to two copies of the Fatty Acid Elongase 1 (FAE1) gene (Fourmann et al. 1998) and that a 4 bp deletion in the coding region of FAE1 confers the low erucic acid trait (Wu et al. 2008).

Oils containing high levels of monounsaturated FAs (oleic, 18:1n9, ω-9) and reduced levels of polyunsaturated FAs (particularly linolenic, 18:3n3, ω-3) exhibit higher thermal stability and are suitable for high temperature cooking and frying. Rapeseed mutants with high oleic (HO) acid and low linolenic (LL) acid seed content have been produced through mutagenesis and characterized (Rakow 1973, Auld et al. 1992). Briefly, one or two major loci corresponding to the *FAD2 (fatty acid desaturase)* genes which code for the enzymes involved in the desaturation of oleic to linoleic acid have been shown to control the HO trait (Schierholt et al. 2001, Falentin et al. 2007). Similarly, two major loci corresponding to the FAD3 genes which code for the enzymes involved in the desaturation of linoleic to linolenic acid have been shown to control the LL trait (Barret et al. 1999, Jourdren et al. 1996a). Several molecular markers have been identified in tightly association with the HO and LL traits (Tanhuanpää et al. 1995, Jourdren et al. 1996a, b, Thormann et al. 1996, Hu et al. 1999, Rajcan et al. 1999, Schierholt et al. 2000). In addition, the *FAD* genes have been mapped to QTL controlling HO (*FAD2*) and LL (*FAD3*) characteristics (Hu et al. 2006).

Stearic acid (18:0) is used as an emulsifier and/or hardener to make candles, plastics, soaps, and candies, among other products. Interestingly, even though stearic acid is a saturated FA, its consumption has little or no effect on cholesterol plasma levels (Grundy 1994, Yu et al. 1995) making it a "healthy" saturated FA. A vegetable source of stearic acid, usually obtained from animal tallow, is currently highly demanded by the food and cosmetic industries. Conventional high stearic acid (HS) lines exist in soybean and sunflower (Graef et al. 1985, Osorio et al. 1995, Perez-Vich et al. 2006). Attempts to obtain *B. napus* lines exhibiting enriched levels of this FA in seed oil have been based on the use of genetic engineering by overexpression of heterologous genes (*Garm FatA1*, Hawkins and Kridl (1998)) or down regulation of the stearoyl-acyl carrier protein desaturase (SAD) gene (Knutzon et al. 1992, Zarhloul et al. 2006).

Oil content and related traits such as fatty acid composition and presence of nutritional (e.g., tocopherols and carotenoids) and antinutritional factors (e.g., glucosinolates and sinapates) are quantitative (polygenic) in nature and are heavily influenced by environmental conditions (Barker et al. 2007). Understanding the complex molecular basis of these key quality traits is critical to the predictive development of *B. napus* cultivars with specific attributes. For example, the discovery of two SNPs in mutant alleles of *FAD2* and *FAD3* genes not only suggested mechanisms that could explain the altered FA compositions observed in the mutant seeds, but also became useful markers for selection of these mutant alleles during marker-assisted trait introgression, greatly accelerating efforts to develop HOLL varieties (Hu et al. 2006). Additionally, detailed understanding of the molecular basis of seed quality traits will provide a list of candidate genes (orthologues) for breeding modified FA content in other increasingly important *Brassica* oilseeds (*Brassica juncea* (L.) Czern and *Camelina sativa* (L.) Crantz).

10.4.2 Oil Content

One of the primary objectives in rapeseed CanOLA breeding has always been increasing seed oil content (%) as a way of maximizing oil production. With that goal in mind, significant progress has been made in the identification and mapping of QTLs associated with oil content and FA composition in *B. napus*. Several QTLs for oil content in seeds have been identified in different mapping populations, all of them exhibiting relatively minor effects (Burns et al. 2003, Ecke et al. 1995, Zhao et al. 2005, Delourme et al. 2006, Qiu et al. 2006, Zhao et al. 2006). Three of these QTLs, located in linkage groups A1, A8, and C3, were consistently revealed across different populations (Delourme et al. 2006). Undoubtedly, the development of a consensus genetic map for the species will allow better comparing QTLs and unifying results across different mapping populations.

The QTL located on linkage groups A8 and C3 coincide with major QTLs for erucic acid content in *B. napus* seeds (Howell et al. 1996, Thormann et al. 1996, Burns et al. 2003) and are closely linked to genes involved in the synthesis of this FA, FAE1.1, and FAE1.2 (Ecke et al. 1995, Zhao et al. 2006, Qiu et al. 2006). This supported the early hypothesis that erucic acid content is per se a major determinant of seed oil content in oilseed rape (Ecke et al. 1995).

Seed oil content is strongly affected by the environment (Zhao et al. 2006, Delourme et al. 2006) and concomitant increases of seed oil content and seed yield have proven to be difficult to achieve in classical breeding programs. Several transgenic approaches aimed at increasing seed oil content in oilseeds have been undertaken. From these, most attempts relied on the overexpression of genes involved in lipid biosynthesis (reviewed by Snydera et al. 2009). An increase of 40% seed oil content, however, was achieved by increasing the levels of glycerol-3-phosphate (Gly3P) in *B. napus* seeds through the overexpression of the yeast Gly3P dehydrogenase gene (Vigeolas et al. 2007). By this means, seeds of three different transgenic lines exhibited increased total FA levels ranging from 57 to 81 mg/g of fresh weight compared to 43 mg/g of fresh weight exhibited by wild-type controls.

10.4.3 Flowering Time Variation: Winter vs. Spring

Understanding the genetic basis of the control of flowering time in *B. napus* is important for plant breeders in order to optimize crop performance in specific environments. The existence of natural flowering time variation has been thought to evolve as part of a mechanism that allowed crucifers to withstand cold winters and high summer temperatures (Schranz and Osborn 2000, Osborn and Lukens 2003, Shindo et al. 2005). This variation has been exploited through domestication of different *B. napus* types for growth in different environments (see section 10.2).

Plant breeders and geneticists have long been studying the variation observed during the transition from the vegetative to reproductive phase of most important crops (Allard 1999). The time that a crop takes to flower is crucial and can affect net seed yields, especially if one takes into account an ever changing climatic scenario. In *B. napus*, flowering time is used to classify rapeseed cultivars into annuals and biannuals types. Both types, however, undergo the same physiological and developmental events in order to flower (e.g., formation of inflorescence meristem from stem meristem, floral formation, bud development, and blooming). Annuals often referred to as spring rapeseed are grown in countries with extremely cold winters, warm short spring and mild summer seasons (e.g., Canada). Spring rapeseed flowers during the first growing season and typically takes between 40 and 60 days to flower. Biannuals often referred to as winter rapeseeds are preferred commercially in temperate climates like those exhibited in northern European Countries and parts of China. Biennials do not flower during the first growing season and they require a period at low temperatures in order to flower (vernalization). This can be achieved by over-wintering in farm fields or artificially by exposing plants to temperatures between 5 and 7°C in cold rooms or environmental chambers. Winter rapeseed usually flowers after 150–185 days from sowing date.

The close evolutionary relationship between *Arabidopsis thaliana* and *Brassica* species allows researchers to conduct detailed genome-wide comparative analysis to identify candidate genes. Flowering time variation observed in *Brassica*

QTL mapping experiments (using annual x biannual crosses) during the late 1990s and early 2000s provided the plant genomic research community with useful case studies. Osborn and Lukens (2003) reviewed the molecular genetic control of flowering time variation in the A and C *Brassica* genomes. On the basis of genome comparative analysis it was stated that genes synthenic to the top of *Arabidopsis* chromosome V explained most of the variation in flowering time in *B. napus* and other *Brassica* crops. Among these genes, two transcription factors were identified as potential candidates underlying flowering time variation, *Flowering Locus C* (*FLC*) and *Constans* (*CO*). *CO* is a regulator of the photoperiod flower promotion pathway (Putterill et al. 1995). Wild-type *Arabidopsis* plants flower later under short days than under long days, however, mutants that have lost *CO* function are insensitive to day length and flower late under both long- and short-day conditions (Suarez-Lopes et al. 2001). Additionally, plants that are late flowering due to a *CO* mutation have very little response to vernalization. Four orthologues of *CO* have been isolated in *B. napus* and one of these copies, designated *BnCOa1*, was able to complement an *Arabidopsis co* mutant (Robert et al. 1998). *FLC* is a key regulator of the autonomous flowering pathway (Michaels and Amasino 1999). Plants with late-flowering alleles have delayed flowering regardless of photoperiod, but vernalization largely negates the effects of these alleles. Levels of *FLC* expression are positively correlated with the delay in flowering, and vernalization diminishes *FLC* transcript levels (Michaels and Amasino 1999). High levels of *FLC* expression require a functional allele at another flowering time locus, *FRIGIDA* (*FRI*), and loss of either *FLC* or *FRI* function causes early flowering (Amasino 2005). The regulation of *FLC* mRNA is under epigenetic control (Sung and Amasino 2004, Sung et al. 2006, Schmitz et al. 2008). Studies on flowering time in *B. napus* also yielded fruitful knowledge on the redundant nature of the *Brassica* genome and its organization (Lagercrantz 1998, Ryder et al. 2001, Osborn and Lukens 2003, Mayerhofer et al. 2005, Parkin et al. 2005). *FLC* orthologues have also been identified in *B. napus* (*BnFLC*) (Tadage et al. 2001). Copies of *BnFLC* were shown to confer winter requirement in *B. napus* and accounted for the major vernalization-responsive flowering time differences in a manner analogous to that of *Arabidopsis* late-flowering ecotypes. In addition, Schranz et al. (2002) concluded that replicated *BnFLC* genes may have a similar function and interact in an additive manner to modulate flowering time variation, in rapeseed, through vernalization.

Although flowering time is a highly heritable trait, and to a certain extent can be Mendelized (i.e., isolation of large effect QTL and subsequent test crossing) (Koornneef et al. 1997) through backcrossing of either late or early alleles (Osborn et al. 1997), days to flower are still quite dependable upon environmental cues (cold exposure, day length, etc.). In spite of this, the majority of experiments on flowering time in *B. napus* have been conducted in a limited set of environments and only dealt with treatments pertinent to the flowering behavior of parental lines (Teutonico and Osborn 1994, Ferreira et al. 1994, Osborn et al. 1997, Butruille et al. 1999b). The number of flowering time QTL detected by these studies ranged from 4 to 9. In all cases, detected QTL explained large phenotypic effects but no minor QTL could be detected efficiently due to the existence of QTL x environment interactions. In order to account for these interactions, Long and colleagues (2007) conducted a QTL study using 11 locations, a DH population, and its derived reconstructed F2 population. Even though flowering time varied greatly in each environment, 60% of the phenotypic variation was explained by genetic effects. A total of 36 major and 6 minor QTL were detected in all environments. In addition, over 50 pairs of mapped loci were found to have interactions with flowering time, 50% of which laid between detected QTL regions. About 28% of candidate genes from *Arabidopsis* were positioned under QTL regions for flowering time in *B. napus*. Importantly, this experiment identified loci that were specific and had major effects in winter (one QTL in A10 -*BnFLC10*-) and spring (a QTL cluster in C6) environments. More recent experiments on flowering time variation also located QTLs in similar genomic regions (Cai et al. 2008, Mei et al. 2009) as described by Long et al. (2007) and others (Teutonico and Osborn 1994, Ferreira et al. 1994, Osborn et al. 1997, Butruille et al. 1999b). However, for these populations the main QTL effects for flowering time, photoperiod sensitivity, and vernalization response were located in different genomic regions. For instance, Cai and coauthors (2008) found a QTL on linkage group C8 affecting days to flower and photoperiod sensitivity that explained the highest percentage of the phenotypic variation in a DH spring population. Mei et al. (2009) found that the main QTL explaining flowering time variation in a semi-winter F2 population was located on C3 and not A10 as described by Long and colleagues (2007). This QTL also showed a high correlation with plant height and collocated with a plant height QTL in linkage group C3.

10.4.4 Hybrids, Population Development, and Seed Yield Improvement

As mentioned earlier, rapeseed CanOLA has gained impressive acceptance by vegetable oil producers. Indeed, the rapid rise of rapeseed as a source of vegetable oil and protein meal has been accompanied by the introduction of hybrid rapeseed. The production of hybrid seed is associated with purity, uniformity, and quality, which grants breeders with the capabilities to act effectively and rapidly when faced with environmental and market changes (Duvick 1984). Not surprisingly, hybrid seed has steadily replaced open pollinated (OP) rapeseed varieties in the farmer's field representing over 90% of the sown seed (Bundessortenamt 2005). *B. napus* is suitable for hybrid seed development, as there are well-defined pollination control systems (Mariani et al. 1990, 1992, Buzza 1995, Fu 1995, Stiewe et al. 1995, Renard et al. 1997). Additionally, heterotic groups within spring and winter types can be used to estimate general combining ability (GCA) and specific combining ability (SCA) of rapeseed inbred lines (Sernyk and Stefansson 1983, Brandle and McVetty 1989, Diers et al. 1995, Starmer et al. 1998, Quijada et al. 2006). For example, estimates of mid-parent heterosis or hybrid vigor (i.e., the positive value in the difference between an F1 and the average of its two parents or the highest parent) between spring European x spring Canadian hybrids ranged between 30 and 60% (Grant and Beversdorf 1985). In a more recent study, Radoev et al. (2008)

conducted a QTL mapping experiment in order to unmask the genetics behind heterotic effects for yield and yield components in rapeseed. Interestingly, 30% of the detected QTL showed significant dominance effects ranging from partial dominance to overdominance. In addition, a number of epistatic interactions (5–12) were found either affecting seed yield or yield-related traits. Radoev et al. (2008) concluded that epistasis, together with partial and complete dominance, and overdominance seemed to be responsible for the expression of heterosis (30% of mid-parent values – i.e., yield average of two parents) in the cross examined.

Researches have used a number of different strategies to improve rapeseed yield. One approach involved the introduction of winter alleles in spring germplasm (Butruille et al. 1999a, Quijada et al. 2006); another was based on the introgression of exotic alleles from unadapted or base germplam sources (Diers and Osborn 1994, Becker et al. 1995, Udall et al. 2005, Li et al. 2006, Chen et al. 2008, Kramer et al. 2009). The introduction of winter alleles into spring rapeseed was investigated by Quijada et al. (2006). Although QTLs were detected in two testcross populations (DH lines by commercial tester) for which the winter alleles increased seed yield (6 QTL), there were instances where QTLs harboring positive winter alleles decreased hybrid rapeseed yield. Interestingly, these winter alleles mapped to genomic regions carrying homoeologous non-reciprocal translocations. This suggested that allelic configurations created by these rearrangements could have contributed to the genetic variation for seed yield traits in rapeseed and hence account for a portion of the heterotic effects in hybrid combinations. These findings show that this approach could not always represent a valuable route to improve hybrid seed yield. Recently, Kramer et al. (2009) showed that the re-evaluation of seed yield QTL described by Butruille et al. (1999b), Quijada et al. (2006), and Udall et al. (2006) using the same genetic background as the original QTL mapping studies and also using different genetic backgrounds not always followed the predicted effect on hybrid seed yield. Interactions between previously detected genomic regions that increased seed yield (donor QTL) and new genetic background (recurrent QTL) may not always result in positive effects. In this context, marker-assisted selection (MAS) could prove effective, for a low heritable trait such as seed yield, however, the analysis of genetic background QTL interactions must be considered. Additionally, the introgression of exotic alleles from either winter ancestry into spring backgrounds or vice versa and/or from unadapted sources may not differ from, or produced significantly lower hybrid seed yield than, elite alleles from similar heterotic pools. Therefore, in order to determine the most informative and predictable alleles to be screened through MAS in a breeding program, comparisons of multiple QTL alleles are recommended. QTL complementation across testers would be required to predict their effects in multiple hybrid combinations. In a different study but still relying on the concept of introgressing exotic or unadapted germplam, Li et al. (2006) used *B. rapa* and *Brassica carinata* A. Braun subgenomic regions to create a partially new *B. napus* synthetic line carrying alleles that confer intergenomic heterotic effects for seed yield. This strategy could also be applied to recruit genomic regions harboring disease-resistance alleles.

Bottleneck events or the loss of genetic variation through selection (natural or artificial) appears to be a common theme during the domestication of horticultural and field crops. One thing to consider while introducing alleles from one background into another is the possibility of fixing unfavorable allele combinations at different loci and eroding genetic variation. For example, an introgression study revealed that a QTL that increased seed yield was linked in coupling phase to a genomic region responsible for high glucosinolate content. This suggested that the transition of "++" into "00" rapeseed could have resulted in the loss of favorable seed yield alleles (Quijada et al. 2006, Osborn et al. 2007).

Population development in rapeseed has been an active area of research. Balanced or bi-parental populations such as first-generation backcross, DH, F2, or recombinant inbred lines (RILs) have been the more prevalent genetic tool used for seed yield QTL analysis in field crops (MacKay 2001). In rapeseed breeding, DH lines have represented the core of population development (Table 10.1) (Quijada et al. 2006, Chen et al. 2007, Radoev et al. 2008, Shi et al. 2009). DH lines are fully homozygous and provide an ideal source from where to detect additive effects. Additionally, DH lines represent immortal populations, which are extremely useful to test the interaction that seed yield QTL have with the environment. Multiple locations and replicated QTL analysis can easily be conducted. This is a key feature for rapeseed breeding programs due to the complex nature of QTL x environment interactions. Very little is known about genes underlying seed yield and their interaction with environmental factors in rapeseed yield improvement. The limitations of its understanding can have profound implications on MAS. Recently and for the first time, a study using 10 natural environments and 2 related populations (DH lines and derived fixed homozygous F2 lines) was conducted to unravel the complex nature of seed yield and yield-related traits in rapeseed (Shi et al. 2009). This study reported an impressive number of 870 identified rapeseed yield QTLs in a single genetic linkage map using QTL meta-analysis (Arcade et al. 2004). From these, 85 QTL were specific to seed yield and 785 for associated yield traits (e.g., flowering time, plant height, maturity). A remarkable finding is that very few QTL were universally detected at all environments tested suggesting that the high proportion of location/environment-specific QTL is indicative of the large effect that abiotic and biotic factors have on genes underlying seed yield in general. This provides breeders with information required to breed lines with specific adaptabilities for specific micro-environments. In addition, this methodology is ideal to identify QTL that showed consistency across environments and can, therefore, be used for MAS for wide target of environments. The feasibility of finding candidate genes through bioinformatics and comparative analysis renders this type of experiment extremely valuable. Detailed meta-QTL analysis also can illustrate the pleiotropic nature of seed yield and yield-related traits. For instance, 55 QTL for seed yield were all dependent on the QTL effect of other yield-associated traits (Shi et al. 2009). Lastly, Shi et al. (2009) demonstrated that meta-analysis of seed yield QTL and yield-associated QTL allows to precisely identify QTL indicators (i.e., trait correlations) for seed yield and yield-associated in order to make effective gains for each specific environment. This approach could also be used to breed for seed quality-related traits.

Advanced and inbred backcross lines, two forms of unbalanced populations, are extremely useful at Mendelizing complex traits while avoiding confounding effects made during the process of line selection or arisen through population structure (e.g., pedigree structures). For example, Butruille et al. (1999b) was able to identify winter alleles from a German cultivar (Ceres) on linkage groups A3 and C4 which improved seed yield (up to 7%) in spring hybrids.

Finally, an active and leading area of research that also affects rapeseed yield is breeding for disease resistance. The amount of information in this area is vast and a comprehensive review on the status of the genetics of disease resistance in *B. napus* will go beyond the scope of this chapter. Other areas of research oriented to seed yield improvement include, for example, breeding for abiotic stress resistance (Zhao et al. 2008) and low incidence of pod shattering (Wang et al. 2007, Gan et al. 2008) and preharvest sprouting (Feng et al. 2009). These traits are all governed by polygenic inheritance and efforts to identify genomic regions explaining phenotypic effects have been concentrated on germplasm evaluation and QTL analysis using similar strategies to those described in this section.

10.4.5 Other Important Oil Quality-Related Traits

B. napus seeds are rich in oil (40–50% v/v). After oil extraction, the produced meal is protein rich (40% dry weight) and constitutes a good alternative for animal feed. Breeding efforts aimed at increasing the quality of the rapeseed meal include increasing essential amino acid content and reducing the amount of antinutritional factors (ANFs) such as glucosinolates, sinapate esters, phytic acid, tannins, and crude fiber (reviewed in Nesi et al. 2008 and Wittkop et al. 2009) as well as increasing other secondary metabolites to aid quality. Breeding for all these traits can be cumbersome due to the existence of pleiotropic effects. Additionally, ANFs play important roles in plant development and stress resistance and ideally, their content reduction should be targeted or limited to the edible parts of the plant.

Protein content and composition is a very important quality trait for the utilization of rapeseed meal in human and animal nutrition. Specifically, high contents of essential amino acids (methione and cysteine) are desired. Three main storage proteins can be found in *B. napus* seeds: cruciferins (12S globulins), napins (2S albumins), and oleosins; with napins exhibiting a higher percentage of essential (sulfuric and aromatic) amino acids. This difference in storage proteins amino acid content was used to evaluate *B. napus* germplasm. As a result, it was shown that the cruciferin (12S)/napin (2S) ratio can range from 0.7 to 2, with the total protein content remaining stable (reviewed in Nesi et al. 2008). Early attempts to modify this ratio through an increase of the napin fraction have been based on the use of genetic engineering (Guerche et al. 1990, Altenbach et al. 1992, Kohno-Murase et al. 1994). As expected, transgenic plants exhibited not only increased levels of napins (and thus essential amino acids) but also decreased levels of cruciferins, which highlighted the tight metabolic controlled exerted on protein content. In addition, it is worth to mention that CanOLA meals exhibit a higher cruciferin/napin ratio than traditional rapeseed meals. This could be due, at least in part, to the reduced glucosinolate content of CanOLA meals (Malabat et al. 2003).

Glucosinolates are secondary metabolites produced in species of the Brassicaceae family; they contain sulfur and nitrogen and are derived from glucose and a particular amino acid. They are called aliphatic if they are derived from methionine, alanine, valine, leucine, or isoleucine, aromatic if they are derived from tyrosine or phenyalanine, and indole-glucosinolates if they are derived from tryptophan. High levels of glucosinolates in rapeseed meal reduce its palatability and acceptance by animals and can also lead to goitrogenic hypertrophy, liver, and kidney problems (Walker and Booth 2001) making them ANF. One spring cultivar (Bronowski) was identified as a low glucosinolate form and together with Liho provided the alleles to create the first double zero spring variety (Tower) exhibiting low erucic acid and low glucosinolate content (8–15 μmol/g) in seeds. Four seed glucosinolate content QTL located on chromosomes A9, C2, C7, and C9 have been recurrently mapped in different *B. napus* studies (Uzunova et al. 1995, Howell et al. 2003, Sharpe and Lydiate 2003, Zhao and Meng 2003, Basunanda et al. 2007). Breeding efforts to further reduce glucosinolate content are ongoing and targeted to reducing specific types of glucosinolates. Interestingly, a recent study associated gene-linked SSR markers to seed glucosinolate content in *B. napus* using comparative genomics with *Arabidopsis* (Hasan et al. 2008). By this means, it will be possible to apply the vast biochemical knowledge developed in this model species to select molecular markers with putative linkage to genes active on specific branches of the glucosinolate biosynthetic pathway and thus facilitate targeting the reduction of specific glucosinolate types (Hasan et al. 2008).

Sinapate esters usually represent 1–2% of the dry rapeseed weight (Bell 1993). They reduce the palatability of the rapeseed meal due to their bitter taste and astringency and also reduce the bioavailability and digestibility of the meal proteins by forming complexes with amino acids (Kozlowska et al. 1990, Shahidi and Naczk 1992). Therefore, sinapate esters are also considered ANF. Classical approaches to reduce sinapate ester seed content have been based in the characterization and utilization of existing genetic variation for this trait, which exhibits high heritabilities (Zum Felde et al. 2006). Recently, four sinapate ester content QTL explaining 53% of the genetic variance were detected, but the strongest QTL mapped within the confidence intervals of the two erucic acid content QTL located on linkage groups A8 and C3 (Amar et al. 2008). In addition, a negative correlation was found between erucic acid content and sinapate ester content, suggesting a pleiotropic effect of the two erucic acid genes on sinapate ester content, with alleles for low erucic acid content increasing sinapate ester content (Amar et al. 2008). Successful transgenic approaches to reduce sinapate ester content have been mainly based on the silencing of seed-specific genes involved in the final metabolic steps of this ANF synthesis, resulting in 70–90% reduction (Hüsken et al. 2005, Weier et al. 2007, Bhinu et al. 2009).

High-fiber and tannin content in rapeseed meals causes digestibility problems. These ANFs are mainly located in the seed coat which represents a significant percentage of the total *B. napus* seed volume. The seed coat of yellow-seeded rapeseed varieties do not contain tannins and do not exhibit the characteristic brown color. In addition, yellow-seeded lines exhibit lower fiber and higher protein content in their seeds (Simbaya et al. 1995, Wittkop et al. 2009). Thus, two strategies could be followed to reduce fiber and tannin content: breeding varieties with a thinner testa (Wittkop et al. 2009) or developing yellow-seeded varieties through the introgression of yellow seed color genes from related species or mutant *B. napus* lines (Nesi et al. 2008). This latter strategy, however, has been hindered by the inability to develop true breeding lines that consistently produce yellow seeds under different environmental conditions (Rahman 2001). This genotype by environment interaction also precluded the use of seed color as a morphological marker for selection of lines with low-fiber seed content. Luckily, the development of near-infrared reflectance spectroscopy (NIRS) calibration equations to determine seed fiber content offered the possibility of readily screening and selecting breeding lines with improved nutritional value (Wittkop et al. 2009). Such screening allowed the identification of previously overlooked phenotypic variation for fiber content in dark-seeded varieties, indicating that improvement of meal quality could be achieved breeding conventional black-seeded *B. napus* (Wittkop et al. 2009). As an added bonus, it is possible that fiber seed content reduction through the selection of a thinner seed coat will also result in reduced sinapate and glucosinolate levels (Wittkop et al. 2009).

B. napus oil quality can also be enhanced by the presence of a series of beneficial hydrophobic compounds including carotenoids, tocochromanols, and phytoesterols. Both carotenoids and tocochromanols are potent antioxidants and essential components of the human diet providing a source of provitamin A (beta-carotene) and E (alpha-tocopherol). Attempts to increase these beneficial compounds in *B. napus* oil include the production of transgenic lines exhibiting elevated levels of total carotenoid (50-fold) and tocopherol (3.6-fold) content in seeds (Shewmaker et al. 1999, Fujisawa et al. 2009, Kumar et al. 2005, Raclaru et al. 2006). Phytoesterols are produced by the isoprenoid biosynthetic pathway and possess LDL cholesterol lowering properties (Best et al. 1954, Law 2000, Nissinen et al. 2002, Trautwein et al. 2003). Recently, three QTL for total phytoesterol content were detected in a winter rapeseed DH population, explaining 60% of the genetic variance for this trait (Amar et al. 2008). Once again, evidence for the existence of pleiotropic effects exerted by the two erucic acid genes was revealed by a significant correlation between phytoesterol and erucic acid content. Fortunately, the existence of this negative correlation indicates that it should be possible to select for high phytoesterol content in CanOLA oil (Amar et al. 2008).

10.5 Conclusions

The current state of knowledge in rapeseed *B. napus* genetics and genomics is vast, and a battery of genetic and genomic tools is available for this crop species. These tools have enabled researchers to investigate the genetic factors governing traits of agronomical and nutritional importance though further work needs to be conducted in order to fully understand the complexity of these quantitative traits. The recent advent of faster and cheaper "next-generation sequencing" technologies and improved bioinformatic and statistical analyses assist in the development of highly dense molecular linkage maps and expedite genome sequencing efforts. Undoubtedly, these technological advances will impact several aspects of *B. napus* breeding and genetics. First, insights into the genomic variation exhibited by *B. napus* will help to clarify its origins. This will prove useful to broaden the narrow genetic pool of commercial rapeseed cultivars. Second, a closer look at the genetic determinants affecting hybrid seed performance will increase our understanding of the molecular basis of heterosis. Third, the dissection of complex quantitative traits such as yield, seed oil content, fatty acid profile, seed nutritional and ANF content, and flowering time will better determine the existence of pleiotropic effects and genotype by environment interactions affecting these traits. This will provide rapeseed breeders with the knowledge required to make effective decisions while breeding for a number of traits that exhibit significant correlations among them as observed for many seed-related traits. Clearly, future nutritional, industrial, and agronomical demands will only be met through a solid understanding of *B. napus* genetics and genomics.

Acknowledgments The authors gratefully acknowledge the Agri aquaculture Nutritional Genomic Center (CGNA) for support on the development of a *B. napus* breeding program and Fondecyt 1090726 and 1100732 for supporting the research related to carotenoid content enhancement and SNP discovery in *B. napus*, respectively.

References

Albert TJ, Molla MN, Muzny DM, Nazareth L, Wheeler D, Song X, Richmond TA, Middle CM, Rodesch MJ, Charles J, Packard CJ, Weinstock GM, Gibbs RA (2007) Direct selection of human genomic loci by microarray hybridization. Nat Methods 4:903–905

Allard RW (1999) Principals of plant breeding, 2nd edn. Wiley, New York, NY

Allender C, Evered C, Lynn J, King G (2005) Tracing the origins of Brassica napus using chloroplast microsatellites. In: Plant and Animal Genomes XIII Conference. Book of Abstract p. 411

Altenbach SB, Kuo CC, Staraci LC, Pearson KW, Wainwright C, Georgescu A, Townsend J (1992) Accumulation of a Brazil nut albumin in seeds of transgenic canola results in enhanced levels of seed protein methionine. Plant Mol Biol 26:1115–1124

Amar S, Ecke W, Becker HC, Möllers C (2008) QTL for phytosterol and sinapate ester content in Brassica napus L. collocate with two erucic acid genes. Theor Appl Genet 116: 1051–1061

Amasino RM (2005) Vernalization and flowering time. Curr Opin Biotechnol 16:154–158

Applequist LA, Ohlson R (1972) Rapeseed. Elsevier, MO, USA

Arcade A, Labourdette A, Falque M, Mangin B, Chardon F, Charcosset A, Joets J (2004) BioMercator: integrating genetic maps and QTL towards discovery of candidate genes. Bioinformatics 20:2324–2326

Auld DL, Heikkinen MK, Erickson DA, Sernyk JL, Romero JE (1992) Rapeseed mutants with reduced levels of polyunsaturated fatty acids and increased levels of oleic acid. Crop Sci 32:657–662

Babula D, Kaczmarek M, Barakat A, Delseny M, Quiros CF, Sadowski J (2003) Chromosomal mapping of Brassica oleracea based on ESTs from Arabidopsis thaliana: complexity of the comparative map. Mol Genet Genomics 268:656–665

Barker GC, Larson TR, Graham IA, Lynn JR, King GJ (2007) Novel insights into seed fatty acid synthesis and modification pathways from genetic diversity and quantitative trait loci analysis of the Brassica C Genome. Plant Phys 144:1827–1842

Barret P, Delourme R, Brunel D, Jourdren C, Horvais R, Renard M (1999) Low linolenic acid level in rapeseed can be easily assessed through the detection of two single base substitution in fad3 genes. In: Proceeding of the 10th International Rapeseed Congress, Canberra, Australia pp. 26–29.

Basunanda P, Spiller TH, Hasan M, Gehringer A, Schondelmaier J, Lühs W, Friedt W, Snowdon RJ (2007) Marker-assisted increase of genetic diversity in a double-low seed quality winter oilseed rape genetic background. Plant Breed 126:581–587

Becker HC, Engqvist GM, Karlsson B (1995) Comparison of rapeseed cultivars and resynthesized lines based on allozyme and RFLP markers. Theor Appl Genet 91:62–67

Bell JM (1993) Factors affecting the nutritional value of canola meal. Can J Anim Sci 73: 679–697

Best MM, Duncan CH, van Loon EJ, Wathens JD (1954) Lowering of serum cholesterol by the administration of plant sterol. Circulation 10:201–206

Bhinu VS, Schäfer UA, Li R, Huang J, Hannoufa A (2009) Targeted modulation of sinapine biosynthesis pathway for seed quality improvement in Brassica napus. Transgenic Res 18:31–44

Boys EF, Roques SE, Evans N, Latunde-Dada AO, West JS, Fitt BDL (2007) Resistance to infection by stealth: Brassica napus (winter oilseed rape) and Pyrenopeziza brassicae (light leaf spot). Eur J Plant Pathol 118:307–321

Bradshaw JE, Griffiths DW (1990) Sugar content of swedes for stock-feeding. J Sci Food Agric 50:167–172

Brandle JE, McVetty PBE (1989) Heterosis and combining ability in hybrids derived from oilseed rape cultivars and inbred lines. Crop Sci 29:1191–1195

Brown G, Formanova N, Jin H, Wargachuk R, Dendy C, Patil P, Laforest M, Zhang J, Cheung WY, Landry BS (2003) The radish Rfo restorer gene of ogura cytoplasmic male sterility encodes a protein with multiple pentatricopeptide repeats. Plant J 35:262–272

Bundessortenamt (2005) Beschreibende Sortenliste 2005: Getreide, Mais, Ölfruchte, leguminosen (Groβkörnig) Hackfrucht (auβer Kartoffeln) Deutscher Landwirtschaffsverlag GmbH, Hannover, Germany.

Burns MJ, Barnes SR, Bowman JG, Clarke MHE, Werner CP, Kearsey MJ (2003) QTL analysis of an intervarietal set of substitution lines in Brassica napus: (i) Seed oil content and fatty acid composition. Heredity 90:39–48

Butruille DV, Guries RP, Osborn TC (1999a) Increasing yield of spring oilseed rape hybrids through introgression of winter germplasm. Crop Sci 39:1491–1496

Butruille DV, Guries RP, Osborn TC (1999b) Linkage analysis of molecular markers and quantitative trait loci in populations of inbred backcross lines of Brassica napus L. Genetics 153:949–964

Buzza GC (1995) Plant breeding. In: Kimber DS, McGregor DI. (eds) Brassica oilseeds: production and utilization. CABI Publishing, Wallingford, CT, pp. 153–175

Cai CC, Tu J, Fu TD, Chen BY (2008) The genetic basis of flowering time and photoperiod sensitivity in rapeseed Brassica napus L. Russ J Genet 44:326–333

Chen X, Li M, Shi J, Fu D, Qian W, Zou J, Zhang C, Meng J (2008) Gene expression profiles associated with intersubgenomic heterosis in Brassica napus. Theor Appl Genet 117: 1031–1040

Chen W, Zhang Y, Xueping Liu X, Baoyuan Chen B, Tu J, Tingdong F (2007) Detection of QTL for six yield-related traits in oilseed rape (Brassica napus) using DH and immortalized F2 populations. Theor Appl Genet 115:849–858

Cheng X, Xu J, Xia S, Gu J, Yang Y, Fu J, Qian X, Zhang S, Wu J (2009) Development and genetic mapping of microsatellite markers from genome survey sequences in Brassica napus. Theor Appl Genet 118:1121–1131

Cheung WY, Champagne G, Hubert N, Landry BS (1997) Comparison of the genetic maps of Brassica napus and Brassica oleracea. Theor Appl Genet 94:569–582

Choi SR, Teakle GR, Prikshit P, Kim JH, Allender CJ, Beynon E, Piao ZY, Soengas P, Han TH, King GJ, Barker GC, Hand P, Lydiate DJ, Batley J, Edwards D, Koo DH, Bang JW, Park B-S, Lim YP (2007) The reference genetic linkage map for the multinational Brassica rapa genome sequencing project. Theor Appl Genet 115:777–792

Cloutier S, Cappadocia M, Landry BS (1995) Study of microspore-culture responsiveness in oilseed rape (Brassica napus L.) by comparative mapping of a F2 population and two microspore-derived populations. Theor Appl Genet 91:841–847

Collins FS, Brooks LD, Chakravarti A (1998) A DNA polymorphism discovery resource for research on human genetic variation. Genome Res 8:1229–1231

Damania AB, Valkoun J, Willcox G, Qualset CO (1997) The origins of agriculture and crop domestication: the harlan symposium. ICARDA, IPGRI, FAO and UC/GRCP Publishers, California, CA

Delourme R, Bouchereau A, Hubert N, Renard M, Landry BS (1994) Identification of RAPD markers linked to a fertility restorer gene for the Ogura radish cytoplasmic male sterility of rapeseed (Brassica napus L.). Theor Appl Genet 88:741–748

Delourme R, Falentin C, Huteau V, Clouet V, Horvais R, Gandon B, Specel S, Hanneton L, Dheu JE, Deschamps M, Margale E, Vincourt P, Renard M (2006) Genetic control of oil content in oilseed rape (Brassica napus L.). Theor Appl Genet 113:1331–1345

Diederichsen E, Frauen M, Linders Enrico GA, Hatakeyama K, Hirai M (2009) Status and perspectives of clubroot resistance breeding in crucifer crops. J Plant Growth Regul 28: 265–281

Diers BW, McVetty PBE, Osborn TC (1995) Relationship between heterosis and genetic distance based on restriction fragment length polymorphism markers in oilseed rape (Brassica napus L.). Crop Sci 36:79–83

Diers BW, Osborn TC (1994) Genetic diversity of oilseed Brassica napus germplasm based on restriction fragment length polimorphisms. Theor Appl Genet 88:662–668

Downey RK, Craig BM (1964) Genetic control of fatty acid biosynthesis in rapeseed (Brassica napus L.). J Am Oil Chem Soc 41:475–478

Duvick DN (1984) Genetic contributions to yield gains of US hybrid maize. In Fehr WR (ed) Genetic contributions to yield gains of five major crop plants. Crop Science Society of America. Madison, WI, American Society of Agronomy, pp 1930–1980

Ecke W, Uzunova M, Weissleder K (1995) Mapping the genome of rapeseed (Brassica napus L.). II Localization of genes controlling erucic acid synthesis and seed oil content. Theor Appl Genet 91:972–977

Falentin C, Brégeon M, Lucas MO, Deschamps M, Leprince F, Fournier MT, Delourme R, Renard M (2007) Identification of fad2 mutations and development of allele-specific markers for high oleic acid content in rapeseed (Brassica napus L.) In: Proceeding of the 12th International Rapeseed Congress, Wuhan, China, pp. 117–119

Feng F, Liu P, Hong D, Yang G (2009) A major QTL associated with preharvest sprouting in rapeseed (Brassica napus L.). Euphytica 169:57–68

Ferreira ME, Williams PH, Osborn TC (1994) RFLP mapping of Brassica napus using doubled haploid lines. Theor Appl Genet 89:615–621

Foisset N, Delourme R, Barret P, Hubert N, Landry BS, Renard M (1996) Molecular-mapping analysis in Brassica napus using isozyme, RAPD and RFLP markers on a doubled haploid progeny. Theor Appl Genet 93:1017–1025

Food and Agriculture Organization of the United Nations (FAO) (2008) Food Outlook June 2008: Oilseeds, Oils and Meals, (ftp://ftp.fao.org/docrep/fao/010/ai466e/ai466e00.pdf).

Food and Drug Administration (1995) Code of Federal Regulations, Title 21, Food and Drugs, Part 184, "Direct Food Substances Affi rmed as Generally Recognized as Safe", Sections 1555, as of April 1, 1995. US Government Printing Office, Washington, DC

Fourmann M, Barret P, Renard M, Pelletier G, Delourme R, Brunel D (1998) The two genes homologous to Arabidopsis FAE1 cosegregate with the two loci governing erucic acid content in Brassica napus. Theor Appl Genet 96:852–858

Fray MJP, Puangsomlee P, Goodrich J, Coupland G, Evans EJ, Arthur AE, Lydiate DJ (1997) The genetics of stamenoid petal production in oils seed rape (Brassica napus) and equivalent variation in Arabidopsis thaliana. Theor Appl Genet 94:731–736

Fu TD (1995) Breeding and utilization of rapeseed hybrid. Hubei Science and Technology Press, Wuhan, pp 42–135

Fu FY, Liu LZ, Chai YR, Chen L, Yang T, Ma AF, Qu CM, Jiang L, Zhang ZS, Li JN (2007) Localization of QTLs for husk proportion and lignin content using a high-density genetic linkage map of Brassica napus. Korean J Genet 29:343–353

Fujisawa M, Takita E, Harada H, Sakurai N, Suzuki H, Ohyama K, Shibata D, Misawa N (2009) Pathway engineering of Brassica napus seeds using multiple key enzyme genes involved in ketocarotenoid formation. J Exp Bot 60:1319–1332

Gaeta RT, Pires JC, Iniguez-Luy F, Leon E, Osborn TC (2007) Genomic changes in resynthesized Brassica napus and their effect on gene expression and phenotypic variation. Plant Cell 19:3403–3417

Gan Y, Malhi SS, Brandt SA, McDonald CL (2008) Assessment of seed shattering resistance and yield loss in five oilseed crops. Can J Plant Sci 88:267–270

Ganal MW, Altmann T, Röder M (2009) SNP identification in crop plants. Curr Opin Plant Biol 12:211–217

Gao M, Li G, Bo Yang B, Qiu D, Farnham M, Quiros C (2007) High-density Brassica oleracea linkage map: identification of useful new linkages. Theor Appl Genet 115:277–287

Gomez-Campo C (1999) Biology of Brassica coenospecies. Elsevier, The Netherlands

Graef GL, Fehr WR, Hammond EG (1985) Inheritance of three stearic acid mutants of soybean. Crop Sci 25:1076–1079

Grant I, Beversdorf WD (1985) Heterosis and combining ability estimates in springplanted oilseed rape (Brassica napus L.). Can J Genet Cytol 27:472–478

Grundy SM (1994) Influence of stearic acid on cholesterol metabolism relative to other long-chain fatty acids. Am J Clin Nutr 60:986–990

Guerche P, De Almeida ER, Schwarztein MA, Gander E, Krebbers E, Pelletier G (1990) Expression of the 2S albumin from Bertholletia excelsa in Brassica napus. Mol Gen Genet 221:306–314

Gupta SK, Pratap A (2007) History, origin and evolution. In: Gupta SK (ed) Advances in botanical research, incorporating advances in plant pathology, rapeseed breeding, vol 45. Academic Press-Elsevier, London, UK, pp 2–17

Harvey BL, Downey RK (1964) The inheritance of erucic acid content in rapeseed (Brassica napus L.). Can J Plant Sci 44:104–111

Hasan V, Friedt W, Pons-Kühnemann J, Freitag NM, Link K, Snowdon RJ (2008) Association of gene-linked SSR markers to seed glucosinolate content in oilseed rape (Brassica napus ssp. napus). Theor Appl Genet 116:1035–1049

Hawkins DJ, Kridl JC (1998) Characterization of acyl-ACP thioesterases of mangosteen (Garcinia mangostana) seed and high levels of stearate production in transgenic canola. Plant J 13: 743–752

He J, Ke L, Hong D, Xie Y, Wang G, Liu P, Yang G (2008) Fine mapping of a recessive genic male sterility gene (Bnms3) in rapeseed (Brassica napus) with AFLP- and Arabidopsis-derived PCR markers. Theor Appl Genet 117:11–18

Hougen FW, Stefansson BR (1983) Rapeseed, in advances in cereal science and technology, vol 5. American Association of Cereal Chemists, St. Paul, pp 261–289

Howell PM, Marshall DF, Lydiate DJ (1996) Towards developing intervarietal substitution lines in Brassica napus using marker-assisted selection. Genome 39:348–358

Howell PM, Sharpe AG, Lydiate DJ (2003) Homoelogous loci control the accumulation of seed glucosinolates in oilseed rape (Brassica napus). Genome 46:454–460

Hu J, Li G, Struss D, Quiros C (1999) SCAR and RAPD markers associated with 18-carbon fatty acids in rapeseed, Brassica napus. Plant Breed 118:145–150

Hu X, Sullivan-Gilbert M, Gupta M, Thompson S (2006) Mapping of the loci controlling oleic and linolenic acid contents and development of fad2 and fad3 allele-specific markers in canola (Brassica napus L.). Theor Appl Genet 113:497–507

Hüsken A, Baumert A, Strack D, Becker HC, Möllers C, Milkowski C (2005) Reduction of sinapate ester content in transgenic oilseed rape (Brassica napus L.) by dsRNAi-based suppression of BnSGT1 gene expression. Mol Breed 16:127–138

Iniguez-Luy FL, Lukens L, Farnham MW, Amasino RM, Osborn TC (2009) Development of public immortal mapping populations, molecular markers and linkage maps for rapid cycling Brassica rapa and Brassica oleracea. Theor Appl Genet 120:31–43

Iniguez-Luy FL, Sass ME, Jung G, Johns MA, Nienhuis J (2006) Development of SCAR markers that distinguish the six cultivated Brassica species and sub-species of the U-triangle. American society of horticultural sciences. J Amer Soc Hort Sci 131:424–432

Iniguez-Luy FL, Van de Voort A, Osborn TC (2008) Development of a set of SSR markers derived from genomic sequence of a rapid cycling Brassica oleracea L. genotype. Theor Appl Genet 117:977–985

International HapMap Consortium, Frazer K, Ballinger DG, Cox DR, Hinds DA, Stuve LL, Gibbs RA, Belmont JW, Boudreau A, Hardenbolo P, et al (2007) A second generation human haplotype map of over 3.1 million SNPs. Nature 449:851–861

Jiang JJ, Zhao XX, Tian W, Li TB, Wang YP (2009) Intertribal somatic hybrids between Brassica napus and camelina sativa with high linolenic acid content. Plant Cell Tiss Organ Cult 99:91–95

Jourdren C, Barret P, Brunel D, Delourme R, Renard M (1996a) Specific molecular marker of the genes controlling linolenic acid content in rapeseed. Theor Appl Genet 93:512–518

Jourdren C, Barret P, Horvais R, Delourme R, Renard M (1996b) Identification of RAPD markers linked to linolenic acid genes in rapeseed. Euphytica 90:351–357

Kelly AL, Sharpe AG, Nixon JH, Evans EJ, Lydiate DJ (1997) Indistinguishable patterns of recombination resulting from male and female meiosis in Brassica napus (oilseed rape). Genome 40:49–56

Kim JS, Chung TY, King GJ, Jin M, Yang T-J, Jin Y-M, Kim H-I, Park B-S (2006) A sequence-tagged linkage map of Brassica rapa. Genetics 174:29–39

Kimber DS, McGregor DI (1995) The species and their origins, cultivation and world production. In: Kimber DS, McGregor DI. (eds) Brassica oilseeds: production and utilization. CABI Publishing, Wallingford, CT, pp 1–9

Knutzon DS, Thompson GA, Radke SE, Johnson WB, Knauf VC, Kridl JC (1992) Modification of Brassica seed oil by antisense expression of a stearoyl-acyl carrier protein desaturase gene. Proc Nat Acad Sci USA 89:2624–2628

Kohno-Murase J, Murase M, Ichiwaka H, Imamura J (1994) Effects of an antisense napin gene on seed storage compounds in transgenic Brassica napus seeds. Plant Mol Biol 26:1115–1124

Kole C, Thormann CE, Karlsson BH, Palta JP, Gaffney P, Yandell B, Osborn TC (2002) Comparative mapping of loci controlling winter survival and related traits in oilseed Brassica rapa and Brassica napus. Mol Breed 9:201–210

Koornneef M, Alonso-Blanco C, Peeters AJM (1997) Genetic approaches in plant physiology. New Phys 137:1–8

Kozlowska H, Naczk M, Shahidi F, Zadernowski R (1990) Phenolic acids and tannins in rapeseed and canola. In: Shahidi F (ed) Canola and rapeseed. production chemistry, nutrition and processing technology. Van Nostrand Reinhold, New York, NY, pp 193–210

Kramer CC, Polewicz H, Osborn TC (2009) Evaluation of QTL alleles from exotic sources for hybrid seed yield in the original and different genetic backgrounds of spring-type Brassica napus L. Mol Breed 24:419–431

Kumar R, Raclaru M, Schsseler T, Gruber J, Sadre R, Lhs W, Zharhloul KM, Friedt W, Enders D, Frentzen M, Weier D (2005) Characterization of plant tocopherol cyclases and their overexpression in transgenic Brassica napus seeds. FEBS Lett 579:1357–1364

Lagercrantz U (1998) Comparative mapping between Arabidopsis thaliana and Brassica nigra indicates that Brassica genomes have evolved through extensive genome replication accompanied by chromosomes fusions and frequent rearrangements. Genetics 150:1217–1228

Lagercrantz U, Lydiate DJ (1996) Comparative mapping in Brassica. Genetics 144:1903–1910

Lagercrantz U, Putterill J, Coupland G, Lydiate DJ (1996) Comparative mapping in Arabidopsis and Brassica, fine scale genome collinearity and congruence of genes controlling flowering time. Plant J 9:13–20

Landry BS, Hubert N, Etoh T, Harada J, Lincoln S (1991) A genetic map for Brassica napus based on restriction fragment length polymorphisms detected with expressed DNA sequences. Genome 34:543–552

Law M (2000) Plant sterol and stanol margarines and health. Br Med J 320:861–864

Li M, Chen X, Meng J (2006) Intersubgenomic heterosis in rapeseed production with a partial new-typed Brassica napus containing subgenome Ar from Brassica rapa and Cc from Brassica carinata. Crop Sci 46:234–242

Li G, Gao M, Yang B, Quiros CF (2003) Gene for gene alignment between the Brassica and Arabidopsis genomes by direct transcriptome mapping. Theor Appl Genet 107:168–180

Li Y, Ma C, Fu T, Yang G, Tu J, Chen Q, Wang T, Zhang X, Li C (2006) Construction of a molecular functional map of rapeseed (Brassica napus L.) using differentially expressed genes between hybrid and its parents. Euphytica 152:25–39

Li G, Quiros CF (2001) Sequence-related amplified polymorphism (SRAP), a new marker system based on a simple PCR reaction: its application to mapping and gene tagging in Brassica. Theor Appl Genet 103:455–461

Lombard V, Delourme R (2001) A consensus linkage map for rapeseed (Brassica napus L): construction and integration of three individual maps from DH populations. Theor Appl Genet 103:491–507

Long Y, Shi J, Qiu D, Li R, Zhang C, Wang J, Hou J, Zhao J, Shi L, Park B-S, Choi SR, Lim YP, Meng J (2007) Flowering time quantitative trait loci analysis of oilseed Brassica in multiple environments and genomewide alignment with arabidopsis. Genetics 177:2433–2444

MacKay T (2001) The genetic architecture of quantitative traits. Annu Rev Genet 35:30–39

Malabat C, Atterby H, Chaudhry Q, Renard M, Guéguen J (2003) Genetic variability of rapeseed protein composition. In: Sorensen H, Sorencen JC, Sorencen S, Bellostas Muguerza N, Bjegegaard C (eds) 11th international rapeseed congress. The Royal Veterinary and Agricultural University, Copenhagen, DNK, pp 205–208

Margulies M, Egholm M, Altman WE, Attiya S, Bader JS, Bemben LA, Berka J, Braverman MS, Chen Y-J, Chen Z, Dewell SB, Du L (2005) Genome sequencing in microfabricated high-density picolitre reactors. Nature 437:376–380

Mariani C, De Beuckeler M, Truettner J, Leemans J, Goldberg RB (1990) Induction of male sterility in plants by a chimaeric ribonuclease gene. Science 347:737–741

Mariani C, Gossele V, De Beuckeler M, De Block M, Golberg RB, De Greef W, Leemans J (1992) A chimaeric ribonuclease inhibitor gene restores fertility to male sterile plants. Nature 357: 384–387

Mayerhofer R, Wilde K, Mayerhofer M, Lydiate D, Bansal V, Good A, Parkin I (2005) Complexities of chromosome landing in a highly duplicated genome: towards map based cloning of a gene controlling blackleg resistance in Brassica napus. Genetics 171: 1977–1988

McCarthy MI, Abecasis GR, Cardon LR, Goldstein DB, Little J, Ioannidis JP, Hirschhorn JN (2008) Genome wide association studies for complex traits: consensus, uncertainty and challenges. Nat Rev Genet 9:356–369

McNaughton IH, Thow RF (1972) Swedes and turnips. Field Crop Abs 25:1–12

Mei DS, Wang HZ, Hu Q, Li YD, Xu YS, Li YC (2009) QTL analysis on plant height and flowering time in Brassica napus. Plant Breed 128:458–465

Michaels SD, Amasino RM (1999) FLOWERING LOCUS C encodes a novel MADS domain protein that acts as a repressor of flowering. Plant Cell 11:949–956

Nesi N, Delourme R, Brégeon M, Falentin C, Renard M (2008) Genetic and molecular approaches to improve nutritional value of *Brassica napus* L. seed. CR Biologies 331:763–771

Nissinen M, Gylling H, Vuoristo M, Miettinen TA (2002) Micellar distribution of cholesterol and phytoesterols after duodenal plant stanol ester infusion. Am J Physiol Gastrointest Liver Physiol 282:1009–1015

Okou DT, Steinberg KM, Middle C, Cutler CD, Albert TJ, Zwick ME (2007) Microarray-based genomic selection for highthroughput resequencing. Nat Methods 4:907–909

Olson M (2007) Enrichment of super-sized resequencing targets from the human genoma. Nat Methods 4:891–892

Osborn TC, Butruille DV, Sharpe AG, Pickering KJ, Parkin IAP, Parker JS, Lydiate DJ (2003a) Detection and effects of a homoeologous reciprocal transposition in Brassica napus. Genetics 165:1569–1577

Osborn TC, Kole C, Parkin IAP, Sharpe AG, Kuiper M, Lydiate DJ, Trick M (1997) Comparison of flowering time genes in Brassica rapa, B. napus and Arabidopsis thaliana. Genetics 146: 1123–1129

Osborn TC, Kramer CC, Graham E, Braun CJ (2007) Insights and innovations from wide crosses: examples from canola and tomato. Crop Sci 47:S228–S237

Osborn TC, Lukens L (2003) The molecular genetic basis of flowering time variation in Brassica species. In: Nagata T, Tabata S (eds) Biotechnology in agriculture and forestry. Brassica and legumes: from gene structure to breeding. Springer-Verlag, Berlin, pp. 69–86

Osborn TC, Pires JC, Birchler JA, Auger DL, Chen ZJ, Lee H-S, Comai L, Madlung A, Doerge RW, Colot V, Martienssen RA (2003b) Understanding mechanisms of novel gene expression in polyploids. Trends Genet 19:141–147

Osorio J, Fernandez-Martinez J, Mancha M, Garces R (1995) Mutant sunflowers with high concentration of saturated fatty acids in the oil. Crop Sci 35:739–742

Parkin IAP, Gulden SM, Sharpe AG, Lukens L, Trick M, Osborn TC, Lydiate DJ (2005) Segmental structure of the Brassica napus genome based on comparative analysis with Arabidopsis thaliana. Genetics 171:765–781

Parkin IAP, Sharpe AG, Keith DJ, Lydiate DJ (1995) Identification of the A and C genomes of amphidiploid Brassica napus (oilseed rape). Genome 38:1122–1131

Perez-Vich B, Leon AJ, Grondona M, Velasco L, Fernandez-Martinez JM (2006) Molecular analysis of the high stearic acid content in sunflower mutant CAS-14. Theor Appl Genet 112:867–875

Pilet ML, Delourme R, Foisset N, Renard M (1998a) Identification of loci contributing to quantitative Weld resistance to blackleg disease, causal agent Leptosphaeria maculans (Desm.) Ces. et de Not., in winter rapeseed (Brassica napus L.). Theor Appl Genet 96:23–30

Pilet ML, Delourme R, Foisset N, Renard M (1998b) Identification of QTL involved in field resistance to light leaf spot (Pyrenopziza brassicae) and blackleg resistance (Leptosphaeria maculans) in winter rapeseed (Brassica napus L.). Theor Appl Genet 97: 398–406

Pilet ML, Duplan G, Archipiano M, Barret P, Baron C, Horvais R, Tanguy X, Lucas MO, Renard M, Delourme R (2001) Stability of QTL for Weld resistance to blackleg across two genetic backgrounds in oilseed rape. Crop Sci 41:197–205

Piquemal J, Cinquin E, Couton F, Rondeau C, Seignoret E, Doucet I, Perret D, Villeger MJ, Vincourt P, Blanchard P (2005) Construction of an oilseed rape (Brassica napus L.) genetic map with SSR markers. Theor Appl Genet 111:1514–1523

Prakash S, Hinata K (1980) Taxonomy, cytogenetics and origin of crop Brassica, a review. Opera Bot 55:1–57

Putterill J, Robson F, Lee K, Simon R, Coupland G (1995) The CONSTANS gene of Arabidopsis promotes flowering and encodes a protein showing similarities to zinc finger transcription factors. Cell 80:847–857

Qiu D, Morgan C, Shi J, Long Y, Liu J, Li R, Zhuang X, Wang Y, Tan X, Dietrich E, Weihmann T, Everett C, Vanstraelen S, Beckett P, Fraser F, Trick M, Barnes S, Wilmer J, Schmidt R, Li J,

Li D, Meng J, Bancroft I (2006) A comparative linkage map of oilseed rape and its use for QTL analysis of seed oil and erucic acid content. Theor Appl Genet 114:67–80

Quijada PA, Udall JA, Lambert B, Osborn TC (2006) Quantitative trait analysis of seed yield and other complex traits in hybrid spring rapeseed (Brassica napus L.): 1. IdentiWcation of genomic regions from winter germplasm. Theor Appl Genet 113:549–561

Raclaru M, Gruber J, Kumar R, Sadre R, Lhs W, Zarhloul KM, Friedt W, Frentzen M, Weier D (2006) Increase of the tocochromanol content in transgenic Brassica napus seeds by overexpression of key enzymes involved in prenylquinone biosynthesis. Mol Breed 18:93–107

Radoev M, Becker H, Ecke W ((2008)) Genetic analysis of heterosis for yield and yield components in rapeseed (Brassica napus L.) by quantitative trait locus mapping. Genetics 179:1547–1558

Rafalski A (2002) Applications of single nucleotide polymorphisms in crop genetics. Curr Opin Plant Biol 5:94–100

Rahman MH (2001) Production of yellow-seeded Brassica napus through interspecific crosses. Plant Breeding 120:463–472

Rajcan I, Kasha K, Kott LS, Beversdorf WD (1999) Detection of molecular markers associated with linolenic and erucic acid levels in spring rapeseed (Brassica napus L.). Euphytica 105: 173–181

Rakow G (1973) Selektion auf Linol- und Linolensäuregehalt in Rapssamen nach mutagener Behandlung. Z Planzen 69:205–209

Ramsay LD, Bradshaw JE, Griffiths DW, Kearsey MJ (2001) The inheritance of quantitative traits in Brassica napus ssp. rapifera (Swedes): augmented triple test cross analyses of production characters. Euphytica 121:65–72

Rana D, Van den Boogaart T, O'Neill CM, Hynes L, Bent E, Macpherson L, Park JY, Lim YP, Bancroft I (2004) Conservation of the microstructure of genome segments in Brassica napus and its diploid relatives. Plant J 40:725–733

Renard M, Delourme R, Pierre J (1997) Market introduction of rapeseed hybrid varieties. GCIRC Bulletin 14:114–119

Robert LS, Robson F, Sharpe A, Lydiate DJ, Coupland G (1998) Conserved structure and function of the Arabidopsis. flowering time gene CONSTANS in Brassica napus. Plant Mol 37:763–772

Ryder CD, Smith LB, Teakle GR, King GJ (2001) Contrasting genome organization: two regions of the Brassica oleracea genome compared with collinear regions of the Arabidopsis thaliana genome. Genome 44:808–817

Rygulla W, Snowdon RJ, Friedt W, Happstadius I, Cheung WY, Chen D (2007) Identification of quantitative trait loci for resistance against verticillium longisporum in oilseed Rape (Brassica napus). Phytopathology 98:215–221

Schierholt A, Becker HC, Ecke W (2000) Mapping a high oleic acid mutation in winter oilseed rape (Brassica napus L.). Theor Appl Genet 101:897–901

Schierholt A, Rcker B, Becker HC (2001) Inheritance of high oleic acid mutations in winter oilseed rape (Brassica napus L.). Crop Sci 41:1444–1449

Schmitz RJ, Sung S, Amasino RM (2008) Histone arginine methylation is required for vernalization-induced epigenetic silencing of FLC in winter-annual Arabidopsis thaliana. Proc Natl Acad Sci USA 105:411–416

Schranz ME, Osborn TC (2000) Novel flowering time variation in the resynthesized polyploid Brassica napus. J Hered 91:242–246

Schranz ME, Quijada P, Sung S-B, Lukens L, Amasino RM, Osborn TC (2002) Characterization and effects of the replicated flowering time gene FLC in Brassica rapa. Genetics 162: 1457–1468

Sebastian RL, Howell EC, King GJ, Marshall DF, Kearsey MJ (2000) An integrated AFLP and RFLP Brassica oleracea linkage map from two morphologically distinct doubled-haploid mapping populations. Theor Appl Genet 100:75–81

Sernyk JL, Stefansson BR (1983) Heterosis in summer rape (Brassica napus L.). Can J Plant Sci 63:407–413

Shahidi F (1990) Rapessed and canola: global production and distribution. In: Sahidi F (ed) Canola and rapeseed: production, chemistry and processing technology. Van Nostrand Reinhold, New York, NY, pp 3–13
Shahidi F, Naczk M (1992) An overview of the phenolics of canola and rapeseed: chemical, sensory and nutritional significance. J Am Oil Chem Soc 69:917–924
Sharpe AG, Lydiate DJ (2003) Mapping the mosaic of ancestral genotypes in a cultivar of oilseed rape (Brassica napus) selected via pedigree breeding. Genome 46:461–468
Sharpe AG, Parkin IAP, Keith DJ, Lydiate DJ (1995) Frequent nonreciprocal translocations in the amphidiploid genome of oilseed rape (Brassica napus). Genome 38:1112–1121
Shewmaker C, Sheehy J, Daley M, Colburn S, Ke D (1999) Seed-specific overexpression of phytoene synthase: increase in carotenoids and other metabolic effects. Plant J 20: 401–412
Shi J, Li R, Qiu D, Jiang C, Long Y, Morgan C, Bancroft I, Zhao J, Meng J (2009) Unraveling the complex trait of crop yield with quantitative trait loci mapping in Brassica napus. Genetics 182:851–861
Shindo C, Aranzana MJ, Lister C, Baxter C, Nicholls C, Nordborg M, Dean C (2005) Role of FRIGIDA and FLOWERING LOCUS C in determining variation in flowering time in Arabidopsis accessions. Plant Physiol 138:1163–1173
Simbaya J, Slominski BA, Rakow G, Campbell LD, Downey RK, Bello JM (1995) Quality characterisitics of yellow seeded Brassica seed meals: protein, carbohydrates, and dietary fiber components. J Agric Food Chem 43:2062–2066
Snowdon RJ, Friedt W (2004) Molecular markers in Brassica oilseed breeding: current status and future possibilities. Plant Breed 123:1–8
Snydera CL, Yurchenkoa OP, Silotoa R, Chena X, Liua Q, Mietkiewskaa E, Weselake RJ (2009) Acyltransferase action in the modification of seed oil biosynthesis. New Biotechnol 26:11–16
Somers DJ, Rakow G, Prabhu VK, Friesen KRD (2001) Identification of a major gene and RAPD markers for yellow seed coat colour in Brassica napus. Genome 44:1077–1082
Song KM, Osborn TC (1992) Polyphylatic origins of B. napus: new evidence based on organelle and nuclear RFLP analysis. Genome 35:992–1001
Song KM, Osborn TC, Williams PH (1988a) Brassica taxonomy based on nuclear restriction fragment length polymorphisms (RFLPs). 1. Genome evolution of diploid and amphidiploid species. Theor Appl Genet 75:784–794
Song KM, Osborn TC, Williams PH (1988b) Brassica taxonomy based on nuclear restriction fragment length polymorphisms (RFLPs). 2. Preliminary analysis of subspecies within B. rapa (syn. campestris) and B. oleracea. Theor Appl Genet 76:593–600
Song KM, Osborn TC, Williams PH (1997) Taxonomy based on nuclear RFLP analysis. In: Kalia HR, Gupta SK (eds) Recent advances in oilseed Brassicas. Kalyani Publishers, New Delhi, pp 12–24
Song KM, Tang KL, Osborn TC (1993) Development of synthetic Brassica amphidiploids by reciprocal hybridization and comparison to natural amphidiploids. Theor Appl Genet 86:811–821
Starmer KP, Brown J, Davis JB (1998) Heterosis in spring canola hybrids grown in Northern Idaho. Crop Sci 38:376–380
Stiewe G, Witt U, Hansen S, Theis R, Abel WO, Röbbelen G (1995) Natural and experimental evolution of CMS for rapeseed breeding. Adv Plant Breeding 18:89–110
Stringam GR, Ripley VL, Love HK, Mitchell A (2003) Transgenic herbicide tolerant canola. The Canadian experience. Crop Sci 43:1590–1593
Suarez-Lopez P, Wheatley K, Robson F, Onouchi H, Valverde F, Coupland G (2001) CONSTANS mediates between the circadian clock and the control of flowering in Arabidopsis. Nature 410:1116–1120
Sun Z, Wang Z, Tu J, Zhang J, Yu F, McVetty PBE, Li G (2007) An ultradense genetic recombination map for Brassica napus, consisting of 13551 SRAP markers. Theor Appl Genet 114:1305–1317

Sung S, Amasino RM (2004) Vernalization in Arabidopsis thaliana is mediated by the PHD finger protein VIN3. Nature 427:159–164

Sung S, He Y, Eshoo TW, Tamada Y, Johnson L, Nakahigashi K, Goto K, Jacobsen SE, Amasino RM (2006) Epigenetic maintenance of the vernalized state in Arabidopsis thaliana requires LIKE HETEROCHROMATIN PROTEIN 1. Nature Genetics 38:706–710

Suwabe K, Tsukazaki H, Iketani H, Hatakeyama K, Kondo M, Fujimura M, Nunome T, Fukuoka H, Hiria M, Matsumoto S (2006) Simple sequence repeat-based comparative genomics between Brassica rapa and Arabidopsis thaliana: the origin of clubroot resistance. Genetics 173:309–319

Tadage M, Sheldon CC, Helliwell CA, Stoutjesdijk P, Dennis ES, Peacock WJ (2001) Control of flowering time by FLC orthologues in Brassica napus. Plant J 28:545–553

Takeda S, Matsuoka M (2008) Genetic approaches to crop improvement: responding to environmental and population changes. Nat Rev Genet 9:444–457

Tang J, Zhang J, Ma C, Tang W, Gao C, Li F, Wang X, Liu Y, Fu T (2009) CAPS and SCAR markers linked to maintenance of selfincompatibility developed from SP11 in Brassica napus L. Mol Breeding 24:245–254

Tanhuanpää P, Vilkki J, Vilkki HJ (1995) Association of a RAPD marker with linolenic acid concentration in the seed oil of rapeseed (Brassica napus L.). Genome 38:414–416

Teutonico RA, Osborn TC (1994) Mapping of RFLP and qualitative trait loci in Brassica rapa and comparison to linkage maps of B. napus, B. oleracea and Arabidopsis thaliana. Theor Appl Genet 89:885–894

Thormann CE, Romero J, Mantet J, Osborn TC (1996) Mapping loci controlling the concentrations of erucic and linolenic acids in seed oil of Brassica napus L. Theor Appl Genet 93: 282–286

Trautwein EA, Duchateau G, Lin Y, Melcnikov SM, Molhuizen H, Ntanios FY (2003) Proposed mechanisms of cholesterol-lowering action of plant sterols. Eur J Lipid Sci Technol 105: 171–185

Trick M, Long Y, Meng J, Bancroft I (2009) Single nucleotide polymorphism (SNP) discovery in the polyploid Brassica napus using Solexa transcriptome sequencing. Plant Biotech J 7: 334–346

U N (1935) Genome analysis in Brassica with special reference to the experimental formation of B. napus and its peculiar mode of fertilization. Jpn J Bot 7:389–452

Udall JA, Quijada PA, Lambert B, Osborn TC (2006) Quantitative trait analysis of seed yield and other complex traits in hybrid spring rapessed (Brassica napus L.): 2. Identification of alleles from unadapted germplasm. Theor Appl Genet 113:597–609

Udall JA, Quijada PA, Osborn TC (2005) Detection of chromosomal rearrangements derived from homeologous recombination in four mapping populations of Brassica napus L. Genetics 169:967–979

United States Department of Agriculture (2009) Oilseeds: world market and trade. Foreign Agricultural Service Circular Series 11–09, November.

Uzunova M, Ecke W, Weissleder K, Röbbelen G (1995) Mapping the genome of rapeseed (Brassica napus L.). I. Construction of an RFLP linkage map and localization of QTLs for seed glucosinolate content. Theor Appl Genet 90:194–204

Vigeolas H, Waldeck P, Zank T, Geigenberger P (2007) Increasing seed oil content in oil-seed rape (Brassica napus L.) by overexpression of a yeast glycerol-3-phophate dehydrogenase under the control of a seed-specific promoter. Plant Biotechnol J 5:431–441

Vignal A, Milan D, Sancristobal M, Eggen A (2002) A review on SNP and other types of molecular markers and their use in animal genetics. Genet Sel Evol 34:275–305

Walker KC, Booth EJ (2001) Agricultural aspects of rape and other Brassica products. Eur J Lipid Sci Technol 103:441–446

Wang R, Ripley VL, Rakow G (2007) Pod shatter resistance evaluation in cultivars and breeding lines of Brassica napus, B. juncea and Sinapis alba. Plant Breed 126:588–595

Warwick SI, Black LD (1993) Molecular relationships in subtribe Brassicinae (Cruciferar, tribe Brassiceae). Can J Botany 71:906–918

Weier D, Mittasch J, Strack D, Milkowski C (2007) The genes BnSCT1 and BnSCT2 from Brassica napus encoding the final enzyme of sinapine biosynthesis: molecular characterization and suppression. Planta 227:375–385

Westermeier P, Wenzel G, Moler V (2009) Development and evaluation of single-nucleotide polymorphism markers in allotetraploid rapeseed (Brassica napus L.). Theor Appl Genet 119:1301–1311

Wheeler DA, Srinivasan M, Egholm M, Shen Y, Chen L, McGuire A, He W, Chen Y-J, Makhijani V, Roth GT, Gomes X, Tartaro K, et al (2008) The complete genome of an individual by massively parallel DNA sequencing. Nature 452:872–876

Wittkop B, Snowdon RJ, Friedt W (2009) Status and perspectives of breeding for enhanced yield and quality of oilseed crops for Europe. Euphytica 170:131–140

Wu G, Wu Y, Xiao L, Li X, Changming L (2008) Zero erucic acid trait of rapeseed (Brassica napus L.) results from a deletion of four base pairs in the fatty acid elongase 1 gene. Theor Appl Genet 116:491–499

Yu S, Derr J, Etherton TD, Kris-Etherton PM (1995) Plasma cholesterol-predictive equations demonstrate that stearic acid is neutral and monounsaturated fatty acids are hypocholesterolemic. Am J Clin Nutr 61:1129–1139

Zarhloul MK, Stoll C, Lhs W, Syring-Ehemann A, Hausmann L, Töpfer R, Friedt W (2006) Breeding high-stearic oilseed rape (Brassica napus) with high- and low-erucic background using optimized promoter-gene constructs. Mol Breed 18:241–251

Zhao J, Becker H, Zhang D, Zhang Y, Ecke W (2005) Oil content in an European x Chinese rapeseed population: QTL with additive and epistatic effects and their genotype-environment interactions. Crop Sci 45:51–59

Zhao J, Becker H, Zhang D, Zhang Y, Ecke W (2006) Conditional QTL mapping of oil content in rapeseed with respect to protein content and traits related to plant development and grain yield. Theor Appl Genet 133:33–38

Zhao J, Meng J (2003) Detection of loci controlling seed glucosinolate content and their association with Sclerotinia resistance in Brassica napus. Plant Breed 122:19–23

Zhao H, Shi L, Duan X, Xu F, Wang Y, Jinling Meng J (2008) Mapping and validation of chromosome regions conferring a new boron-efficient locus in Brassica napus. Mol Breed 22:495–506

Zum Felde T, Becker HC, Möllers C (2006) Genotype x environment interactions, heritability and trait correlations of sinapate ester content in winter rapeseed (*Brassica napus* L.). Crop Sci 46:2195–2199

Chapter 11
Genetics of *Brassica juncea*

Akshay K. Pradhan and Deepak Pental

Abstract *Brassica juncea* (AABB) is a natural allopolyploid of *Brassica rapa* (AA) and *Brassica nigra* (BB). *B. juncea* group contains both oilseed types and vegetable types. Extensive phenotypic and molecular marker-based studies on oilseed types of *B. juncea* have identified two major divergent gene pools, the Indian gene pool and the east European gene pool. Hybrids between the lines of these two gene pools are heterotic for yield, and such hybrids have now been commercialized in India. The two gene pools are significantly contrasting in major agronomic traits like seed size, pod size, pod number, height and oil content. A major future challenge in heterosis breeding is to transfer superior traits from one gene pool to the other while retaining the overall genetic divergence. This would allow further increases in yield. Recent work has provided reasonable number of markers for molecular tagging of both qualitative and quantitative traits and comparative mapping with model crucifer *Arabidopsis* and other *Brassica* species. Current work involves development of "00" canola-quality lines and their agronomic trials, disease resistance and QTL introgression. As the crop holds great promise for dryland agriculture in India and many other such areas around the globe, a concerted international effort is required on germplasm exchange, genetics, genomics and breeding of *B. juncea*.

Keywords Mustard · Molecular markers · Gene tagging · QTL · Disease resistance

Abbreviations

AFLP	Amplified fragment length polymorphism
CAPS	Cleaved amplified polymorhic sequence
cM	centi-Morgan
DH	Doubled haploid
FAE1	Fatty acid elongase 1
IP	Intron polymorphism

A.K. Pradhan (✉)
Department of Genetics, Centre for Genetic Manipulation of Crop Plants, University of Delhi South Campus, New Delhi 110021, India
e-mail: pradhancgmcp@gmail.com

R. Schmidt, I. Bancroft (eds.), *Genetics and Genomics of the Brassicaceae*, Plant Genetics and Genomics: Crops and Models 9, DOI 10.1007/978-1-4419-7118-0_11,

QTL	Quantitative trait loci
RAPD	Random amplification of polymorphic DNA
RFLP	Restriction fragment length polymorphism
RIL	Recombinant inbred line
RSB	Recurrent selection backcross
RuBisCO	Ribulose-1,5-bisphosphate carboxylase oxygenase
SCAR	Sequence characterized amplified region
var.	Variety

Contents

11.1 Introduction 324
11.2 Available Variability 326
11.3 Genome Mapping in *B. juncea* 327
11.4 Genetics and Mapping of Important Traits 331
11.4.1 Erucic Acid Content and Oil Content 332
11.4.2 Glucosinolates and the Importance of Context 334
11.4.3 Seed Coat Colour 336
11.4.4 Agronomic and Yield Traits 337
11.4.5 Disease Resistance 338
11.5 Future Prospects 340
References 341

11.1 Introduction

Brassica juncea is a natural allopolyploid (AABB), i.e. it contains genomes of two diploid species, *Brassica rapa* (AA) and *Brassica nigra* (BB). This relationship was first proposed by U (1935) in the form of famous U's traingle showing the relationship of diploid species *B. rapa* (AA), *B. nigra* (BB) and *Brassica oleracea* (CC) with the three allopolyploid species *B. napus* (AACC), *B. juncea* (AABB) and *Brassica carinata* (BBCC), the six most economically important species of family *Brassicaceae*. The allopolyploid nature of *B. juncea* ($2n = 36$) has been confirmed by artificial resynthesis (Prakash and Chopra 1991), and analysis of RuBisCO (Uchimiya and Wildman 1978) and RFLP of nuclear DNA (Song et al. 1988). Analysis of chloroplast DNA has shown *B. rapa* to be the donor of cytoplasm to *B. juncea* (Palmer et al. 1983, Erickson et al. 1983).

The genome size of *B. juncea* has been estimated to be 1,068 Mbp, that of *B. rapa* 529 Mbp and that of *B. nigra* 632 Mbp. Compared to this the genome size of the model crucifer species *Arabidopsis thaliana* is around 157 Mbp (Johnston et al. 2005). Diploid *Brassica* species are hypothesized to contain derivatives of three genomes based on the conclusions of Schranz et al. (2006). Consequently, *B. juncea* can have six paralogous/homoeologous regions, three from the genome of each progenitor species.

B. juncea, commonly known as mustard, is cultivated as an oilseed and a condiment, as well as a vegetable in some parts of the world. Various phenotypes of mustard evolved spontaneously in nature and were further improved under domestication. These include root mustard (*B. juncea* var. *napiformis* Paill. et Bois.), stem mustard (*B. juncea* var. *tsatsai* Mao), leaf mustard (*B. juncea* var. *rugosa* Bailey), seed stalk mustard (*B. juncea* var. *utilis* Li) and seed mustard (*B. juncea* var. *juncea*) (Gladis and Hammer 1992). Whereas seed mustard is grown extensively in India and to a lesser extent in Canada, Australia, Russia and Ukraine as edible oilseed crop, the vegetable types are mainly grown in China. Seed mustard is also grown in Canada and Europe on a limited acreage for condiment purposes.

The centre of origin of various *B. juncea* types is unclear. However, cytogenetical (Olsson 1960, Prakash 1973), biochemical (Vaughan 1977) and DNA polymorphism analyses (Song et al. 1988) show that *B. juncea* has polyphyletic origin, and that variable diploid types of *B. rapa* and *B. nigra* have been involved in the evolution of allopolyploid *B. juncea* types with distinct phenotypes. On the basis of the sympatric distribution of diploid progenitor species *B. rapa* and *B. nigra*, two main centers of origin have been suggested: (a) The Middle East and Indian region and (b) east European and Chinese region. The Middle East and Indian types evolved in the direction of oil-yielding types, while Chinese forms mainly evolved in the direction of leafy vegetables (Gomez-Campo and Prakash 1999).

As hybridization among the diploids and between the diploid and the tetraploid *Brassica* species is fairly common (Prakash et al. 1999, Ford et al. 2006), there is a strong case for multiple origins of allopolyploid *B. juncea*. However, such hybridizations followed by chromosome duplications might have faced a major limitation. As most of the *B. juncea* types are annual, it may be presumed that only those allopolyploid events that had minimal homoeologous pairing and consequently high sexual fecundity would have survived. All the three allopolyploid species of U's triangle including *B. juncea* behave as functional diploids; there is no pairing between homoeologous chromosomes. However, artificially resynthesized *Brassica* allopolyploids, including those of *B. juncea*, show multivalent formation and laggards, depressed sexual fecundity, genome instability (Prakash et al. 1999) and low yield.

India has around six million hectares of land under oilseed mustard. The crop is mostly grown in the north-western dryland areas of India during the winter season (October–March). It is sown at the end of the rainy season on residual moisture. Due to its better drought hardiness, mustard is being worked upon as a replacement crop for *B. napus* in Canada and Australia, particularly in the drought-prone areas of these countries. Canada is presently a distant second to India in production and acreage under mustard.

In terms of research and development efforts, oilseed mustard has been studied far less intensively than *B. napus*, which is a crop primarily of the developed world. However, since 1980s, efforts have been underway on systematic molecular genetic studies in *B. juncea* due to its increasing international acceptance as an important oilseed crop. The availability of extensive DNA sequence data and marker data from related *Brassica* species and the model species *A. thaliana* and the availability of techniques of transformation and doubled haploid production have opened up enormous possibilities for carrying out meaningful studies on the genetics of agronomically important traits in oilseed mustard.

However, it will be prudent not to think of working on the genetics of *B. juncea à la A. thaliana*. Although the genome of one of the diploid parents, *B. rapa,* is being sequenced, it may be some time yet before laboratories venture to sequence a polyploid genome such as that of *B. juncea*. Currently, it is considered more important to adopt practical and achievable goals of understanding genetics of major agronomic traits in this crop. In this endeavour, ascertaining the location(s) of the trait-controlling genes is more important and takes precedence over understanding the mechanism of expression of the traits. Instances of traits tagged in *B. juncea* and discussed here amply prove that for all practical purposes, genetics of localization through markers is far more critical for diversification and crop improvement.

11.2 Available Variability

Genetic variability is the key to effective breeding. One of the most significant findings in this regard in *B. juncea* was delineating the existence of two major gene pools, namely the Indian and the east European gene pools. Pradhan et al. (1993) studied genetic divergence in *B. juncea* at the morphological level using 25 accessions of Indian and east European origins and some resynthesized *B. juncea* lines produced in an earlier study (Prakash 1973). The accessions were grouped into different clusters based on nine agronomic and yield traits. Maximum divergence was observed between the Indian and the east European types. The existence of two gene pools was subsequently confirmed through marker-based phylogenetic analysis (Jain et al. 1994, Srivastava et al. 2001, Burton et al. 2004). These studies revealed that the Indian gene pool is constituted of a very narrow base and is significantly divergent from the east European gene pool. Jain et al. (1994) used RAPD markers to study genetic variation among 12 Indian and 11 exotic (east European) *B. juncea* accessions. Cluster analysis identified two groups, group A comprised only of the exotic genotypes and group B comprised of all the Indian and four of the exotic genotypes. Srivastava et al. (2001) using AFLP markers studied diversity among 21 natural *B. juncea* accessions originating from India, Russia, Australia, and Canada and nine resynthesisized lines developed in India. Three distinct diversity groups were identified based on the similarity coefficient. All the Indian types formed one group, the resynthesized lines formed a separate cluster and the lines from Australia, Canada and Russia (all east European types) formed the third cluster. Burton et al. (2004) used AFLP markers to assess the genetic diversity of *B. juncea* consisting of 77 canola-quality breeding lines derived from the Canadian and the Australian breeding programmes and 15 non-canola-quality genotypes from India, Russia, Canada and China. Clustering of the genotypes on the basis of AFLP polymorphism separated the Indian types from the rest of the genotypes forming two major clusters. The divergence of the two gene pools is also reflected at the biochemical level and exemplified by the difference in the seed erucic acid content and types of aliphatic glucosinolates. All Indian germplasm contain around 46% erucic acid whereas east European lines contain around 20% erucic acid (Kirk and Hurlstone

1983). As for glucosinolates, Indian types contain both propyl and butyl glucosinolates whereas east European mustard lines contain mainly propyl glucosinolates (Gland et al. 1981, Sodhi et al. 2002).

Hybrids between the two pools have been shown to be heterotic (hybrid vigour) for yield. Pradhan et al. (1993) studied heterosis in 10 × 10 diallel crosses where the parents were selected on the basis of genetic divergence based on the yield-influencing component traits. Maximum heterosis was recorded between the genetically diverse Indian × east European crosses. These observations were confirmed by Jain et al. (1994). Crosses made between genetically divergent Indian × east European lines showed significant level of heterosis; on the other hand, crosses made within the Indian or the east European genotypes recorded either negative or insignificant yield heterosis. These observations have been used to produce first ever *B. juncea* hybrid DMH-1 based on a cross between an Indian gene pool line, Pusabold, and an east European gene pool line, EH-2 (Sodhi et al. 2006).

The east European lines, however, are ill adapted to the Indian agronomic conditions, and Indian types are not productive enough for the summer season growth in Canada. Therefore, the genetic improvement of the two gene pools with introgression of desirable characters from one divergent pool to the other is the most important overall objective in oilseed mustard improvement programme. East European lines contain many desirable traits such as resistance to white rust (*Albugo candida*), low erucic acid and low glucosinolates, higher oil content, yellow seed coat and yield components such as branching, pod density and number (Pradhan et al. 2003). In comparison, Indian types display many desirable traits such as early flowering, shorter height, better pod length, bolder seeds and better resistance to pod shattering (Ramchiary et al. 2007b). Most of these traits are complex in inheritance, and the transfer of these traits from one to the other gene pool while retaining the overall variability of the two gene pools is a major challenge. Maintaining the overall divergence of the two pools is necessary to exploit the possibility of heterosis breeding in *B. juncea*. Therefore, a systematic genetic analysis through the use of molecular markers and subsequent marker-assisted precision introgressions need to be undertaken for the manipulation of these complex traits.

11.3 Genome Mapping in *B. juncea*

A review on genome mapping and molecular breeding of *B. juncea* was published by Edwards et al. (2007) which described different maps published till 2003. Since then significant progress has been made in molecular mapping in *B. juncea*, the latest being a comparative map describing the genome synteny between *B. juncea* and other *Brassica* species (Panjabi et al. 2008). In this review we shall describe some of the salient features of the maps published after 2003. The basic information on mapping in *B. juncea* has been summarized, indicating the parents, the nature of the population and the markers used, in Table 11.1.

Table 11.1 *Brassica juncea* mapping studies: summary of linkage maps developed and traits investigated

Parents of the mapping population	Population size and type	Number and type of markers	Map length (cM)	Traits investigated	References
J90-4317 × J90-2733	119 F_1DH	343 RFLP	2,073	Erucic acid content, linolenic acid content, resistance to white rust, oil content, seed glucosinolate content	Cheung et al. (1997, 1998a, b)
(J-o-7DH1 × J-o-3DH1) × J-o-3DH1	120 BC_1	183 RFLP	1,266	–	Axelsson et al. (2000)
(J-o-7DH1 × BjRS) × J-o-3DH1	60 Test cross	269 RFLP	1,041		
Varuna × BEC144	94 RILs	114 RAPD	790.4	Oleic acid content, white rust resistance	Sharma et al. (2002a), Mukherjee et al. (2001), Varshney et al. (2004)
BJ-70 × BJ-99	131 F_1DH	264 AFLP, 9 RAPD	1,641	Oil content, palmitic acid content, stearic acid content, oleic acid content, linoleic acid content, linolenic acid content, eicosanoic acid content, erucic acid content, seed coat colour, glucosinolate content, days to flower, plant height, 1,000-seed weight	Lionneton et al. (2002, 2004)
Varuna × Heera	123 F_1DH	1,297 AFLP, 72 RFLP, 13 gene markers, 486 IP markers	1,992.2	Erucic acid content, seed coat colour, plant height, days to flowering, number of primary and secondary branches, main shoot length, number of siliques on a plant, silique density, silique length, number of seeds in a silique, 1,000-seed weight, oil content, seed glucosinolate content	Pradhan et al. (2003), Gupta et al. (2004), Padmaja et al. (2005), Ramchiary et al. (2007a, b), Panjabi et al. (2008), Bisht et al. (2009)

Table 11.1 (continued)

Parents of the mapping population	Population size and type	Number and type of markers	Map length (cM)	Traits investigated	References
RLM-514 × canola-quality inbred	111 F_1DH	280 RFLP	1,564	Erucic acid content, oleic acid content, linoleic acid content, linolenic acid content, seed aliphatic glucosinolate content, seed coat colour, silique length, number of seeds per silique, number of siliques per main raceme, 1,000-seed weight, flowering time, maturity duration	Mahmood et al. (2003a, b, 2005a, b, 2006, 2007)
"AC Vulcan" × UM 3122	140 F2	323 Microsatellite, RFLP	1,577.3	Blackleg resistance	Christianson et al. (2006)

A map of *B. juncea* genome based on microsatellite and RFLP markers was constructed for mapping two loci controlling resistance to blackleg disease by Christianson et al. (2006). A F2 population of 140 individuals derived from a cross between blackleg-resistant *B. juncea* line "AC Vulcan" and blackleg-susceptible line "UM3132" was used for the construction of the linkage map. A total of 323 loci were distributed on 18 linkage groups covering 1,577.3 cM. Using a set of 75 common markers, the map was observed to be almost perfectly collinear with the previously published map of *B. juncea* (Axelsson et al. 2000) except for four pairs of markers.

The linkage map of *B. juncea* based primarily on AFLP markers (Pradhan et al. 2003) was further enriched with more AFLP, RFLP, microsatellites and gene-based markers by Ramchiary et al. (2007b). Microsatellite markers used in this study were of *B. rapa*, *B. napus* and *B. nigra* origin (www.brassica.info/). This *B. juncea* map included 1,448 markers comprising 1297 AFLP, 72 RFLP, 69 microsatellites and 10 gene-based markers covering a total length of 1,840.1 cM.

The first comprehensive comparative linkage map, establishing gene-to-gene synteny between *A. thaliana* and *B. juncea*, was published by Panjabi et al. (2008). Unlike other comparative maps that made use of common RFLP markers, this map was developed by the use of PCR-based markers based on polymorphism in the intronic regions. Single-copy genes from *A. thaliana* physically located at an approximate distance of 100–200 kb were used to design PCR primers spanning introns. In cases where large genomic regions were devoid of single-copy genes, primers were designed from multiple-copy genes. Primers for PCR amplification were designed from exon sequences which showed high nucleotide conservation between *A. thaliana* and the corresponding EST or GSS sequence described for *Brassica* species. A total of 486 intron polymorphic (IP) loci of *A. thaliana* covering all the five chromosomes were mapped relative to the earlier framework map of *B. juncea* (Pradhan et al. 2003). The markers were distributed to all the 18 linkage groups of *B. juncea* covering 1,992.2 cM.

The comparative map of Panjabi et al. (2008) was used to study homoeologous relationships, diversification and evolution of the A, B and C genomes of *Brassica* species. Use of gene-based makers allowed comparative mapping between *B. juncea* and the other allotetraploid and diploid *Brassica* species of U's triangle. This was done by comparing the syntenic blocks of the *B. juncea* (AABB) map with the earlier published comparative maps of *B. napus* (AACC) (Parkin et al. 2005) and *B. nigra* (BB) (Lagercrantz 1998). The maps were converted to common *A. thaliana* blocks following the genomic block nomenclature of Schranz et al. (2006), and the block arrangements of A, B and C genomes in the three different species were compared. This comparison leads to the identification of homoeologous groups between A, B and C genomes of *Brassica* species (Fig. 11.1). The comparison of the three genomes showed a high level of conserved macro-level collinearity between the A genome of *B. juncea* and *B. napus* and between the B genome of *B. juncea* and *B. nigra*, signifying an absence of large-scale perturbation during the formation of the allopolyploid species. It also allowed the prediction of possible macro-level chromosomal changes that led to the diversification of the three diploid *Brassica* genomes. The conclusion in this chapter on the absence of large-scale genomic upheaval during the process of formation of allopolyploid *Brassica* species is in

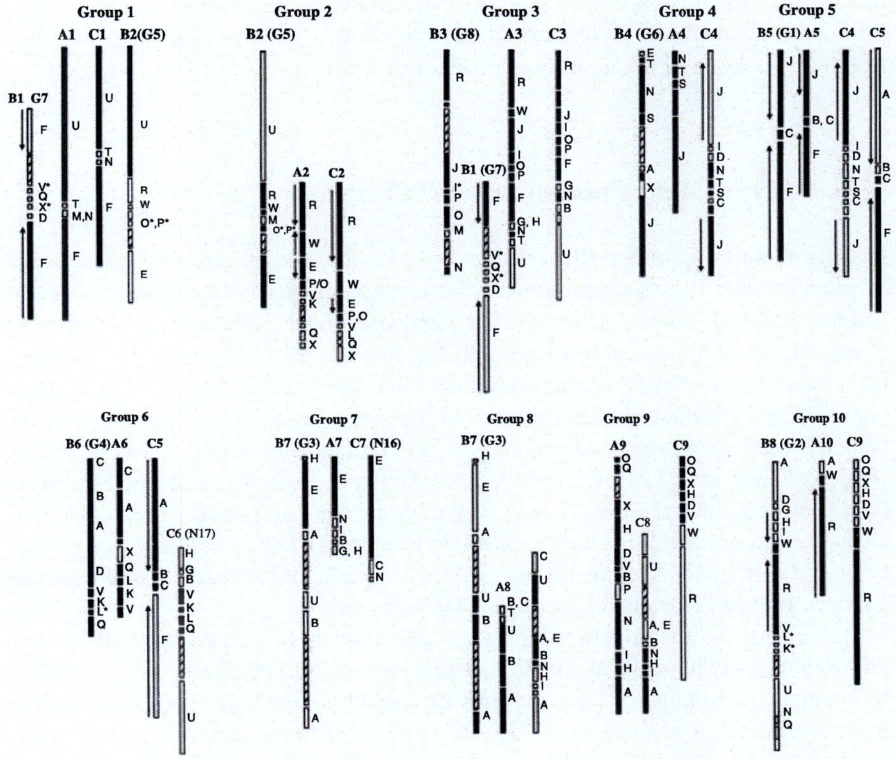

Fig. 11.1 The block arrangements in the A and B genomes are based on the consensus block arrangement of the A genome of *Brassica juncea* (A1–A10; Panjabi et al. 2008) and *Brassica napus* (N1–N10) (Parkin et al. 2005) and B genome of *B. juncea* (A11–A18; Panjabi et al. 2008) and *B. nigra* (G1–G8) (Lagercrantz 1998). The C genome is based on the *B. napus* map (N11–N19) (Parkin et al. 2005). The original nomenclature of the 2 LGs of C genomes (N16 and N17) and all the LGs of B genome (G1–G8) have been shown in *parenthesis* with the redesignated nomenclature. Certain single-gene insertions were considered as putative blocks (marked with *asterisk*) if a similar block was found present at the corresponding region in the homoeologous chromosome. *Filled bars* represent the common blocks shared between all the three members of the group. Large gaps ($\geq$10 cM regions devoid of any markers) in the LGs have been depicted by *hatched boxes*. *Arrows* represent the orientation of the gene order (within the block) with respect to the corresponding regions in *Arabidopsis thaliana* (adapted from fig. 7 in Panjabi et al. 2008)

consonance with the earlier contention of Axelsson et al. (2000) in this regard. However, these observations are contrary to some earlier studies suggesting rapid and extensive genome rearrangements in resynthesized *B. napus* (Song et al. 1995, Osborn 2004, Pires et al. 2004).

11.4 Genetics and Mapping of Important Traits

Genetic studies of qualitative traits pertaining to leaf, stem, flower, pod and seed morphology have been reviewed (Banga 1996). These studies predominantly showed digenic inheritance of different traits. In recent years, advances in the molecular marker technologies along with genomics resources from *A. thaliana* and related *Brassica* species have accelerated research on agronomically important traits in *B juncea*. Some of the findings are discussed here.

11.4.1 Erucic Acid Content and Oil Content

Genetic variations with less than 2% erucic acid in the seed oil and less than 20 μmol of glucosinolate in seed meal are the basic requirements for the development of canola-quality *B. juncea*. These quality parameters have already been achieved in *B. napus*. Work on breeding of zero erucic acid mustard was initiated following the identification of low erucic acid lines Zem 1 and Zem 2 by Kirk and Oram (1981). The inheritance of the erucic acid trait in *B. juncea* was shown to be controlled by two genes with additive effects, zero erucic being recessive in expression (Kirk and Hurlstone 1983, Bhat et al. 2002, Gupta et al. 2004). Kirk and Hurlstone (1983) reported that the Indian gene pool of *B. juncea* containing around 49% erucic acid contains alleles for high erucic acid level at two loci, whereas the gene pool from China–Eastern Europe–Canada containing around 25% erucic acid contains an allele for high erucic acid content at one locus only.

Earlier studies of mapping the erucic acid trait in *B. juncea* used RFLP and AFLP markers (Cheung et al. 1998b, Lionneton et al. 2002, Mahmood et al. 2003b). Using mapping populations derived from crosses involving high erucic acid parents from the Indian gene pool, two QTL for erucic acid were identified. These QTL mapped to two different linkage groups with unequal contributions (Cheung et al. 1998b, Mahmood et al. 2003b). However, mapping of the erucic acid trait involving a high erucic acid parent from the east European gene pool identified only one QTL (Lionneton et al. 2002). Some of the above studies also predicted the involvement of *FAE1* genes (Cheung et al. 1998b, Lionneton et al. 2002) for controlling the erucic acid content in *B. juncea*, as has been shown in the case of *B. napus* (Jourden et al. 1996, Fourmann et al. 1998). However, RFLP and AFLP markers are not suitable for high-throughput applications in plant breeding programmes.

Further refinement in genetic analysis of the erucic acid trait in *B. juncea* was undertaken by Gupta et al. (2004). The aim of this study was to develop markers tagged to erucic acid that could be conveniently used in the marker-assisted selection of low and high erucic acid lines in a backcross breeding programme. Two *FAE1* genes, *FAE1.1* and *FAE1.2*, cloned from low and high erucic acid mustard lines revealed the presence of four and three single nucleotide polymorphisms (SNPs) in *FAE1.1* and *FAE1.2*, respectively. Using these SNPs, the *FAE1.1* and *FAE1.2* genes were mapped to the *B. juncea* map described by Pradhan et al. (2003) (Table 11.1). Their map positions coincided with the two QTL for erucic acid located on two different linkage groups, one belonging to the A genome and the other belonging to the B genome. The QTL in the A genome explained 60% and the QTL in the B genome explained 38% of the phenotypic variance. Genotype–phenotype association showed that all seven SNPs could distinguish low from high erucic acid types, and the heterozygotes were found to be intermediate in phenotype. Validation study showed wide applicability of these SNP markers in marker-assisted selection.

Some of the above-mentioned studies also undertook QTL analyses for the content of fatty acids in seed oil of *B. juncea* other than erucic acid (Cheung et al. 1998b, Sharma et al. 2002a, Lionneton et al. 2002, Mahmood et al. 2003b). Almost all the studies showed association of erucic acid QTL with QTL detected for contents of other fatty acids such as palmitic acid (16:0), stearic acid (18:0), oleic acid (18:1), linoleic acid (18:2) and linolenic acid (18:3). The contents of these fatty acids also exhibit negative correlation with the content of erucic acid. It seems that *FAE1* is an efficient enzyme converting the common substrate oleic acid to erucic acid thus limiting the availability of the same substrate for the desaturation and other branching pathways. Similar findings were reported in *B. rapa* (Tanhuanpaa et al. 1996, 1998) and *B. napus* (Chen and Beversdorf 1990). In a genetic network model where a major gene(s) exerts pleiotropic effects on the expression of other genes, true genes/loci contributing to the affected traits remain undetected. Hence, for meaningful genetic dissection of these affected traits contributing to natural variations, one would require special genetic stocks which lack the genes that exert pleiotropic effects.

The most significant finding of the genetic analysis of erucic acid trait is the negative association between low erucic acid and high oil content in *Brassica* species. This correlation has been shown earlier in *B. napus* (Ecke et al. 1995, Qiu et al. 2006). Cheung et al. (1998b) identified two QTL for oil content coinciding with the two QTL for erucic acid mapping to the locations of the *FAE1* genes. Mahmood et al. (2006) showed significant positive correlation between erucic acid content and oil content and of the five QTL detected for oil content, one QTL co-mapped to an erucic acid QTL region. Recently, Ramchiary et al. (2007b) dissected the oil content trait using an AFLP-based map of *B. juncea*. The study identified seven QTL for oil content, of which two major QTL mapped to the locations of the genes for *FAE1.1* and *FAE1.2*. It has been suggested that the erucic acid content of the seed oil is itself a major determinant of seed oil content in *Brassica* species since the increase in the molecular mass during the elongation of oleic acid (C18:1) to erucic acid (C22:1) could explain higher oil content (Ecke et al. 1995, Burns et al. 2003).

The only exception to the above phenomenon was reported by Lionneton et al. (2002) who observed no correlation between oil content and erucic acid in *B. juncea*. The two QTL detected for oil content mapped to linkage groups different from those where the two QTL for the erucic acid content were located. It could be noted that the two parental lines used for developing the mapping population in this study (Table 11.1) were not highly contrasting for both erucic acid and oil content traits, although the DH mapping population showed high positive correlation between oil content and eicosanoic acid (C20:1) content. Similar observation has also been reported in *B. napus*, in which no QTL for oil content was detected in the positions of erucic acid in a population segregating for erucic acid content whereas some QTL for oil content were detected in erucic acid positions in another population not segregating for erucic acid (Delourme et al. 2006). From the latter study the authors inferred that the genes responsible for oil content in the erucic acid regions might be linked but not correspond to them. In cases where deleterious association is due to tight linkage, marker-assisted disruption through fine mapping of the target regions could be followed for the elimination of linkage drag. In case of pleiotropic situation as has been shown in *B. napus* (Ecke et al. 1995, Burns et al. 2003, Qiu et al. 2006) and *B. juncea* (Cheung et al. 1998b, Mahmood et al. 2003b, Ramchiary et al. 2007b), the goal of breeding for high oil and zero erucic acid could be achieved through screening the low erucic acid germplasm for the compensatory genes for oil content and pyramiding these genes.

11.4.2 Glucosinolates and the Importance of Context

The pioneering work of Canadian canola breeders provided the necessary building blocks for the breeding of low glucosinolate *B. juncea* (Love et al. 1990b). The first low glucosinolate *B. juncea* line BJ 1058 was developed from an interspecific cross between an Indian *B. juncea* line having non-allyl, 3-butenyl glucosinolate with low glucosinolate *B. rapa*, followed by one backcross to the non-allyl, 3-butenyl glucosinolate *B. juncea* parent and selfing of backcrossed plants (Love et al. 1990b). This line has been subsequently used in various breeding programme for the development of low glucosinolate *B. juncea*.

Methionine-derived aliphatic glucosinolates are the major glucosinolates present in *Brassica* species. Previous studies have shown that the aliphatic glucosinolate profiles of east European mustard lines are less complex than the glucosinolate profiles of the Indian types. The former contain mostly 2-propenyl glucosinolate while the latter contain both 2-propenyl and 3-butenyl glucosinolates (Gland et al. 1981, Love et al. 1990a, b, Sodhi et al. 2002). The difference in the glucosinolate profile is also reflected at the genetic level; aliphatic glucosinolate in the east European lines have been shown to be controlled by two loci (Love et al. 1990a) while the cross involving low glucosinolate east European and high glucosinolate Indian type lines revealed contribution of six to eight loci to the glucosinolate content (Stringam and Thiagarajah 1995, Sodhi et al. 2002).

Cheung et al. (1998b) identified two and three QTL, respectively, for 2-propenyl and 3-butenyl glucosinolates. Mahmood et al. (2003a) undertook a more extensive QTL analysis of seed aliphatic glucosinolates and identified five QTL for total aliphatic glucosinolates, four QTL for 2-propenyl glucosinolate and three QTL for 3-butenyl glucosinolate in different environments. Of the six common QTL identified with the individual glucosinolates, two minor QTL were found to be inconsistent across environments. On the basis of the consistency across the environments and high phenotypic variance explained by different QTL, the study suggested four QTL (GSL-A2a, GSL-A2b, GSL-F and GSL-B3) for marker-assisted selection for canola breeding programme in Canada. A study by Lionneton et al. (2004) identified two major QTL for both 2-propenyl and 3-butenyl glucosinolates by using an AFLP map (Table 11.1). Both the parents were high in aliphatic glucosinolates but were reciprocally contrasting for 2-propenyl and 3-butenyl glucosinolate content, one being higher for 2-propenyl glucosinolate and the other higher for 3-butenyl glucosinolate.

Genetics of glucosinolate biosynthesis has been studied in considerable detail in *A. thaliana*. In comparison, lesser but significant details are available in *B. napus* and *B. oleracea*. Many of the candidate genes involved in glucosinolate biosynthetic pathway have been characterized (see review by Halkier and Gershenzon 2006). Detailed studies on mapping of the loci involved in the biosynthesis of aliphatic glucosinolates have been reported in *B. juncea* recently (Ramchiary et al. 2007a, Bisht et al. 2009). This was accomplished through the use of DNA markers based on the functional genomics information available from allied *Brassica* species and *A. thaliana*. Study by Ramchiary et al. (2007a) provided an insight into the complexity of this trait in *B. juncea* showing involvement of context-dependent loci (effects of QTL alleles differing in magnitude and direction in different genetic backgrounds) in the expression of the trait. Following recurrent selection backcross (RSB) method with a DH generation interspersing the backcross generations, the low glucosinolate trait was introgressed from canola-quality east European line Heera to a major Indian cultivar Varuna. Comparison of phenotypic data for various glucosinolates between early (F_1DH) and advanced (BC_4DH) backcross generations revealed a shift in the mean values and percentage of transgressive segregants with the advancement of backcrossing indicating a change in the selective values of the alleles with change in the genetic background due to the existence of epistasis and context dependency. This phenomenon was confirmed by comparing the QTL data obtained from F_1DH and BC_4DH generations. QTL analysis for various glucosinolates detected 10 loci in the F_1DH mapping population compared to 6 loci detected in the BC_4DH mapping population. Three consistent QTL across the backcross generations showed significant change in the additive effects and R^2 values. The most significant observation was the disappearance of a major QTL in the BC_4DH that was detected in the F_1DH generation and appearance of a new major QTL in the BC_4DH generation. Six "true" QTL, four major (*J2Gsl1*, *J3Gsl2*, *J9Gsl3* and *J16Gsl4*) and two minor QTL (*J17Gsl5* and *J3Gsl6*), were identified, and their suitability in the marker-assisted introgression to the Indian germplasm was validated.

The above-mentioned study has led to the identification of "true" QTL for marker-assisted introgression of low glucosinolate trait from an east European donor to Indian cultivars. Two major conclusions from the study were that (a) the QTL data of balanced mapping population such as F_1DH and RILs may not be informative for dissecting "true" loci for complex traits exhibiting epistasis and context dependency and (b) the low glucosinolate loci identified in this study for marker-assisted introgression to Indian germplasm may not be applicable to the east European germplasm due to the difference in the glucosinolate profiles of the two gene pools.

In a follow-up paper, Bisht et al. (2009) undertook fine mapping of the QTL identified earlier by Ramchiary et al. (2007a) for efficient marker-assisted introgression of low glucosinolate trait into the germplasm of Indian gene pool. Based on the DNA sequences from *A. thaliana* and *B. oleracea* for the different genes involved in the aliphatic glucosinolate biosynthesis, candidate genes were cloned from high and low glucosinolate *B. juncea* lines Varuna and Heera, respectively. Seventeen paralogues belonging to six gene families were mapped in *B. juncea*. Candidate gene *BjuA.GSL-ELONG.a* mapped to the QTL interval of *J2Gsl1*, *BjuA.GSL-ELONG.c*, *BjuA.GSL-ELONG.d* and *BjuA.Myb28.a* mapped to QTL interval of *J3Gsl2*, *BjuA.GSL-ALK.a* mapped to QTL interval of *J3Gsl6* and *BjuB.Myb28.a* mapped to QTL interval of *J17Gsl5*. No candidate gene could be mapped to QTL *J9Gsl3* and *J16Gsl4*. The functionality and contribution of different candidate genes/QTL were assessed by allelic variation study using phenotypic data of 785 BC_4DH lines. It was observed that *BjuA.Myb28.a* and *J9Gsl3* contributed significantly to the base-level glucosinolate production while *J16Gsl4* (probably *GSL-PRO*), *BjuA.GSL-ELONG.a* and *BjuA.GSL-ELONG.c* contributed to the C3, C4 and C5 elongation pathways, respectively. Three A genome QTL, *J2Gsl1* harbouring *BjuA.GSL-ELONG.a*, *J3Gsl2* harbouring both *BjuA.GSL-ELONG.c* and *BjuA.Myb28.a* and *J9Gsl3*, were identified as most important loci for breeding low glucosinolate *B. juncea*. The two-step genetic control postulated from the study is (a) an overall decrease of glucosinolate by *BjuB.Myb28.a* and *J9Gsl3* and (b) control of elongation process by *BjuA.GSL-ELONG.a* and *BjuA.GSL-ELONG.c*.

In terms of practical breeding, the studies of Ramchiary et al. (2007a, b) observed negative associations between the trait-enhancing alleles of some yield component QTL and low glucosinolate alleles of QTL *J2Gsl1* and *J3Gsl2* in these two linkage groups. With the development of candidate gene markers in these loci, marker-assisted precision breeding would be possible for the development of productive low glucosinolate *B. juncea* cultivars belonging to the Indian gene pool.

11.4.3 Seed Coat Colour

In *B. juncea*, both yellow and brown seed occur in nature. Most of the east European types are yellow seeded, while majority of the Indian cultivars are brown seeded. The quality of canola mustard can be further improved through the incorporation of yellow seed coat colour trait as the yellow seeds have higher oil content (Shirzadegan and Röbbelen 1985) and yield aesthetically appealing bright yellow oil that adds to its market value. Besides, yellow-seeded mustard contains less fibre and more protein and hence enhances the dietary feed value of canola-quality meal (Simbaya et al. 1995, Slominski et al. 1999).

The seed coat colour in *B. juncea* has been shown to be controlled by two independently segregating genes, with either of the dominant alleles producing brown seed coat. Furthermore, there is an influence of the maternal genotype on the expression of the trait (Vera et al. 1979, Vera and Woods 1982, Anand et al. 1985, Thiagarajah and Stringam 1993, Choudhary and Solanki 2007). Due to the recessive nature of the yellow seed coat colour coupled with maternal influence, it is important to find molecular markers linked to the seed coat colour loci for application in backcross breeding.

An attempt to tag the seed coat colour trait was made by Negi et al. (2000). Of the three AFLP markers, namely AFLP5, AFLP8 and AFLP29, showing linkage with seed coat colour trait, the marker AFLP8 showing least recombination was converted to a locus-specific SCAR marker after genome walking and cloning. In a follow-up work from the same laboratory (Sabharwal et al. 2004) two loci controlling the seed coat colour were mapped by AFLP markers. Following association mapping approach, 11 AFLP markers linked to seed coat colour were mapped to two linkage groups in which the earlier markers AFLP5 and AFLP29 were mapped to one linkage group and the marker AFLP8 for which the SCAR marker was developed was mapped to another linkage group.

Lionneton et al. (2004) also identified two loci for the seed coat colour mapping to two different linkage groups using an AFLP map (Table 11.1). Mahmood et al. (2005a) mapped the two loci of seed coat colour by QTL analysis using a RFLP map (Table 11.1). Phenotyping of the trait was done quantitatively based on light reflectance. Two QTL, SC-B4 and SC-A6, tightly linked to RFLP markers *wg7b6cNM* and *wg5a1a*, respectively, were mapped to two different linkage groups.

The mapping of seed coat colour in *B. juncea* by PCR-based microsatellite markers was reported by Padmaja et al. (2005). Using three different mapping populations, three microsatellite markers, Ra2-A11, Na10-A08 and Ni4-F11, showed tight linkage with two loci of seed coat colour, *BjSc1* and *BjSc2*, respectively. Markers Ra2-A11 and Na10-A08 along with *BjSc1* mapped to a linkage group in the A genome, and the marker Ni4-F11 along with *BjSc2* mapped to another linkage group in the B genome. Validation study of the identified microsatellite markers (Na10-A08 and Ni4-F11) was undertaken in a number of yellow-seeded eastern European and brown-seeded Indian germplasm of *B. juncea*, and it was concluded that these two co-dominant microsatellite markers could be effectively used in marker-aided introgression programme. Yan et al. (2009) mapped the A genome seed coat colour locus in *B. juncea* using a BC_6F_2 mapping population. Two RAPD markers designated as SCM57 and SCM1078 were found to be linked to the gene for brown seed coat, and these were converted to SCAR markers. A third SCAR marker SZ1 amplifying 331 bp was developed using information from Negi et al. (2000). These three markers along with Ra2-A11, previously shown to be linked to the A genome seed coat colour locus (Padmaja et al. 2005), were found to be linked to the trait.

11.4.4 Agronomic and Yield Traits

Most of the agronomically important and yield-related traits are multigenic, and most of these are highly influenced by the environment and therefore are difficult to dissect into their component loci through conventional genetic analysis. Following different mating designs, the overall genetic architecture of these complex traits can be classified primarily into additive and non-additive genetic components. Banga (1996) reviewed the genetics of yield and its component traits in *B. juncea* and observed prevalence of both additive and non-additive gene actions for most of the yield-related traits. The recent molecular map-based QTL analyses (Lionneton et al. 2004, Mahmood et al. 2005b, Ramchiary et al. 2007b) (Table 11.1) provide an opportunity to genetically dissect these complex traits into their component loci and locate their tentative positions in the genome. This kind of analysis could help locate the favourable alleles and determine their organization and, consequently, help breeders in devising strategies for transferring these loci from one gene pool to the other.

Lionneton et al. (2004) undertook the QTL mapping in *B. juncea* for four agronomic and yield-related traits using an AFLP map (Table 11.1). The mapping population was evaluated for 2 years in the field for days to flowering, plant height, 1,000-seed weight and seed oil content. A total of 12 QTL were detected in a 2-year study. For each trait three QTL were identified. In one linkage group, the QTL for days to flower overlapped with the QTL for 1,000-seed weight.

QTL dissection of yield and yield components was carried out by Mahmood et al. (2005b) using a RFLP map (Table 11.1). Quantitative data from the mapping population for the number of siliques on the main raceme, number of seed/siliqua, silique length (mm), 1,000-seed weight (g) and seed yield (kg/hectare) were recorded from the replicated field trials conducted over 2 years in four locations. QTL analysis revealed no stable QTL for seed yield. Four QTL for silique length, one for number of seeds/silique, five for number of siliques on the main raceme and one for 1,000-seed weight were identified. The effects of the majority of QTL were consistent across trials/environments.

Ramchiary et al. (2007b) carried out QTL analysis of yield-influencing traits in *B. juncea* (Table 11.1). The mapping population was grown in three different environments and phenotyped for 12 quantitative traits. A total of 65 QTL spreading over 13 linkage groups were identified from the three environments. QTL analysis revealed major contribution of A genome towards the domestication of *B. juncea* as an oilseed crop. The organization of QTL in the *B. juncea* genome revealed the clustering of major QTL in a few linkage groups, particularly in A7 and A10 of the A genome. These genetic intervals with QTL hotspots in the linkage groups A7 and A10 contained QTL of different traits having agronomically antagonistic allelic effects. Some of the major QTL of yield components pertaining to seed size, number of seeds in a silique, silique density and number were mapped to these genetic intervals. QTL analysis also identified some well-separated QTL which could be readily transferred between the two gene pools.

11.4.5 Disease Resistance

Among the different fungal diseases in *B. juncea*, Alternaria blight (*Alternaria brassicae*), white rust (*A. candida*) and stem rot (*Sclerotinia sclerotiorum*) cause considerable yield losses. In India yield loss up to 70% due to infection of Alternaria blight infection (Saharan and Chand 1988) and up to 92% due to stem rot infection (Shivpuri et al. 1999) have been reported. No source of resistance for Alternaria blight and stem rot has so far been reported in the *B. juncea* germplasm (Saharan 1992).

The white rust disease caused by *A. candida* is one of the destructive diseases of mustard in India and Canada. Most of the released Indian cultivars are highly susceptible to white rust. Thirteen races of this pathogen have been reported from different *Brassica* species (Verma et al. 1999), the predominant race of white rust infecting *B. juncea* being identified as race-2 (Petrie 1988, Rimmer et al. 2000). Sources of resistance to this race are available in the east European gene pool of *B. juncea*. Genetic analysis in *B. juncea* revealed that resistance to race-2 is controlled by a single dominant gene (Tiwari et al. 1988, Sachan et al. 1995).

Genetic analysis of white rust resistance in *B. juncea* has been undertaken at the molecular level to locate the gene(s) and to identify markers for MAS applications. Mapping work on white rust resistance has been reviewed by Edwards et al. (2007). All the earlier studies have identified a single locus for white rust resistance in *B. juncea*. Cheung et al. (1998a) mapped the resistance locus *Acr* with three RFLP markers X140a, X42 and X83 using a DH mapping population derived from a cross between a white rust-susceptible cultivar (J90-4317) and a white rust-resistant cultivar (J90-2733). Following bulk segregant analysis, the white rust resistance locus $Ac2_1$ was mapped by two RAPD markers WR2 and WR3 flanking the gene. These markers were shown to be specific to the Russian source of resistance (Prabhu et al. 1998). Mukherjee et al. (2001) undertook molecular mapping of a locus *Ac2(t)* conferring resistance to white rust by RAPD markers using a resistance source (BEC-144) from the east European gene pool through bulk segregant analysis. This work was further extended by Varshney et al. (2004) who converted one of earlier identified linked RAPD markers to a cleaved amplified polymorphic sequence (CAPS) marker and also developed a tightly linked AFLP marker for the same white rust resistance gene.

Somers et al. (2002) mapped the resistance to another race 2 variant 2V in *B. juncea* for which no natural resistance is available in *B. juncea*. Resistance against the race 2V was introgressed from *B. napus* to *B. juncea* through interspecific hybridization. A BC3F2 population segregating for resistance to race 2V was used to map the resistance locus by AFLP markers using bulk segregant analysis. Eight *B. napus*-specific AFLP markers were identified to be linked to white rust resistance.

Although all the above studies identified one locus for the gene conferring resistance to white rust, it would be interesting to determine whether the loci governing the resistance identified in independent studies are the same or not. This is currently not possible due to lack of common set of markers in different mapping populations.

There is no high degree of resistance to Alternaria blight among the cultivated *Brassica* species. However, sources of resistance to this fungal disease have been identified in wild crucifers such as *Sinapis alba* (Burns et al. 1987), *Eruca sativa* (Tiwari 1991), *Camelina sativa* and *Capsella bursa-pastoris* (Conn et al. 1988). Although attempts have been made to transfer the resistance genes from these wild sources to cultivated *Brassica* species by both sexual and somatic cell hybridization (Sharma et al. 2002b), tangible results are yet to come out.

Partial resistance to Sclerotinia stem rot has been reported in *B. napus* cv. Zhong You 821 (Buchwaldt et al. 2003). Recently, a new source exhibiting outstanding resistance to Sclerotinia stem rot has been identified in *B. napus* genotype, ZY006 (Li et al. 2009).

11.5 Future Prospects

There are three facets of genetic analysis – inheritance pattern, mapping and study of genic interactions. Inheritance of some of the qualitative traits, which have been studied in some detail, follows digenic models of inheritance confirming that *B. juncea* is a recent allopolyploid, and neofunctionalisation or subfunctionalisation of the paralogues in the two parental genomes AA (from *B. rapa*) and BB (from *B. nigra*) has not taken place.

A significant breakthrough in genetics and breeding of *B. juncea* is heterosis breeding based on the availability of two divergent gene pools. A CMS system (Sodhi et al. 2006) and transgenic method of pollination control (Jagannath et al. 2001, 2002) are available. Hybrids based on these systems have shown increase in yield by 20–30% over the pure line check varieties in the mustard-growing areas of India. The challenge now is to increase the yield even further. The two gene pools, the east European and the Indian, need to be further improved by introgression of useful traits from one gene pool to the other. However, the overall genetic architecture of the two pools has to be maintained to exploit heterosis. The availability of markers will allow quicker and more precise introgression of quality traits (mostly recessive) and avoid serious problems associated with linkage drag.

Quantitative traits like seed size, seed number, silique density and oil content require deeper understanding. Fortunately, these traits barring the oil content have high heritability (Ramchiary et al. 2007b), and therefore strong phenotypic selection is possible. Earlier mapping work in *B. juncea* mostly used RFLP, RAPD and AFLP markers. More recent work with gene-specific markers will allow navigation between the genomes of *Brassica* species with that of *A. thaliana*. Comparative mapping will help in providing clues to the candidate genes of QTL involved with yield-influencing traits like seed size, silique length and number and may allow further advances in yield. Consolidation of different maps within *B. juncea* and comparative maps across *Brassica* species will also allow study of conserved QTL which work across the lines and species. However, introgression of these traits will require better understanding of genetic and molecular interactions behind these traits. Fine mapping and expression studies will be required to understand the genetic and molecular interactions underlying these traits.

An interesting and a useful area would be introgression of genetic/genomic information from basic diploids and allopolyploid species into *B. juncea* to further diversify its genetic base. As has been shown in rice (Brar and Khush 1997) and tomato (Lippman et al. 2007), such inputs from the secondary gene pool may improve both the yield and the resistance to disease. A fine example is the use of *B. nigra* and *B. juncea* resources to confer resistance to blackleg disease in *B. napus* (Dixelius 1999). Wide hybridization, even beyond U's triangle species, may be required to confer resistance to diseases like Alternaria blight and stem rot. Fortunately, *Brassica* species can be crossed with many members of the *Brassica* coenospecies quite readily (Prakash et al. 1999) and introgression could be monitored by markers, at least for the background selection. Although wide hybridization does not seem so exciting in the parlance of modern biology, it is a very effective tool for diversification and disease resistance.

B. juncea is a drought hardy crop. Given the specter of global warming, water scarcity and shortage of quality edible oil, it is important to launch a global R&D effort on such a crop. If the experience of *Arabidopsis* is anything to go by, open source and knowledge sharing can significantly strengthen research in *B. juncea* to resolve intricate genetics questions. *Brassica* geneticists, breeders, pathologists and molecular biologists need to share germplasm, markers and other resources for rapid advances in research and development.

References

Anand IJ, Reddy WR, Rawat DS (1985) Inheritance of seed coat colour in mustard. Indian J Genet 45:34–37

Axelsson T, Bowman CM, Sharpe AG, Lydiate DJ, Lagercrantz U (2000) Amphidiploid *Brassica juncea* contains conserved progenitor genomes. Genome 43:679–688

Banga SS (1996) Genetics and breeding. In: Chopra VL, Prakash S (eds) oilseed and vegetable Brassicas: Indian perspective, pp 50–76. Oxford and IBH publishing Co Pvt Ltd, New Delhi

Bhat MA, Gupta ML, Banga SK, Raheja RK, Banga SS (2002) Erucic acid heredity in *Brassica juncea* – some additional information. Plant Breed 121:456–469

Bisht NC, Gupta V, Ramchiary N, Sodhi YS, Mukhopadhyay A, Arumugam N, Pental D, Pradhan AK (2009) Fine mapping of loci involved with glucosinolate biosynthesis in oilseed mustard (*Brassica juncea*) using genomic information from allied species. Theor Appl Genet 118: 413–421

Brar DS, Khush GS (1997) Alien introgression in rice. Plant Mol Biol 35:35–47

Buchwaldt L, Yu FQ, Rimmer SR, Hegedus DJ (2003) Resistance to *Sclerotinia sclerotiorum* in a Chinese *Brassica napus* cultivar. Proc 8th Int Congr of Plant Pathol, Christchurch, p 289

Burns MJ, Barnes SR, Bowman JG, Clarke MHE, Werner CP, Kearsey MJ (2003) QTL analysis of an intervarietal set of substitution lines in *Brassica napus:* (i) seed oil content and fatty acid composition. Heredity 90:39–48

Burns H, Plessis J, Renard M (1987) Resistance to some crucifers to *Alternaria brassicae* (Berk.) Sacc. Proc 7th Int Rapeseed Congr Poznan, Poland, pp 1222–1227

Burton W, Ripley V, Potts D, Salisbury P (2004) Assessment of genetic diversity in selected breeding lines and cultivars of canola quality *Brassica juncea* and their implications for canola breeding. Euphytica 136:181–192

Chen JL, Beversdorf WD (1990) Fatty acid inheritance in microspore-derived populations of spring rapeseed (*Brassica napus* L.). Theor Appl Genet 80:465–469

Cheung WY, Friesen LJ, Rakow G, Seguin-Swartz G, Landry BS (1997) An RFLP based linkage map in mustard *Brassica juncea* (L.) Czern. and Coss. Theor Appl Genet 94:841–851

Cheung WY, Gugel RK, Landry BS (1998a) Identification of RFLP markers linked to white rust resistance gene (Acr) in muatard (*Brassica juncea* (L.) Czern. & Coss.). Genome 41:626–628

Cheung WY, Landry BS, Raney P, Rakow GFW (1998b) Molecular mapping of seed quality traits in *Brassica juncea* L. Czern., and Coss. Acta Hort 459:139–147

Choudhary BR, Solanki ZS (2007) Inheritance of siliqua locule number and seed coat colour in *Brassica juncea.* Plant Breed 126:104–106

Christianson JA, Rimmer SR, Good AG, Lydiate DJ (2006) Mapping genes for resistance to *Leptosphaeria maculans* in *Brassica juncea.* Genome 49:30–41

Conn KL, Tewari JP, Dahiya JS (1988) Resistance to *Alternaria brassicae* and phytoalexin-elicitation in rapeseed and other crucifers. Plant Sci 56:21–25

Delourme R, Falentin C, Huteau V, Clouet V, Horvais R, Gandon B, Specel S, Hanneton L, Dheu JE, Deschamps M, Margale E, Vincourt P, Renard M (2006) Genetic control of oil content in oilseed rape (*Brassica napus* L.). Theor Appl Genet 113:1331–1345

Dixelius C (1999) Inheritance of the resistance to *Leptosphaeria maculans* of *Brassica nigra* and *B. juncea* in near-isogenic lines of *B. napus*. Plant Breed 118:151–156

Ecke W, Uzunova M, Weissleder K (1995) Mapping the genome of rapeseed (*Brassica napus* L.) II. Localization of genes controlling erucic acid synthesis and seed oil content. Theor Appl Genet 91:972–977

Edwards D, Salisbury PA, Burton WA, Hopkins CJ, Batley J (2007) Indian mustard. In: Cole C (ed) Genome mapping and molecular breeding in plants Oilseeds, pp 179–210. Springer, Berlin, Heidelberg

Erickson LR, Straus NA, Beversdorf WD (1983) Restriction patterns reveal origins of chloroplast genomes in *Brassica* amphidiploids. Theor Appl Genet 65:201–206

Ford CS, Allainguillaume J, Grilli-Chantler P, Cuccato G, Allender CJ, Wilkinson MJ (2006) Spontaneous gene flow from rapeseed (*Brassica napus*) to wild *Brassica oleracea*. Proc R Soc B 273:3111–3115

Fourmann M, Barret P, Renard M, Pelletier G, Delourme R, Brunel D (1998) The two genes homologous to *Arabidopsis FAE1* co-segregate with the two loci governing erucic acid content in *Brassica napus*. Theor Appl Genet 96:852–858

Gladis T, Hammer K (1992) The *Brassica* collection in gatersleben: *Brassica juncea*, *Brassica napus*, *Brassica nigra*, and *Brassica rapa*. Feddes Rep 103:467–507

Gland A, Röbbelen G, Thies W (1981) Variation of alkenyl glucosinolates in seeds of *Brassica* species. Z. Pflanzenzüchtg 87:96–110

Gomez-Campo C, Prakash S (1999) Origin and domestication. In: Gomez-Campo C (ed) Biology of Brassica coenospecies, pp 33–58. Elsevier, Amsterdam

Gupta V, Mukhopadhyay A, Arumugam N, Sodhi YS, Pental D, Pradhan AK (2004) Molecular tagging of erucic acid trait in oilseed mustard (*Brassica juncea*) by QTL mapping and single nucleotide polymorphisms in FAE1 gene. Theor Appl Genet 108:743–749

Halkier BA, Gershenzon J (2006) Biology and biochemistry of glucosinolates. Annu Rev Plant Biol 57:303–333

Jagannath A, Arumugam N, Gupta V, Pradhan AK, Burma PK, Pental D (2002) Development of transgenic barstar lines and identification of a male sterile (barnase)/restorer(barstar) combination for heterosis breeding in Indian oilseed mustard (*Brassica juncea*). Curr Sci 82:46–52

Jagannath A, Bandyopadhyay P, Arumugam N, Gupta V, Burma PK, Pental D (2001) The use of spacer DNA fragment insulates the tissue specific expression of a cytotoxic gene (*barnase*) and allows high frequency generation of transgenic male sterile lines in *Brassica juncea* L. Mol Breeding 8:11–23

Jain A, Bhatia S, Banga SS, Prakash S, Lakshmikumaran M (1994) Potential use of random amplified polymorphic DNA (RAPD) technique to study the genetic diversity in Indian mustard (*Brassica juncea*) and its relationship to heterosis. Theor Appl Genet 88:116–122

Johnston JS, Pepper AE, Hall AE, Chen ZJ, Hodnett G, Drabek J, Lopez R, Price HJ (2005) Evolution of genome size in Brassicaceae. Ann Bot 95:229–235

Jourdren C, Barret P, Horvais R, Foisset N, Delourme R, Renard M (1996) Identification of RAPD markers linked to the loci controlling erucic acid level in rapeseed. Mol Breed 2:61–71

Kirk JTO, Hurlstone CJ (1983) Variation and inheritance of erucic acid content in *Brassica juncea*. Z Pflanz 90:331–338

Kirk JTO, Oram RN (1981) Isolation of erucic acid free lines of *Brassica juncea*: Indian mustard now a potential oilseed crop in India. J Aust Inst Agric Sci 47:51–52

Lagercrantz U (1998) Comparative mapping between *Arabidopsis thaliana* and *Brassica nigra* indicates that Brassica genomes have evolved through extensive genome replication accompanied by chromosome fusions and frequent rearrangements. Genetics 150:1217–1228

Li CX, Liu SY, Sivasithamparam K, Barbetti MJ (2009) New sources of resistance to *Sclerotinia sclerotiorum* in Chinese and Australian *Brassica napus* and *B. juncea* germplasm screened under Western Australian conditions. Australasian Plant Pathol 38:149–152

Lionneton E, Aubert G, Ochatt S, Merah O (2004) Genetic analysis of agronomic and quality traits in mustard (*Brassica juncea*). Theor Appl Genet 109:792–799

Lionneton E, Ravera S, Sanchez L, Aubert G, Delourme R, Ochatt S (2002) Development of an AFLP-based linkage map and the localization of QTLs for seed fatty acid content in condiment mustard (*Brassica juncea*). Genome 45:1203–1215

Lippman ZB, Semel Y, Zamir D (2007) An integrated view of quantitative trait variation using tomato interspecific introgression lines. Curr Opin Genet Dev 17:545–552

Love HR, Rakow G, Raney JP, Downey RK (1990a) Development of low glucosinolate mustard. Can J Plant Sci 70:419–424

Love HR, Rakow G, Raney JP, Downey RK (1990b) Genetic control of 2-propenyl and 3-butenyl glucosinolate synthesis in mustard. Can J Plant Sci 70:425–429

Mahmood T, Ekuere U, Yeh F, Good AG, Stringam GR (2003a) Molecular mapping of seed aliphatic glucosinolates in *Brassica juncea*. Genome 46:753–760

Mahmood T, Ekuere U, Yeh F, Good AG, StringamG R (2003b) RFLP linkage analysis and mapping genes controlling the fatty acid profile of *Brassica juncea* using reciprocal DH populations. Theor Appl Genet 107:283–290

Mahmood T, Rahman MH, Stringam GR, Raney JP, Good AG (2005a) Molecular markers for seed colour in *Brassica juncea*. Genome 48:755–760

Mahmood T, Rahman MH, Stringam GR, Yeh F, Good AR (2005b) Molecular markers for yield components in *Brassica juncea* – do these assist in breeding for high seed yield? Euphytica 144:157–167

Mahmood T, Rahman MH, Stringam GR, Yeh F, Good AG (2006) Identification of quantitative trait loci (QTL) for oil and protein contents and their relationships with other seed quality traits in *Brassica juncea*. Theor Appl Genet 113:1211–1220

Mahmood T, Rahman MH, Stringam GR, Yeh F, Good AG (2007) Quantitative trait loci for early maturity and their potential in breeding for earliness in *Brassica juncea*. Euphytica 154: 101–111

Mukherjee AK, Mohapatra T, Varshney A, Sharma R, Sharma RP (2001) Molecular mapping of a locus controlling resistance to *Albugo candida* in Indian mustard. Plant Breed 120:483–487

Negi MS, Devic M, Delseny M, Lakshmikumaran M (2000) Identification of AFLP fragments linked to seed coat colour in *Brassica juncea* and conversion to a SCAR marker for rapid selection. Theor Appl Genet 101:146–152

Olsson G (1960) Species crosses within the genus *Brassica* I. artificial *Brassica juncea* coss. Hereditas 46:171–222

Osborn TC (2004) The contribution of polyploidy to variation in *Brassica* species. Physiol Plant 121:531–536

Padmaja KL, Arumugam N, Gupta V, Mukhopadhyay A, Sodhi YS, Pental D, Pradhan AK (2005) Mapping and tagging of seed coat colour and the identification of microsatellite markers for marker-assisted manipulation of the trait in *Brassica juncea*. Theor Appl Genet 111:8–14

Palmer JD, Shields CR, Cohen DB, Orton TJ (1983) Chloroplast DNA evolution and the origin of amphiploid *Brassica* species. Theor Appl Genet 65:181–189

Panjabi P, Jagannath A, Bisht NC, Lakshmi Padmaja K, Sharma S, Gupta V, Pradhan AK, Pental D (2008) Comparative mapping of *Brassica juncea* and *Arabidopsis thaliana* using Intron Polymorphism (IP) markers: homeologous relationships, diversification and evolution of the A, B and C Brassica genomes. BMC Genomics 9:113

Parkin IAP, Gulden SM, Sharpe AG, Lukens L, Trick M, Osborn TC, Lydiate DJ (2005) Segmental structure of the *Brassica napus* genome based on comparative analysis with *Arabidopsis thaliana*. Genetics 171:765–781

Petrie GA (1988) Races of *Albugo candida* (white rust and staghead) on cultivated Cruciferae in Saskatchewan. Can J Plant Pathol 10:142–150

Pires JC, Zhao J, Schranz ME, Leon EJ, Quijada PA, Lukens LN, Osborn TC (2004) Flowering time divergence and genomic rearrangements in resynthesized *Brassica* polyploids (Brassicaceae). Biol J Linnean Soc 82:675–688

Prabhu KV, Somers DJ, Rakow G, Gugel RK (1998) Molecular markers linked to white rust resistance in mustard *Brassica juncea*. Theor Appl Genet 97:865–870

Pradhan AK, Gupta V, Mukhopadhyay A, Arumugam N, Sodhi YS, Pental D (2003) A high density linkage map in *Brassica juncea* (Indian mustard) using AFLP and RFLP markers. Theor Appl Genet 106:607–614

Pradhan AK, Sodhi YS, Mukhopadhyay A, Pental D (1993) Heterosis breeding in Indian mustard (*Brassica juncea* L. Czern. & Coss.): analysis of component characters contributing to heterosis for yield. Euphytica 69:219–229

Prakash S (1973) Artificial *Brassica juncea* Coss. Genetica 44:249–260

Prakash S, Chopra VL (1991) Cytogenetics of crop Brassicas and their allies. In: Tsuchiya T, Gupta PK (eds) Chromosome engineering in plants: Genetics, Breeding, Evolution, Part B, pp 161–180. Elsevier, Amsterdam

Prakash S, Takahata Y, Kirti PB, Chopra VL (1999) Cytogegetics. In: Gomez-Campo C (ed) Biology of Brassica coenospecies, pp 59–106. Elsevier, Amsterdam

Qiu D, Morgan C, Shi J, Long Y, Liu J, Li R, Zhuang X, Wang Y, Tan X, Dietrich E, Weihmann T, Everett C, Vanstraelen S, Beckett P, Fraser F, Trick M, Barnes S, Wilmer J, Schmidt R, Li J, Li D, Meng J, Bancroft I (2006) A comparative linkage map of oilseed rape and its use for QTL analysis of seed oil and erucic acid content. Theor Appl Genet 114:67–80

Ramchiary N, Bisht NC, Gupta V, Mukhopadhyay A, Arumugam N, Sodhi YS, Pental D, Pradhan AK (2007a) QTL analysis reveals context-dependent loci for seed glucosinolate trait in the oilseed *Brassica juncea*: importance of recurrent selection backcross scheme for the identification of 'true' QTL. Theor Appl Genet 116:77–85

Ramchairy N, Padmaja KL, Sharma S, Gupta V, Sodhi YS, Mukhopadhyay A, Arumugam N, Pental D, Pradhan AK (2007b) Mapping of yield influencing QTL in *Brassica juncea*: implications for breeding of a major oilseed crop of dryland areas. Theor Appl Genet 115:807–817

Rimmer S, Mathur S, Wu C (2000) Virulence of isolatesof *Albugo candida* from western Canada to *Brassica* species. Can J Plant Pathol 22:229–235

Sabarwal V, Negi MS, Banga SS, Lakshmikumaran M (2004) Mapping of AFLP markers linked to seed coat colour loci in *Brassica juncea* (L.) Czern. Theor Appl Genet 109: 160–166

Sachan J, Kolte S, Singh B (1995) Genetics of resistance to white rust (*Albugo candida* race 2) in mustard (*Brassica juncea* (L.) Czern & Coss.). In: GCIRC 9th Int Rapeseed Congr, Cambridge, UK, pp1295–1297

Saharan GS (1992) Disease resistance. In: Labana KS, Banga S, Banga SK (eds) Breeding oilseed Brassicas, pp 181–205. Narosa, New Delhi

Saharan G, Chand J (1988) Diseases of oilseed crops. Haryana Agricultural University, Hisar, India, pp 268

Schranz ME, Lysak MA, Mitchell-Olds T (2006) The ABC's of comparative genomics in the Brassicaceae: building blocks of crucifer genomes. Trends Plant Sci 11:535–542

Sharma R, Aggarwal RAK, Kumar R, Mohapatra T, Sharma RP (2002a) Construction of an RAPD linkage map and localization of QTLs for oleic acid level using recombinant inbreds in mustard (*Brassica juncea*). Genome 45:467–472

Sharma G, Dinesh Kumar V, Haque A, Bhat SR, Prakash S, Chopra VL (2002b) Brassica coenospecies: a rich reservoir for genetic resistance to leaf spot caused by *Alternaria brassicae*. Euphytica 125:411–417

Shirzadegan M, Röbbelen G (1985) Influence of seed colour and hull proportions on quality properties of seeds in *Brassica napus* L. Fette Seifen Anstrichm 87:235–237

Shivpuri A, Sharma K, Chipa H (1999) Some studies on the stem rot (*Sclerotinia sclerotiorum*) disease of rapeseed/mustard in Rajasthan, India. Proc 10th Int Rapeseed Congr Canberra, 26–29 Sept (CD-ROM)

Simbaya J, Slominski BA, Rakow G, Campbell LD, Downey RK, Bell JM (1995) Quality characteristics of yellow seeded *Brassica* seed meal: protein, carbohydrates and dietary fibre components. J Agric Food Chem 43:2062–2066

Slominski BA, Simbaya J, Campbell LD, Rakow G, Guenter W (1999) Nutritive value for broilers of meal derived from newly developed varieties of yellow seeded canola. Anim Feed Sci Technol 78:249–262

Sodhi YS, Chandra A, Verma JK, Arumugam N, Mukhopadhyay A, Gupta V, Pental D, Pradhan AK (2006) A new cytoplasmic male sterility system for hybrid seed production in Indian oilseed mustard *Brassica juncea*. Theor Appl Genet 114:93–99

Sodhi YS, Mukhopadhyay A, Arumugam N, Verma JK, Gupta V, Pental D, Pradhan AK (2002) Genetic analysis of total glucosinolate in crosses involving a high glucosinolate Indian variety and a low glucosinolate line of *Brassica juncea*. Plant Breed 121:508–511

Somers DJ, Rakow G, Rimmer SR (2002) *Brassica napus* DNA markers linked to white rust resistance in *Brassica juncea*. Theor Appl Genet 104:1121–1124

Song K, Lu P, Tang K, Osborn TC (1995) Rapid genome change in synthetic polyploids of *Brassica* and its implications for polyploidy evolution. Proc Natl Acad Sci 92:7719–7723

Song KM, Osborn TC, Williams PH (1988) *Brassica* taxonomy based on nuclear restriction fragment length polymorphism (RFLPs). I. Genone evolution of diploid and amphidiploid species. Theor Appl Genet 75:784–794

Srivastava A, Gupta V, Pental D, Pradhan AK (2001) AFLP-based genetic diversity assessment amongst agronomically important natural and some newly synthesized lines of *Brassica juncea*. Theor Appl Genet 102:193–199

Stringam GR, Thiagarajah MR (1995) Inheritance of alkenyl glucosinolates in traditional and microspore-derived doubled haploid populations of *Brassica juncea* L. Czern. and Coss. Proc 9th Int Rapeseed Congr, Cambridge. pp: 804–806

Tanhuanpää PK, Vikki JP, Vihinen M (1998) Mapping and cloning of FAD2 gene to develop allele-specific PCR for oleic acid in spring turnip rape (*Brassica rapa* ssp. *oleifera*). Mol Breed 4: 543–550

Tanhuanpää PK, Vikki JP, Vikki HJ (1996) Mapping of a QTL for oleic acid concentration in spring turnip rape (*Brassica rapa* ssp. *oleifera*). Theor Appl Genet 92:952–956

Thiagarajah MR, Stringam GR (1993) A comparison of genetic segregationin traditional and microspore-derived populations of *Brassica juncea* L. Czern. and Coss. Plant Breed 111:330–334

Tiwari JP (1991) Structural and biochemical bases of the black spot diseases of crucifers. Adv Struc Biol 1:25–34

Tiwari A, Petrie G, Downey R (1988) Inheritance of resistance to *Albugo candida* race 2 in mustard *Brassica juncea* (L.) Czern. Can J Plant Sci 68:297–300

U N (1935) genome analysis in Brassica with special reference to the experimental formation of *B. napus* and peculiar mode of fertilization. Jpn J Bot 7:389–452

Uchimiya H, Wildman SG (1978) Evolution of fraction I protein in relation to origin of amphidiploid *Brassica* species and other member of cruciferae. J Heredity 69:299–303

Varshney A, Mohapatra T, Sharma RP (2004) Development and validation of CAPs and AFLP markers for white rust resistance gene in *Brassica juncea*. Theor Appl Genet 109:153–159

Vaughan JG (1977) A multidisciplinary study of fraction I protein in relation to origin of amphidiploid *Brassica* species and other member of *cruciferae*. J Heredity 69:299–303

Vera CL, Woods DL (1982) Isolation of independent gene pairs at two loci for seed coat colour in *Brassica juncea*. Can J Plant Sci 62:47–50

Vera CL, Woods DL, Downey RK (1979) Inheritance of seed coat colour in *Brassica juncea*. Can J Plant Sci 59:635–637

Verma PR, Saharan GS, Bartaria AM, Shivpuri A (1999) Biological races of *Albugo candida* on *Brassica juncea* and *B. rapa* var. *toria* in India. J Mycol Plant Pathol 29:75–82

Yan M, Liu Z, Guan C, chen S, Yuan M, Liu X (2009) Inheritance and molecular markers for the seed coat color in *Brassica juncea*. Front Agric China 3:1–6

Chapter 12
Arabidopsis lyrata Genetics

Outi Savolainen and Helmi Kuittinen

Abstract *Arabidopsis lyrata* is a wild predominantly outcrossing perennial relative which diverged from *Arabidopsis thaliana* about 6 million years ago (MYA). The two species differ at 12% of synonymous nucleotide sites. *A. lyrata* has become a model organism for population and ecological genetics of outcrossing species. It has been used for studies of local adaptation and herbivore resistance. The genetics and evolution of the self-incompatibility system have been studied in detail. The complex demography and consequent highly diverged population structure set limits on the use of sequence variation and association studies for functional genomics. The full genome sequence has recently become available and will significantly broaden the opportunities for molecular evolutionary and functional genomic studies.

Keywords Population structure · Association mapping · Nucleotide diversity · Systematics · Adaptation

Contents

12.1 Introduction 348
12.2 Systematics and Distribution 348
12.3 *A. lyrata* Genome 350
12.4 *A. lyrata* Is Self-Incompatible and Has Inbreeding Depression 351
12.5 The Mating System Influences Genome Evolution 353
12.6 Population Genetic Diversity in Individual Populations 354
12.7 Disjunct Populations Are Highly Differentiated 358
12.8 Genetics of Local Adaptation 361
12.9 Perspectives for *A. lyrata* for Functional and Population Genomics 365
References 366

O. Savolainen (✉)
Department of Biology and Biocenter, University of Oulu, Oulu FIN-90014, Finland
e-mail: outi.savolainen@oulu.fi

R. Schmidt, I. Bancroft (eds.), *Genetics and Genomics of the Brassicaceae*, Plant Genetics and Genomics: Crops and Models 9, DOI 10.1007/978-1-4419-7118-0_12,

12.1 Introduction

Arabidopsis lyrata L. O'Kane and Al-Shehbaz is a perennial and mostly self-incompatible, diploid close relative of *Arabidopsis thaliana*. It diverged from *A. thaliana* about 5 MYA (million years ago) (Koch and Matschinger 2007), and the two species differ at 12% of synonymous nucleotide sites (Barrier et al. 2003). The whole genome sequence of an *A. lyrata* ssp. *lyrata* individual has been made available by the Joint Genome Institute (http://genome.jgi-psf.org/Araly1/Araly1.home.html).

The interest in *A. lyrata* stems from multiple reasons (Mitchell-Olds 2001). The increasing information on *A. thaliana* suggested early on that this resource could also be used for studying other closely related species. *A. lyrata* itself is becoming a model species for population genetics of outcrossing plant populations, where advanced population genomic analyses have been used (Ross-Ibarra et al. 2008). Further, these two related species with different mating systems have provided opportunities to compare the population genetic structure and genome evolution of selfing and outcrossing species (Savolainen et al. 2000, Charlesworth and Wright 2001, Wright et al. 2003, Lockton et al. 2008).

The availability of the *A. thaliana* sequence and functional information has made it attractive to study the genetics and evolution of traits related to adaptation, such as disease resistance, life history evolution, or herbivore resistance (Clauss and Koch 2006). Combining the increasing genetic resources with tractable field studies has already allowed demonstrating selection effects at individual loci in the wild (Kivimäki et al. 2007). It is also possible to study the targets and action of selection in such populations (Sandring et al. 2007, Sandring and Agren 2009).

Here we discuss genetic studies of *A. lyrata*, mainly with reference to the potential of using the populations for functional genomic studies. In many ways, *A. lyrata* studies can complement and add to *A. thaliana*, but it also provides many interesting opportunities for functional genomics in its own right, as an outcrossing species, especially now that the full genome sequence is available.

12.2 Systematics and Distribution

The systematic position of *A. lyrata* L. O'Kane and Al-Shehbaz 1997 (O'Kane and Al-Shehbaz 1997, 2003) was recently reviewed, and two taxa, *Arabis lyrata* L. and *Cardaminopsis petraea* (L.) Hiit., were considered to belong to this same species. Molecular systematic studies have shown that *A. lyrata* is a close relative of *A. thaliana* (Koch et al. 1999). Nuclear and cytoplasmic sequences suggest a divergence time from *A. thaliana* of 3–5.8 MYA (Koch et al. 2000, 2001, Koch et al. 2008). Later studies have described several evolutionary lineages and subspecies of *A. lyrata* (Koch and Matschinger 2007, Koch et al. 2008). The non-Arctic *A. lyrata* ssp. *lyrata* occurs mainly in eastern North America and in eastern Russia and *A. lyrata* ssp. *petraea* in Europe and possibly also in eastern Russia. A third evolutionary lineage, subspecies *A. lyrata* ssp. *kamchatica*, has been described from eastern Russia, but a similar type is found, e.g., in Japan and Korea. This taxon also occurs in Alaska (Koch et al. 2008, Schmickl et al. 2008). It has recently been shown that this lineage *A. lyrata* ssp. *kamchatica* is an allopolyploid hybrid of *Arabidopsis halleri* (ssp. *gemmifera*) and *A. lyrata*, which has arisen multiple times in Asia and in North America (Shimizu et al. 2005, Shimizu-Inatsugi et al. 2009). Shimizu et al. (2005) have considered this a separate species, with two subspecies, *kamchatica* and *kawasakiana*. There are still many open issues on the systematics of *A. lyrata*, as a species-wide analysis of populations has not yet been published; e.g., large parts of Russia and some parts of the USA have not been included in the analyses (but see Koch et al. 2008).

The overall distribution of *A. lyrata* consists of highly disjunct populations in the northern hemisphere (Fig. 12.1) (Jalas and Suominen 1994, Hoffman 2005). The populations occur in various low competition habitats (see Section 12.8 for more details). In Europe, ssp. *petraea* is quite common at least in Iceland, in Norway, and in a limited coastal area of Sweden (Ericson and Mascher 1978, Jalas and Suominen 1994). *A. lyrata* ssp. *lyrata* occurs in eastern USA, possibly also in the Far East. The *kamchatica* lineage occurs in western North America and in eastern Asia. Hoffman (2005) has analyzed the changes of distribution in the genus *Arabidopsis*. *A. lyrata*

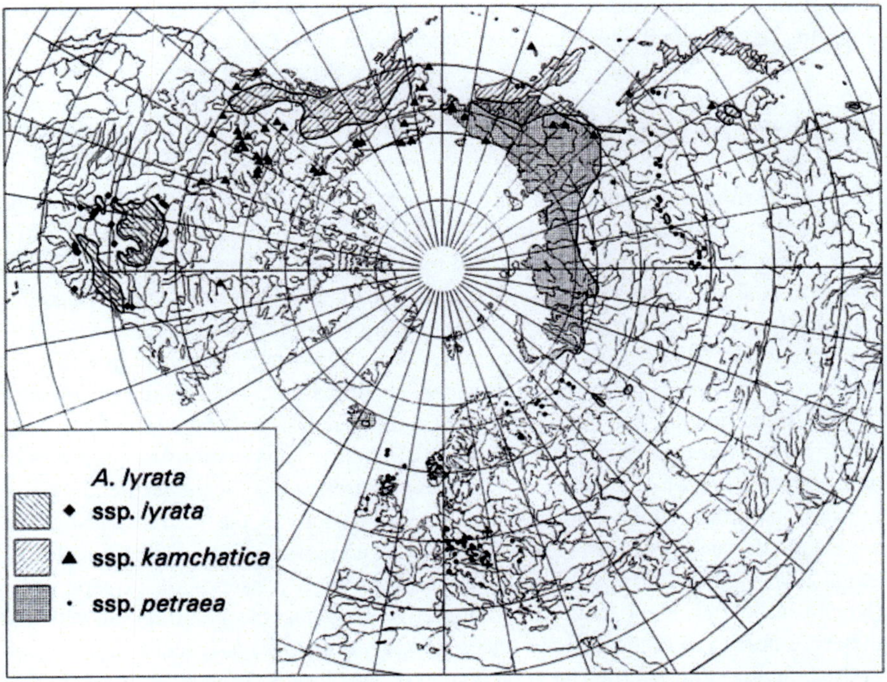

Fig. 12.1 Distribution map of *Arabidopsis lyrata* subspecies. Modified from Hoffman (2005)

ssp. *petraea* has moved to more northern areas compared to the core area of the genus. The *kamchatica* lineage seems to occupy the broadest area, and it has the widest climatic niche (Hoffman 2005).

There are two other species closely related to *A. lyrata* (and *A. thaliana*). *A. lyrata* (as a maternal parent) can be crossed with *A. halleri* to yield F1 hybrids. The F1 has been used to produce backcross progeny, with *A. lyrata* as maternal parent (Willems et al. 2007). The seed set is good when *A. lyrata* is used as the maternal parent, but very low in the other reciprocal. Koch and Matschinger (2007) recently showed that both *A. halleri* and *Arabidopsis arenosa* share cpDNA haplotypes with *A. lyrata*. Population genetic evidence has shown traces of introgression from *A. halleri* at nuclear genes of *A. lyrata* (Ramos-Onsins et al. 2004). There is also evidence of extensive sharing of self-incompatibility alleles between the two species (Castric et al. 2008).

A. lyrata has been crossed with the more distant *A. thaliana* (Mesicek 1967, Redei 1974) to yield sterile homoploid hybrids with 13 chromosomes. Later, Nasrallah et al. (2000) have produced further hybrids using *A. thaliana* as the female. For many traits, the hybrids were intermediate between the two parents. The hybrids were pollen sterile and could not be selfed. Hybrids could be used as female parents in backcrosses to either parent. There was no crossing over or recombination within chromosomes in the hybrid; only whole chromosomes recombined. The backcross progeny to *A. thaliana* produced pollen sterile plants (with developed flower parts), but the backcrosses to *A. lyrata* produced hardly any plants with stamen-bearing flowers. The F1 progeny could yield self-fertile plants upon doubling of the chromosome number (Nasrallah et al. 2000).

12.3 *A. lyrata* Genome

The base chromosome number in *A. lyrata* is 8, which is the ancestral chromosome number in the Brassicaceae (Chapter 1 by Lysak and Koch, this volume). The first genetic maps were obtained recently (Kuittinen et al. 2004, Yogeeswaran et al. 2005, Hansson et al. 2006). Comparison of orthologous genes between *A. thaliana* and *A lyrata* using these sparse maps demonstrated high conservation of gene order. The gene order and chromosome number differences between *A. thaliana* ($n = 5$) and *A. lyrata* ($n = 8$) were most parsimoniously accounted for by two reciprocal translocations, three fusions, and three inversions in the *A. thaliana* lineage (Kuittinen et al. 2004, Yogeeswaran et al. 2005, Hansson et al. 2006). The positions of centromeres and synteny between *A. lyrata* and *A. thaliana* chromosomes were verified by chromosome painting, which also revealed that the fusions in the *A. thaliana* lineage were actually due to pericentromeric inversions and reciprocal translocations (Lysak et al. 2006). *Capsella rubella* and *A. lyrata* have high conservation in gene order (Koch and Kiefer 2005), and the comparative chromosome painting showed that they resemble closely the inferred ancestral karyotype of Brassicaceae (Lysak et al. 2006, Schranz et al. 2006). The genetic map of the close relative *A. halleri* is highly similar to that of *A. lyrata* (Willems et al. 2007). The high synteny with *A. thaliana* has made it possible to use positional information from *A. thaliana* when choosing candidate genes underlying QTLs in *A. lyrata* (Heidel et al. 2006) and *A. halleri* (Courbot et al. 2007).

The positions of centromeres as well as the arrangement of interphase chromosome have been conserved between *A. thaliana* and *A. lyrata*, but there are some changes in satellite repeat type between species (Berr et al. 2006). The evolution of centromere satellite families in *A. lyrata* and *A. halleri* has been studied further by Kawabe and Charlesworth (2007b), who found evidence for exchange between repeat family members and recent rapid spread of one variant type in *A. lyrata*.

Natural *A. lyrata* populations are considered to be mostly diploid, but there are also polyploid plants in Central Europe (Polatchek 1966, Mesicek 1970, see Mable 2004 and references therein). *A. lyrata ssp. kamchatica* is likely to be allotetraploid between *A. lyrata* and *A. halleri* (Shimizu et al. 2005, Shimizu-Inatsugi et al. 2009). Synthetic allopolyploids between *A. thaliana* and *A. lyrata* have been created by spontaneous somatic doubling of F1 amphidiploid shoots in order to study the changes in genome structure and expression occurring after polyploidization (Beaulieu et al. 2009).

The genome sequencing of *C. rubella* and *A. lyrata* will allow more in-depth studies in genome evolution using *C. rubella* as an outgroup and comparing the changes in *A. thaliana* and *A. lyrata* (http://genome.jgi-psf.org/Araly1/Araly1.home.html). The *A. lyrata* sequence is available, but has not yet been fully annotated. The *C. rubella* genome is currently being sequenced.

12.4 *A. lyrata* Is Self-Incompatible and Has Inbreeding Depression

Outcrossing species in the Brassicaceae family characteristically have a sporophytic self-incompatibility system (Chapter 14 by Nasrallah, this volume). The genes coding for SI specificity in pollen (S-locus cysteine-rich SCR) and pistils (S-locus receptor kinase, SRK) are found in self-recognition haplotypes. These haplotypes can be maintained because there is very little recombination between the loci (Awadalla and Charlesworth 1999). The SI system of *A. lyrata* has been well described from a functional point of view (Kusaba et al. 2001 and Chapter 14 by Nasrallah, this volume). The Brassica SI system in general has mostly been studied in cultivated species, but *A. lyrata* and the related *A. halleri* allow studying a naturally evolving system (Schierup et al. 2001, Mable et al. 2003).

There are a large number of alleles, several dominance classes, and additional complicating factors such as nonreciprocal compatibility differences (Mable et al. 2003). Frequency-dependent selection favoring low-frequency alleles is likely to govern the polymorphism. The SI alleles have high diversity and low differentiation between populations, as predicted by models of frequency-dependent selection (Schierup et al. 2006, 2008). The *A. lyrata* SI locus is part of a multigene family, where other loci of the family do not show as high diversity or as low differentiation (Charlesworth et al. 2003b). The SI alleles are to a large degree shared between

A. lyrata and *A. halleri* (Castric et al. 2008). Introgression and selection for rare alleles have given rise to this higher level of shared polymorphism at the SI loci than between species at other loci (Ramos-Onsins et al. 2004, Castric et al. 2008).

Most populations of *A. lyrata* are nearly completely outcrossing due to the SI system. They are pollinated by various small insects, such as bees and flies (Clauss and Koch 2006). The flowers of *A. lyrata* are much larger than those of the selfing *A. thaliana*. Pollinators are attracted by benzenoid compounds emitted by the flowers. The volatile floral compounds are different from those in the selfing *A. thaliana* (Peer and Murphy 2003, Abel et al. 2009).

The SI system has broken down in North America in populations of *A. lyrata* in Ontario and Michigan (Mable et al. 2005). The self-compatible populations have high homozygosity and reduced variation at neutral marker loci, concordant with the selfing mating system (Mable and Adam 2007). Diversity at the SRK locus is not reduced in these populations, nor can the self-compatibility be associated with any specific haplotype of the SRK locus. Thus, some other factor may have caused the loss of self-incompatibility (Mable et al. 2005). The sequence of events leading to self-compatible populations still remains to be fully understood.

The original Brassica SI system has broken down in the selfing *A. thaliana*, in a complex set of events, estimated to have taken place about 1 million years ago (Tang et al. 2007, Shimizu et al. 2008). The *A. lyrata* SI-locus alleles transformed to *A. thaliana* render that species self-incompatible (Nasrallah et al. 2002), but as for *A. lyrata*, the details of the evolution of selfing are still not known.

Plants from self-incompatible *A. lyrata* populations can be artificially selfed using bud pollination. Inbreeding depression is an inherent feature of most outcrossing species (Charlesworth and Charlesworth 1987), and *A. lyrata* displays considerable inbreeding depression upon selfing (Kärkkäinen et al. 1999). There is some evidence that the genetic basis of the inbreeding depression may be due to deleterious recessive alleles, but there was also a suggestion for the role of overdominance (Kärkkäinen et al. 1999).

The outcrossing mating system, extensive heterozygosity, and inbreeding depression complicate many genetic studies. In some outcrossing important crop species it has been possible to generate inbred lines (through extensive crossing efforts), but such lines are at least not yet available in *A. lyrata*. The self-fertile populations of North America are a good starting point for studies where selfing and high homozygosity are an advantage. For instance, a selfed progeny with lower heterozygosity than in outcrossing populations was chosen for the whole genome sequencing.

In outcrossing species genetically variable populations in common garden or laboratory studies result in higher error variances, and there is often a need for larger sample sizes than for genetically homogeneous materials. QTL mapping experiments are best conducted in three-generation pedigrees, where four grandparents are used to generate two unrelated F1s which are crossed to generate F2s (e.g., Heidel et al. 2006). So far, no recombinant inbred lines or near isogenic lines are available. It has been feasible to use vegetatively propagated or tissue-cultured F2 individuals in multiple mapping studies (Kuittinen et al. 2004, Heidel et al. 2006).

While outcrossing, heterozygous plants may be more complicated organisms for many genetic studies than the selfing *A. thaliana*; however, outcrossing is the most frequent mating system of flowering plants. Thus, even with the complications, such species merit attention. Further, there are also some advantages to the study of outcrossing organisms. The extensive population genetics theory can be more easily applied, without needing to be concerned about the variable rate of selfing (Hein et al. 2005). Further, in outcrossing populations individual loci are more likely to be independent of each other, an advantageous feature in some studies, such as association mapping. Methods of analysis used in *Drosophila* or in human studies can easily be adopted for outcrossing plant species.

12.5 The Mating System Influences Genome Evolution

Selfing has evolved from outcrossing in the lineage leading to *A. thaliana* after the species have diverged, and *A. lyrata–A. thaliana* species pair is a good model to study the effects of mating system on genome evolution. Breeding system affects the organization of genetic variation in populations. In inbreeding populations the frequency of homozygotes is increased. This affects both the nature of diversity within species and the accumulation of differences between species, i.e., genome evolution (Charlesworth and Wright 2001).

Due to increased homozygosity, inbreeding populations have smaller effective sizes and less effective recombination, which has two consequences. First, as the level of neutral genetic variation depends on the effective population size, less neutral variability is expected in inbreeding compared with otherwise similar but outbreeding populations. Second, selection is less effective in small populations; thus weaker purifying selection is expected in predominantly inbreeding species. This should be seen as an elevated frequency of nonsynonymous low-frequency variants in inbreeding species and as more rapid accumulation of nonsynonymous variants in inbreeding compared with outcrossing lineages (Charlesworth and Wright 2001). In spite of this expectation, Wright et al. (2002) did not find evidence for a significant excess of amino acid substitutions or reduction in codon bias in the lineage leading to *A. thaliana* in comparison with *A. lyrata*. There was similar codon preference in the two species, driven by the tRNA abundance, and the rate of protein evolution between the species was negatively correlated with expression level (Wright et al. 2004). Foxe et al. (2008) studied the patterns of divergence and diversity in *A. lyrata* populations and found that *A. lyrata* showed an excess of low-frequency polymorphisms consistent with weak purifying selection, which was a similar pattern to what has been found in *A. thaliana* (Bustamante et al. 2002).

It is expected that in the regions of low recombination there is reduced diversity due to negative (background selection) and positive (selective sweeps) selection of linked variants, which decreases the effective size in that region. Empirical data from Drosophila support this, but data from plants are conflicting, e.g., in *A. thaliana* there is no strong correlation between recombination and diversity in genome-wide samples (Nordborg et al. 2005). While the genome-wide sampling in *A. thaliana* may complicate the patterns, this relationship can be studied using diversity data within individual populations in *A. lyrata*. However, in contrast to the expectations, the diversity was not reduced at low-recombination centromeric areas. This was attributed to more frequent occurrence of selective sweeps in the chromosomal arms than in centromeric areas (Wright et al. 2006, Kawabe et al. 2008).

Transposable element (TE) families have spread recently in genomes of *A. thaliana* and *A. lyrata*, enabling comparison of TE dynamics in species with contrasting breeding system (Lockton et al. 2008). TE abundance in natural populations is controlled by transposition increasing the copy number and purifying selection removing insertions from populations. Host breeding system is expected to affect TE dynamics in a rather complex way through its effect on the frequency of homozygotes, effective population size, and recombination (see Wright et al. 2001 and references therein). Polymorphic element insertions occurred at lower frequencies in *A. lyrata* than in *A. thaliana* which was compatible with weak purifying selection in *A. lyrata* and relaxed selection in *A. thaliana* (Wright et al. 2001). Patterns of transposable element occurrence varied among different *A. lyrata* populations suggesting that their demographic histories have a significant impact on the efficacy of selection (Lockton et al. 2008).

A. thaliana and *A. lyrata* have been used as a model to study the evolution of an imprinted seed growth gene, *MEDEA*, which is assumed to be under intergenomic conflict. The parental genomic conflict hypothesis assumes that there is a conflict of interest between genomes inherited from mother and father over the allocation of resources from mother to seeds (Wilkins and Haig 2003). The optimum offspring size is larger for paternal side than the maternal side in outcrossing species, but in inbreeding species genomic conflict is not expected because both the egg and the pollen come from the same parent. At the *MEDEA* gene, the patterns of sequence variation in *A. lyrata* populations and divergence from *A. thaliana* support an arms race type of evolution in the outcrossing *A. lyrata* but not in *A. thaliana,* as expected from theory (Kawabe and Charlesworth 2007a, Miyake et al. 2007, Spillane et al. 2007).

12.6 Population Genetic Diversity in Individual Populations

Population genetics tools are being used to detect DNA variants that govern important adaptive traits. There are two main approaches. First, analysis of sequence variation can provide signals of the effects of natural selection and can suggest loci for further functional studies (Charlesworth et al. 2003a, Nordborg and Weigel 2008). Searching for these signals is complicated by the signals due to demographic history. Second, population-level studies of association between DNA variation and phenotypic variation can provide direct evidence of the role of polymorphisms in governing trait variation (Zhu et al. 2008). The interpretation of association genetic studies requires that the effects of family structure, population structure, and the extent of linkage disequilibrium are taken into account (Yu et al. 2006). Thus, understanding the genetic structure of *A. lyrata* populations is important not only for evolutionary studies but also for functional genomics. Here we first deal with the genetic structure of individual populations and then consider population divergence in the next section.

The distribution of plants in Eurasia has been influenced by repeated glaciations and interglacial periods and by the associated colonizations and range reductions (Comes and Kadereit 1998). North American species also have been influenced by glaciations. Many northern plant species thus have a complicated demographic history. *A. lyrata* consists of widely separated groups of populations (Jalas and Suominen 1994, Hoffman 2005). The detailed history of the disjunct *A. lyrata* populations across the range still remains to be studied (Koch et al. 2008), but the Central European populations have likely been the source of at least Icelandic and Scandinavian populations (Wright et al. 2003, Muller et al. 2008), and the North American populations are likely due to more recent colonizations (Wright et al. 2003, Schmickl et al. 2008). This history of colonizations and range changes has a large influence on the genetic diversity of populations, as described below. Currently, individual *A. lyrata* populations vary in size, from the small remnant populations (e.g., in Scotland or Wales) (Kunin et al. 2009) to the extensive populations found, e.g., in Norway (Gaudeul et al. 2007) or in Iceland (Schierup 1998).

As described above, because of the SI system, most populations are outcrossing. Genotypes of neutral markers in individual *A. lyrata* populations are close to Hardy–Weinberg frequencies (e.g., Jonsell et al. 1995, Van Treuren et al. 1997, Clauss and Mitchell-Olds 2006). This is consistent with the prevailing close to random mating outcrossing and rather efficient pollen flow by pollinators (Clauss and Koch 2006). Some North American populations, which have lost their self-incompatibility system and contain individuals capable of setting seeds without cross-pollination, have a different genetic structure (Mable et al. 2005). Some of these populations have outcrossing rates as low as 0.20 (Mable et al. 2005, Mable and Adam 2007). These populations also have higher homozygosity than H-W-expectation. Icelandic and German populations have shown evidence of inbreeding due to relatedness of the parents (biparental inbreeding) (Schierup 1998, Clauss and Mitchell-Olds 2006). There is also evidence that seeds within a fruit may often share a father (correlated paternity within fruits) (Schierup 1998). Biparental inbreeding and correlated paternity are still minor deviations from random mating.

The populations have a low level of spatial structure due to relatedness. In Central Europe, groups of relatives do not extend beyond a few meters (Clauss and Mitchell-Olds 2006). The related patches seem to be due to limited seed movement. In Norwegian mountains, spatial relatedness extends to 0.5 m due to clonal growth (Jonsell et al. 1995, Gaudeul et al. 2007, Lundemo et al. (2010)). Thus, samples from the natural populations are likely to show a low level of family structure.

Genetic diversity has been studied at multiple levels in individual *A. lyrata* populations. We first compare diversity in different populations within a larger region. The northern populations of Iceland, Norway, Sweden, and Russia all had about equal average levels of within-population diversity at 13 isozyme loci (Jonsell et al. 1995). Swedish populations on average had about 30% more diversity than the

Table 12.1 Microsatellite diversity (as expected heterozygosity, He) (Muller et al. 2008) and nucleotide polymorphism (θ_π) at 78 loci (Ross-Ibarra et al. 2008) in individual populations of *Arabidopsis lyrata*

	He microsat	θ_π (mean)	θ_π (median)
Germany	0.562	0.0135	0.0209
Iceland	0.364	0.0129	0.0083
Norway	0.305	–	–
Sweden	0.314	0.0097	0.0045
Russia	0.202	0.0071	0.0025
USA Ind/NC	0.210	0.0060	0.0013
Canada	–	0.0055	0.0012

Norwegian populations (Gaudeul et al. 2007). In Germany and Austria, diversity levels at microsatellite loci ranged between 0.45 and 0.58, with an average of 0.52 (Clauss and Mitchell-Olds 2006). Within the regions, the diversity of individual populations varied less than twofold, much less than is common in populations of selfing plants such as in *A. thaliana* (Schoen and Brown 1991, Stenøien et al. 2005, Bakker et al. 2006a), where some populations can be monomorphic and others highly polymorphic.

Microsatellite diversity in individual populations from many different regions is compared in Table 12.1 (Muller et al. 2008). The German population had the highest variation, and Iceland about 70% of that in Germany. Similar findings were made on comparing Germany and Iceland with a different set of loci (Clauss and Mitchell-Olds 2006). The Scandinavian populations had about half the variation of the Central European population. The Russian population (Karhumäki, Karelia) and a North American population from North Carolina had much reduced microsatellite diversity (Table 12.1).

Studies on small sets of nuclear loci showed that silent (and synonymous) and total nucleotide diversity follow similar patterns (Wright et al. 2003, Balana-Alcaide et al. 2006, Kuittinen et al. 2008). Recently, Ross-Ibarra et al. (2008) studied nearly 80 loci in six different populations. The means and medians of the estimated nucleotide diversity in each population are also shown in Table 12.1. On average, the German population was most variable, with mean nucleotide heterozygosity of 0.0135. This can be compared to the genome-wide *A. thaliana* estimate of 0.0083 in 12 accessions representing worldwide diversity (Schmid et al. 2005). Thus an individual population contained more diversity than *A. thaliana* in a wide set of accessions. There was much less diversity in the other populations, with lowest levels in North America. Even if not exactly the same populations were included in these studies of microsatellite and nucleotide diversity, the general results are similar – high diversity in Central Europe and less in northern Europe and North America (Table 12.1).

The northern (and other) marginal populations, which are currently practically isolated from the central populations, could have lower diversity because of a present low effective population size, where less genetic variation can be maintained at equilibrium. On the other hand, the diversity could still be reduced due to effects of old bottlenecks. In a random mating, constant size, large equilibrium populations, and the frequency distribution of the individual sequence variants should conform to the expectations derived from the standard neutral equilibrium model (SNE) (Hudson 1990, Hein et al. 2005). When a population has lost genetic variation in a bottleneck, it takes a very long time to recover the genetic variation, even if population grows to its original size (Charlesworth et al. 2003a). Soon after a bottleneck, the number of sequence variants at intermediate frequencies will be above that expected for a population at mutation-drift equilibrium (because the rare variants have been eliminated in the bottleneck).

The Central European *A. lyrata* populations did not show evidence of bottlenecks in a microsatellite study (Clauss and Mitchell-Olds 2006). The northern European populations had less variation than the German ones, but they also did not show traces of bottlenecks. The rapidly evolving neutral microsatellite markers have reached an equilibrium (Clauss and Mitchell-Olds 2006, Muller et al. 2008). Coding DNA sequences have much lower mutation rates than the repetitive microsatellites and take much longer to reach equilibrium. Thus, many *A. lyrata* populations still show traces of bottlenecks, as indicated by genome-wide excess of intermediate frequency variants in the northern European and North American populations (Wright et al. 2003, Ross-Ibarra et al. 2008). The Central European population (Plech), however, seemed to be at an equilibrium even as measured with the slowly evolving sequence variation (Clauss and Mitchell-Olds 2003, Wright et al. 2006, Ross-Ibarra et al. 2008).

The bottlenecks and population expansions are expected to influence not just nucleotide diversity at individual loci but also correlations between nucleotide sites and linkage disequilibrium. When variants at two nucleotide sites are statistically independent, there is no linkage disequilibrium (LD). Linkage disequilibrium can be generated (and maintained) in small populations by genetic drift. Another demographic effect, admixture between differentiated populations, also results in LD. The LD generated by genetic drift in small populations is maintained for very long times, as for instance in human populations, where LD ranges over at least tens of kilobases (Hein et al. 2005, Hinds et al. 2005). In *A. lyrata*, the extent of LD has not yet been extensively studied over long genomic regions, but an estimate of the ratio of recombination/diversity from Germany is 20 times higher than that in *A. thaliana* (Kim et al. 2007, Ross-Ibarra et al. 2008). The Central European population recombination rate is also much higher and lower LD than in other populations of *A. lyrata* (Wright et al. 2003, Ross-Ibarra et al. 2008). LD is quite variable between genomic regions and between these individual populations, but in Germany LD starts decreasing within individual genes. As a comparison, in *A. thaliana*, in worldwide collections, LD extends about 10 kb (Kim et al. 2007), but in local populations may extend over whole chromosomes (Nordborg et al. 2002).

The above (genome-wide) departures from the standard neutral model are likely due to the demographic history of the species. However, natural selection also causes departures from the SNE model – balancing selection causes excess intermediate frequencies, and strong directional selection causes loss of variability and an excess of rare variants (Hein et al. 2005). Thus, patterns of nucleotide diversity can be used to search for the effects of natural selection (e.g. Wright et al. 2005). When the populations have a strong signal of demography, it can, however, be difficult to distinguish between the effects of selection and demography. Given the described history, it is clearly not sufficient to show that the sequence data in a population depart from the standard neutral equilibrium model. Preferably, the tests of selection should be conducted against a model including the demographic history, including the effects of range expansion (Excoffier and Ray 2008). This will be especially important in species where the demographic history is as complex as in *A. lyrata* or *A. thaliana* (Nordborg et al. 2005, Ross-Ibarra et al. 2008). For *A. lyrata*, it may be easiest to detect signatures due to selection in populations close to demographic equilibrium, such as the Central European populations (Clauss and Mitchell-Olds 2006). On the other hand, populations in recently colonized areas may have undergone periods of strong selection for local adaptation (Leinonen et al. 2009), but the genetic effects of selection may be difficult to detect due to the genome-wide genetic consequences of the colonization.

Individual *A. lyrata* populations can be used for association genetic studies. The low level of family structure and the rather low level of linkage disequilibrium make it feasible to examine associations between phenotypic traits and candidate genes. An insertion–deletion polymorphism (of 14 amino acids) at the *FRI* gene was found to be associated with flowering time (of non-vernalized plants) in both Swedish and Norwegian populations (Kuittinen et al. 2008). In this case, transformations to *A. thaliana* of the two variants confirmed that this gene and not just linked variants was responsible for the effect. For genome-wide studies, the low LD sets a requirement for a very dense set of SNPs for detecting the associations. On the other hand, once an association with a marker is detected, then the causative locus is likely to be close by.

Within-population association studies would not be able to detect the genetic basis of differences between populations if the loci are not variable within populations. Other loci, which are fixed for alternate alleles within populations, could contribute to differences between populations. Further, the strong demographic signals and the departures from the SNE model could also influence the association studies within populations, as has been found in examining human populations (Slatkin 1999). Once these features are known, it is possible to statistically correct for them (Yu et al. 2006, Zhao et al. 2007).

12.7 Disjunct Populations Are Highly Differentiated

A. lyrata has colonized many different areas and environments. The currently disjunct populations have been fully isolated or have very reduced gene flow for considerable times. Bottlenecks during migrations and genetic drift during isolation have resulted in genetically very distinct populations. The differentiation caused by this kind of genetic factors influences all loci in the genome, including neutral loci that have no influence on the fitness of the individual. Before searching for

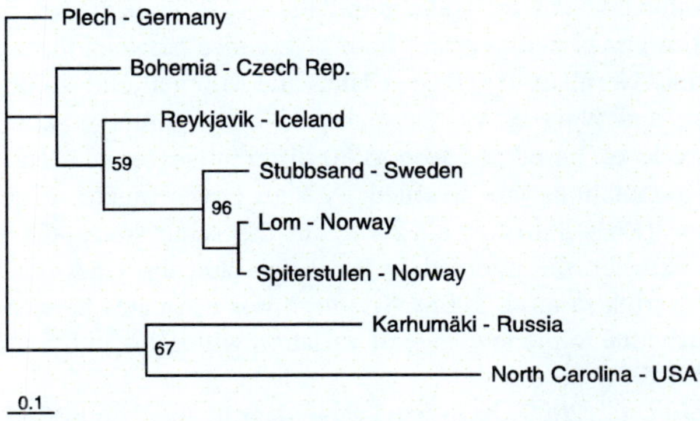

Fig. 12.2 Neighbor joining tree of *Arabidopsis lyrata* populations based on Nei's genetic distances of microsatellite data of Muller et al. (2008). *Numbers* give bootstrap support of branches

genetic differences due to differential natural selection in different environment, it is important to be aware of the general genome-wide level of differentiation between populations. Here we discuss this level of differentiation, and the next section concerns studying local adaptations.

Genetic differentiation between populations of a species is often measured by F_{ST}, the proportion of variation due to differences between populations (Wright 1943, 1951). Several studies have examined the differentiation of individual populations within a region. In northern Europe, populations within a region (Iceland, Sweden) were more similar to each other than to populations from other regions (Jonsell et al. 1995, Gaudeul et al. 2007). Within Norway and Sweden, there was in each case a clear isolation by distance effect, with higher divergence in Norwegian than in Swedish populations (average F_{ST} 0.23 and 0.18, respectively) (Gaudeul et al. 2007). A hierarchical structure of variability has also been found in studies of microsatellites in Central Europe. Most variation was within populations, 17% between close populations within a region, and F_{ST} 0.05 between regions of Germany and Austria (Clauss and Mitchell-Olds 2006). These data indicate limited gene flow between these populations. Thus, even in these rather limited spatial scales, the populations show rather high neutral divergence. There is a strong geographically related hierarchical genetic structure (Fig. 12.2), much stronger than is found in *A. thaliana* in these same areas (Nordborg et al. 2005, Stenøien et al. 2005).

Most interest has focused on divergence between a few European populations and North America. Isozyme and microsatellite variation showed that Swedish and Norwegian populations were quite closely related, whereas the Russian population from Karelia was most diverged (Jonsell et al. 1995, Van Treuren et al. 1997). The North American populations of subspecies *lyrata* were most diverged from the European ones based on microsatellites (Van Treuren et al. 1997). DNA sequence studies of several nuclear loci showed that the subspecies *lyrata* was highly diverged from the European populations, but there were few fixed differences between the subspecies, suggesting rather recent divergence (Wright et al. 2003).

The Scandinavian and Icelandic populations are evidently derived from the Central European ones, with current high divergence between them at microsatellite loci (Iceland–Germany F_{ST} 0.25, Clauss and Mitchell-Olds 2006, Muller et al. 2008; Sweden and Norway vs. Germany 0.3–0.4, Muller et al. 2008). At the DNA sequence level, based on close to 80 different sequenced loci, the Icelandic and Swedish populations are diverged with an F_{ST} estimate of about 0.2 from Central Europe (Ross-Ibarra et al. 2008). In the same study, the F_{ST} estimates were highest between the populations of Russia and the USA and Canada (F_{ST} about 0.6) (Ross-Ibarra et al. 2008). The high F_{ST} estimates between these populations are partly due to the low level of variation within the individual populations (Charlesworth 1998).

Ross-Ibarra et al. (2008) have used their data to fit a simplified demographic model. In the model, all populations originate from the Central European German population. This was motivated because the pairwise divergence with German and the other populations was lower than between other pairs, and there were many shared variants between the other populations and the German one. The estimates for the pairwise divergence times from the German population were recent, less than 50,000 years. These estimates suggest that genetic drift may have produced the divergence at the neutral level very rapidly. The population history is likely to be considerably more complex, and especially the North American population history is not yet well understood (Schmickl et al. 2008). This kind of model including demography provides an important starting point for analyzing selection (see Section 12.8).

Wright et al. (2008) found much higher divergence in chloroplast DNA than at nuclear loci, most variation was between populations, with F_{ST} 0.85. The North American, Swedish, and Russian populations had no variation within populations. The patterns of variation compared to nuclear loci suggest that much of this variation is neutral or else influenced by very similar rapid selective episodes that may occur at nuclear loci. The lower diversity and higher differentiation are consistent with the lower effective size and a low mutation rate of the organelle genomes (about 10% of the nuclear genome mutation rate).

The high divergence level and complex demographic history set some limits on how the genetic divergence data are used for detecting local adaptation. A higher differentiation between populations at a specific locus than the neutral genome-wide differentiation can support the role of the locus in local adaptation (Beaumont and Nichols 1996, Beaumont and Balding 2004). The background divergence of *A. lyrata* is so high that it can be difficult to detect the adaptive divergence from the overall high divergence.

An analysis for selection in the total set of highly differentiated populations was conservative (for several statistical reasons) (Ross-Ibarra et al. 2008). However, in a more geographically restricted setting, with lower genome-wide differentiation, it may be easier to apply this kind of approach. The very high differentiation of *A. lyrata* populations also implies that the populations cannot easily be combined in association analyses. When the phenotypes are genetically differentiated between populations, spurious associations may arise (Yu et al. 2006). When these are corrected, the correction may also remove the genetic differences due to local adaptation. Thus, other methods need to be used to analyze the between-population differences.

12.8 Genetics of Local Adaptation

A. lyrata occurs in scattered populations in open habitats along a wide geographical range. The environments occupied vary from dolomitic or gypsum rock outcrops in Central Europe to lake shores, river banks, sand dunes, gravel, cliffs, and sometimes serpentine in North America and North Europe (Ericson and Mascher 1978, Schierup 1998, Kärkkäinen et al. 1999, Clauss and Koch 2006). Although *A. lyrata* is intolerant to competition it tolerates well abiotic stress like salt spray in coastal areas, serpentine soil, flood at river banks, or summer drought. The species is found in a range of different altitudes, from coastal lowland to high mountains. The habitats can span a wide range of altitudes even with a small geographic area, as in closely situated Norwegian lowland and mountain populations. Consequently, there are large climatic differences among the sites, e.g., in winter and summer temperatures (Kunin et al. 2009).

The populations are relatively stable and often large due to the perenniality and outcrossing mode of reproduction. The current populations likely have a long history of geographical isolation (Muller et al. 2008). These life history features and occurrence of *A. lyrata* populations in ecologically variable, isolated habitats are expected to provide opportunities for local adaptation. In each habitat traits evolve in the direction that confers high fitness in that habitat, leading to diversifying selection in heterogenous environments in the absence of strong gene flow. With the possibility to utilize molecular and functional information from the close relative *A. thaliana*, *A. lyrata* has become a promising study object for studies of functional and genetic aspects of adaptation.

There is ample variation in various traits that may confer adaptation to the ecologically variable environments both within and among populations. In *A. lyrata*, among most intensively studied traits of potential ecological significance are those relating to plant defence (Kärkkäinen and Ågren 2002, Kärkkäinen et al. 2004, Clauss et al. 2006, Kivimäki et al. 2007, Loe et al. 2007, Jorgensen and Emerson 2009), life history traits (Riihimäki and Savolainen 2004, Riihimäki et al. 2005, Kuittinen et al. 2008), and reactions to soil conditions (Vergeer et al. 2008). The differences have often been quantified in common garden experiments to demonstrate a genetic component in differentiation. Many of the genetic differences in ecologically important traits between populations may be due to adaptation to local conditions. However, some changes may be due to other processes, like pleiotropic effects of underlying genes or drift associated with possible bottlenecks during colonization (as explained above). The most direct evidence for local adaptation comes from reciprocal transplant experiments, where the local populations should have higher survival and fecundity in their own environments compared with populations from other sites (Kawecki and Ebert 2004). This kind of evidence for adaptive differentiation has been obtained for populations originating from Germany, Norway, and Sweden where differences in several life history characters contributed to fitness differences among populations (Leinonen et al. 2009). It was suggested that divergence in these traits were adaptations to the local conditions of early plant establishment; e.g., differences in winter survival could be due to differential cold tolerance, mediated by differences in cessation of growth and development of cold tolerance in response to photoperiod (Leinonen et al. 2009).

The focus of research on adaptive variation has been in evolutionary genetics. One major question is the genetic architecture of adaptive variation (Orr 1998). Adaptations are sometimes under a simple genetic control, e.g., governed by polymorphisms at individual genes. However, most adaptive traits are likely to be quantitative, i.e., they show more or less continuous variation and are controlled by many genes together with environment. In this case, it is of interest to know the distribution of individual locus effects, interactions between loci, interactions between genes and environment, and the extent of pleiotropy. The nature of actual mutations – whether they are regulatory or coding mutations – is a question to be answered (Wray 2007). It is also interesting to which extent variation in orthologous genes is responsible for similar adaptations in closely related species (Hoekstra and Coyne 2007).

Traits relating to plant defence, tolerance, or resistance against pathogens and insects have been the focus of much research in *A. lyrata*. A long-lived perennial plant encounters a variety of different pathogens and insects and is likely to have developed a variety of defence traits. These include trichomes on leaves and other parts of the plant, production of secondary metabolites like glucosinolates, and detection of microbial signals and activation of plant resistance by resistance proteins.

Trichome density on the leaf surface is related to defence against insects, UV light, and drought (Dalin et al. 2009). The protective effect of trichomes against herbivore damage was demonstrated in Swedish and Norwegian *A. lyrata* populations (Kivimäki et al. 2007, Loe et al. 2007). There is variation in the ability to produce trichomes not only between *A. lyrata* populations but also between individuals within populations (Kärkkäinen et al. 2004). In a Swedish population polymorphism at one gene determined the presence or absence of trichomes. Glabrousness was recessive (Kärkkäinen and Ågren 2002). The lack of trichomes was later associated with coding region mutations changing the GL1 protein (Kivimäki et al. 2007). In a Russian population of *A. lyrata,* glabrousness was associated with another mutation in the same gene *GL1* (Hauser et al. 2001), showing that glabrousness has evolved independently in lineages leading to the Russian and Swedish *A. lyrata* populations, based on different mutations in the same gene. The gene causing glabrousness was more differentiated among populations that were neutral isozyme loci, suggesting that hairiness was under differential selection in Swedish *A. lyrata* populations (Kärkkäinen et al. 2004).

Glucosinolates are secondary metabolites showing structural variation between and within Brassicaceous species (Windsor et al. 2005). Insect resistance was influenced by the level of various glucosinolates in *A. lyrata* ssp. *petraea* (Clauss et al.

2006). Glucosinolates were suggested to be responsible for discriminating *A. lyrata* ssp. *petraea* populations in the analysis of metabolomic fingerprints by PCA (Davey et al. 2008).

Glucosinolate content in *A. lyrata* is a continuously varying, complex trait, potentially affected by many genes. Genetic architecture of glucosinolate variation was studied by QTL mapping in a cross between two European populations, and it was found to be governed by a few major QTL (Heidel et al. 2006). One of the QTL co-localized with a candidate glucosinolate biosynthetic locus, *MAM*. Significant association between polymorphism at *MAM* with variation in glucosinolate phenotype within a natural German population suggested that this gene may control the glucosinolate phenotype (Heidel et al. 2006). This study showed that the same gene was responsible for both within- and between- population variation.

Plant disease resistance genes (*R*-genes) are important in pathogen recognition and defence. Natural variation in resistance in *A. thaliana* populations is known to be accounted for by several *R*-genes, and selection contributes to the maintenance of *R*-gene variation (Bakker et al. 2006b). Variation in the life history and population dynamics between selfing *A. thaliana* and outcrossing *A. lyrata* may cause differences in the evolution of *R*-genes. Jorgensen and Emerson (2009) studied sequence variation in *A. lyrata* populations in an orthologue of the *RPW8* locus known to be mainly responsible for variation in powdery mildew resistance in *A. thaliana*. They found that haplotypes coding truncated RPW8 proteins due to many different loss-of-function mutations were frequent in all studied populations, and these were significantly associated with susceptibility to powdery mildew. Neutrality tests found evidence for recent selective sweeps within some of the populations. The results would fit with a model of conditional neutrality where loss-of-function mutations accumulate in the absence of pathogen, and the functional allele is selected and persists in populations in response to infrequent infections (Jorgensen and Emerson 2009).

Sequence variation in strong candidate genes in natural populations of *A. lyrata* can be analyzed to infer whether there are signs of positive directional, balancing, or divergent selection, even without phenotypic information (Clauss and Mitchell-Olds 2003). Ross-Ibarra et al. (2008) scanned genome-wide sequence divergence between populations using a set of 80 genes. A NBS-LRR gene, a candidate for disease resistance, showed an excess of population differentiation compared with other genes, which was indicative of positive selection due to local adaptation. A more detailed study focusing on sequence variation in this and another neighboring NBS-LRR gene showed, however, no evidence for divergence between populations due to positive selection or balancing selection within populations. It was suggested that conditional neutrality model may explain the elevated level of truncated alleles (Gos and Wright 2008).

Another group of potentially adaptive traits of interest are those relating to life history. Populations are likely to have different optimal timing of various developmental transitions as they encounter different climatic conditions. In contrast to *A. thaliana*, *A. lyrata* is perennial and reproduces also vegetatively by ramets; thus it has the possibility to choose whether it invests in sexual reproduction or growth in any growth season. Common garden studies designed to reveal variation in response to environmental cues to induce flowering have revealed genetic differences between European populations (Riihimäki and Savolainen 2004, Riihimäki et al. 2005, Kuittinen et al. 2008). Southern populations had higher probability to flower, and they flowered earlier than northern populations. In experiments with short (14 h) and long (20 h) days and different vernalization treatments, the ability to flower was promoted by long days and vernalization in northern populations but not in southern populations. Long photoperiod and vernalization reduced the time to flowering among those plants that flowered, more in northern than in southern populations (Riihimäki and Savolainen 2004, Riihimäki et al. 2005, Kuittinen et al. 2008). Variation was thus due to differences in photoperiodic responses and requirements for vernalization among populations. This kind of latitudinal variation is indicative of adaptation to climatic variation. More direct evidence for adaptive significance of flowering was demonstrated by Sandring et al. (2007), as early flowering was found to confer higher fitness in an alpine Norwegian population in 1 year. Selection for optimal floral display and flowering time may be mediated by pollinators, in addition to climatic factors (Sandring and Agren 2009). The *FRIGIDA* gene has a major role in determining flowering time in *A. thaliana* ecotypes (Le Corre et al. 2002, Gazzani et al. 2003, Le Corre 2005). To find out whether the same gene was responsible for flowering time differences in *A. lyrata*, Kuittinen et al. (2008) studied sequence variation in *FRIGIDA* in three European populations. An in-frame insertion–deletion polymorphism in the coding region was polymorphic and associated with flowering time variation within two northern populations, suggesting that it (or a linked polymorphism) was involved in flowering time variation (Kuittinen et al. 2008). However, since the frequency of the variant conferring late flowering was high in southern populations, this gene did not explain the flowering time differences between *A. lyrata* populations, and other loci must be involved.

Traits related to adaptation to abiotic environment, such as soil conditions, are a third group of traits that are under active research. These include tolerance to serpentine soil and response to nutrients. Ecological adaption to serpentine has been studied in many plant species. Low calcium to magnesium ratio, high heavy metal content, and low nutrient and moisture content make serpentine soils a high stress environment for a plant. Some *A. lyrata* populations can be found in serpentine soil. By hybridizing genomic DNA from *A. lyrata* to *A. thaliana* Affymetrix tiling arrays, Turner et al. (2008) were able to identify genes with high differentiation between closely situated *A. lyrata* populations occurring either in granitic or in serpentine soil. Some of these genes may be involved in adaptation to serpentine, but verification requires further work (Turner et al. 2008). Although some populations can be found on serpentine soil, *A. lyrata* generally tolerates poor soils containing heavy metals. A close relative, *A. halleri*, instead has constitutive heavy metal tolerance and associated hyperaccumulation. *A. halleri* has some variation between populations in tolerance to Zn, but there are no Cd-sensitive plants. To circumvent the lack of useful variation within species, a cross between these two species was successfully used to map major QTLs for Zn and Cd tolerance and accumulation (Courbot et al. 2007, Willems et al. 2007).

Vergeer et al. (2008) studied the response of *A. lyrata* plants to different nitrogen levels. Populations originating from regions with naturally low nitrogen deposition levels grew faster and had higher leaf turnover rates and shorter times to flowering, and their growth rate responded more strongly to added nitrogen than populations originating from high deposition areas. This indicated adaptation to local nitrogen levels and a loss of plastic response to low nitrogen availability in areas with high current rates of natural nitrogen deposition (Vergeer et al. 2008).

12.9 Perspectives for *A. lyrata* for Functional and Population Genomics

A. lyrata has been of interest to geneticists only for about 10 years, and in this short time it has become a promising model species for ecological and population genetics. It is among the first plant species to be fully sequenced. Its role can increase now that the full sequence of the species is available. Furthermore, the sequence of another relative, *C. rubella*, will also soon be available by the JGI, and other relatives, such as *A. halleri*, *A. arenosa*, and the more distant salt tolerant *Thellungiella*, and the selfing *Boechera* are also likely to be sequenced soon.

The whole genome sequences of several sets of closely related species have proved to be a valuable resource for both evolutionary and functional studies. Detailed genetic comparisons between closely related species have provided important insights for genome evolution in humans and other hominids or in studies of 12 species of sequenced Drosophila (Drosophila 12 Genomes Consortium 2007). The set of two sequenced *Arabidopsis* species and *Capsella* will allow determining which evolutionary changes are novelties in each of the *A. thaliana* and *A. lyrata* lineages, and what are ancestral states. Comparisons of the evolutionary consequences of the selfing and outcrossing mating system will continue to be an important part of these studies, e.g., involving the nature of selection in selfing and outcrossing populations or on features of molecular evolution (Charlesworth and Wright 2001). The details of the loss of self-incompatibility both in this species and in *A. thaliana* can be examined in more depth (Mable et al. 2005, Tang et al. 2007).

Aside from the full genome sequence, the *Arabidopis lyrata* genetic resources are so far quite limited. There are no publicly available inbred lines or sets of recombinant inbred lines. Publicly available population samples are also quite limited, compared, e.g., to *A. thaliana*. This situation can improve rapidly, as studies are extended to new populations, and more genetic resources, such as inbred lines, can be generated.

A. lyrata represents the large majority of flowering plants that are outcrossing. The population and ecological genetics will be of special interest, because the extensive genomic and functional information on *A. thaliana* can be applied in related species. Combining these advanced tools with the field studies will also allow researchers to examine the fitness consequences of the variants.

Other closely related species will also be of immediate interest to functional genomics. So far many studies of *A. thaliana* relatives have been on *A. lyrata*, but many other relatives are also now being studied, such as *Arabidopsis suecica* and *A. arenosa* for polyploidy (Pontes et al. 2004), *Arabis alpina* for the basis of perenniality (Wang et al. 2009), *A. halleri* for heavy metal tolerance (Willems et al. 2007). A more remote relative, *Thlaspi*, is also important for studies of heavy metal tolerance (Lombi et al. 2000). The genus *Boechera* has received attention for the genetics of apomixis (Schranz et al. 2005) and also for studies of drought tolerance (Knight et al. 2006)

Recent work on *A. lyrata* population genetics has uncovered patterns of diversity and demonstrated the complex demographic history of the species. For functional genomics, one advantage is that in the individual wild populations it will be possible to obtain association mapping samples with very low relatedness. This avoids complications of association mapping in highly related breeding populations of cultivated plants. Further, selection can be efficiently analyzed in some close to equilibrium populations. The high differentiation of populations sets some limitations on the use of sets of these populations in searching for the traces of selection. It will be critical that the population structure is taken into account when analyzing sequence variation and in association studies.

Acknowledgments We acknowledge the financial support of the Biocenter, University of Oulu, the Biosciences and Environment Research Council of Finland, the ERA-Net Plant Genomics Program ARelatives for financial support of our research, and our Oulu University plant genetics group members and David Remington for comments on the manuscript.

References

Abel C, Clauss M, Schaub A, Gershenzon J, Tholl D (2009) Floral and insect-induced volatile formation in *Arabidopsis lyrata* ssp *petraea*, a perennial, outcrossing relative of *A. thaliana*. Planta 230:1–11

Awadalla P, Charlesworth D (1999) Recombination and selection at Brassica self-incompatibility loci. Genetics 152:413–425

Bakker EG, Stahl EA, Toomajian C, Nordborg M, Kreitman M et al (2006a) Distribution of genetic variation within and among local populations of Arabidopsis thaliana over its species range. Mol Ecol 15:1405–1418

Bakker EG, Toomajian C, Kreitman M, Bergelson J (2006b) A genome-wide survey of R gene polymorphisms in Arabidopsis. Plant Cell 18:1803–1818

Balana-Alcaide D, Ramos-Onsins SE, Boone Q, Aguade M (2006) Highly structured nucleotide variation within and among *Arabidopsis lyrata* populations at the FAH1 and DFR gene regions. Mol Ecol 15:2059–2068

Barrier M, Bustamante CD, Yu J, Purugganan MD (2003) Selection on rapidly evolving proteins in the Arabidopsis genome. Genetics 163:723–733

Beaulieu J, Jean M, Belzile F (2009) The allotetraploid *Arabidopsis thaliana-Arabidopsis lyrata* subsp *petraea* as an alternative model system for the study of polyploidy in plants. Mol Genet Genomics 281:421–435

Beaumont MA, Balding DJ (2004) Identifying adaptive genetic divergence among populations from genome scans. Mol Ecol 13(4):969–980

Beaumont MA, Nichols RA (1996) Evaluating loci for use in the genetic analysis of population structure. Proc R Soc Lond 263:1619–1626

Berr A, Pecinka A, Meister A, Kreth G, Fuchs J et al (2006) Chromosome arrangement and nuclear architecture but not centromeric sequences are conserved between *Arabidopsis thaliana* and *Arabidopsis lyrata*. Plant J 48:771–783

Bustamante CD, Nielsen R, Sawyer SA, Olsen KM, Purugganan MD et al (2002) The cost of inbreeding in *Arabidopsis*. Nature 416:531–534

Castric V, Bechsgaard J, Schierup MH, Vekemans X (2008) Repeated adaptive introgression at a gene under multiallelic balancing selection. PLoS Genet 4:9

Charlesworth B (1998) Measures of divergence between populations and the effect of forces that reduce variability. Mol Biol Evol 15:538–543

Charlesworth D, Charlesworth B (1987) Inbreeding depression and its evolutionary consequences. Annu Rev Ecol Syst 18:237–268

Charlesworth B, Charlesworth D, Barton NH (2003a) The effects of geographic and genetic structure on neutral genetic variation. Annu Rev Ecol Syst 34:99–125

Charlesworth D, Mable BK, Schierup MH, Bartolome C, Awadalla P (2003b) Diversity and linkage of genes in the self-incompatibility gene family in *Arabidopsis lyrata*. Genetics 164: 1519–1535

Charlesworth D, Wright SI (2001) Breeding system and genome evolution. Curr Opin Genet Dev 11:685–690

Clauss MJ, Dietel S, Schubert G, Mitchell-Olds T (2006) Glucosinolate and trichome defenses in a natural *Arabidopsis lyrata* population. J Chem Ecol 32:2351–2373

Clauss MJ, Koch MA (2006) Poorly known relatives of *Arabidopsis thaliana*. Trends Plant Sci 11:449–459

Clauss MJ, Mitchell-Olds T (2003) Population genetics of tandem trypsin inhibitors in *Arabidopsis* species with contrasting ecology and life history. Mol Ecol 12:1287–1299

Clauss MJ, Mitchell-Olds T (2006) Population genetic structure of *Arabidopsis lyrata* in Europe. Mol Ecol 15:2753–2766

Comes HP, Kadereit JW (1998) The effect of Quaternary climatic changes on plant distribution and evolution. Trends Plant Sci 3:432–438

Courbot M, Willems G, Motte P, Arvidsson S, Roosens N et al (2007) A major quantitative trait locus for cadmium tolerance in *Arabidopsis halleri* colocalizes with HMA4, a gene encoding a heavy metal ATPase. Plant Physiol 144:1052–1065

Dalin P, Ågren J, Björkman C, Huttunen P, Kärkkäinen K (2009) Leaf trichome formation and plant resistance to herbivory. In: Schaller A (ed) Induced plant resistance to herbivory. Springer, Heidelberg

Davey MP, Burrell MM, Woodward FI, Quick WP (2008) Population-specific metabolic phenotypes of *Arabidopsis lyrata* ssp. *petraea*. New Phytol 177:380–388

Drosophila 12 Genomes Consortium (2007) Evolution of genes and genomes on the Drosophila phylogeny. Nature 450:203–218

Ericson L, Mascher JW (1978) *Cardaminopsis petraea*, strandtrav, i Ångermanland. Svensk Botanisk Tidsrift 71:415–418

Excoffier L, Ray N (2008) Surfing during population expansions promotes genetic revolutions and structuration. Trends Ecol Evol 23:347–351

Foxe JP, Dar VUN, Zheng H, Nordborg M, Gaut BS et al (2008) Selection on amino acid substitutions in *Arabidopsis*. Mol Biol Evol 25:1375–1383

Gaudeul M, Stenoien HK, Agren J (2007) Landscape structure, clonal propagation, and genetic diversity in Scandinavian populations of *Arabidopsis lyrata* (Brassicaceae). Am J Bot 94: 1146–1155

Gazzani S, Gendall AR, Lister C, Dean C (2003) Analysis of the molecular basis of flowering time variation in *Arabidopsis* accessions. Plant Physiol 132:1107–1114

Gos G, Wright SI (2008) Conditional neutrality at two adjacent NBS-LRR disease resistance loci in natural populations of Arabidopsis lyrata. Mol Ecol 17:4953–4962

Hansson B, Kawabe A, Preuss S, Kuittinen H, Charlesworth D (2006) Comparative gene mapping in *Arabidopsis lyrata* chromosomes 1 and 2 and the corresponding *A. thaliana* chromosome 1: recombination rates, rearrangements and centromere location. Genet Res 87:75–85

Hauser MT, Harr B, Schlotterer C (2001) Trichome distribution in *Arabidopsis thaliana* and its close relative *Arabidopsis lyrata*: molecular analysis of the candidate gene GLABROUS1. Mol Biol Evol 18:1754–1763

Heidel AJ, Clauss MJ, Kroymann J, Savolainen O, Mitchell-Olds T (2006) Natural variation in MAM within and between populations of *Arabidopsis lyrata* determines glucosinolate phenotype. Genetics 173:1629–1636

Hein J, Schierup MH, Wiuf C (2005) Gene genealogies, variation and evolution. Oxford University Press, New York, NY

Hinds DA, Stuve LL, Nilsen GB, Halperin E, Eskin E et al (2005) Whole-genome patterns of common DNA variation in three human populations. Science 307:1072–1079

Hoekstra HE, Coyne JA (2007) The locus of evolution: evo devo and the genetics of adaptation. Evolution 61:995–1016

Hoffman MH (2005) Evolution of the realized niche in the genus *Arabidopsis* (Brassicaceae). Evolution 59:1425–1436

Hudson RR (1990) Gene genealogies and the coalescent process. In: Futuyma D, Antonovics J (eds) Oxford surveys in evolutionary biology, pp 1–44. Oxford University Press, New York, NY

Jalas J, Suominen J (1994) Atlas Florae Europaeae. Cruciferae (Sisymbrium to Aubrieta). Helsinki University Printing House, Helsinki

Jonsell B, Kustås K, Nordal I (1995) Genetic variation in *Arabis petraea*, a disjunct species in northern Europe. Ecography 18:321–332

Jorgensen TH, Emerson BC (2009) RPW8 and resistance to powdery mildew pathogens in natural populations of *Arabidopsis lyrata*. New Phytol 182:984–993

Kärkkäinen K, Ågren J (2002) Genetic basis of trichome production in *Arabidopsis lyrata*. Hereditas 136:219–226

Kärkkäinen K, Kuittinen H, van Treuren R, Vogl C, Oikarinen S et al (1999) Genetic basis of inbreeding depression in *Arabis petraea*. Evolution 53:1354–1365

Kärkkäinen K, Loe G, Ågren J (2004) Population structure of *Arabidopsis lyrata*: evidence for divergent selection on trichome production. Evolution 58:2831–2836

Kawabe A, Charlesworth D (2007a) Diversity patterns of the Medea gene in Arabidopsis lyrata. Genes Genet Syst 82:528–528

Kawabe A, Charlesworth D (2007b) Patterns of DNA variation among three centromere satellite families in *Arabidopsis halleri* and *lyrata*. J Mol Evol 64(2):237–247

Kawabe A, Forrest A, Wright SI, Charlesworth D (2008) High DNA sequence diversity in pericentromeric genes of the plant *Arabidopsis lyrata*. Genetics 179:985–995

Kawecki TJ, Ebert D (2004) Conceptual issues in local adaptation. Ecol Lett 7:1225–1241

Kim S, Plagnol V, Hu TT, Toomajian C, Clark RM et al (2007) Recombination and linkage disequilibrium in *Arabidopsis*. Nat Genet 39:1151–1155

Kivimäki M, Karkkainen K, Gaudeul M, Loe G, Agren J (2007) Gene, phenotype and function: GLABROUS1 and resistance to herbivory in natural populations of *Arabidopsis lyrata*. Mol Ecol 16:453–462

Knight CA, Vogel H, Kroymann J, Shumate A, Witsenboer H et al (2006) Expression profiling and local adaptation of Boechera holboellii populations for water use efficiency across a naturally occurring water stress gradient. Mol Ecol 15:1229–1237

Koch M, Bishop J, Mitchell-Olds T (1999) Molecular systematics and evolution of *Arabidopsis* and *Arabis*. Plant Biol 1:529–537

Koch MA, Haubold B, Mitchell-Olds T (2000) Comparative evolutionary analysis of chalcone synthase and alcohol dehydrogenase loci in *Arabidopsis*, *Arabis*, and related genera (Brassicaceae). Mol Biol Evol 17:1483–1498

Koch M, Haubold B, Mitchell-Olds T (2001) Molecular systematics of the Brassicaceae: evidence from coding platid *MatK* and nuclear *CHS*. Am J Bot 88:534–544

Koch MA, Kiefer M (2005) Genome evolution among cruciferous plants: a lecture from the comparison of the genetic maps of three diploid species – *Capsella rubella*, *Arabidopsis lyrata* subsp. *petraea*, and *A. thaliana*. Am J Bot 92:761–767

Koch MA, Matschinger M (2007) Evolution and genetic differentiation among relatives of *Arabidopsis thaliana*. Proc Natl Acad Sci USA 104:6272–6277

Koch MA, Wernisch M, Schmickl R (2008) *Arabidopsis thaliana's* wild relatives: an updated overview on systematics, taxonomy and evolution. Taxon 57:933–943

Kuittinen H, de Haan AA, Vogl C, Oikarinen S, Leppälä J et al (2004) Comparing the linkage maps of the close relatives *Arabidopsis lyrata* and *Arabidopsis thaliana.* Genetics 168:1575–1584

Kuittinen H, Niittyvuopio A, Rinne P, Savolainen O (2008) Natural variation in *Arabidopsis lyrata* vernalization requirement conferred by a FRIGIDA indel polymorphism. Mol Biol Evol 25:319–329

Kunin WE, Vergeer P, Kenta T, Davey MP, Burke T et al (2009) Variation at range margins across multiple spatial scales: environmental temperature, population genetics and metabolomic phenotype. Proc R Soc Lond B Biol Sci 276:1495–1506

Kusaba M, Dwyer K, Hendershot J, Vrebalov J, Nasrallah JB et al (2001) Self-incompatibility in the genus *Arabidopsis*: characterization of the S locus in the outcrossing *A. lyrata* and its autogamous relative *A. thaliana.* Plant Cell 13:627–643

Le Corre V (2005) Variation at two flowering time genes within and among populations of *Arabidopsis thaliana*: comparison with markers and traits. Mol Ecol 14:4181–4192

Le Corre V, Roux F, Reboud X (2002) DNA polymorphism at the FRIGIDA gene in *Arabidopsis thaliana*: extensive nonsynonymous variation is consistent with local selection for flowering time. Mol Biol Evol 19:1261–1271

Leinonen P, Sandring S, Quilot B, Clauss MJ, Mitchell-Olds T et al (2009) Local adaptation in European populations of *Arabidopsis lyrata* (Brassicaceae). Am J Bot 96:1129–1137

Lockton S, Ross-Lbarra J, Gaut BS (2008) Demography and weak selection drive patterns of transposable element diversity in natural populations of *Arabidopsis lyrata.* Proc Natl Acad Sci USA 105:13965–13970

Loe G, Torang P, Gaudeul M, Agren J (2007) Trichome production and spatiotemporal variation in herbivory in the perennial herb *Arabidopsis lyrata.* Oikos 116:134–142

Lombi E, Zhao FJ, Dunham SJ, McGrath SP (2000) Cadmium accumulation in populations of Thlaspi caerulescens and Thlaspi goesingense. New Phytol 145:11–20

Lundemo S, Savolainen O, Stenoien HK (2010) Investigating the effects of topography and clonality on genetic structuring within a large Norwegian population of *Arabidopsis lyrata.* Ann Bot 106:243–254

Lysak MA, Berr A, Pecinka A, Schmidt R, McBreen K et al (2006) Mechanisms of chromosome number reduction in Arabidopsis thaliana and related Brassicaceae species. Proc Natl Acad Sci USA 103:5224–5229

Mable BK (2004) Polyploidy and self-compatibility: is there an association? New Phytol 162: 803–811

Mable BK, Adam A (2007) Patterns of genetic diversity in outcrossing and selfing populations of *Arabidopsis lyrata.* Mol Ecol 16:3565–3580

Mable BK, Robertson AV, Dart S, Di Berardo C, Witham L (2005) Breakdown of self-incompatibility in the perennial *Arabidopsis lyrata* (Brassicaceae) and its genetic consequences. Evolution 59:1437–1448

Mable BK, Schierup MH, Charlesworth D (2003) Estimating the number, frequency, and dominance of S-alleles in a natural population of *Arabidopsis lyrata* (Brassicaceae) with sporophytic control of self-incompatibility. Heredity 90:422–431

Mitchell-Olds T (2001) *Arabidopsis thaliana* and its wild relatives: a model system for ecology and evolution. Trends Ecol Evol 16:693–700

Miyake T, Takebayashi N, Salter D, Wolf ED (2007) Natural selection on imprinted genes in outcrossing *Arabidopsis lyrata* and self-fertilizing *A. thaliana.* Genes Genet Syst 82: 543–543

Muller MH, Leppala J, Savolainen O (2008) Genome-wide effects of postglacial colonization in *Arabidopsis lyrata.* Heredity 100:47–58

Msiček J (1967) The chromosome morphology of *Arabidopsis thaliana* (L.) Heynh. and some remarks on the problem of *Hylandra suecica* (Fr.) Love. Folia Geobotanica Phytotaxonomica 2:433–436

Msiček J (1970) Chromosome counts in *Cardamine arenosa* agg (Cruciferae). Preslia 42:225–248
Nasrallah ME, Liu P, Nasrallah JB (2002) Generation of self-incompatible *Arabidopsis thaliana* by transfer of two S locus genes from A. lyrata. Science 297:247–249
Nasrallah ME, Yogeeswaran K, Snyder S, Nasrallah JB (2000) Arabidopsis species hybrids in the study of species differences and evolution of amphiploidy in plants. Plant Physiol 124: 1605–1614
Nordborg M, Borewitz JO, Bergelson J, Berry C, Chory J et al (2002) The extent of linkage disequilibrium in *Arabidopsis thaliana*. Nat Genet 30:190–193
Nordborg M, Hu TT, Ishino Y, Juahveri J, Toomajian C et al (2005) The pattern of polymorphism in *Arabidopsis thaliana*. PLoS Biol 3:1289–1299
Nordborg M, Weigel D (2008) Next-generation genetics in plants. Nature 456:720–723
O'Kane SL, Al-Shehbaz IA (1997) A synopsis of *Arabidopsis* (Brassicaceae). Novon 7:323–327
O'Kane SL, Al-Shehbaz IA (2003) Phylogenetic position and generic limits of *Arabidopsis* (Brassicaceae) based on sequences of nuclear ribosomal DNA. Ann Mo Bot Gard 90:603–612
Orr HA (1998) The population genetics of adaptation: the distribution of factors fixed during adaptive evolution. Evolution 52:935–949
Peer WA, Murphy AS (2003) Floral scent of *Arabidopsis lyrata* (Brassicaceae). Biochem Syst Ecol 31:1193–1195
Polatchek A (1966) Cytotaxonomische Beiträge zur Flora der Ostalpenländer I. Österreichische Botanische Zeitschrift 113:1–44
Pontes O, Neves N, Silva M, Lewis MS, Madlung A et al (2004) Chromosomal locus rearrangements are a rapid response to formation of the allotetraploid Arabidopsis suecica genome. PNAS 101:18240–18245
Ramos-Onsins SE, Stranger BE, Mitchell-Olds T, Aguadé M (2004) Multilocus analysis of variation and speciation in the closely related species *Arabidopsis halleri* and *A. lyrata*. Genetics 166:373–388
Redei GP (1974) Is *Hylandra* an amphidiploid of *Arabidopsis* and *Cardaminopsis arenosa*? Arabidopsis Inf Serv 11:5
Riihimäki M, Kuittinen H, Podolsky RH, Koelewijn H, Savolainen O (2005) Studying genetics of adaptive variation in model organisms: flowering time variation in *Arabidopsis lyrata*. Genetica 123:63–74
Riihimäki M, Savolainen O (2004) Environmental and genetic effects on flowering differences between northern and southern populations of *Arabidopsis lyrata* (Brassicaceae). Am J Bot 91:1036–1045
Ross-Ibarra J, Wright SI, Foxe JP, Kawabe A, DeRose-Wilson L et al (2008) Patterns of polymorphism and demographic history in natural populations of *Arabidopsis lyrata*. PLoS One 3:e2411
Sandring S, Agren J (2009) Pollinator-mediated selection on floral display and flowering time in the perennial herb *Arabidopsis lyrata*. Evolution 63:1292–1300
Sandring S, Riihimäki MA, Savolainen O, Ågren J (2007) Selection on flowering time and floral display in an alpine and a lowland population of Arabidopsis lyrata. J Evol Biol 20:558–567
Savolainen O, Langley CH, Lazzaro B, Freville H (2000) Contrasting patterns of nucleotide variation at the alcohol dehydrogenase locus in the outcrossing *Arabidopsis lyrata* and the selfing *Arabidopsis thaliana*. Mol Biol Evol 17:645–655
Schierup MH (1998) The effect of enzyme heterozygosity on growth in a strictly outcrossing species, the self-incompatible *Arabis petraea* (Brassicaceae). Hereditas 128:21–31
Schierup MH, Bechsgaard JS, Christiansen FB (2008) Selection at work in self-incompatible *Arabidopsis lyrata*. II. Spatial distribution of S haplotypes in Iceland. Genetics 180:1051–1059
Schierup MH, Bechsgaard JS, Nielsen LH, Christiansen FB (2006) Selection at work in self-incompatible *Arabidopsis lyrata*: mating patterns in a natural population. Genetics 172: 477–484
Schierup MH, Mable BK, Awadalla P, Charlesworth D (2001) Identification and characterization of a polymorphic receptor kinase gene linked to the self-incompatibility locus of *Arabidopsis lyrata*. Genetics 158:387–399

Schmickl R, Jorgensen MH, Brysting AK, Koch MA (2008) Phylogeographic implications for the North American boreal-arctic Arabidopsis lyrata complex. Plant Ecol Divers 1:245–254
Schmid KJ, Ramos-Onsins S, Ringys-Beckstein H, Weisshaar B, Mitchell-Olds T (2005) A multi-locus sequence survey in Arabidopsis thaliana reveals a genome-wide departure from a neutral model of DNA sequence polymorphism. Genetics 169:1601–1615
Schoen DJ, Brown AHD (1991) Intraspecific variation in population gene diversity and effective population size correlates with the mating system in plants. Proc Natl Acad Sci USA 88: 4494–4497
Schranz ME, Dobes C, Koch MA, Mitchell-Olds T (2005) Sexual reproduction, hybridization, apomixis, and polyploidization in the genus Boechera (Brassicaceae). Am J Bot 92: 1797–1810
Schranz ME, Lysak MA, Mitchell-Olds T (2006) The ABC's of comparative genomics in the Brassicaceae: building blocks of cruciefer genomes. Trends Plant Sci 11:535–542
Shimizu-Inatsugi R, Lihová J, Iwanaga H, Kudoh H, Marhold K et al (2009) The allopolyploid *Arabidopsis kamchatica* originated from multiple individuals of *A. lyrata* and *A. halleri*. Mol Ecol 18(19):4024–4048
Shimizu KK, Fujii S, Marhold K, Watanabe K, Kudoh H (2005) *Arabidopsis kamchatica* (Fish. ex EC) K. Shimizu Kudoh and *A. kamchatica* subsp. *kawasakiana* (Makino) K. Shimizu Kudoh, new combinations. Acta Phytotox Geobot 56:165–174
Shimizu KK, Shimizu-Inatsugi R, Tsuchimatsu T, Purugganan MD (2008) Independent origins of self-compatibility in Arabidopsis thaliana. Mol Ecol 17:704–714
Slatkin M (1999) Disequilibrium mapping of a quantitative-trait locus in an expanding population. Am J Hum Genet 64:1765–1773
Spillane C, Schmid KJ, Laoueille-Duprat S, Pien S, Escobar-Restrepo J-M et al (2007) Positive darwinian selection at the imprinted MEDEA locus in plants. Nature 448:349–352
Stenøien HK, Fenster CB, Tonteri A, Savolainen O (2005) Genetic variability in natural populations of Arabidopsis thaliana in northern Europe. Mol Ecol 14:137–148
Tang CL, Toomajian C, Sherman-Broyles S, Plagnol V, Guo YL et al (2007) The evolution of selfing in *Arabidopsis thaliana*. Science 317:1070–1072
Turner TL, von Wettberg EJ, Nuzhdin SV (2008) Genomic analysis of differentiation between soil types reveals candidate genes for local adaptation in *Arabidopsis lyrata*. PLoS One 3:e3183
Van Treuren R, Kuittinen H, Kärkkäinen K, Baena-Gonzalez E, Savolainen O (1997) Evolution of microsatellites in *Arabis petraea* and *A. lyrata*, outcrossing relatives of *Arabidopsis thaliana*. Mol Biol Evol 14:220–229
Vergeer P, van den Berg LLJ, Bulling MT, Ashmore MR, Kunin WE (2008) Geographical variation in the response to nitrogen deposition in *Arabidopsis lyrata petraea*. New Phytol 179:129–141
Wang RH, Farrona S, Vincent C, Joecker A, Schoof H et al (2009) PEP1 regulates perennial flowering in *Arabis alpina*. Nature 459:423–427
Wilkins JF, Haig D (2003) What good is genomic imprinting: the function of parent-specific gene expression. Nat Rev Genet 4:359–368
Willems G, Drager DB, Courbot M, Gode C, Verbruggen N et al (2007) The genetic basis of zinc tolerance in the metallophyte Arabidopsis halleri ssp halleri (Brassicaceae): an analysis of quantitative trait loci. Genetics 176:659–674
Windsor AJ, Reichelt M, Figuth A, Svatos A, Kroymann J et al (2005) Geographic and evolutionary diversification of glucosinolates among near relatives of *Arabidopsis thaliana* (Brassicaceae). Phytochemistry 66:1321–1333
Wray GA (2007) The evolutionary significance of cis-regulatory mutations. Nat Rev Genet 8: 206–216
Wright S (1943) Isolation-by-distance. Genetics 28:114–138
Wright S (1951) The genetical structure of populations. Ann Eugen 15:323–354
Wright SI, Bi IV, Schroeder SG, Yamasaki M, Doebley JF et al (2005) The effects of artificial selection on the maize genome. Science 308:1310–1314

Wright SI, Foxe JP, DeRose-Wilson L, Kawabe A, Looseley M et al (2006) Testing for effects of recombination rate on nucleotide diversity in natural populations of *Arabidopsis lyrata*. Genetics 174:1421–1430

Wright SI, Lauga B, Charlesworth D (2002) Rates and patterns of molecular evolution in inbred and outbred *Arabidopsis*. Mol Biol Evol 19:1407–1420

Wright S, Lauga B, Charlesworth D (2003) Subdivision and haplotype structure in natural populations of *Arabidopsis lyrata*. Mol Ecol 12:1247–1263

Wright SI, Le QH, Schoen DJ, Bureau TE (2001) Population dynamics of an Ac-like transposable element in self- and cross-pollinating arabidopsis. Genetics 158:1279–1288

Wright SI, Nano N, Foxe JP, Dar VU (2008) Effective population size and tests of neutrality at cytoplasmic genes in Arabidopsis. Genet Res 90:119–128

Wright SI, Yau CBK, Looseley M, Meyers BC (2004) Effects of gene expression on molecular evolution in *Arabidopsis thaliana* and *Arabidopsis lyrata*. Mol Biol Evol 21:1719–1726

Yogeeswaran K, Frary A, York TL, Amenta A, Lesser AH et al (2005) Comparative genome analyses of Arabidopsis spp.: inferring chromosomal rearrangement events in the evolutionary history of *A. thaliana*. Genome Res 15:505–515

Yu JM, Pressoir G, Briggs WH, Bi IV, Yamasaki M et al (2006) A unified mixed-model method for association mapping that accounts for multiple levels of relatedness. Nat Genet 38:203–208

Zhao K, Aranzana MJ, Sung K, Lister C, Shindo C et al (2007) An *Arabidopsis* example of association mapping in structured samples. PLoS Genet 3:e4

Zhu C, Gore M, Buckler ES, Yu J (2008) Status and prospects of association mapping in plants. Plant Genome 1:5–20

Chapter 13
The Genetics of *Capsella*

Günter Theißen

Abstract The genus *Capsella* comprises three well-defined species, *C. grandiflora*, *C. rubella* and *C. bursa-pastoris* (the widely known Shepherd's purse). Together with some intraspecific varieties, however, these species show remarkable differences in ploidy level, breeding systems, habitat range, and some morphological features such as leaf structure and floral architecture. Similar differences are also found in many other plant groups, thus understanding their genetic bases is of great interest for developmental and evolutionary biologists alike. Since *Capsella* is closely related to the major model plant *Arabidopsis thaliana* numerous molecular genetic tools are available for respective studies, and for one species (*C. rubella*) even sequencing of the whole genome is currently being pursued. Therefore, we currently see a renaissance of research interest in *Capsella* addressing some key issues of developmental and evolutionary biology such as speciation, adaptation, the developmental genetic basis of plant form, and whether evolution can proceed in a saltational way, that is generate profound changes within just a few or even only one generation of organisms. Some highlights of recent investigations are briefly presented.

Keywords Capsella · Disomic inheritance · Genome evolution · Flowering time · Fruit structure · Homeotic mutant · Leaf development · Saltational evolution · Self-incompatibility · Shepherd's purse · Tetraploidy

Abbreviations

A.	*Arabidopsis*
C.	*Capsella*
SC	Self-compatibility
SCR	S-locus cysteine-rich protein
SI	Self-incompatibility
SRK	S-locus receptor kinase

G. Theißen (✉)
Department of Genetics, Friedrich Schiller University Jena, Jena D-07743, Germany
e-mail: guenter.theissen@uni-jena.de

R. Schmidt, I. Bancroft (eds.), *Genetics and Genomics of the Brassicaceae*, Plant Genetics and Genomics: Crops and Models 9, DOI 10.1007/978-1-4419-7118-0_13,

Contents

13.1 Introduction 374
13.2 Speciation 375
13.2.1 On the Ancestry of *C. grandiflora* 375
13.2.2 On the Origin of *C. rubella* 376
13.2.3 On the Origin of *C. bursa-pastoris* 376
13.3 Genome and Chromosome Evolution 377
13.4 Evolution and Development of Phenotypic Traits 378
13.4.1 Leaf Development 378
13.4.2 Flowering Time 378
13.4.3 Floral Structure and Function 379
13.4.4 Fruit Structure 383
13.5 Outlook 384
References 384

13.1 Introduction

Capsella is a small genus within the mustard family (Brassicaceae) and is closely related to *Arabidopsis*; the lineages that led to the two genera separated less than 8 million years ago (German et al. 2009, Franzke et al. 2009). Quite a number of *Capsella* species have been described in the past, but these are now mainly of historical interest. Current taxonomy distinguishes just three well-defined species, *Capsella grandiflora* (Fauché & Chaub.) Boiss., *Capsella rubella* Reut. and *Capsella bursa-pastoris* (L.) Medik., which, however, show remarkable differences in habitat range, ploidy level, and breeding systems (Hurka and Neuffer 1997, Zunk et al. 1999, Hurka et al. 2005). While *C. grandiflora* and *C. rubella* are diploid ($2n = 2x = 16$), the well-known weed *C. bursa-pastoris* (Shepherd's purse) is a tetraploid species with disomic inheritance ($2n = 4x = 32$). *C. grandiflora* is obligatory outbreeding due to a sporophytic self-incompatibility (SI) system and grows in a quite restricted habitat in western Greece, Albania and Northern Italy. It has relatively large, fragrant flowers with showy petals to attract pollinators. In contrast, *C. rubella* shows a high rate of self-fertilization and has much smaller flowers. This species grows around the Mediterranean Sea, but has also expanded into Middle Europe, Northern Africa, Australia and the Americas, often by following European settlers (Hurka et al. 1989, Hurka and Neuffer 1997, Paetsch et al. 2006). The predominantly selfing *C. bursa-pastoris* is one of the most widely distributed flowering plants on earth (Hurka et al. 2003). One typically finds it in disturbed, "man-made" habitats, like arable soils, the margins of agricultural fields or ruderal habitats all over the world, except in the hot and humid tropics (Hurka and Neuffer 1997).

The species-specific differences found within the genus *Capsella* exist also in quite a number of other plant groups, including close relatives of *Capsella* such as *Arabidopsis* (Barrett 2002, Mable et al. 2005). They are the results of evolutionary transitions that are of great biological interest. A shift from outcrossing to self-pollination, for example, is one of the most prevalent evolutionary changes in flowering plants (Stebbins 1950, Shimizu et al. 2004). Within the small genus

Capsella we can analyze a whole spectrum of transitions leading from a diploid, self-incompatible, obligatory outcrossing species with comparatively large and attractive flowers but a quite restricted area of distribution (*C. grandiflora*), to a tetraploid, self-compatible, predominantly selfing species with relatively small flowers but a colonization success almost all around the globe (*C. bursa-pastoris*).

Moreover, in case of *C. bursa-pastoris* a homeotic variety (i.e., a variety in which proper organs develop at a position where usually organs of another identity develop) occurs in quite stable populations in the wild-a rare and remarkable phenomenon which is potentially of great evolutionary importance (Nutt et al. 2006, Hintz et al. 2006). Populations of the homeotic variety have been established at several places in Europe and provide an excellent opportunity to investigate the origin and performance of a floral homeotic mutant in the wild, with potential implications for a better understanding of sympatric speciation (i.e. speciation without prior geographic separation) and evolutionary novelties.

Thus the genus *Capsella* has much to offer in terms of interesting traits to students of development, microevolution (intraspecific evolution and speciation) and macroevolution (evolution above the species level, especially events that generate evolutionary novelties or key innovations). It is also a quite tractable system; due to the close relationship between *Capsella* and *Arabidopsis* numerous genetic and molecular tools developed for the one can be utilized also for the other and many more are being developed.

After the closely related *Arabidopsis thaliana* had been selected by the late 1980s as one of the major model systems to study plant features by molecular biology, it rapidly outperformed all other plants also on the field of genetics, even including traditional "supermodels" ranging, e.g., from Mendel's pea (*Pisum sativum*) to snapdragon (*Antirrhinum majus*) and maize (*Zea mays*). No wonder, therefore, that it is almost forgotten that also members of the genus *Capsella* served as important models during early days of traditional genetics. For instance, not long after the rediscovery of Mendel's laws, *C. bursa-pastoris* provided an early example of deviation from "Mendelian segregation ratios" when following the inheritance of a morphological character (fruits shape), thus revealing a polygenetic effect (see Section 13.4.4). Likewise, *C. grandiflora* was used as an early pioneer in studies on self-incompatibility (termed "self-sterility" at the time) (Riley 1936), now a major research topic within *Capsella* and far beyond (see Section 13.4.3.3).

As we will see, *Capsella* genetics has come a long way since these pioneering days.

13.2 Speciation

13.2.1 On the Ancestry of C. grandiflora

Even though the genus *Capsella* is only very small, and the relationships between its three well-defined species have been under investigation for quite some time, the phylogeny of the genus could be clarified only very recently employing molecular data, and is still not fully resolved. In any case, comparisons with closely related species from other genera strongly suggests that *C. grandiflora* represents the most ancestral character states concerning reproductive biology and ploidy level (Hurka et al. 2005).

13.2.2 On the Origin of C. rubella

Comparative sequence analysis of orthologous genes recently confirmed that *C. rubella* originated from *C. grandiflora* and that both species are very closely related (Foxe et al. 2009, Guo et al. 2009). The available data and analyses suggest that *C. rubella* is monophyletic, i.e. originated from *C. grandiflora* only once. On evolutionary timescales both lineages separated very recently; in two different studies estimates range from less than 25,000 years ago to 30,000–50,000 years ago (Foxe et al. 2009, Guo et al. 2009). A low degree of genetic diversity indicates that *C. rubella* went through an extreme population bottleneck, and may have originated by a single, selfing individual of the ancestral *C. grandiflora*, probably living in Greece (Guo et al. 2009).

13.2.3 On the Origin of C. bursa-pastoris

The endeavour to understand the origin of *C. bursa-pastoris* started earlier and was therefore less straight forward than attempts to understand the origin of *C. rubella*. Pioneering isozyme electrophoresis and isoelectric focusing experiments of Rubisco suggested that *C. bursa-pastoris* is an allopolyploid of *C. rubella* and *C. grandiflora* (Hurka et al. 1989, Mummenhoff and Hurka 1990). Additional data from restriction enzyme site variation in the chloroplast genome, however, were then taken as evidence that *C. bursa-pastoris* is an autopolyploid of *C. grandiflora* (Hurka and Neuffer 1997). Recently, phylogenetic analyses based on data obtained with chloroplast and nuclear DNA markers did not find support for either of these scenarios, even though some accessions of *C. bursa-pastoris* were found to share alleles with *C. rubella* at nuclear loci (Slotte et al. 2006, 2008). According to these analyses, *C. bursa-pastoris* is probably neither an autopolyploid of *C. rubella* or *C. grandiflora* nor was either of these species the maternal parent in an allopolyploidization event (Slotte et al. 2006). *C. bursa-pastoris* has disomic inheritance and a level of divergence between nuclear homoeologous loci of about 1.4–4.3%, suggesting an allopolyploid rather than autopolyploid origin (Slotte et al. 2006). Thus *C. bursa-pastoris* possibly originated by allopolyploidization with *C. rubella* as paternal ancestor and with a diploid ancestor as maternal parent, which is extinct or has not been investigated yet. However, both chloroplast and nuclear gene data are more compatible with the alternative scenario that *C. bursa-pastoris* originated completely from extinct diploid ancestors followed by repeated hybridization and backcrossing with *C. rubella*, leading to the introgression of *C. rubella* alleles into the *C. bursa-pastoris* genome (Slotte et al. 2006, 2008). In line with this, Slotte et al. (2008) found evidence for unidirectional gene flow from *C. rubella* to *C. bursa-pastoris* in western Eurasia, where both species live in sympatry in some habitats, but not in eastern Eurasia (China), where *C. rubella* does not occur.

A low level of intraspecific polymorphism in chloroplast DNA sequences indicates that the maternal effective population size of *C. bursa-pastoris* is small, possibly because the species had a single maternal origin, which is in contrast to many other polyploid species (Ceplitis et al. 2005, Slotte et al. 2006). Estimates for the most recent common ancestor (MRCA) range between 43,000 and 430,000 years, which makes it possible that *C. bursa-pastoris* arose after the last Pleistocene glaciation (Slotte et al. 2006). As in case of *C. rubella*, the origin of *C. bursa-pastoris* constituted a severe bottleneck, after which the species underwent a remarkable population expansion (Slotte et al. 2008).

Tetraploidization during the origin of *C. bursa-pastoris* was accompanied or followed by a shift to disomic inheritance, which led to "fixed heterozygosity" and thus increased the intraspecific genetic diversity. This may have helped to avoid inbreeding depression when outcrossing is rare, and could be one of the reasons why this species has a much higher colonization potential and weediness than *C. rubella* (Hintz et al. 2006).

13.3 Genome and Chromosome Evolution

Mapping populations of *C. grandiflora* × *C. rubella* and of *Arabidopsis* species, and the availability of the *A. thaliana* genome sequence enabled important insights into the structure and evolution of *Capsella* genomes and chromosomes (Acarkan et al. 2000, Boivin et al. 2004, Koch and Kiefer 2005). It turned out that the order, orientation and sequence of genes is very similar in *A. thaliana* and *C. rubella*, with more than 90% sequence identity within exons (Acarkan et al. 2000, Koch and Kiefer 2005). This allows the identification of genes within *Capsella* with the help of the *A. thaliana* genome. The eight *C. rubella* linkage groups (representing chromosomes) are completely collinear with the eight linkage groups of *A. lyrata*. This enabled reconstruction of a parsimonious scenario of chromosomal rearrangements during evolution, suggesting that the existence of eight chromosomes in these species represents the ancestral state, while the five chromosomes in *A. thaliana* represent a derived state (Koch and Kiefer 2005, Yogeeswaran et al. 2005). In line with this, there is evidence that rates of chromosomal changes (inversions, reciprocal translocations, fusions) have been very different in different lineages. While chromosomal mutation rates have been very low in the lineages that led to *C. rubella* and *A. lyrata*, most chromosomal rearrangements occurred in the lineage that led to *A. thaliana* after the lineage that led to *A. lyrata* had already branched off about 5.3 million years ago (Koch and Kiefer 2005, Yogeeswaran et al. 2005). This is also true for the genomic regions of pericentromeres, in which *A. thaliana* has undergone recent, significant expansions, in some cases measuring hundreds of kilobases (Hall et al. 2006). The reason for this extensive genome and chromosome reshuffling in the *A. thaliana* lineage is neither known nor obvious. It is thus clear, however, that *C. rubella* and *A. lyrata* are much better models to study basic chromosome structures in Brassicaceae than *A. thaliana*. Meanwhile 24 chromosomal building blocks have been defined in a unified comparative genomics framework of Brassicaceae based on their positions in a proposed ancestral karyotype ($n = 8$) very similar to those of *C. rubella* and *A. lyrata*, but somewhat different from the reduced and reshuffled genome of *A. thaliana* ($n = 5$) (Lysak and Koch, Chapter 1, this issue; Schranz et al. 2006).

13.4 Evolution and Development of Phenotypic Traits

Capsella species, especially *C. bursa-pastoris*, show extensive variability for quite a number of morphological traits such as leaf and fruit shape, and life-history traits such as flowering time. This makes them good model systems for developmental genetics trying to find out how genes map onto phenotypes, and how differences in developmental genetics affect evolution. Some traits are heavily affected by a single genetic locus. Many traits of interest, however, are affected by both segregation of alleles at many loci and the environment. Since these traits typically vary in a quantitative rather than qualitative way, they are called quantitative traits, and the loci contributing to them, quantitative trait loci (QTL).

13.4.1 Leaf Development

Four main rosette leaf types have been described in *C. bursa-pastoris*, termed *heteris*, *rhomboidea*, *tenuis*, and *simplex*, which differ in leaf shape and lobing. Two Mendelian loci, each with two alleles, affect this trait. Gene A influences the elongation of the primary lobes, and gene B the division of the lobes, in a way that *heteris* with dominant alleles at both loci (A-B-) develops elongated primary lobes that divide the leaf to the midrib, while the double recessive *simplex* (aabb) forms lobes that show little elongation and do not divide the leaf to the midrib (Neuffer and Linde 1999, and literature cited therein). Accordingly, *rhomboidea* (aaB-) and *tenuis* (A-bb) show the dominant trait only in the one or the other character. It was shown that the frequency of some leaf types correlates with environmental factors; for example, the frequency of the B allele increases with elevation above sea level in the Alps (Neuffer and Bartelheim 1989). The adaptive value of leaf shape (if any) is not clear yet. Molecular cloning of genes A and B may open the door for more sophisticated studies but has not been reported yet. The genetic basis of differences in leaf shape is currently successfully being investigated using *Cardamine hirsuta* (dissected leaf form) in comparison with *A. thaliana* (simple leaf form) (Hay and Tsiantis 2006, Barkoulas et al. 2008). By demonstrating the importance of Class I KNOTTED1-like homeobox proteins and of PINFORMED1 auxin efflux transporter for the development of dissected leaves these studies may have paved the way for similar investigations in *Capsella*.

13.4.2 Flowering Time

Flowering time (or "time to flower") is an important life-history trait contributing significantly to plant fitness, especially in annual plants (Roux et al. 2006). Different flowering strategies have evolved in response to local growth conditions (Mitchell-Olds and Schmitt 2006). The genetic basis of differences in flowering time has been intensively studied in *A. thaliana*, where several dozens of crucial genes have been identified by mutant analysis. Here, four main pathways, termed the autonomous, photoperiod, vernalization and gibberellin pathway, have been defined that help the plant to control the floral transition dependent on favourable inner states, day-length, temperature/season and hormonal status, respectively. Integrator genes and cross-talk between pathways fine-tune the transition from vegetative to reproductive development (Mouradov et al. 2002).

As in *Arabidopsis*, studies on phenotypic variation in *C. bursa-pastoris* detected diverse accessions and ecotypes with striking differences in flowering time (Neuffer and Bartelheim 1989, Neuffer and Meyer-Walf 1996, Hurka and Neuffer 1997). Employing an F_2 population derived from a cross of two accessions from two different climatic regions in California, mapping involving molecular markers revealed three major QTL for the onset of flowering, but the genetic loci underlying these QTL remained unknown (Linde et al. 2001).

A recent gene expression study employing *A. thaliana* microarrays showed that early- and late-flowering accessions differ in the photoperiod and in the gibberellin pathway. It turned out that gibberellins biosynthesis genes are down-regulated in late flowering accessions, whereas circadian-clock genes in the photoperiodic pathway, such as *CIRCADIAN CLOCK-ASSOCIATED1* (*CCA1*) and *TIMING OF CAB EXPRESSION1* (*TOC1*), are differentially expressed between early- and late-flowering accessions (Slotte et al. 2007). These genes are thus promising candidates for the evolution of adaptive flowering time variation in *C. bursa-pastoris* (Slotte et al. 2007).

Interestingly, even though the orthologue of the *A. thaliana* key flowering time gene *FLOWERING LOCUS C* (*FLC*) seems to play a conserved role in the vernalization response also in *C. bursa-pastoris*, it is not differentially expressed prior to vernalization between late- and early-flowering accessions and thus does not play a role in generating natural variation in flowering time (Slotte et al. 2007). This is in strong contrast to the situation in *A. thaliana*, where variation at the *FLC* locus, and especially multiple independent mutations at its activator *FRIGIDA* (*FRI*), can explain a great deal of genetic variation in flowering time (Caicedo et al. 2004, Slotte et al. 2007, and references cited therein).

13.4.3 Floral Structure and Function

13.4.3.1 Floral Size

The flowers of the selfing *C. rubella* and *C. bursa-pastoris* show the typical "selfing syndrome", characterized by a considerable reduction in flower size compared to the outcrossing *C. grandiflora* (what's in a name!) in addition to a breakdown of the self-incompatibility (SI) system (see below). This remarkable morphological change very likely occurred during less than 50,000 years, i.e. on a very short evolutionary timescale, probably due to strong natural selection on reproductive assurance in a colonizing species (Foxe et al. 2009). It will be interesting to determine the developmental genetic mechanisms underlying the reduction in flower size. Studies on a very similar difference in flower size between the selfing *A. thaliana* and the obligate outcrossing *A. lyrata* are ongoing (Stransfeld and Lenhard 2007) and may provide candidate genes for changes underlying this important aspect of the selfing syndrome in *Capsella*.

13.4.3.2 Saltational Change in Floral Architecture

In addition to all its merits as an experimental system outlined above, the genus *Capsella* provides also an almost unprecedented opportunity to study saltational evolution in a population of extant organisms not long after the critical mutational event happened.

Whether all evolutionary changes are gradual, as suggested by Charles Darwin, or whether also saltational changes occur, remains a highly controversial topic (Bateman and DiMichele 2002, Theißen 2006, 2009). In the tradition of the predominant "Neodarwinism" and the "Synthetic Theory of Evolution" the majority of evolutionary biologists have usually rejected saltational evolution with arguments taken from population genetics, seemingly demonstrating that the establishment of any major mutant in the wild is extremely unlikely. To reject saltational evolution, however, oversimplistic arguments are often put forward that may not reflect realistic scenarios in natural habitats (for a typical example considering flowering plants, see Crepet and Niklas 2009).

Homeotic mutants, in which proper organs develop at a position where usually organs of another identity develop, are reasonable representatives of saltational change, and several lines of evidence suggest that homeotic changes played a considerable role during plant evolution (Theißen 2006, 2009). Studying the fitness of homeotic mutants in natural habitats thus may tell us a great deal about the modes and mechanisms of saltational evolution (Bateman and DiMichele 2002, Vergara-Silva 2003, Dietrich 2003, Theißen 2006, 2009). A floral homeotic mutant of *C. bursa-pastoris* represents a promising model system along these lines (Nutt et al. 2006, Hintz et al. 2006). The mutant variety is termed *Stamenoid petals* (*Spe*), because it has petals completely replaced by stamens, while all other organs are unchanged (see Fig. 13.1). In contrast to most other floral homeotic mutants known, *Spe* exists in some natural habitats at least for decades, so one can conclude that its fitness must be quite similar to that of the wild type. How is that possible despite the drastic change of floral architecture?

Recently, Ziermann et al. (2009) have shown that visitation by potential pollinators (see Fig. 13.1), such as hoverflies, wild bees and thrips, is about twice as high in wild-type plants as in *Spe* plants. Important reasons could be not only optical cues provided by petals, but also attractive volatiles such as monoterpenes and 3,4-dimethylbenzaldehyde, which are also very likely only produced by petals (Ziermann et al. 2009). Nevertheless, the number of seeds per fruits was about the same in wild type and mutant. Thus flower structure and floral visitation are obviously only of minor importance for *C. bursa-pastoris*, very likely because it

Fig. 13.1 Curious about *Capsella*. The pictures show insects visiting inflorescences of a *C. bursa-pastoris* wild type (**a**) and a *Stamenoid petals* (*Spe*) variety (**b**). The *Spe* variety is a naturally occurring floral homeotic mutant in which all petals are replaced by stamens. Note also the characteristic heart-shaped fruits ("Shepherd's purse") visible in both a and b

is mainly a self-pollinating plant species. However, because of differences in plant habit (especially branching pattern), wild-type plants produced more flowers, fruits and seeds than *Spe*-plants, whereas germination capacity of *Spe* seeds was higher than that of the wild type. Whether the observed linkages between floral phenotype, plant architecture and seed germination rate also hold at the population genetic level in natural habitats is currently unknown. Also the underlying genetic components have not been identified yet (Ziermann et al. 2009), but it appears unlikely that all the differences between wild type and *Spe* plants are caused by mutation of the *Spe* locus alone.

The reported findings led to the conclusion that the similar fitness of *Spe* and wild-type *C. bursa-pastoris* in the field represents a kind of evolutionary "stalemate" in which differences in plant architecture and germination capacity compensate each others (Ziermann et al. 2009).

In addition to a high rate of selfing, considerable differences in the flowering time have very likely contributed to the persistence of *Spe* in sympatry with wild-type plants (Hameister et al. 2009). A lower genetic variation of the *Spe* subpopulation compared to the wild-type indicates a recent local origin or recent introduction in one habitat in vineyards in southwest Germany (Hameister et al. 2009).

The phenotype of *Spe* is caused by a co-dominant mutant allele at a single locus conferring stamen identity in one of the two disomically inherited genomes of *C. bursa-pastoris* (Nutt et al. 2006). Development of the *Spe* phenotype is probably brought about by ectopic expression of a class C floral organ identity gene, specifying stamens when co-expressed with a class B gene, in the organs of the second floral whorl (Nutt et al. 2006). Cloning of the relevant gene and understanding of its lesion in the mutant are well underway (Pia Nutt, Janine Ziermann and Günter Theißen, unpublished data).

13.4.3.3 Self-Incompatibility

Many angiosperms are obligate outcrossers because of self-incompatibility (SI). The inability to self-fertilize is often determined by a single self-incompatibility (S)-locus. SI originated independently about a dozen times during angiosperm evolution, but loss of SI is even more common. By enabling reproduction via selfing and thus reproductive assurance breakdown of SI is thought to provide short-term advantages when pollinators or mates are rare (Barrett 2002).

In the genus *Capsella*, as well as in the whole of the Brassicaceae, sporophytic SI is very likely the ancestral condition, and the transition from outcrossing to selfing has occurred repeatedly, e.g. in *Arabidopsis* as well as in *Capsella* (Fobis-Loisy et al. 2004). The S-locus of the Brassicaceae consists of two determinant genes which are usually not separated by recombination. One gene, *SRK*, encodes a transmembrane receptor kinase that is expressed at the stigmatic surface of the female reproductive organs (carpels); the other gene, *SCR*, encodes a small soluble ligand that is deposited in the pollen wall of pollen produced by male reproductive organs (stamens). When the SCR protein binds to SRK from the same haplotype, an SI response occurs, preventing self-pollination through a cascade of downstream events (Fobis-Loisy et al. 2004, Rea and Nasrallah 2008, Nasrallah, Chapter 14, this issue).

It has long been assumed that the self-compatible (SC) breeding systems of *C. rubella* and *C. bursa-pastoris* were caused by breakdown of the SI system still active in *C. grandiflora* (Hurka and Neuffer 1997, Hurka et al. 2005). Characterization of multilocus patterns of neutral DNA sequence diversity across the genome revealed that in case of *C. rubella*, the transition to selfing and subsequent geographic expansion have taken place during the past 20,000 years, involving a near-complete bottleneck causing a strong reduction in effective population size (Foxe et al. 2009). In a parallel study it was estimated that a single S allele in *C. rubella* was fixed at least 27,000 years ago and that coalescence of this allele and the corresponding allele of *C. grandiflora* occurred 30,000–60,000 years ago (Guo et al. 2009). Selfing was probably favoured during colonization as new habitats emerged after the most recent glaciations and the intensification of human agriculture. It could well be, therefore, that breakdown of SI played a causal role in founding the *C. rubella* species, even though it cannot be excluded that speciation predated loss of SI (Guo et al. 2009).

The S-locus of *C. grandiflora* comprises at least 38 haplotypes and is very polymorphic because of strong frequency-dependent selection (Paetsch et al. 2006, Nasrallah et al. 2007). In *C. rubella*, *SCR* apparently evolved into a pseudogene by loss of all but the first two exons, whereas functional versions of *SRK* may have persisted, even though several independent loss of function mutations have occurred as well (Guo et al. 2009). This is in interesting contrast to evolutionary changes in the genus *Arabidopsis*, where the different S-locus haplotypes of the selfer *A. thaliana* that have been investigated so far all contain *SRK* pseudogenes, whereas *SCR* shows different states ranging from gene loss, pseudogene formation, to retention of function (Kusaba et al. 2001, Tang et al. 2007, Sherman-Broyles et al. 2007).

13.4.4 Fruit Structure

Fruit shape and structure vary a lot within Brassicaceae, providing interesting characters for developmental genetic studies. Many Brassicaceae develop dry, dehiscent elongated fruits termed siliques, while *Capsella* forms the characteristic heart-shaped siliculas (see Fig. 13.1). About 100 years ago, however, Shull (1914) discovered a variant of *C. bursa-pastoris* which developed elongated siliques similar to the fruits of *Arabidopsis*. When this variant, termed "*C. heegeri*", was crossed with wild-type *C. bursa-pastoris*, all fruits were heart shaped in the F_1 generation, indicating that the formation of wild-type fruits is dominant over that of *heegeri* fruits. Surprisingly, in the F_2 progeny, the plants developing mutant fruits segregated about 1:15 rather than 1:3. This can be explained by the assumption that two unlinked genes (C, D) control the development of fruit shape and that both of them have to be homozygous mutant for recessive alleles (ccdd) for the mutant phenotype to be expressed. It suffices that one locus has at least one dominant wild-type allele (such as Ccdd or ccDd) for heart-shaped fruits to develop. These findings suggest that loss of function of two unlinked, functionally redundant genes was responsible for expression of the *heegeri* character. This redundancy may be caused by equivalent gene pairs that reflect the tetraploidy of the *C. bursa-pastoris* genome. This way, the studies by Shull (1914) provided an early example of deviation from "Mendelian segregation ratios" revealing a polygenetic effect on a morphological character.

Unfortunately, the *heegeri* variant is no longer available, so that molecular cloning and analysis of the affected genes are not possible. However, the heart-shaped fruits of *Capsella* remain an attractive system to study the development genetics of fruit shape.

Differences between fruit types reflect differential growth during fruit development. Little is known about genes that control such processes. Interestingly, constitutive expression of *CYP78A9*, a gene encoding a cytochrome P450 enzyme, converts the silique of *A. thaliana* into a silicula-like structure by lateral expansion; this effect is stronger in an *apetala2-1* mutant background than in a wild-type background (Ito and Meyerowitz 2000). How differential growth during fruit development is influenced by the encoded proteins remains unclear. CYP78A9 could well be involved in the synthesis of a secondary compound that acts as a signalling molecule, but APETALA2 (AP2) very likely functions as a transcription factor, so their mode of interaction is probably indirect and not immediately obvious. Also expression of the MADS-domain transcription factor FRUITFULL (FUL) fused to a strong transcriptional activation domain of the virus VP16 (FUL::VP16) generates heart-shaped fruits in *A. thaliana* (Cristina Ferrándiz, personal communication). Like in case of leaf development studies in *A. thaliana* thus have provided clues or even candidate genes that may be used as starting points for understanding an interesting morphogenetic feature of *Capsella*, its characteristic fruits (Shepherd's purse!).

13.5 Outlook

We have seen that the genus *Capsella* is a treasure trove of interesting biological phenomena that deserve being studied more intensively in the future. Some technical advantages compared to many other plant systems support respective studies. Like *A. thaliana*, *Capsella* species are easy to cultivate and propagate. Their life cycles are fairly short, allowing the growth of more than one generation per year under optimized growth conditions. One of the species, *C. rubella*, is both diploid and self-compatible, two important prerequisites to develop it into an efficient system for standard forward genetic approaches (Bowman 2006). In addition, genome sequencing of *C. rubella* is currently underway (Joint Genome Institute, United States Department of Energy) and will tremendously boost comparative genomics within Brassicaceae and beyond.

At least in case of *C. bursa-pastoris* it has already been shown for a *Capsella* species that it can be transformed by the "floral dip" method (Bartholmes et al. 2008). Thus transgenes can be introduced quite effectively, facilitating studies on gene function.

As we have seen, *Capsella* genetics has come a long way since the pioneering days during the first half of the twentieth century. Clearly, it is a flourishing field of research now. However, with all its interesting biological features and more and more efficient experimental tools being developed, I am confident that the best years for *Capsella* genetics are yet to come.

Acknowledgements I am indebted to Barbara Neuffer (Osnabrück), Renate Schmidt (Gatersleben) and Ian Bancroft (Norwich) for valuable comments that helped to improve the manuscript. Many thanks to Cristina Ferrándiz (Valencia) for a personal communication, and to Janine Ziermann (Jena) for the original photographs used in Fig. 13.1. Work on the *Capsella bursa-pastoris Spe* variety in the author's laboratory has been funded by grants TH 417/4-1 and -2 from the Deutsche Forschungsgemeinschaft (DFG).

References

Acarkan A, Roßberg M, Koch M, Schmidt R (2000) Comparative genome analysis reveals extensive conservation of genome organisation for *Arabidopsis thaliana* and *Capsella rubella*. Plant J 23:55–62

Barkoulas M, Hay A, Kougioumoutzi E, Tsiantis M (2008) A developmental framework for dissected leaf formation in the *Arabidopsis* relative *Cardamine hirsuta*. Nat Genet 40:1136–1141

Barrett SCH (2002) The evolution of plant sexual diversity. Nat Rev Genet 3:274–284

Bartholmes C, Nutt P, Theißen G (2008) Germline transformation of Shepherd's purse (*Capsella bursa-pastoris*) by the floral dip method as a tool for evolutionary and developmental biology. Gene 409:11–19

Bateman RM, DiMichele WA (2002) Generating and filtering major phenotypic novelties: neoGoldschmidtian saltation revisited. In: Cronk QCB, Bateman RM, Hawkins JA (eds) Developmental genetics and plant evolution, pp. 109–159. Taylor & Francis, London

Boivin K, Acarkan A, Mbulu RS, Clarenz O, Schmidt R (2004) The *Arabidopsis* genome sequence as a tool for genome analysis in Brassicaceae. a comparison of the *Arabidopsis* and *Capsella rubella* genomes. Plant Physiol 135:735–744

Bowman JL (2006) Molecules and morphology: comparative developmental genetics of the Brassicaceae. Plant Syst Evol 259:199–215

Caicedo AL, Stinchcombe JR, Olsen KM, Schmitt J, Purugganan MD (2004) Epistatic interaction between *Arabidopsis FRI* and *FLC* flowering time genes generates a latitudinal cline in a life history trait. Proc Natl Acad Sci USA 101:15670–15675

Ceplitis A, Su YT, Lascoux M (2005) Bayesian inference of evolutionary history from chloroplast microsatellites in the cosmopolitan weed *Capsella bursa-pastoris* (Brassicaceae). Mol Ecol 14:4221–4233

Crepet WL, Niklas KJ (2009) Darwin's second "abominable mystery": why are there so many angiosperm species? Am J Bot 96:366–381

Dietrich MR (2003) Richard Goldschmidt: hopeful monsters and other 'heresies'. Nat Rev Genet 4:68–74

Fobis-Loisy I, Miege C, Gaude T (2004) Molecular evolution of the S locus controlling mating in the Brassicaceae. Plant Biol 6:109–118

Foxe JP, Slotte T, Stahl EA, Neuffer B, Hurka H, Wright SI (2009) Recent speciation associated with the evolution of selfing in *Capsella*. Proc Natl Acad Sci USA 106:5241–5245

Franzke A, German D, Al-Shehbaz IA, Mummenhoff K (2009) *Arabidopsis* family ties: molecular phylogeny and age estimates. Taxon 58:425–432

German DA, Friesen N, Neuffer B, Al-Shehbaz IA, Hurka H (2009) Contribution to ITS phylogeny of the Brassicaceae with special reference to some Asian taxa. Plant Syst Evol 283:33–56

Guo Y-L, Bechsgaard JS, Slotte T, Neuffer B, Lascoux M, Weigel D, Schierup MH (2009) Recent speciation of *Capsella rubella* from *Capsella grandiflora*, associated with loss of self-incompatibility and extreme bottleneck. Proc Natl Acad Sci USA 106:5246–5251

Hall AE, Kettler GC, Preuss D (2006) Dynamic evolution at pericentromeres. Genome Res 16: 355–364

Hameister S, Neuffer B, Bleeker W (2009) Genetic differentiation and reproductive isolation of a naturally occurring floral homeotic mutant within a wild-type population of *Capsella bursa-pastoris* (Brassicaceae). Mol Ecol 18:2659–2667

Hay A, Tsiantis M (2006) The genetic basis for differences in leaf form between *Arabidopsis thaliana* and its wild relative *Cardamine hirsuta*. Nat Genet 38:942–947

Hintz M, Bartholmes C, Nutt P, Ziermann J, Hameister S, Neuffer B, Theißen G (2006) Catching a 'hopeful monster': shepherd's purse (*Capsella bursa-pastoris*) as a model system to study the evolution of flower development. J Exp Bot 57:3531–3542

Hurka H, Bleeker W, Neuffer B (2003) Evolutionary process associated with biological invasions in the Brassicaceae. Biol Invasions 5:281–292

Hurka H, Freundner S, Brown AHD, Plantholt U (1989) Aspartat aminotransferase isozymes in the genus *Capsella* (Brassicaceae): subcellular location, gene duplication and polymorphism. Biochem Genet 27:77–90

Hurka H, Neuffer B (1997) Evolutionary processes in the genus *Capsella* (Brassicaceae). Plant Syst Evol 206:295–316

Hurka H, Paetsch M, Bleeker W, Neuffer B (2005) Evolution within the Brassicaceae. Nova Acta Leopoldina NF92 342:113–127

Ito T, Meyerowitz EM (2000) Overexpression of a gene encoding a cytochrome P450, *CYP78A9*, induces large and seedless fruit in *Arabidopsis*. Plant Cell 12:1541–1550

Koch MA, Kiefer M (2005) Genome evolution among cruciferous plants: a lecture from the comparison of the genetic maps of three diploid species – *Capsella rubella*, *Arabidopsis lyrata* ssp. *petraea*, and *A. thaliana*. Am J Bot 92:761–767

Kusaba M, et al. (2001) Self-incompatibility in the genus *Arabidopsis*: characterization of the S locus in the outcrossing *A. lyrata* and its autogamous relative *A. thaliana*. Plant Cell 13:627–643

Linde M, Diel S, Neuffer B (2001) Flowering ecotypes of *Capsella bursa-pastoris* (L.) Medik. (Brassicaceae) analysed by a cosegregation of phenotypic characters (QTL) and molecular markers. Annals Bot 87:91–99

Mable BK, Robertson AV, Dart S, di Berardo C, Witham L (2005) Breakdown of self-incompatibility in the perennial *Arabidopsis lyrata* (Brassicaceae) and its genetic consequences. Evolution 59:1437–1448

Mitchell-Olds T, Schmitt J (2006) Genetic mechanisms and evolutionary significance of natural variation in *Arabidopsis*. Nature 441:947–952

Mouradov A, Cremer F, Coupland G (2002) Control of flowering time: interacting pathways as a basis for diversity. Plant Cell 14:S111–S130

Mummenhoff K, Hurka H (1990) Evolution of the tetraploid *Capsella bursa-pastoris* (Brassicaceae): isoelectric focusing analysis of Rubisco. Plant Syst Evol 172:205–213

Nasrallah JB, Liu P, Sherman-Broyles S, Schmidt R, Nasrallah ME (2007) Epigenetic mechanisms for breakdown of self-incompatibility in inter-specific hybrids. Genetics 175:1965–1973

Neuffer B, Bartelheim S (1989) Gen-ecology of *Capsella bursa-pastoris* from an altitudinal transsect in the alps. Oecologia 81:521–527

Neuffer B, Linde M (1999) *Capsella bursa-pastoris* – colonisation and adaptation: a globe-trotter conquers the world. Plant evolution in man-made habitats (van Ramsdonk LWD, den Nijs JCM, Eds.). Proc 7th Symp IOPB, Amsterdam 1998

Neuffer B, Meyer-Walf M (1996) Ecotype variation in relation to man made habitats in *Capsella*: field and trampling area. Flora 191:49–57

Nutt P, Ziermann J, Hintz M, Neuffer B, Theißen G (2006) *Capsella* as a model system to study the evolutionary relevance of floral homeotic mutants. Pl Syst Evol 259:217–235

Paetsch M, Mayland-Quellhorst S, Neuffer B (2006) Evolution of the self-incompatibility system in the Brassicaceae: identification of the S-locus receptor kinase (SRK) in self-incompatible *Capsella grandiflora*. Heredity 97:283–290

Rea AC, Nasrallah JB (2008) Self-incompatibility systems: barriers to self-fertilization in flowering plants. Int J Dev Biol 52:627–636

Riley HP (1936) The genetics and physiology of self-sterility in the genus Capsella. Genetics 21:24–39

Roux F, Touzet P, Cuguen J, Le Corre V (2006) How to be early flowering: an evolutionary perspective. Trends Plant Sci 11:375–381

Schranz ME, Lysak MA, Mitchell-Olds T (2006) The ABC's of comparative genomics in the Brassicaceae: building blocks of crucifer genomes. Trends Plant Sci 11:535–542

Sherman-Broyles S, Boggs N, Farkas A, Liu P, Vrebalov J, Nasrallah ME, Nasrallah JB (2007) S locus genes and the evolution of self-fertility in *Arabidopsis thaliana*. Plant Cell 19:94–106

Shimizu K, Cork JM, Caicedo AL, Mays CA, Moore RC, Olsen KM, Ruzsa S, Coop G, Bustamante CD, Awadalla P, Purugganan MD (2004) Darwinian selection on a selfing locus. Science 306:2081–2084

Shull GH (1914) Duplicate genes for capsule-form in *Capsella bursa-pastoris*. Z Abst u Vererbl 12:97–149

Slotte T, Ceplitis A, Neuffer B, Hurka H, Lascoux M (2006) Intrageneric phylogeny of *Capsella* (Brassicaceae) and the origin of the tetraploid *C. bursa-pastoris* based on chloroplast and nuclear DNA sequences. Am J Bot 93:1714–1724

Slotte T, Holm K, McIntyre LM, Lagercrantz U, Lascoux M (2007) Differential expression of genes important for adaptation in *Capsella bursa-pastoris* (Brasicaceae). Plant Physiol 145:160–173

Slotte T, Huang H, Lascoux M, Ceplitis A (2008) Polyploid speciation did not confer instant reproductive isolation in *Capsella* (Brassicaceae). Mol Biol Evol 25:1472–1481

Stebbins GL (1950) Variation and evolution in plants. Columbia Univ Press, New York, NY
Stransfeld L, Lenhard M (2007) Plant organ size control. encyclopedia of life sciences. Wiley, Hoboken, NJ. DOI: 10.1002/9780470015902.a0003363
Tang C, Toomajian C, Sherman-Broyles S, Plagnol V, Guo Y-L, Hu TT, Clark RM, Nasrallah JB, Weigel D, Nordborg M (2007) The evolution of selfing in *Arabidopsis thaliana*. Science 317:1070–1072
Theißen G (2006) The proper place of hopeful monsters in evolutionary biology. Theory Biosci 124:349–369
Theißen G (2009) Saltational evolution: hopeful monsters are here to stay. Theory Biosci 128: 43–51
Vergara-Silva F (2003) Plants and the conceptual articulation of evolutionary developmental biology. Biol Philos 18:249–284
Yogeeswaran K, Frary A, York TL, Amenta A, Lesser AH, Nasrallah JB, Tanksley SD, Nasrallah ME (2005) Comparative genome analyses of *Arabidopsis* spp.: inferring chromosomal rearrangement events in the evolutionary history of *A. thaliana*. Genome Res 15:505–515
Ziermann J, Ritz MS, Hameister S, Abel C, Hoffmann MH, Neuffer B, Theißen G (2009) Floral visitation and reproductive traits of *Stamenoid petals*, a naturally occuring floral homeotic variant of *Capsella bursa-pastoris* (Brassicaceae). Planta 230:1239–1249
Zunk K, Mummenhoff K, Hurka H (1999) Phylogenetic relationship in tribe Lepidieae (Brassicaceae) based on chloroplast DNA restriction site variation. Can J Bot 77:1504–1512

Chapter 14
Self-Incompatibility in the Brassicaceae

June B. Nasrallah

Abstract Flowering plants have evolved several mechanisms for controlling pollination. Genetic self-incompatibility is one of the most elaborate and "smart" systems known to date. In the Brassicaceae, recognition of "self" pollen in the self-incompatibility response is based on highly specific interactions between matched stigma surface receptors and pollen coat ligands encoded by haplotypes of the *S* locus, which triggers arrest of pollen tube development. This chapter presents a brief historical account of the analysis of SI in the Brassicaceae, an overview of our current understanding of the recognition and response phases of SI, and a summary of progress made in elucidating the genetic basis of loss of SI and switches to self-fertility in various lineages, with an emphasis on knowledge gained from analysis of a recently developed transgenic *Arabidopsis thaliana* self-incompatible model.

Keywords Self-incompatibility · Receptor · Ligand · Signaling · Mating system · Self-fertility · *Arabidopsis thaliana*

Abbreviations

ARC1	Arm-repeat containing protein 1
CVR	C-terminal variable region
EGF	Epidermal growth factor
eSRK	Soluble form of the extracellular domain of SRK
hvI, hvII, hvIII	Hypervariable regions I, II, III
KAPP	Kinase-associated protein phosphatase
MLPK	*M* locus protein kinase
RLK	Receptor-like kinase
RLP	Receptor-like protein
SCR	*S*-locus cystine-rich protein
SI	Self-incompatibility
SLG	*S*-locus glycoprotein

J.B. Nasrallah (✉)
Department of Plant Biology, Cornell University, Ithaca, NY 4853, USA
e-mail: jbn2@cornell.edu

R. Schmidt, I. Bancroft (eds.), *Genetics and Genomics of the Brassicaceae*, Plant Genetics and Genomics: Crops and Models 9, DOI 10.1007/978-1-4419-7118-0_14,

S locus	Self-incompatibility locus
SRK	*S*-locus receptor kinase
THL	Thioredoxin h-like
U-box	A sequence motif characteristic of a family of E3 ligases

Contents

14.1 Introduction 390
14.2 Genetics of Self-Incompatibility 391
14.3 Mechanism of Recognition and Inhibition of "Self" Pollen 392
14.3.1 Cytological Responses 392
14.3.2 Molecular Studies 393
14.4 *S* haplotype Structure, Suppressed Recombination, and Diversification 400
14.4.1 Diversification of SRK and SCR 400
14.5 Mating-Type Dimorphism in the Brassicaceae: Loss of SI and the Switch From an Outbreeding to a Self-Fertile Mode of Mating 403
14.5.1 Analysis of Self-Fertility in Non-model Members of the Brassicaceae . . 403
14.5.2 Analysis of Self-Fertility in the Model Plant *A. thaliana* 404
14.6 Future Prospects 405
References 406

14.1 Introduction

It is estimated that approximately half of the species of the Brassicaceae are self-sterile due to the operation of a self-incompatibility (SI) system that triggers arrest of pollen tubes after self-pollination. SI is an intra-specific barrier to self-fertilization that ensures high rates of out-crossing. In the Brassicaceae, SI is a pollen–stigma recognition system controlled by a single and highly polymorphic genetic locus known as the *S* locus. In self-incompatible members of the family, the stigma can discriminate between pollen grains on the basis of the *S*-locus variant they express. It recognizes pollen grains as "self" pollen if they express the same *S*-locus variant as it expresses (whether they are derived from the same flower, the same plant, or other plants expressing the same variant). These "self" pollen grains trigger a response in stigma epidermal cells, which leads to inhibition of pollen hydration, germination, and tube growth. In contrast, pollen grains that express a different *S*-locus variant from that expressed in the stigma (designated as "non-self") do not trigger the inhibition response. Rather, they elaborate tubes that grow into the pistil through the extracellular matrix of the stigma, stylar transmitting tract, and ovary, where they are targeted to the ovules and effect fertilization and seed production.

The phenomenon of self-incompatibility has intrigued plant researchers with varied interests, from those delving into basic biological questions, such as geneticists, biochemists, developmental biologists, ecologists, and evolutionary biologists, to those concerned with developing breeding programs for hybrid seed production and crop improvement. Several reviews of plant SI systems and of Brassicaceae SI in particular have been published in the last few years (deNettancourt 2001, Franklin-Tong 2008, Nasrallah 2005, Takayama and Isogai 2005, Samuel et al. 2008, Sherman-Broyles and Nasrallah 2008). This chapter presents an historical perspective on the study of SI in the Brassicaceae, followed by a brief and by no means exhaustive review of our current understanding of the molecular mechanism underlying "self" pollen recognition and rejection and of the genetic basis of switches in mating system from out-breeding to self-fertile.

14.2 Genetics of Self-Incompatibility

The earliest report of self-incompatibility (SI) in the Brassicaceae appeared in 1920 by Stout who observed it in radish (*Raphanus sativus;* Stout 1920). But it was only in the 1950s that rigorous genetic analyses were carried out by Bateman (Bateman 1954) who, working in *Iberis amara* and interpreting previous results from *Raphanus,* first concluded that SI in the Brassicaceae is controlled by a single multi-allelic locus, with sporophytic control of pollen SI phenotype (i.e., pollen SI phenotype is determined by the diploid genotype of the pollen-producing parent). This mode of genetic control was later confirmed in several other taxa, including *Cardamine, Capsella, Raphanus raphanistrum,* and *Brassica oleracea* (Bateman 1955, Sampson 1957 and 1964, Thompson 1957), and more recently in extensive studies of several *Brassica* species (see below), *Arabidopsis lyrata* (Charlesworth et al. 2003, Kusaba et al. 2001, Mable 2003, Schierup et al. 2001, Bechsgaard et al. 2004), and *Capsella grandiflora* (Nasrallah et al. 2007). In parallel with these genetic studies, two advances were made that greatly facilitated analysis of SI. First, visualization of pollen tube growth in pollinated pistils by UV fluorescence microscopy was established as a rapid and more effective method than seed set for scoring SI phenotype (Kho and Baer 1968). Second, developmental analyses demonstrated that the SI response is regulated during stigma maturation: stigmas are initially compatible with self pollen and only acquire the ability to reject self pollen in conjunction with anther dehiscence 1–2 days before flower opening or anthesis (Nasrallah 1974, Shivanna et al. 1978). This feature allows the production and maintenance of *S*-locus homozygotes by manual self-pollination of the immature stigmas of young buds before the onset of SI.

The number of *S*-locus variants in a particular species can be quite large: for example, over 50 variants exist in *B. olearacea* (Ockendon 1982) and ~100 variants in *B. rapa* (Nou et al. 1993). In all Brassicaceae species analyzed, and in view of sporophytic control of pollen SI phenotype, *S*-locus alleles were found to exhibit complex genetic interactions, not only in stigma but also in pollen (Bateman 1954, Sampson 1964, Thompson and Taylor 1966). While many allelic combinations exhibit co-dominance in pollen and stigma, other allelic combinations can exhibit mutual weakening or dominant/recessive interactions, forming non-linear dominance series. Interestingly, allelic interactions can differ in stigma and pollen: for example, a particular allelic combination may exhibit co-dominance in the stigma but a dominant/recessive interaction in pollen. This non-concordance of allelic interactions in pollen and stigma was initially puzzling, but is now easily explained by molecular data (see below).

14.3 Mechanism of Recognition and Inhibition of "Self" Pollen

14.3.1 Cytological Responses

Sporophytic control of pollen SI phenotype is the only one feature that distinguishes Brassicaceae SI from other molecularly well-characterized SI systems (i.e., those of the Solanaceae, Plantaginaceae, Rosaceae, and Papaveraceae), all of which exhibit gametophytic control of pollen SI phenotype. Other features distinctive of the Brassicaceae include the site in the pistil where inhibition of pollen occurs (stigma surface rather than sub-epidermal cells or style), the site where the primary cellular response to pollen–pistil recognition takes place (the stigma epidermis rather than the pollen tube), and the ultimate outcome of self-recognition (biostatic inhibition of pollen rather than cell death).

The stigma surface in Brassicaceae is of the "dry" type (Heslop-Harrison 1975) and in a successful pollination, pollen grains obtain water for hydration and germination from stigma epidermal cells, and pollen tubes grow within the epidermal cell wall before accessing internal regions of the pistil. Therefore, it is of no surprise that the crucifer stigma, and more specifically each of the highly differentiated stigma epidermal cells known as papillar cells, functions as a highly selective sieve that determines the success or failure of both intra-specific and inter-specific pollinations. It is at the stigma surface that discrimination between self and non-self pollen occurs, with recognition of self resulting in the disruption of the very early events of pollen activation: pollen grains fail to either hydrate or germinate, and if tubes are formed at all, they fail to grow into the stigma epidermal cell wall. Consequently, inhibition of pollen does not involve the killing of pollen tubes which occurs in later-acting SI systems, where cell death is required to arrest pollen tubes that have already grown into the pistil.

An individual stigma epidermal cell can discriminate between a self and a non-self pollen grain (Dickinson et al. 2000), indicating that this cell responds to pollen grains in a spatially restricted manner. In support of this notion, microscopic observations have shown that incompatible and compatible pollen produce different cytological changes just beneath the site of pollen–stigma contact (Dickinson and Lewis 1973). Among these are callose deposition in incompatible, but not in compatible pollinations (Dickinson 1995, Dearnaley et al. 1997, Elleman and Dickinson 1999), as well as specific changes in the vacuolar network and in actin filaments involving actin reorganization and possibly degradation in incompatible pollinations and actin polymerization in compatible pollinations (Iwano et al. 2007). Although it is tempting to conclude that these changes are relevant for the outcome of pollination, it has been shown that at least callose deposition is not required for SI (Singh and Paolillo 1990, Sulaman et al. 1997).

14.3.2 Molecular Studies

14.3.2.1 Identification of the Stigma and Pollen Determinants of SI: From Immunogenetics, Protein Electrophoresis, to Molecular Cloning

As in all SI systems, it was evident early on that the most direct approach to elucidate the mechanism of self-recognition in SI was through identification of molecules encoded by the highly polymorphic *S* locus. Hence, the approach was to search for molecules that exhibited *S* allele-associated polymorphisms. The first identification of such a molecule was achieved in the 1960s in a novel approach using immunogenetic analysis of stigma extracts to identify polymorphic proteins encoded by the *Brassica S* locus (Nasrallah and Wallace 1967a and 1967b). The novelty of this approach was not in the search for polymorphic molecules or in the use of immunological methods, as these had previously been used to study SI in several other families. Rather, the novelty was in a simple shift in experimental strategy, from one based on analysis of pollen, which had been the focus of previous studies, to one based on analysis of the stigma, which had been previously largely ignored. The premise underlying these studies was that analysis of stigma extracts from plants expressing different SI specificities could lead to the identification of polymorphic *S* locus-associated molecules, a goal that had not been achieved by repeated analysis of pollen extracts. This novel approach proved to be a critical turning point in studies of SI. It sets the stage for subsequent visualization of *S* allele-associated polymorphic stigma proteins by electrophoresis (Nasrallah et al. 1970 and 1972, Nishio and Hinata 1982), as well as eventual molecular cloning of *S*-locus genes and characterization of the *S* locus, first in *Brassica* (Nasrallah et al. 1985), and subsequently in *Nicotiana* and Papaver (reviewed in Zhang and Xue 2008, Franklin-Tong 2008). In all of these SI systems, it was the pistil component of SI that was first isolated, while the pollen component was only reported many years later in the Brassicaceae (Schopfer et al. 1999), Solanaceae (reviewed in Zhang and Xue 2008), and *Papaver* (Wheeler et al., 2009).

Molecular analysis of SI in the Brassicaceae began in earnest in the early 1980s using the genetically well-characterized *S* alleles of *Brassica oleracea* and culminated in the molecular cloning of the first *S* locus-linked gene in any SI system (Nasrallah et al. 1985). This gene provided a molecular anchor within the *S* locus that enabled subsequent identification of the stigma and pollen determinants of SI.

The Stigma Determinant of SI Specificity

The first *S*-locus gene to be identified in the Brassicaceae was the *S*-Locus Glycoprotein (*SLG*) gene (Nasrallah et al. 1985). *SLG* was identified by differential screening of a stigma cDNA library and shown to be a stigma-specific gene that is tightly linked to the *S* locus and exhibits *S* allele-specific polymorphisms. Its protein product is an abundant glycoprotein expressed specifically in the stigma epidermal cell wall (Kandasamy et al. 1989), and it corresponds to the stigma-specific polymorphic protein antigens identified by the immunological and electrophoretic studies described earlier. However, even though *SLG* exhibits several of the characteristics expected for the stigma determinant of SI specificity and is found in the majority of *Brassica* and *Raphanus S* loci, this gene turned out not to have this function. Indeed, *SLG* is absent from a small number of self-incompatible *Brassica* strains (Suzuki et al. 2000) as well as from self-incompatible *A. lyrata* (Kusaba et al. 2001) and *C. grandiflora* (Nasrallah et al. 2007).

In any case, it was realized shortly after the isolation of *SLG*, that this gene is a member of a relatively large gene family, one of which was identified as another *S*-locus linked gene, the *S*-locus Receptor Kinase (*SRK*) gene (Stein et al. 1991). *SRK* encodes a single-pass transmembrane kinase with an extracellular domain that exhibits extensive sequence similarity to SLG. This SRK proved to be the stigma determinant of SI specificity. Analysis of spontaneous or induced loss-of-function *Brassica* mutants (Goring et al. 1993, Nasrallah et al. 1994 and 2000) and gain-of-function transgenic experiments (Takasaki et al. 2000) demonstrated that SRK is both necessary and sufficient for determining SI specificity in stigmas, but not in pollen. Consistent with this function, *SRK* is highly polymorphic and is expressed primarily in stigma epidermal cells, although at much lower levels (~200-fold lower) than *SLG* (Stein et al. 1991).

SRK and SLG are the prototypical members of a distinct class of receptor-like kinases (RLKs) and receptor-like proteins (RLPs) in plants that is defined by a unique extracellular "S" (for S locus) domain. In SRK and SLG, this S domain is glycosylated and contains 12 conserved cysteine residues. Similar to other S-domain RLKs, the SRK gene consists of seven exons (Fig. 14.1): exon 1 encodes the signal peptide and the extracellular S domain, exon 2 encodes the transmembrane domain, and exons 3–7 encode the cytoplasmic kinase domain (Stein et al. 1991). Based on computer-generated 3D models, the extracellular S domain of SRK (hereafter eSRK) consists of several structural modules that include sequentially from the amino terminus to the carboxy terminus (Fig. 14.1): two lectin-like domains (usually, but not always, indicative of carbohydrate-binding activity), an epidermal growth factor (EGF)-like domain containing the first six conserved cysteines, and a PAN_APPLE domain (typically a protein interaction domain) containing the remaining six conserved cysteines (Naithani et al. 2007).

The Pollen Determinant of SI Specificity

Identification of the pollen determinant of SI proved to be a difficult task. Success was eventually achieved through construction of large-insert libraries, isolation of *S* locus-derived clones by hybridization with *SLG* or *SRK* sequences, and painstakingly searching these clones for a gene having the characteristics expected for a pollen determinant. In particular, the criteria used for gene identification were that the gene had to be expressed (likely specifically) in anthers in a pattern consistent with sporophytic control of SI, it had to exhibit *S* allele-associated polymorphisms and, most importantly, it had to confer a modified SI specificity in pollen of transgenic plants.

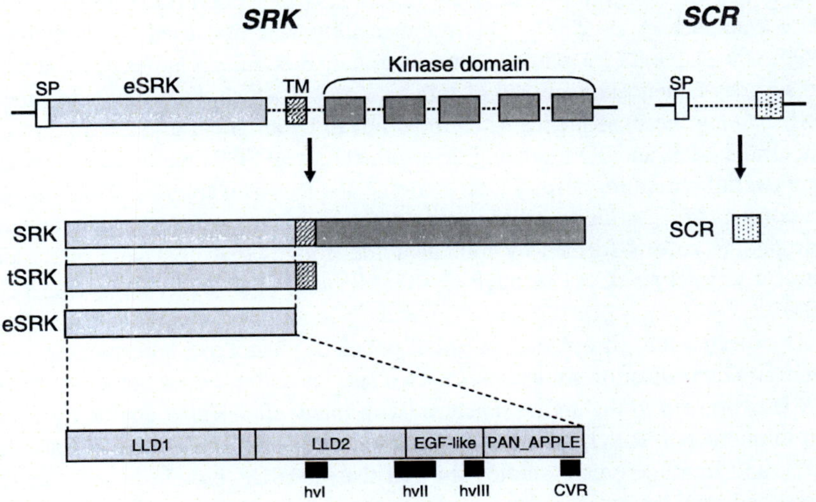

Fig. 14.1 The *S* locus-encoded stigma and pollen determinants of specificity in the self-incompatibility response. The *top* diagrams show the structures of the *SRK* and *SCR* genes. Exons are depicted by *rectangles* and introns by *dashed lines* to indicate that the size of introns can vary dramatically between alleles. Below the genes are the structures of the mature SRK isoforms and SCR proteins lacking their signal peptides (SP). The full-length SRK consists of an extracellular S domain or eSRK (*light gray*), followed by a transmembrane domain (TM; *cross-hatch*), and a serine-threonine kinase domain (*dark gray*). The tSRK is a truncated protein lacking the kinase domain, and the eSRK is a soluble form of the extracellular domain. The *vertical lines* within the eSRK diagram at bottom delineate the signal peptide and the various structural domains predicted by 3D modeling: LLD1: lectin-like domain 1; LLD2: lectin-like domain2; EGF-like; PAN_APPLE. The four *black boxes* show the location of the four hypervariable regions (hvI, hvII, hvIII, and CVR)

All three criteria were fulfilled by a gene that was isolated from a bacterial artificial chromosome (BAC) clone derived from the *S8* locus of *Brassica rapa* (Schopfer et al. 1999). This gene was expressed specifically in anthers and, importantly, it conferred the expected SI specificity in transgenic pollen. Sequence analysis showed the presence of two exons (Fig. 14.1), the first encoding a signal peptide and the second encoding a small protein approximately 50 amino acids in length that contained eight cysteine residues. Accordingly, the gene was named the *S*-locus *C*ysteine-*R*ich (*SCR*) gene (Schopfer et al. 1999). Furthermore, comparison of the *B. rapa SCR8* allele with two *SCR* alleles isolated from *B. oleracea* (Schopfer et al. 1999) revealed the extreme polymorphism of SCR variants. Only a small number of residues were found to be conserved among SCRs: eight cysteine residues, a glycine residue in the GlyxCys2 motif, and an aromatic residue in the Cys3xxxTyr/Phe motif.

Subsequent studies confirmed the results of this first analysis of SCR function and described additional features of the gene that are consistent with known characteristics of SI in the Brassicaceae. Alleles of this gene were isolated from several self-incompatible strains and species. In the process, the gene was given a second designation, SP-11 (S pollen #11), by researchers in Japan (Takayama and Isogai 2005, Watanabe et al. 2000). This nomenclature was first used by Suzuki et al. (1999) who sequenced a genomic clone spanning a *B. rapa S* haplotype and identified several sequences expressed in anthers, none of which were shown to function in SI. Among these, sequence #11 turned out to be an allele of *SCR*. In any case, structural studies and 3D modeling determined that all SCRs analyzed assumed the same overall structure, despite their extreme sequence divergence. This 3D structure consists of a cystine-stabilized αβ fold similar to defensins, in which the buried hydrophobic core is formed by four disulfide bridges that link pairs of the eight conserved cysteines (Cys1 through Cys8) (Mishima et al. 2003, Chookajorn et al. 2004).

As expected for sporophytic control of pollen SI phenotype, reporter gene assays and in situ hybridization studies in *Brassica* species and *A. lyrata* demonstrated that *SCR* is expressed in the anther tapetum (with many alleles also exhibiting expression in microspores; Schopfer and Nasrallah 2000, Shiba et al. 2001, Kusaba et al. 2002), and immunological studies showed that the SCR protein is secreted from tapetal cells and becomes incorporated into the outer pollen coat (Shiba et al. 2001, Kachroo et al. 2001, Takayama et al. 2001). Furthermore, expression studies of *SCR* in *S*-locus homozygotes and heterozygotes provided an explanation for the allelic interactions of dominance/recessiveness exhibited by some combinations of *S* haplotypes in pollen. It was shown that *SCR* alleles derived from pollen-recessive *S* haplotypes are silenced in the presence of a pollen-dominant *S* allele (Kakizaki et al. 2003, Kusaba et al. 2002, Shiba et al. 2002). Unexpectedly, this silencing of recessive *SCR* alleles is not effected by transcripts of the dominant "suppressing" *SCR* allele: it does not occur when transcripts of a dominant allele are expressed from a transgene in a plant homozygous for a recessive allele (Schopfer et al. 1999) and it does not even require that the "suppressing" allele be transcribed (Fujimoto et al. 2006a). Like other silenced genes, a recessive *SCR* allele exhibits increased promoter methylation (Shiba et al. 2006), but the specific features in pollen-recessive *SCR* alleles that make these genes susceptible to silencing are not known.

14.3.2.2 The S Haplotype and Control of Recognition Specificity

The fact that SI specificity is determined by two tightly linked genes rather than one, means that the classical designation of "*S* allele" is no longer valid. Rather, we use the more appropriate term "*S* haplotype" to designate a particular variant of the *S* locus containing specific *SRK* and *SCR* alleles that together determine a unique SI specificity. For example, the *B. oleracea S6* haplotype contains the *SRK6* and *SCR6* alleles and determines *S6* recognition specificity.

S haplotypes containing closely linked *SRK* and *SCR* genes have been identified in all self-incompatible crucifer species examined to date, including *Brassica* and *Raphanus* species, *A. lyrata*, and *C. grandiflora*. However, the chromosomal context in which the *S* locus is found can differ between taxa due to the large-scale chromosomal rearrangements that differentiate these taxa (Conner et al. 1998, Kusaba et al. 2001). Additionally, for at least one set of *SRK/SCR* alleles in most of the genera analyzed, transgenic experiments have been performed, which demonstrate that these two genes determine SI specificity in stigma and pollen, as first demonstrated in *Brassica* species. By extension, it may be safely concluded that *SRK* and *SCR* genes arranged in close linkage within an *S* haplotype must also determine SI specificity across the Brassicaceae.

Shortly after the isolation of *SRK* and *SCR,* it was realized that the tight genetic linkage and highly polymorphic nature of these genes and the location of their protein products at the stigma surface and in the pollen coat, respectively, were consistent with the notion that these proteins function as "matched" pairs of receptor and ligand that determine the outcome of pollination through their *S* haplotype-specific interactions (Fig. 14.2). Studies of SRK and SCR proteins were therefore launched to test this prediction.

SRK was found to produce in stigmas, not only the full-length SRK receptor kinase shown to be an integral plasma membrane protein (Stein et al. 1996), but also two shorter products: a soluble version of the extracellular domain, designated

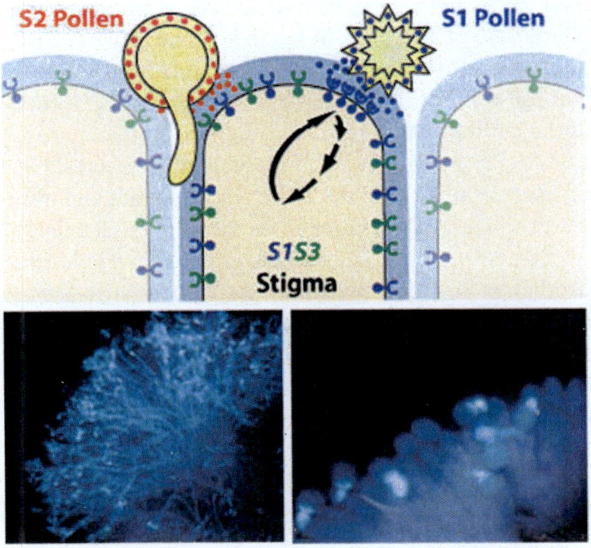

Fig. 14.2 Receptor–ligand interactions and recognition of "self" pollen at the stigma epidermal cell surface. The diagram shows the behavior of "self" and "nonself" pollen tubes at the surface of a single epidermal cell in a self-incompatible S1S3 heterozygote. The SCR ligand is located in the pollen coat and is delivered to the epidermal cell surface when the grain lands on the stigma. The SCR1 ligand from pollen grains produced by plants expressing the S1 haplotype binds and activates the SRK1 receptor (*blue*) in the S1S3 stigma cell, triggering a signal transduction cascade that leads to inhibition of pollen hydration, germination, and tube growth. In contrast, a pollen grain derived from a plant that expresses neither the S1 nor the S3 haplotypes, e.g., the S2 haplotype (*red*) produces an SCR2 ligand that fails to bind and activate SRK and pollen tube growth proceeds unimpeded. The micrographs below the diagram show successful pollen tube development in a "non-self" (cross) pollination (*left*) and the inhibition of pollen tube growth in a "self" pollination (*right*). Modified from Nasrallah (2005)

eSRK (Giranton et al. 2000), and a membrane spanning protein, designated tSRK, which is produced by post-translational cleavage and consists of the extracellular, transmembrane, and juxtamembrane domains but lacks the kinase domain (Shimosato et al. 2007) (Fig. 14.1). SRK was shown to form oligomers in unpollinated stigmas (Giranton et al. 2000) and to interact physically with SCR via its extracellular domain, both in vitro and in vivo (Kachroo et al. 2001, Takayama et al. 2001, Chookajorn et al. 2004). Importantly, the SRK–SCR interaction is *S* haplotype specific, such that a particular SCR variant will only bind and activate the SRK variant encoded by the same S haplotype but not an SRK variant encoded by a different S haplotype. This *S* haplotype-specific SRK–SCR interaction explains the exquisite specificity in the recognition of self pollen that is the hallmark of the SI response. Furthermore, the rapid activation of the SRK kinase (Cabrillac et al. 2001, Takayama et al. 2001, Shimosato et al. 2007) triggered by this interaction is consistent with the rapid inhibition of self pollen at the stigma surface.

A detailed description of the SRK–SCR recognition complex and its molecular composition is lacking, however. In particular, the role of each of the various SRK isoforms is not well understood. A recent study has shown that the membrane-tethered SRK and tSRK, but not the soluble eSRK, exhibit high-affinity binding to SCR in vivo (Shimosato et al. 2007). What then is the role of eSRK? Is it required for SI and if so, in what capacity? The eSRK is predicted to be secreted into the stigma epidermal cell wall, similar to SLG. It is possible, therefore, that the two proteins might fulfill a similar function, which might explain why SLG is dispensable in some *Brassica* strains and in other self-incompatible species. Clues to this function derive from analysis of SLG in the *B. rapa scf1* mutant and transgenic tobacco. The *scf1* mutant exhibits breakdown of SI in the stigma, drastically reduced SLG transcript and protein levels, and lack of SRK protein despite normal levels of *SRK* transcripts (Nasrallah et al. 1992, Dixit et al. 2000). Furthermore, in tobacco leaves expressing *SRK* under control of the Cauliflower Mosaic Virus 35S promoter, the SRK protein exhibits aberrant oligomerization when expressed alone but not when co-expressed with SLG (Dixit et al. 2000). These results suggest that SLG and possibly eSRK might be required for the stabilization or proper maturation of the SRK full-length receptor.

14.3.2.3 Signal Transduction

Also poorly understood is how the SRK–SCR interaction and subsequent activation of the SRK kinase culminate in inhibition of self pollen. A number of proteins have been identified on the basis of their interaction with the kinase domain of *Brassica* SRK in the yeast two-hybrid system. These include the *A*rmadillo *r*epeat-*c*ontaining protein ARC1 (Gu et al. 1998), the thioredoxin h-like proteins THL1 and THL2 (Bower et al. 1996), the kinase-associated protein phosphatase KAPP, calmodulin, and a sorting nexin (Braun et al. 1997, Vanoosthuyse et al. 2003). While a role in SI for the latter three proteins remains to be established, evidence indicates that THL1 and THL2 function as negative effectors that maintain SRK in an inactive state in the absence of SCR (Cabrillac et al. 2001, Haffani et al. 2004). In contrast,

ARC1, a stigma-specific member of the plant U-box family that exhibits E3 ligase activity, is thought to function as a positive effector of SI (Stone et al. 1999 and 2003). Antisense suppression of *ARC1* in *B. napus* was associated with partial loss of SI. Furthermore, ARC1 is phosphorylated by SRK and it co-localizes with the proteasome in the presence of an active SRK in tobacco cell cultures (Stone et al. 2003). It has been suggested that subsequent to SRK activation in the stigma, ARC1 is phosphorylated by SRK and then effects the ubiquitination and degradation of specific stigma proteins that function either as inhibitors of SI or as factors required for successful pollen tube growth (Stone et al. 1999 and 2003). It is interesting to note that ARC1 shuttles between cytosol and nucleus in tobacco suspension cells (Stone et al. 2003). However, it is not known if ARC1 behaves in this manner in the stigma and if such behavior might be important for SI. In any case, a yeast 2-hybrid screen recently identified a protein related to Exo70 as an ARC1 interactor (Samuel et al. 2008). Exo70 is a conserved component of the exocyst, a complex that functions in polarized secretion. In the context of the stigma, this protein appears to function in compatible pollen tube growth, suggesting that during the SI response, ARC1 would suppress the function of this protein, causing pollen rejection (Samuel et al. 2008).

Another positive effector of SI has been identified by map-based cloning of the *m* (*modifier*) mutation, a spontaneous mutation that causes breakdown of SI in the stigma of *B. rapa* (Murase et al. 2004). This protein is a cytoplasmic plasma membrane-tethered kinase designated *M-locus protein kinase. MLPK* is thought to function in a complex with SRK because it interacts with SRK in tobacco protoplasts and can complement the *m* mutation in transient assays of stigma epidermal cells (Murase et al. 2004, Kakita et al. 2007a and 2007b). It should be noted, however, that a role for MLPK in SI has not as yet been established by complementation of the *m* mutation by stable transformation of *Brassica* plants with a wild-type MLPK allele.

Elucidation of the SRK signal transduction pathway clearly holds the key to understanding how "self" pollen is inhibited at the stigma surface. Thus, the response phase of SI will be a major focus of study in the future. As had been repeatedly demonstrated for other biological phenomena, progress in this area of research is expected to result from application of forward genetic approaches involving screens for mutants that exhibit defects in the SI response and subsequent map-based cloning of the affected genes. Attempts have been made to mutagenize self-incompatible *Brassica* plants (Nasrallah et al. 2000) and self-fertile mutants have been isolated. However, such mutant screens have been hampered by difficulties in implementing standard seed mutagenesis schemes in self-incompatible plants that do not set seed. Similarly, target gene isolation and validation are complicated by the highly duplicated genomes and inefficient transformation methods in *Brassica* species. These hurdles have been recently overcome by transfer of the SI trait into the model crucifer *A. thaliana* (Nasrallah et al. 2002 and 2004), which has many known advantages as a molecular genetic experimental system, not least of which are established and efficient mutagenesis and transformation methods. The normally highly self-fertile *A. thaliana* was shown to express SI when transformed with functional *SRK* and *SCR* genes isolated from the *Sb* haplotype of *A. lyrata,* the closest self-incompatible relative of *A. thaliana* (Nasrallah et al. 2002 and 2004). This transgenic *A. thaliana* self-incompatible model system promises to facilitate the design of efficient screens for mutants exhibiting loss of SI and cloning of genes required for SI.

14.4 *S* haplotype Structure, Suppressed Recombination, and Diversification

The integrity of the *S* haplotype and the absolute genetic linkage of matched alleles of SRK and SCR are clearly critical for persistence of SI. Self-incompatible plants are typically heterozygous at the *S* locus, and recombination events between *SRK* and *SCR* would produce recombinant *S* haplotypes having mismatched *SRK* and *SCR* alleles whose products could not interact to effect inhibition of self pollen. Therefore, some mechanism(s) must act to maintain the linkage of these genes over time. In fact, several studies have demonstrated that the *S*-locus region exhibits suppressed recombination. In all genetic studies performed over more than half a century in various species, segregation ratios in populations of plants segregating for S haplotype were consistent with the inheritance of a single genetic locus. In none of these studies were recombinants isolated, even when large populations were analyzed (Casselman et al. 2000) and even for *S* haplotypes that extend over several hundred kilobases (Boyes et al. 1997, Casselman et al. 2000). Furthermore, linkage disequilibrium has been estimated to extend over 400–600 kilobases in the *A. lyrata S*-locus region (Hagenblad et al. 2006, Kamau and Charlesworth 2005). Several factors likely underlie the low frequency of recombination between *SRK* and *SCR*. In some *S* haplotypes, recombination may be rare because of the close physical proximity of the two genes, such as in the *B. rapa S8, A. lyrata Sa, and C. grandiflora S7* haplotypes (Fig. 14.3; Schopfer et al. 1999, Kusaba et al. 2001, Nasrallah et al. 2007). However, the size of *S* haplotypes can vary extensively (Boyes et al. 1997, Kusaba et al. 2001), suggesting that other factors must contribute to low recombination frequencies. Indeed, different *S* haplotypes within a species have been shown to be structurally heteromorphic, i.e., they exhibit highly rearranged gene order, with *SRK* and *SCR* occupying different locations relative to each other and to flanking markers (Fig. 14.3; Boyes et al. 1997, Kusaba et al. 2001). Furthermore, similar to other chromosomal regions that exhibit suppressed recombination due to long-term independent evolution, *S* haplotypes contain haplotype-specific sequences that are often related to transposons (Boyes et al. 1997, Fujimoto et al. 2006b, Sherman-Broyles et al. 2007).

14.4.1 Diversification of SRK and SCR

Understanding the basis of specificity in the SRK–SCR interaction, how the two proteins co-evolve, and how new SI specificities were generated is a major goal of future studies of SI. No doubt, the answer to these questions must lie in the

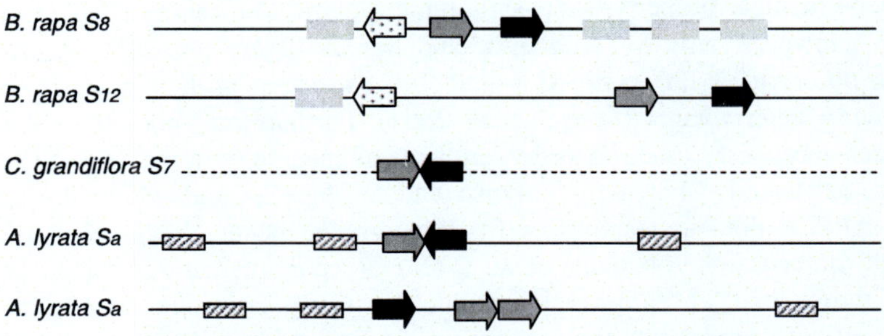

Fig. 14.3 Intra-specific and inter-specific structural heteromorphism of *S* haplotypes. In each *S* haplotype, the 5' -> 3' gene orientation is depicted by *black arrows* for *SRK*, *dark gray arrows* for *SCR,* and *stippled arrows* for *SLG* (which is absent from the *S* locus of *Arabidopsis* and *Capsella* species). *S* locus-linked genes that do not function in SI are shown as *rectangles*. Note that the genomic context of the *S* locus differs between *Brassica* species and *Arabidopsis* species; thus, the genes that flank *SRK* and *SCR* in the two genera are different, as indicated by the different shading of these rectangles (*gray* for *B. rapa* and *cross-hatched* for *A. lyrata*). The *dashed line* in the *C. grandiflora S7* haplotype indicates that flanking genes have not been identified in this species. The drastic differences in the orientations of the *S*-locus genes relative to each other and to flanking genes and the varying distances separating the genes contribute to reduced recombination in the *S*-locus region

highly polymorphic sequences of the eSRK and SCR. In all species investigated to date, SRKs exhibit high levels of intra-specific sequence divergence (e.g., Stein et al. 1991, Kusaba et al. 2001, Paetsch et al. 2006). SRK variants can exhibit as much as 35 and 51% amino acid sequence divergence within *Brassica* species (Stein et al. 1991) and *A. lyrata* (Schierup et al. 2001), respectively. SCR variants exhibit even greater intra-specific polymorphisms, and amino acid sequence divergence in excess of 60% is often observed between SCR variants (Schopfer et al. 1999, Watanabe et al. 2000, Kusaba et al. 2001). Importantly, alleles of both *SRK* and *SCR* exhibit trans-specific and even trans-generic polymorphisms, i.e., alleles from one species may share higher sequence identity with alleles from a different species or genus than with other alleles in the same species (Schierup et al. 2001, Bechsgaard et al. 2006, Kimura et al. 2002, Sato et al. 2003, Nasrallah et al. 2007). This feature indicates that *SRK* and *SCR* polymorphisms are very old and are subject to diversifying selection (Bechsgaard et al. 2006, Charlesworth et al. 2006).

In the eSRK, polymorphic residues are scattered throughout the entire sequence, with concentrations in four regions (Fig. 14.1): three designated "hypervariable" (hv) regions hvI, hvII, and hvIII, and a fourth called the "C-terminal Variable Region" or CVR (Kusaba et al. 1997, Nasrallah et al. 1987). These regions are thought to function as determinants of SI specificity because they often, though not always, exhibit ratios of non-synonymous to synonymous substitutions (K_a/K_s) greater than 1, which is indicative of diversifying selection rather than relaxed constraint. These regions also contain the majority of amino acid residues having a high posterior probability of being under selective pressure to change in physico-chemical property, such as volume, polarity, or charge (Sainudiin et al. 2005). However, only hvI and hvII, and not hvIII, are perfectly conserved in SRKs that exhibit the same SI specificity, consistent with a potential role as specificity determinants (Miege et al. 2001). Furthermore, not all polymorphic residues necessarily function as determinants of specificity in the SRK–SCR interaction, as recently shown for polymorphic residues located within the CVR in PAN_APPLE domain responsible for ligand-independent eSRK self-association or self-dimerization (Naithani et al. 2007). Recent interaction studies in yeast have shown that 11 amino acids within the CVR contribute to specificity in SRK self-association (i.e., the propensity of a particular eSRK to exhibit stronger self-interaction with itself than with other eSRK variants) (Naithani et al. 2007).

Thus, identification of amino acid residues or regions in the eSRK that determine SI specificity cannot be achieved by sequence inspection alone, but rather requires empirical evidence from domain swapping or site-directed mutagenesis experiments. Such evidence was recently obtained (Boggs et al. 2009a) using a newly developed transgenic self-incompatible *A. thaliana* model system (Nasrallah et al. 2002 and 2004) that allows efficient and relatively rapid *in planta* functional analysis of large numbers of engineered eSRK chimeras and site-directed mutants. After confirming that a fragment encompassing the hvI, hvII, and hvIII regions is sufficient to determine a particular SRK specificity, a detailed analysis of the hvI–hvIII region from two SRK variants determined that substitutions at the majority of polymorphic sites in this region, even those showing signals of positive selection had no effect on SRK function (Boggs et al. 2009a). In fact, only a few of the ~100 polymorphic sites were found to be essential for the activation of SRK by its cognate SCR (Boggs et al. 2009a). These essential sites are clustered in two non-contiguous regions, one located toward the C-terminal end of hvI and the other located in the C-terminal half of hvII (Boggs et al. 2009a). A high-resolution 3D structure of the eSRK in its ligand-bound and unbound forms is required to determine if these two amino acid clusters are surface-exposed regions that are brought into close proximity in a 3D structure to form part of an SCR-binding pocket.

In the case of SCR, sequence alignments are not helpful for focusing attention on potential specificity-determining residues because the entire sequence is polymorphic. Nevertheless, such residues have been identified for at least two SCR alleles because of the availability of an in vivo pollination assay that allows rapid functional analysis of site-directed mutants of SCR or chimeric proteins generated by swapping domains between two different SCR variants (Chookajorn et al. 2004, Sato et al. 2004). In this assay, an SCR protein expressed in bacteria or produced synthetically is manually applied to a stigma expressing the cognate SRK, after which the activation of this SRK is assayed by pollinating with non-self pollen. For example, saturating an *S13* stigma with SCR13 protein will cause inhibition of pollen grains that are normally compatible with *S13* stigmas, such as *S2* and *S6* pollen. Use of this assay determined that the region between Cys5 and Cys6, which is predicted to form a surface-exposed loop that might be involved in binding eSRK, contains specificity determinants in two SCRs analyzed (Chookajorn et al. 2004, Sato et al. 2004). However, it has not been possible to extrapolate these results to other SCR variants (Chookajorn et al. 2004), possibly because other regions function as specificity determinants in other SCRs.

14.5 Mating-Type Dimorphism in the Brassicaceae: Loss of SI and the Switch From an Outbreeding to a Self-Fertile Mode of Mating

The obligate out-crossing mode of mating conferred by SI is clearly advantageous. At the population level, it produces high levels of polymorphisms and increased capacity to adapt to a variety of environmental challenges. It also confers a distinct advantage on individual plants that express a new SI specificity because of increased number of potential mating partners. Despite these advantages, however, SI has been repeatedly lost in self-incompatible lineages (Igic et al. 2006), including the Brassicaceae (Foxe et al. 2009, Guo et al. 2009, Nasrallah et al. 1992, 1994, and 2007, Mable et al. 2005, Okamoto et al. 2007) and almost half the taxa in the Brassicaceae are self-compatible (Bateman 1955). Of course, the establishment of self-fertile strains and species resulting from loss of SI must be accompanied by purging of the recessive deleterious mutations (i.e., the genetic load) and associated inbreeding depression that typically characterizes obligate out-crossers (Barrett 1988). In any case, determining how SI was lost in different lineages is important from an evolutionary perspective because the mating system adopted by a plant species has profound consequences for levels of polymorphism in populations and for the mode and tempo of evolutionary change. Additionally, elucidating the genetic basis of the switch to self-fertility has the potential to identify novel factors required for the operation of SI.

14.5.1 Analysis of Self-Fertility in Non-model Members of the Brassicaceae

The genetic basis of loss of SI has been investigated in two ways: (1) by analysis of self-compatible strains that arose spontaneously or were experimentally induced in self-incompatible strains, and (2) by investigating self-fertile inter-specific hybrids generated by crossing a self-incompatible and a self-compatible species. These analyses have shown the occurrence of mutations at loci outside the *S* locus (called SI "modifier" loci) that disrupt the SI response without reducing expression of *S*-locus genes (e.g., the *B. rapa m* mutation mentioned earlier; Murase et al. 2004), which likely disrupt components of the *SRK* signal transduction pathway. However, *S*-locus genes are by far the major targets of natural or experimental selection for self-fertility. Indeed, loss of SI frequently occurs by mutations in the *S*-locus genes themselves (Goring et al. 1993, Nasrallah et al. 1992, 1994, and 2000, and 2007, Schopfer et al. 1999, Okamoto et al. 2007). Additionally, the *S* locus can be targeted indirectly by mutations at modifier loci that affect expression of its genes, such as the *B. rapa scf1* mutation described earlier (Nasrallah et al. 1992).

Disrupted *S*-locus gene expression was also found to underlie self-fertility in inter-specific hybrids, as shown in the self-fertile homoploid hybrids generated by crossing a self-incompatible strain of *C. grandiflora* carrying the *S7* haplotype with a self-fertile strain of *Capsella rubella* that harbors a non-functional *S* (*S0*) locus (Nasrallah et al. 2007). Analysis of F1 and F2 progenies derived from this cross demonstrated that self-fertility segregates as a dominant pollen-specific trait and is due to silencing of the *C. grandiflora SCR* allele in *S7S0* heterozygous anthers, similar to the silencing of recessive *SCR* alleles in plants heterozygous for two functional *S* haplotypes (Nasrallah et al. 2007).

14.5.2 Analysis of Self-Fertility in the Model Plant A. thaliana

Recent studies of naturally occurring transitions from out-breeding to selfing in the Brassicaceae have focused on the model plant *A. thaliana.* This small self-fertile plant and its closest self-incompatible relatives, *A. lyrata* and *A. halleri,* are thought to have diverged from a common ancestor approximately 5 million years ago (Ramos-Onsins et al. 2004). Therefore, parallel studies of *A. thaliana* and its self-incompatible sister species promise to address many questions regarding the differences in mating system, life history, and ecology of these species. Of course, a major strength of these studies lies in the many advantages of *A. thaliana* as a molecular genetic model, including a sequenced and annotated genome (Arabidopsis Genome Initiative 2000), a short life cycle well suited to genetic studies, ample resources for map-based cloning and gene identification, availability of over 750 accessions collected from various geographical locations and of a large number of knock-out mutants, and ease of transformation that allows efficient functional analyses. Importantly for analysis of out-crossing and selfing, these well-known advantages were further augmented by successful transfer of the SI trait to *A. thaliana* by transformation with *SRK–SCR* gene pairs isolated from functional *S* haplotypes of *A. lyrata* (Boggs et al. 2009b, Nasrallah et al. 2002 and 2004).

Initial studies of self-fertility in *A. thaliana* focused on the *S* locus and its *SRK* and *SCR* sequences, first in the reference Col-0 accession and subsequently in other accessions. In all cases investigated so far, the *S* locus was found to contain mutated *SRK* or *SCR* sequences and is therefore referred to as the pseudo-*S* (ΨS) locus (Boggs et al. 2009c, Kusaba et al. 2001, Nasrallah et al. 2002 and 2004, Sherman-Broyles et al. 2007, Shimizu et al. 2004 and 2008, Tang et al. 2007). Interestingly, *A. thaliana* exhibits substantial *S*-locus polymorphisms, with different accessions harboring distinct ΨS haplotypes that differ in gene content and organization. Species wide, three classes of ΨS haplotypes were identified, which are orthologous to three distinct functional *S* haplotypes of *A. lyrata* and *A. halleri* (Bechsgaard et al. 2006). Each of these classes comprises ΨS haplotypes that differ in the extent of decay of their ΨSRK and ΨSCR sequences, with evidence for partial or complete deletion and rearrangement of these sequences. Further, ΨS haplotypes can differ significantly in their overall organization, either because they belong in classes descended from distinct functional *S* haplotypes or as a result of subsequent rearrangements or inter-haplotypic recombination (Sherman-Broyles et al. 2007).

The availability of the transgenic self-incompatible *A. thaliana SRK–SCR* model has shown that the species also harbors cryptic mutations at SI modifier loci. This conclusion is based on the observation that transformation of different accessions with functional *A. lyrata SRK* and *SCR* genes (hereafter *AlSRK–SCR*) results in SI phenotypes that vary in strength and stability, as manifested by pollen tube counts and amount of seed set (Boggs et al. 2009c, Nasrallah et al. 2002 and 2004). For example, *AlSRK–SCR* transformants of the C24 accession exhibit an SI phenotype identical to that of self-incompatible *A. lyrata*, both in its strength and its stability over the course of stigma development. In contrast, *AlSRK–SCR* transformants of several accessions, including Col-0, exhibit transient SI or pseudo-compatibility, whereby robust SI is observed in young mature stigmas but is lost at later stages of development. These differences indicate that, while both the C24 and the Col-0 *S* haplotypes are non-functional, all other genes required for SI are functional in C24 but not in Col-0 (Nasrallah et al. 2004). Crosses between *AlSRK–SCR* plants of different accessions established a genetic basis for these differences and demonstrated the dominance of stable SI over transient SI with several contributing loci (Boggs et al. 2009c, Nasrallah et al. 2002, Liu et al. 2007). Map-based cloning identified *PUB8 (PLANT U-BOX8)*, a gene located at one boundary of the *S* locus, as a modifier that causes transient SI (Liu et al. 2007). *PUB8* encodes a previously uncharacterized Arm repeat- and U box-containing protein that appears to regulate *SRK* transcript levels at late stages of stigma development (Liu et al. 2007). Accessions that exhibit transient SI contain a *PUB8* allele that is expressed at lower levels than the *PUB8* allele of C24 plants and this reduced expression is associated with reduced *SRK* transcript levels. This result indicates that transient SI is largely due to reduced *SRK* transcripts and protein at late stages of stigma development (Liu et al. 2007).

Together, these studies of *S*-locus polymorphism studies and of cryptic polymorphisms at SI modifier loci have suggested that the switch to self-fertility in *A. thaliana* did not occur by a single event, such as by selective sweep of a single mutation as initially suggested (Shimizu et al. 2004). Rather, it appears to have involved multiple mutational events, possibly occurring in distinct small populations, where a gene that promotes selfing would have a strong selective advantage (Sherman-Broyles et al. 2007, Tang et al. 2007).

14.6 Future Prospects

Much progress has been made in the past few years toward elucidating the mechanism of SI in the Brassicaceae. The *SRK* receptor and its *SCR* ligand have been identified as the stigma and pollen determinants of specificity in the SI response and the puzzle of how the stigma can discriminate between "self" and "non-self" pollen has been solved. The molecular characterization of the *S* locus has also set the stage for investigating aspects of the evolution of SI. However, much remains to be learned about both the recognition and the response phases of SI in this family. One of the most challenging tasks is to understand how SRK and SCR co-evolve to maintain their interaction and how the large repertoire of SI specificities was generated. In the more immediate future, studies of SI must provide a detailed description of the SRK–SCR interaction and of the 3D structure of the receptor in its ligand-bound and unbound forms and define the composition of the active recognition complex. Similarly, it will be important to elucidate the SI signal transduction pathway and determine how the haplotype-specific SRK–SCR interactions at the stigma surface ultimately lead to arrest of pollen tube development.

Addressing many of these mechanistic issues is likely to be facilitated by the availability of the recently established transgenic *A. thaliana* self-incompatible model. The expectation is that this model will allow saturation mutagenesis of the SI response and generation of a collection of mutants defective in SI signaling, efficient in vivo functional analysis of receptor/ligand chimeras and site-directed mutants to identify amino acid residues that determine SI specificity, and genetic analyses aimed at understanding the basis of dominant–recessive interactions of *SRK* and *SCR* alleles. In addition, genome-sequencing projects currently underway in several Brassicaceae species, including *Brassica* species, *A. lyrata,* and *C. rubella,* will allow genome-wide comparisons between species with similar and different modes of mating. These comparisons and parallel analyses of cryptic natural variation for expression of SI in *A. thaliana* will no doubt provide insight into the origin of SI and its genes and illuminate the evolution of breeding systems in the Brassicaceae.

Acknowledgment Research in the author's laboratory is supported by grants from the United States National Science Foundation and Department of Agriculture.

References

Arabidopsis Genome Iniative (2000) Analysis of the genome sequence of the flowering plant Arabidopsis thaliana. Nature 408:796–815

Barrett SCH (1988) The evolution, maintenance, and loss of self-incompatibility systems. In: Lovett Doust J and Lovett Doust L (eds) Plant reproductive ecology: patterns and strategies, pp 98–124. Oxford University Press, Oxford

Bateman AJ (1954) Self-incompatibility systems in angiosperms. II. *Iberis amara.* Heredity 8: 305–332

Bateman AJ (1955) Self-incompatibility systems in angiosperms III. cruciferae. Heredity 9:53–68

Bechsgaard J, Bataillon T, Schierup MH (2004) Uneven segregation of sporophytic self-incompatibility alleles in *Arabidopsis lyrata.* J Evol Biol 17:554–561

Bechsgaard J, Castric V, Charlesworth D, Vekemans V, Schierup MH (2006) The transition to self-compatibility in *Arabidopsis thaliana* and evolution within S-haplotypes over 10 million years. Mol Biol Evol 23:1741–1750

Boggs NA, Dwyer KG, Nasrallah ME, Nasrallah JB (2009a) In vivo detection of residues required for ligand-selective activation of the S-locus receptor in arabidopsis. Curr Biol 19:786–791

Boggs NA, Dwyer KG, Shah P, McCulloch AA, Nasrallah ME, Nasrallah JB (2009b) Expression of distinct self-incompatibility specificities in *Arabidopsis thaliana.* Genetics 182:1313–1321

Boggs NA, Nasrallah ME, Nasrallah JB (2009c) Independent S-locus mutations caused self-fertility in *Arabidopsis thaliana.* PLoS Genet 5:e1000426

Bower MS, Matias DD, Fernandes-Carvalho E, Mazzurco M, Gu T, Rothstein SJ, Goring D (1996) Two members of the thioredoxin-h family interact with the kinase domain of a *Brassica* S locus receptor kinase. Plant Cell 8:1641–1650

Boyes DC, Nasrallah ME, Vrebalov J, Nasrallah JB (1997) The self-incompatibility (*S*) haplotypes of *Brassica* contain highly divergent and rearranged sequences of ancient origin. Plant Cell 9:237–247

Braun DM, Stone JM, Walker JC (1997) Interaction of the maize and arabidopsis kinase interaction domains with a subset of receptor-like kinase: implications for transmembrane signaling in plants. Plant J 12:83–95

Cabrillac D, Cock JM, Dumas C, Gaude T (2001) The S-locus receptor kinase is inhibited by thioredoxins and activated by coat proteins. Nature 410:220–223

Casselman AL, Vrebalov J, Conner JA, Singhal A, Giovannoni J, Nasrallah ME, Nasrallah JB (2000) Determining the physical limits of the *Brassica S* locus by recombinational analysis. Plant Cell 12:23–33

Charlesworth D, Kamau E, Hagenblad J, Tang C (2006) *Trans*-specificity at loci near the self-incompatibility loci in *Arabidopsis*. Genetics 172:2699–2704

Charlesworth D, Mable BK, Schierup MH, Bartolome C, Awadalla P (2003) Diversity and linkage of genes in the self-incompatibility gene family in *Arabidopsis lyrata*. Genetics 164: 1519–1535

Chookajorn T, Kachroo A, Ripoll DR, Clark AG, Nasrallah JB (2004) Specificity determinants and diversification of the *Brassica* self-incompatibility pollen ligand. Proc Natl Acad Sci USA 101:911–917

Conner JA, Conner P, Nasrallah ME, Nasrallah JB (1998) Comparative mapping of the *Brassica S* locus region and its homeolog in arabidopsis: implications for the evolution of mating systems in the brassicaceae. Plant Cell 10:801–812

Dearnaley J, Levine N, Lew R, Heath I, Goring D (1997) Interrelationships between cytoplasmic Ca^{2+} peaks, pollen hydration, plasma membrane conductances during compatible and incompatible pollinations of *Brassica napus* papillae. Plant Cell Physiol 38: 985–999

deNettancourt D (2001) Incompatibility and incongruity in wild and cultivated plants. Springer, Berlin, Heidelberg, New York, NY

Dickinson H (1995) Dry stigmas, water and self-incompatibility in *Brassica*. Sex Plant Reprod 8:1–10

Dickinson HG, Elleman CJ, Doughty J (2000) Pollen coatings – chimaeric genetics and new functions. Sex Plant Reprod 12:302–309

Dickinson HG, Lewis D (1973) Cytochemical and ultrastructural differences between intraspecific compatible and incompatible pollinations in *Raphanus*. Proc R Soc Lond B 183:21–28

Dixit R, Nasrallah ME, Nasrallah JB (2000) Post-transcriptional maturation of the S receptor kinase of *Brassica* correlates with co-expression of the S-locus glycoprotein in the stigmas of two brassica strains and in transgenic tobacco plants. Plant Physiol 124:297–311

Elleman CJ, Dickinson HG (1999) Commonalities between pollen/stigma and host/pathogen interactions: calcium accumulation during stigmatic penetration by *Brassica oleracea* pollen tubes. Sex Plant Reprod 12:194–202

Foxe JP, Slotte T, Stahl EA, Neuffer B, Hurka H, Wright SI (2009) Recent speciation associated with the evolution of selfing in capsella. Proc Natl Acad Sci USA 106(13):5241–5245

Franklin-Tong VE (2008) Self-incompatibility in flowering plants. Springer, Berlin, Heidelberg, New York, NY

Fujimoto R, Okazaki K, Fukai E, Kusaba M, Nishio T (2006b) Comparison of the genome structure of the self-incompatibility (*S*) locus in interspecific pairs of *S* haplotypes. Genetics 173: 1157–1167

Fujimoto R, Sugimura T, Fukai E, Nishio T (2006a) Suppression of gene expression of a recessive SP11/SCR allele by an untranscribed SP11/SCR allele in brassica self-incompatibility. Plant Mol Biol 61:577–587

Giranton J, Dumas C, Cock JM, Gaude T (2000) The integral membrane S-locus receptor kinase of brassica has serine/threonine kinase activity in a membranous environment and spontaneously forms oligomers in planta. Proc Natl Acad Sci USA 97:3759–3764

Goring DR, Glavin TL, Schafer U, Rothstein SJ (1993) An S receptor kinase gene in self-compatible *Brassica napus* has a 1-bp deletion. Plant Cell 5:531–539

Gu T, Mazzurco M, Sulaman W, Mathias DD, Goring D (1998) Binding of an arm repeat protein to the kinase domain of the S-locus receptor kinase. Proc Natl Acad Sci USA 95:382–387

Guo YL, Bechsgaard JS, Slotte T, Neuffer B, Lascoux M, Weigel D, Schierup MH (2009) Recent speciation of capsella rubella from capsella grandiflora, associated with loss of self-incompatibility and an extreme bottleneck. Proc Natl Acad Sci USA 106:5246–5251

Haffani YZ, Gaude T, Cock JM, Goring DR (2004) Antisense suppression of thioredoxin h mRNA in *Brassica napus* cv. westar pistils causes a low level constitutive pollen rejection response. Plant Mol Biol 55:619–630

Hagenblad J, Bechsgaard J, Charlesworth D (2006) Linkage disequilibrium between incompatibility locus region genes in the plant *Arabidopsis lyrata*. Genetics 173:1057–1073

Heslop-Harrison J (1975) Incompatbility and the pollen stigma interaction. Annu Rev Plant Physiol 26:403–425

Igic B, Bohs L, Kohn JR (2006) Ancient polymorphism reveals unidirectional breeding system transitions. Proc Natl Acad Sci USA 103:1359–1363

Iwano M, Shiba H, Matoba K, Miwa T, Funato M, Entani T, Nakayama P, Shimosato H, Takaoka A, Isogai A, Takayama S (2007) Actin dynamics in papilla cells of *Brassica rapa* during self- and cross- pollination. Plant Physiol 144:72–81

Kachroo A, Schopfer CR, Nasrallah ME, Nasrallah JB (2001) Allele-specific receptor-ligand interactions in Brassica self-incompatibility. Science 293:1824–1826

Kakita M, Murase K, Iwano M, Matsumoto T, Watanabe M, Shiba H, Isogai A, Takayama S (2007a) Two distinct forms of M-locus protein kinase localize to the plasma membrane and interact directly with S-locus receptor kinase to transduce self-incompatibility signaling in *Brassica rapa*. Plant Cell 19:3961–3973

Kakita M, Shimisato H, Murase K, Isogai A, Takayama S (2007b) Direct interaction between the *S*-locus receptor kinase and the *M*-locus protein kinase involved in *Brassica* self-incompatibility signaling. Plant Biotechnol 24:185–190

Kakizaki T, Takada Y, Ito A, Suzuki G, Shiba H, Takayama S, Isogai A, Watanabe M (2003) Linear dominance relationship among four class-II S haplotypes in pollen is determined by the expression of SP11 in *Brassica* self-incompatibility. Plant Cell Physiol 44:70–75

Kamau E, Charlesworth D (2005) Balancing selection and low recombination affect diversity near the self-incompatibility loci of the plant *Arabidopsis lyrata*. Curr Biol 15:1773–1778

Kandasamy MK, Paolillo DJ, Faraday CD, Nasrallah JB, Nasrallah ME (1989) The S-locus specific glycoproteins of *Brassica* accumulate in the cell wall of developing stigma papillae. Dev Biol 134:462–472

Kho YO, Baer J (1968) Observing pollen tubes by means of fluorescence. Euphytica 17:298–303

Kimura R, Sato K, Fujimoto R, Nishio T (2002) Recognition specificity of self-incompatibility maintained after the divergence of *Brassica oleracea* and *Brassica rapa*. Plant J 29:215–223

Kusaba M, Dwyer KG, Hendershot J, Vrebalov J, Nasrallah JB, Nasrallah ME (2001) Self-incompatibility in the genus *Arabidopsis*: characterization of the S locus in the outcrossing A. *lyrata* and its autogamous relative A. *thaliana*. Plant Cell 13:627–643

Kusaba M, Nishio T, Satta Y, Hinata K, Ockendon D (1997) Striking sequence similarity in inter- and intra-specific comparisons of class I *SLG* alleles from *Brassica oleracea* and *Brassica campestris*: implications for the evolution and recognition mechanism. Proc Natl Acad Sci USA 94:7673–7678

Kusaba M, Tung CW, Nasrallah ME, Nasrallah JB (2002) Monoallelic expression and dominance interactions in anthers of self-incompatible *Arabidopsis lyrata*. Plant Physiol 128:17–20

Liu P, Sherman-Broyles S, Nasrallah ME, Nasrallah JB (2007) A cryptic modifier causing transient self-incompatibility in *Arabidopsis thaliana*. Curr Biol 17:734–740

Mable BK (2003) Estimating the number, frequency, and dominance of S-alleles in a natural population of *Arabidopsis lyrata* (Brassicaceae) with sporophytic control of self-incompatibility. Heredity 90:422–431

Mable BK, Robertson AV, Dart S, DiBerardo C, Witham L (2005) Breakdown of self-incompatibility in the perennial *Arabidopsis lyrata* (Brassicaceae) and its consequences. Evolution 59:1437–1448

Miege C, Ruffio-Chable V, Schierup MH, Cabrillac D, Dumas C, Gaude T, Cock JM (2001) Intrahaplotype polymorphism at the *Brassica* S locus. Genetics 159:811–822

Mishima M, Takayama S, Sasaki K, Jee J, Kojima C, Isogai A, Shirakawa M (2003) Structure of the male determinant factor for *Brassica* self-incompatibility. J Biol Chem 278: 36389–36395

Murase K, Shiba H, Iwano M, Che F, Watanabe M, Isogai A, Takayama S (2004) A membrane-anchored protein kinase involved in *Brassica* self-incompatibility signaling. Science 303: 1516–1519

Naithani S, Chookajorn T, Ripoll DR, Nasrallah JB (2007) Structural modules for receptor dimerization in the S-locus receptor kinase extracellular domain. Proc Natl Acad Sci USA 104:12211–12216

Nasrallah ME (1974) Genetic control of quantitative variation of self incompatibility proteins detected by immunodiffusion. Genetics 76:45–50

Nasrallah JB (2005) Recognition and rejection of self in plant self-incompatibility: comparisons to animal histocompatibility. Trends Imm 26:412–418

Nasrallah ME, Barber JT, Wallace DH (1970) Plant self-incompatibility proteins–detection, genetic determination and possible mode of action. Heredity 25:23–27

Nasrallah ME, Kandasamy MK, Chang M-C, Stadler Z, Lim S, Nasrallah JB (2000) Identifying genes for pollen-stigma recognition in crucifers. Annals Bot 85(Suppl A):125–132

Nasrallah ME, Kandasamy MK, Nasrallah JB (1992) A genetically defined trans-acting locus regulates S-locus function in *Brassica*. Plant J 2:497–506

Nasrallah JB, Kao TH, Chen CH, Goldberg ML, Nasrallah ME (1987) Amino-acid sequence of glycoproteins encoded by three alleles of the S locus of *Brassica oleracea*. Nature 326: 617–619

Nasrallah JB, Kao TH, Goldberg ML, Nasrallah JB (1985) A cDNA clone encoding an S locus specific glycoprotein from *Brassica oleracea*. Nature 318:263–267

Nasrallah ME, Liu P, Nasrallah JB (2002) Generation of self-incompatible *Arabidopsis thaliana* by transfer of two *S* locus genes from *A. lyrata*. Science 297:247–249

Nasrallah ME, Liu P, Sherman-Broyles S, Boggs N, Nasrallah JB (2004) Natural variation in expression of self-incompatibility in *Arabidopsis thaliana*: implications for the evolution of selfing. Proc Natl Acad Sci USA 101:16070–16074

Nasrallah JB, Liu P, Sherman-Broyles S, Schmidt R, Nasrallah ME (2007) Epigenetic mechanisms for breakdown of self-incompatibility in interspecific hybrids. Genetics 175:1965–1973

Nasrallah JB, Rundle SJ, Nasrallah ME (1994) Genetic evidence for the requirement of *Brassica* S-locus receptor kinase gene in the self-incompatibility response. Plant J 5:373–384

Nasrallah ME, Wallace DH (1967a) Immunochemical detection of antigens in self-incompatibility genotypes of cabbage. Nature 213:700–701

Nasrallah ME, Wallace DH (1967b) Immunogenetics of self-incompatibility in *Brassica oleracea* L. Heredity 22:519–527

Nasrallah ME, Wallace DH, Savo RM (1972) Genotype, protein, phenotype relationships in self-incompatibility of *Brassica*. Genet Res 20:151–160

Nishio T, Hinata K (1982) Comparative studies on S-glycoproteins purified from different S-genotypes in self-incomatpible *Brassica* species. I. purification and chemical properties. Genetics 100:641–647

Nou IS, Watanabe M, Isogai A, Hinata K (1993) Comparison of *S*-alleles and *S*-glycoproteins between two wild populations of *Brassica* Turkey and Japan. Sex Plant Reprod 6:79–86

Ockendon DJ (1982) An S-allele suvey of cabbage (*Brassica oleracea* var *capitata*). Euphytica 31:325–331

Okamoto S, Odashima M, Fujimoto R, Sato Y, Kitashiba H, Nishio T (2007) Self-compatibility in *Brassica napus* is caused by independent mutations in *S*-locus genes. Plant J 50: 391–400

Paetsch M, Mayland-Quellhorst S, Neuffer B (2006) Evolution of the self-incompatibility system in the brassicaceae: identification of S-locus receptor kinase (SRK) in self-incompatible *Capsella grandiflora*. Heredity 97:283–290

Ramos-Onsins SE, Stranger BE, Mitchell-Olds T, Aguade M (2004) Multilocus analysis of variation and speciation in closely related species *Arabidopsis halleri* and *A. lyrata*. Genetics 166:373–388

Sainudiin R, Shuk WWW, Yogeeswaran K, Nasrallah JB, Yang Z, Nielsen R (2005) Detecting site-specific shysiochemical selective sressures: applications to the class I HLA of the human major histocompatibility complex and the SRK of the plant sporophytic self-incompatibility system. J Mol Evol 60:315–326

Sampson DR (1957) The genetics of self-incompatibility in the radish. J Hered 48:26–29

Sampson DR (1964) A one-locus self-incompatibility system in *raphanis raphanistrum*. Can J Genet Cytol 6:435–445

Samuel MA, Yee D, Haasen KE, Goring DR (2008) 'Self' pollen rejection through the intersection of two cellular pathways in the brassicaceae: self-incompatibility and the compatible pollen response. In: Franklin-Tong VE (ed) Self-Incompatibility in Flowering Plants, pp 173–191. Springer, Berlin

Sato Y, Fujimoto R, Toriyama K, Nishio T (2003) Commonality of self-recognition specificity of S haplotypes between *Brassica oleracea* and *Brassica rapa*. Plant Mol Biol 52:617–626

Sato Y, Okamoto S, Nishio T (2004) Diversification and alteration of recognition specificity of the pollen ligand SP11/SCR in self-incompatibility of *Brassica* and *Raphanus*. Plant Cell 16: 3230–3241

Schierup MH, Mable BK, Awadalla P, Charlesworth D (2001) Identification and characterization of a polymorphic receptor kinase gene linked to the self-incompatibility locus of *Arabidopsis lyrata*. Genetics 158:387–399

Schopfer CR, Nasrallah JB (2000) Self-incompatibility. prospects for a novel putative peptide-signaling molecule. Plant Physiol 124:935–940

Schopfer CR, Nasrallah ME, Nasrallah JB (1999) The male determinant of self-incompatibility in *Brassica*. Science 286:1697–1700

Sherman-Broyles S, Boggs N, Farkas A, Liu P, Vrebalov J, Nasrallah ME, Nasrallah JB (2007) *S*-locus genes and the evolution of self-fertility in *Arabidopsis thaliana*. Plant Cell 19:94–106

Sherman-Broyles S, Nasrallah JB (2008) Self-incompatibility and the evolution of mating systems in the brassicaceae. In: VE Franklin-Tong (ed) Self-Incompatibility in Flowering Plants, pp123–pp147. Springer, Berlin

Shiba H, Iwano M, Entani T, Ishimoto K, Shimosato H, Che F, Satta Y, Ito A, Takada Y, Watanabe M, Isogai A, Takayama S (2002) The dominance of alleles controlling self-incompatibility in *Brassica* pollen is regulated at the RNA level. Plant Cell 14:491–504

Shiba H, Kakizaki T, Iwano M, Tarutani Y, Watanabe M, Isogai A, Takayama S (2006) Dominance relationships between self-incompatibility alleles controlled by DNA methylation. Nat Genet 38:297–299

Shiba H, Takayama S, Iwano M, Shimosato H, Funato M, Nakagawa T, Che F, Suzuki G, Watanabe M, Hinata K, Isogai A (2001) A pollen coat protein, SP11/SCR, determines the pollen S-specificity in the self-sncompatibility of *Brassica* species. Plant Physiol 125:2095–2103

Shimizu KK, Cork JM, Caicedo AL, Mays CA, Moore RC, Olsen KM, Ruzsa S, Coop G, Bustamante CD, Awadalla P, Purugganan MD (2004) Darwinian selection on a selfing locus. Science 306:2081–2084

Shimizu KK, Shimizu-Inatsugi R, Tsuchimatsu T, Purugganan MD (2008) Independent origins of self-compatibility in *Arabidopsis thaliana*. Mol Ecol 17:704–714

Shimosato H, Yokota N, Shiba H, Iwano M, Entani T, Che F, Watanabe M, Isogai A, Takayama S (2007) Characterization of the SP11/SCR high affinity binding site involved in self/nonself recognition in brassica self incompatibility. Plant Cell 19:109–117

Shivanna KR, Heslop-Harrison Y, Heslop-Harrison J (1978) The pollen-stigma interaction: bud pollination in the cruciferae. Acat Bot Neerl 27:107–119

Singh A, Paolillo DJ (1990) Role of calcium in the callose response of self-pollinated brassica stigmas. Amer J Bot 77:128–133

Stein J, Dixit R, Nasrallah ME, Nasrallah JB (1996) SRK, the stigma-specific S locus receptor kinase of *Brassica*, is targeted to the plasma membrane in transgenic tobacco. Plant Cell 8: 429–445

Stein JC, Howlett B, Boyes DC, Nasrallah ME, Nasrallah JB (1991) Molecular cloning of a putative receptor protein kinase gene encoded at the self-incompatibility locus of *Brassica oleracea*. Proc Natl Acad Sci USA 88:8816–8820

Stone SL, Anderson EM, Mullen RT, Goring DR (2003) *ARC1* is an E3 ubiquitin ligase and promotes the ubiquitination of proteins during the rejection of self-incompatible *Brassica* pollen. Plant Cell 15:885–898

Stone SL, Arnoldo M, Goring DR (1999) A breakdown of *Brassica* self-incompatibility in ARC1 antisense transgenic plants. Science 286:1729–1731

Stout AB (1920) Further experimental studies on self-incompatibility in hermaphroditic plants. J Genet 9:85–129

Sulaman W, Arnoldo M, Yu K, Tulsieram L, Rothstein SJ, Goring DR (1997) Loss of callose in the stigma papillae does not affect the *Brassica* self-incompatibility phenotype. Planta 203: 327–331

Suzuki G, Kai N, Hirose T, Fukui K, Nishio T, Takayama D, Isogai A, Watanabe M, Hinata K (1999) Geomic organization of the *S*-locus: identification and characterization of genes in the *SLG/SRK* region of *S9* haplotype of (*Brassica campestris* syn. *rapa*). Genetics 153: 391–400

Suzuki T, Kusaba M, Matsushita M, Okazaki K, Nishio T (2000) Characterization of brassica S-haplotypes lacking S-locus glycoprotein. FEBS Lett 482:102–108

Takasaki T, Hatakeyama K, Suzuki G, Watanabe M, Isogai A, Hinata K (2000) The S receptor kinase determines self-incompatibility in *Brassica* stigma. Nature 403:913–916

Takayama S, Isogai A (2005) Self-incompatibility in plants. Annu Rev Plant Biol 56:467–489

Takayama S, Shimosato H, Shiba H, Funato M, Che F, Watanabe M, Iwano M, Isogai A (2001) Direct ligand-receptor complex interaction controls *Brassica* self-incompatibility. Nature 413:534–538

Tang C, Toomajian C, Sherman-Broyles S, Plagnol V, Guo Y, Hu TT, Clark RM, Nasrallah JB, Weigel D, Nordborg M (2007) The evolution of selfing in *Arabidopsis thaliana*. Science 317:1070–1072

Thompson KF (1957) Self-incompatibility in marrow-stem kale, *Brassica oleracea* var *acephala*. I. demonstration of a sporophytic system. J Genet 55:45–60

Thompson KF, Taylor JP (1966) Non-linear dominance relationships between *S* alleles. Heredity 21:345–362

Vanoosthuyse V, Tichtinsky G, Dumas C, Gaude T, Cock JM (2003) Interaction of calmodulin, a sorting nexin and kinase associated protein phosphatase with *Brassica oleracea* S locus receptor kinase. Plant Physiol 133:919–929

Watanabe M, Ito Y, Takada Y, Ninomiya C, Kakizaki T, Takahata Y, Hatakeyama K, Hinata K, Suzuki G, Takasaki T, Satta Y, Shiba H, Takayama S, Isogai A (2000) Highly divergent sequences of the pollen self-incompatiblity (S) gene in class-I S haplotypes of *Brassica campestris* (syn. *rapa*) L. FEBS Lett 473:139–144

Wheeler MJ, de Graaf BHJ, Hadjiosif N, Perry RM, Poulter NS, Osman K, Vatovec S, Harper A, Franklin CH, Franklin-Tong VE (2009) Identification of the pollen self-incompatibility determinant in *Papaver ehoeas*. Nature 459:992–995

Zhang Y, Xue Y (2008) Molecular biology of S-RNase-based self-incompatibility. In: Franklin-Tong VE (ed) Self-Incompatibility in Flowering Plants, pp 193–215. Springer, Berlin

Chapter 15
Sequencing the Gene Space of *Brassica rapa*

Jeong-Hwan Mun and Beom-Seok Park

Abstract The Brassicaceae includes some of the world's most important dicotyledonous crop species, including vegetables, forages, condiments, and oil crops. Among the *Brassica* crops, *Brassica rapa* is considered as an excellent model for genomic research on the *Brassica* species. *B. rapa* has a relatively compact diploid genome, a large collection of cultivars, and a broad array of available genomic resources. Because of these desirable features, there is growing consensus that *B. rapa* is an ideal candidate to act as a reference species for *Brassica* genomic studies. Already, various genomic studies have been conducted, including the construction of a BAC-based physical map, >150,000 expressed sequence tags, and ongoing genome sequencing. Sequencing and comparative analysis of the selected seed BACs have suggested that the triplicated *B. rapa* genome contains only approximately twice the number of genes in *Arabidopsis*. The imminent *B. rapa* genome sequence will offer novel insights into the organization and evolution of *Brassica* as well as the similarities and differences between the genomes of *A. thaliana* and other plant model species. In parallel, the transfer of knowledge from *B. rapa* to other *Brassica* crops is highly expected.

Keywords *Brassica rapa* · Genome · Genomic resources · Sequencing

Contents

15.1 *B. rapa* as a Reference for the *Brassica* "A" Genome 414
15.2 Genome Structure of *B. rapa* 415
15.2.1 Cytogenetic Study of the *B. rapa* Genome 415
15.2.2 Repetitive Sequences of *B. rapa* 416
15.2.3 Triplicated Nature of the *B. rapa* Genome 419
15.3 Genomic Resources for *B. rapa* 420

J.-H. Mun (✉)
Genomics Division, Department of Agricultural Biotechnology, National Academy of Agricultural Science, Suwon 441-707, Korea
e-mail: munjh@rda.go.kr

R. Schmidt, I. Bancroft (eds.), *Genetics and Genomics of the Brassicaceae*, Plant Genetics and Genomics: Crops and Models 9, DOI 10.1007/978-1-4419-7118-0_15,

15.3.1 BAC Libraries and BAC-end Sequences 420
15.3.2 Genetic Map 422
15.3.3 Physical Map 422
15.3.4 Expressed Sequence Tags and Transcriptome Analysis 424
15.3.5 Information Resources 427
15.4 Progress of Genome Sequencing 427
15.4.1 Sequencing of Euchromatic Regions Based on the Clone-by-Clone Strategy 427
15.4.2 Seed BAC Selection 428
15.4.3 Characteristics of the Seed BAC Sequences 430
15.4.4 Sequencing Process 432
15.5 Perspective 433
References 434

15.1 *B. rapa* as a Reference for the *Brassica* "A" Genome

There are approximately 3,700 species of Brassicaceae with diverse characteristics, many of which are of agronomic importance as vegetables, condiments, and oil crops (Beilstein et al. 2006). Economically, *Brassica* species contribute approximately 10% of the world's vegetable crop produce and approximately 12% of the worldwide edible oil supplies (Economic Research Service and USDA 2008). *Brassica* crops are characterized by diverse morphologies including, in some cases, enlarged vegetative and floral meristems. The morphological diversity among the *Brassica* species may be linked to the genomic changes associated with polyploidy (Lukens et al. 2004). These characteristics have long been the targets of breeding programs worldwide, and it is in this regard that the study of genomics could be most beneficial.

The genetic relationships among the *Brassica* species, characterized by U's triangle, with the chromosome numbers and estimated genome sizes are represented in Chapter 2 (Johnston et al. 2005, U 1935). Of the six widely cultivated *Brassica* species, *B. rapa* (AA, $2n = 20$, 529 Mbp), *B. nigra* (BB, $2n = 16$, 632 Mbp), and *B. oleracea* (CC, $2n = 18$, 696 Mbp) are monogenomic diploids. The interspecific breeding between these three diploid species results in the creation of three new species of allotetraploid hybrids, namely *B. juncea* (AABB, $2n = 36$, 1,068 Mbp), *B. napus* (AACC, $2n = 38$, 1,132 Mbp), and *B. carinata* (BBCC, $2n = 34$, 1,284 Mbp) (Johnston et al. 2005). Thus, investigation of the *Brassica* genomes provides substantial opportunities to study the divergence of gene function and genome evolution associated with polyploidy, extensive duplication, and hybridization.

Based on the close phylogenetic relationship between the *Brassica* species and the model plant, *Arabidopsis thaliana*, knowledge transfer from the results of studies on *Arabidopsis* for *Brassica* crop improvement should be straightforward. However, the complex genome organization of the *Brassica* species as a result of multiple rounds of polyploidy and genome hybridization renders the identification of orthologous relationships of genes between the genomes highly difficult. In particular, the comparative genomic studies of *B. rapa* and *B. oleracea* with *A. thaliana* revealed that the *Brassica* genome triplicated very soon after divergence from the *Arabidopsis* lineage. Subsequent extensive interspersed gene-loss or gene-gain events and large-scale chromosomal rearrangements including segmental duplications or deletions in the *Brassica* lineage complicate the orthologous relationships of the loci between the two genomes (Mun et al. 2009, Town et al. 2006, Yang et al. 2006).

The genomes of several *Brassica* crop species have been characterized in detail over the past few years. *B. rapa* is native to Europe and East Asia, where the existence of a large native *B. rapa* population has provided an important resource for the breeding program. There are wide morphological variations in *B. rapa,* including the leafy type (Chinese cabbage and PakChoi), turnip type (vegetable turnip), and oil type (yellow sarson). Several *B. rapa* species are of regional agricultural importance either as vegetable or oil crops. As a consequence of its native distribution and agronomic usage, *B. rapa* has great potential for use as a model for the study of both basic and applied aspects of plant biology. In particular, *B. rapa* ssp. *pekinensis* (Chinese cabbage, cv. *Chiifu*), one of the most widely cultivated annual vegetable crops in northeast Asia, exhibits characteristics that are useful for the study of genome characteristics such as diploidy and small genome size (529 Mbp) (Yang et al. 2005). In response to the need for a simple genetic system with favorable genetic attributes for research among *Brassica* species, *B. rapa* has played a central role as a model species representing the *Brassica* "A" genome and is the focus of multinational genome projects. Genomic studies on *B. rapa* ssp. *pekinensis* cv. *Chiifu* were initiated in 2003 when a bacterial artificial chromosome (BAC) library was constructed (Park et al. 2005) and a collection of BAC-end sequences was initiated. Shortly afterward, a cytogenetic study, generation of DNA markers, construction of high-density genetic (Kim et al. 2006) and physical maps (Mun et al. 2008), and eventually large-scale genome sequencing based on the clone-by-clone strategy (http://www.brassica.info) was initiated. The information and genomic resources produced during the course of investigations are useful for understanding the genetic system of *B. rapa*. Moreover, these resources will be beneficial for *Brassica* crop breeding because they enable comparative genomic studies and subsequent transfer of knowledge from *B. rapa* to other crop *Brassica* species. In this chapter, we summarize the current state of information and the genomic resources pertaining to the *B. rapa* genome and the progress on *B. rapa* genome sequencing conducted mainly by the National Academy of Agricultural Science (NAAS), RDA, Korea.

15.2 Genome Structure of *B. rapa*

15.2.1 Cytogenetic Study of the B. rapa *Genome*

Identification of the karyotype is the first step in the analysis of the genome structure of *B. rapa*. Most of the cytogenetic analyses of the *B. rapa* genome have been performed on mitotic metaphase chromosomes (Lim et al. 2005, 2007). These analyses have shown that the *B. rapa* genome is organized in relatively small, compact chromosomes with centromeric repeat sequences and rDNA sequences arranged as tandem arrays in the heterochromatin. The individual chromosome size ranges from 2.1 to 4.5 μm, with a total chromosome length of 32.6 μm. An alternative cytogenetic map based on a pachytene DAPI (4′, 6-diamidino-2-phenylindole dihydrochloride) and fluorescence in situ hybridization (FISH) karyogram was also constructed. On DAPI analysis, the mean lengths of 10 pachytene chromosomes were revealed to range from 23.7 to 51.3 μm, with a total chromosome length of 385.3 μm. Thus, chromosomes in the meiotic prophase stage are 12 times longer than those in the mitotic metaphase and display a well-differentiated pattern of brightly fluorescing heterochromatin segments (Koo et al. 2004).

The relative positions of 33 sequence tagged site (STS) makers and 5 repetitive sequences on the metaphase chromosomes were examined by FISH, and the genetic and cytogenetic correspondences between the linkage groups and chromosomes were determined (Kim et al. 2006, Lim et al. 2005). As of 2008, more than 100 gene-containing BAC clones have been analyzed by FISH (Mun et al. 2009) and the distribution of five major centromeric and pericentromeric repeats have been determined. By the cytogenetic method, all chromosomes can be identified based on their lengths, centromere positions, heterochromatin patterns, and positions of various repeat sequences (Fig. 15.1a). In the chromosomal structure in the mitotic metaphase, centromeric satellites were estimated to encompass approximately 30% of the total chromosomes, particularly in the core centromeric blocks of all the chromosomes. Moreover, in the pachytene FISH, the total length of the pericentromeric heterochromatin regions was estimated to be 38.2 μm, which is approximately 10% of the total chromosome length (Koo et al. 2004). Thus, overall, heterochromatin comprises approximately 40% of the total *B. rapa* genome. All gene-containing BACs are localized to the euchromatin (Mun et al. 2009). Thus we can estimate the size of the euchromatin as ≈60% of the whole genome size, i.e., ≈330 Mb. The relationship between cytogenetic pachytene FISH and sequence contig distance at several locations in the euchromatin has been estimated to range from 400 to 500 kbp/μm (Fig. 15.1b). The FISH karyotype has created a rational basis for integrating the molecular, genetic, and cytogenetic maps of *B. rapa*. More importantly, it provides a basis for targeting the gene space of *B. rapa* and for strategic clone-by-clone genome sequencing that is likely to enable the discovery of most genes.

15.2.2 Repetitive Sequences of B. rapa

The centromere is a dynamic and rapidly evolving structure consisting largely of highly repetitive sequences such as tandem satellite repeats and retrotransposons. The centromere repeats characterized in plant genomes often extend over several millions of nucleotides with 150–180 motifs, including the pAL1 satellite in *Arabidopsis* (Copenhaver et al. 1999). CentO in rice (Zhang et al. 2004), CentC in maize (Ananiev et al. 1998), and MtRs in *Medicago truncatula* (Kulikova et al. 2004). The composition of the repetitive sequences in the *B. rapa* genome has been surveyed by a similarity search of 10,204 BAC-end sequences with a

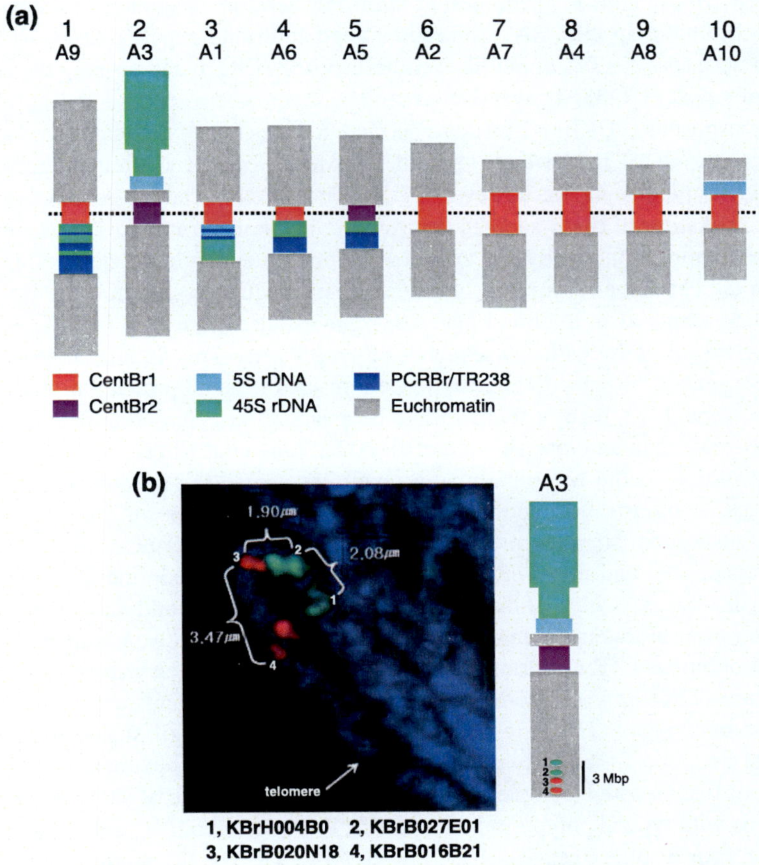

Fig. 15.1 Karyotype of *B. rapa* ssp. *pekinensis* cv. *Chiifu*. (**a**) Idiogram of *B. rapa* chromosomes represented by euchromatic arms and heterochromatic repetitive sequence blocks. Each chromosome was assigned to the sequencing center of the *B. rapa* genome sequencing consortium. (**b**) An example to estimate the physical length of a euchromatin arm using the relationship between pachytene FISH and sequence contig distance in chromosome A3; 1 μm in pachytene FISH is equivalent to 400–500 kbp of nucleotide length

previously reported tandem repeat that contains a *Hin*dIII site at both ends (Harrison and Heslop-Harrison 1995). Approximately 30% of them showed a high similarity. This fraction contained two different centromeric satellite repeat classes named CentBr1 and CentBr2. They are 176 bp tandem repeats with 82% sequence similarity with each other (Lim et al. 2005). From FISH data, CentBrs were revealed to be centromeric and chromosome specific, with CentBr1 on eight chromosomes (A1, A2, A4, and A6-A10) and CentBr2 on only two chromosomes (A3 and A5; Fig. 15.1a). The CentBr repeats were also present in other *Brassica* species and were particularly abundant in the *B. oleracea* whole-genome shotgun sequences, indicating that these repeats are the major components of the centromeric sequences of *Brassica* "A" and "C" genomes, at least. Furthermore, sequence analysis of the heterochromatin-specific BAC clones identified additional repeat classes, including centromere-specific Ty1-copia-like retrotransposon (CRB), 238 bp-long degenerate tandem repeat (TR238) arrays, rDNAs, and pericentromere-specific Ty3-gypsy-like retrotransposons (PCRBr) (Lim et al. 2007). CRB was one of the major components of all centromeres in the three diploid *Brassica* species and their three reciprocal allotetraploid hybrids; however, TR238 and PCRBr were A-genome specific. Characterization of these specific centromeric or pericentromeric repeat elements may be important in identifying the heterochromatin/euchromatin borders. In addition to the centromeric or pericentromeric repeat sequence searches, the BAC-end sequence search also identified many *B. rapa*-specific sequences (≈50% of BES) that had no similarity with any sequence of *Arabidopsis*. This finding suggested that *B. rapa* genome expansion appears to be a result of the amplification of *B. rapa*-specific sequences, many of which may be arranged as heterochromatic blocks of transposons or tandem repeats (Lim et al. 2005, Yang et al. 2006).

In addition to the heterochromatic repetitive sequences, retrotransposons also comprise a significant fraction of the euchromatin region of plant genomes. Retrotransposons are classified into two classes: long terminal repeat (LTR) retrotransposons and non-LTR retrotransposons. LTRs include the Ty1-copia and Ty3-gypsy types, which show distinct sequence features and the order of the encoded gene products (Kumar and Bennetzen 1999). Recently, a unique group of non-autonomous LTR retrotransposons, namely terminal-repeat retrotransposons in miniature (TRIM), was reported from *B. rapa*. TRIM elements are characterized by terminal repeats (TR) ranging from 100 to 250 bp in length that encompass an internal domain of ≈300 bp and create 5 bp target site duplications. The internal sequence begins with a complement of the primer-binding site of tRNA-methionine and ends with typical polypurine tract motifs. From 96 Mbp BAC-end sequences of *B. rapa*, four distinct lineages of TRIMs (Br1-Br4 TRs), with lengths ranging from 364 to 1,311 bp, were identified (Yang et al. 2007). The estimated copy number of Br TRs was more than six times greater in the *Brassica* species than in *Arabidopsis*, suggesting that various TRIM elements were inserted into the *Brassica* genome after divergence from the *Arabidopsis* lineage. Although the distribution of the Br TRs in the *B. rapa* genome is unknown, many of them appear to be located in the euchromatic region, based on the evidence of the distribution of *Arabidopsis* TRIM elements in the euchromatic gene space and the frequent occurrence of TRIMs in many *B. rapa* ESTs. The abundant TRIMs in the euchromatin of the *B. rapa* genome are expected to play an important role not only in the reconstruction of the host genome but also in the modification of the gene features by insertion of promoter or terminator sequences residing inside the elements. This modification may act as the driving force for gaining new functions even among the duplicated genes in the *Brassica* genome.

The euchromatic distribution and higher insertion polymorphism of the Br TRs present a useful application of these elements as DNA markers to distinguish the various *Brassica* crop species. Kwon et al. (2007) developed a transposon-display system using Br1 and Br2 TRs. The TRIM display system successfully accessed the genetic diversity in the Brassicaceae family and effectively identified 16 commercial F1 hybrids of *B. rapa* and other *Brassica* crops. Thus, transposons in the euchromatic regions can be a valuable source of applied genomics.

15.2.3 Triplicated Nature of the B. rapa Genome

With the progress of whole-genome sequencing of model genomes, one of the important challenges in plant genomics is to redirect the knowledge gained from the study of model genomes into the important biological and agronomical questions among crop species. Comparative structural genomic studies as a strategy for applied agriculture is well established in several plant families such as cereals and legumes (Cannon et al. 2006, Choi et al. 2004, Paterson et al. 2003). Because *Brassica* species are closely related to *A. thaliana*, comparative genomic studies are aimed at the *Brassica* genome organization and evolution with the *A. thaliana* genome as the basis for comparison. One of the profound findings from the comparative analysis was the recognition of the triplicated nature of the *Brassica* genome, indicating whole-genome triplication after the split from the *A. thaliana* lineage (Lagercrantz 1996, Lagercrantz and Lydiate 1996, Lysak et al. 2005, O'Neill and Bancroft 2000, Town et al. 2006, Yang et al. 2006).

The triplicated nature of the *B. rapa* genome was revealed by comparative genetic and physical mapping and sequence-level synteny comparison. The foundation of comparative mapping has been an RFLP marker-based genetic map of cv. *Jangwon* (Kim et al. 2006). Kim et al. used 545 genetic markers based on *B. rapa* ESTs and *Arabidopsis* genes. Among the 520 RFLP markers, a majority of the markers detected >1 locus, and only 62 EST markers showed a single locus anchor. Interchromosomal comparison of the linkage groups identified many homologous blocks in the genome. Comparative physical mapping between *Arabidopsis* and *B. rapa* further revealed triplicated homologous blocks including flowering locus C (*FLC*) regions in the *B. rapa* genome with collinearity of genes in the blocks (Park et al. 2005). These findings collectively suggested that the diploid *B. rapa* genome has been derived from a hexaploid ancestor. Sequence-level analysis of the triplicated *FLC* regions of *B. rapa* revealed the diploidization process in the triplicated genome. Yang et al. (2006) compared four paralogous BAC clones covering the *FLC* loci and the homologous 124 kbp segment of *Arabidopsis* chromosome 5. They estimated the time elapsed since the divergence of the paralogous and homologous lineages and reported that the three paralogous subgenomes of *B. rapa* triplicated around 13–17 MYA, very soon after the *Arabidopsis* and *Brassica* divergence occurred around 17–18 MYA. In addition, the *B. rapa* genome underwent segmental duplication around 0.8 MYA. Moreover, the *B. rapa* genome segments showed extensive interspersed gene loss and insertion of specific transposons such as TRIMs relative to the *Arabidopsis* genome segment. The diploidization process was estimated to reduce 40% of the triplicated genome length; thus, the *B. rapa* genome would contain only approximately a 1.7-fold higher number of genes as compared to *Arabidopsis*.

The duplicated genes retained in the *B. rapa* genome may have a selective advantage owing to their gaining new functions (neofunctionalization) or partitioning of the original function (subfunctionalization) between the two duplicates. Genome triplication and following diploidization in the *Brassica* lineage would affect the functional diversification of redundant genes because these events could permit mutations in the ancestor genome loci that are normally under tight selective constraints; the resulting subfunctionalization or neofunctionalization of the duplicated genes might lead to phenotypic diversification of *Brassica.* A study on the functional diversification of triplicated genes would provide an insight into the role of polyploidization in the *Brassica* crop genomes.

15.3 Genomic Resources for *B. rapa*

15.3.1 BAC Libraries and BAC-end Sequences

To accomplish a genomic study of any crop species, development of various genomic resources is indispensable. The early fruits of investigation of the *B. rapa* genome are evident in the recent advances in our understanding of the *Brassica* "A" genome structure and evolution. The genomic resources available for the *B. rapa* Genome Sequencing Project (BrGSP) are summarized in Table 15.1. A successful structural genomic study of the *B. rapa* genome relies on the quality and availability of detailed large-insert genomic libraries. As of 2008, five large-insert BAC libraries of *B. rapa* ssp. *pekinensis* cv. *Chiifu* are publicly available, providing a 53-fold genome coverage overall. These libraries were constructed using the restriction enzymes *Eco*RI, *Bam*HI, *Hin*dIII, and *Sau*3AI.

Using the public BAC libraries, a total of 260,637 BAC-end sequences (BES) have been generated from 146,688 BAC clones (≈203 Mbp) as a collaborative outcome of the multinational BrGSP community. Analysis of BES combined with BAC sequence surveys enabled the outlining of the features of the whole-genome structure of *B. rapa.* BLAST search of BES identified that up to 30% of BES were estimated to contain centromeric or pericentromeric repetitive sequences. An additional 10% of BES matched with transposons and other repeat sequences. Based on this data and FISH analysis (see Section 15.2.1), the heterochromatic region was postulated to occupy ≈40% of the *B. rapa* genome, while the euchromatic space was postulated to constitute ≈60% of the *B. rapa* genome (Lim et al. 2007, Yang et al. 2005). Comparison of BES with the *B. rapa* ESTs and *Arabidopsis* coding sequences (CDS) led to the recognition of approximately 11% of BES as protein-coding genes. Comparison of BES with the *Arabidopsis* genome proposed an efficient application of the BES data set to select the seed BAC clones for genome sequencing (Yang et al. 2005). *In silico* comparative sequence matching of 91,000 *B. rapa* BES on the *Arabidopsis* chromosome sequences identified approximately 50% BES showing significant sequence similarity along with overall collinearity with counterpart *Arabidopsis* chromosomal regions. Based on the comparative

Table 15.1 Summary of the genomic resources available for the *B. rapa* genome sequencing project

Resources	Material	Characteristics	Number[a]	References[b]
BAC library	5 libraries		234,544 clones (53.4×)	KBGP
KBrB	cv. *Chiifu/BamHI*	avr. insert size 120 kbp	55,296 clones (12.5×)	
KBrE	cv. *Chiifu/EcoRI*	avr. insert size 139 kbp	23,040 clones (6×)	
KBrH	cv. *Chiifu/HindIII*	avr. insert size 120 kbp	56,448 clones (12.8×)	
KBrS1	cv. *Chiifu/Sau3AI*	avr. insert size 100 kbp	55,296 clones (10.5×)	
KBrS2	cv. *Chiifu/Sau3AI*	avr. insert size 132 kbp	46,464 clones (11.6×)	
BAC end sequence	146,688 clones	both end single-pass sequence	260,637 reads (203 Mbp)	KBGP, NCBI
KBrB	55,296 clones		97,912 reads (75 Mbp)	KBGP, NCBI
KBrE	23,040 clones		43,168 reads (33 Mbp)	KBGP
KBrH	50,688 clones		88,951 reads (73 Mbp)	KBGP, NCBI
KBrS1	6,144 clones		8,117 reads (5 Mbp)	KBGP, NCBI
KBrS2	11,520 clones		22,489 reads (17 Mbp)	KBGP
Genetic map	3 reference maps			
JWF3p	cv. *Jangwon*	$F_{2:3}$ population 134 lines	1,287 cM, 545 markers	Kim et al. (2006)
CK	cv. *Chiifu* × cv. *Kenshin*	doubled haploid F_2 population 78 lines	1,182 cM, 556 markers	Choi et al. (2007)
VCS	cv. *VC1* × cv. *SR5*	doubled haploid F_2 population 91 lines	837 cM, 355 markers	KBGP
Physical map				
Build 2	67,468 BAC fingerprints	HICF map	1,428 contigs (717 Mbp)	Mun et al. (2008)
Expressed sequence tag	33 cDNA libraries of cv. *Chiifu*	cDNA single-pass sequence	152,253 ESTs (91 Mbp)	KBGP
Microarray	2 Nimble gen chips			
KBGP-24 K	24,000 unigenes	60mer, 6 probes/gene		Lee et al. (2008)
KBGP-50 K	32,000 unigenes, 17,000 CDS	60mer, 7 probes/gene		KBGP
BAC sequence	KBrB, KBrE, KBrH, KBrS clones	BAC shotgun sequence	On-going[c]	KBGP, NCBI

[a]Total sequence length or genome coverage is represented in parenthesis. Genome coverage was estimated based on the haploid genome equivalent of *B. rapa* as 529 Mbp.
[b]KBGP, Korea *B. rapa* genome project (http://www.brassica-rapa.org). NCBI, the national center for biotechnology information.
[c]As of December 2008, 886 BAC sequences have been deposited in NCBI.

genetic map and microcollinearity between the two genomes, the BAC clones allocated on the *Arabidopsis* genome can be chosen as seed BAC clones even before a complete physical map is established (see Section 15.4.2).

15.3.2 Genetic Map

Several genetic maps of *B. rapa* have thus far been reported using various mapping populations. Most of them were constructed using hybridization-based markers. Recently, three high-density genetic maps of *B. rapa* ssp. *pekinensis* have been reported as references to guide the on-going genome sequencing. One of them is an intercultivar map based on a cv. *Chiifu* × cv. *Kenshin* cross (Choi et al. 2007). The CK map used 78 doubled haploid lines of the F2 generation, thus providing direct and accurate genetic information for *Chiifu* genome sequencing. For this reason, the BrGSP community chose the CK map as the reference map for the on-going genome sequencing project. This map consists of a total of 556 markers over a length of 1,182 cM; however, the number of anchored BACs is limited. The second map is based on an intracultivar F2:3 linkage map based on the cultivar *Jangwon* cross (Kim et al. 2006). The Jangwon map was first developed using 520 RFLP markers and 25 PCR-based markers over a length of 1,287 cM. Although this map does not represent perfect information for *Chiifu* sequencing, it plays a role in determining the positioning of the sequenced BAC clones relative to the *B. rapa* linkage map. Currently, most of the RFLP markers of the Jangwon map have been replaced by PCR-based STS markers, including gene-targeting or SSR markers designed on the BAC sequences. As of 2008, a total of 513 sequenced BACs have been anchored on the Jangwon map, which contains 905 genetic loci over a length of 1,311 cM. Additionally, a third genetic map has been established from a cross between cv. *VC1* and cv. *SR5* (http://www.brassica-rapa.org). The VCS map additionally provides the positions of 225 BACs that are not included in the Jangwon map. In addition, this map provides enhanced genetic resolution in those regions in which suppression of recombination or distorted segregation was shown in the previous maps. Integration of these three maps using a set of anchored STS markers is underway and an integrated map will soon be available to the sequencing community.

The availability of the genome sequence, DNA markers, and high-density genetic maps has significantly accelerated the identification and isolation of genes that are important for crop quality and development, for example, the genes pertaining to disease resistance, flowering time, or biosynthesis of valuable phytochemicals. In addition, high levels of genome collinearity between *B. rapa* and other *Brassica* species offers the possibility of using positional cloning guided by conserved synteny as a basis for isolating orthologous genes of interest in other crops of the Brassicaceae family.

15.3.3 Physical Map

One of the crucial components for the success of a genome sequencing activity using the clone-by-clone strategy is the availability of a genome-wide, sequence-ready physical map. A physical map not only makes it possible to identify clones for genome sequencing with comprehensive coverage with reduced sequencing redundancy but also enables one to simplify the sequence assembly by arranging the sequence contigs in order. Thus far, the utility of physical maps has been reported by major genome sequencing projects on human (International Human Genome Sequencing Consortium 2001). *A. thaliana* (Marra et al. 1999), rice (Chen et al. 2002), and *M. truncatula* (Mun et al. 2006). These physical maps were constructed using a combination of techniques including restriction enzyme-digested BAC fragment fingerprinting on agarose gels and assembly of the fingerprints using the FPC software package (Soderlund et al. 2000). The agarose method has been successful, but has limited throughput (Nelson et al. 2005). Alternatively, fluorescence-labeled fingerprinting methods using a DNA sequencing gel or capillary electrophoresis have been developed to increased throughput (Ding et al. 2001, Gregory et al. 1997, Luo et al. 2003, Xu et al. 2004). Fluorescence-labeled capillary electrophoresis methods include the three-enzyme method and the High-Information Content Fingerprinting (HICF) methods, which use type IIS restriction enzyme or the SNaPshot labeling technique. The first genome-wide plant HICF physical map was constructed for maize (Nelson et al. 2005).

The first genome-wide BAC-based physical map of *B. rapa* was constructed by the SNaPshot method (Mun et al. 2008). To create a robust sequence-ready physical map, a total of 99,456 BAC clones from the three independent BAC libraries (≈22.5× coverage) were fingerprinted by digestion with combinations of five restriction enzymes (*Eco*RI, *Bam*HI, *Xba*I, *Xho*I, and *Hae*III) followed by SNaPshot reagent labeling of four colors at the 3′-ends of the restriction fragments and sizing on an ABI 3730×1 capillary sequencer. Of these fingerprints, a total of 93,689 clones (94.2%) were successfully fingerprinted to be used for contig assembly. From the initial data set, 26,221 BAC clones containing heterochromatic repetitive sequences were removed from the contig assembly. This process significantly enriches the euchromatic contigs in the resulting build. The physical contig map was assembled using 67,468 high-quality heterochromatic repeat-free BAC fingerprints from the initial data set. These BAC clones represent 15.2× coverage of the *B. rapa* genome; they were condensed into 1,417 contigs and the resulting contigs were manually edited to validate reliability. With the results of contig evaluation, manual editing of the initial contig build yielded 1,428 contigs with an average length of 512 kbp spanning 717 Mb, 1.3× coverage of the genome. An unsatisfactory aspect of this assembly is its large number of questionable (Q) clones. The Q clones in this assembly corresponded to 15% of the clones. However, three specifically deep contigs contributed to ≈48.3% of all the Q clones in the build. Thus, when these deep contigs of the initial build were excluded due to false-positive overlaps, the Q clones in the remaining contigs correspond to 7.7% of all clones. This ratio is similar to the ratios reported in catfish (7.3%) (Quiniou et al. 2007) and maize (11%) HICF maps (Nelson et al. 2005).

The contigs produced in the course of the fingerprinting work were tagged with 315 anchored genetic markers. The practically important aspect of this was that the integration of a physical map into a genetic map enabling the positioning of 242 gene-rich contigs to specific locations on 10 chromosomes, thereby providing seeds

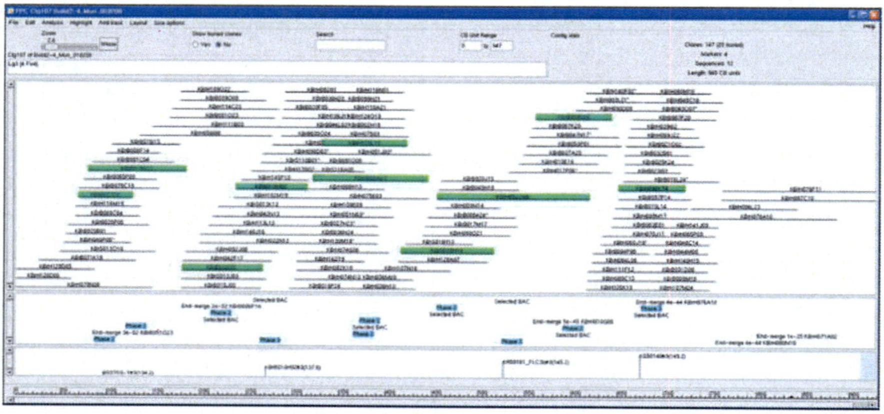

Fig. 15.2 An example of a physical contig used as a platform in BAC selection for genome sequencing. This contig consists of 147 BAC clones from three BAC source libraries and is estimated to cover approximately 1.1 Mbp. This contig was anchored to the region around 134–150 cM of the linkage group A3 using two gene-target markers, KBrB037I10-1 and KBrB010H02, one SSR marker, KS50140, and one RFLP marker, KR50161_FLC3b. All the highlighted BAC clones are in the genome sequencing pipeline and the sequencing phases that they are in have been indicated. The clones prefixed with KBrH are from the *Hin*dIII library, those prefixed with KBrB are from the *Bam*HI library, and those prefixed with KBrS are from the *Sau*3AI library (Table 15.1)

for the genome sequencing effort. An example of such a contig is shown in Fig. 15.2. The extent of the contigs associated with genetic loci is ≈160.7 Mb or 30% of the total genome. However, the total coverage of the physical contigs suggests that most contigs do not have sufficient overlaps and the gaps between the contigs need to be filled by additional fingerprinting. To improve the map, additional fingerprinting of approximately 30,000 clones of two recently constructed BAC libraries (KBrE and KBrS2) is being performed. These data will be merged into the current build to continue the refinement of the physical map. In parallel, linkage analysis of simple sequence repeats (SSRs) and single nucleotide polymorphisms (SNPs) in the seed BACs and BES will be carried out in order to provide even more anchoring points for the physical map. The details of these BACs and associated markers are presently available from the Korea *B. rapa* Genome Project (KBGP) website of NAAS (http://www.brassica-rapa.org). This information is playing an important role in identifying the BAC clones for sequencing.

15.3.4 Expressed Sequence Tags and Transcriptome Analysis

As of December 2008, a total of 152,253 *B. rapa* ESTs, obtained from 33 cDNA libraries representing a variety of organs and development stages, have been deposited in public databases (Table 15.2). These ESTs were clustered into 39,095 unique sequences (unigenes), including 16,898 tentative consensus sequences. *In silico* mapping of the 39,095 unigenes on the *Arabidopsis* genome identified 85% of unigenes covering 75% of the overall *Arabidopsis* counterpart coding sequences. It was found that the remaining 15% of the *B. rapa* unigenes were not

Table 15.2 Summary of *B. rapa* ssp. *pekinensis* cv. *Chiifu* ESTs from the libraries

Library	Tissue source	Number of ESTs	NCBI accession
KBAY	Anther, young anther	1, 859	Ex015357–Ex017215
KBCD	Whole plant, cold treated	6, 732	Ex017216–Ex023947
KBCG	Callus, developing callus	11, 847	Ex023948–Ex035794
KBFL	Floral bud, >2 mm in size	10, 332	Ex035795–Ex046126
KBFS	Floral bud, <2 mm in size	9,102	Ex046127–Ex055228
KBLS	Whole plant, salt treated	6, 894	Ex055229–Ex062122
KBLW	Non-photosynthetic mature leaf	2, 379	Ex062123–Ex064501
KBPL	Mature pollen	3, 587	Ex064502–Ex068088
KBPS	Seedling, 1-week old	4, 174	Ex068089–Ex072262
KBRT	Root, 1-month old	4, 682	Ex072263–Ex076944
KBSP	Silique, 1–10 days after pollination	6, 226	Ex076945–Ex083170
KBSQ	Silique, 10–25 days after pollination	933	Ex083171–Ex084103
KBST	Floral stem, bolting	1, 577	Ex084104–Ex085680
KCOV	Ovule, before pollination	421	Ex115626–Ex116046
KCOW	Ovule, 5–10 days after pollination	1, 225	Ex116047–Ex127271
KFFB	Floral bud, open flower	8, 771	Ex085681–Ex094451
KFPC	Leaf, *Pectobacterium carotoborum* infected	3, 708	Ex094452–Ex098159
KFRT	Root, mixture of 1, 3, and 7-weeks old	3, 555	Ex098160–Ex101714
KFSD	Mature seed and 2 days old germinating seed	4, 689	Ex101715–Ex106403
KFYP	Young plant, 3-weeks old	8, 408	Ex106404–Ex114811
KHCT	Cotyledon, in greening stage	2, 733	Ex117272–Ex120004
KHLD	Defected leaf	2, 037	Ex120005–Ex122041
KHLM	Mature green leaf	3, 423	Ex122042–Ex125464
KHLW	Non-photosynthetic mature leaf	2, 541	Ex125465–Ex128005
KHOS	Ovule and silique	2, 390	Ex128006–Ex130395
KHRT	Root, mixed stage and treatment	8, 265	Ex130396–Ex138660
KLPS	Seedling, 1-week old, etiolated	3, 840	Ex138661–Ex142500
NRFB	Floral bud and open flower, normalized	814	Ex114812–Ex115625
KFRC	Root, *Plasmodiophora brassicae* infected	5,169	
KFXT	Whole plant, heat treated	6, 481	
KFFO	Floral organ, mixed stages	8, 654	
KFFC	Carpel, regular library	2, 409	
KFFA	Anther, regular library	2, 396	
Total		152, 253	127,144

homologous with any gene in *Arabidopsis*, representing novel *Brassica*-specific genes (http://www.brassica-rapa.org/BrEMD). The functional category of ESTs based on gene ontology analysis did not show any significant overrepresentation of any specific functional category of genes in the *B. rapa* genome as compared to the *Arabidopsis* genome, as summarized in Fig. 15.3. Additional analysis of the EST collections identified approximately 10,000 putative full-length cDNA sequences containing a complete open reading frame with poly-A tail (unpublished data). It is anticipated that these sequences will contribute to the formation of a *B. rapa*-specific set of sequences not only to train gene prediction programs for *B. rapa* and other *Brassica* species but also to evaluate structure and alternative splicing of genes.

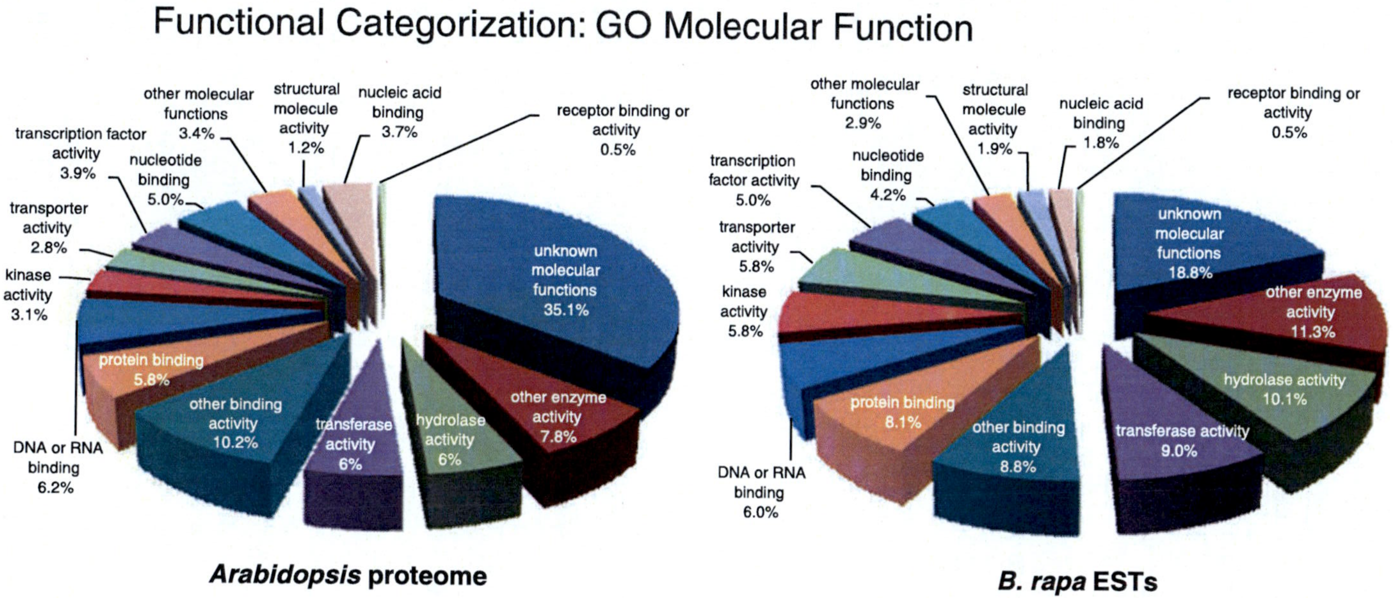

Fig. 15.3 Functional categories of the *B. rapa* expressed sequence tags compared to those of *Arabidopsis* proteome. The *graph* shows the proportion of clones in the EST collection corresponding to various predicted molecular functions based on gene ontology analysis

The corresponding cDNA clones have been used to construct microarrays for expression profiling during development or under various biotic- and abiotic-stress conditions. Two microarrays, namely KBGP-24K and KBGP-50K, were developed using the NimbleGen platform. Both microarrays included the 6 (KBGP-24K) or 7 (KBGP-50K) 60-nucleotide-long probes per gene. The 24K chip covered approximately 24,000 unigenes clustered by 127,144 ESTs from 20 cDNA libraries, whereas the 50K chip doubled up the gene contents by including an additional 8,500 unigenes plus 17,500 genes predicted from the seed BAC sequences in the genome sequencing pipeline. These microarrays enabled the examination of the changes in the genome-wide gene expression of *B. rapa*, which reflected its own transcriptional changes. Using the KBGP-24K chip, genome-wide transcriptome analysis in response to three abiotic stresses, namely salt, cold, and drought (which significantly affect the productivity of *Brassica* crops) was conducted (Lee et al. 2008). This analysis successfully identified stress-related genes along with novel transcription factor genes, suggesting the existence of a *B. rapa*-specific signaling pathway that works together with the common stress–response pathway under abiotic stress conditions. Information on the EST, unigene, and microarray experiments can be retrieved from the *B. rapa* EST and Microarray Database (BrEMD) website (http://www.brassica-rapa.org/BrEMD).

15.3.5 Information Resources

The online database on the KBGP website (http://www.brassica-rapa.org) provides easy access to the information generated by KBGP. This web interface provides the nucleotide sequences of the BAC clones with tentative annotation of the predicted genes, genetic maps along with the DNA markers, a physical map that can be viewed on WebFPC, cDNA sequences, and various transcriptome profiling data obtained by KBGP-24K microarray analysis. The BAC and cDNA libraries and clones can even be ordered through this site. The Korea *Brassica* Genome Resource Bank (KBGRB, http://www.brassica-resource.org) is a material resource database of *B. rapa* and is supported by the Korea National Plant Research Resource Center. Resources, including the seeds of various *B. rapa* wild accessions and cultivars, RI lines, mapping populations, and BAC and cDNA libraries, can be obtained through this database.

15.4 Progress of Genome Sequencing

15.4.1 Sequencing of Euchromatic Regions Based on the Clone-by-Clone Strategy

The most important single genomic resource for any species is its nuclear genome sequence, and despite the central role of *Brassica* in plant biology and world agriculture, the genomes of none of the *Brassica* species have been fully sequenced thus far. Considering the importance of the *Brassica* species and the value of *B. rapa* as a model *Brassica*, sequencing nearly all of its gene spaces by focusing on the euchromatic regions will be significantly beneficial. In general, two well-established genome sequencing approaches, namely clone-by-clone sequencing and whole-genome shotgun (WGS) sequencing, can be used to investigate the whole-genome structure of the species. However, the best approach to sequencing of crop genomes remains debatable because both the above approaches have their own strengths and weaknesses (Paterson 2006). The clone-by-clone strategy typically involves a "minimum tiling path" of large-insert ($\approx$100 kbp) clones such as BACs of a known order, which is determined using a combination of genetic, physical, and/or cytogenetic mapping. One of the risks of this method is the high cost of assembling large-insert libraries and arranging the clones. In addition, failure to sample highly repetitive regions could cause insufficient resolution of the genome. In contrast, the WGS approach is based on gathering enough sequences so that each nucleotide in the genome is covered, i.e., an average coverage equivalent to 7-genome coverage by paired-end sequence combination for randomly chosen clones of short insert (3–4 kbp) and fosmid ($\approx$36 kbp) clones. This method is more cost-effective than the clone-by-clone method and is even able to capture the gene-poor heterochromatic regions. WGS samples the entire genome, but sometimes fails to provide contiguity across elements that are found at high copy numbers. Polyploidy and a higher content and more recent origin of repetitive DNA are particularly serious issues in most plants. Both methods benefit from genetically anchored STS markers that help to build up aligned sequence scaffolds.

Cytogenetic studies based on extensive FISH analysis of both metaphase and pachytene chromosomes as described above have provided a detailed insight into the organization of heterochromatic and euchromatic regions. This work demonstrated that the genome of *B. rapa* is organized into distinct regions of pericentromeric heterochromatin, which are rich in repetitive sequences, and gene-rich euchromatin, which are not highly interspersed with heterochromatin. Moreover, sequencing of several BAC clones, chosen because of preliminary evidence that they were gene-rich, has confirmed that the gene density in *B. rapa* is relatively high on the order of 1 gene per 3–4 kbp (Yang et al. 2006). Each of the gene-rich BAC clones examined so far by FISH (>100 BACs) was found to be localized to the visible euchromatic region of the genome (Mun et al. 2009). At the same time, a whole-genome shotgun pilot sequencing of *B. oleracea* with 0.44-fold genome coverage generated transposable element-rich-sequences (Ayele et al. 2005, Zhang and Wessler 2004). Taken together, these data indicate a genome organization where the overwhelming majority of the *B. rapa* euchromatic space can be sequenced in a highly efficient manner by the clone-by-clone strategy.

15.4.2 Seed BAC Selection

Clone-by-clone sequencing generally starts from the defined seed sites and grows outward. The selection of seed BACs evenly scattered on the gene-rich euchromatic regions of the genome is a strong challenge for the clone-by-clone genome sequencing approach. The fingerprint-based physical map, combined with BES and genetic anchoring data, provides a basis for selecting seed BACs and for creating a draft tiling path. Alternatively, comparative approaches using an already sequenced closely related model genome, named as the comparative tiling method, can be used as a backbone for *in silico* clone validation of seed BACs even before the availability of a complete physical map; for example, in-depth comparative sequence analyses of several sequenced *B. rapa* BAC clones revealed overall collinearity of genes with a homologous region of the *Arabidopsis* counterpart. Thus, if we allocate *B. rapa* BAC clones on the *Arabidopsis* euchromatin using BES matches and scatter them onto the *B. rapa* chromosomes, selection of gene-rich BAC clones will be straightforward even without the need for using physical map information (Fig. 15.4). Based on this idea, Yang et al. (2005) compared the BES of 46,000 *B. rapa* BAC clones with *Arabidopsis* sequences using BLASTZ and allocated the BAC clones on the five *Arabidopsis* chromosomes by *in silico* matching based on unique, significant (<E^{-6}), and directional matches: one BAC end forward and

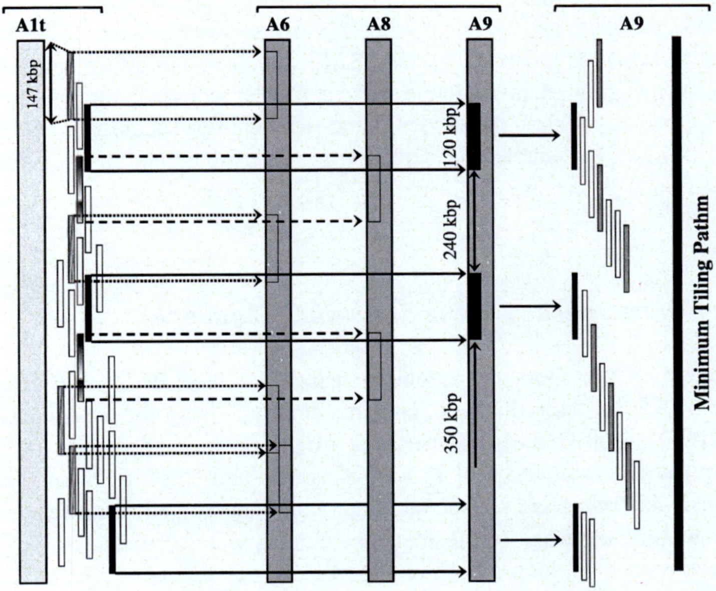

Fig. 15.4 Schematic representation of the comparative tiling method to select seed BAC clones from the euchromatic regions of *B. rapa* using *in silico* matching of BESs with known *Arabidopsis* sequences. A total of 4,317 *B. rapa* BAC clones with an average insert size of 120 kbp were mapped *in silico* on the euchromatic regions of the *Arabidopsis* genome. On *in silico* comparative allocation, 589 minimally overlapping BAC clones were selected and sequenced as seed BACs and then scattered on the euchromatic regions of the 10 *B. rapa* chromosomes by a combination of genetic and cytogenetic mapping. Generally, BAC clones covering the same *Arabidopsis* counterpart region were distributed over three *B. rapa* chromosomes due to the triplicated nature of the *Brassica* genome. The physical gaps between the seed BACs will be filled with additional BAC sequencing based on the minimum tiling path obtained from physical map information

the other end reverse direction, with a complement match within the 30–500 kbp interval. As a result, approximately 50% of BES showed significant sequence similarity with unique *Arabidopsis* sequences, and a total of 4,317 BAC clones were allocated on the *Arabidopsis* chromosomes by significant matching with both ends within the 30–500 kbp interval. A single *B. rapa* BAC clone was calculated to span an average of 147 kbp of the *Arabidopsis* counterpart sequence. These *B. rapa* BAC clones spanned 93 Mbp of the *Arabidopsis* sequences, representing ≈78% of the total *Arabidopsis* genome. BAC-FISH and STS mapping using BES of select BAC clones positioned on the counterpart *Arabidopsis* chromosomes showed the real euchromatic locations of the BAC clones scattered on the chromosomes. Therefore, if minimally overlapping BAC clones allocated on the *Arabidopsis* genomes were selected and scattered onto the *B. rapa* chromosomes, they could provide a seed point for bidirectional outward genome sequencing. As of July, 2008, 589 minimum tiled *B. rapa* BAC clones spanning 75 Mbp of the *Arabidopsis* euchromatic genome were selected from the *in silico* allocation and sequenced in phase 3 (finished sequences) or phase 2 (sequences that are fully oriented and ordered, but contain some small sequence gaps and low-quality regions). Each BAC clone was distributed onto the 10 *B. rapa* chromosomes by STS mapping, FISH analysis, and physical mapping information. All the sequenced BAC clones were provided to BrGSP as seed BACs and used as starting points for chromosome sequencing (data available on the KBGP website). Integration of seed BACs with the physical map provides "gene-rich" contigs spanning ≈160 Mbp (see Section 15.3.3). This enabled the selection of clones to extend the initial sequence contigs.

15.4.3 Characteristics of the Seed BAC Sequences

Construction of the sequence assembly using 589 seed BAC sequences generated the 65.8 Mbp nonredundant sequence contigs. Using these sequences, Mun et al. (2009) reported the characteristics of the euchromatic space of *B. rapa*. Based on the BLAT analysis of *B. rapa* unigenes and TBLASTN comparison of *Arabidopsis-M. truncatula* conserved single-copy genes against the sequence contigs, the genome coverage of the sequence contigs was calculated to be ≈30% of euchromatin; thus, the euchromatic region of *B. rapa* is estimated to be at least ≈220 Mbp, 42% of the whole genome when considering the genome size for *B. rapa* as 529 Mbp. The structures of the potential protein-coding genes in the seed BAC sequences have been predicted using a variety of ab initio, similarity-based, and ESTs/full-length cDNAs-based methods. Gene annotation of the sequence contigs resulted in the construction of 15,762 gene models. Taken together with the genome coverage of *B. rapa* sequences, the overall number of protein-coding genes in the *B. rapa* genome would be at least 52,000–53,000, which is higher than those of any other plant genomes sequenced thus far. However, the estimated total number of genes in the *B. rapa* genome is only twice that of *Arabidopsis*.

Table 15.3 Comparison of overall composition of the annotated protein coding genes in the *B. rapa* seed BAC sequence contigs and the euchromatic counterpart of the *A. thaliana* genome

Feature	*B. rapa*	*A. thaliana*[a]
Total sequence length (Mbp)	65.8	75.3
Transposons (%)	6	3
Number of protein coding genes	15,762	19,639
Number of exons per gene	4.7	5.5
Intron size (bp)	141	162
Exon size (bp)	225	230
Average gene size (kbp)	1.6	2.2
Average gene density (kbp/gene)	4.2	3.8
Overall G/C content (%)	35.2	35.8
Exons	46.3	44.6
Introns	32.6	32.0
Intergenic regions	31.3	31.8

[a]Statistics of *A. thaliana* is based on the annotation of TAIR7 version from the *Arabidopsis* information resource website (http://www.arabidopsis.org/portals/genAnnotation). Information for the *B. rapa* genome was based on the report of Mun et al. (2009)

The statistics of the gene structure and density are shown in Table 15.3. Although the structural features of the protein-coding genes in *B. rapa* are almost similar to those of the *B. oleracea* genes reported previously (Town et al. 2006), the former show a smaller average gene length (1.6 kbp) as compared to *Arabidopsis* counterparts (2.2 kbp). This difference appears to amount to almost one lesser exon per gene (4.7 and 5.5 exons per gene in *B. rapa* and *Arabidopsis*, respectively) along with a shorter exon (225 bp in *B. rapa* and 230 bp in *Arabidopsis*, respectively) and intron length (141 bp in *B. rapa* and 162 bp in *Arabidopsis*, respectively) in *B. rapa*. The base composition was found to be very similar in *B. rapa* and *Arabidopsis*. The average gene density of one per 4.2 kbp in the seed BAC sequences was slightly lower than that in *Arabidopsis* (one per 3.8 kbp). Thus, the *Arabidopsis/B. rapa* ratio of gene density was 0.90, indicating the slightly less compact organization of the *B. rapa* euchromatic region. A similarity search of the protein-coding genes of *B. rapa* against the public databases indicated that approximately 18% of the predicted genes showed no significant similarity with any of the genes reported. This result is roughly consistent with the results of previous studies conducted using EST analysis (See Section 15.3.4) and comparative analysis of *FLC* regions (Yang et al. 2006).

Repetitive sequence analysis revealed that 6% of the seed BAC sequences are composed of transposons, twofold higher proportions than those identified in the counterpart *Arabidopsis* euchromatic genome, presumably due to the increased number of LTRs and long interspersed elements (LINEs) (Table 15.4). In addition, low complexity repetitive sequences were quite abundant in the *B. rapa* euchromatic region, representing *B. rapa*-specific expansion of repetitive sequences. The distribution of repetitive sequences and transposons along the chromosomes was not uneven. It has previously reported that *B. oleracea* (696 Mbp) has a significantly

Table 15.4 Comparison of repetitive sequences identified in the *B. rapa* seed BAC sequence contigs and the euchromatic counterpart of the *A. thaliana* genome

Family	Genome coverage (%)[a]	
	B. rapa	*A. thaliana*
DNA transposon	2.2	1.6
LTR	2.4	0.8
LINE	1.3	0.3
SINE	0.1	0.0
Satellite	0.4	0.0
Low complexity repetitive sequence	4.4	1.0
Other[b]	0.4	0.1
Total	11.2	3.8

[a]Genome coverage was calculated using 65.8 Mbp for *B. rapa* and 75.3 Mbp for the euchromatic counterpart of *A. thaliana* as a whole sequence size.
[b]Other family includes simple sequence repeats and short tandem repeats. Information for the *B. rapa* genome was based on the report of Mun et al. (2009).

higher proportion of both class I and class II transposons in the partial draft genome shotgun sequences than *Arabidopsis* (Zhang and Wessler 2004) Taken together with the previous reports (Ayele et al. 2005, Lim et al. 2007), this result suggests that transposons were partly responsible for genome expansion in the *Brassica* lineage with predominant accumulation in the heterochromatic regions of *B. rapa*.

15.4.4 Sequencing Process

Extensive cytogenetic and sequencing data indicate that the *B. rapa* genome is organized into distinct gene-rich euchromatic regions. Thus, the BrGSP has jointly decided that the clone-by-clone approach to *B. rapa* genome sequencing will entail the best use of the available resources (http://brassica.bbsrc.ac.kr/brassica_genome_sequencing_concept.htm). The effort is being focused on the euchromatic arms of all 10 chromosomes, with sequencing distributed across teams in Korea, Australia, UK, China, Canada, and Japan, as summarized in Fig. 15.1. The project aims initially to produce a "phase 2" sequence, with accessible trace files to aid completion of specific regions as required. As compared to phase 3 sequencing, this will significantly reduce the sequencing cost and time involved. BAC clones are sequenced by shotgun and standard assembly methods, and then, gene annotation is conducted by a combination of a semiautomatic method and manual editing. STS markers including SSR or SNP markers generated from the BAC sequences will localize BACs onto the reference genetic map.

Sequencing technology is a dynamic field, and it is possible that opportunities that will enable the centers to sequence more efficiently may arise during the course of the project. Advances in sequencing technology using next-generation sequencing machines could lower the sequencing costs. In fact, the sequencing center of

NAAS tested a combination of GS-FLX and ABI 3730 sequencing machines and could reduce the sequencing costs by almost 20%; however, the lower resolution of the poly-mononucleotide tracks in the pyrosequencing data required additional validation.

As of December, 2008, a total 107 Mbp from 886 BAC clones (401 phase 3, 445 phase 2, and 40 phase 1 clones) have been sequenced and deposited in the HTGS database of NCBI by three sequencing centers, Korea (681 BACs), UK/China (200 BACs), and Australia (5 BACs). The other participating sequencing center will soon increase its capacity to accelerate sequencing. Besides the reported BAC clones, NAAS sequenced an additional 1,173 BAC clones as a result of sequencing of the chromosomes A3 and A9. Using 884 BAC clones, the sequence scaffolds for the two chromosomes were constructed. In case of A3, 535 BAC sequences generated seven sequence scaffolds spanning 34.9 Mbp; however, in case of A9, 349 BAC clones generated 15 scaffolds comprising 28.5 Mbp. Based on the estimated size of the *B. rapa* euchromatic region, the assemblies thus far constitute 90 and 85%, respectively, of the entire euchromatic chromosome arms with residual euchromatic gaps of 4 and 12.5 Mbp for A3 and A9, respectively (unpublished data). Additional effort continues to attempt to fill the gaps between the scaffolds.

15.5 Perspective

The most important beneficiaries of *B. rapa* sequencing will be *Brassica* crop researchers, ranging from breeders to plant biologists. *B. rapa* will provide an excellent reference genome sequence representing most *Brassica*-specific properties as well as valuable information to understand the genetic systems in the crop species that is positioned for evolutionary comparisons with *Arabidopsis*. The large number of *B. rapa* EST collections along with full-length cDNA sets and the genome sequences will enable the identification and isolation of the genes of interest that are important to agriculture. High-density genetic maps together with STS markers and associated sequence information have accelerated genome mapping and map-based cloning. Furthermore, the comparative genomics approach will benefit genomic investigation of closely related *Brassica* crops, including *B. oleracea* and *B. napus*, by transferring the knowledge on *B. rapa* along with the enhanced breeding efforts. In fact, quantitative trait loci or association mapping of valuable phytochemical-related genes in *B. napus* has taken the advantage of the available *B. rapa* genome sequence data and resources (personal communication from Prof. J. Meng). In addition, genome sequencing of other *Brassica* crops, particularly the construction of sequence assemblies and scaffolds of *B. napus*, will be improved owing to the presence of information on the *B. rapa* genome. Even next generation sequencing techniques can be applied with the *B. rapa* genome as a reference. There is no doubt that increasing genomic resources will facilitate the molecular genetic studies on *B. rapa* and will eventually contribute to improving *Brassica* crops.

Acknowledgment We thank the many participants of the Korea *Brassica rapa* Genome Project. This work was supported by the National Academy of Agricultural Science (PJ006759 and PJ006691), Rural Development Administration, Korea.

References

Ananiev EV, Phillips RL, Rines HW (1998) Chromosome-specific molecular organization of maize (*Zea mays* L.) centromeric regions. Proc Natl Acad Sci USA 95:13073–13078

Ayele M et al (2005) Whole genome shotgun sequencing of *Brassica oleracea* and its application to gene discovery and annotation in *Arabidopsis*. Genome Res 15:487–495

Beilstein MA, Al-Shehbaz IA, Kellogg EA (2006) Brassicaceae phylogeny and trichome evolution. Am J Bot 93:607–619

Cannon SB et al (2006) Legume genome evolution viewed through the *Medicago truncatula* and *Lotus japonicus* genomes. Proc Natl Acad Sci USA 103:14959–14964

Chen M et al (2002) An integrated physical and genetic map of the rice genome. Plant Cell 14: 537–545

Choi HK et al (2004) Estimating genome conservation between crop and model legume species. Proc Natl Acad Sci USA 101:15289–15294

Choi SR et al (2007) The reference genetic linkage map for the multinational *Brassica rapa* genome sequencing project. Theor Appl Genet 115:777–792

Copenhaver GP et al (1999) Genetic definition and sequence analysis of *Arabidopsis* centromeres. Science 286:2468–2474

Ding Y et al (2001) Five-color-based high-information-content fingerprinting of bacterial artificial chromosome clones using type IIS restriction endonucleases. Genomics 74:142–154

Economic Research Service, USDA (2008) Vegetables and melons outlook. http://www.ers.usda.gov/Publications/VGS/Tables/World.pdf.

Gregory SG, Howell GR, Bentley DR (1997) Genome mapping by fluorescent fingerprinting. Genome Res 7:1162–1168

Harrison GE, Heslop-Harrison JS (1995) Centromeric repetitive DNA sequences in the genus *Brassica*. Theor Appl Genet 90:157–165

International Human Genome Sequencing Consortium (2001) A physical map of the human genome. Nature 409:934–941

Johnston JS et al (2005) Evolution of genome size in Brassicaceae. Ann Bot 95:229–235

Kim JS et al (2006) A sequence-tagged linkage map of *Brassica rapa*. Genetics 174:29–39

Koo DH et al (2004) A high-resolution karyotype of *Brassica rapa* ssp. *pekinensis* revealed by pachytene analysis and multicolor fluorescence *in situ* hybridization. Theor Appl Genet 109:1346–1352

Kulikova O et al (2004) Satellite repeats in the functional centromere and pericentromeric heterochromatin of *Medicago truncatula*. Chromosoma 113:276–283

Kumar A, Bennetzen JL (1999) Plant retrotransposons. Annu Rev Genet 33:479–532

Kwon SJ et al (2007) Terminal repeat retrotransposon in miniature (TRIM) as DNA markers in *Brassica* relatives. Mol Genet Genomics 278:361–370

Lagercrantz U (1996) Comparative mapping in *Arabidopsis* and *Brassica*, fine scale genome collinearity and congruence of genes controlling flowering time. Plant J 9:13–20

Lagercrantz U, Lydiate DJ (1996) Comparative genome mapping in *Brassica*. Genetics 144: 1903–1910

Lee SC et al (2008) Transcriptome analysis in Brassica rapa under the abiotic stresses using *Brassica* 24 K oligo microarray. Mol Cells 26:595–605

Lim KB et al (2005) Characterization of rDNAs and tandem repeats in the heterochromatin of *Brassica rapa*. Mol Cells 19:436–444

Lim KB et al (2007) Characterization of the centromere and peri-centromere retrotransposons in *Brassica rapa* and their distribution in related *Brassica* species. Plant J 49:173–183

Lukens LN et al (2004) Genome redundancy and plasticity within ancient and recent *Brassica* crop species. Biol J Linn Soc Lond 82:665–674
Luo MC et al (2003) High-throughput fingerprinting of bacterial artificial chromosomes using the SNaPshot labeling kit and sizing of restriction fragments by capillary electrophoresis. Genomics 82:378–389
Lysak MA et al (2005) Chromosome triplication found across the tribe Brassiceae. Genome Res 15:516–525
Marra M et al (1999) A map for sequence analysis of the *Arabidopsis thaliana* genome. Nat Genet 22:265–270
Mun J-H et al (2006) Distribution of microsatellites in the genome of *Medicago truncatula*: a resource of genetic markers that integrate genetic and physical maps. Genetics 172:2541–2555
Mun J-H et al (2008) The first generation of a BAC-based physical map of *Brassica rapa*. BMC Genomics 9:280
Mun J-H et al (2009) Genome-wide comparative analysis of the *Brassica rapa* gene space reveals genome shrinkage and differential loss of duplication events after whole genome triplication. Genome Biol (10:R111)
Nelson WM et al (2005) Whole-genome validation of high-information-content fingerprinting. Plant Physiol 139:27–38
O'Neill CM, Bancroft I (2000) Comparative physical mapping of segments of the genome of *Brassica oleracea* var. *alboglabra* that are homoelogous to sequenced regions of chromosomes 4 and 5 of *Arabidopsis thaliana*. Plant J 23:233–243
Park JY et al (2005) Physical mapping and microsynteny of *Brassica rapa* ssp. *pekinensis* genome corresponding to a 222 kbp gene-rich region of *Arabidopsis* chromosome 4 and partially duplicated on chromosome 5. Mol Genet Genomics 274:579–588
Paterson AH et al (2003) Structure and evolution of cereal genomes. Curr Opin Genet Dev 13: 644–650
Paterson AH (2006) Leafing through the genomes of our major crop plants: strategies for capturing unique information. Nat Rev Genet 7:174–184
Quiniou SMA, Waldbieser GC, Duke MV (2007) A first generation BAC-based physical map of the channel catfish. BMC Genomics 8:40
Soderlund C et al (2000) Contigs built with fingerprints, markers, and FPC V4.7. Genome Res 11:934–941
Town CD et al (2006) Comparative genomics of *Brassica oleracea* and *Arabidopsis thaliana* reveal gene loss, fragmentation, and dispersal after polyploidy. Plant Cell 18:1348–1359
U N (1935) Genome analysis in *Brassica* with special reference to the experimental formation of *B. napus* and peculiar mode of fertilization. Jpn J Bot 7:389–452
Xu Z et al (2004) Genome physical mapping with large-insert bacterial clones by fingerprint analysis: methodologies, source clone genome coverage, and contig map quality. Genomics 84:941–951
Yang TJ et al (2005) The Korea *Brassica* Genome Projects: a glimpse of the *Brassica* genome based on comparative genome analysis with *Arabidopsis*. Comp Funct Genomics 6:138–146
Yang TJ et al (2006) Sequence-level analysis of the diploidization process in the triplicated *FLOWERING LOCUS C* region of *Brassica rapa*. Plant Cell 18:1339–1347
Yang TJ et al (2007) Characterization of terminal-repeat retrotransposon in miniature (TRIM) in *Brassica* relatives. Theor Appl Genet 114:627–636
Zhang Y et al (2004) Structural features of the rice chromosome 4 centromere. Nucleic Acid Res 32:2023–2030
Zhang X, Wessler SR (2004) Genome-wide comparative analysis of the transposable elements in the related species *Arabidopsis thaliana* and *Brassica oleracea*. Proc Natl Acad Sci USA 101:5589–5594

Chapter 16
Germplasm and Molecular Resources

Emma M. Knee, Luz Rivero, Deborah Crist, Erich Grotewold, and Randy Scholl

Abstract A wide range of genetic diversity exists in various members of the Brassicaceae, notably in the genera *Arabidopsis* and *Brassica*. Comprehensive seed collections include mutants, transgenic lines, chromosomal variants, mapping populations, Targeting Induced Local Lesions in Genomes (TILLING), and natural accessions, collected from the wild. Many molecular resources have been developed and are widely utilized in research. These include expressed sequence tags (ESTs); full length, sequence-validated cDNA clones of many genes; bacterial artificial chromosome (BAC) libraries; and vectors with multiple applications. Diverse resources are maintained by numerous institutions and specialized centers. Various public and private databases with comprehensive information about the collections are also available. This chapter focuses on the resources for the members of this family, with emphasis on collections that are publicly available.

Keywords Accession · Clone collection · Diversity fixed foundation set · Forward genetic resource · Gene bank · Genomic library · Germplasm collection · Mapping population · Mutant collection · Natural accession · Recombinant inbred line · Resource Center · Reverse genetic resource · Seed collection · Stock Center · Teaching tool

Abbreviations

ABRC	*Arabidopsis* Biological Resource Center
AGI	*Arabidopsis* genome initiative
AGRIKOLA	*Arabidopsis* Genomic RNAi Knock-Out Line Analysis
amiRNA	artificial microRNA
ATIDB	*Arabidopsis thaliana* integrated database
BAC	Bacterial artificial chromosome

R. Scholl (✉)
Department of Plant Cellular and Molecular Biology, Arabidopsis Biological Resource Center, The Plant Biotechnology Center, The Ohio State University, Columbus, OH 43210, USA
e-mails: scholl.1@osu.edu; abrc@arabidopsis.org

Emma M. Knee and Luz Rivero are co-first authors

R. Schmidt, I. Bancroft (eds.), *Genetics and Genomics of the Brassicaceae*, Plant Genetics and Genomics: Crops and Models 9, DOI 10.1007/978-1-4419-7118-0_16,

BBSRC	Biological Sciences Research Council
BCCM	Belgian coordinated collections of micro-organisms
BiBAC	Binary BAC
BRC	Bioresource Center
CATMA	Complete *Arabidopsis* transcriptome microarray
CEPH	Centre d'Etude du polymorphisme humain
CIC	CEPH INRA CNRS
CNRGV	Centre National de Resource Genomique Vegetables
CNRS	Centre National de Recherche Scientifique
CSHL	Cold Spring Harbor Laboratory
CUGI	Clemson University Genome Institute
DFFS	Diversity fixed foundation sets
DH	Double haploid
DNA	Deoxyribonucleic acid
ECPGR	European Cooperative Programme for Plant Genetic Resources
EMS	Ethylmethanesulfonate
EST	Expressed sequence tag
EU RESGEN	European Union genetic resources project
GetCID	Gene transfer, clone identification, and distribution service
GRIN	Germplasm resources information network
GST	Gene-specific tag
HRI	Horticultural Research International
INRA	Institut National de la Recherche Agronomique
JGI	Joint Genomes Institute
MATDB	Munich Information Center for Protein Sequences *Arabidopsis thaliana* Database
MBGP	Multinational *Brassica* genome project
MBrGSP	Multinational *Brassica rapa* genome sequencing project
MTA	Material transfer agreement
NASC	Nottingham *Arabidopsis* Resource Center
NCBI	National Center for Biotechnology Information
NIL	Near-Isogenic line
NPGS	National Plant Germplasm System
ORF	Open reading frame
PCR	Polymerase chain reaction
PGML	Plant Genome Mapping Laboratory
QTL	Quantitative trait loci
RI	Recombinant inbred
RIL	Recombinant inbred line
RNA	Ribonucleic acid
RNAi	RNA interference
SIGnAL	Salk Institute Genomic Analysis Laboratory
TILLING	Targeting Induced Local Lesions in Genomes

Contents

16.1 Introduction 439
16.2 Germplasm Resources 440
16.2.1 *A. thaliana* 445
16.2.2 *Brassica* Species 447
16.3 Molecular Resources 451
16.3.1 Genomic Library/Clone Resources for *A. thaliana* 456
16.3.2 Genomic Library/Clone Resources for Members of the Brassicaceae . . . 459
16.3.3 Other Molecular Resources for *Arabidopsis* 460
16.3.4 New Resources for the Brassicaceae 463
16.4 Conclusions 463
References 466

16.1 Introduction

The Brassicaceae contains the model species *Arabidopsis thaliana* and a diverse array of cultivated types including oilseed, mustard, vegetable (leaf, root, and stem), and fodder (leaf and root) crops. It is the focus of plant breeding and genetic research including classical cytogenetics, and molecular biology. Wide genetic diversity has been generated through traditional breeding, mutagenesis and molecular techniques, and many samples have been collected from the wild. The Brassica family plays an important role in the field of plant genomics. Several genome sequencing projects have been completed or are presently underway. Partially or completely sequenced species include *A. thaliana*, *Arabidopsis lyrata*, *Capsella rubella*, *Thellungiella halophila*, and *Brassica rapa*, and comprehensive comparative sequencing of *A. thaliana* natural accessions is underway (http://borevitzlab.uchicago.edu). Planned sequencing projects at the U.S. Department of Energy Joint Genome Institute (DOE JGI, http://www.jgi.doe.gov) involve *Arabidopsis arenosa*, *Arabidopsis halleri*, *Boechera divericarpa*, *Boechera holboellii*, and *Boechera stricta*. The existing genomic sequence of *A. thaliana* (Arabidopsis Genome Initiative 2000) represents the reference point for evaluation of other plant sequences as these are developed across the family as well as in other plant families (Lysak and Lexer 2006). Comparative genetic mapping has found significant synteny and conservation of genomic blocks across a number of species (Schranz et al. 2007a, b). Mapping and quantitative genetics have been aided by the development of double haploid (DH)-derived populations, and Targeting Induced Local Lesions in Genomes (TILLING) has been conducted in several species. Comprehensive molecular resources for *A. thaliana* have been developed. Many are publicly available and are widely utilized in research. Included are expressed sequence tags (ESTs); full-length, sequence-validated cDNA clones of many genes; bacterial artificial chromosome (BAC) libraries; and diverse vectors with multiple applications. BAC libraries and ESTs are currently the primary molecular resources developed in other species of Brassicaceae.

The many germplasm and molecular resources that have been developed in the Brassicaceae are held and distributed by numerous institutions and specialized centers. This chapter will survey and assess the resources available for the members of this group with focus on collections that are publicly available.

16.2 Germplasm Resources

Most of the Brassicaceae are conserved as seeds. Resource institutions have the mission to collect, document, evaluate, enhance, preserve, and distribute seeds with diverse genetic backgrounds, procuring accessions through international and domestic exchanges and developing in situ maintenance programs for crop plants and their wild relatives. In situ conservation is particularly desirable for wild species, which may be difficult to regenerate ex situ (Boukema and van Hintum 1999). Large and small collections of biological resources for important members of the Brassicaceae are maintained by several institutions, including university programs, state-owned or private organizations, and national or regional gene banks. Large national collections, such as those in the USA, China, and India, are generally maintained as a network of active regional collections. For safety reasons, some collections are duplicated in separate in situ or ex situ locations. Biological resources in most public institutions are freely accessible, while access to unique resources in private institutions may often require a material transfer agreement (MTA) or similar contract that governs the transfer of tangible research materials between organizations, defining the rights of the provider and the recipient with respect to the materials and any derivatives. Since most *Brassica* crop species are out-crossing and biennial, it is very important to follow strict procedures in maintaining the genetic integrity of the germplasm during regeneration (Boukema and van Hintum 1999). Good storage conditions are important to reduce the regeneration frequency. The major factors influencing seed longevity are storage temperature and seed moisture content (the higher the value of either, the shorter the lifespan of the seeds). A sealed or any moisture-proof container, where seeds have the optimal 3–7% moisture content, is recommended (Boukema and van Hintum 1999, Rivero-Lepinckas et al. 2006). Short-to medium-term storage conditions are suitable for active collections since these are accessed often. A convenient temperature is approximately 4°C (standard refrigerator temperature). For long-term base collections, which are infrequently accessed, seed viability can be maintained for decades or centuries at –20°C (Rivero-Lepinckas et al. 2006).

The majority of germplasm collections have been developed since the 1970s, including the establishment of public and private databases with comprehensive and accessible information about the collections. These efforts greatly enhance the management and use of germplasm and genomic resources worldwide. These collections, including URLs for obtaining detailed information, are summarized in Table 16.1. Additional information about selected resource centers is provided below, by species.

Table 16.1 Primary resource centers, holdings, and contact information

Continent, country	Name of institution/location	No. of accessions	Nature/scope of collection	URL[a]
Africa	–	–	–	–
Ethiopia	Institute of Biodiversity Conservation (IBC), Addis Ababa	1,332	*B. carinata, B. nigra*	http://ibc–et.org/ibc/dbase
Asia	–	–	–	–
China	Institute of Crop Germplasm Resources (ICGR)/Chinese Academy of Agricultural Sciences (CAAS), Beijing	21,000	*B. alboglabra, B. carinata, B. chinensis, B. campestris, B. juncea, B. napus* var. *napobrassica, B. nigra, B. oleracea, B. rapa, Capsella* sp., *Raphanus* sp.	http://icgr.caas.net.cn
India	National Bureau of Plant Genetic Resources (NBPGR)/New Delhi	4,530	*B. juncea, B. nigra, B. rapa, B. tournefortii*	http://www.nbpgr.ernet.in
–	All India Coordinated Research Project (AICRP) on Rape & Mustard/Haryana University, Hisar, Haryana	7,314	*B. juncea, B. rapa*	–
–	Indian Agricultural Research Institute (IARI)/Pusa Campus, New Delhi	1,200	*B. juncea*	–
Japan	National Institute of Agrobiological Sciences (NIAS)/Tsukuba, Ibaraki	2,168	*B. campestris, B. carinata, B. juncea, B. napus, B. nigra, B. oleracea, B. parachinensis, B. pekinensis, B. rapa, Raphanus sativus*	http://www.gene.affrc.go.jp
–	RIKEN BRC/Tsukuba-shi, Ibaraki	80,000	*A. thaliana* and close relatives	http://www.brc.riken.jp
Korea	Korea Brassica Genome Resource Bank (KBGRB)/Chungnam National University, Daejeon	5,507	*B. rapa, Brassica* sp., wild types, mutants, recombinant inbred lines, TILLING lines	http://brassica-resource.org

Table 16.1 (continued)

Continent, country	Name of institution/location	No. of accessions	Nature/scope of collection	URL[a]
Russia	N.I. Vavilov Research Institute of Plant Industry (VIR)/St. Petersburg	3,503	*B. botrytis, B. capitata, B. chinensis, B. gongylodes, B. japonica, B. juncea, B. napus, B. nigra, B. oleracea, B. pekinensis, B. sabauda*	http://vir.nw.ru
Taiwan	Asian Vegetable Research and Development Center (AVRDC)/Shanhua	1,386	*B. campestris, B. carinata, B. juncea, B. napus, B. rapa*	http://avrdc.org
Europe	–	–	–	–
Czech Republic	Oseva/Research Institute of Oilseed Crops, Opava	922	*B. carinata, B. juncea, B. napus, B. nigra, B. rapa*	http://oseva.cz
France	Institut National de la Recherche Agronomique (INRA)	69,000 1,200	*A. thaliana, B. oleracea*	http://dbsgap.versailles.inra.fr/vnat
–	Unite Experimentale d'angers Geves/Brion	1,396	*B. chinensis, B. napus, B. oleracea*	–
Germany	Institute of Plant Genetics and Crop Plant Research (IPK)/Gatersleben	4,284	*B. carinata, B. hybrida, B. juncea, B. napus, B. nigra, B. oleracea, B. rapa,* wild species	http://www.ipk-gatersleben.de
Spain	Department of Plant Biology/Polytechnic University of Madrid, Madrid	1,900	*Brassica* sp., numerous cruciferous genera, wild types	
The Netherlands	Centre for Genetic Resources, the Netherlands (CGN)/Wageningen University	1,750	*B. carinata. B. napus, B. oleracea, B. rapa, R.* sativus	http://cgn.wur.nl

Table 16.1 (continued)

Continent, country	Name of institution/location	No. of accessions	Nature/scope of collection	URL[a]
United Kingdom	Nottingham *Arabidopsis* Stock Centre (NASC)/University of Nottingham, Loughborough	400,000	*A. thaliana, A. arenosa, A. lyrata, A. korshinsky, A. shokei, A. suecica, C. rubella, Crucihimalaya lasiocarpa, Olimarabidopsis pumila, Sisymbrium irio, Thellungiella salsuginea, Thellungiella parvula*	http://arabidopsis.info
–	Warwick Horticulture Research International (Warwick HRI)/University of Warwick, Wellesbourne	6,915	*B. juncea, B. napus, B. oleracea, B. rapa, Raphanus, Eruca, Sinapsis, Crambe, Rorippa*	http://www2.warwick.ac.uk/fac/sci/whri/research/gru
North America	–	–	–	–
Canada	Plant Gene Resources of Canada (PGRC)/Saskatoon Research Center, Saskatoon	3,917	*B. barrelieri, B. carinata, B. juncea, B. napus, B. nigra, B. rapa, B. oleracea, B. tournefortii*	http://pgrc3.agr.gc.ca
USA	*Arabidopsis* Biological Resource Center (ABRC)/The Ohio State University, Columbus, Ohio	400,000	*A. thaliana, A. arenosa, A. lyrata, A. korshinsky, A. shokei, A. suecica, C. rubella, C. lasiocarpa, O. pumila, S.irio, T. salsuginea, T. parvula, B. napus, B. oleracea, B. rapa*	http://arabidopsis.org

Table 16.1 (continued)

Continent, country	Name of institution/location	No. of accessions	Nature/scope of collection	URL[a]
–	The National Plant Germplasm System (NPGS)	5,338	*Brassica* sp., *Raphanus* sp., *Arabidopsis* sp., *Arabis* sp., *Boechera* sp., *Capsella* sp., *Sisymbrium* sp.	http://www.ars-grin.gov
Oceania	–	–	–	–
Australia	Australian Temperate Field Crops Collection (ATFCC)/Dept. of Primary Industries, Horsham, Victoria	3,599	*B. carinata, B. cretica, B. elongate, B. fruticulosa, B. incana, B. insularis, B. juncea, B. macrocarpa, B. montana, B. napus, B. nigra, B. oleracea, B. oxyrrhinia, B. rapa, B. rupestris, B. souliei, Brassica* sp., *B. tournefortii, B. villosa*	–

[a] Additional information at http://bioversityinternational.org

16.2.1 A. thaliana

The substantial diversity of *A. thaliana* resources is primarily maintained in resource centers. Two of these, the *Arabidopsis* Biological Resource Center (ABRC) at The Ohio State University, USA, and the Nottingham *Arabidopsis* Stock Centre (NASC) at the University of Nottingham, UK, are comprehensive in the breadth and depth of available materials. ABRC and NASC have an ongoing mutual interchange of stocks, so that their seed collections largely mirror each other. Their collections include a wide range of *A. thaliana* resources, as well as a number of related resources, such as natural accessions or synthetic hybrids of various species of *Arabidopsis*, domesticated *Brassica* crops, and other related species. All of these resources are now popular for studies of molecular biology, genetics, plant physiology, developmental biology, functional and comparative genomics.

A. thaliana seed holdings at ABRC and NASC consist of over 400,000 lines including mutants, transgenic lines, TILLING lines, chromosomal variants, natural accessions, and other species (Table 16.2). Seeds enter ABRC and NASC by donation from researchers. Backups of stocks are maintained, typically as duplicates in separate facilities, to ensure preservation. The resources and data are accessible to the scientific community through The *Arabidopsis* Information Resource (TAIR, http://arabidopsis.org) and the NASC database (http://arabidopsis.info). The mutant collection consists primarily of characterized lines, many of which are described in detail in publications. The natural accessions represent approximately 1,300 unique lines and have recently been fingerprinted by PCR marker analysis in the laboratory of J. Borevitz (http://borevitzlab.uchicago.edu). An array of recombinant inbred populations is available, generated by hybridization of diverse accessions. Large numbers of single T-DNA and transposon insertion lines, which are progeny lines of independent transformants or transposants, are also available and for many flanking

Table 16.2 Types and sources of *Arabidopsis* seed stocks

Type of resource	Number of lines	Source
Characterized mutants	4,200	ABRC, NASC
Transgenes	700	ABRC, NASC
Recombinant inbred	28 populations	ABRC, NASC
	30 populations	INRA
	17 populations	Miscellaneous
Mapping lines	400	ABRC, NASC
T-DNA insertion	615,000	ABRC, NASC
	55,000	INRA (FLAG_FST)
	50,000	RIKEN BRC
	900	Biological Research Center (Hungary)
Transposon insertion	70,000	ABRC, NASC
	18,000	RIKEN BRC
Fox hunting	15,000	RIKEN BRC
TILLING	11,000	ABRC, NASC
RNA interference	21,000	ABRC, NASC
Natural accessions	1,300	ABRC, NASC
	500	INRA

sequence has been determined so that their insertions are anchored on the genome sequence. Comprehensive numbers of pooled T-DNA lines are accessible mainly for phenotypic screening/mutant tagging. Because of the size of these populations, at least one and often two independent insertion alleles exist in the collection for the great majority of genes (more than 26,000). The efforts of several laboratories contributed to these collections including (1) Salk Institute, USA, (2) John Innes Centre, UK, (3) Max Planck Institute for Plant Breeding, Germany, (4) University of Bielefeld, Germany, (5) University of Wisconsin, USA, (6) Syngenta, (7) Cold Spring Harbor Laboratory, USA. Details of these populations are summarized in Table 16.3. In addition, substantial forward genetic insertional resources are available from ABRC and NASC (Table 16.4). These have been used in the past for gene tagging. Various insertional resources, mostly sequence-tagged lines, are also available from a number of other institutions (Table 16.2).

The RIKEN Bio Resource Center (BRC, http://www.brc.riken.jp), Japan, preserves and distributes *A. thaliana* seed stocks including transposon-tagged, T-DNA-tagged, and full-length cDNA over-expressing (Fox hunting) lines. Diverse mutant and wild-type stocks formerly distributed by The Sendai *Arabidopsis* Seed Stock Center in Japan are now preserved and distributed by BRC, which replaces the Sendai Center and has substantially expanded collections and functions (Table 16.2). The Institute National de la Recherche Agronomic (INRA) in France is responsible for germplasm and genomic resources for *A. thaliana*, among other plants. The *Arabidopsis thaliana* Resource Centre for Genomics has made the following three collections available to the scientific community: a collection of 55,000 independent insertion mutants, more than 500 natural accessions, and nearly 30 recombinant inbred line (RIL) populations consisting of

Table 16.3 Reverse genetic seed resources of *A. thaliana* available from ABRC/NASC

Resource name	Type	Number of lines
AGRIKOLA	RNAi knockout	21,000
Bancroft/Dean	Transposon insertion	200
Chromatin Charting	Transposon insertion/launch pad/transposase	250
CSHL	Transposon gene trap and enhancer trap	2,000
GABI-Kat	T-DNA insertion	80,000
Haseloff/U. Cambridge	T-DNA enhancer trap	250
IMA	Transposon insertion	500
JIC activate	Transposon activation trap	1,000
JIC SM	Transposon (Ds-Spm) insertion	26,000
Poethig/U. Penn	T-DNA enhancer trap	300
Syngenta (SAIL/GARLIC/TMRI)	T-DNA insertion	57,000
Salk	T-DNA insertion	153,000
Salk (confirmed)	T-DNA insertion	25,000
TAMARA	Transposon-mediated activation tag	9,000
TILLING	EMS mutagenized	11,000
Wisconsin Ds-Lox	T-DNA/launch pad (Ds Lox) insertion	10,000+
Miscellaneous	Transposon insertion/launch pad	550

Table 16.4 Forward genetic seed resources of *A. thaliana* available from ABRC/NASC

Resource name	Type	Number of lines
Alonso, Crosby, and Ecker	T-DNA simple insert (sets of pools and individual lines)	40,000
Bressan, Yokoi, and Koiwa	T-DNA activation tag (sets of pools)	23,000
Calyx	T-DNA transactivation enhancer trap (individual lines)	7,600
Feldmann	T-DNA simple insert (sets of pools and individual lines)	10,000
Gallois	T-DNA enhancer trap (set of lines)	250
INRA	T-DNA promoter trap (set of pools)	10,000
Jack	T-DNA enhancer trap (set of pools)	11,000
JIC gene trap (EXOTIC)	Transposon gene trap (individual lines)	23,000
JIC enhancer trap	Transposon enhancer trap (individual lines)	2,000
JIC GMT (MT/MET)	Transposon enhancer trap (individual lines)	2,000
LeClere and Bartel	T-DNA over-expression (set of pools)	33,000
Loake	T-DNA activation tag (individual lines)	600
Scheible and Somerville	T-DNA activation tag (set of pools)	62,000
SLAT	Transposon insertion (individual lines)	4,000
Soll/Johnson	Transposon insertion (individual lines)	150
Sussman and Amasino	T-DNA simple insert (set of pools)	60,000
Syngenta (Zeneca-Mogen)	T-DNA promoter trap (individual lines)	1,000
Weigel	T-DNA activation tag (set of pools)	63,000
Wisconsin, KO	T-DNA insertion (set of pools)	60,000
Miscellaneous insertion	T-DNA and transposon insertion (individual lines)	950

approximately 450 lines each (http://dbsgap.versailles.inra.fr/vnat) (Table 16.2). ABRC and NASC coordinate activities with this organization and the functions of the three organizations are largely complementary.

Lehle Seeds (http://arabidopsis.com) is a private company in Texas, USA, and distributes *A. thaliana* seeds for wild types, EMS M_2, fast neutron M_2, and gamma-ray and fast neutron M_1 populations. Lehle Seeds provides these specific services, which can be supported in a for-profit operational atmosphere, to *Arabidopsis* researchers, and the public stock centers maintain the broader based genetic collections for the community.

16.2.2 **Brassica** *Species*

Traditionally, most germplasm, including the material of existing collections, was based on local cultivated selections or landraces developed over generations. Core collections have been developed (Boukema et al. 1997) more recently for *Brassica* species, organized principally by The European Union Genetic Resources Project (EU RESGEN) and other programs. These collections are designed to represent the genetic diversity of a crop species and its relatives with a minimum of repetition and are typically held in public genetic resource collections. Information on germplasm characteristics and ordering locations for the collections is maintained in databases, such as Bioversity International (http://www.bioversityinternational.org).

The web site http://www.brassica.info has been established as the common portal for compiling information about *Brassica* species, crops, and their genomes. The European Cooperative Programme for Plant Genetic Resources (ECPGR) in the Netherlands maintains the ECPGR *Brassica* database (http://documents.plant.wur.nl/cgn/pgr/brasedb) and compiles information on genetic resource accessions held in public collections. Requests for the more than 24,000 accessions are directed to the collection holders of the correspondent gene banks in different countries such as Austria, Azerbaijan, Belgium, Bulgaria, Czech Republic, Germany, Estonia, France, Georgia, Greece, Hungary, Italy, Lithuania, Poland, Portugal, Romania, Russia, Serbia, Slovakia, Slovenia, Spain, Sweden, Switzerland, the Netherlands, Turkey, Ukraine, and the UK. Since there are over 130 collections of *Brassica* species distributed across almost 60 countries worldwide, only a selection of large or otherwise important collections will be described (see also Table 16.1).

The ABRC holds and distributes a small collection of *B. napus* self-compatible allopolyploids, which were developed by crossing the homozygous parents *B. rapa* and *B. oleracea* followed by colchicine doubling. The progeny were then advanced by single seed descent to the S5 generation.

The National Plant Germplasm System (NPGS) in the USA is a cooperative effort between public and private organizations to preserve the genetic diversity of plants including members of the Brassicaceae. The NPGS, through the National Genetic Resources Program, aids scientists by acquiring, preserving, evaluating, documenting, and distributing crop germplasm. Since many important crop species originate outside the USA, the first steps toward diversity are acquisition and introduction by donation from foreign cooperators or international germplasm collections. Two active regional centers (Northeast Regional Plant Introduction Station, Cornell University, Geneva, New York, and North Central Regional Plant Introduction Station, Iowa State University, Ames, Iowa) hold the majority of the Brassicaceae represented in germplasm collections. Many of these are duplicated in long-term storage at the National Center for Genetic Resources Preservation in Fort Collins, Colorado. The Germplasm Resources Information Network (GRIN) provides comprehensive germplasm information (http://www.ars-grin.gov). The Brassicaceae collection includes a total of 5,347 accessions, where the genus *Brassica* represents the majority, with 4,562 accessions, corresponding to 85 species from 83 countries, followed by *Raphanus*, with 761 accessions representing 4 species from 31 countries. Other genera in this collection are *Arabidopsis*, *Arabis*, *Capsella*, *Boechera*, and *Sisymbrium*, totaling 24 accessions.

The Canadian Plant Germplasm System, as part of Plant Gene Resources Canada (http://pgrc3.agr.gc.ca), is a national institution devoted to the free and unrestricted exchange of germplasm with all nations. Access to Canadian collections is permitted by any person with a justified use. It has a total of 3,917 accessions of the genus *Brassica*.

Warwick Horticulture Research International (Warwick HRI), through the Genetic Resources Unit, maintains a collection of 6,915 accessions mostly *B. oleracea, B. napus, B. rapa, and B. juncea,* but with other genera such as *Eruca, Sinapsis, Crambe, Rorippa*, and *Raphanus* available for pure and applied scientific research and educational purposes throughout the world. Requestors are required to sign an MTA. The only other qualifier is a request to return information for inclusion in the database to benefit other users. In addition, two reference collections are maintained and distributed, the *Brassica* S-allele collection and the European Clubroot Differential Series for *Brassica*. Warwick HRI also distributes the Diversity Fixed Foundation Sets (DFFS http://www.brassica.info) which are informative sets of genetically fixed lines representing a structured sample of diversity across a gene pool.

The Center for Genetic Resources in the Netherlands is part of Wageningen University and maintains a crucifer collection that comprises more than 40 crop groups, including 1,750 accessions available to the public. A standard MTA is required for access to this collection. The collection contains mainly cultivars (1,080), plus landraces (294) and research material (181). Only 14 accessions are documented as wild populations and the population type of 181 accessions is not known. *B. oleracea* is represented with the highest number of accessions, followed by *B. rapa*, *R. sativus*, *B. napus*, and *B. carinata*.

China holds several regional collections, consisting of more than 21,000 accessions for various *Brassica*, *Raphanus*, and *Capsella* species. The Institute of Crop Germplasm Resources of the Chinese Academy of Agricultural Sciences in Beijing is responsible for the collection and preservation of these resources and the Chinese Crop Germplasm Information System (http://icgr.caas.net.cn) is the central repository for genetic resources information. The Korea *Brassica* Genome Resource Bank is a *Brassica*-related seed and DNA stock center, which supplies plant materials for structural and functional genomics of *Brassica* species. It has developed mapping populations, recombinant inbred lines, and mutants of *Brassica* species and holds 4,643 accessions and 864 TILLING lines. Moreover, seeds of species and/or cultivars showing wild-type and/or unusual characteristics have been collected and multiplied. These genetic resources are the basal materials for the Multinational *Brassica* Genome Project (MBGP) and its subproject, the Multinational *Brassica rapa* Genome Sequencing Project (MBrGSP).

The National Bureau of Plant Genetic Resources in New Delhi, India, is the nodal organization for exchange, quarantine, collection, conservation, evaluation, and the systematic documentation of plant genetic resources. It holds 4,530 accessions of *B. juncea*, *B. nigra*, *B. tournefortii*, *B. rapa*, and other *Brassica* species. The All India Coordinated Research Project maintains rape and mustard stocks at the Haryana University, Hisar, including 7,314 accessions of *B. juncea*, *B. rapa*, other *Brassica* species. The Indian Agricultural Research Institute, Pusa Campus, New Delhi, maintains 1,200 accessions of *B. juncea*.

The Institute of Biodiversity Conservation in Ethiopia was established in 1976 with the objectives of collection and preservation of crops. The collection consists of a number of traditional cultivars/landraces of *Brassica*.

There are several other important collections worldwide. The N. I. Vavilov Research Institute of Plant Industry in Russia holds a global collection encompassing 3,503 accessions of *Brassica*. The Polytechnic University of Madrid in Spain maintains a collection of 1,900 closely related genera of the Brassicaceae including wild types of *Brassica* and related species. There are two major resource centers in France, the Unite Experimentale d'Angers Geves holding 1,396 accessions of *B. chinensis*, *B. napus*, and *B. oleracea* and the Institut National de la Recherche Agronomique (INRA) having more than 1,200 accessions of *Brassica*. The Institute of Plant Genetics and Crop Plant Research in Germany has a total of 4,284 accessions corresponding to various *Brassica* crop and wild species.

16.2.2.1 Diversity Fixed Foundation Sets (DFFS)

Recent molecular analyses have identified considerable genotypic variation within individual extant accessions of *Brassica*. This is present both as within-line heterogeneity and heterozygosity. As such, any coordinated trait and genetic studies will be very difficult to interpret unless material with a high degree of homozygosity is available. To address this issue and provide researchers with relevant seed and DNA resources, Diversity Fixed Foundation Sets (DFFS) are being constructed for *B. napus, B. oleracea, B. rapa*, and non-crop *Brassica* C genome species. The latest are diploid species ($2n = 18$) including *B. cretica, B. incana, B. macrocarpa*, and *B. villosa*, which are able to intercross with *B. oleracea* (C genome $2n = 18$). The DFFS are designed to represent the majority of natural allelic variation within respective gene pools in a form suitable for long-term exploitation by researchers and end users. Cumulative data will allow long-term comparative analysis and enable experimental trials to be established with replicate plants. For *B. napus*, development of the diversity sets was initiated with a collection of accessions based on the EU RESGEN core collection (http://documents.plant.wur.nl/cgn/pgr/brasedb/brasresgen.htm), which now contains genetically fixed lines, double haploid (DH) or inbred lines, mapping population parents, accessions with extended geographic origins, crop types, and wild types. Similarly, the *B. oleracea* DFFS represents crop types and mapping parents and includes a number of genetically fixed lines. The *Brassica* C genome species DFFS was initiated with accessions of wild relatives of *B. oleracea* which have not been domesticated and thus may possess adaptations to survival in the wild that have been lost by the crop species. Sets and public availability can be found at http://www.brassica.info.

16.2.2.2 Mapping Populations

Numerous DH and some recombinant inbred line (RIL) populations of *B. juncea, B. napus, B. oleracea*, and *B. rapa* were generated by crossing important cultivars and have been used to construct linkage maps comprising genetic and phenotypic markers. These populations of "immortal" homozygous lines are useful resources for genetic mapping, building, and integrating reference maps of different species and identification of molecular markers linked to important agronomic traits, quantitative trait loci (QTL) analysis, and marker-assisted selection in breeding programs

(Pink et al. 2008). To increase the available locus resolution prior to generating near-isogenic lines (NILs), a series of substitution line populations have been generated through recurrent backcross to a common parent. These reference populations and the associated key data sets (genetic linkage map, provenance, and quality assurance validation) are available through the Crop Store relational database (http://www.brassica.info/CropStore), which has been developed to address the need for low maintenance, explicit curation, and management of integrated data sets for crop plant genetics. Seeds are provided in the form of quality-assured starter packs on a cost of replacement basis. Specific information on these populations can be found at http://www.brassica.info/resource/plants/mapping-populations.php.

16.2.2.3 Emerging Resources

One of the functional genomics resources being developed for *Brassica* species is TILLING (Wang et al. 2008). A number of projects have been initiated and are benefiting from the higher mutant load achievable in *Brassica*, due to gene duplications. Information about the ongoing programs and available resources is collated via the Multinational *Brassica* Genome Project (MBGP, http://www.brassica.info/research/activities/tilling.php). This features details of available screening services in order to identify mutants in specific genes. Another advanced resource being developed is an RNAi pipeline (http://brassica.info/ukbrc/advab) for downregulation of multiple gene copies; this will be established in *B. oleracea, B. napus*, and *B. rapa*.

16.2.2.4 Educational Resources

Numerous educational resources and outreach programs have been developed using *A. thaliana* and different species of the rapid cycling *Brassica* (*B. rapa, B. nigra, B. oleracea, B. juncea, B. napus*, and *B. carinata*). These successful teaching tools are used worldwide, providing opportunities for students to perform scientific research in the classrooms of schools and colleges. These resources and programs represent the efforts by private or educational institutions to introduce students to concepts in plant biology, plant physiology, genetics, molecular biology, genomics, biotechnology, and bioinformatics. Plants grow in compact areas and have short generation cycles, making them convenient for classroom studies. Students of all ages can explore a wide variety of questions and learn by engaging in genuine scientific inquiry. The germplasm and molecular resources are part of the educational materials that are included in creative experiments and kits. Resources, descriptions, and ordering information are available online (Table 16.5).

16.3 Molecular Resources

The foundation of genomic resources in the Brassicaceae is an array of large insert bacterial artificial chromosome (BAC) libraries which have been used to derive genome sequence for *A. thaliana* and continue to be utilized in sequencing

Table 16.5 Publicly available resources for education

Institution/company	URL
Arabidopsis Biological Resource Center (ABRC)	http://abrc.osu.edu
Carolina Biological Supply	http://www.carolina.com
EDVOTEK	http://www.edvotek.com
Fast plants	http://www.fastplants.org
Greenomes	http://www.greenomes.org
Nottingham Arabidopsis Stock Centre (NASC)	http://www.arabidopsis.info/CollectionInfo?id=49
Partnership for Research and Education in Plants (PREP)	http://www.prep.biotech.vt.edu
The Polyploidy Portal	http://polyploidy.org/index.php/MapPlants
Other teaching resources	http://www.arabidopsis.org/portals/education/programs.jsp

Brassica genomes and in comparative genomic studies. Yeast artificial chromosome (YAC) and cosmid libraries have also proven useful in this regard, but the BACs afford compelling advantages. The complete genome sequence of *A. thaliana* was assembled in annotation units derived largely from BAC clones. Many BAC libraries of *A. thaliana* and other members of the Brassicaceae are available (Tables 16.6–16.8). EST and full-length cDNA clones have been important in establishing experimental evidence for transcription of predicted gene models in *A. thaliana* and are being put to the same use in other members of the Brassicaceae. The next phase of genomic clone resources in *A. thaliana* is coming into play in the form of Gene-Specific Tag (GST) clones and RNA interference (RNAi) constructs that can be used for targeted disruption of individual genes or gene families in functional genomic studies. Multifunctional expression vectors designed for use with the numerous cDNA and open reading frame (ORF) clone resources are also proving useful in transcriptomic and functional genomic studies.

Various storage formats are possible for molecular resources. However, for frequently accessed collections, most clones are maintained in an *E. coli* host as a glycerol stock stored at –70°C. Many are distributed as bacterial stab cultures which are easy to prepare, maintain viability during shipping, and are easy to grow upon receipt. For distribution of single clones, individual glycerol stocks in cryovials, stored at –70°C, are the most robust format for clone preservation. This allows access to individual clones without thawing and with minimal risk of cross contamination. Many libraries and large collections of clones are stored as glycerol stocks in 96- or 384-well microtiter plates. Strict protocols for replicating and accessing clones are required to minimize cross contamination, when either manual or robotic procedures are employed. Storage of plasmid DNA on a paper matrix such as Whatmann Clonesaver™ minimizes the storage space required for large numbers of clones and represents a good alternative in specific situations.

Table 16.6 *A. thaliana* BAC and P1 libraries utilized in the *Arabidopsis* genome initiative sequencing project and available from ABRC

Library	Vector	Average insert size (kb)	Cloning enzyme	Format available	Restrictions	Other sources
TAMU	pBeloBAC11	100	*Hind*III	Single clones, colony blot filters	–	–
IGF (AT_BBb)	pBeloBAC-Kan	100	*Eco*RI	Single clones, colony blot filters	–	CUGI
TAC (K, At_CTa)	pYLTAC7	80	*Hind*III	Library, single clones, filters	Basic or academic research only, MTA if obtained from GetCID or CUGI	CUGI, GetCID (library, single clones, filters)
PAC (Mi/P1)	pAd10-SacBII	80	*Sau*3AI	Library, single clones, filters	Basic or academic research only	–
Choi (At_CBa)	pBeloBAC11	187.5	*Hind*III	Single clones	MTA if obtained from CUGI	CUGI (library, single clones, filters)

Table 16.7 Other *A. thaliana* large insert libraries

Library	Natural accession	Vector	Average insert size (kb)	Cloning enzyme	Format available	Source
JAtY	Columbia	pYLTAC17	80	*Hind*III	Library, single clones, filters	GetCID
AT	Unspecified	pECBAC1	127	*Bam*HI	Library, filters	Amplicon Express
CD4-35	Landsberg	pBiBAC2	162	*Bam*HI	Library	ABRC
CD4-11	Wassilewskija	POCA18-hyg	25	*Cla*I	Single and pooled clones	ABRC
BC	Columbia	pCLD04541	17	*Bam*HI	Library, single clones, filters	GetCID
JAtC	Columbia	pCLD04541	80	*Bam*HI	Library, single clones, filters	GetCID
JLerB	Landsberg	pCLD04541	17	*Bam*HI	Library, single clones, filters	GetCID
H51	Stockholm × Limburg	pCLD04541	–	*Bam*HI	Library, single clones, filters	GetCID

Table 16.8 BAC libraries derived from members of the Brassicaceae

Species	Name/number	Vector	Insert size (kb)	Cloning enzyme	Format available	Source
A. arenosa	AER	pECBAC1	115	*Bam*HI	Library, single clones, filters	ABRC/Amplicon Express
A. cebennensis	–	pIndigoBAC536	–	*Hind*III	Library	ABRC
A. halleri	–	pIndigoBAC536	–	*Hind*III	Library	ABRC
A. lyrata	ALY	pECBAC1	130	*Bam*HI	Library, single clones, filters	ABRC/Amplicon Express
A. lyrata	–	pIndigoBAC536	–	*Hind*III	Library	ABRC
B. juncea	BJM	pECBAC1	135	*Bam*HI	Library, filters	Amplicon Express
B. napus	Bna-B	pIndigoBAC5	200	*Hind*III	Library, single clones	CNRGV
B. napus	JBnB	pBAC/SACB1	145	*Hind*III	Library, single clones, filters	GetCID
B. napus	JBnY	pYLTAC7	85	*Hind*III	Library, single clones, filters	GetCID
B. oleracea	JBo	pBiBAC2	145	*Bam*HI	Library, single clones, filters	GetCID
B. oleracea	BOCIG	pIndigoBAC5	90	*Hind*III	Library	PGML
B. oleracea	BOBRO	pECSBAC4	100	*Hind*III	Library	PGML
B. oleracea	BOTO1	pIndigoBAC5	105	*Hind*III	Library	PGML
B. oleracea	BoB	pBeloBAC11	110	*Hind*III	Library, single clones	Warwick HRI
B. rapa	JBr	pBiBAC2	130	*Bam*HI	Library, single clones, filters	GetCID
B. rapa	KBrH	pCUGIBACI	115	*Hind*III	Library, single clones, filters	KBGRBGetCID
B. rapa	KBrB	pCUGIBACI	125	*Hind*III	Library, single clones, filters	KBGRBGetCID
C. rubella	CR	pECBAC1	100	*Bam*HI	Library, single clones, filters	ABRC/Amplicon Express
C. rubella	–	pIndigoBAC536	–	*Hind*III	Library	ABRC
O. pumila	OPM	pECBAC1	115	*Bam*HI	Library, single clones, filters	ABRC/Amplicon Express
R. sativus	RS_Dba	pBeloBACII45	–	*Hind*III	Library, single clones, filters	CUGI
R. sativus	Rsa-B	pIndigoBAC536	–	*Hind*III	Library, single clones	CNRGV
S. irio	SIR	pECBAC1	140	*Bam*HI	Library, single clones, filters	ABRC/Amplicon Express

16.3.1 *Genomic Library/Clone Resources for* A. thaliana

16.3.1.1 Resources Utilized by the *Arabidopsis* Genome Initiative

The *Arabidopsis* Genome Initiative (AGI) generated the *A. thaliana* genome sequence from the accession Columbia-0 (*Arabidopsis* Genome Initiative 2000). The sequence was largely completed in the year 2000 and serves as a reference for other natural accessions of *A. thaliana*, members of the Brassicaceae, and higher plants in general. Most of the large insert clones utilized in assembling the AGI sequence are available, without restrictions, from the ABRC (Table 16.6) and can be identified and ordered through The *Arabidopsis* Information Resource (TAIR, http://arabidopsis.org). The genome sequence is publicly available from a number of sources with varying degrees of annotation and integration of related data: TAIR, The National Center for Biotechnology Information (NCBI), The J. Craig Venter Institute (JCVI), the Munich Information Center for Protein Sequences (MIPS) *Arabidopsis thaliana* database (MATDB), *Arabidopsis thaliana* Ensembl, the Salk Institute Genomic Analysis Laboratory (SIGnAL) database, and the *Arabidopsis thaliana* Integrated Database (ATIDB). The URLs for these sites are provided in Table 16.9.

Table 16.9 Web resources for *A. thaliana* genome sequence and annotation

Name	Abbreviation	URL
The *Arabidopsis* Information Resource	TAIR	http://arabidopsis.org
The National Center for Biotechnology Information	NCBI	http://ncbi.nlm.nih.gov
The Munich Information Center for Protein Sequences *Arabidopsis thaliana* Database	MATDB	http://mips.helmholtz-muenchen.de/plant/athal/index.jsp
Arabidopsis thaliana Ensembl	AtEnsembl	http://atensembl.arabidopsis.info/index.html
The J. Craig Venter Institute	JCVI	http://blast.jcvi.org/er-blast/index.cgi?project=ath1
The Salk Institute Genomic Analysis Laboratory	SIGnAL	http://signal.salk.edu
The *Arabidopsis Thaliana* Integrated Database	ATIDB	http://atidb.org

16.3.1.2 Bacterial Artificial Chromosome and P1 Libraries and Clones Used to Generate Genome Sequence

Most of the *A. thaliana* genome sequence was derived from BAC or P1 clones developed at Texas A&M, USA (library name abbreviated as TAMU, TBAC, or T) (Choi et al. 1995), Institute fur Genbiologische Forschung, Germany (library name abbreviated as IGF) (Mozo et al. 1998), Mitsui, Japan (library name abbreviated as MI P1) (Liu et al. 1995), and Kazusa, Japan (library name abbreviated as TAC/K)

(Liu et al. 1999). BAC or P1 sequence was assembled into annotation units that comprise the genome sequence. Each annotation unit is derived primarily from sequence of a single BAC or P1 clone and that sequence has been deposited in GenBank under the name of the clone. However, sequence from adjacent clones was added to one or both ends of a significant number of annotation units, meaning that there is not a precise correspondence between the clone and the annotation unit bearing the same. Researchers must therefore exercise caution when selecting a BAC or P1 clone for PCR amplification of genomic sequence based on the sequence of the annotation unit and should consult any notes placed in the GenBank record describing whether sequence was added from an adjacent clone, the length of the sequence, and from which adjacent clone it was derived. A small number of annotation units were derived from cosmid clones or sub-cloned segments of BAC and yeast artificial chromosome (YAC) clones.

16.3.1.3 Sources of Agi BAC and P1 Libraries, Filters, and Clones

The BAC and P1 libraries utilized in the AGI sequencing project (TAMU, IGF, MI P1, and TAC/K) have been deposited with ABRC in 384-well format. Individual clones from these libraries can be ordered through the TAIR web site. They are also available as colony blot filters for hybridization-based screening and the entire TAC/K and MI P1 libraries may be ordered in 384-well microtiter plate format. Only a subset of the BAC clones in these libraries were fully sequenced for use in assembling the AGI sequence. However, end sequence is available for many more clones allowing them to be anchored on the *A. thaliana* genome. These clones may be visualized using the TAIR map viewer or searched based on end sequence available in the public databases. Individual BAC clones comprising the bulk of the AGI sequence have been isolated from the original libraries, and single cultures have been established in cryovials at ABRC to facilitate preparation of orders for single clones. The above AGI resources comprise the bulk of BAC clone distribution by ABRC. The AGI BAC clones have also been arrayed in 96-well microtiter plates at ABRC and can be ordered as a whole genome set or as sets representing individual chromosomes. Other sources of individual AGI BAC clones, filters, and entire libraries include the Clemson University Genomics Institute (CUGI), USA, and the John Innes Genome Laboratory Gene Transfer, Clone Identification and Distribution Service (GetCID), UK (Table 16.6). Some cosmid clones utilized to fill in sequence in specific regions of the genome where BAC coverage was incomplete are also available from ABRC.

16.3.1.4 Other *Arabidopsis* Genomic Clone Resources

Yeast Artificial Chromosome Libraries and Clones

Early physical mapping of the *A. thaliana* genome, performed before the genome sequencing, involved creation of a tiling path of YACs. A subset of YAC clones from the CEPH INRA CNRS (CIC) (Creusot et al. 1995) library have been positioned on the *A. thaliana* physical map. These have been established as a collection of individual cultures at ABRC that can be ordered through the TAIR web site. This resource is now essentially archival and is rarely distributed. Additional clones from the CIC and other YAC libraries have been archived, but are visible on the TAIR sequence or map viewers and available from ABRC upon request.

Transformable BAC Libraries and Clones

Plant-transformable BAC clones can potentially be used for direct complementation of a mutant phenotype. The TAC/K library utilized in the AGI sequencing project is a transformable library. End sequence is available for some clones from this library. Complete sequence is available for a limited number of clones which were used to generate the AGI sequence for chromosomes 3 and 5. For the remainder, no sequence is available. This limits the utility of the library for complementation studies. However, colony blot filters are available from ABRC and CUGI to allow for identification of potential clones by hybridization. Clones from the JAtY transformable BAC library have been end sequenced and anchored on the *A. thaliana* genome. These are available from GetCID.

Large Insert Libraries from Various *Arabidopsis* Accessions

Additional large insert *A. thaliana* libraries not utilized in the AGI sequencing project are available from various sources. Several were developed from accessions other than Columbia-0 and may be useful in studies of comparative genomics. ABRC distributes a transformable binary BAC (BiBAC) library prepared from the common strain Landsberg *erecta* (Chang et al. 2003) and a cosmid library from natural accession Wassilewskija (Table 16.7) (Schulz et al. 1998). Large insert libraries are also available from GetCID, which provides binary cosmid and BiBAC libraries from Columbia, a binary cosmid library from Landsberg *erecta*, and a binary cosmid library from an inbred line derived from a cross between Stockholm and Limburg (Table 16.7). These are available as entire libraries, filters, and individual clones via a screening service. Amplicon Express, USA, also provides an *A. thaliana* BAC library from an unspecified natural accession as the entire library or as filters (Table 16.7).

16.3.1.5 Utilization of Large Insert Genomic Libraries

A. thaliana BAC and YAC clones have been utilized in ongoing efforts to sequence pericentromeric regions not sequenced as part of the initial release of the AGI sequence. In addition, BAC clones representing subsections of chromosomes with high homology to *Brassica* species have been utilized in comparative genomic studies. ABRC is predominantly distributing *A. thaliana* large insert clones for use as a template for amplification of genomic DNA, and demand for these clones is not as high as it was in the early days after completion of the genome sequence. Clones from other *A. thaliana* natural accessions and related species might be expected to be in demand for use in comparative genomics and genome sequencing projects as well as for complementation of mutant phenotypes, particularly if transformable BAC clones are available.

16.3.2 Genomic Library/Clone Resources for Members of the Brassicaceae

The well-annotated *A. thaliana* genome sequence serves as a reference for other plants, especially members of the Brassicaceae. *Brassica* genomes have undergone multiple rounds of hybridizations/duplications. Polyploid genomes of recent origin exist and have been studied in depth. The *B. napus* genome is predicted to encode approximately 100,000 genes (Love et al. 2006) in comparison to approximately 33,000 (according to the TAIR9 genome release, June 2009) in *A. thaliana*. The *B. rapa* genome sequencing project is ongoing and new Brassica sequence continues to become available. Currently much of the sequence deposited in GenBank is derived from whole BACs, BAC ends, and ESTs of *B. rapa*, *B. oleracea*, *B. napus*, *B. nigra*, *B. juncea*, and *B. carinata*. There has also been a significant amount of sequence generated from small insert genomic clones of the TIGR *Brassica oleracea* Clone Collection (Ayele et al. 2005, Katari et al. 2005). The *Brassica rapa* Genome Sequencing Project (BrGSP) is being carried out by laboratories in Korea and the UK utilizing publicly available BAC libraries (see Chapters 6 and 15). The A genome of *B. rapa* can also be found in *B. napus* and *B. juncea*, both of which are major oil-producing crops. There are a number of online repositories of *Brassica* sequence and other information. These include the *Brassica* Genome Gateway site (http://brassica.bbsrc.ac.uk/), brassica.info, BrassicaDB, CropStore, Ensembl with *Brassica* tracks, physical mapping (UK), TAIR, and ATIDB (see Chapter 22).

16.3.2.1 Resources Associated with *Brassica* Sequencing Projects

BAC libraries from members of the Brassicaceae other than *A. thaliana* are available from ABRC, CUGI, GetCID, Amplicon Express, the French Plant Genomic Resource Center, Centre National de Ressource Genomiques Vegetales (CNRGV), the Plant Genome Mapping Laboratory (PGML), University of Georgia, USA, The Korea *Brassica* Genome Resource Bank (KBGRB), and Warwick Horticultural Research International (HRI), UK (Table 16.8). Sequence information for clones from many of these libraries is limited. Construction of physical maps of the *Brassica* A and C genomes and genome sequencing is being carried out in Korea and in the UK utilizing BAC library resources developed from *B. rapa* variety Chiifu-401 (Mun et al. 2008). These libraries are available from KBGRB (brassicagenome.org) and from GetCID (KBrH and KBrB, see Table 16.8). Partial physical mapping of the genome of *B. napus* is being conducted in Canada and in the European Union. PGML is conducting hybridization-based physical mapping in *B. oleracea* and distributes clones from three libraries utilized in this effort. *B. rapa* genomic survey sequence from the KBrH BIBAC library has been integrated as a G-Browse track in the TAIR database, in AtEnsembl, and in ATIDB in reference to the *A. thaliana* genome sequence to facilitate identification of clones of interest. Individual clones can then be ordered from GetCID. The Biotechnology and Biological Sciences Research Council (BBSRC) *Brassica* IGF project utilized a *B. oleracea* BAC library available from GetCID (JBo) and has generated contigs for the C genome that were anchored on the *A. thaliana* genome using *Arabidopsis* GST clones. The small insert genomic clones of the TIGR *Brassica oleracea* Clone Collection are available from Warwick HRI and sequence has been deposited in GenBank and also aligned to the *A. thaliana* genome in a number of online databases. *A. lyrata, T. halophila*, and *C. rubella* sequencing projects are underway at the US Department of Energy Joint Genome Institute (JGI).

16.3.3 *Other Molecular Resources for* Arabidopsis

With the completion of the *A. thaliana* genome sequence, the focus has shifted from structural to functional genomic studies. According to the 2009 annual report of the Multinational Coordinated *Arabidopsis thaliana* Functional Genomics Project, approximately 25,000 loci in *A. thaliana* are represented by at least one expressed sequenced tagged (EST) clone. Approximately 21,000 are represented by a full-length cDNA or open reading frame (ORF) clone. However, not all of these resources are publicly available. EST and full-length cDNA clones have been important in establishing experimental evidence for transcription of predicted gene models. In other members of the Brassicaceae they are beginning to be put to the same use. A large number of *A. thaliana* full-length cDNA clones and ORF clones are available, although a comprehensive set covering the entire genome has still not been established. Many ORFs are now being cloned into expression constructs for determination of expression patterns and for probing gene function. Techniques for rapid and efficient transfer of ORFs to new vectors and the availability of multifunctional vectors have been very important in this regard. Emerging genomic resources for *A. thaliana* include multifunctional vectors, gene-specific tag (GST) clones, and RNA interference (RNAi) constructs.

16.3.3.1 Expressed Sequence Tags and cDNA Clones

Some of the earliest publicly available *A. thaliana* ESTs were developed by Tom Newman, Chris Somerville (Newman et al. 1994), and Christoph Benning (White et al. 2000) at Michigan State University, USA, and by a consortium at Centre National de la Recherche Scientifique (CNRS), France (Cooke et al. 1996). These have been available through ABRC since the early 1990s. The original phage libraries used to generate the Newman and Benning ESTs are also available from ABRC as are a number of other cDNA libraries distributed as phage suspension. A small subset of ESTs available from ABRC has been experimentally verified and fully sequenced by the research community. These are identified in TAIR and can be ordered from ABRC. A subset of the MSU ESTs is also available from CUGI. The RIKEN BioResource Center (BRC) in Japan has generated and also distributes large numbers of end sequenced cDNAs. Many of these have now been fully sequenced in a joint venture with the Salk Institute, Stanford, Plant Gene Expression Center (SSP) consortium increasing their value as a resource (Yamada et al. 2003). This is described in the next section (see Table 16.10 for a summary of publicly available EST and cDNA resources).

Table 16.10 *Arabidopsis* EST and full-length cDNA/ORF resources available from ABRC and other sources

Resource	Vector	Format available	Restrictions	Source
Newman EST	pZL1, pSHLOX1, pBluescript	Individual clones, partial set in 96-well format	MTA if obtained from CUGI	ABRC CUGI (partial set only)
CNRS EST	pBluescript	Individual clones	None	ABRC
Benning EST	pBluescript	Individual clones	None	ABRC
RIKEN cDNA	pBluescript derived	Individual clones	MTA	RIKEN BRC
Genoscope GSLT cDNA	Gateway™ entry	Individual clones	MTA, Invitrogen limited use license	CNRGV
REGIA cDNA	Gateway™ entry	Individual clones	Invitrogen limited use license	ImaGenes GmbH
ATOME ORF	Gateway™ entry	Individual clones	Invitrogen limited use license	CNRGV
SSP and Salk ORF and cDNA	Univector	Individual clones, partial set in 96-well format	None	ABRC
Salk ORF	Gateway™ entry	Individual clones, set in 96-well format	Invitrogen limited use license	ABRC
TIGR ORF and cDNA	Gateway™ entry	Individual clones, set in 96-well format	Invitrogen limited use license	ABRC
Peking–Yale ORF	Gateway™ entry	Individual clones, set in 96-well format	Invitrogen limited use license	ABRC
Arabidopsis Membrane Interactome Project ORF	Gateway™ entry or TopoTA™	Individual clones	Invitrogen limited use license	ABRC
Judy Callis	Gateway™ entry	Individual clones	Invitrogen limited use license	ABRC

16.3.3.2 Sequenced Full-Length cDNA and ORF Clones in Entry Vectors

The bulk of the cDNA and ORF resources currently being generated are in Gateway™ (Invitrogen™) entry vectors to facilitate transfer of the ORF/cDNA to a number of available destination vectors for further study. Genoscope has generated large numbers of fully sequenced cDNA clones in a Gateway™ entry vector (Castelli et al. 2004) that are distributed by CNRGV. The largest collection of full-length ORF clones was generated by the SSP consortium (Yamada et al. 2003) by sub-cloning and fully sequencing cDNAs generated at RIKEN. The original cDNA clones are available from RIKEN BRC and sub-cloned sequence-verified ORF clones are available from ABRC. The bulk of the ORF clones in this collection is available in both an entry vector for Cre-Lox recombinational cloning (pUNI51) and a Gateway™ entry vector. J. Ecker at the Salk Institute, USA, continues to generate full-length, fully sequenced ORF clones in the pUNI51 vector, adding to the ORF Clone Collection at ABRC. Another collection of sequence-validated ORF and cDNA clones primarily targeting genes of unknown function previously not supported by a full-length cDNA clone was developed at TIGR. These are available from ABRC and complement the SSP and Salk collections. ABRC has received smaller collections of ORF clones focused on specific gene families or functional groupings from The Peking–Yale (PY) consortium (Gong et al. 2004), The *Arabidopsis* Membrane Interactome Project, and Judy Callis (Stone et al. 2005). Several smaller projects have also generated ORF clones that are made available through various centers summarized in Table 16.10. A number of the cDNA and ORF clones have also been transferred to expression vectors and are available from ABRC and other sources. These are not listed here as many are derived from starting material summarized in Table 16.10.

16.3.3.3 Gene-Specific Tag and RNA Interference Clones

GST clones from the Complete *Arabidopsis* Transcriptome MicroArray (CATMA) consortium covering most of the *A. thaliana* genome are available from ABRC and NASC. These clones were used to generate the CATMA arrays and also formed the starter material for RNAi constructs developed by the *Arabidopsis* Genomic RNAi Knock-Out Line Analysis (AGRIKOLA) consortium (Hilson et al. 2004). AGRIKOLA RNAi clones are available from NASC and it is anticipated that they will also be made available through ABRC. A limited number of verified CATMA and AGRIKOLA RNAi clones are also available from Belgian Coordinated Collections of Micro-organisms (BCCM) and researchers may request verification of a specific clone through this service. Approximately 8,000 artificial microRNA (amiRNA) clones, representing as many genes, from Cold Spring Harbor Laboratory (CSHL), USA (Schwab et al. 2006), are available from Open Biosystems and ABRC. This is an ongoing project in which three RNAi constructs targeting each gene in *A. thaliana* will be developed.

16.3.3.4 Multifunctional Vectors

A wide array of vectors, for expression in plants, *E. coli,* yeast, and others, are available from ABRC (Table 16.11). These vectors are designed for an array of advanced functions, including monitoring of expression of different genes by multiple wavelength fluorescence microscopy. These vectors also typically allow production of proteins with versatile affinity and/or antigenic tags. The vectors are useful in *A. thaliana* and other plant species. Plant transformation and RNAi vectors are also available. In addition, ABRC distributes a range of expression vectors designed for use with the UNI-vector system (Liu et al. 1998) allowing transfer of a clone from an entry vector into a destination vector designed for expression in bacteria, yeast, insect, mammalian, or plant cells. N- or C-terminal tagging with C-MYC or His is also possible. These vectors can be used with the original sequence-validated ORF clones. Numerous vectors which can be recombined with the Gateway™ ORF clones are also commercially available.

16.3.4 New Resources for the Brassicaceae

Brassica EST and cDNA clones are fast growing resources that have the potential to be useful in a number of areas of study including establishment of experimental evidence for transcription of predicted gene models. There is no centralized repository for *Brassica* EST and cDNA clones and many of the clones from which sequence has been derived have not been made publicly available. A handful of available resources are listed here, but numerous stocks would likely materialize if resource repositories were developed. CNRGV distributes two *B. napus* EST unigene sets, GOPT6 and RFO2. Warwick HRI is coordinating the *B. oleracea* EST sequencing program and will generate and sequence more than 15,000 clones. The libraries from which the clones are sequenced will be generated from different tissues and subjected to different growing conditions. They plan to make the libraries available for screening and/or other sequencing projects. Shinhan Shiu at Michigan State University, USA, has deposited *R. raphanistrum* and *R. sativus* EST sequence in GenBank and has donated seven EST collections utilized in this project to ABRC. There is also the potential for utilization of vectors developed for use with *A. thaliana* and other systems and for use of *A. thaliana* clones in *Brassica* systems.

16.4 Conclusions

Diverse biological resources exist for the Brassicaceae. The primary resources are seed lines, genomic clones, and cDNA clones. The greatest development of genetic resources exists for *A. thaliana*. The seed collections of *A. thaliana* include a vast array of mutants, transgenic lines, natural accessions, and resources developed to study natural variation. All of these materials are available at public stock centers.

Table 16.11 Multifunctional vectors available from ABRC

Vector name	Stock number	Donor(s)	Description
pEarleyGate series	CD3-683 to CD3-696 and CD3-724	Craig Pikaard and Keith Earley	Gateway^{TM} plant expression vectors for creating C- and N-terminal fusions with fluorescent proteins, HA, AcV5, MYC, FLAG, and TAP
pMN20GW	CD3-680	David Jackson and Amitabh Mohanty	Gateway^{TM} plant expression vector with tetramerized CaMV enhancers in the T-DNA region
pORE series	CD3-920 to CD3-934	Dwayne Hegedus	Vectors for plant expression, promoter analysis, and general plant transformation featuring both monocot and dicot promoters
PKYLX-myc9-loxP	CD3-677	Joseph Ecker	Plant expression vector for use with the univector system
pMDC series	CD3-737 to CD3-762	Mark Curtis and Ueli Grossniklaus	Gateway^{TM} plant expression vectors for functional analysis of genes
pFGC and pGSA series	CD3-446 to CD3-458	Natalie Doetsch and Richard Jorgensen	Plant expression vectors based on pCAMBIA1200 for dsRNA silencing
pSY series	CD3-818 to CD3-821	Nir Ohad and Shaul Yalovsky	BiFC vectors for live protein interaction in plant cells
pDEST-GADT7	CD3-763	Pascale Rossignol, Sarah Collier, John Doonan, and Peter Shaw	Gateway^{TM} yeast expression vector based on pGADT7
pDEST-GBKT7	CD3-764	Pascale Rossignol, Sarah Collier, John Doonan, and Peter Shaw	Gateway^{TM} yeast expression vector based on pGBKT7
pYL435	CD3-678	Savithramma Dinesh-Kumar	Gateway^{TM} yeast expression vector for generation of C-terminal TAP fusion
LIC6	CD3-1009	Savithramma Dinesh-Kumar and Michael Snyder	Gateway^{TM} and ligation-independent cloning (LIC)-compatible plant expression vector
pYL436 (pC-TAPa)	CD3-679	Savithramma Dinesh-Kumar and Xing-Wang Deng	Gateway^{TM} plant expression vector for generation of C-terminal TAP fusion
pN-TAPa	CD3-696	Savithramma Dinesh-Kumar, Xing-Wang Deng, and Vicente Rubio	Gateway^{TM} plant expression vector for generation of N-terminal alternative tandem affinity purification (TAPa) fusion

Table 16.11 (continued)

Vector name	Stock number	Donor(s)	Description
pHB series	CD3-590 to CD3-595	Stephen Elledge	*E. coli* expression vectors for the generation of fusion proteins tagged with HA, MYC, GST, and FLAG for use with the univector system
pHI series	CD3-603 to CD3-599	Stephen Elledge	Baculovirus expression vectors for generation of fusion proteins tagged with GluGlu, FLAG, GST, HA, His6, and MYC for use with the univector system
pHM series	CD3-613 to CD3-615	Stephen Elledge	Mammalian expression vectors for generation of fusion proteins tagged with HA, MYC, and FLAG for use with the univector system
pHY series	CD3-604 to CD3-612	Stephen Elledge	Yeast expression vectors for generation of untagged proteins and fusion proteins tagged with HA and MYC for use with the univector system

The seed resources of other members of the Brassicaceae are highlighted by natural variation and domesticated strains of a vast number of species. Many such species are available from local collections. Genetic resources such as TILLING, RI, and double haploid populations exist for various species or are being developed. Genomic clones, some utilized for whole-genome sequencing, have been developed for a number of species. Resources derived from cDNA have also been developed. *A. thaliana* sequence-validated clones are available as entry and expression clones in vectors designed for a variety of applications. RNAi clones are also being developed, notably in *Arabidopsis*. The *A. thaliana* DNA resources are primarily available from ABRC, and various organizations distribute clones for other members of the Brassicaceae.

References

Arabidopsis Genome Initiative (2000) Analysis of the genome sequence of the flowering plant *Arabidopsis thaliana*. Nature 408(6814):796–815

Ayele M, Haas BJ, Kumar N et al (2005) Whole genome shotgun sequencing of *Brassica olearacea* and its application to gene discovery and annotation in *Arabidopsis*. Genome Res 15:487–495

Boukema IW, van Hintum TJL, Astley D et al (1997) Creation and composition of the *Brassica oleracea* core collection. Plant Genet Resour Newsl 111:29–32

Boukema IW, van Hintum TJL (1999) Genetic Resources. In: Gomez-Campo C (ed) Developments in plant genetics and breeding 4. Biology of *Brassica* coenospecies. Elsevier, The Netherlands

Castelli V, Aury JM, Jaillon O et al (2004) Whole sequence comparisons and "full-length" cDNA sequences: a combined approach to evaluate and improve *Arabidopsis* genome annotation. Genome Res 14:406–413

Chang YL, Henriquez X, Preuss D et al (2003) A plant transformation-competent BIBAC library prepared from the *Arabidopsis thaliana* Landsberg ecotype for functional and comparative genomics. Theor Appl Genet 106:269–276

Choi S, Creelman RA, Mullet JE, Wing RA et al (1995) Construction and characterization of a bacterial artificial chromosome library for *Arabidopsis thaliana*. Weeds World 2:17–20

Cooke R, Raynal M, Laudié M et al (1996) Further progress towards a catalogue of all *Arabidopsis* genes: analysis of a set of 5000 non-redundant ESTs. Plant J 9(1):101–124

Creusot F, Fouilloux E, Dron M et al (1995) The CIC library: a large insert YAC library for genome mapping in *Arabidopsis thaliana*. Plant J 8(5):763–770

Gong Z, Dong CH, Lee H et al (2004) Genome wide ORFeome cloning and analysis of transcription factor genes. Plant Physiol 135(2):773–782

Hilson P, Allemeersch J, Altmann T et al (2004) Versatile gene-specific sequence tags for *Arabidopsis* functional genomics: transcript profiling and reverse genetics applications. Genome Res 14:2176–2189

Katari MS, Balija V, Wilson RK et al (2005) Comparing low coverage random shotgun sequence data from *Brassica oleracea* and rice genome sequence for their ability to add to the annotation of *Arabidopsis thaliana*. Genome Res 15:487–495

Liu Y-G, Mitsukawa N, Vazquez-Tello A, Whittier RF et al (1995) Generation of a high-quality P1 Library of *Arabidopsis* suitable for chromosome walking. Plant J 7(2):351–358

Liu Q, Li MZ, Leibham D et al (1998) The univector plasmid fusion system, a method for rapid construction of recombinant DNA without restriction enzymes. Curr Biol 8:1300–1309

Liu YG, Shirano Y, Fukaki H et al (1999) Complementation of plant mutants with large genomic DNA fragments by a transformation competent artificial chromosome vector accelerates positional cloning. Proc Natl Acad Sci 96:6535–3540

Love C, Logan E, Erwin T et al (2006) Analysis of the *Brassica* A and C genomes and comparison with the genome of *Arabidopsis thaliana*. Acta Horticulturae (ISHS) 706:99–104

Lysak MA, Lexer C et al (2006) Towards the era of comparative evolutionary genomics in Brassicaceae. Plant Syst Evol 259:175–198

Mozo T, Fischer S, Shizuya H, Altmann T et al (1998) Construction and characterization of the IGF *Arabidopsis* BAC library. Mol Gen Genet 258(5):562–570

Mun JH, Kwon SJ, Yang TJ et al (2008) The first generation of a BAC-based physical map of *Brassica rapa*. BMC Genomics 9:280–291

Newman T, deBruijn FJ, Green P et al (1994) Genes galore: a summary of methods for accessing results from large-scale partial sequencing of anonymous *Arabidopsis* cDNA clones. Plant Physiol 106(4):1241–1255

Pink D, Bailey L, McClement S et al (2008) Double haploids, markers and QTL analysis in vegetable *Brassicas*. Euphytica 164:509–514

Rivero-Lepinckas L, Christ D, Scholl R (2006) Growth of plants and preservation of seeds. In Salinas J, Sanchez-Serrano J (eds) Methods in molecular biology. *Arabidopsis* protocols. Humana Press, Totowa, NJ

Schranz ME, Song BH, Windsor AJ, Mitchell-Olds T (2007a) Comparative genomics in the Brassicaceae: a family-wide perspective. Curr Opin Plant Biol 10(2):168–175

Schranz ME, Windsor AJ, Song BH et al (2007b) Comparative genetic mapping in *Boechera stricta*, a close relative of *Arabidopsis*. Plant Physiol 144(1):286–298

Schulz B, Bennett MJ, Dilkes P, Feldmann KA et al (1998) T-DNA tagging in *Arabidopsis thaliana*: cloning by gene disruption. In Gelvin S, Schilperoort R (eds) Plant molecular biology manual K3. Kluwer Academic Publishers, Belgium

Schwab R, Ossowski S, Riester M et al (2006) Highly specific gene silencing by artificial MicroRNAs in *Arabidopsis*. Plant Cell May 18(5):1121–1133

Stone SL, Hauksdóttir H, Troy A et al (2005) Functional analysis of the RING-type ubiquitin ligase family of *Arabidopsis*. Plant Physiol 137(1):13–30

Wang N, Wang Y, Tian F et al (2008) A functional genomics resource for *Brassica napus*: development of an EMS mutagenized population and discovery of FAE1 point mutations by TILLING. New Phytol 180(4):751–765

White JA, Tood J, Newman T et al (2000) A new set of *Arabidopsis* expressed sequence tags from developing seeds. The metabolic pathway from carbohydrates to seed oil. Plant Physiol 124(4):1582–1594

Yamada K, Lim J, Dale JM et al (2003) Empirical analysis of transcriptional activity in the *Arabidopsis* genome. Science 302:842–846

Chapter 17
Resources for Metabolomics

Christoph Böttcher, Edda von Roepenack-Lahaye, and Dierk Scheel

Abstract Metabolomics is developing toward an integral component of functional genomics approaches. The large structural diversity of plant metabolites requires different analytical techniques for broad metabolite analysis. In addition, new bioinformatics tools and databases are necessary for data analysis and storage. This chapter describes the resources available for comprehensive analysis of plant secondary metabolites focusing on *Arabidopsis thaliana* and *Brassica* species. In particular, a platform for non-targeted profiling of semi-polar plant metabolites based on liquid chromatography coupled to mass spectrometry is described.

Keywords *Arabidopsis thaliana* · Flavonoids · Glucosinolates · Lipids · Liquid chromatography-mass spectrometry · Metabolite profiling · Phenylpropanoids · Secondary metabolism

Abbreviations

APCI	Atmospheric pressure chemical ionization
API	Atmospheric pressure ionization
APPI	Atmospheric pressure photoionization
CID	Collision-induced dissociation
DGDG	Digalactosyldiacylglycerol
ESI	Electrospray ionization
FT-ICR	Fourier-transform ion cyclotron resonance
GC	Gas chromatography
LC	Liquid chromatography
MS	Mass spectrometry
MGDG	Monogalactosyldiacylglycerol
NMR	Nuclear magnetic resonance
PC	Phosphatidylcholine

C. Böttcher (✉)
Department of Stress and Developmental Biology, Leibniz Institute of Plant Biochemistry, Weinberg 3, 06120 Halle/Saale, Germany
e-mail: cboettch@ipb-halle.de

R. Schmidt, I. Bancroft (eds.), *Genetics and Genomics of the Brassicaceae*, Plant Genetics and Genomics: Crops and Models 9, DOI 10.1007/978-1-4419-7118-0_17,

PE	Phosphatidylethanolamine
PG	Phosphatidylglycerol
PI	Phosphatidylinositol
ppb	Parts per billion
ppm	Parts per million
PS	Phosphatidylserine
QTOF	Quadrupole-time-of-flight
SQDG	Sulfoquinovosyldiacylglycerol
UPLC	Ultra-performance liquid chromatography

Contents

17.1 Introduction 470
17.2 Non-targeted Profiling of Semi-polar Plant Metabolites Using UPLC/ESI-QTOF-MS 472
17.2.1 Experimental Design and Sampling – General Considerations 472
17.2.2 Sample Preparation 473
17.2.3 Data Acquisition 476
17.2.4 Data Extraction in Non-targeted Analysis of Metabolite Profiles 481
17.2.5 Elucidation of Molecular Structures 483
17.3 Compound Classes Amenable for LC/API-MS-Based Profiling Approaches . . . 488
17.3.1 Secondary Metabolites in *Arabidopsis* 488
17.3.2 Lipids 493
17.4 Conclusion and Outlook 495
References 496

17.1 Introduction

A plant's metabolome comprises all small molecule metabolites of cells, tissues, and organs during its entire lifetime. Plants contain primary metabolites, which are required for growth, development, survival, and propagation, and secondary metabolites that are not necessary for these basic processes. However, they may be important for adaptation to variable environmental conditions, defense against pathogens and herbivores, and protection against abiotic stresses (Kliebenstein 2004). Originating from primary metabolism, secondary metabolites undergo an enormous structural diversification, which is the basis for their functional diversity and divergent biological activities. Plants contain a huge number of metabolites, in particular, secondary metabolites. Estimates for a given species range from 4,000 to 20,000 metabolites (Fernie et al. 2004) and more than 100,000 known plant secondary metabolites are believed to represent only a small portion of the total number (Schwab 2003).

Metabolomics is the comprehensive quantitative and qualitative analysis of all metabolites of an organism (Fiehn et al. 2000), which has not been achieved yet for any plant species and will in principle be a difficult task given the structural diversity and diverse concentration ranges of metabolites, as well as their spatial and temporal patterns of appearance. Metabolomics also implies the integration with genomics, transciptomics, proteomics, and enzymatic data sets in a functional genomics and systems biology approach (Oksman-Caldentey and Saito 2005). Metabolite profiling represents the quantitative analysis of a subset of cellular metabolites and allows the comparison of profiles from wild type and mutant plants, different ecotypes or cultivars and specific organs, as well as monitoring of changes in metabolite profiles during development or in response to environmental factors (Fernie et al. 2004). It can be performed in a non-targeted or targeted way, the latter focusing on specific classes of metabolites, such as amino acids, lipids, phenylpropanoids.

Due to their structural diversity, plant metabolites have diverse physicochemical properties and therefore, different analytical techniques are required to detect, quantify, and identify all types of compounds. The technologies employed in metabolomics studies are mostly based on mass spectrometry (MS) and nuclear magnetic resonance (NMR) spectroscopy (Last et al. 2007). Because of its low sensitivity, NMR has been less frequently used in metabolomics. However, two-dimensional NMR, in particular, provides exact structural information and is therefore an excellent synergetic complementary technique to MS. In addition, NMR has been very successfully employed for metabolite fingerprinting; for example, of Arabidopsis ecotypes (Ward et al. 2003) and several cultivars of *Brassica rapa* and their metabolic response to fungal infection (Abdel-Farid et al. 2007). The number of metabolites identified in these studies usually ranges between 20 and 40 and is much lower than with MS techniques. MS can detect a broad range of metabolites with high sensitivity and excellent accuracy (Last et al. 2007). Therefore, less biological materials are required and also allows tissue- and organ-specific approaches, in order to detect spatial differences. Traditionally, MS is coupled with gas chromatography (GC), more recently, capillary electrophoresis and liquid chromatography (LC) are being used for pre-fractionation. In particular recent developments in LC, such as ultra-performance liquid chromatography (UPLC), have improved LC/MS coupling almost to a level comparable to GC/MS. MS instruments with high mass resolution, such as Fourier-transform ion cyclotron resonance (FT-ICR) mass spectrometers have successfully been employed to analyze plant extracts without prior pre-fractionation (Aharoni et al. 2002; Hirai et al. 2005; Ohta et al. 2007; Tohge et al. 2005).

Metabolomic analyses demonstrate their impact particularly in an integral functional genomics approach. For example, the integral analysis of transcriptome and metabolome data sets of *Arabidopsis thaliana* have the potential to identify novel gene functions and metabolic networks (Saito et al. 2008). Furthermore, profiling of 162 *Arabidopsis* lines for different abundances of transcript, proteins, and metabolites allowed QTL mapping of more than 40,000 molecular and 139 phenotypic traits and the identification of genome hot spots (Fu et al. 2009). The use of selected *Arabidopsis* knock-out lines for metabolomic analysis, in particular in combination with feeding of labeled or tagged precursors, has successfully been employed for elucidation of biosynthetic pathways (Böttcher et al. 2009a) and identification of novel metabolites in *A. thaliana* (Böttcher et al. 2008).

Novel bioinformatic tools and data bases have been generated for data analysis and interpretation, as well as for integrative analysis of large sets of unrelated data, many of which are publicly available (Last et al. 2007).

We focus here primarily on the description of the platform we developed for non-targeted profiling of semi-polar plant metabolites by LC/MS and its optimization for analysis of extracts from *A. thaliana* and *Brassica* species.

17.2 Non-targeted Profiling of Semi-polar Plant Metabolites Using UPLC/ESI-QTOF-MS

17.2.1 Experimental Design and Sampling – General Considerations

In order to draw valid and objective conclusions from metabolomic data sets, each metabolomic study should base on a well thought-out experimental design. Metabolite levels are highly variable and depend on numerous environmental parameters, which cannot be perfectly controlled and reproduced even under laboratory conditions. Therefore, factorial experimental designs are preferably used in metabolomic studies, in which several parameters are simultaneously varied across the experiment, whereas others are kept as constant as possible. The effect of parameters and their interactions can then be assessed by comparison of study design classes. Typical factorial designs of metabolomic studies can be generalized as genotype × environment interactions. To reduce the impact of bias on data structure and statistical analyzes, the experimental design implicates establishment of strictly randomized procedures at any stage of the study, be it plant growth, sampling, sample preparation, or data acquisition. In addition, the optimal number of replicates within study design classes has to be determined in order to achieve adequate statistical power. This implies an estimation of biological variability of individuals and technical variability of the analytical protocols in pre-study experiments. Especially when analyzing plant material originating from field trials, increased biological variability of metabolite levels in comparison to plants grown under controlled laboratory conditions has to be considered. Consequently, high numbers of individual analyses per study design class are required for statistically sound detection of metabolic differences (Catchpole et al. 2005). To reduce the effects of biological variability, pooling of a sufficient number of individuals can be performed, in particular, when primary interest is not at the individual level, but rather on differences between study design classes. The reduction of variability in a pooled experimental design becomes more effective, the larger biological variability is relative to technical variability (Kendziorski et al. 2005). If this applies, the benefit of pooling is a decrease of the replicate number at constant statistical power or an increase of power at a constant number of replicates.

Another important point is the establishment of robust sampling procedures. When analyzing plant organs that can hardly be sampled as a whole due to their high biomass, attention has to be paid to the possible presence of metabolite gradients within the sample matrix. For instance, significant radial gradients were detected for several amino acids, hexoses, and organic acids in potato tubers (Shepherd et al. 2007). To obtain a representative metabolic profile of a whole tuber and to avoid gradient-induced bias during sampling, freeze-drying, and pooling of two diametrically opposed segments was suggested. When metabolite profiling is performed on photosynthetic tissue it should be considered that levels of most metabolites are subject to diurnal rhythm. To ensure comparability of metabolite profiling studies, sampling has to be carried out at the same time point within the light and dark period. Measuring metabolites requires further the immediate inactivation of metabolism, because the turnover of metabolites can be extremely rapid, as compared to DNA or RNA. Typically, metabolic conversions could be arrested by flash-freezing in liquid nitrogen immediately after harvest or alternatively by careful freeze-drying. It should be noted that minimum reporting standards for plant biology context information in metabolomics studies have been suggested (Fiehn et al. 2007).

17.2.2 Sample Preparation

17.2.2.1 Extraction

The quality of sample preparation is a crucial factor for any metabolite profiling approach. Ideally, it provides quantitative extraction of preferably all the metabolites compatible to the applied analytical technique from a complex biological matrix, while removing matrix components potentially interfering with the analysis. In addition, the extract is pre-concentrated in order to detect a maximal number of metabolites within their dynamic range. Important parameters that need to be evaluated or optimized for every new sample preparation procedure and sample matrix, are repeatability, recovery rate, as well as the number of reproducibly detected endogenous features (unique *m/z*-retention time pairs) and metabolites. A simple but effective mean to evaluate sample preparation quality in terms of recovery rate, are pre- and post-extraction spiking experiments performed on a set of non-endogenous compounds with diverse physicochemical properties similar to that of the putative metabolites (t'Kindt et al. 2008). Using this approach, metabolome coverage of a sample preparation procedure can also be estimated. Common solvent mixtures used for extraction of semi-polar plant metabolites comprise methanol/water (von Roepenack-Lahaye et al. 2004), which is by far the most popular, acetonitrile/water (t'Kindt et al. 2008), isopropanol (Glauser et al. 2008b), methanol/water/formic acid (De Vos et al. 2007), or methanol/water/chloroform (Giavalisco et al. 2008). To rapidly inactivate enzymes, pre-cooled extraction solvent is typically added to frozen plant tissue homogenate and immediately mixed. Given the fact that certain metabolites, such as cyanogenic glycosides or glucosinolates, are rapidly degraded by endogenous glycosyl hydrolases upon tissue disruption, fast quenching of enzyme activities is crucial, in particular when analyzing secondary metabolites. For the sake of compound stability, extraction at room temperature is preferred in most cases and frequently accelerated by sonication. In order to evaluate overall technical performance within large metabolomics studies, pre-extraction spiking of a set of internal standards is advisable.

17.2.2.2 Fractionation

Because the aim of non-targeted analytical approaches is the detection of as many metabolites as possible, sample preparation procedures for metabolite profiling are kept as simple and universal as possible. In comparison to the recent developments of analytical platforms in terms of resolution and sensitivity, relatively little research efforts have therefore been spent on fractionation schemes for crude plant extracts. With regard to the broad concentration range within a metabolome, establishment of enrichment and depletion strategies for selected compound classes will be of prime importance to obtain better metabolome coverage. Common techniques for fractionation are liquid–liquid extraction and solid-phase extraction. Because of the commercial availability of a broad range of silica and polymer-based adsorbents with different retention mechanisms, solid-phase extraction has a high potential in metabolomics applications. Mixed mode adsorbents with multiple retention mechanisms allow separation of analytes with different physicochemical properties using the same solid phase. Furthermore, the fascinating concept of molecularly imprinted polymers could be applied, for instance, for depletion of selected high abundant metabolites. Current procedures frequently use reversed-phase solid-phase extraction to remove analytes and matrix components of extremely high polarity, such as saccharides and inorganic salts, or of extremely low polarity, such as lipids and hydrophobic pigments (Grata et al. 2008). However, also ion-exchange solid-phase extraction was employed for fractionation of crude *A. thaliana* leaf extracts (Giavalisco et al. 2008) and for depletion of major metabolite classes, such as glucosinolates and phenolic choline esters from *A. thaliana* seed extracts (Böttcher et al. 2008).

Phenolic choline esters accumulate to considerable amounts in seeds of the Brassicaceae. Sinapoylcholine constitutes the major member of this compound class in *A. thaliana* and *Brassica napus* and comprises about 1–2% of dry matter (Bouchereau et al. 1991). It is accompanied by a large number of structurally related compounds (Böttcher et al. 2009a). Phenolic choline esters are characterized by excellent ionization properties upon electrospray ionization (ESI) in positive ion mode, which, in combination with their abundance, seriously complicates LC/MS-based metabolite profiling of seed extracts. In order to reliably quantify both phenolic choline esters as well as metabolites of lower abundance and ionization efficiency, a two-step metabolite profiling approach was developed (von Roepenack, unpublished). In this approach, crude methanolic seed extracts (50 mg fresh weight/ml) were diluted ninefold allowing analysis of phenolic choline esters within the dynamic range of the mass spectrometer. For profiling of other semi-polar metabolites, phenolic choline esters were depleted from total extracts using weak cation-exchange solid-phase extraction. After this fractionation step, the non-retained analytes could be concentrated by a factor of 15 (75 mg fresh weight/ml) compared to the diluted total extract. Depletion and concentration results in detection of about 400 additional novel features that were below the detection limit in the diluted total extracts (Fig. 17.1).

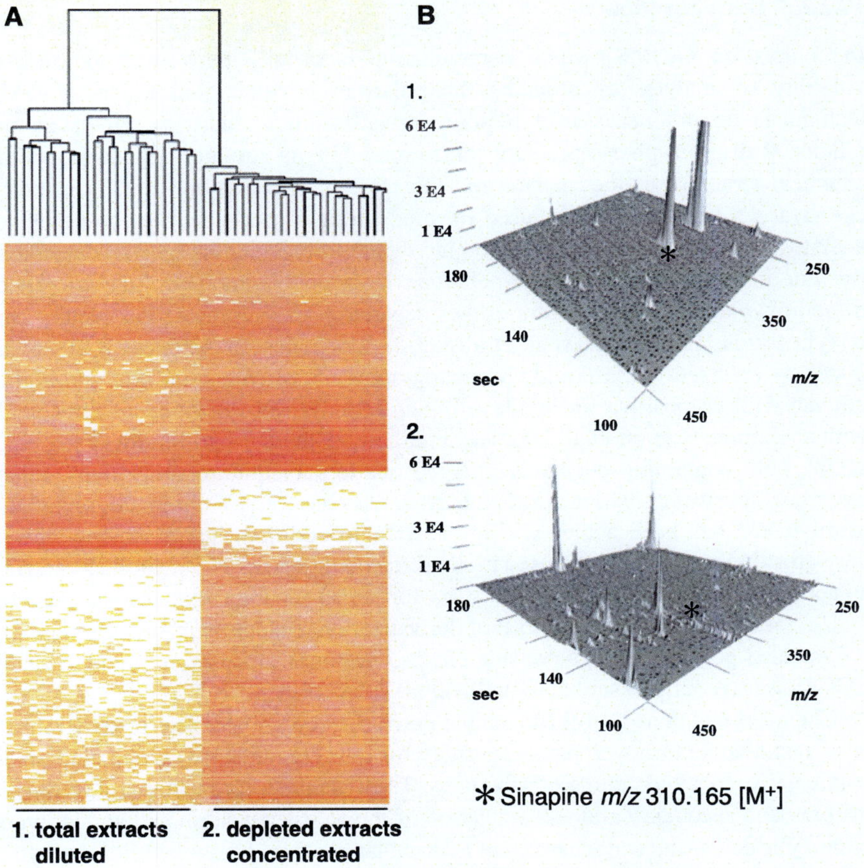

Fig. 17.1 The effect of depletion of phenolic choline esters from *B. napus* seed extracts and subsequent extract concentration on the number and intensity distribution of reliably detected features (unique *m/z*-retention time pairs) in LC/MS metabolite profiles. Twenty-five individual seeds were extracted using 100 μl of aqueous methanol. The resulting samples were diluted threefold. The sixth part of the extracts were again diluted by a factor of three and directly analyzed by UPLC/ESI(+)-QTOF-MS. The other part of the sample was subjected to weak cation-exchange solid-phase extraction. Compounds, such as choline esters possessing a quarternary ammonium group and hence a permanent charge were retained on the column. The non-retained fractions were concentrated up to 15 times in comparison to the diluted total extract and subsequently analyzed by UPLC/ESI(+)-QTOF-MS. Profiles of total extracts and depleted extracts were arranged in two sample classes for data processing. After peak picking and alignment using the XCMS package a hierarchical cluster analysis was performed on features appearing reliably in all profiles of at least one sample class. The heatmap (**a**), in which *white* colouring indicates low abundant or absent features and *red* colouring high-abundant ones, represents the intensity distribution of 1093 features in the two sample classes consisting of 25 files each. The cluster analysis shows that about 421 features (∼40%) detected in depleted and concentrated samples cannot be detected anymore in the diluted total extracts. After solid-phase extraction samples lost about 174 features (∼16%) and showed only in 18 (∼2%) features, a reduction in intensity. The *right part* of the figure shows the complete removal of the major phenolic choline ester sinapine after fractionation using weak-cation solid-phase extraction (**b**)

17.2.2.3 Derivatization

In contrast to GC/MS, where derivatization is usually performed to increase volatility of analytes by reducing the polarities of functional groups, LC/MS techniques do not necessarily require derivatization of analytes. However, the efficiency of atmospheric pressure ionization (API) is strongly dependent on the chemical structure of the analyte and therefore highly non-uniform. Principally, chemical derivatization can be used for modulating physicochemical properties of analytes to improve their chromatographic separability and mass spectral detectability. For instance, addition of pre-charged hydrophobic moieties like tris(2,4,6-trimethoxyphenyl) phosphonium allows separation of low molecular weight carboxylic acids, amines (Leavens et al. 2002), aldehydes, or ketones (Barry et al. 2003) by reversed-phase liquid chromatography and provides uniform and highly efficient ESI in positive ion mode (Fig. 17.2A). By derivatization of fatty acids with cholamine (Lamos et al. 2007) or 3-carbinol-1-methyl-pyridinium (Yang et al. 2007), both containing positively charged quarternary ammonium groups, sensitive mass spectrometric detection after reversed-phase liquid chromatography and positive ESI has been achieved (Fig. 17.2B). In addition, heavy isotope-labeled derivatization reagents can be used in rapid screening for novel metabolites carrying an appropriate functional group, and allow further an accurate relative quantification by isotope dilution. On basis of a set of derivatization reagents carrying differentially ^{13}C-labeled polyethylene glycol side chains, multiplex analysis of amino acids by LC/MS was recently described (Abello et al. 2008) (Fig. 17.2C).

Chemical reactions useful for analyte derivatization are characterized by robustness, quantitative turnover, high selectivity for a single functional group, and broad applicability to a wide range of substrates. These include, for instance, acylation of primary and secondary amines, alkylation of thiols, esterification and amidation of carboxylic acids, and condensation of hydroxylamines or hydrazines with aldehydes or ketones. Besides the alteration of physicochemical properties for improvement of separation and detection, derivatization reagents with additional functional aspects have been designed. Incorporation of, e.g., chlorine results in an isotope signature, which is typically not observed among most metabolites and can, therefore, be used for fast screening (Carlson and Cravatt 2007b). Furthermore, cationic or anionic moieties introduced by derivatization can be used as purification tags using ion-exchange solid-phase extraction. Recently, solid-phase-bound chemoselective probes were developed for enrichment of amines, carboxylic acids, thiols, and aldehydes/ketones from biological matrices (Fig. 17.2D). Following proteolytic release of the captured analytes via a trypsin cleavage site, selected chemical classes of metabolites have been tagged and analyzed by LC/MS-based metabolite profiling (Carlson and Cravatt 2007a).

17.2.3 Data Acquisition

When optimizing analytical conditions for LC/MS, the correct interplay between separation, ionization, and mass spectral detection requires special attention. For

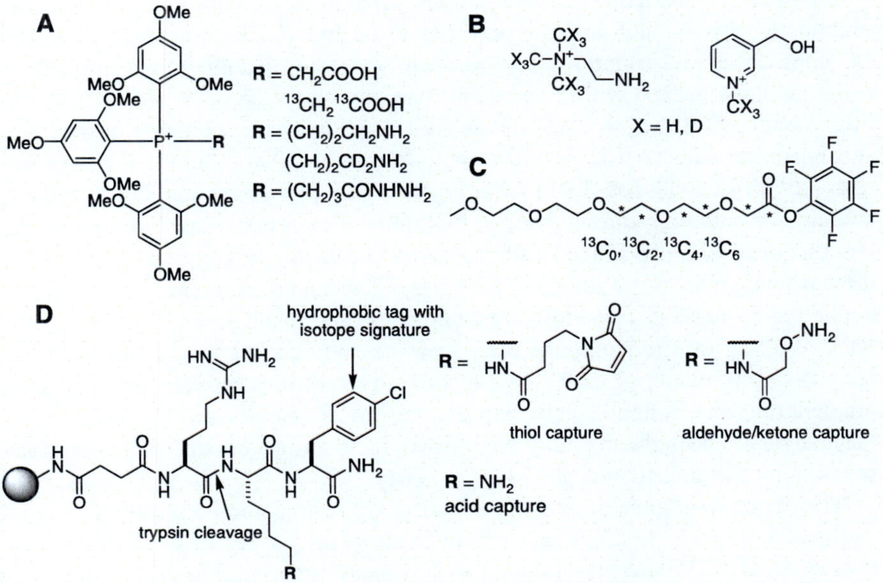

Fig. 17.2 Derivatization reagents for LC/ESI-MS. Tris-(2,4,6-trimethoxyphenyl)-phosphonium-based derivatization reagents were developed for labeling of primary and secondary amines, carboxylic acids as well as aldehydes and ketones (**a**). Cholamine and 3-carbinol-1-methyl-pyridinium as well as their ^{2}H-labeled derivatives were applied for derivatization and isotope dilution analysis of fatty acids (**b**). Pentafluorophenyl-activated esters of different ^{13}C-containing poly(ethylene glycol) chains were used for derivatization and multiplex analysis of amino acids and glutathione (**c**). Metabolite enrichment by tagging and proteolytic release (METPR) (**d**)

instance, ion sources for LC/MS applications are restricted to certain flow rate regimes, which require in case of incompatibility post-column splitting or addition prior to ionization. Additives and pH of the mobile phase have to assure chromatographic resolution but need to be compatible with the applied ionization method. If necessary, post-column addition of modifiers can be used to adjust the properties of the mobile phase in order to allow efficient analyte ionization. When using ESI, its susceptibility to ion suppression and enhancement has to be considered and good chromatographic separation is necessary to minimize such effects induced by co-elution of analytes or non-volatile matrix components (Böttcher et al. 2007). ESI-MS acts as concentration-sensitive detector within flow rate regimes typically applied in narrow-bore, micro-bore, and capillary HPLC in contrast to atmospheric pressure chemical ionization (APCI), which is a mass-flow sensitive detector. Principally, concentration sensitivity of ESI and the limited flow rate ranges of corresponding ion sources implicate miniaturization of chromatographic columns (Tomer et al. 1994). When performing fast chromatographic separations resulting in chromatographic peak widths of a few seconds or less, the mass spectrometer has to be capable to acquire a sufficient number of data points across a chromatographic peak to allow accurate quantification and deconvolution.

For optimization of any chromatographic separation, a compromise between resolution, sensitivity, and analysis time has to be found. The efficiency of several LC approaches and column types has been evaluated, including conventional LC using columns packed with 5 and sub-2 μm particles ($\Delta p_{max} = 400$ bar), monolithic columns ($\Delta p_{max} = 200$ bar), as well as UPLC using columns packed with sub-2 μm particles ($\Delta p_{max} = 1000$ bar) (Grata et al. 2008). Among them, combination of UPLC with sub-2 μm particles stationary phases appeared to be the best strategy for both high-throughput and high-resolution separations. Further parameters that need to be optimized within a given chromatographic system are column dimensions, gradient run time and flow rate. Elevated column temperatures can be applied to decrease viscosity of the mobile phase and to increase diffusivity of analytes resulting in a reduction of back pressure and acceleration of mass transfer, respectively (Grata et al. 2009). A functional relationship between peak capacity, gradient run time, column length, particle size, linear velocity, and the diffusivity of the analyte has been theoretically established, facilitating assessment of appropriate separation conditions (Neue and Mazzeo 2001).

The chromatographic system we are adopting for metabolite profiling of crude methanolic extracts originating from different plant tissues of *A. thaliana* consists of an UPLC platform equipped with a micro-bore column (1.0 × 100 mm) packed with 1.8-μm particles. Chromatography was performed at a flow rate of 150 μl/min, which is compatible with the applied ESI interface without splitting and results in a maximal system pressure of 550 bar when the separation is carried out at 40°C. Typically, a binary linear gradient of water and acetonitrile acidified with 0.1% formic acid is applied. To provide separation of a wide range of metabolites within a single analysis, initial and final gradient compositions were set to 5 and 95% acetonitrile, respectively. A gradient run time of 15 min results in a moderate-efficient separation with a peak capacity of 250. Due to a total cycle time of 20 min including washing and equilibration steps, about 60 samples per day and machine can be processed. Although, maximal system pressure and ion source would allow a separation at flow rates of up to 250 μl/min, in practice, the observed gain in resolution could not compensate the reduction of sensitivity using a concentration sensitive detector, such as ESI-MS. Likewise, only slightly higher peak capacities were reached upon use of a longer column (1.0 × 150 mm), which has the disadvantage of increased system pressures and longer equilibration times. Crude plant extracts have to be injected in solvent mixtures containing sufficient amounts of organic solvent to dissolve hydrophilic as well as hydrophobic metabolites. In combination with a comparatively large injection volume (2 μl) a relatively broad injection profile is obtained, which might limit further improvement of chromatographic resolution by increasing flow rate and column length.

For the separation of crude plant extracts a modified C_{18} stationary phase was used (Waters HSS T3), which provides balanced retention of polar and nonpolar metabolites. Application of 0.1% formic acid as mobile phase additive results in symmetric peak shapes for the major secondary metabolite classes found in *Arabidopsis* including glucosinolates, benzenoids, phenylpropanoids, flavonoids, indolics, and fatty acid derivatives (D'Auria and Gershenzon 2005). The chromatographic system is further characterized by an excellent reproducibility, which is of major importance for precise alignment of LC/MS profiles. The observed maximal retention time deviation across large-scale metabolomics studies consisting of some hundred injections is in the order of a chromatographic peak half-width (2–3 s). Column performance is maintained for 2,500–3,000 injections of crude plant extracts. To maximize metabolome coverage, mass-spectral detection is routinely performed after ESI in both positive and negative ion mode. Due to the use of 0.1% formic acid as mobile-phase additive, which is less suited for the negative ion mode, the number of features detected in methanolic extracts from different *Arabidopsis* tissues is in general by a factor of two to three lower after negative than after positive ion ESI. However, major compound classes, such as glucosinolates, flavonol glycosides, and certain phenolic compounds, can be sensitively detected under these conditions in negative ion mode. Besides putative higher metabolome coverage, combined acquisition of LC/MS profiles in positive and negative ion mode has further advantages. In case of the detection of a metabolite in both positive and negative ion mode, the relative quantification between study design classes can be validated and the annotation of features is considerably facilitated due to the formation of complementary quasi-molecular and fragment ions. Unfortunately, the impact of different ionization techniques, ion modes, and mobile-phase additives on metabolome coverage has not been systematically evaluated yet for any plant species. However, initial studies on human serum extracts suggest that metabolome coverage can be highly increased by application of multiple ionization approaches (Nordstrom et al. 2008). While being reported for plant metabolomics (Aharoni et al. 2002; Thiocone et al. 2008), the potential of APCI providing a higher dynamic range, less susceptibility to matrix effects, and ionization of more nonpolar analytes in comparison to ESI is widely unexploited. It remains to be seen, if technical problems in terms of robustness and reproducibility associated with nano-ESI can be overcome to capitalize on dramatic higher ionization efficiencies and strongly reduced ion suppression effects of this ionization technique compared to ESI (Gangl et al. 2001; Schmidt et al. 2003).

Among the multitude of mass analyzers (Dunn 2008), time-of-flight mass spectrometry (TOF-MS) is particularly suitable for combination with UPLC platforms. TOF-MS provides a high spectral acquisition rate allowing adequate sampling of data points across chromatographic peaks, moderate to high mass resolution and a sensitive detection of ions over a wide mass range. In our metabolite profiling platform ions are analyzed by quadrupole-time-of-flight mass spectrometry (QTOF-MS), which combines the advantages of TOF-MS with the tandem-MS functionalities of triple–quadrupole MS (Chernushevich et al. 2001). The instrument used is characterized by spectral acquisition rates of up to 20 Hz and a mass spectral resolution of about 10,000 resulting in mass accuracies of 10 ppm following external calibration and 5 ppm, if internal recalibration is applied (Table 17.1). The mass spectrometer is equipped with an 8-bit analog-to-digital converter, which provides a higher dynamic range compared to time-to-digital converter-based ion recording systems. In addition, mass peak centroids are nearly independent of the

Table 17.1 Accuracy of mass and relative isotope abundances using state-of-the-art QTOF-MS. A mixture of 15 natural products was repetitively analyzed by UPLC/ESI(+)-QTOF-MS (external mass calibration, N=9). For most of the analyzed ions a mass accuracy of ±5 ppm is reached together with an excellent reproduction of the isotope pattern

Compound	Elemental composition	Ion type	Mass accuracy ($N = 9$) Measured m/z mean ± SD×10^3	Δm (ppm)	RIA accuracy ($N = 9$) 1st isotopic rel. intensity [%] mean ± SD (calc.)	2nd isotopic rel. intensity [%] mean ± SD (calc.)
Kinetin	$C_{10}H_9N_5O$	$[M+H]^+$	216.0867 ± 0.5	–6.1	13.4 ± 0.4(12.8)	1.0 ± 0.3(1.0)
Chelidonine	$C_{20}H_{19}NO_5$	$[M+H]^+$	354.1333 ± 1.6	–0.7	22.4 ± 0.8(22.4)	3.4 ± 0.1(3.4)
Xanthohumol	$C_{21}H_{22}O_5$	$[M+H]^+$	355.1514 ± 1.1	–7.4	22.8 ± 0.7(23.2)	4.2 ± 0.3(3.6)
Laudanosin	$C_{21}H_{27}NO_4$	$[M+H]^+$	358.2011 ± 1.8	–0.5	24.0 ± 0.4(23.6)	3.4 ± 0.1(3.4)
Bicuculline	$C_{20}H_{17}NO_6$	$[M+H]^+$	368.1125 ± 1.5	–1.0	22.4 ± 0.5(22.4)	3.6 ± 0.1(3.6)
Methylumbelliferylglucoronide	$C_{16}H_{16}O_9$	$[M+Na]^+$	375.0665 ± 0.9	–5.9	18.2 ± 1.0(17.8)	3.3 ± 0.1(3.4)
Rotenone	$C_{23}H_{22}O_6$	$[M+H]^+$	395.1479 ± 1.3	–2.6	24.5 ± 0.5(25.4)	4.0 ± 0.1(4.3)
Noscapine	$C_{22}H_{23}NO_7$	$[M+H]^+$	414.1558 ± 2.0	+2.8	25.1 ± 0.9(24.7)	4.2 ± 0.1(4.4)
Aloin	$C_{21}H_{22}O_9$	$[M+Na]^+$	441.1157 ± 1.4	+0.3	22.1 ± 1.1(23.3)	5.2 ± 0.7(4.4)
Armentoflavone	$C_{30}H_{18}O_{10}$	$[M+H]^+$	539.0965 ± 1.7	–1.4	31.1 ± 1.2(33.1)	7.2 ± 0.3(7.3)
Vitexin-2-O-ŕhamnoside	$C_{27}H_{30}O_{14}$	$[M+H]^+$	579.1686 ± 1.5	–3.8	29.5 ± 0.8(30.1)	6.8 ± 0.1(7.2)
Reserpine	$C_{33}H_{40}N_2O_9$	$[M+H]^+$	609.2828 ± 3.1	+3.5	37.7 ± 1.2(37.2)	8.3 ± 0.3(8.6)
Rutin	$C_{27}H_{30}O_{16}$	$[M+H]^+$	611.1582 ± 1.7	–4.3	30.6 ± 0.9(30.2)	8.4 ± 0.9(7.7)
Erythromycin	$C_{37}H_{67}NO_{13}$	$[M+H]^+$	734.4684 ± 3.0	–0.1	42.2 ± 0.7(41.7)	10.5 ± 0.2(11.1)
Rifampicin	$C_{43}H_{58}N_4O_{12}$	$[M+H]^+$	823.4117 ± 2.9	–0.8	48.5 ± 1.0(49.1)	13.8 ± 0.5(14.3)

RIA, relative isotope abundance; SD, standard deviation; calc., calculated peak intensities and do not need to be corrected as in case of time-to-digital converters (Chernushevich et al. 2001). This characteristic results in accurate masses over a wide dynamic range. Assuming sufficient peak intensities and the absence of underlying interferences, a further hallmark of the instrument is its ability to precisely measure the isotopic pattern (Table 17.1), which facilitates the reduction of candidate elemental compositions calculated on basis of accurate mass measurements (Kind and Fiehn 2006). For many metabolites relevant to secondary metabolism of *Arabidopsis*, detection limits of about 1 pmol are achievable using UPLC/ESI-QTOF-MS. For compounds possessing functional groups with high proton affinities (e.g., alkaloids) or intrinsic charges (e.g., choline esters) detection limits of about 100 fmol are characteristic for positive ion mode.

In order to monitor chromatographic and mass spectral performance of the UPLC/ESI-QTOF-MS platform a mixture containing a set of eight standards is used comprising acidic, basic, and neutral compounds of different polarities. For rapid evaluation, scripts allowing a fully automated determination of retention times, accurate masses, and peak areas, as well as a graphical representation of analytical data over a certain time period were developed. The standard mixture is repeatedly analyzed in positive and negative ion mode at the beginning of a sample batch to assess instrument performance and, if necessary, to adjust detector sensitivity. In addition, it is injected within a sample batch after each 10 injections. Typically, a sample batch comprises injection of 60 samples and 6 standard mixtures. Thereafter, the ESI interface has to be cleaned and the chromatographic column flushed with a high organic eluent mixture for a longer period. Large-scale metabolomic studies consisting of several hundred analyses require consecutive processing of numerous sample batches. To reduce the impact of instrument drift, samples need to be strictly randomized within these batches. To evaluate overall process efficiency a subset of five out of eight standards is spiked pre-extraction to each sample. Prior to any kind of further analysis, intensities and retention times of spiked standards are assessed and outlier excluded from further analysis.

17.2.4 Data Extraction in Non-targeted Analysis of Metabolite Profiles

In general, two approaches for analysis of metabolite profiles obtained from a non-biased data acquisition can be distinguished. In multi-target approaches a subset of features related to pre-selected analytes is quantified and statistically evaluated across study design classes. In contrast, non-targeted approaches analyze all features detectable without any pre-selection and prior knowledge about their relation to a certain analyte. Therefore, they provide the opportunity to discover potentially novel metabolic biomarkers. A variety of methods and software tools have been developed for non-targeted analysis of hyphenated mass spectrometry data originating from metabolomic and proteomic experiments (Katajamaa and Oresic 2005). Common freely available software platforms, which are capable to process high-resolution

LC/MS data, include metAlign (Lommen 2009), mzMine (Katajamaa et al. 2006), OpenMS (Sturm et al. 2008), and XCMS (Smith et al. 2006). The open source software XCMS has been continuously developed further and offers besides its primary functionalities an additional highly sensitive feature detection algorithm (Tautenhahn et al. 2008), mass-spectral deconvolution, and annotation tools (Tautenhahn et al. 2007), as well as algorithms to automatically search tandem mass spectrometry data against a reference library (Benton et al. 2008). Usually, the metabolomic data processing workflow comprises pre-processing, feature detection and quantification, alignment, and normalization. Generally, this workflow results in a transformation of a set of raw data files into a matrix of intensities of the form [feature × sample], which can be evaluated after further transformation by univariate or multivariate statistical approaches. In case of XCMS, pre-processing includes conversion of raw data files from the vendor-specific file format to netCDF, mzData, or mzXML and centroidization of mass spectra depending on the peak picking algorithm to be used. Furthermore, samples have to be assigned to separate sample classes. Usually, study design classes are defined as sample classes and populated by biological and/or technical replicates. Applying the centWave algorithm for feature detection, chromatographic peak half-width, signal-to-noise ratio (*S/N*), and absolute peak intensity threshold, as well as tolerable mass deviation of consecutive scans across a chromatographic peak, need to be specified (Tautenhahn et al. 2008). Typically, the centWave algorithm detects about 5.000 features with *S/N* > 3 in UPLC/ESI(+)-QTOF-MS profiles of methanolic leaf extracts of *A. thaliana*. The alignment algorithm implemented in XCMS is based on iterative cycles of feature grouping and subsequent retention time correction, which can be graphically evaluated (Smith et al. 2006). An essential parameter for alignment is the maximal retention time deviation, which can be estimated, for instance, from retention times of spiked standard compounds. Due to the high reproducibility of UPLC, retention times correction can be omitted in the majority of cases. Furthermore, the XCMS group function allows rejection of features, which are only partially detected within the replicates of a sample class. In addition to the intensity matrix, separate columns containing information related to the presence of a feature within a sample class are generated and have proven useful in further filtering steps. It should be noted, that each alignment has to be evaluated in terms of quality prior to any further analysis. Unfortunately, computational methods for assessing alignment quality are lacking to date. Nevertheless, manual inspection of aligned features related to spiked internal standards allows a rough estimate of alignment quality. In principal, two normalization approaches are available to reduce technical variability while retaining biologically relevant variability related to the experimental design. Statistical approaches, which are mostly adapted from microarray data analysis, calculate optimal scaling factors to normalize, for instance, the unit-norm or the median of feature intensities of each sample within a sample set. Additionally, normalization by a single or by multiple internal standards has been proposed. The chemical diversity of metabolites leading to variability in recovery during extraction and to different responses during ionization renders single standard normalization approaches for LC/MS profiling unfeasible. However, combination of pre-extraction spiking of five internal standards related to the major classes of lipid conjugates and a recently developed multiple internal standard normalization approach proved superior over commonly utilized normalization approaches in reduction of technical variability in lipid profiling of mouse liver (Sysi-Aho et al. 2007). Nevertheless, systematic evaluation of normalization procedures and establishment of internal standard sets allowing multiple standard normalization approaches will be needed in the future.

Before extracting biologically relevant information, the data structure of the intensity matrix has to be adapted to the applied statistical method. Frequently, a logarithmic transformation is used to render the feature intensity distribution more symmetrically, to correct for variance inhomogeneity and to convert multiplicative models into additive ones. However, several other data pretreatment methods were described for metabolomics data and shown to strongly affect the outcome of data analysis (van den Berg et al. 2006). In general, for analysis of metabolomic data sets unsupervised and supervised statistical methods can be applied. Unsupervised methods aim at sample classification, while prior class membership of samples is unknown. Popular methods for exploratory data analysis are, for instance, hierarchical cluster analysis, principal component analysis, and independent component analysis (Scholz et al. 2004; Scholz and Selbig 2007; Steuer et al. 2007). It should be noted, that these tools can also be used to evaluate the quality of technical replicates. In contrast, the aim of supervised methods is the identification of features discriminating study design classes (Broadhurst and Kell 2006). Here, common statistical tools like ANOVA or the multitudes of statistical tests developed for binary classification can be used.

17.2.5 Elucidation of Molecular Structures

At the current state of analytical technology and LC/MS databases, generation and confirmation of structural hypotheses are by far the most laborious and time-consuming step of metabolomics experiments. Consequently, structure elucidation is performed at a very late stage of the experimental workflow, typically, only for features discriminating study design classes. For MS-based de novo structural elucidation, MS platforms providing fragmentation capabilities and a high spectral resolution are mandatory. Besides the more expensive Fourier transform MS platforms, QTOF-MS has proven to possess a particular potential for structural elucidation of unknown metabolites.

In general, API results in formation of cluster, adduct, and fragment ions. Therefore, a given metabolite is represented by a group of features that are characterized by high chromatogram correlation. Assignment of discriminating features to such correlation groups and reconstruction of individual compound mass spectra represent the initial step of the structure elucidation procedure (Werner et al. 2008a). Next, charge states of each ion of a compound mass spectra are determined and ions annotated as cluster, quasi-molecular, or fragment ion. Reliable identification and annotation of quasi-molecular ions allowing the postulation of a molecular mass is particularly important for any further step of de novo structure elucidation. The co-occurrence of several quasi-molecular ions, such as $[M+H]^+$, $[M+Na]^+$, and $[M+K]^+$, or cluster ions, such as $[2M+H]^+$ and $[2M+H+K]^{2+}$, with different susceptibilities to fragmentation (sodium and potassium adducts are generally less prone to collision-induced dissociation (CID)), strongly facilitates identification of quasi-molecular ions. Likewise, complementary quasi-molecular ions formed under positive and negative ionization can also be very useful for this purpose. Furthermore, having identified putative fragment ions, the corresponding precursor ions can be searched for by precursor ion scanning, a scanning technique not only adapted from triple quadrupole-MS but also available on some QTOF-MS platforms. Identification and annotation of feature correlation groups are supported within the XCMS framework by the recently released library CAMERA (http://bioconductor.org/packages/bioc/html/CAMERA.html).

Following annotation, targeted CID-MS/MS experiments are performed on quasi-molecular ions using QTOF-MS. Depending on the isolation width of the first quadrupole, only the monoisotopic peak or the whole isotope cluster can be subjected to CID. In the latter case, relative isotope abundance information can be used for calculating elemental compositions of fragment ions. However, a major drawback of using broader isolation widths is co-fragmentation of co-eluting ions within the isolation window resulting in spectra of low purity. In addition, acquisition of CID-mass spectra without any selection in the first quadrupole was suggested (MS^E) to simultaneously fragment each precursor ion during a single analysis. Powerful deconvolution software is needed to assign precursor and product ions to unique spectra. However, the quality and purity of spectra obtained by this approach are questionable in particular for complex samples, such as plant extracts in which incomplete separation and co-elution of metabolites are frequently observed. In addition, data-dependent acquisition (auto-MS^2) of a QTOF-MS platform was used to create a comprehensive library of CID-mass spectra of precursor ions detected in extracts of several *Arabidopsis* tissues (Matsuda et al. 2009). Without further spectra acquisition, 50% of the features detected in these extracts could be tagged by CID-mass spectra speeding up downstream structural analysis. It should be noted that acquisition of CID-mass spectra from different quasi-molecular ions in both positive and negative ion mode is advantageous because they often contain complementary information that can be used for validation of structure hypotheses. Furthermore, CID-mass spectra of fragment ions generated by in-source fragmentation can be acquired to analyze the relationship of fragment ions. Results from such pseudo-MS^3 experiments can be clearly represented for further analyses in form of a triangular matrix as shown in Fig. 17.3B.

Fig. 17.3 Determination of putative elemental compositions of a protonated molecular ion detected at *m/z* 736 by UPLC/ESI(+)-QTOF-MS in a *A. thaliana* flower bud extract. (**a**): On basis of the accurate monoisotopic mass and a pre-defined set of elements 2017 putative elemental compositions were calculated using an instrument specific mass accuracy of 10 ppm. By application of the nitrogen rule and restriction of the number of ring and double bond equivalents (RDBE) 1782 formulas can be filtered out. Analysis of relative isotope abundancies (RIA) allows rejection of further 183 candidate formulas by restriction of the number of ^{12}C and ^{32}S.

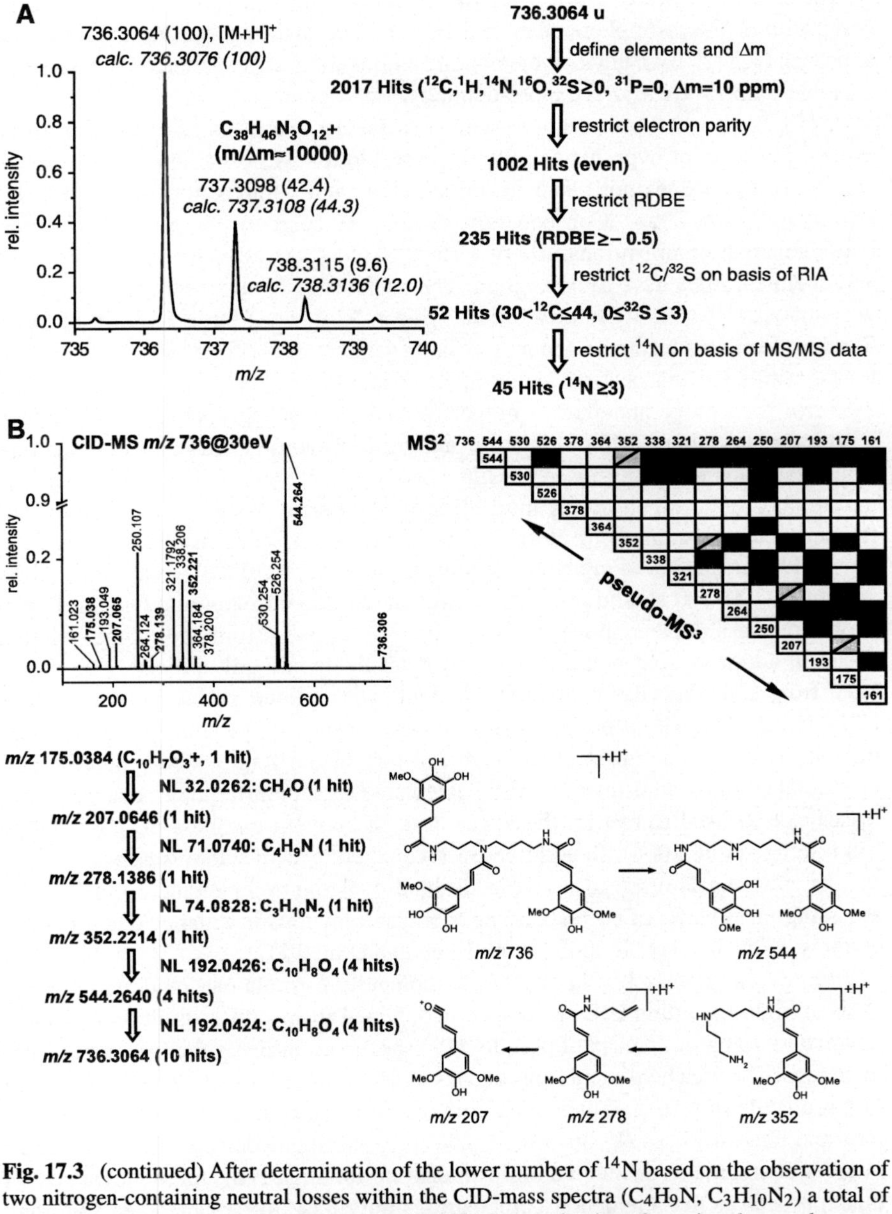

Fig. 17.3 (continued) After determination of the lower number of ^{14}N based on the observation of two nitrogen-containing neutral losses within the CID-mass spectra (C_4H_9N, $C_3H_{10}N_2$) a total of 45 putative elemental compositions were retained. (**b**): Determination of putative elemental compositions by analysis of the CID-mass spectra obtained from the protonated molecular ion. To analyze the filiations of the fragment ions a total of 15 pseudo-MS^3 experiments were carried out. On basis of these experiments a triangular matrix was constructed, in which the occurrence of fragment ions within a certain pseudo-MS^3 experiment (row) is marked by a black box. Afterward, the precursor sum formula can be reconstructed using this matrix by adding neutral losses starting from low-mass fragment ions with preferably unique elemental compositions. Possible decompositions are, for instance, 161→193→250→321→530→736 and 161→193→264→352→544→736. For the decomposition 175→207→278→352→544→736 a total of 10 putative elemental compositions could be deduced for the precursor ion due to combinations of multiple hits for the neutral loss of 192.042

The next step toward a structure hypothesis is the determination of the elemental composition. Two complementary strategies can be distinguished here. In the first approach (Fig. 17.3A), putative elemental compositions are calculated on basis of an accurate monoisotopic mass, a pre-defined set of elements (typically ^{12}C, ^{1}H, ^{14}N, ^{16}O, ^{32}S, ^{31}P) and the instrument-specific mass accuracy. As calculated recently, even application of hypothetical ultra-high-resolution MS capable of mass accuracies of 100 ppb does not result in unique elemental compositions above 500 Da (Kind and Fiehn 2006). Consequently, in order to decrease the number of candidate elemental compositions, filtering steps based on relative isotope abundances and chemical rules have to be applied. The analysis of relative isotope abundances was shown to be a particularly efficient mean to reject a large fraction of putative sum formulas fitting an accurate mass (Ojanpera et al. 2006; Stoll et al. 2006). It was estimated that a mass spectrometer capable of 3 ppm mass accuracy and 2% error for isotope abundances outperforms a mass spectrometers with less than 1 ppm mass accuracy without isotope pattern information (Kind and Fiehn 2006). Chemical rules frequently used for filtering include the restriction of the parity of the number of nitrogen atoms in relation to the parity of the nominal mass and the parity of the number of electrons (nitrogen rule) (Pellegrin 1983), as well as thresholds for ring and double bond equivalents and molecular element ratios (Kind and Fiehn 2007). It should be noted, however, that the nitrogen rule becomes erroneous for mass defects above 0.5 Da leading to an overestimation of the nominal mass. In addition, ring and double bond equivalents can only precisely be calculated from sum formulas containing elements with unique valence states, such as C, H, N, and O. Due to the fact that each sum formula of a fragment ion and neutral loss has to be a subset of the sum formula of the fragmented precursor ion, elemental compositions unequivocally calculated for small fragment ions or neutral losses can be used to restrict the lower limit of element numbers of, e.g., ^{14}N or ^{32}S (Grange et al. 2006). In addition, in vivo labeling with stable isotopes, such as ^{13}C or ^{15}N, can be used to derive additional constraints for element numbers by assessing mass shifts of corresponding monoisotopic ions in unlabeled and labeled extracts (Giavalisco et al. 2008; 2009; Hegeman et al. 2007).

In a second approach (Fig. 17.3B), a molecular formula can be determined on basis of high-resolution CID-mass spectra obtained from a quasi-molecular ion and the related triangular matrix of pseudo-MS^3 spectra comprising the filiation of fragment ions. The idea behind this approach is to decompose the elemental composition of a molecule in a sum of elemental compositions calculated for derived fragment ions and neutral losses (Konishi et al. 2007). Assuming occurrence of an adequate number of fragment ions with a uniform distribution in *m/z*, the major advantage of this approach is that elemental compositions only have to be calculated for small monoisotopic masses or neutral losses resulting in less or unique solutions, even with mass accuracies obtainable by QTOF-MS.

Having determined putative elemental compositions, compound databases can be queried including general chemical databases (Chemical Abstract Service Database, Beilstein, ChemSpider [http://www.chemspider.com/], PubChem [http://pubchem.ncbi.nlm.nih.gov/]), natural product databases (*Dictionary of Natural*

Products, KNApSAcK [http://kanaya.naist.jp/KNApSAcK/]), metabolic databases with pathway information (KEGG [http://www.genome.jp/kegg/ligand.html], AraCyC [http://www.arabidopsis.org/biocyc/index.jsp]), or specialized compound databases, for instance, for lipid conjugates (LIPID Maps [http://www.lipidmaps.org], LIPIDBANK [http://www.lipidbank.jp]), or flavonoids (flavonoid viewer [http://www.metabolome.jp/software/FlavonoidViewer]). A detailed overview of currently available compound databases can be found in the recent literature (Tohge and Fernie 2009; Werner et al. 2008b). Assuming one or multiple hits after database queries, molecular structures that do not explain the experimental CID-mass spectra can be excluded from further analysis. In order to speed up this process, it was suggested to match the experimental CID-mass spectra with theoretical spectra calculated for compounds retrieved from databases searches (Hill et al. 2008). Despite their currently limited content, API mass spectral databases, such as MASSBANK (http://www.massbank.jp/) or METLIN (http://metlin.scripps.edu/), can be queried. In addition, hydrogen–deuterium-exchange experiments, derivatization reactions or basic chemical transformations can be carried out to probe the presence of certain functional groups in order to further reduce the number of structure hypotheses. According to the Metabolomics Standards Initiative, at least two orthogonal types of data relative to an authentic standard have to fit under identical experimental conditions for proper identification (annotation level 1) of a compound (Sumner et al. 2007). If the identity of a compound is derived by comparison to published physicochemical and spectral data without a chemical reference annotation level 2 is reached, which implies putative annotation of a compound.

In case database queries do not result in structural hypotheses, the careful interpretation of the fragmentation pattern is often the key to determine the compound class and to recognize molecular substructures that can then be used for refinement of a database search and to generate structure hypotheses. For the analysis of CID-mass spectra, sum formulas have to be calculated for fragment ions and neutral losses and a fragmentation tree has to be postulated. For the validation of selected edges being important for the interpretation of the fragmentation tree, pseudo-MS^3 experiments can be carried out on a QTOF-MS platform. If possible, additional fragmentation experiments by ion trap multistage MS can be performed. Structural information can be derived by identification of characteristic fragment ions, neutral losses, and sub-spectra that are related to specific structural elements. Furthermore, identification of similar spectra of already identified compounds can give a clue toward structure hypotheses. Because of the current state of API mass spectral databases and the low reproducibility of CID-mass spectra between different instruments and fragmentation methods (tandem-in-time, tandem-in-space) (Hopley et al. 2008), the interpretation of CID-mass spectra of non-peptidic, low molecular weight compounds requires specific experience. As described above, CID experiments can be complemented by basic chemical reactions to determine the number of exchangeable protons or the presence of a certain functional group. Degradation reactions can be performed and resulting products analyzed, optionally with orthogonal analytical techniques. In addition, heavy isotope-labeled precursor feeding has been applied to identify derived metabolites by screening for pairs of labeled and non-labeled features (Malitsky et al. 2008). Likewise, a fluorinated precursor has been used for this purpose (Böttcher et al. 2009b). A further method, which is particularly feasible for *Arabidopsis* is the analysis of biosynthetic mutants carrying a block in a given metabolic pathway. By simple comparison of wild-type and mutant profiles the spectrum of metabolites downstream of this block can be identified and structurally elucidated. This approach has been used to identify products of flavonoid (Böttcher et al. 2008; Yonekura-Sakakibara et al. 2008), phytoalexin (Böttcher et al. 2009b), and oxylipin biosynthesis (Glauser et al. 2008a).

In general, de novo structural elucidation without any prior knowledge solely on the basis of MS techniques is unfeasible. However, biochemical knowledge on the general compound classes and possible enzymatic transformation occurring in a plant species allows in several cases deduction of correct structure hypotheses for unknown metabolites. It should be noted, that regio- and stereochemical aspects of a molecular structure could only be addressed by MS in extremely rare cases. Therefore, in-depth structure elucidation of an unknown compound requires isolation and subsequent NMR analysis. As demonstrated recently for low abundant wound-induced jasmonate derivatives in *Arabidopsis*, combination of a preparative LC/MS approach with sensitive NMR techniques (capillary NMR) provided structure elucidation of these metabolites already at the microgram level (Glauser et al. 2008b).

17.3 Compound Classes Amenable for LC/API-MS-Based Profiling Approaches

17.3.1 Secondary Metabolites in Arabidopsis

Due to the pronounced interest in *Arabidopsis* as plant model system and the progress in analytical technology, the number of reported secondary metabolites has grown rapidly over the last two decades from ~40 (Chapple et al. 1994) to ~170 (D'Auria and Gershenzon 2005) to, at present, approximately 300–400 (personal estimation). In particular, GC and LC/MS-based metabolite profiling approaches have strongly contributed in exploring the chemical diversity within certain compound classes during the last decade. The fact that 15–20% of the annotated genes in *Arabidopsis* are predicted to encode enzymes participating in secondary metabolism and a large fraction of these enzymes has not yet been functionally characterized suggests that many secondary metabolites in *Arabidopsis* may await discovery. The accumulation of secondary metabolites is characterized by a spatial and temporal variability and has to be triggered in many cases by environmental stimuli. The recent discoveries of novel secondary metabolites in *Arabidopsis*, such as hydroxycinnamoyl spermidine conjugates accumulating in flower buds (Fellenberg et al. 2008) and seeds (Böttcher et al. 2008), demonstrates that sensitive and careful analyses of all plant organs under numerous environmental conditions at different developmental stages are required to catalogue the metabolome of *Arabidopsis*.

In particular, mass spectrometry-based analytical approaches in combination with techniques allowing sampling at tissue or cell level (Moco et al. 2009) will play a major role for the discovery of novel compounds and the determination of their spatial distribution.

The set of secondary metabolites characterized so far from *Arabidopsis* can be grouped in seven major biosynthetic classes including glucosinolates, indole and indole-sulfur compounds, phenylpropanoids, benzenoids, flavonoids, terpenes, and fatty acid derivatives (Fig. 17.4) (D'Auria and Gershenzon 2005). Glucosinolates are considered the most characteristic class of secondary metabolites within the Brassicaceae. A total of 36 glucosinolates were reported to occur in *A. thaliana* (Reichelt et al. 2002) of which methionine-derived glucosinolates constitute the biggest and structurally most diverse subclass. Further precursor amino acids used for glucosinolate biosynthesis in *Arabidopsis* are tryptophan, phenylalanine, and leucine. Glucosinolates were detected in all organs of the plant and show highly variable accumulation patterns among different organs and developmental states (Brown et al. 2003; Petersen et al. 2002). Targeted glucosinolate analysis frequently employs group separation using anion exchanger, desulfation, and quantification by reversed-phase LC with UV detection. Being sulfate monoesters, glucosinolates can be easily ionized in particular under negative ion API conditions. Therefore, reversed-phase LC or capillary electrophoresis coupled to ESI-MS has been applied for sensitive detection and identification of intact glucosinolates (Bringmann et al. 2005; Cataldi et al. 2007; Tian et al. 2005). Due to the uniform fragmentation behavior of the deprotonated molecular ions upon CID, tandem MS-based screening approaches have been developed (Mellon et al. 2002; Rochfort et al. 2008). Upon tissue damage glucosinolates are cleaved by separately compartmented myrosinases resulting in formation of a wide array of hydrolysis products, which act as defense compounds against herbivores or pathogens. These hydrolysis products include, for instance, isothiocyanates and nitriles, which are typically separated by GC due to their low polarity and volatility. However, in particular hydrolysis products of aromatic glucosinolates and of the higher homologs of the aliphatic glucosinolates are also amenable to LC/ESI-MS analysis.

A further hallmark of many Brassicaceae is the formation of tryptophan-derived indolic phytoalexins. These compounds show a high structural diversity, but most of them are substituted at the 3-position with sulfur- or nitrogen-containing side chains (Pedras et al. 2000). The formation of indole-3-carbaldehyde, indole-3-carboxylic acid, as well as camalexin as major indolic phytoalexin is characteristic for *Arabidopsis* (Hagemeier et al. 2001; Tan et al. 2004; Tsuji et al. 1992). In addition, the accumulation of methyl and glucosyl derivatives of indole-3-carboxylic acid has been reported (Bednarek et al. 2005). Hydroxy derivatives of indole-3-carboxylic acid and camalexin were identified as aglycones in several glycosides and malonyl glycosides (Bednarek et al. 2005; Böttcher et al. 2009b). As demonstrated by comprehensive mass spectral analysis of 23 indolic phytoalexins, reversed-phase LC/positive ion ESI-MS represents a valuable tool for separation and sensitive detection of this compound class (Pedras et al. 2006). The published CID

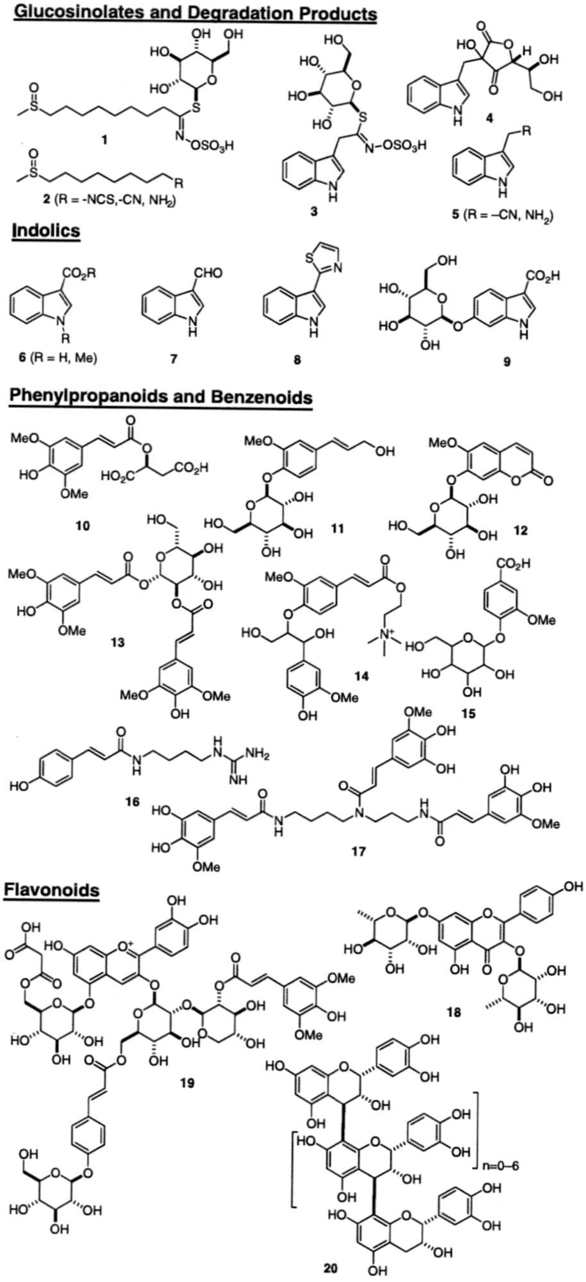

Fig. 17.4 (continued)

mass spectra should facilitate rapid identification of already known representatives and discovery of novel ones.

Like many other Brassicaceae, *Arabidopsis* accumulates, in addition to flavonoids, a set of structurally diverse phenylpropanoids. Among them, esters and amides of substituted hydroxycinnamic acids, coumarins (Kai et al. 2008; 2006), as well as mono- and oligolignols (Bednarek et al. 2005; Pauwels et al. 2008) have been identified. Sinapoylmalate is the dominating phenylpropanoid ester in leaf tissue, whereas sinapoylcholine accumulates exclusively in seeds. Both metabolites are accompanied by sinapoylglucose being their biosynthetic precursor. Besides sinapoylcholine a set of 26 phenolic choline esters has been characterized from *Arabidopsis* seeds (Böttcher et al. 2009a). Interestingly, malate and choline esters of hydroxycinnamic acids oxidatively coupled to monolignols via 4-*O*-8′ and 5-8 linkages have been found in inflorescences and seeds, respectively (Böttcher et al. 2008; Rohde et al. 2004). Except *N*-coumaroylaspartate (Mock et al. 1993), the presence of hydroxycinnamoyl amides has only been recently discovered in several organs of *Arabidopsis* including bis- and tris-substituted spermidine conjugates in seeds (Böttcher et al. 2008), roots (Luo et al. 2009), and flowers (Fellenberg et al. 2008), as well as agmatine conjugates accumulating upon pathogen attack in leaf tissue (Muroi et al. 2009). Due to the number of hetero atoms present in phenylpropanoid conjugates, which allow adduct formation with protons or alkali metal ions, most members within this compound class can be detected by positive ion ESI-MS, in particular nitrogen-containing compounds like choline esters, spermidine and agmatine conjugates. In addition, negative ion ESI can be successfully applied for glucose and malate esters. Many representatives exhibit relatively simple and intuitive fragmentation patterns permitting structure elucidation by means of tandem-MS techniques (Matsuda et al. 2009).

The most prominent flavonoids present in *Arabidopsis* are flavonol glycosides accumulating constitutively in many organs of the plant (Kerhoas et al. 2006; Yonekura-Sakakibara et al. 2008). Comprehensive flavonol profiling revealed more than 30 distinct structures with kaempferol, quercetin, and isorhamnetin as aglycones. Glycosylation occurs at position 3-*O* and 7-*O*. Typically, monosaccharides,

Fig. 17.4 Major classes of secondary metabolites and selected representatives found in *A. thaliana* and amenable for analysis by reversed-phase LC/ESI-MS. 8-Methylsulfinyloctylglucosinolate (**1**) and related degradation products (**2**); indolyl-3-methylglucosinolate (**3**) and related degradation products (**4, 5**); indole-3-carboxylic acid and methyl derivates (**6**); indole-3-carbaldehyde (**7**); camalexin (**8**); 6-β-D-glucopyranosyloxyindole-3-carboxylic acid (**9**); 2-*O*-sinapoyl-L-malate (**10**); coniferin (**11**); scopolin (**12**); 1,2-di-*O*-sinapoyl-β-D-glucopyranose (**13**); feruloylcholine 4-*O*-8′-coupled to coniferyl alcohol (**14**); 4-*O*-glycosylated vanillic acid (**15**); coumaroylagmatine (**16**); N^1,N^5,N^{10}-tris-(5-hydroxyferuloyl)spermidine (**17**); kaempferol 3-*O*-α-L-rhamnopyranoside-7-*O*-α-L-rhamnopyranoside (**18**); cyanidin 3-*O*-{2-*O*-[2-*O*-(sinapoyl)-β-D-xylopyranosyl]-6-*O*-(4-β-D-glucopyranosyloxycinnamoyl)-β-D-glucopyranoside}-5-*O*-[6-*O*-(malonyl)-β-D- glucopyranoside]; epicatechin-type proanthocyanidin oligomers (**19**). The lack of stereochemical information reflects the current annotation level of some metabolites such as rhamnose and glucose, are linked to position 7-*O* of the aglycone, whereas at position 3-*O* the sugar moieties variably consist of mono-, di- and trisaccharides (Stobiecki et al. 2006; Veit and Pauli 1999). Like many other flavonoids, flavonol glycosides can be detected in both positive and negative ion mode by ESI-MS with high sensitivity. Glycoside composition, as well as the structure of the aglycones, can easily be determined by tandem-MS techniques (Cuyckens and Claeys 2004; Kachlicki et al. 2008; Wolfender et al. 2000). Besides flavonol glycosides a set of anthocyanins has been reported to occur in leaves and stems of *Arabidopsis* (Bloor and Abrahams 2002). In-depth structural analysis of these compounds revealed cyanidine as aglycone linked at position 3-*O* to a disaccharide and at position 5-*O* to a monosaccharide being esterified with different hydroxycinnamic acids or malonic acid, respectively. After induction of the flavonoid biosynthetic pathway by overexpression of the MYB transcription factor gene, *PAP1* (PRODUCTION OF ANTHOCYANIN PIGMENT 1), 11 cyanidine glycosides have been identified from their UV–visible absorption spectra and structurally annotated by analysis of the fragmentation patterns (Tohge et al. 2005). Proanthocyanidins, the third class of flavonoids present in *Arabidopsis*, are formed by oligomerization of flavanol-type monomers, such as epicatechin (Dixon et al. 2005). They accumulate in the seed coat and have been detected by LC/MS in seed extracts (Routaboul et al. 2006). The fragmentation pattern of oligomeric proanthocyanindins has been studied in detail and sequence analysis of this compound class is achievable by tandem-MS (Li and Deinzer 2007).

About 30 genes encoding terpene synthases have been identified within the *Arabidopsis* genome. Therefore, it is not surprising that a multitude of mono- and sesquiterpenes have been identified as volatiles being emitted from several *Arabidopsis* organs (Chen et al. 2003; Steeghs et al. 2004; Tholl et al. 2005). In addition, the presence of numerous genes encoding putative oxidosqualene cyclases suggests formation of higher terpenes in *Arabidopsis*, which, however, have not yet been detected *in planta*. Because of their physicochemical properties, mono- and sesquiterpenes are analyzed by GC-based techniques and not amenable for LC/MS analysis. Likewise, higher terpenoids with a low number of functional groups are not well covered by ESI-MS techniques due to their low ionization efficiency. However, given a certain degree of functionalization of the carbon skeleton as found, for instance, in triterpene saponins, LC/MS-based approaches can be applied for profiling of such compound classes (Huhman and Sumner 2002).

Except terpenes and fatty acid derivatives, five out of seven biosynthetic classes present in the secondary metabolism of *Arabidopsis* are well covered by reversed-phase LC/ESI-MS. It remains to be elucidated whether by application of different ionization techniques, such as APCI and atmospheric pressure photoionization (APPI) or multi-mode ionization allowing fast switching between ESI and APCI within the same analysis, a better coverage of less polar metabolites can be achieved to enable efficient profiling and discovery of novel metabolites within this polarity range.

17.3.2 Lipids

Lipid research or nowadays lipidomics has been a rapidly expanding research field over the last decade. This has been indicated not only by the ever growing amount of relevant publications but also by the instalment of, for example, the European Lipidomics Initiative (http://www.lipidomics.net/) or the Kansas Lipidomics Research Center (http://www.k-state.edu/lipid/lipidomics/index.htm), the latter more focusing on plant research. In addition, it is also mirrored by the increasing number of databases available to the general public (http://www.lipidmaps.org; http://www.lipidlibrary.co.uk; http://www.cyberlipid.org; http://www.lipidat.ul.ie; http://www.lipidbank.jp; http://www.byrdwell.com/).

Lipids are as such not only a class of compounds of relatively high structural diversity. They also seem to be involved in a staggering array of cellular processes that include photosynthesis, signal transduction, vesicle trafficking, secretion, cytoskeletal rearrangement, seed germination, organ differentiation, pollination, and responses to biotic and abiotic stresses (reviewed in Farmer et al. 2003; Isaac et al. 2007; Roberts et al. 2008; Wallis and Browse 2002; Wang 2002; Welti and Wang 2004). As lipids are such an important compound class in plants in general and *A. thaliana* being "the" model plant of the last two decades it is not asthonishing that there has been a lot of research in that particular field of interest. The lipid profile of *Arabidopsis* is dominated by glycerolipids presenting 56% of the total lipid fraction accompanied by lower amounts of wax, cutin monomers, chlorophyll, plastoquinone, carotenoids, tocopheroids, sterols, and sphingolipids. Being a seed plant the glycerolipid fraction of *Arabidopsis* consists for the most part of galactolipids (monogalactosyldiacylglycerol (MGDG) and digalactosyldiacylglycerol (DGDG)) followed by phosphatidylglycerol (PG), phosphatidylcholine (PC), phosphatidylethanolamine (PE), phosphatidylinositol (PI), sulfoquinovosyldiacylglycerol (SQDG), and phosphatidylserine (PS). The fatty acid composition of the different *Arabidopsis* tissues prevalently consists of linolenic (18:3) and linoleic acid (18:2). Figure 17.5 as well as Tables 17.2 and 17.3 shows an overview of the lipid composition in *Arabidopsis* adapted from a number of publications of the last 15 years. A very comprehensive analysis of lipid species identified in *Arabidopsis* leaves has been published in Devaiah et al. (2006).

Advancement in the area of lipid research has been driven by the development of analytical technologies in the last decade in particular in the field of chromatography coupled to mass spectrometry. This chapter does not intend to represent a comprehensive overview of the advantages and disadvantages of the different analytical methods applied in lipidomic research, for the simple reason that there have been written excellent reviews concerning this topic in the recent years (Isaac et al. 2007; Roberts et al. 2008; Welti and Wang 2004). It will merely focus on those techniques, which are currently predominantly used in the analysis of plant lipids. GC/MS is a powerful, well-established technique for separation, quantification, and identification of fatty acids and fatty acid-derived metabolites in lipidomic research. However, compounds to be analyzed by GC/MS have to be volatilized prior to analysis, which

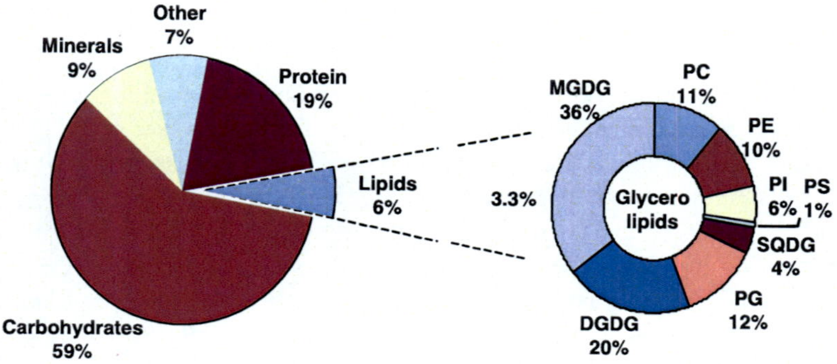

Fig. 17.5 Composition of *A. thaliana* (ecotype Wassilewskija) leaves (distribution of dry weight in percent of total; distribution of glycerolipids in mol%). MGDG, monogalactosyldiacylglycerol; DGDG, digalactosyldiacylglycerol; PG, phosphatidylglycerol; PC, phosphatidylcholine; PE, phosphatidylethanolamine; PI, phosphatidylinositol; SQDG, sulfoquinovosyldiacylglycerol; PS, phosphatidylserine. Data adapted from Browse and Somerville (1994)

requires in many cases derivatization. A frequently applied derivatization method for fatty acids, which are typically released from complex lipids by hydrolysis, is methylation to give fatty acid methyl esters. In contrast to this destructive approach only allowing the measurement of total fatty acid composition, ESI-MS techniques facilitate direct introduction of the sample without hydrolysis and derivatization and permit detection of intact molecular species. Hence, in-depth chemical characterization of complex lipids, such as glycerolipids, by ESI-MS gained a growing importance in lipidomics during the last decade. Two methodological approaches have been mainly described using ESI-MS as technique of choice: targeted and non-targeted lipid profiling. In the targeted option direct infusion-ESI/tandem-MS

Table 17.2 Glycerolipid composition of different organs and plastids of *A. thaliana*

Lipid source Lipid class	Leaves	Chloroplast	Extra- chloroplast	Roots	Seeds
PC	11.2	12	47.8	45.4	48.1
PE	10	–	36.5	27.5	22.1
PI + PS	6.4	–	10.9	12.9	18.9
SQDG	4.3	3.9	–	–	–
PG	12	9.5	4.4	3.8	4.6
DGDG	20.2	20.9	–	2.0	3.3
MGDG	35.5	53.7	–	3.4	3.0
Total lipid mg/g f.w.	3.2	2.3	0.9	1.5	350

Values shown are mol%; f.w., fresh weight. Data adapted from Bonaventure et al. (2003), Browse and Somerville (1994), Li et al. (2006). Extra-chloroplast defines all membranes not belonging to the chloroplast. MGDG, monogalactosyldiacylglycerol; DGDG, digalactosyldiacylglycerol; PG, phosphatidylglycerol; PC, phosphatidylcholine; PE, Phosphatidylethanolamine; PI, phosphatidylinositol; SQDG, sulfoquinovosyldiacylglycerol; PS, phosphatidylserine.

Table 17.3 Fatty acid (FA) composition of different organs of *A. thaliana*

Tissue source FA composition	Leaf	Flower	Root	Seeds
Palmitic acid 16:0	17.5	27.4	25.3	8.3
Palmitoleic acid 16:1(9)	0.6	1.7	0.9	0.3
Palmitoleic acid 16:1(3)	3.7	–	–	–
Hexadecatrienoic acid 16:3	11.8	1.2	–	–
Stearic acid 18:0	1.5	2.4	6.6	3.5
Oleic acid 18:1(9)	2.6	3.2	3.6	13
Linoleic acid 18:2	14.4	30.6	28.8	27.5
Linolenic acid 18:3	47.7	31.3	21.3	18.6
Other[a]	–	–	13.2	28.6

Values shown are mol%.
[a]Mostly consisting of eicosenoic acid (20:1).
Data adapted from Bonaventure et al. (2003), and Li et al. (2006).

strategies are applied to address specific classes of lipids via precursor ion or neutral loss scanning. For example, PCs can be measured by scanning for the phosphoric acid choline ester fragment at *m/z* 184 using precursor ion scanning in positive ion mode. Using different scanning techniques for a variety of fragment ions, each independent measurement allows the coverage of many lipid classes. It also reduces the complexity of the sample and allows to omit prior chromatographic separation (Isaac et al. 2007; Welti and Wang 2004). In contrast to direct-infusion experiments, non-targeted profiling experiments are predominantly conducted using LC/MS and LC/tandem-MS platforms. Here, liquid chromatography is directly coupled to a high-resolution mass spectrometer scanning a complete mass range rather than for a particular mass value. Due to the preceding sample separation, even very complex compound mixtures can be analyzed without loosing scarce substances to matrix effects occurring during ionization (Böttcher et al. 2007; Harrabi et al. 2009). Depending on the equipment of the laboratory analyzing the lipid composition of a sample either the targeted or the non-targeted profiling option is chosen. Laboratories with MS platforms of lower resolution, such as triple quadrupole or ion trap instruments, tend to go for targeted approaches. Higher resolution instruments, such as QTOF-MS, are mainly used for the non-targeted profiling option. Both profiling options have advantages and disadvantages and both are successfully applied in lipid research.

17.4 Conclusion and Outlook

Since the introduction of LC/MS-based metabolite profiling approaches tremendous progress has been achieved in this analytical discipline. On the one hand, both liquid chromatography and mass spectrometry platforms have been technically improved. In particular, introduction of UPLC on sub-2 μm particles allowing high-throughput separation represents an important milestone, since metabolite profiling of large genetic plant populations is now possible. On the other hand, development of software packages facilitating peak picking, alignment and deconvolution now enables reliable large-scale comparative analyses of metabolite profiles. Further hallmarks are the development of standard mass spectrometry data formats (mzData, mzXML, mzML) and reporting standards for metabolomics experiments. However, relative little efforts have been spent on the evaluation and optimization of wet chemistry aspects and ionization techniques. Therefore, to fully exploit the potential of LC/MS-based profiling approaches, in particular in terms of better metabolome coverage, depletion and enrichment strategies for selected compound classes have to be developed. Furthermore, mass spectrometric detectability across the whole polarity range using additional ionization techniques needs to be established.

At present, a major bottleneck of any metabolomics experiments is the annotation and identification of metabolites. To partially solve this issue the development of large natural product mass spectral databases, as well as algorithms for automatic analysis of tandem-MS data is required. Furthermore, inter-laboratory comparability of profiling studies has to be evaluated and strategies for data normalization developed in order to enable construction of profile databases as known for microarray data.

The recent achievements in LC/MS-based high-throughput quantitative metabolite profiling will have a major impact in integrated functional genomics approaches. The possibility to profile large plant populations in a non-targeted manner will allow a new view into natural metabolic variation, which can be used in the genetic dissection of quantitative traits. First GC/MS-based studies that link growth-related phenotypes to metabolic signatures have already been successfully performed (Keurentjes et al. 2008; Meyer et al. 2007) and a comprehensive study profiling transcripts, proteins, and metabolites in a recombinant inbred line population included also the LC/MS-based profiling (Fu et al. 2009). Furthermore, the integral analysis of metabolomics data sets with transcriptomics data can identify novel gene functions and metabolic networks (Saito et al. 2008). Finally, the use of genetic resources, such as knock-out lines in metabolomic studies is helpful in elucidating biochemical pathways and identifying novel metabolites (Böttcher et al. 2008; 2009b).

References

Abdel-Farid IB, Kim HK, Choi YH, Verpoorte R (2007) Metabolic characterization of *Brassica rapa* leaves by NMR spectroscopy. J Agric Food Chem 55:7936–7943

Abello N, Geurink PP, van der Toorn M, van Oosterhout AJM, Lugtenburg J, van der Marel GA, Kerstjens HAM, Postma DS, Overkleeft HS, Bischoff R (2008) Poly(ethylene glycol)-based stable isotope labeling reagents for the quantitative analysis of low molecular weight metabolites by LC-MS. Anal Chem 80:9171–9180

Aharoni A, Ric de Vos CH, Verhoeven HA, Maliepaard CA, Kruppa G, Bino R, Goodenowe DB (2002) Nontargeted metabolome analysis by use of Fourier-transform ion cyclotron mass spectrometry. Omics 6:217–234

Barry SJ, Carr RM, Lane SJ, Leavens WJ, Manning CO, Monte S, Waterhouse I (2003) Use of S-pentafluorophenyl tris(2,4,6-trimethoxyphenyl) phosphonium acetate bromide and

(4-hydrazino-4-oxobutyl) [tris(2,4,6-trimethoxyphenyl)]phosphonium bromide for the derivatization of alcohols, aldehydes and ketones for detection by liquid chromatography/electrospray mass spectrometry. Rapid Commun Mass Spectrom 17:484–497

Bednarek P, Schneider B, Svatos A, Oldham NJ, Hahlbrock K (2005) Structural complexity, differential response to infection, and tissue specificity of indolic and phenylpropanoid secondary metabolism in Arabidopsis roots. Plant Physiol 138:1058–1070

Benton HP, Wong DM, Trauger SA, Siuzdak G (2008) XCMS2: Processing tandem mass spectrometry data for metabolite identification and structural characterization. Anal Chem 80:6382–6389

Bloor SJ, Abrahams S (2002) The structure of the major anthocyanin in *Arabidopsis thaliana*. Phytochemistry 59:343–346

Bonaventure G, Salas JJ, Pollard MR, Ohlrogge JB (2003) Disruption of the FATB gene in Arabidopsis demonstrates an essential role of saturated fatty acids in plant growth. Plant Cell 15:1020–1033

Böttcher C, von Roepenack-Lahaye E, Schmidt J, Clemens S, Scheel D (2009a) Analysis of phenolic choline esters from seeds of *Arabidopsis thaliana* and *Brassica napus* by capillary liquid chromatography/electrospray-tandem mass spectrometry. J Mass Spectrom 44: 466–476

Böttcher C, von Roepenack-Lahaye E, Schmidt J, Schmotz C, Neumann S, Scheel D, Clemens S (2008) Metabolome analysis of biosynthetic mutants reveals a diversity of metabolic changes and allows identification of a large number of new compounds in Arabidopsis. Plant Physiol 147:2107–2120

Böttcher C, von Roepenack-Lahaye E, Willscher E, Scheel D, Clemens S (2007) Evaluation of matrix effects in metabolite profiling based on capillary liquid chromatography electrospray ionization quadrupole time-of-flight mass spectrometry. Anal Chem 79:1507–1513

Böttcher C, Westphal L, Schmotz C, Prade E, Scheel D, Glawischnig E (2009b) The multifunctional enzyme CYP71B15 (PHYTOALEXIN DEFICIENT3) converts cysteine-indole-3-acetonitrile to camalexin in the indole-3-acetonitrile metabolic network of *Arabidopsis thaliana*. Plant Cell 21:1830–1845

Bouchereau A, Hamelin J, Lamour I, Renard M, Larher F (1991) Distribution of sinapine and related compounds in seeds of Brassica and allied genera. Phytochemistry 30:1873–1881

Bringmann G, Kajahn I, Neususs C, Pelzing M, Laug S, Unger M, Holzgrabe U (2005) Analysis of the glucosinolate pattern of *Arabidopsis thaliana* seeds by capillary zone electrophoresis coupled to electrospray ionization-mass spectrometry. Electrophoresis 26:1513–1522

Broadhurst DI, Kell DB (2006) Statistical strategies for avoiding false discoveries in metabolomics and related experiments. Metabolomics 2:171–196

Brown PD, Tokuhisa JG, Reichelt M, Gershenzon J (2003) Variation of glucosinolate accumulation among different organs and developmental stages of *Arabidopsis thaliana*. Phytochemistry 62:471–481

Browse J, Somerville CR (1994) Glycerolipids. In: Meyerowitz EM, Somerville CR (eds) Arabidopsis, pp 881–912. Cold Spring Harbor Laboratory Press, Cold Spring Habor

Carlson EE, Cravatt BF (2007a) Chemoselective probes for metabolite enrichment and profiling. Nat Methods 4:429–435

Carlson EE, Cravatt BF (2007b) Enrichment tags for enhanced-resolution profiling of the polar metabolome. J Am Chem Soc 129:15780–15782

Cataldi TRI, Rubino A, Lelario F, Bufo SA (2007) Naturally occurring glucosinolates in plant extracts of rocket salad (Eruca sativa L.) identified by liquid chromatography coupled with negative ion electrospray ionization and quadrupole ion-trap mass spectrometry. Rapid Commun Mass Spectrom 21:2374–2388

Catchpole GS, Beckmann M, Enot DP, Mondhe M, Zywicki B, Taylor J, Hardy N, Smith A, King RD, Kell DB, Fiehn O, Draper J (2005) Hierarchical metabolomics demonstrates substantial compositional similarity between genetically modified and conventional potato crops. Proc Natl Acad Sci U S A 102:14458–14462

Chapple CCS, Shirely BW, Zook M, Hammerschmidt R, Somerville SC (1994) Secondary metabolism in Arabidopsis. In: Meyerowitz EM, Somerville CR (eds) Arabidopsis, pp 989–1030. Cold Spring Harbor Laboratory Press, Cold Spring Habor

Chen F, Tholl D, D'Auria JC, Farooq A, Pichersky E, Gershenzon J (2003) Biosynthesis and emission of terpenoid volatiles from Arabidopsis flowers. Plant Cell 15:481–494

Chernushevich IV, Loboda AV, Thomson BA (2001) An introduction to quadrupole-time-of-flight mass spectrometry. J Mass Spectrom 36:849–865

Cuyckens F, Claeys M (2004) Mass spectrometry in the structural analysis of flavonoids. J Mass Spectrom 39:1–15

De Vos RC, Moco S, Lommen A, Keurentjes JJ, Bino RJ, Hall RD (2007) Untargeted large-scale plant metabolomics using liquid chromatography coupled to mass spectrometry. Nat Protoc 2:778–791

Devaiah SP, Roth MR, Baughman E, Li MY, Tamura P, Jeannotte R, Welti R, Wang XM (2006) Quantitative profiling of polar glycerolipid species from organs of wild-type Arabidopsis and a PHOSPHOLIPASE D alpha 1 knockout mutant. Phytochemistry 67:1907–1924

Dixon RA, Xie DY, Sharma SB (2005) Proanthocyanidins - a final frontier in flavonoid research? New Phytol 165:9–28

Dunn WB (2008) Current trends and future requirements for the mass spectrometric investigation of microbial, mammalian and plant metabolomes. Phys Biol 5. doi:10.1088/1478-3975/5/1/011001

D'Auria JC, Gershenzon J (2005) The secondary metabolism of *Arabidopsis thaliana*: growing like a weed. Curr Opin Plant Biol 8:308–316

Farmer EE, Almeras E, Krishnamurthy V (2003) Jasmonates and related oxylipins in plant responses to pathogenesis and herbivory. Curr Opin Plant Biol 6:372–378

Fellenberg C, Milkowski C, Hause B, Lange PR, Bottcher C, Schmidt J, Vogt T (2008) Tapetum-specific location of a cation-dependent O-methyltransferase in *Arabidopsis thaliana*. Plant J 56:132–145

Fernie AR, Trethewey RN, Krotzky AJ, Willmitzer L (2004) Metabolite profiling: from diagnostics to systems biology. Nat Rev Mol Cell Biol 5:763–769

Fiehn O, Kopka J, Dormann P, Altmann T, Trethewey RN, Willmitzer L (2000) Metabolite profiling for plant functional genomics. Nat Biotechnol 18:1157–1161

Fiehn O, Sumner LW, Rhee SY, Ward J, Dickerson J, Lange BM, Lane G, Roessner U, Last R, Nikolau B (2007) Minimum reporting standards for plant biology context information in metabolomic studies. Metabolomics 3:195–201

Fu J, Keurentjes JJB, Bouwmeester H, America T, Verstappen FWA, Ward JL, Beale MH, de Vos RCH, Dijkstra M, Scheltema RA, Johannes F, Koornneef M, Vreugdenhil D, Breitling R, Jansen RC (2009) System-wide molecular evidence for phenotypic buffering in Arabidopsis. Nat Genet 41:166–167

Gangl ET, Annan M, Spooner N, Vouros P (2001) Reduction of signal suppression effects in ESI-MS using a nanosplitting device. Anal Chem 73:5635–5644

Giavalisco P, Hummel J, Lisec J, Inostroza AC, Catchpole G, Willmitzer L (2008) High-resolution direct infusion-based mass spectrometry in combination with whole ^{13}C metabolome isotope labeling allows unambiguous assignment of chemical sum formulas. Anal Chem 80:9417–9425

Giavalisco P, Kohl K, Hummel J, Seiwert B, Willmitzer L (2009) ^{13}C Isotope-labeled metabolomes allowing for improved compound annotation and relative quantification in liquid chromatography-mass spectrometry-based metabolomic research. Anal Chem 81:6546–6551

Glauser G, Grata E, Dubugnon L, Rudaz S, Farmer EE, Wolfender JL (2008a) Spatial and temporal dynamics of jasmonate synthesis and accumulation in Arabidopsis in response to wounding. J Biol Chem 283:16400–16407

Glauser G, Guillarme D, Grata E, Boccard J, Thiocone A, Carrupt PA, Veuthey JL, Rudaz S, Wolfender JL (2008b) Optimized liquid chromatography-mass spectrometry approach for the isolation of minor stress biomarkers in plant extracts and their identification by capillary nuclear magnetic resonance. J Chromatogr A 1180:90–98

Grange AH, Zumwalt MC, Sovocool GW (2006) Determination of ion and neutral loss compositions and deconvolution of product ion mass spectra using an orthogonal acceleration time-of-flight mass spectrometer and an ion correlation program. Rapid Commun Mass Spectrom 20:89–102

Grata E, Boccard J, Guillarme D, Glauser G, Carrupt PA, Farmer EE, Wolfender JL, Rudaz S (2008) UPLC-TOF-MS for plant metabolomics: a sequential approach for wound marker analysis in *Arabidopsis thaliana*. J Chromatogr B Analyt Technol Biomed Life Sci 871: 261–270

Grata E, Guillarme D, Glauser G, Boccard J, Carrupt PA, Veuthey JL, Rudaz S, Wolfender JL (2009) Metabolite profiling of plant extracts by ultra-high-pressure liquid chromatography at elevated temperature coupled to time-of-flight mass spectrometry. J Chromatogr A 1216: 5660–5668

Hagemeier J, Schneider B, Oldham NJ, Hahlbrock K (2001) Accumulation of soluble and wall-bound indolic metabolites in *Arabidopsis thaliana* leaves infected with virulent or avirulent *Pseudomonas syringae* pathovar tomato strains. Proc Natl Acad Sci U S A 98:753–758

Harrabi S, Herchi W, Kallel H, Mayer PM, Boukhchina S (2009) Liquid chromatographic-mass spectrometric analysis of glycerophospholipids in corn oil. Food Chem 114:712–716

Hegeman AD, Schulte CF, Cui Q, Lewis IA, Huttlin EL, Eghbalnia H, Harms AC, Ulrich EL, Markley JL, Sussman MR (2007) Stable isotope assisted assignment of elemental compositions for metabolomics. Anal Chem 79:6912–6921

Hill DW, Kertesz TM, Fontaine D, Friedman R, Grant DF (2008) Mass spectral metabonomics beyond elemental formula: chemical database querying by matching experimental with computational fragmentation spectra. Anal Chem 80:5574–5582

Hirai MY, Klein M, Fujikawa Y, Yano M, Goodenowe DB, Yamazaki Y, Kanaya S, Nakamura Y, Kitayama M, Suzuki H, Sakurai N, Shibata D, Tokuhisa J, Reichelt M, Gershenzon J, Papenbrock J, Saito K (2005) Elucidation of gene-to-gene and metabolite-to-gene networks in Arabidopsis by integration of metabolomics and transcriptomics. J Biol Chem 280:25590–25595

Hopley C, Bristow T, Lubben A, Simpson A, Bul E, Klagkou K, Herniman J, Langley J (2008) Towards a universal product ion mass spectral library - reproducibility of product ion spectra across eleven different mass spectrometers. Rapid Commun Mass Spectrom 22:1779–1786

Huhman DV, Sumner LW (2002) Metabolic profiling of saponins in *Medicago sativa* and *Medicago truncatula* using HPLC coupled to an electrospray ion-trap mass spectrometer. Phytochemistry 59:347–360

Isaac G, Jeannotte R, Esch SW, Welti R (2007) New mass-spectrometry-based strategies for lipids. Genet Eng (N Y) 28:129–157

Kachlicki P, Einhorn J, Muth D, Kerhoas L, Stobiecki M (2008) Evaluation of glycosylation and malonylation patterns in flavonoid glycosides during LC/MS/MS metabolite profiling. J Mass Spectrom 43:572–586

Kai K, Mizutani M, Kawamura N, Yamamoto R, Tamai M, Yamaguchi H, Sakata K, Shimizu B (2008) Scopoletin is biosynthesized via ortho-hydroxylation of feruloyl CoA by a 2-oxoglutarate-dependent dioxygenase in *Arabidopsis thaliana*. Plant J 55:989–999

Kai K, Shimizu B, Mizutani M, Watanabe K, Sakata K (2006) Accumulation of coumarins in *Arabidopsis thaliana*. Phytochemistry 67:379–386

Katajamaa M, Miettinen J, Oresic M (2006) MZmine: toolbox for processing and visualization of mass spectrometry based molecular profile data. Bioinformatics 22:634–636

Katajamaa M, Oresic M (2005) Processing methods for differential analysis of LC/MS profile data. BMC Bioinformatics 6. doi:10.1186/1471–2105–6–179

Kendziorski C, Irizarry RA, Chen KS, Haag JD, Gould MN (2005) On the utility of pooling biological samples in microarray experiments. Proc Natl Acad Sci U S A 102:4252–4257

Kerhoas L, Aouak D, Cingoz A, Routaboul JM, Lepiniec L, Einhorn J, Birlirakis N (2006) Structural characterization of the major flavonoid glycosides from *Arabidopsis thaliana* seeds. J Agric Food Chem 54:6603–6612

Keurentjes JJB, Koornneef M, Vreugdenhil D (2008) Quantitative genetics in the age of omics. Curr Opin Plant Biol 11:123–128

Kind T, Fiehn O (2006) Metabolomic database annotations via query of elemental compositions: mass accuracy is insufficient even at less than 1 ppm. BMC Bioinformatics 7. doi:10.1186/1471–2105–7–234

Kind T, Fiehn O (2007) Seven Golden Rules for heuristic filtering of molecular formulas obtained by accurate mass spectrometry. BMC Bioinformatics 8. doi:10.1186/1471–2105–8–105

Kliebenstein DJ (2004) Secondary metabolites and plant/environment interactions: a view through *Arabidopsis thaliana* tinged glasses. Plant Cell Environ 27:675–684

Konishi Y, Kiyota T, Draghici C, Gao JM, Yeboah F, Acoca S, Jarussophon S, Purisima E (2007) Molecular formula analysis by an MS/MS/MS technique to expedite dereplication of natural products. Anal Chem 79:1187–1197

Lamos SM, Shortreed MR, Frey BL, Belshaw PJ, Smith LM (2007) Relative quantification of carboxylic acid metabolites by liquid chromatography-mass spectrometry using isotopic variants of cholamine. Anal Chem 79:5143–5149

Last RL, Jones AD, Shachar-Hill Y (2007) Towards the plant metabolome and beyond. Nat Rev Mol Cell Biol 8:167–174

Leavens WJ, Lane SJ, Carr RM, Lockie AM, Waterhouse I (2002) Derivatization for liquid chromatography/electrospray mass spectrometry: synthesis of tris(trimethoxyphenyl)phosphonium compounds and their derivatives of amine and carboxylic acids. Rapid Commun Mass Spectrom 16:433–441

Li Y, Beisson F, Pollard M, Ohlrogge J (2006) Oil content of Arabidopsis seeds: the influence of seed anatomy, light and plant-to-plant variation. Phytochemistry 67:904–915

Li HJ, Deinzer ML (2007) Tandem mass spectrometry for sequencing proanthocyanidins. Anal Chem 79:1739–1748

Lommen A (2009) MetAlign: interface-driven, versatile metabolomics tool for hyphenated full-scan mass spectrometry data preprocessing. Anal Chem 81:3079–3086

Luo J, Fuell C, Parr A, Hill L, Bailey P, Elliott K, Fairhurst SA, Martin C, Michael AJ (2009) A novel polyamine acyltransferase responsible for the accumulation of spermidine conjugates in Arabidopsis seed. Plant Cell 21:318–333

Malitsky S, Blum E, Less H, Venger I, Elbaz M, Morin S, Eshed Y, Aharoni A (2008) The transcript and metabolite networks affected by the two clades of Arabidopsis glucosinolate biosynthesis regulators. Plant Physiol 148:2021–2049

Matsuda F, Yonekura-Sakakibara K, Niida R, Kuromori T, Shinozaki K, Saito K (2009) MS/MS spectral tag-based annotation of non-targeted profile of plant secondary metabolites. Plant J 57:555–577

Mellon FA, Bennett RN, Holst B, Williamson G (2002) Intact glucosinolate analysis in plant extracts by programmed cone voltage electrospray LC/MS: performance and comparison with LC/MS/MS methods. Anal Biochem 306:83–91

Meyer RC, Steinfath M, Lisec J, Becher M, Witucka-Wall H, Törjék O, Fiehn O, Eckardt Ä, Willmitzer L, Selbig J, Altmann T (2007) The metabolic signature related to high plant growth rate in *Arabidopsis thaliana*. Proc Natl Acad Sci U S A 104:4759–4764

Mock HP, Wray V, Beck W, Metzger JW, Strack D (1993) Coumaroylaspartate from cell-suspension cultures of *Arabidopsis thaliana*. Phytochemistry 34:157–159

Moco S, Schneider B, Vervoort J (2009) Plant micrometabolomics: the analysis of endogenous metabolites present in a plant cell or tissue. J Proteome Res 8:1694–1703

Muroi A, Ishihara A, Tanaka C, Ishizuka A, Takabayashi J, Miyoshi H, Nishioka T (2009) Accumulation of hydroxycinnamic acid amides induced by pathogen infection and identification of agmatine coumaroyltransferase in *Arabidopsis thaliana* . Planta 230:517–527

Neue UD, Mazzeo JR (2001) A theoretical study of the optimization of gradients at elevated temperature. J Sep Sci 24:921–929

Nordstrom A, Want E, Northen T, Lehtio J, Siuzdak G (2008) Multiple ionization mass spectrometry strategy used to reveal the complexity of metabolomics. Anal Chem 80:421–429

Ohta D, Shibata D, Kanaya S (2007) Metabolic profiling using Fourier-transform ion-cyclotron-resonance mass spectrometry. Anal Bioanal Chem 389:1469–1475

Ojanpera S, Pelander A, Pelzing M, Krebs I, Vuori E, Ojanpera I (2006) Isotopic pattern and accurate mass determination in urine drug screening by liquid chromatography/time-of-flight mass spectrometry. Rapid Commun Mass Spectrom 20:1161–1167

Oksman-Caldentey KM, Saito K (2005) Integrating genomics and metabolomics for engineering plant metabolic pathways. Curr Opin Biotechnol 16:174–179

Pauwels L, Morreel K, De Witte E, Lammertyn F, Van Montagu M, Boerjan W, Inze D, Goossens A (2008) Mapping methyl jasmonate-mediated transcriptional reprogramming of metabolism and cell cycle progression in cultured Arabidopsis cells. Proc Natl Acad Sci U S A 105: 1380–1385

Pedras MS, Adio AM, Suchy M, Okinyo DP, Zheng QA, Jha M, Sarwar MG (2006) Detection, characterization and identification of crucifer phytoalexins using high-performance liquid chromatography with diode array detection and electrospray ionization mass spectrometry. J Chromatogr A 1133:172–183

Pedras MS, Okanga FI, Zaharia IL, Khan AQ (2000) Phytoalexins from crucifers: synthesis, biosynthesis, and biotransformation. Phytochemistry 53:161–176

Pellegrin V (1983) Molecular formulas of organic compounds-the nitrogen rule and degree of unsaturation. J Chem Educ 60:626–633

Petersen BL, Chen SX, Hansen CH, Olsen CE, Halkier BA (2002) Composition and content of glucosinolates in developing *Arabidopsis thaliana*. Planta 214:562–571

Reichelt M, Brown PD, Schneider B, Oldham NJ, Stauber E, Tokuhisa J, Kliebenstein DJ, Mitchell-Olds T, Gershenzon J (2002) Benzoic acid glucosinolate esters and other glucosinolates from *Arabidopsis thaliana*. Phytochemistry 59:663–671

Roberts LD, McCombie G, Titman CM, Griffin JL (2008) A matter of fat: an introduction to lipidomic profiling methods. J Chromatogr B Analyt Technol Biomed Life Sci 871:174–181

Rochfort SJ, Trenerry VC, Imsic M, Panozzo J, Jones R (2008) Class targeted metabolomics: ESI ion trap screening methods for glucosinolates based on MSn fragmentation. Phytochemistry 69:1671–1679

Rohde A, Morreel K, Ralph J, Goeminne G, Hostyn V, De Rycke R, Kushnir S, Van Doorsselaere J, Joseleau JP, Vuylsteke M, Van Driessche G, Van Beeumen J, Messens E, Boerjan W (2004) Molecular phenotyping of the *pal1* and *pal2* mutants of *Arabidopsis thaliana* reveals far-reaching consequences on phenylpropanoid, amino acid, and carbohydrate metabolism. Plant Cell 16:2749–2771

Routaboul JM, Kerhoas L, Debeaujon I, Pourcel L, Caboche M, Einhorn J, Lepiniec L (2006) Flavonoid diversity and biosynthesis in seed of *Arabidopsis thaliana*. Planta 224:96–107

Saito K, Hirai MY, Yonekura-Sakakibara K (2008) Decoding genes with coexpression networks and metabolomics - 'majority report by precogs'. Trends Plant Sci 13:36–43

Schmidt A, Karas M, Dulcks T (2003) Effect of different solution flow rates on analyte ion signals in nano-ESI MS, or: when does ESI turn into nano-ESI? J Am Soc Mass Spectrom 14: 492–500

Scholz M, Gatzek S, Sterling A, Fiehn O, Selbig J (2004) Metabolite fingerprinting: detecting biological features by independent component analysis. Bioinformatics 20:2447–2454

Scholz M, Selbig J (2007) Visualization and analysis of molecular data. In: Weckwerth W (ed) Metabolomics: methods and protocols, pp 87–104. Humana Press, Totowa, NJ

Schwab W (2003) Metabolome diversity: too few genes, too many metabolites? Phytochemistry 62:837–649

Shepherd T, Dobson G, Verrall SR, Conner S, Griffiths DW, McNicol JW, Davies HV, Stewart D (2007) Potato metabolomics by GC-MS: what are the limiting factors? Metabolomics 3: 475–488

Smith CA, Want EJ, O'Maille G, Abagyan R, Siuzdak G (2006) XCMS: processing mass spectrometry data for metabolite profiling using nonlinear peak alignment, matching, and identification. Anal Chem 78:779–787

Steeghs M, Bais HP, de Gouw J, Goldan P, Kuster W, Northway M, Fall R, Vivanco JM (2004) Proton-transfer-reaction mass spectrometry as a new tool for real time analysis of root-secreted volatile organic compounds in Arabidopsis. Plant Physiol 135:47–58

Steuer R, Morgenthal K, Weckwerth W, Selbig J (2007) A gentle guide to the analysis of metabolomic data. In: Weckwerth W (ed) Metabolomics: Methods and Protocols, pp 105–126. Humana Press, Totowa, NJ

Stobiecki M, Skirycz A, Kerhoas L, Kachlicki P, Muth D, Einhorn J, Mueller-Roeber B (2006) Profiling of phenolic glycosidic conjugates in leaves of *Arabidopsis thaliana* using LC/MS. Metabolomics 2:197–219

Stoll N, Schmidt E, Thurow K (2006) Isotope pattern evaluation for the reduction of elemental compositions assigned to high-resolution mass spectral data from electrospray ionization fourier transform ion cyclotron resonance mass spectrometry. J Am Soc Mass Spectrom 17:1692–1699

Sturm M, Bertsch A, Gropl C, Hildebrandt A, Hussong R, Lange E, Pfeifer N, Schulz-Trieglaff O, Zerck A, Reinert K, Kohlbacher O (2008) OpenMS - an open-source software framework for mass spectrometry. BMC Bioinformatics 9. doi:10.1186/1471–2105–9–163

Sumner LW, Amberg A, Barrett D, Beale MH, Beger R, Daykin CA, Fan TWM, Fiehn O, Goodacre R, Griffin JL, Hankemeier T, Hardy N, Harnly J, Higashi R, Kopka J, Lane AN, Lindon JC, Marriott P, Nicholls AW, Reily MD, Thaden JJ, Viant MR (2007) Proposed minimum reporting standards for chemical analysis. Metabolomics 3:211–221

Sysi-Aho M, Katajamaa M, Yetukuri L, Oresic M (2007) Normalization method for metabolomics data using optimal selection of multiple internal standards. BMC Bioinformatics 8. doi:10.1186/1471–2105–8–93

t'Kindt R, De Veylder L, Storme M, Deforce D, Van Bocxlaer J (2008) LC-MS metabolic profiling of *Arabidopsis thaliana* plant leaves and cell cultures: optimization of pre-LC-MS procedure parameters. J Chromatogr B Analyt Technol Biomed Life Sci 871:37–43

Tan J, Bednarek P, Liu J, Schneider B, Svatos A, Hahlbrock K (2004) Universally occurring phenylpropanoid and species-specific indolic metabolites in infected and uninfected *Arabidopsis thaliana* roots and leaves. Phytochemistry 65:691–699

Tautenhahn R, Bottcher C, Neumann S (2008) Highly sensitive feature detection for high resolution LC/MS. BMC Bioinformatics 9. doi:10.1186/1471–2105–9–504

Tautenhahn R, Böttcher C, Neumann S (2007) Annotation of LC/ESI-MS mass signals. In: Hochreiter S (ed) Bioinformatics and research and development, pp 371–380. Springer, Heidelberg

Thiocone A, Farmer EE, Wolfender JL (2008) Screening for wound-induced oxylipins in *Arabidopsis thaliana* by differential HPLC-APCI/MS profiling of crude leaf extracts and subsequent characterisation by capillary-scale NMR. Phytochem Anal 19:198–205

Tholl D, Chen F, Petri J, Gershenzon J, Pichersky E (2005) Two sesquiterpene synthases are responsible for the complex mixture of sesquiterpenes emitted from Arabidopsis flowers. Plant J 42:757–771

Tian QG, Rosselot RA, Schwartz SJ (2005) Quantitative determination of intact glucosinolates in broccoli, broccoli sprouts, brussels sprouts, and cauliflower by high-performance liquid chromatography-electrospray ionization-tandem mass spectrometry. Anal Biochem 343:93–99

Tohge T, Fernie AR (2009) Web-based resources for mass-spectrometry-based metabolomics: a user's guide. Phytochemistry 70:450–456

Tohge T, Nishiyama Y, Hirai MY, Yano M, Nakajima J, Awazuhara M, Inoue E, Takahashi H, Goodenowe DB, Kitayama M, Noji M, Yamazaki M, Saito K (2005) Functional genomics by integrated analysis of metabolome and transcriptome of Arabidopsis plants over-expressing an MYB transcription factor. Plant J 42:218–235

Tomer KB, Moseley MA, Deterding LJ, Parker CE (1994) Capillary liquid-chromatography mass-spectrometry. Mass Spectrom Rev 13:431–457

Tsuji J, Jackson EP, Gage DA, Hammerschmidt R, Somerville SC (1992) Phytoalexin accumulation in *Arabidopsis thaliana* during the hypersensitive reaction to *Pseudomonas syringae* pv *syringae*. Plant Physiol 98:1304–1309 van den Berg RA, Hoefsloot HC, Westerhuis JA, Smilde AK, van der Werf MJ (2006) Centering, scaling, and transformations: improving the biological information content of metabolomics data. BMC Genomics 7. doi:10.1186/1471–2164–7–142

Veit M, Pauli GF (1999) Major flavonoids from *Arabidopsis thaliana* leaves. J Nat Prod 62: 1301–1303

von Roepenack-Lahaye E, Degenkolb T, Zerjeski M, Franz M, Roth U, Wessjohann L, Schmidt J, Scheel D, Clemens S (2004) Profiling of Arabidopsis secondary metabolites by capillary liquid chromatography coupled to electrospray ionization quadrupole time-of-flight mass spectrometry. Plant Physiol 134:548–559

Wallis JG, Browse J (2002) Mutants of Arabidopsis reveal many roles for membrane lipids. Prog Lipid Res 41:254–278

Wang X (2002) Phospholipase D in hormonal and stress signaling. Curr Opin Plant Biol 5:408–414

Ward JL, Harris C, Lewis J, Beale MH (2003) Assessment of 1H NMR spectroscopy and multivariate analysis as a technique for metabolite fingerprinting of *Arabidopsis thaliana*. Phytochemistry 62:949–957

Welti R, Wang X (2004) Lipid species profiling: a high-throughput approach to identify lipid compositional changes and determine the function of genes involved in lipid metabolism and signaling. Curr Opin Plant Biol 7:337–344

Werner E, Croixmarie V, Umbdenstock T, Ezan E, Chaminade P, Tabet JC, Junot C (2008a) Mass spectrometry-based metabolomics: accelerating the characterization of discriminating signals by combining statistical correlations and ultrahigh resolution. Anal Chem 80:4918–4932

Werner E, Heilier JF, Ducruix C, Ezan E, Junot C, Tabet JC (2008b) Mass spectrometry for the identification of the discriminating signals from metabolomics: current status and future trends. J Chromatogr B Analyt Technol Biomed Life Sci 871:143–163

Wolfender JL, Waridel P, Ndjoko K, Hobby KR, Major HJ, Hostettmann K (2000) Evaluation of Q-TOF-MS/MS and multiple stage IT-MSn for the dereplication of flavonoids and related compounds in crude plant extracts. Analusis 28:895–906

Yang WC, Adamec J, Regnier FE (2007) Enhancement of the LC/MS analysis of fatty acids through derivatization and stable isotope coding. Anal Chem 79:5150–5157

Yonekura-Sakakibara K, Tohge T, Matsuda F, Nakabayashi R, Takayama H, Niida R, Watanabe-Takahashi A, Inoue E, Saito K (2008) Comprehensive flavonol profiling and transcriptome coexpression analysis leading to decoding gene-metabolite correlations in Arabidopsis. Plant Cell 20:2160–2176

Chapter 18
Transformation Technology in the Brassicaceae

Penny A.C. Sparrow, Cassandra M.P. Goldsack, and Lars Østergaard

Abstract With the accelerating advances in Brassicaceae genetics and genomics, transformation technologies are now routinely being exploited to elucidate gene function as well as contributing to the development of novel enhanced crops. *Agrobacterium*-mediated transformation remains the most broadly used approach for the introduction of transgenes into Brassicaceae. For *Arabidopsis thaliana*, *in planta* transformation is now routinely employed using the relatively low-tech approach of floral dipping. The relative ease of producing independent transgenic lines using this approach has been exploited to create T-DNA insertion mutants or knockout lines for most *Arabidopsis* genes. In *Brassica*, transformation relies mainly on in vitro transformation methods, and yet despite the significant progress made towards enhancing transformation efficiencies, some genotypes remain recalcitrant to transformation. Advances in our understanding of the genetics behind transformation have enabled researchers to identify more readily transformable genotypes for use in routine high-throughput systems. These developments open up exciting new avenues to exploit model *Brassica* genotypes as resources for understanding gene function in complex genomes. Although many other Brassicaceae have served as model species for improving plant regeneration and transformation systems, this chapter focuses on the recent technologies employed for both *Arabidopsis* and *Brassica* transformation.

Keywords Brassica · Brassicaceae · Transformation · Testing gene function

Contents

18.1 Introduction 506
18.2 *Agrobacterium* Transformation Methods 508
    18.2.1 *Agrobacterium tumefaciens* 508
    18.2.2 *Agrobacterium rhizogenes* 510

P.A.C. Sparrow (✉)
John Innes Centre, Norwich Research Park, Norwich NR4 7UH, UK
e-mail: penelope.sparrow@bbsrc.ac.uk

R. Schmidt, I. Bancroft (eds.), *Genetics and Genomics of the Brassicaceae*, Plant Genetics and Genomics: Crops and Models 9, DOI 10.1007/978-1-4419-7118-0_18,

18.3 Direct Uptake Transformation Methods 511
18.4 Chloroplast Transformation 511
18.5 Bacterial Strains and Plasmids 511
18.6 Shoot Regeneration 512
18.6.1 The Genetic Basis of In Vitro Shoot Regeneration 512
18.6.2 Intolerance to In Vitro Conditions 513
18.6.3 Choice of Explant and Tissue Culture Media 513
18.6.4 Shoot Elongation and Rooting In Vitro 514
18.7 Hyperhydricity and Tissue Necrosis: Use of Ethylene Inhibitors 514
18.8 Floral Dipping/Microinjection 515
18.9 Selection of Transgenics 516
18.10 Transformation as a Research Tool 517
18.11 Concluding Remarks 518
References 518

18.1 Introduction

As detailed in other chapters, researchers now have access to a wide range of genomic and genetic resources that are greatly enhancing our knowledge and the speed at which we are able to elucidate gene function (reviewed by Østergaard and Yanofsky 2004, King 2006 and Trick et al. 2007). Furthermore, the close relationship that exists between *Arabidopsis* and *Brassica* can be exploited to extrapolate our findings from the model to the crops. This has been made possible by the completion of the *Arabidopsis* genome sequencing project, the production and screening of *Brassica* BAC libraries to aid identification of orthologous genes, the creation of mapping populations and linkage maps, and the alignment between the genomes of members of the Brassicaceae. This combined with the efficient and developing transformation technologies is set to continue to help in our understanding of gene function in both *Arabidopsis* and *Brassica*.

Transformation, as a research tool, initially focused on the overexpression of gene(s) of interest using genes isolated from the models and then later homologous crop genes, or the use of reporter genes to test specific endogenous promoters (to target specific areas). Most of the common reverse genetics approaches (the ability to knock out or silence genes – see also Chapters 19 and 20) are also based on transformation techniques such as T-DNA insertion (Alonso et al. 2003), RNAi (Horiguchi 2004), artificial miRNA (Schwab et al. 2006), antisense (Ecker and Davies 1986), or the newly developed "target mimicry" technique (for knocking down miRNA function) (Franco-Zorrilla et al. 2007). All of these techniques have been shown to work in *Arabidopsis thaliana* and are also now starting to be applied in *Brassica* species (see Section 18.11). *Brassica* transformation is now commonly used as a powerful research tool to test gene function (Sparrow et al. 2007). Areas of particular interest include biotic and abiotic stress tolerance (Purity et al. 2008), oil synthesis (Mietkiewska et al. 2008), and plant architecture (Østergaard et al. 2006).

The transformation of *A. thaliana* has focused on a small number of key genotypes, mainly Columbia (Col), C24, Landsberg *erecta* (Ler), and Wassilewskija (Ws), although most *Arabidopsis* genotypes can also be transformed by the floral dip method (detailed in Section 18.9) at a rate similar to (within tenfold of) those stated for Columbia (Bent 2006). The routine transformation of *Brassica*, unlike *Arabidopsis*, is still hindered by genotype restrictions, with some genotypes remaining recalcitrant to transformation. Advances in our understanding of the genetics behind transformation have enabled researchers to identify more readily transformable genotypes for use in high-throughput systems for testing gene function (Sparrow et al. 2004a, 2006a). However, the routine high-throughput transformation is still likely to be limited to a number of key genotypes; for example, the commonly reported Spring variety Westar for *Brassica napus* transformation (Moloney et al. 1989, Cardoza and Stewart 2006).

However, there may be occasions when a particular *Brassica* genotype needs to be transformed and where only a few transgenic plants are required. Under these circumstances, high transformation efficiency is less critical. Lessons can be learnt from the range of published papers that have added to our knowledge of tissue culture conditions favoured by different *Brassica* genotypes and species. In this chapter we review the advances made to *Brassica* and *Arabidopsis* transformation methods, particularly of the main *Brassica* species, and discuss how this technology can be further exploited to better understand gene function.

As a commercial crop, the diverse array of different vegetable morphologies and oilseed crops within the *Brassica* genus has been the result of years of hybridization, accompanied by intense selection, between and within the *Brassica* species. The application of conventional breeding has led to the development of many elite cultivars within this genus. With increasing knowledge of the function of genes, and the development of techniques for plant transformation, the potential for further improvement of these species is considerable, be that as a Genetic Modification (GM) product or via conventional breeding once gene function has been elucidated. To date, GM *B. napus* is the only *Brassica* species to gain commercial regulatory approval. GM herbicide-tolerant *B. napus* (canola) was the fourth most planted GM crop in 2007 (James 2007). The other major, so-called first-generation trait introduced into *Brassica* is the *Bacillus thuringiensis* toxin (*Bt*) to alleviate insect attack. Experimental *Bt*-producing *B. napus* lines have been developed (Halfhill et al. 2002, Stewart et al. 1996, 1997, and Ramachandran et al. 1998), and in *Brassica oleracea* the vegetable crops including broccoli (Cao et al. 1999, Metz et al. 1995), cabbage (Jin et al. 2000), and Chinese cabbage (Cho et al. 2001, Xiang et al. 1999) using the *Bt cry1Ac* gene designed to control diamondback moth. Work has also reached field trial stages by Crop and Food Research, New Zealand (www.crop.cri.nz) (unpublished data) and the "Collaboration on Insect Management for Brassicas in Asia and Africa (CIMBAA)". The latter is an organization whose aim is to engineer sustainable insect resistance traits into locally adapted cabbage and cauliflower varieties, free of any technology fees and taken through the regulatory process of countries which would benefit strongly from this material (http://www.cimbaa.org).

However, public perception of GM technology still hinders its global advancement, especially in Europe. In the developed world it is the so-called next generation traits, such as modified oils containing increased Omega-3 fatty acids or broccoli with increased levels of antioxidants, and other such consumer-targeted traits that are likely to gain public acceptance.

18.2 *Agrobacterium* Transformation Methods

Agrobacterium-mediated transformation (either *Agrobacterium tumefaciens* or *Agrobacterium rhizogenes*) still remains the favoured delivery approach for the introduction of transgenes into most dicotyledonous plant species and is fast becoming a common practice for an expanding range of monocots (Smith and Hood 1995, Opabode 2006, Bartlett et al. 2008).

18.2.1 Agrobacterium tumefaciens

Interest in transforming *Arabidopsis* followed the adoption of *A. thaliana* as a model plant species in the 1980s (reviewed by Somerville and Koorneef 2002), with tissue culture-based *A. tumefaciens*-mediated transformation first reported in 1986 by Lloyd et al. (1986) and An et al. (1986). Further improvements to these methods were made by Schmidt and Willmitzer (1988) and Valvekens et al. (1988) using an in vitro root transformation method (which has recently been adapted for *Arabidopsis lyrata* (Fobis-Loisy et al. 2007)). Early *in planta* methods were also developed including direct gene transfer into mesophyll protoplast cells (Damm et al. 1989) and seed transformation (Feldman and Marks 1987); however, these techniques were outcompeted when Bechtold et al. (1993) published *Arabidopsis* transformation using vacuum infiltration of whole plants. These latter approaches moved away from tissue culture-dependent methods, making the process far more applicable to a wide range of non-specialist laboratories. Further improvements to this method demonstrated that vacuum infiltration was not necessary and potted plants could be dipped whole into a suspension of *Agrobacterium* – the so-called floral dip method (Clough and Bent 1998) which has since become the transformation method of choice for *A. thaliana* (discussed later). The floral dip approach has also more recently been successfully reported for *Capsella bursa-pastoris* transformation (Bartholmes et al. 2008).

In *Arabidopsis* and other diploid plant species, it is often critical for further genetic analysis to maintain diploidy in the resulting transgenic plants. However, it was previously shown that aberrant ploidy levels are identified in *Arabidopsis* transformants (Altmann et al. 1994). Depending on the transformation procedure used, 13% (root explant transformation), 33% (cotyledon explant transformation), or 38% (direct gene transfer) of the transformants showed aberrant ploidy levels. Pollen size was found to correlate with ploidy of the plant and can thus be used as a marker for ploidy level by analysis under a light microscope.

The first reports of successful in vitro transformation of *Brassica* emerged in the late 1980s/early 1990s for all six of the major economically important species:

Brassica juncea (Barfield and Pua 1991), *B. napus* (Moloney et al. 1989), *Brassica rapa* (Radke et al. 1992), *B. oleracea* (de Block et al. 1989), *Brassica nigra* (Gupta et al. 1993), and *Brassica carinata* (Narasimhulu et al. 1992). Many publications followed, reporting improvements to tissue culture conditions and the use of reporter genes to determine transformation efficiencies (discussed later); yet, the major limiting factor still remained poor tissue culture response and insensitivity to *Agrobacterium* in many recalcitrant genotypes. Progress was later made to identify and develop protocols that worked for a wider range of genotypes and *Brassica* species (Bhalla and Smith 1998, Sparrow et al. 2004a, 2006a, Bhalla and Singh 2008), as well as to identify genotypes for use in routine transformation studies to test gene function (Cardoza and Stewart 2006, Gasic and Korban 2006, Sparrow et al. 2006b) discussed in detail later.

Knowledge of the genetic variation for in vivo *A. tumefaciens* susceptibility (causing crown gall disease) had previously been observed in a wide range of plant species including *Prunus* (Bliss et al. 1999), soybean (Bailey et al. 1994, Mauro et al. 1995), and grape (*Vitis* sp.) (Szegedi and Kozma 1984). In *A. thaliana* (Nam et al. 1997) susceptibility to *Agrobacterium* was shown to be a dominant and heritable trait, and in *B. oleracea*, QTL associated with susceptibility to wild-type *A. rhizogenes* (hairy root formation) and *A. tumefaciens* (crown galls formation) have also been identified (Cogan et al. 2004, Sparrow et al. 2004b). Screening substitution lines associated with these mapping populations confirmed the significance of these QTL and demonstrated that susceptibility to *Agrobacterium* was indeed a heritable trait and could be introduced into recalcitrant lines. Crossing genotypes with low *Agrobacterium* susceptibility to genotypes with high susceptibility resulted in an intermediate response in the hybrid plants (Sparrow et al. 2004b).

The simplest approaches for *Brassica* are those described by Moloney et al. (1989) for *B. napus* cv. "Westar" (efficiency 55%), Bhalla and Singh (2008) for a range of *B. napus* and *B. oleracea* Australian cultivars (efficiency between 7 and 68%), Babic et al. (1998) for *B. carinata* breeding line C90-1163 (efficiency 30–50%), and Sparrow et al. (2004a, 2006b) for *B. oleracea* genotype AGDH 1012 (efficiency 15–25%; see also www.bract.org for a tutorial guide to this method). All these methods use cotyledonary petioles dipped into a suspension of *A. tumefaciens* and co-cultivated for 72 h on selection-free media before transferring to a basic MS basal medium supplemented with between 2 and 4 mg/l 6-benzyl-aminopurine (BAP). Explants are then sub-cultured to fresh media after 2–3 weeks. Emerging shoots are isolated and rooted on a hormone-free medium (sometimes the addition of an auxin is required).

Hypocotyl methods have also been applied to *B. napus* at a reported efficiency of 25% (for an updated method see Cardoza and Stewart 2006) and *B. juncea* with an efficiency of 16%[1] (Gasic and Korban 2006). These approaches use either

[1]N.B the efficiencies reported in this section are based on either the percentage of explants giving rise to viable transgenic shoots or the percentage of explants giving rise to transgenic rooted plants. All methods successfully generated fertile transgenic plants.

pre-cultured hypocotyl segments (following a short callus induction period) or newly isolated hypocotyl segments, immersed in a suspension of *A. tumefaciens* for 30–60 min. Explants are then washed and transferred to co-cultivation medium for 48 h before transferring to selection medium. Both methods employ different media for callus induction, shoot induction, shoot elongation, and rooting stages (see also Section 18.7.3).

B. rapa remains the most recalcitrant of the cultivated *Brassica* species to transform. However, a limited number of successes have been reported; Radke et al. (1992) using hypocotyl sections, Zhang et al. (2000) and Wahlroos et al. (2003) using cotyledonary petioles, Yang et al. (unpublished data) using cotyledonary leaf sections (as described in Yang et al. 2004), Kuvshinov et al. (1999) using internodes from glasshouse-grown *B. rapa*, and more recently *A. tumefaciens* infiltration of whole plants by Qing et al. (2000) and microinjection of *A. tumefaciens* directly into flower buds by Yan et al. (2003, 2004) (see also Section 18.8)

With the sequencing of the *B. rapa* genome currently underway (www.brassica.info.org) it is likely that efforts and interest in *B. rapa* transformation will rise over the next few years. This will provide excellent resources to optimize the information arising from the Multinational *B. rapa* Genome Sequencing Project for *Brassica* functional genomics as it becomes available.

18.2.2 Agrobacterium rhizogenes

A. rhizogenes (a soil bacterium responsible for the development of hairy root disease of dicotyledonous plants) has, in its modified form, been used to transform over 79 plant species (reviewed in Christey 2001). It has been successfully used to transform *Brassica*, where the focus has been on *B. oleracea* and *B. napus* (Christey and Sinclair 1992, David and Tempe 1988, Puddephat et al. 2001). Overall, transformation rates were low, but in some cases, transgenic *Brassica* plants were more efficiently obtained via *A. rhizogenes*-mediated transformation than *A. tumefaciens* (Christey et al. 1997). The main disadvantage of using *A. rhizogenes*, however, was that *rol* genes were often transferred and expressed in plants regenerated from hairy roots and often exhibiting the associated altered phenotype such as wrinkled leaves, shortened internodes, reduced fertility, and plagiotropic roots. However, these traits often segregated in the next generation facilitating the recovery of normal transgenic plants (Christey et al. 1999, Puddephat et al. 2001).

In *Arabidopsis, A. rhizogenes* root transformation of wild-type shoots (composite plants) is now being used for "in root" gene testing of transgenes and plant–microbe interaction studies (Taylor et al. 2006, Veena and Taylor 2007). Due to the small size of *Arabidopsis* the procedure is carried out in vitro – but for larger plants non-tissue culture conditions can be used. Composite plant production in *Brassica* has been reported by Collier et al. (2005) and Taylor et al. (2006) in which transgene expression of reporter genes in composite plant roots has been demonstrated.

18.3 Direct Uptake Transformation Methods

Direct gene transfer of mesophyll protoplasts using PEG-mediated transfer was demonstrated in *Arabidopsis* by Damm et al. (1989) and in *Brassica* using either electroporation or PEG-mediated transfer for cauliflower protoplasts by Mukhopadhyay et al. (1991) (using hypocotyls protoplasts) and Eimert and Siegemund (1992) (for mesophyll protoplasts). Radchuk et al. (2002) studied a range of factors effecting PEG-mediated transformation in *Brassica* using both kanamycin and hygromycin selection, and Nugent et al. (2006) demonstrated successful transformation in cauliflower mesophyll protoplasts using a *gus* reporter gene and hygromycin selection (see also Section 18.5). Regeneration from protoplasts remains the main limitation of this approach. Another direct approach is biolistics or microparticle bombardment where DNA-coated beads are fired at high speed into plant cells. These approaches again offer an alternative transformation method for genotypes that are not susceptible to *Agrobacterium* infection and are a useful approach for transient expression studies (Puddephat et al. 1999).

18.4 Chloroplast Transformation

Plastid transformation offers a number of potential advantages over nuclear transformation (Maliga 2004). The high number of plastids per plant cell (approximately 10^5 copies per plant cell) means higher expression levels can potentially be achieved than with nuclear transformation. This is particularly desirable for product-based transformation, where high protein yields are desirable (Dhinrga et al. 2004). Inheritance of the introduced transgene(s) will also be maternal and therefore offers containment of the transgene due to lack of gene flow through pollen (Danieli 2002); although observations of gene transfer from chloroplast to nuclear genomes have been reported (Stegemann et al. 2003) frequencies are extremely low. Plastid transformants have been produced in a number of species including *A. thaliana* (Sikdar et al. 1998); and in *Brassica* chloroplast transformation by particle bombardment has been reported in both *B. oleracea* (Hou et al. 2003) and *B. napus* (Liu et al. 2007) and also in *B. napus* using a PEG-mediated approach (Nugent et al. 2006).

18.5 Bacterial Strains and Plasmids

A number of different strains of *A. tumefaciens* have been successfully used to transform Brassicaceae. The most frequently used for *Arabidopsis* is GV3101 (Koncz and Schell 1986), while other disarmed strains of the C58 background can often be substituted with similar results. LBA4404 (Hoekema et al. 1983), an octopine strain, is also commonly used but generally works less well than GV3101; this lower efficiency is sometimes desirable for obtaining a greater proportion of transgenics with low copy number (Bent 2006). LBA4404 is also routinely used in *Brassica*, as are the nopaline strains C58 (Sciaky et al. 1978) and derivatives AGL1 (Lazo et al. 1991) and EHA101 and EHA105 (Hood et al. 1986) which have all been used successfully and routinely. The use of GV3101, the preferred strain for *Arabidopsis* transformation, has also been reported for *Brassica* (De Block et al. 1989, Mehra et al. 2000), although its use is far from routine in the crops. It is likely that over the next few years the ability to use plasmids and *Agrobacterium* strains routinely used for *Arabidopsis* transformation will be highly desirable as researchers move findings from the model species into crops such as *Brassica*.

The type of plasmid used is thought to be less critical than bacterial strain, although choice of promoters and selectable markers is important. Commonly used plasmids for *Brassica* have been the modified pBIN19 (Bevan 1984), the SLJ vectors (Jones et al. 1992) SLJ1714 and SLJ1711, the pCAMBIA vectors in particular pCAM2200 (www.cambia.org), and pGreen (Hellens et al. 2000, www.pgreen.ac.uk). The above examples are based on the 35S promoter driving *nptII*, the favoured selectable marker gene for *Brassica*, but other selectable markers are available (see Section 18.9 Selection of Transgenics).

18.6 Shoot Regeneration

With the exception of the floral dip method (discussed below) the above transformation methods all rely on having a robust regeneration system in place for transformation success (recovery of transgenic plants). Many factors affect the successful regeneration of shoots in vitro, namely genotype, media conditions, and explant type and age. In this section we focus on factors effecting in vitro regeneration systems for *A. tumefaciens*-mediated transformation; however, a number of the points discussed also apply to other transformation methods.

18.6.1 The Genetic Basis of In Vitro Shoot Regeneration

Extensive screening of genotypes and tissue culture conditions have improved the frequency of shoot regeneration for most *Brassica* species. Despite these advances, some genotypes remain highly recalcitrant to in vitro regeneration. The genotype-dependent nature of in vitro shoot regeneration, both within and among the *Brassica* species, was first reported by Murata and Orton (1987) and later by Narasimhulu and Chopra (1988a, 1988b) who suggested shoot regeneration to be a heritable trait. Ono and Takahata (2000) and Sparrow et al. (2004c) looked at the genetic control of shoot regeneration from cotyledonary petioles using diallel crosses in *B. napus* and *B. oleracea*, respectively. They showed that shoot regeneration was under strong genetic control and associated with additive and dominant gene effects, with additive gene effects accounting for the majority of the variation. The ability to introduce or increase the in vitro shoot regeneration potential of a genotype, by conventional breeding, may help overcome restrictions to routine transformation programmes where efficient shoot regeneration is a critical prerequisite. QTL associated with shoot regeneration efficiency in *Arabidopsis* have also been identified (Schiantarelli et al. 2001, Lall et al. 2004), and the genetic regulation of gene expression during shoot development is currently being studied intensively.

18.6.2 Intolerance to In Vitro Conditions

In theory all cells that contain a nucleus are totipotent and retain the genetic information required to regenerate a whole plant. However, some genotypes are simply unable to tolerate in vitro conditions and thus regenerate shoots. Intolerance to in vitro conditions has also been observed in *B. oleracea* and *B. napus* (Sparrow et al. 2004a, c, 2006b) and shown to be a dominant heritable trait. In these papers, cotyledonary petioles from a range of genotypes were screened for regeneration potential in the absence of *Agrobacterium*. A number of genotypes exhibited extreme tissue culture blackening to the petiole base. Genotypes that demonstrated tissue culture blackening failed to regenerate shoots or regenerated a low number of small shoots direct from the petiole base (without a callus phase). These shoots were often non-viable as they failed to develop further and could not be isolated. These genotypes made poor candidates for transformation success. By contrast, genotypes that regenerated multiple shoots, a response associated with a small callus phase, and no blackening were considerably more favourable to *Agrobacterium*-mediated transformation than genotypes that regenerated a small number of shoots direct from the petiole base.

18.6.3 Choice of Explant and Tissue Culture Media

In vitro shoot regeneration of *Brassica* has been achieved from an array of different explant types: cotyledonary petioles (Moloney et al. 1989, Ono et al. 1994), cotyledonary sections (Yang et al. 2004), hypocotyls (Yang et al. 1991), peduncles (Christey and Earle 1991), leaf sections (Alaska-Kennedy et al. 2005), anthers, microspores (Keller and Armstrong 1977, Litcher 1982), thin cell layers of epidermal and subepidermal cells (Klimazewska and Keller, 1985), roots (Xu et al. 1982), and protoplasts (Glimelius 1984, Barsby et al. 1986). However, it is the seedling explants (cotyledons and hypocotyls) that remain favourites for transformation. Seeds can be surface sterilized and germinated in vitro to achieve sterile explant tissue. The age of the explant is also critical, with many researchers finding 3–4-day-old seedlings give optimal results. While it is often the age of the explant that is referred to, it is actually the size of the explant that is more critical. A 3-day-old seedling in one culture room under one light regime may be different (bigger or smaller) than the same seedling grown under different growth room conditions. For hypocotyl sections, older explants (8–10 days) have also been used (Cardoza and Stewart (2006) for *B. napus*) allowing a larger number of explants to be obtained per seedling. Gasic and Korban (2006) on the other hand found hypocotyls from

3–4-day-old seedlings gave optimal results in *B. juncea*. What appears critical in both these systems, however, is the length of the hypocotyl section. Small sections were optimal, i.e. 5–10-mm sections (where smaller explants did better). Long hypocotyl sections had the tendency to curl and therefore loose contact with the culture media.

Not only is in vitro shoot regeneration genotype dependent but regeneration success is also affected by the choice of transformation protocol, hormone, and other media additive levels used. The ability to regenerate from one tissue type does not guarantee regeneration success from another explant source, and indeed it is always advisable to carry out a regeneration study with the genotype and explant of choice before selecting the transformation approach. Hypocotyl and leaf sections often require a callus phase prior to shoot regeneration (Yang et al. 2004, Gasic and Korban 2006) using cotyledonary leaf sections. This is normally achieved using a high cytokinin (usually BAP) to low auxin (e.g. naphthalene acetic acid (NAA)) ratio. Some transformation protocols also exploit a short callus induction phase prior to transformation, a so-called preconditioning stage (Cardoza and Stewart 2006).

18.6.4 Shoot Elongation and Rooting In Vitro

Once shoot regeneration has been achieved from the desired genotype, another stumbling block can be the isolation of viable shoots. This is partially true for *B. rapa*, one of the more recalcitrant *Brassica* species (Sparrow and Goldsack, unpublished data). Small shoots are often prone to undesirable morphology, hyperhydricity, and failure to elongate and root in vitro. A shoot elongation step, where cytokinin levels are lowered but not removed, can often help shoots to elongate (Cardoza and Stewart 2006, Bhalla and Singh 2008). Often hyperhydricity is overcome once shoots have elongated, but occasionally other supplements need to be explored (see Section 18.7). Rooting is often achieved by simply removing the cytokinin (Sparrow et al. 2006b); however in some cases it may also require the addition of an auxin (such as NAA or indole-3-butyric acid (IBA)) Cardoza and Stewart (2006), Verma and Singh (2007), Bhalla and Singh (2008).

18.7 Hyperhydricity and Tissue Necrosis: Use of Ethylene Inhibitors

Hyperhydricity (formerly termed "vitrification") and tissue necrosis can present a serious problem for plant tissue culturists. However, a better understanding of the underlying mechanism of hyperhydricity and its control in vitro can significantly contribute to improved tissue culture success (Ziv 1991). In *B. rapa* and *B. juncea*, Chi and Pua (1989) and Chi et al. (1990) demonstrated that higher regeneration frequencies could be achieved if explants were given the appropriate media and environmental conditions, thus removing/reducing hyperhydricity.

Factors found to effect hyperhydricity and tissue necrosis are accumulation of ethylene and high humidity in culture vessels (de Block et al. 1989), excessively rich media (Ziv 1991), *Agrobacterium* overgrowth/sensitivity (Jin et al. 2000), and high doses of exogenous cytokinin and/or auxin (Ketaeva et al. 1991, Kamal et al. 2007). In *B. napus* (Cardoza and Stewart 2003) and in *B. rapa* (Goldsack and Sparrow, unpublished data) it was observed that increasing the percentage of gelling agent (from 0.8% phytagar to 1.2%) in the shoot elongation medium reduced the relative humidity of the culture vessel and reduced hyperhydricity.

Ethylene is another key factor in optimizing tissue culture conditions for some *Brassica* species (Cardoza and Stewart 2004). Silver nitrate has been used to reduce hyperhydricity in a range of *Brassica* species such as *B. rapa* (Kuvshinov et al. 1999, Xiang et al. 2000, Yang et al. 2004) and *B. napus* (Tang et al. 2003). In particular, *B. rapa*, one of the most recalcitrant *Brassica* species to culture in vitro, responds positively to the addition of silver nitrate (Palmer 1992, Radke et al. 1992). It has also been reported that excluding silver nitrate from tissue culture media can drastically reduce regeneration frequency in some genotypes of *B. napus* (Phogat et al. 2000). Other ethylene inhibitors, which can be used in combination or as an alternative, include silver thiosulphate (Eapen and George 1996, 1997) and aminoethoxyvinylglycine used by Chi et al. (1990) for *B. rapa* and *B. juncea*, Pua and Chi (1993) for *B. juncea*, and Burnett et al. (1994) for *B. rapa*.

18.8 Floral Dipping/Microinjection

The ability to bypass the sometimes complex tissue culture phase associated with the above methods could, potentially, overcome some of the barriers to transformation success observed in *Brassica. In planta* transformation of *Arabidopsis* is now common practice and involves immersing intact inflorescences in suspensions of *A. tumefaciens* (Bechtold et al. 1993, Clough and Bent 1998, Kojima et al. 2006). The *Agrobacterium* targets the ovules for the transformation event (Ye et al. 1999), and therefore species in which the ovary remains open for an extended developmental period may be good candidates for successful *in planta* transformation (Desfeux et al. 2000). Transformation by infiltration of adult *Brassica* plants with A*grobacterium* has been reported for *B. rapa* (Pakchoi) by Liu et al. (1998) and Xu et al. (1998) and Qing et al. (2000). In the latter study two transgenic plants were obtained from 20,000 seeds arising from the dipping of 30–50 plants. While this efficiency is low, it does demonstrate the potential to apply the method to *Brassica*. In *B. napus*, Wang et al. (2003) reported a success rate of 0.18% (approximately 11 putative transgenics arising from four dipped plants) using a double infiltration approach.

Recently, Zhandong et al. (2007) have reported a transformation rate of 2.35% for Chinese cabbage (*B. rapa*) using the method of Liu et al. (1998). In this study the gene of interest was a Turnip mosaic virus resistance gene, and infection with TuMV was used to identify 43 transgenic plants out of 1,831 seeds. A simplified floral dip method has also been reported for *B. carinata* and *B. napus* by Verma et al. (2008).

At present reports of *in planta* transformation of *Brassica* are limited, and further studies on the parameters of the system will be necessary before it can be exploited as a routine transformation method. It should be noted that the efficiencies reported to date for *Brassica* are not dramatically different to early reports in the model plant *Arabidopsis*. However, floral dipping may be more amenable to *Arabidopsis* due to its size and thus ease of handling, faster life cycle, and smaller seed, which lend itself better to subsequent downstream screening for positive transgenics.

As an alternative to floral dipping, microinjection of the bacteria directly into the flower bud (effectively flooding the bud cavity) before fertilization has been investigated. This technique has been successfully applied in *B. rapa* (Chinese cabbage) by Yan et al. (2003, 2004). Efficiencies of up to 0.56% were obtained when the floral stage at which the microinjection was carried out, as well as the concentration of *Agrobacterium*, sucrose, and surfactant used was optimized. This efficiency was based on injecting approximately 50 flower buds with a size of 2–3 mm, from which they obtained 500–800 seeds yielding on average three to four transgenics. Therefore, microinjection also provides an alternative method for *Brassica* transformation, especially where facilities for tissue culture-based techniques are unavailable.

18.9 Selection of Transgenics

Selectable marker genes can be classed as either positive- or negative-based selection. Positive selectable marker genes are defined as those that promote the growth of transformed tissue, where the majority of work has centered on the use of modified sugars such as mannose (Reed et al. 2001), disaccharide, cellobiouronic acid (CbA) (developed by the CAMBIA group, www.cambia.org), or glucose using the trehalose-6-phosphate synthase (TPS) system (Leyman et al. 2006); these types of selectable markers are likely to face less regulatory hurdles if the end goal is commercial transgenic plants. Negative selection systems using toxic agents, such as antibiotics, herbicides, or drugs, were the first to be developed and exploited. Kanamycin is the most commonly used negative selectable marker for most plant transformation work (Miki and McHugh 2004). The level of antibiotic used will depend on both the genotype and the transformation method used and has been reported for kanamycin in the range of 5 mg/l to over 200 mg/l. Moloney et al. (1989) used 15 mg/l kanamycin for the selection of transgenic shoots arising from cotyledonary explants of the *B. napus* cultivar Westar, while Cordoza and Stewart (2006) used 200 mg/l kanamycin when Westar hypocotyl explants were used. Hygromycin has also been successfully used in *Brassica*. Cao et al. (1999) and Lee et al. (2004) found hygromycin to be a more effective selection agent than kanamycin in the *Brassica* genotypes they tested, with very few escapes coming through the system. However, as hygromycin is a much harsher selective agent than kanamycin it may not be suitable for all genotypes. From a regulatory point of view, hygromcin as a selectable marker is unlikely to gain approval for field release, unlike Kanamycin which has now achieved GRAS status (generally regarded as safe)

(EU directive 2001/18).[2] In such cases the ability to produce marker-free transgenics is desirable, i.e. where the selectable marker is later removed (Komari et al. 1996). Basta or glufosinate (herbicide resistance) has also been successfully used in *Brassica* (de Block et al. 1989), although less desirable for cotyledonary-based transformation methods due to Basta mode of action in targeting photosynthetic material. Basta is frequently used as a selectable marker for floral dipping or microinjection transformation methods, where large numbers of seeds can be soil sown and seedlings sprayed with Basta to select for positive transgenics.

18.10 Transformation as a Research Tool

The *Arabidopsis/Brassica* relationship

The relative ease of transformation and access to complete genomic sequences in *Arabidopsis* have led to an array of forward and reverse genetic approaches that provide great insights into gene function. In the previous chapters, it has been discussed that the genomes of diploid *Brassica*s have undergone a triplication event of an ancestral genome similar to that of *Arabidopsis*. It is therefore often the case that three paralogous genes in a diploid *Brassica* genome exist for each gene in *Arabidopsis* (and thus up to six in the amphidiploid *Brassicas* where genomes are combined). This level of gene redundancy may complicate the elucidation of gene function through characterization of classical loss of function mutants. However, verification of conservation of gene function between *Arabidopsis* and *Brassica* can still be elucidated via transgenic approaches, such as complementation of an *Arabidopsis* mutant phenotype with a *Brassica* orthologue or mimicking overexpression phenotypes in both *Arabidopsis* and *Brassica*. Due to the multiple copies of each "*Arabidopsis gene*" that may be present in *Brassica*, the phenotypes observed in *Arabidopsis* transgenics may not always be observed in *Brassica* species. It is thought that RNAi approaches to knock down paralogues with one construct may help to better understand gene function (see also Chapters 19 and 20 on reverse genetics). One such example of RNAi transformation in *Brassica* is that of Byzova et al. (2004) who used an RNAi approach to modify petals into sepaloid organs in *Arabidopsis* and oilseed rape. Silencing of the BPI gene family resulted in the transgenic plants producing male fertile flowers in which the petals were converted into sepals (in *Arabidopsis*) or into sepaloid petals (in *B. napus*). These novel flower phenotypes were shown to be both stable and heritable in both species. Another example is the work of Eason et al. (2005) who demonstrated the downregulated BoCP5, a cysteine protease upregulated during harvest-induced senescence, in broccoli. Post-harvest floret senescence (yellowing) was delayed in the transgenic lines produced. In addition the florets were found to contain significantly greater levels of chlorophyll during post-harvest storage at 20°C when compared to wild-type plants. In 2008, Yu et al. also successfully demonstrated the use of RNAi to enhance

[2]This was the case as of 2008 within the EU, but still remains a contentious issue.

the carotenoid content of *Brassica* napus seeds by downregulating lycopene epsilon cyclase.

18.11 Concluding Remarks

A range of transformation techniques have been used to introduce a wealth of agronomically useful traits into *Brassica* such as insect/disease/bacterial/fungal/viral and herbicide resistance, those involved in flowering control, post-harvest attributes, altered stress tolerance, and altered health benefits such as altering anthocynanin/sulphur/vitamin or amino acid levels (reviewed by Christey and Braun 2007), and we can expect to see this continue at an even higher rate in the years to come. The knowledge that has now been obtained in *Arabidopsis* and other plant species creates exciting opportunities for testing out, in crop plants such as *Brassica*, whether this knowledge can usefully be exploited for improving yield.

There have already been reports to suggest that transferring knowledge and technology from *Arabidopsis* to *Brassica* will be feasible in many cases (Chandler et al. 2005, Østergaard et al. 2006, Lee et al. 2007). For example, pod shattering is a major problem for oilseed rape farmers with average annual losses of 11–25% experienced due to unsynchronized seed dispersal (Price et al. 1996). The extensive knowledge on how fruit development in *Arabidopsis* is regulated showed that it is possible to produce pod shatter-resistant *Brassica* fruits by overexpressing the MADS box gene *FRUITFULL* from *Arabidopsis* in *B. juncea* (Østergaard et al. 2006, 2007). This manipulation resulted in loss of the highly specified valve margin tissue in fruits and consequently to pod shatter resistance, as was also observed in *Arabidopsis* (Ferrándiz et al. 2000).

Pod shattering is just one example of an important trait that can be manipulated based on previous knowledge from *Arabidopsis*. Since oilseed rape is a relatively young crop, in comparison to wheat, barley, rice, and maize, there is plenty of room for improvement of other traits. These include flowering time, branching, canopy architecture, fatty acid composition, overall seed oil production, and disease resistance to name just a few. Transformation of *Brassica* is likely to play a prominent role in obtaining the goals for crop improvement, and it is therefore important to keep optimizing and refining the current protocols as well as developing new approaches.

References

Alaska-Kennedy Y, Yoshida H, Takahata Y (2005) Efficient plant regeneration from leaves of rapeseed (*Brassica napus* L.): the influence of $AgNO_3$ and genotype. Plant Cell Rep 24:649–654

Alonso JM, Stepanova AN, Leisse TJ (2003) Genomewide insertional mutagenesis of *Arabidopsis thaliana*. Science 301:653–657

Altmann T, Damm B, Frommer WB, Martin T, Morris PC, Schweizer D, Willmitzer L, Schmidt R (1994) Easy determination of ploidy level in *Arabidopsis thaliana* plants by means of pollen size measurement. Plant Cell Rep 13: 652–656

An, G, Watsen BD, Chiang CC (1986) Transformation of tobacco, tomato, potato, and *Arabidopsis thaliana* using a binary Ti vector system. Plant Physiol 81:301–305

Babic V, Datla RS, Scoles GJ, Keller WA (1998) Development of an efficient *Agrobacterium*-mediated transformation system for *Brassica carinata*. Plant Cell Rep 17:183–188

Bailey MA, Boerma HR, Parrott WA (1994) Inheritance of *Agrobacterium tumefaciens*-induced tumorigenesis of soybean. Crop Sci 34:514–519

Barfield DG, Pua EC (1991) Gene transfer in plants of *Brassica juncea* using *Agrobacterium tumefaciens*-mediated transformation. Plant Cell Rep 10:308–314

Barsby TL, Yarrow SA, Shepard JF (1986) A rapid and efficient alternative procedure for the regeneration of plants from hypocotyl protoplasts of *Brassica napus*. Plant Cell Rep 5:101–103

Bartholmes C, Nutt P, Theiβen G (2008) Germline transformation of Shepherd's purse (Capsella bursa-pastoris) by the 'floral dip' method as a tool for evolutionary and developmental biology. Gene 409:11–19

Bartlett JG, Alves SC, Smedley M, Snape JW, Harwood WA (2008) High throughput agrobacterium-mediated barley transformation. Plant Methods 4:22

Bechtold N, Ellis J, Pelletier G (1993) *In planta Agrobacterium* mediated gene transfer by infiltration of adult *Arabidopsis thaliana* plants. Comptes Rendus de l'Académie des Sciences, Sciences de la Vie 316:1194–1199

Bent A (2006) *Arabidopsis thaliana* floral dip transformation method. In: Wang K (ed) *Agrobacterium* Protocols, 2nd edn. Humana press, Totowa, NJ

Bevan M (1984) Binary *Agrobacterium* vectors for plant transformation. Nucleic Acids Res 12:8711–8721

Bhalla PL, Singh M (2008) Agrobacterium-mediated transformation of Brassica napus and Brassica oleracea. Nat Protoc 2:181–189

Bhalla PL, Smith N (1998) *Agrobacterium*-mediated transformation of Australian cultivars of cauliflowers, *Brassica oleracea* var botrytis. Mol Breed 4:531–541

Bliss FA, Almehdi AA, Dandekar AM, Schuerman PL, Bellaloui N (1999) Crown gall resistance in accessions of 20 *Prunus* species. HortScience 34:206–209

Burnett L, Arnoldo M, Yarrow S, Huang B (1994) Enhancement of shoot regeneration from cotyledon explants of *Brassica rapa* ssp. oleifera through pretreatment with auxin and cytokinin and use of ethylene inhibitors. Plant Cell Tissue Organ Cult 37:253–256

Byzova M, Verduyn C, de Brouwer D, de Block M (2004) Transforming petals into sepaloid organs in *Arabidopsis* and oilseed rape: implementation of the hairpin RNA mediated gene silencing technology in an organ-specific manner. Planta 218:379–387

Cao J, Tang JD, Strizhov N, Shelton AM, Earle ED (1999) Transgenic broccoli with high levels of *Bacillus thuringiensis* Cry1C protein control diamondback moth larvae resistant to Cry1A or Cry1C. Mol Breed 5:131–141

Cardoza V, Stewart N (2003) Increased *Agrobacterium*-mediated transformation and rooting efficiencies in canola (*Brassica napus* L.) from hypocotyl segment explants. Plant Cell Rep 21:599–604

Cardoza V, Stewart N (2004) Invited review: *Brassica* biotechnology: progress in cellular and molecular biology. In Vitro Cell Dev Biol Plant 40:542–551

Cardoza V, Stewart N (2006) Canola (*Brassica napus* L.). In: Wang K (ed) Agrobacterium protocols, 2nd edn. Methods in molecular biology 343, vol 1. Humana Press, Totowa, NJ

Chandler J, Corbesier L, Spielmann P, Dettendorfer J, Stahl D, Apel K, Melzer S (2005) Modulating flowering time and preventing pod shatter in oilseed rape. Mol Breed 15:87–94

Chi GL, Barfield DG, Sim GE, Pua EC (1990) Effect of $AgNO_3$ and aminovinylglycine on in vitro shoot and root organogenesis from seedling explants of recalcitrant *Brassica* genotypes. Plant Cell Rep 9:195–198

Chi GL, Pua EC (1989) Ethylene inhibitors enhanced de novo shoot regeneration from cotyledons of Brassica campestris ssp. chinensis (Chinese cabbage) in vitro. Plant Sci 64:243–250

Cho H, Cao J, Ren J, Earle E (2001) Control of lepidopteran insect pests in transgenic Chinese cabbage (*Brassica rapa* ssp. *Pekinensis*) transformed with a synthetic bacillus thuringiensis cry1C gene. Plant Cell Rep 20:1–7

Christey MC (2001) Use of RI-mediated transformation for production of transgenic plants. In Vitro Cell Dev Biol Plant 37:687–700
Christey MC, Braun RH (2007) Vegetable *Brassicas*. In: Kole C, Hall TC (eds) A compendium of transgenic crop plants. Vegetable crops, vol 7. Wiley-Blackwell
Christey MC, Braun RH, Reader JK (1999) Field performance of transgenic vegetable *Brassicas* (*B. oleracea* and *B. rapa*) transformed with *Agrobacterium* rhizogenes. Sabrao. J Breed Genet 31:93–108
Christey MC, Earle ED (1991) Regeneration of *Brassica oleracea* from peduncle explants. HortScience 26:1069–1072
Christey MC, Sinclair BK (1992) Regeneration of transgenic kale (*Brassica oleracea* var. *acephala*), rape (*B. napus*) and turnip (*B. campestris var. rapifera*) plants via *Agrobacterium rhizogenes* mediated transformation. Plant Sci 82:161–192
Christey MC, Sinclair BK, Braun RH, Wyke L (1997) Regeneration of transgenic vegetable *brassicas* (*Brassica oleracea* and *B. campestris*) via Ri-mediated transformation. Plant Cell Rep 16:587–593
Clough SJ, Bent AF (1998) Floral dip: a simplified method for *Agrobacterium* mediated transformation of *Arabidopsis thaliana*. Plant J 16:735–743
Cogan NOI, Newbury HJ, Oldacres AM, Lynn JR, Kearsey MJ, King GJ, Phuddephat IJ (2004) Identification and characterization of QTL controlling *Agrobacterium* mediated transient and stable transformation of *Brassica oleracea*. Plant Biotechnol J 2:59–69
Collier R, Fuchs B, Walter N, Lutke WK, Taylor CG (2005) *Ex vitro* composite plants: an inexpensive, rapid method for root biology. Plant J 43:449–457
Damm B, Scmidt R, Willmitz L (1989) Efficient transformation of *Arabidopsis thaliana* using direct gene transfer to protoplasts. Mol Gen Genet 217:6–12
Daniell H (2002) Molecular strategies for gene containment in transgenic crops. Nat Biotechnol 20:581–586
David C, Tempe J (1988) Genetic transformation of cauliflower (*Brassica oleracea* L. var. *Botrytis*) by *Agrobacterium rhizogenes*. Plant Cell Rep 7:88–91
De Block M, Tenning P, de Brouwer D (1989) Transformation of *Brassica napus* and *Brassica oleracea* using *Agrobacterium tumefaciens* and the expression of the *bar* and *neo* genes in the transgenic plants. Plant Physiol 91:694–701
Desfeux C, Clough SJ, Bent AF (2000) Female reproductive tissues are the primary target of *Agrobacterium*-mediated transformation by the *Arabidopsis* floral-dip method. Plant Physiol 123:895–904
Dhinrga A, Portis AR, Daniell H (2004) Enhanced translation of a chloroplast expressed RbcS gene restores small subunit levels and photosynthesis in nuclear RbcS antisense plants. Proc Natl Acad Sci USA 101(6315):6320
Eapen S, George L (1996) Enhancement in shoot regeneration from leaf discs of *Brassica juncea* L. Czern and Coss by silver nitrate and silver thiosulfate. Physiol Mol Biol Plants 2:83–86
Eapen S, George L (1997) Plant regeneration from peduncle segments of oil seed Brassica species: influence of silver nitrate and silver thiosulfate. Plant Cell Tissue Organ Cult 51:228–232
Eason JR, Ryan DJ, Watson LM, Hedderley D, Christey MC, Braun RH, Coupe SA (2005) Suppression of the cysteine protease, aleurain, delays floret and leaf senescence in *Brassica oleracea*. Plant Mol Biol 57:645–657
Ecker JR, Davis RW (1986) Inhibition of gene-expression in plant-cells by expression of antisense RNA. Proc Natl Acad Sci USA 83:5372–5376
Eimert K, Siegemund F (1992) Transformation of cauliflower (*Brassica oleracea L. var. botrytis*) – an experimental survey. Plant Mol Biol 19:485–490
Feldman KA, Marks MD (1987) *Agrobacterium* mediated transformation of germinating seeds of *Arabidopsis thaliana:* a non-tissue culture approach. Mol Gen Genet 208:1–9
Ferrándiz C, Liljegren SJ, Yanofky MF (2000) Negative regulation of the *SHATTERPROOF* genes by *FRUITFULL* during *Arabidopsis* fruit development. Science 289:436–438
Fobis-Loisy I, Chambrier P, Gaude T (2007) Genetic transformation of *Arabidopsis lyrata*: specific expression of the green fluorescent protein (GFP) in pistil tissues. Plant Cell Rep 26:745–753

Franco-Zorrilla JM, Valli A, Todesco M, Mateos I, Puga MI, Rubio-Somoza I, Leyva A, Weigel D, Garcia JA, Paz-Ares J (2007) Target mimicry provides a new mechanism for regulating microRNA activity. Nat Genet 39:1033–1037

Gasic K, Korban SS (2006) Indian mustard [*Brassica juncea* (L.) Czern]. In: Wang K (ed) *Agrobacterium* Protocols, 2nd edn. Humana Press, Totowa, NJ

Glimelius K (1984) High growth rate and regeneration capacity of hypocotyl protoplasts in some *Brassicaceae*. Physiol Plant 61:38–44

Gupta V, Lakshmi Sita G, Shaila MS, Jagannathan V (1993) Genetic transformation of *Brassica nigra* by *Agrobacterium* based vector and direct plasmid uptake. Plant Cell Rep 12:418–421

Halfhill M, Millwood RJ, Raymer PL, Stewart C Jr (2002) Bt-transgenic oilseed rape hybridization with its weedy relative, *Brassica rapa*. Environ Biosafety Res 1:19–28

Hellens R, Mullineaux P, Klee H (2000) Technical focus: a guide to *Agrobacterium* binary Ti vectors. Trends Plant Sci 5(10):446–451

Hoekema A, Hirsch P, Hooykaas P, Schilperoort R (1983) A binary plant vector strategy based on separate *vir* and T region of the *Agrobacterium tumefaciens* Ti plasmid. Nature 303: 179–180

Hood EE, Helmer GL, Fraley RT, Chilton MD (1986) The hypervirulence of *Agrobacterium tumefaciens* A281 is encoded in a region of pTiBo542 outside of T DNA. J Bacteriol 168:1291–1301

Horiguchi G (2004) RNA silencing in plants: a shortcut to functional analysis. Differentiation 72:65–73

Hou B, Zhou Y, Wan L, Zhang Z, Shen G, Chen Z, Hu, Z (2003) Chloroplast transformation in oilseed rape. Transgenic Res 12:111–114

James C (2007) Global status of commercialized biotech/GM crops: Brief No. 37. ISAAA, Ithaca, NY

Jin RG, Liu YB, Tabashnik BE, Borthakur D (2000) Development of transgenic cabbage (*Brassica oleracea* var. *capitata*) for insect resistance by *Agrobacterium tumefaciens*-mediated transformation. In Vitro Cell Dev Biol Plant 36:231–237

Jones JDG, Shlumukov L, Carland F, English J, Scofield SR, Bishop GJ, Harrison K (1992) Effective vectors for transformation, expression of heterologous genes, and assaying transposon excision in transgenic plants. Transgenic Res 1:285–297

Kamal GB, Lllich KG, Asadollah A (2007) Effects of genotype, explant type and nutrient medium components on canola (*Brassica napus* L.) shoot in vitro organogenesis. Afr J Biotechnol 6:861–867

Keller WA, Armstrong KC (1977) Embryogenesis and plant regeneration in *Brassica napus* anther cultures. Can J Bot 55:1383–1388

Ketaeva NV, Alexandrova IG, Butenko RG, Dragavtceva EV (1991) Effect of applied and internal hormones on vitrification and apical necrosis of different plants cultured in vitro. Plant Cell Tissue Organ Cult 27:149–154

King GJ (2006) Utilization of *Arabidopsis* and *Brassica* genomic resources to underpin genetic analysis and improvement of Brassica crops. In: Varshney RK, Koebner RMD (eds) Model plants: crop improvement. CRC Press, Boca Raton, FL

Klimazewska K, Keller WA (1985) High frequency plant regeneration from thin cell layer explants of *Brassica napus*. Plant Cell Tissue Organ Cult 4:83–197

Kojima M, Sparthana P, Teixeira da Silva JA, Nogawa M (2006) Development of *in planta* transformation methods using *Agrobacterium tumefaciens*. In: Teixeira da Silva JA (ed) Floriculture, ornamental and plant biotechnology: advances and topical issues, vol II, 1st edn. Global Science Books, Isleworth, UK

Komari T, Hiei Y, Saito Y, Murai N, Kumashiro T (1996) Vectors carrying two separate T-DNAs for co-transformation of higher plants mediated by Agrobacterium tumefaciens and segregation of transformants free from selection markers. Plant J 10:165–174

Koncz C, Schell J (1986) The promoter of T_L-DNA gene *5* controls the tissue-specific expression of chimeric genes carried by a novel type of *Agrobacterium* binary vector. Mol Gen Genet 204:383–396

Kuvshinov V, Koivu K, Kanera A, Perhu E (1999) *Agrobacterium tumefaciens* mediated transformation of greenhouse-grown *Brassica rapa* ssp. Oleifera. Plant Cell Rep 18:733–777

Lall S, Nettleton D, Decook R, Che P, Howell S (2004) Quantitative trait loci associated with adventitious shoot formation in tissue culture and the program of shoot development in Arabidopsis. Genetics 167:1883–1892

Lazo GR, Stein PA, Ludwig RA (1991) A DNA transformation—competent Arabidopsis genomic library in Agrobacterium. Biotechnology 9:963–967

Lee MK, Kim HS, Kim JS, Kim SH, Park YD (2004) Agrobacterium-mediated transformation system for large-scale production of transgenic Chinese cabbage (*Brassica rapa* L. ssp. *pekinensis*) plants for insertional mutagenesis. J Plant Biol 47:300–306

Lee JH, Park SH, Lee JS, Ahn JH (2007) A conserved role of SHORT VEGETATIVE PHASE (SVP) in controlling flowering time of Brassica plants. Biochim Biophys Acta 1769:455–461

Leyman B, Avonce N, Ramon M, Van Dijck P, Iturriaga G, Thevelein JM (2006) Trehalose-6-phosphate synthase as an intrinsic selection marker for plant transformation. J Biotechnol 121:309–317

Litcher R (1982) Induction of haploid plants from isolated pollen of Brassica napus. Plant Physiol (formally Z. Pflanzenphysiol) 105:427–434

Liu F, Cao MQ, Yao L, Robaglia C, Tourneur C (1998) In Planta transformation of pakchoi (Brassica campestris L. ssp. chinensis) by infiltration of adult plants with Agrobacterium. Acta Hortic 467:187–192

Liu C, Lin C, Chen JJW, Tseng M (2007) Stable chloroplast transformation in cabbage (*Brassica oleracea* L. var. *capitata* L.) by particle bombardment. Plant Cell Rep 26:1733–1744

Lloyd AM, Barnason AR, Rogers SG, Byrne MC, Fraley RT, Horsch RB (1986) Transformation of *Arabidopsis thaliana* with *Agrobacterium tumefaciens*. Science 234:464–466

Maliga P (2004) Plastid transformation in higher plants. Annu Rev Plant Biol 55:289–313

Mauro AO, Pfeiffer TW, Collins GB (1995) Inheritance of soybean susceptibility to Agrobacterium tumefaciens and its relationship to transformation. Crop Sci 35:1152–1156

Mehra S, Pareek A, Bandyopadhyay P, Sharma P, Burma PK, Pental D (2000) Development of transgenics in Indian oilseed mustard (*Brassica juncea*) resistant to herbicide phosphinothricin. Curr Sci 78:1358–1364

Metz T, Roush R, Tang J, Shelton A, Earle E (1995) Transgenic broccoli expressing a *Bacillus thuringiensis* insecticidal crystal protein: implications for pest resistance management strategies. Mol Breed 1:309–317

Mietkiewska E, Hoffman TL, Brost JM (2008) Hairpin-RNA mediated silencing of endogenous FAD2 gene combined with heterologous expression of crambe abyssinica FAE gene causes an increase in the level of erucic acid in transgenic *Brassica carinata* seeds. Mol Breed 22: 619–627

Miki B, McHugh S (2004) Selectable marker genes in transgenic plants: applications, alternatives and biosafety. J Biotechnol 107:193–232

Moloney MM, Walker JM, Sharma KK (1989) High-efficiency transformation of *Brassica napus* using *Agrobacterium* vectors. Plant Cell Rep 8:238–242

Mukhopadhyay A, Topfer R, Pradhan AK, Sodhi YS, Steinbiss HH, Schell J, Pental D (1991) Efficient regeneration of *Brassica oleracea* hypocotyl protoplasts and high frequency genetic transformation by direct DNA uptake. Plant Cell Rep 10:375–379

Murata M, Orton TJ (1987) Callus initiation and regeneration capacities in Brassica species. Plant Cell Tissue Organ Cult 11:111–123

Nam J, Matthysse AG, Gelvin SB (1997) Diferences in susceptibility of Arabidopsis ecotypes to crown gall disease may result from a deficiency in T-DNA integration. Plant Cell 9:317–333

Narasimhulu SB, Chopra VL (1988a) Species specific shoot regeneration response of cotyledonary explants of Brassicas. Plant Cell Rep 7:104–106

Narasimhulu SB, Kirti PB, Mohapatra T, Prakash S, Chopra VL (1992) Shoot regeneration in stem explants and its amenability to Agrobacterium tumefaciens mediated gene transfer in Brassica carinata. Plant Cell Rep 11:359–362

Narasimhulu SB, Prakash S, Chopra VL (1988b) Comparative shoot regeneration responses of diploid brassicas and their synthetic amphidiploid products. Plant Cell Rep 7:525–527

Nugent GD, Coyne S, Ngyuen TT, Kavanagh TA, Dix PJ (2006) Nuclear and plastid transformation of Brassica oleracea var. botrytis (cauliflower) using PEG-mediated uptake of DNA into protoplasts. Plant Sci 170:135–142

Ono Y, Takahata Y (2000) Genetic analysis of shoot regeneration from cotyledonary explants in Brassica napus. Theor Appl Genet 100:895–898

Ono Y, Takahata Y, Kaizuma N (1994) Effect of genotype on shoot regeneration from cotyledonary explants of rapeseed (*Brassica napus* L). Plant Cell Rep 14:13–17

Opabode JT (2006) Agrobacterium-mediated transformation of plants: emerging factors that influence efficiency. Biotechnol Mol Biol 1:12–20

Østergaard L, Kempin SA, Bies D, Klee HJ, Yanofsky MF (2006) Pod shatter resistant fruit produced by ectopic expression of the FRUITFULL gene in Brassica juncea. Plant Biotechnol 4:45–51

Østergaard L, Yanofsky MF (2004) Establishing gene function by mutagenesis in Arabidopsis thaliana. Plant J 39:682–696

Palmer CE (1992) Enhanced shoot regeneration from Brassica campestris by silver nitrate. Plant Cell Rep 11:541–545

Phogat SK, Burma PK, Pental D (2000) High frequency regeneration of *Brassica napus* varieties and genetic transformation stocks containing fertility restorer genes for two cytoplasmic male sterility systems. J Plant Biochem Biotechnol 9:73–79

Price JS, Hobson RN, Neale MA, Bruce DM (1996) Seed losses in commercial harvesting of oilseed rape. J Agric Eng 65:83–191

Pua EC, Chi GL (1993) De novo shoot morphogenesis and plant growth of mustard (Brassica juncea) in vitro in relation to ethylene. Physiol Plant 88:467–474

Puddephat IJ, Robinson HT, Fenning TM, Barbara DJ, Morton A, Pink DAC (2001) Recovery of phenotypically normal transgenic plants of *Brassica oleracea* upon *Agrobacterium* rhizogenes-mediated co-transformation and selection of transformed hairy roots by GUS assay. Mol Breed 7:229–242

Puddephat IJ, Thompson N, Robinson HT Sandhu P, Henderson J (1999) Biolistic transformation of broccoli (*Brassica oleracea* var. *Italica*) for transient expression of the β-glucuronidase gene. J Hortic Sci Biotechnol 74:714–720

Purity RS, Gautam K, Singla-Pareek SL (2008) Towards salinity tolerance in *Brassica*: an overview. Physiol Mol Biol Plants 14:39–49

Qing CM, Fan L, Lei Y, Bouchez D, Tourneur C, Yan L, Robaglia C (2000) Transformation of Pakchoi (*Brassica rapa L.* ssp. *chinensis*) by *Agrobacterium* infiltration. Mol Breed 6: 67–72

Radchuck V, Ryschka U, Schumann G, Klocke E (2002) Genetic transformation of cauliflower (*Brassica oleracea* var. *botrytis*) by direct DNA uptake into mesophyll protoplasts. Physiol Plant 114:429–438

Radke SE, Turner JC, Facciotti D (1992) Transformation and regeneration of *Brassica rapa* using *Agrobacterium tumefaciens*. Plant Cell Rep 11:499–505

Ramachandran S, Buntin G, All J, Raymer P, Stewart C Jr (1998a) Greenhouse and field evaluations of transgenic canola against diamondback moth, *Plutella xylostella*, and corn earworm, *Helicoverpa zea*. Entomol Exp Appl 88:17–24

Reed J, Privalle L, Luann Powell M, Meghji M, Dawson J, Dunder E, Sutthe J, Wenck A, Launis K, Kramer C, Chang YF, Hansen G, Wright M (2001) Phosphomannose isomerase: an efficient selectable marker for plant transformation. In Vitro Cell Dev Biol Plant 37: 127–132

Schiantarelli E, De la Pena A, Candela M (2001) Use of recombinant inbred lines (RILs) to identify, locate and map major genes and quantitative trait loci involved with in vitro regeneration ability in Arabidopsis thaliana. Biomed Life Sci 102:335–341

Schmidt R, Willmitzer L (1988) High efficiency *Agrobacterium tumefaciens*-mediated transformation of *Arabidopsis thaliana* leaf and cotyledon explants. Plant Cell Rep 7:583–586

Schwab R, Ossowski S, Riester M, Warthmann N, Weigel D (2006) Highly specific gene silencing by artificial micoRNAs in *Arabidopsis*. Plant Cell 18:1121–1133

Sciaky D, Montoya AL, Chilton MD (1978) Fingerprints of *Agrobacterium* Ti plasmids. Plasmid 1:238–253

Sikdar S, Serino G, Chaudhuri S, Maliga P (1998) Plastid transformation in Arabidopsis thaliana. Plant Cell Rep 18:20–24

Smith RH, Hood EE (1995) *Agrobacterium tumefaciens* transformation of monocotyledons. Crop Sci 35:301–309

Somerville C, Koorneef M (2002) A fortunate choice: the history of *Arabidopsis* as a model plant. Nat Rev Genet 3:883–889

Sparrow PAC, Dale PJ, Irwin JA (2004a) The use of phenotypic markers to identify *Brassica oleracea* genotypes for routine high-throughput *Agrobacterium*-mediated transformation. Plant Cell Rep 23:64–70

Sparrow PAC, Dale PJ, Irwin JA (2006b) Brassica oleracea. In: Wang K (ed) *Agrobacterium* Protocols, 2nd edn. Methods in molecular biology 343, vol 1. Humana Press, Totowa, NJ

Sparrow PAC, Irwin JA, Goldsack CM, Østergaard L (2007) Brassica transformation: commercial application and powerful research tool. Transgenic Plant J 1:330–339

Sparrow PAC, Snape JW, Dale PJ, Irwin JA (2006a) The rapid identification of *B. napus* genotypes, for high-throughput transformation, using phenotypic tissue culture markers. Acta Hortic 706:239–247

Sparrow PAC, Townsend T, Dale PJ, Irwin JA (2004b) Genetic analysis of *Agrobacterium tumefaciens* susceptibility in *Brasssica oleracea*. Theor Appl Genet 108:664–650

Sparrow PAC, Townsend T, Morgan CL, Arthur AE, Dale PJ, Irwin JA (2004c) Genetic analysis of in vitro shoot regeneration from cotyledonary petioles of *Brassica oleracea*. Theor Appl Genet 108:1249–1255

Stegemann S, Hartmann S, Ruf S, Bock R (2003) High-frequency gene transfer from the chloroplast genome to the nucleus. Proc Natl Acad Sci USA 100:8828–8833

Stewart C Jr, Adang M, All J, Raymer P, Ramachandran S, Parrott W (1996) Insect control and dosage effects in transgenic canola containing a synthetic *Bacillus thuringiensis cryIAc* gene. Plant Physiol 112:115–120

Stewart CN Jr, All JN, Raymer PL, Ramachandran S (1997) Increased fitness of transgenic insecticidal rapeseed under insect selection pressure. Mol Ecol 6:773–779

Szegedi E, Kozma P (1984) Studies on the inheritance of resistance to crown gall disease of grapevine. Vitis 23:121–126

Tang GX, Zhou WJ, Li HZ, Mao BZ, He ZH, Yoneyama K (2003) Medium, explant and genotype factors influencing shoot regneration in oilseed *Brassica* spp. J Agron Crop Sci 189:351–358

Taylor CG, Fuchs B, Collier R, Lutke WK (2006) Generation of composite plants using *Agrobacterium rhizogenes*. In: Wang K (ed) *Agrobacterium* Protocols, vol 1, 2nd edn. Humana press, Totowa, NJ

Trick M, Bancroft I, Lim Y-P (2007) The *Brassica rapa* genome sequencing initiative. Genes Genomes Genomics 1:35–39

Valvekens D, Van Montagu, M, Van Lijsebettens M (1988) *Agrobacterium tumefaciens*-mediated transformation of *Arabidopsis thaliana* root explants by using kanamycin selection. Proc Natl Acad Sci USA 85:5536–5540

Veena V, Taylor CG (2007) *Agrobacterium rhizogenes*: recent developments and promising applications. In Vitro Cell Dev Biol 43:383–403

Verma SS, Chiinnusarny V, Bansal KC (2008) A simplified floral dip method for transformation of *Brassica* napus and B-carinata. J Plant Biochem Biotechnol 17:197–200

Verma R, Singh RR (2007) Regeneration and in vitro flowering in *Brassica Campestris* (L.) Var. *Bhavani*. Our Nature 5:21–24

Wahlroos T, Susi P, Tylkina L, Malyshenko S, Zvereva S, Korpela T (2003) *Agrobacterium*-mediated transformation and stable expression of the green fluorescent protein in *Brassica rapa*. Plant Physiol Biochem 41:733–778

Wang WC, Menon G, Hansen G (2003) Development of a novel *Agrobacterium* mediated transformation method to recover transgenic *Brassica napus* plants. Plant Cell Rep 22:274–281

Xiang Y, Wong WKR, Ma MC, Wong RSC (1999) Agrobacterium-mediated transformation of Brassica campestris ssp. parachinensis with synthetic Bacillus thuringiensis cry1Ab and cry1Ac genes. Plant Cell Rep 19:251–256

Xiang Y, Wong WKR, Ma MC, Wong RSC (2000) *Agrobacterium*-mediated transformation of *Brassica campestris* ssp. *Parachinensis* with synthetic *Bacillus thuringiensis* cry1Ab and cry1Ac genes. Plant Cell Rep 19:251–256

Xu ZH, Davey MR, Cocking EC (1982) Plant regeneration from root protoplasts of Brassica. Plant Sci Lett 24:117–121

Xu H, Wang X, Zhao H, Liu F (1998) An intensive understanding of vacuum infiltration transformation of pakchoi (*Brassica rapa* ssp. *chinensis*). Plant Cell Rep 27:1369–1376

Yan JY, He YK, Cao JS (2004) Factors affecting transformation efficiency by micro-injecting *Agrobacterium* into flower bud of Chinese cabbage. Agric Sci China 3:44–51

Yan JY, HeY K, Cao JS (2003) Transformation of Chinese cabbage (*Brassica rapa* L. ssp. *pekinensis*) by Agrobacterium micro-injection into flower bud. Agric Sci China 2:906–911

Yang MZ, Jia SR, Pua EC (1991) High frequency of plant regeneration from hypocotyl explants of *Brassica carinata A*. Br. Plant Cell Tissue Organ Cult 24:79–82

Yang ZH, Jin H, Plaha P, Woong BT, Jiang GB, Woo JG, Yun HD, Lim YP, Lee HY (2004) An improved regeneration protocol using cotyledonary explants from inbred lines of Chinese cabbage (*Brassica rapa* ssp. *Pekinensis*). J Plant Biotechnol 6:235–239

Ye GN, Stone D, Pang SZ, Creely W, Gonzalez K, Hinchee M (1999) *Arabidopsis* ovule is the target for *Agrobacterium* in planta vacuum infiltration transformation. Plant J 19:249–257

Yu B, Lydiate DJ, Young LW, Scha¨fer UA, Hannoufa A (2008) Enhancing the carotenoid content of *Brassica napus* seeds by down regulating lycopene epsilon cyclase. Transgenic Res 17: 573–585

Zhandong Y, Shuangyi Z, Qiwei H (2007) High level resistance to *Turnip mosaic virus* in Chinese cabbage (*Brassica campestris* ssp. *pekinensis* (Lour) Olsson) transformed with the antisense NIb gene using marker-free *Agrobacterium tumefaciens* infiltration. Plant Sci 172:920–929

Zhang FL, Takahata Y, Watanabe M, Xu JB (2000) *Agrobacterium*-mediated transformation of cotyledonary explants of Chinese cabbage (*Brassica campestris* L. ssp. *pekinensis*). Plant Cell Rep 19:569–575

Ziv M (1991) Quality of micropropagated plants—vitrification. In Vitro Cell Dev Biol Plant 27: 64–69

Chapter 19
Resources for Reverse Genetics Approaches in *Arabidopsis thaliana*

Bekir Ülker and Bernd Weisshaar

Abstract Having many characteristics of an ideal experimental system, *Arabidopsis thaliana* became a very important model system for flowering plants. Its completed genome sequence data provided scientists the first fundamental tool towards understanding its genome structure and genes that it possess. There are more than 33,000 predicted genes in *Arabidopsis* and this number is increasing as novel methods develop and our understanding of genome organization and regulation expands. Reverse genetics that aim to reveal the functions of all *Arabidopsis* genes and the related resources were next most important tools that plant scientists needed. Numerous consortia were formed to supply scientists with such resources and tools necessary to determine the functions of *Arabidopsis* genes. Thanks to these international community efforts, now there are around 426,000 independent T-DNA/transposon insertion lines representing near saturation of all genes in *Arabidopsis* available for the research community. Besides insertion lines, several other sophisticated technologies and resources crucial for large-scale gene function studies in *Arabidopsis* were also developed. In this chapter, we discuss most of these important reverse genetics resources for gene function analysis.

Keywords Reverse genetics · *Arabidopsis* · T-DNA · Transposon · Activation tagging · Knockout · T-DNA insertion collections · T-DNA vectors · *Agrobacterium* · TILLING · EcoTILLING · Deleteagene · Zink finger nuclease · Homologous recombination · Gene silencing · Antisense · RNAi · MicroRNA · miRNA · Microarray · Gene trap · Promoter trap · Enhancer trap · Overexpression · Mutant · EMS mutagenesis · Fast neutron bombardment · SALK · Gabi-Kat · FLAGdb · SAIL

B. Ülker (✉)
Plant Molecular Engineering Group, IZMB (Institute of Cellular and Molecular Botany), University of Bonn, D-53115 Bonn, Germany
e-mail: ulker@uni-bonn.de

R. Schmidt, I. Bancroft (eds.), *Genetics and Genomics of the Brassicaceae*, Plant Genetics and Genomics: Crops and Models 9, DOI 10.1007/978-1-4419-7118-0_19,

Abbreviations

Ac/Ds	Activator/dissociation
AchrDNA	*Agrobacterium* chromosomal DNA
AGI	*Arabidopsis* genome initiative
AGRICOLA	*Arabidopsis* genomic RNAi knock-out line analysis
AMAZE/ZIGIA	*Arabidopsis En-1* transposon insertion lines from Max-Planck-institute for plant breeding, Cologne
amiRNAs	artificial microRNAs
CaMV	Cauliflower mosaic virus
BLAST	Basic local alignment search tool
DHPLC	High-performance liquid chromatography
dsRNA	double-stranded RNA
EMS	Ethylmethanesulfonate
En/Spm	Enhancer/supressor–mutator
FLAGdb	Functional analysis of the *Arabidopsis* genome database, also known as INRA/Versailles lines
FST	Flanking sequence tag
GABI-Kat	Genomanalyse im biologischen System Pflanze *Arabidopsis* T-DNA lines
GFP	Green fluorescent protein
GUS	β-glucuronidase
LUC	Luciferase protein
INDELS	Insertions/deletions
MASC	Multinational *Arabidopsis* steering committee
MIPS	The Munich Institute for Protein Sequences
miRNAs	microRNAs
NASC	Nottingham *Arabidopsis* stock centre
PCR	Polymerase chain reaction
SNPs	Single nucleotide polymorphisms
SLAT	Sainsbury laboratory *Arabidopsis thaliana* transposants
SAIL	Syngenta *Arabidopsis* insertion library, formerly known as GARLIC (Gilroy *Arabidopsis* reverse lethal insertion collection)
SALK	SALK institute T-DNA insertion lines
siRNA	small interfering RNA
TAIR	The *Arabidopsis* information resource
TAIL PCR	Thermal asymmetric interlaced polymerase chain reaction
TAMARA	Transposable element-mediated activation tagging mutagenesis in *Arabidopsis*
taRNAs	Transacting RNAs
T-DNA	Transferred DNA
TILLING	Targeting-Induced Local Lesions In Genomes
tiling arrays	Arrays that cover the whole genome
Ti plasmid	Tumour inducing plasmid
TT1	*Arabidopsis* transparent testa 1 protein
uidA	Gene coding for β-glucuronidase
ZFN	Zinc finger nuclease

Contents

19.1 Introduction 529
19.2 Gene Function Analyses 529
19.2.1 Similarity to Other Known Proteins 530
19.2.2 Expression Analyses 531
19.2.3 Mutation Analyses 533
19.2.4 Over-Expression/Activation-Mediated Functional Assays 547
19.2.5 Gene Silencing-Mediated Functional Analysis 548
19.3 Outlook 552
References 553

19.1 Introduction

The demand to produce food for human consumption and feed for animals is ever increasing. Modern genetics and plant breeding may have much to contribute to meeting these challenges. Plants have a great potential of adaptation to adverse conditions, and they carry an enormous biosynthetic potential. However, improving the adaptation of crops to adverse conditions and fully harnessing the biosynthetic potential of plants requires a deep understanding of the molecular basis by which plants control growth and development and respond to environmental signals and stresses. Therefore, understanding of plant gene function is a prerequisite to accelerated crop improvement.

Arabidopsis thaliana is playing a major role in plant biology research since its selection as the model for dicotyledonous flowering plants. Two decades of investment in *Arabidopsis* have generated a fully sequenced and annotated genome (*Arabidopsis* Genome Initiative 2000), as well as a wealth of tools and resources. Until the year 2000, functions of only a few hundred of the about 33,000 (TAIR9 Genome Release, June 19, 2009) predicted genes were known. Thanks to strong community efforts and due to the intense efforts into development and use of resources which are the subject of this chapter, these numbers are currently rapidly increasing. Besides gene function assignment, there is also progress in discovering how pathways of gene regulation or signalling are regulated. Basic research and gene function analysis in *Arabidopsis* is not only enhancing our understanding of plant biology and generating scientific knowledge, but also laying foundations for an improvement of sustainable food and feed production.

19.2 Gene Function Analyses

Traditionally, forward genetics approaches relying on phenotypic observations or the selection of a desired phenotype have been used in determining gene functions in *Arabidopsis*. However, these strategies have elucidated the functions of only a small number of all *Arabidopsis* genes (Meinke and Scholl 2003, Alonso and Ecker 2006). Therefore, several novel methods have been developed to help determining the function(s) of genes that are known only from DNA sequence information.

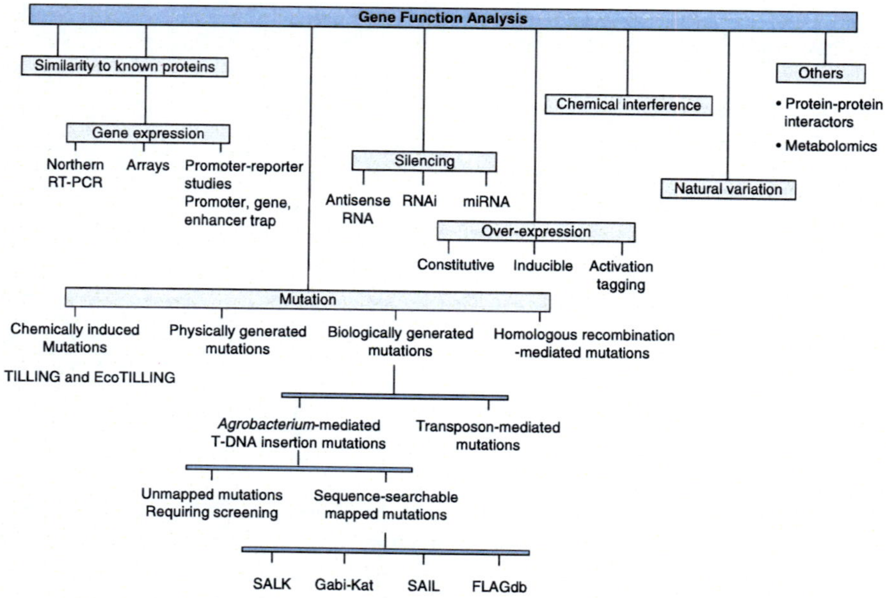

Fig. 19.1 Schematic summary of strategies and methods available for determining the function(s) of genes in *Arabidopsis* with an emphasis on reverse genetic strategies

These diverse approaches are outlined schematically in Fig. 19.1. Reverse genetics approaches that aim to mutate every gene in the genome, and the targeted determination of their functions, are one of the most often used strategy and will be discussed below in detail. The generation of plant resources required for these strategies is expensive and time consuming. Hence, quite a number of consortia have been formed around the world to help in generating such material for the science community. These community resources and databases will be described in this chapter.

19.2.1 Similarity to Other Known Proteins

Approximately 70% of the predicted proteins in every organism bear similarity to proteins in other distantly related organisms, but about 30% are more or less unique in a given organism (Rubin et al. 2000). Like many researchers in biological sciences, plant scientists have been using the information gained from research in model organisms such as *Escherichia coli* and yeast as well as other organisms to predict functions of genes found in plants based on sequence similarities. Researchers were able to assign putative functions to 70% of the *Arabidopsis* gene products on the basis of sequence similarity to proteins of known function in other organisms ("*Arabidiopsis* Genome Initiative", 2000).

A study showed that 80% of the *Arabidopsis* genes have a homolog in rice and 40% of the rice genes in *Arabidopsis* (Rensing et al. 2005). Another study concluded that 71% of predicted rice proteins were reasonably similar to *Arabidopsis* proteins (Paterson et al. 2005). The high similarities suggest that basic cellular and biochemical functions of many crop genes can be inferred from results produced with *Arabidopsis*.

19.2.2 Expression Analyses

Clues obtained from expression analysis of a particular gene can help to come up with hypotheses as to what the function of a given gene could be. For example, such analyses can determine where and when the gene of interest is expressed, how its activity is regulated in response to developmental signals, abiotic stresses (such as drought, cold, or salt), biotic stresses (such as pathogens, herbivores, or pests), phytohormones, photoperiod, and other changes in environment.

However, inferring function of a gene product from mRNA level is indirect and the correlation between mRNA changes and changes in protein level in a cell can be weakly correlated (Steinmetz and Davis 2004). Similarly, reliability and reproducibility must be considered before expression data can be considered for hypotheses.

Gene expression data can be collected from RNA gel-blot analyses, quantitative RT-PCR experiments, and microarray analyses. Transient or stable expression of fluorescently tagged gene products in cells could be used to determine their subcellular locations. Finally, transgenic lines containing promoter–reporter constructs could be used to determine spatial and temporal expression patterns of the studied genes.

19.2.2.1 Array-Related Expression Data (Microarrays)

Thousands of microarray experiments have been conducted with *Arabidopsis* which provide a large amount of quantitative data on gene expression in different tissues and in response to different treatments and experimental conditions. Moreover, such microarray analyses have been performed in numerous mutants. Most of these microarray data are deposited in public databases such as NASC (Nottingham *Arabidopsis* Stock Centre), TAIR (The *Arabidopsis* Information Resource), and ArrayExpress (http://www.ebi.ac.uk/microarray-as/ae/). NASCArray, for example, currently holds 3,727 publicly available microarray hybridizations (May 08, MASC, Multinational *Arabidopsis* Steering Committee report 2008). Researchers can access the data directly from these resources or can analyse the data using GENEVESTIGATOR (Zimmermann et al. 2004) or MAPMAN tools (Thimm et al. 2004). The ability to analyse gene expression with tiling arrays (arrays that are not biased by gene annotation but cover the whole genome) makes this technology superior to previous microarray technologies (Yamada et al. 2003, Laubinger et al. 2008).

These compilations of gene expression data provide insights into gene regulation and furthermore into the signalling pathways that control complex responses. Therefore, such gene expression databases offer a very powerful opportunity to assign functional information to genes of otherwise unknown function.

19.2.2.2 Expression Analyses Using Promoter:Reporter Gene Fusions and Promoter, Gene, and Enhancer Trap Lines

Promoter–reporter gene fusions provide another way of studying gene expression and conditions that might regulate genes. This is done typically by cloning a few kilobases upstream region from start codon (ATG) of the gene and fusing it to a reporter like *uidA* (*ß-glucuronidase, GUS*), *green fluorescent protein* (*GFP*), or *luciferase* (*LUC*). The constructs carrying such fusions are transformed into plants using either *Agrobacterium*-mediated transformation or direct delivery methods such as micro-projectile bombardment. Transgenic plants containing the reporter construct are then analysed for reporter gene expression. However, regulatory sequences of genes flanking the insertion site of the reporter construct may in some instances influence the expression pattern of the introduced promoter–reporter gene fusion and such promoters might miss other important regulatory regions in its native location. Therefore researchers typically need to analyse several independent lines for each construct and need to interpret the results obtained from such analyses more cautiously.

Promoter, gene, and enhancer traps are reporter genes that are not normally expressed unless they are integrated near or within a gene (Fig. 19.2) (Martienssen 1998). This system was initially developed to be used in bacterial genetics for generating random insertional fusions of *lacZ* reporter gene (Casadaban and Cohen 1979). The applicability of promoter/gene fusion technologies in plants was first

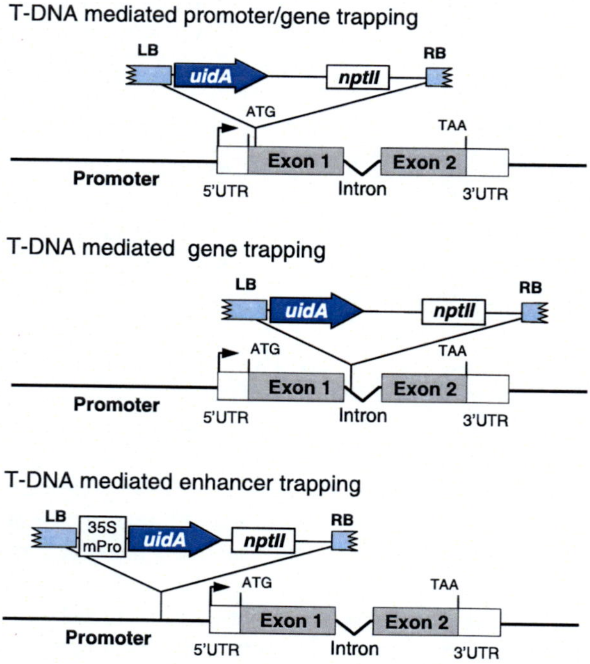

Fig. 19.2 Promoter, gene, and enhancer trapping strategy

demonstrated in tobacco and *Arabidopsis* using a promoterless kanamycin phosphotransferase gene (Teeri et al. 1986, Koncz et al. 1989). Follow-up studies made use of more versatile reporters like *GUS* (Fobert et al. 1991, Kertbundit et al. 1991, Topping et al. 1991), *LUC* (Alvarado et al. 2004), and *GFP* (Kiegle et al. 2000).

Small collections of promoter, gene, and enhancer trap lines, each carrying an independent reporter gene insertion somewhere in the genome, have been established in *Arabidopsis* (Topping et al. 1994, Sundaresan et al. 1995, Campisi et al. 1999, Alvarado et al. 2004). The Cold Spring Harbor Gene and Enhancer Trap lines (Sundaresan et al. 1995, Martienssen 1998) and the GFP insertion lines of Jim Haselhoff (Kiegle et al. 2000) are publicly available for the research community (Table 19.1). FLAGdb T-DNA insertion line collection is another relatively large resource (currently contains over 37,000 sequence indexed lines) for tagging with *GUS* (Samson et al. 2002). The pGKB5 vector used to generate this population contains a promoterless *GUS* gene at the right border (RB). Therefore, in some FLAGdb lines in-frame fusions of GUS and the flanking coding sequence may have occurred (Bouchez et al. 1993a, Menand et al. 2002).

19.2.3 Mutation Analyses

Mutant analysis typically provides reliable means to assign gene function (Meinke et al. 1998). Several thousand mutants of *Arabidopsis* defective in almost every aspect of plant growth and development have been identified over the past 30 years. The most widely used mutagens in plant genetics are chemicals, ionizing radiation, and T-DNA/transposon insertions. Deletions, insertions, and rearrangements are more likely to result in loss-of-function alleles, whereas point mutations can lead to a broader range of effects such as alleles of reduced, enhanced, or novel gene function.

19.2.3.1 Chemically Generated Mutants

Mutagenesis with ethylmethanesulfonate (EMS) has most often been used for forward genetic screens in *Arabidopsis* due to high level of mutations that are generated by this method. EMS is an alkylating agent and can form adducts with nucleotides, causing them to "pair" with non-complementary bases, thus introducing base changes after replication. EMS typically produces transition mutations (G/C to A/T) because it alkylates G residues and the alkylated G residue pairs with T instead of the conventional base pairing with C (Koornneef et al. 1982, Greene et al. 2003).

EMS-induced mutations are randomly distributed (Greene et al. 2003). Due to such random base substitutions in proteins, EMS mutagenesis provides allelic series as well as true knockouts. However, the frequency of null alleles, for example mutations that cause an early stop codon in the open reading frame of a given gene, is quite low. This strategy can yield insights into protein function by providing data on the role of specific amino acid residues. Such allelic series can also provide useful information for understanding the function of essential genes by generating weak

Table 19.1 T-DNA insertion collections and number of lines with mapped insertions[a]

T-DNA insertion collection	Ecotype	References	Number of lines with mapped insertions	Data collection date
SALK T-DNA http://signal.salk.edu	Columbia	Alonso et al. (2003)	165, 851	23.06.2009
SALK T-DNA homozygous http://signal.salk.edu	Columbia	Alonso et al. (2003)	33, 468	23.06.2009
Other T-DNA homozygous	Columbia		2, 379	23.06.2009
GABI-Kat http://www.gabi-kat.de Available from GABI-Kat	Columbia	Rosso et al. (2003)	64, 058	08.07.2009
GABI-Kat confirmed, donated to NASC http://Arabidopsis.info/	Columbia	Rosso et al. (2003)	6, 676	20.07.2009
SAIL, the database moved to NASC http://Arabidopsis.info/	Columbia[b]	Sessions et al. (2002)	61, 930	23.06.2009
Wisconsin FST available from ABRC	Columbia	Sussman et al. (2000)	17, 190	23.06.2009
FLAGdb http://urgv.evry.inra.fr/projects/FLAGdb++/HTML/index.shtml best searched at http://signal.salk.edu/cgi-bin/tdnaexpress Available from INRA, France	Wassilevskija	Balzerque et al. (2001), Samson et al. (2002)	37, 142	23.06.2009
JIC SM FST http://arabidopsis.info/CollectionInfo?id=33	Landsberg	See NASC website	24, 203	23.06.2009
RIKEN FST (Ds transposon) http://rarge.gsc.riken.go.jp/dsmutant/index.pl Available from RIKEN BRC	Nössen	Ito et al. (2002, 2005), Kuromori et al. (2004)	18, 622	23.06.2009
CSHL FST (Ds transposon) http://genetrap.cshl.org/	Columbia	Sundaresan et al. (1995), Martienssen (1998)	21, 682	23.06.2009

Table 19.1 (continued)

T-DNA insertion collection	Ecotype	References	Number of lines with mapped insertions	Data collection date
IMA FST Institute of molecular agrobiology (IMA) lines from V. Sundaresan and colleagues. Modified maize transposable dissociation (Ds) elements inserted into different positions within the genome. Available from ABRC Summary page at TAIR: http://www.arabidopsis.org/portals/mutants/findmutants.jsp	Landsberg	Parinov et al. (1999)	805	23.06.2009
Other FST			175	23.06.2009
Total			411,662	20.07.2009

[a]Data obtained from the SIGnAL-Salk Institute Genomic Analysis Laboratory website (http://signal.salk.edu) on 07.08.2009

[b]It is important to note that the SAIL lines transformed with the pCSA110 vector are in the background of homozygous *qrt* (quartet, pollen tetrad defect) mutation but the pDAP101 is in wild-type Columbia background (Preuss et al. 1994, McElver et al. 2001). The pCSA110 vector contains GUS reporter gene driven by the tomato LAT52 pollen-specific promoter (Eyal et al. 1995) and was transformed into qrt/qrt plants to facilitate segregation analysis in pollen for related projects

nonlethal alleles. However, given that EMS produces a high number of independent mutations in a genome, unrelated mutations must be removed by several rounds of backcrossing to the parental non-mutagenized line before the phenotype of a mutant of interest can be analysed in detail. Another disadvantage of the EMS mutagenesis technique has been the long time and extensive effort that it usually takes to determine a particular mutated locus because there is no molecular tag (e.g. insertion) that would allow the direct cloning of a mutated gene of interest. Map-based cloning, often called positional cloning, has been the most effective strategy to isolate the gene that corresponds to a chemically or radiation-derived mutant (Ostergaard and Yanofsky 2004). Map-based cloning is labour intensive and time consuming, hence several groups tried to develop alternative methods to identify induced mutations. For example, McCallum et al. demonstrated that mutations induced by EMS in *A. thaliana* can be easily identified by analysing heteroduxplex DNA derived from wild-type and mutant gene via denaturing high-performance liquid chromatography (DHPLC) (McCallum et al. 2000). The authors have named this technique TILLING (*Targeting-Induced Local Lesions IN Genomes)* and it is discussed in detail in the following section. A less expensive and faster modification of the TILLING protocol was published by Colbert et al. (2001).

TILLING and EcoTILLING

TILLING is a reverse-genetic strategy to identify SNPs (single nucleotide polymorphisms) and/or INDELS (insertions/deletions) in a gene of interest from a mutagenized population. It has been developed to identify mutations at an *Arabidopsis* locus, but TILLING is suitable for essentially any organism that can be mutagenized or displays natural variation between accessions. During the TILLING protocol, DNA is isolated from pooled M2 plants that are chemically mutagenized, the region of interest is amplified using PCR and fluorescent primers (Fig. 19.3). PCR products are denatured and re-annealed to form heteroduplexes between a mutated sequence and its wild-type counterpart. The resulting pool contains a mixture of homo- and heteroduplexes. Mismatched base pairs in the heteroduplex DNA fragments are cleaved by single-strand specific mismatch endonucleases such as CEL1 from celery (Oleykowski et al. 1998). Following cleavage, products are analysed on denaturing polyacrylamide gels and visualized by fluorescence detection systems. A mutation in a region of interest will generate fragments of cleaved heteroduplexes whose additive sizes should equal the total length of the entire product. Once the cleaved fragments and their respective polymorphic site are identified, the region of interest is sequenced in all plants making up the pool in which the mutation was detected to identify and verify the induced mutation. Comparison of the CEL1 nuclease with other single-strand-specific nucleases by Till and colleagues revealed that the endonuclease activity of the CEL1 does not differ significantly from that of *Aspergillus* S1 nuclease or from mung bean nuclease if the digestion conditions are optimized for each enzyme (Till et al. 2004).

TILLING has been successfully used for the study of functional genomics in *Arabidopsis* (Colbert et al. 2001). The numerous mutations in *A. thaliana* that

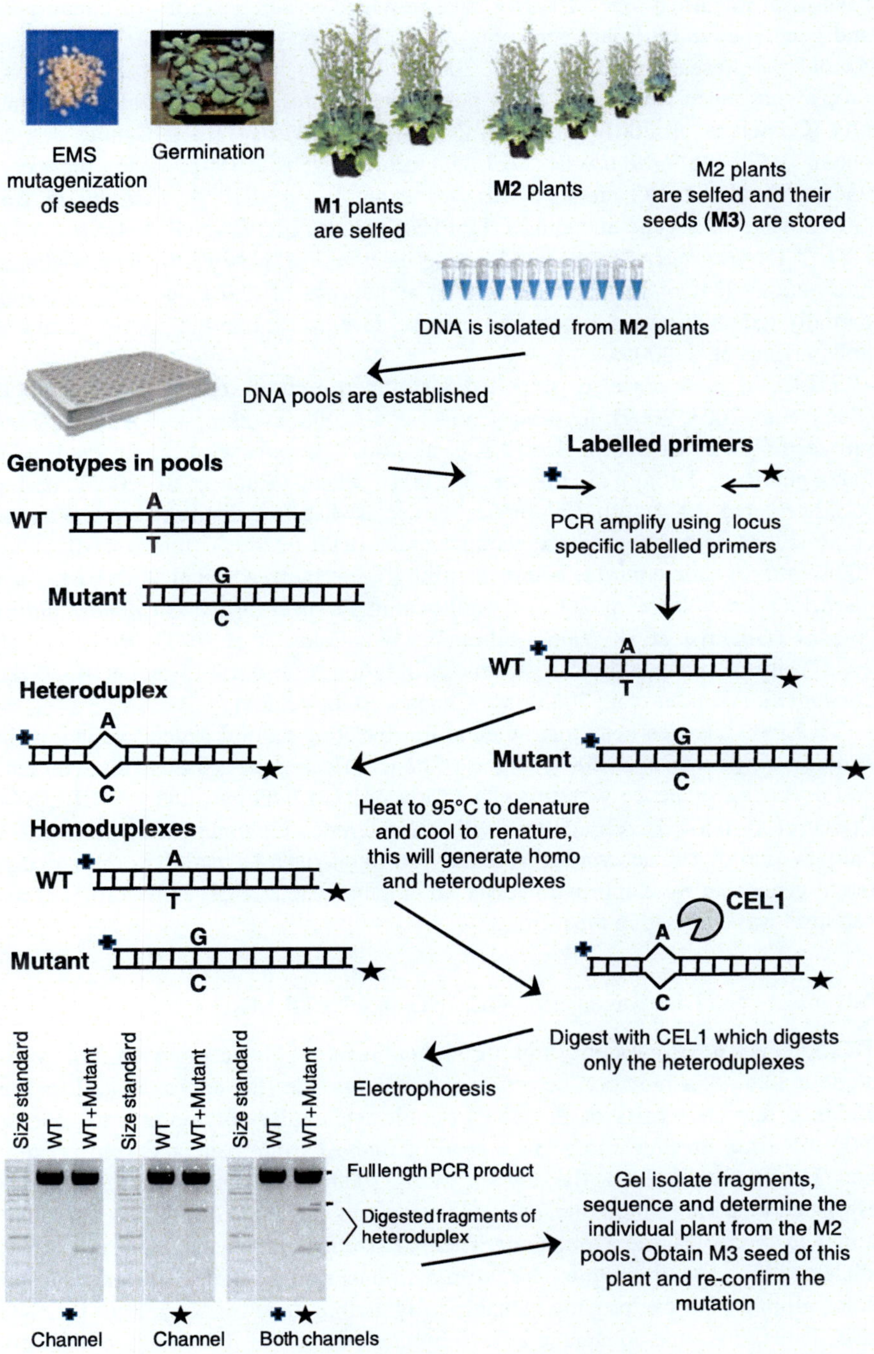

Fig. 19.3 TILLING method and its application to mutation discovery in *Arabidopsis* plants

have been identified via TILLING have provided an allelic series of phenotypes and genotypes to elucidate gene and protein function throughout the genome for *Arabidopsis* researchers (Till et al. 2003a). In 2003, the *Arabidopsis* TILLING Project has been introduced as a public service for the *Arabidopsis* community (Colbert et al. 2001, Till et al. 2003b). *Arabidopsis* TILLING lines can be obtained via the *Arabidopsis* TILLING project website (http://tilling.fhcre.org:9366). The TILLING strategy was, for example, used in *A. thaliana* to isolate mutants of a gene-encoding CTP:PHOSPHORYLETHANOLAMINE CYTIDYLYLTRANSFERASE, a rate-limiting enzyme in phosphatidylethanolamine biosynthesis (Mizoi et al. 2006), as well as mutants of *GSL1* and *GSL5* playing partially redundant roles in both sporophytic development and the development of pollen (Enns et al. 2005).

TILLING lines are also generated for numerous other plants including lotus (Perry et al. 2003, 2009), soybean (Cooper et al. 2008, Dierking and Bilyeu 2009), rapeseed (Wu et al. 2008), rice (Till et al. 2007) (http://www.tilling.ucdavis.edu/index.php/Rice_Tilling) (Suzuki et al. 2008), wheat (Slade et al. 2005), barley (Caldwell et al. 2004) (http://bioinf.scri.sari.ac.uk/distilling/distilling.html) (Talame et al. 2008) (http://www.distagenomics.unibo.it/TILLMore/), maize (Till et al. 2004) (http://genome.purdue.edu/maizetilling/), sorghum (Xin et al. 2008), *Brassica oleracea* (Himelblau et al. 2009) as well as animals species such as *Caenorhabditis elegans* (Gilchrist et al. 2006), zebrafish (Wienholds et al. 2003, Moens et al. 2008) (https://webapps.fhcrc.org/science/tilling/index.php), rat (Smits et al. 2004), *Drosophila* (Winkler et al. 2005) (see Chapter 20 for details).

TILLING has also been instrumental for detecting natural genetic variations in *Arabidopsis* (Comai et al. 2004). In this setting, the technique is used to discover natural occurring sequence variation. This form of TILLING is called EcoTILLING. Comai et al. pooled each ecotype with the reference Columbia ecotype in a 1:1 ratio. A total of 192 accessions were assayed to uncover 55 haplotypes in five different genes that were approximately 1 kb in length each. A large proportion of the variation was detected in the introns.

Advantages and Disadvantages of TILLING and EcoTILLING

TILLING is a non-transgenic, high-throughput reverse genetic approach. This technique unlike other mutation detection methods provides the approximate location within a few base pairs of the SNPs (single nucleotide polymorphisms) and/or INDELS (insertions/deletions) in a gene of interest from a mutagenized population (Henikoff et al. 2004, Barkley and Wang 2008). McCallum et al. (2000) have estimated that 5% of mutations introduce stop codons, about 65% missense mutations and about 30% silent mutations. Therefore through mutagenesis one can obtain alleles with partial or complete loss of function as well as alleles with novel functions, all of which can provide valuable insights into gene function. EcoTILLING also shares this same advantage as a technique except that it focuses on naturally occurring variation as opposed to induced variation as in TILLING. The disadvantages of these methods are the requirement for locus-specific PCR products (difficult for gene families with very similar sequences and in polyploids) and the inability to detect mutations near SSRs (because of the flare caused by polymerase slippage-induced deletions).

19.2.3.2 Mutants Generated by Physical Agents

In plants, fast neutron bombardment is one of the best known examples of mutations generated by physical agents and has been shown to be a very effective mutagen (Koornneef et al. 1982, Li et al. 2001). Fast neutron bombardment induces deletion mutations and rearrangements (Sun et al. 1992, Salmeron et al. 1996, Li et al. 2001). Li and co-workers have demonstrated that deletion mutants can be identified for targeted plant genes by screening fast neutron-mutagenized populations via an efficient PCR screening procedure called "Deleteagene" (Li et al. 2002b). Their screening strategy requires locus-specific primers amplifying both the wild-type and the mutant gene carrying a deletion. In order to detect a mutant in a mixture of a large number of wild-type lines, the PCR extension time is shortened so that amplification of the wild-type fragment is suppressed but amplification of the deletion mutant fragment is enhanced. Using this strategy, in a population of 51,840 fast neutron lines, Li et al. have found deletion mutants for more than 80% of the 25 loci tested (Li et al. 2001). Based on these data, they estimate that a population size of 130,397 will yield a 99% probability of success in isolating a deletion in any target locus.

An alternative and high-throughput strategy for identifying genomic deletions is the use of whole-genome tiling arrays. If a genomic deletion is responsible for a particular mutant phenotype, then hybridization of genomic DNA from the mutant line to a set of probes covering the entire genome can be used to rapidly identify the probe or probes corresponding to a genomic deletion that is present in the mutant line. This method was used to successfully identify a 523-bp deletion in the *HKT1*-gene of *A. thaliana*, the deletion mutant showed a sodium over-accumulation phenotype (Gong et al. 2004, Wang et al. 2004).

One potential limitation for using deletion mutants obtained with fast neutron mutagenesis is that some deletions can be quite large and affect more than one gene. However, large deletions are beneficial if deletion of two to three copies of tandem arranged homologous genes are desired (Li et al. 2001, Zhang et al. 2003). A drawback of this method is that it requires genomic sequence information for a relatively large region around the gene of interest.

19.2.3.3 Biologically Generated Insertional Mutants

Unlike yeast, *Drosophila*, mouse, or *Physcomitrella* the efficiency of targeted gene mutations in *Arabidopsis* and in higher plants in general is very low (Puchta and Hohn 2005). Therefore, the plant community lacked for a long while the valuable genome saturating mutant collections the researches of other organisms readily obtained. However, thanks to the development of simple *Arabidopsis* transformation methods using *Agrobacterium tumefaciens* (Bechtold et al. 1993, Clough and

Bent 1998), and due to several projects that generated T-DNA insertion mutation collections, large resources are available for *Arabidopsis*.

A. tumefaciens is a soil-borne bacterial pathogen of plants. In nature, *Agrobacterium* transfers a defined segment of tumour-inducing Ti plasmid (T-DNA) into the host leading to the formation of crown gall tumours controlled by the T-DNA-encoded oncogenes (Zupan et al. 2000, Gelvin 2003). This unique DNA transfer mechanism has been successfully exploited for the introduction of transgenes into plants, a methodology that has revolutionized studies in molecular plant biology and agricultural biotechnology (Van Montagu 2003). *Agrobacterium*-mediated DNA transfer has also been successfully used in transforming other organisms such as yeast, fungi, and even human cells (Lacroix et al. 2006).

The natural T-DNA is located on the Ti (tumour inducing) plasmid and contains genes for biosynthesis of opines, a class of nutrient molecules used by *Agrobacterium*. The borders of the T-DNA are marked by two imperfect 25 bp long DNA sequence repeats, designated LB and RB (left and right border). Beside the T-DNA, the Ti plasmid also carries genes encoding enzymes for the catabolism of opines and virulence (vir) genes encoding proteins involved in transfer of T-DNA into plants. After perception of signals from wounded plants, expression of the vir genes is induced and the T-DNA is cleaved at LB and RB from the Ti plasmid and subsequently transferred into a plant cell's nucleus (Gelvin 2003). The genes within the T-DNA are dispensable for transfer and can be replaced by any given DNA and thereby transferred to plants. The simplicity and high efficiency of the *Agrobacterium*-mediated transformation compared to other DNA delivery methods is the reason for its broad use both in dicots and monocots (Birch 1997, Van Montagu 2003).

T-DNA insertional mutagenesis has so far been one of the most common ways to produce plant materials for reverse genetics. Besides mutagenesis, T-DNA insertions were used to generate activation tagging, promoter, gene, and enhancer trapping lines (Fig. 19.4). All of these methods have been very instrumental in elucidating functions and regulations of *Arabidopsis* genes (Alonso and Ecker 2006).

Agrobacterium T-DNA-Mediated Insertional Mutations

The ease of introducing foreign DNA with *Agrobacterium* using the floral dip method (Bechtold et al. 1993, Clough and Bent 1998) allowed the generation of thousands of transformants and made systematic analysis of gene function feasible. Although the distribution of isolated T-DNA insertions appears not to be fully random (Schneeberger et al. 2005, Li et al. 2006, Gelvin and Kim 2007, Zhang et al. 2007), there are currently over 400,000 mapped T-DNA insertions and one or more insertion mutants have been obtained for more than 95% of annotated *Arabidopsis* genes (Table 19.1).

Four of the most widely used T-DNA insertion line collections are discussed in detail in this chapter. These are GABI-Kat lines (Genomanalyse in biologischen System Pflanze Arabidopsis T-DNA lines, (Rosso et al. 2003)); SAIL (Syngenta Arabidopsis Insertion Library), formerly known as GARLIC (Gilroy Arabidopsis Reverse Lethal Insertion Collection, (Sessions et al. 2002)); FLAGdb (functional analysis of the Arabidopsis genome database), also known as INRA/Versailles lines

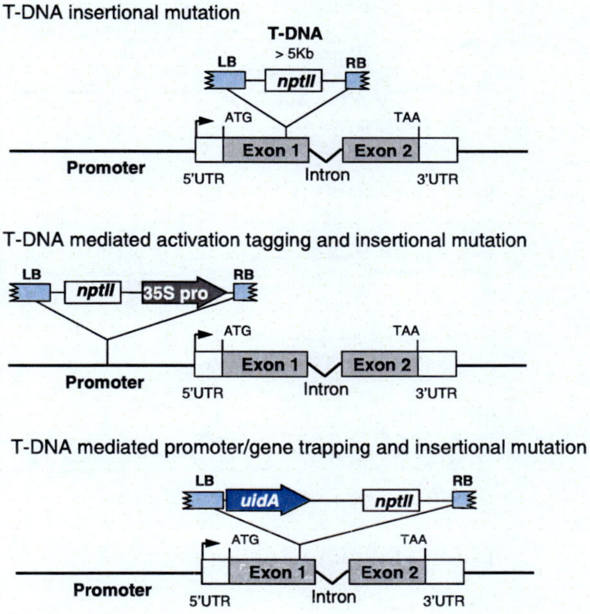

Fig. 19.4 Approaches using T-DNA insertions into *Arabidopsis* genome to elucidate gene function

(Samson et al. 2002) and SALK Institute T-DNA insertion lines (Alonso et al. 2003). The details of the T-DNA constructs used in generating these four T-DNA insertion populations are shown in Fig. 19.5.

As depicted in Fig. 19.6, these T-DNA insertion lines are generated by transferring the corresponding T-DNA constructs to either the Columbia or Wassilewskija accession of *Arabidopsis* using the *Agrobacterium* floral dip method (Bechtold et al. 1993, Clough and Bent 1998). T_1 transformants are selected using a method appropriate for the respective resistance gene used. For example, SALK lines were selected with kanamycin (the *nptII* gene confers resistance) while GABI-Kat were selected on sulfadiazine (the *sul* gene confers resistance)-containing media. DNA is isolated from these surviving T_1 plants (or their offspring) and used to determine the position of T-DNA insertion in the *Arabidopsis* genome by obtaining sequences flanking the T-DNA borders. Various methods are used in determining the T-DNA flanking sequences. These methods are typically known as FST (flanking sequence tag) recovery methods. Adapter ligated PCR (Alonso et al. 2003), TAIL PCR (Liu et al. 1995, Sessions et al. 2002), or inverse PCR (Thomas et al. 1994) are the most frequently used FST recovery methods. PCR fragments isolated from such recovery methods are sequenced and the sequences are deposited to databases. These sequences are then compared with *Arabidopsis* genome sequence via BLAST (Basic Local Alignment Search Tool) and the ones having significant matches are determined (Fig. 19.6).

Depending on T-DNA insertion collection, these sequences and T2 and later generation offspring seeds are stored by the institution involved in generating these material and made publicly available or deposited to stock centres where users can access sequence data and obtain seeds of such insertion lines for a small fee. The websites hosting such T-DNA insertion sequences and lines are given in Table 19.1.

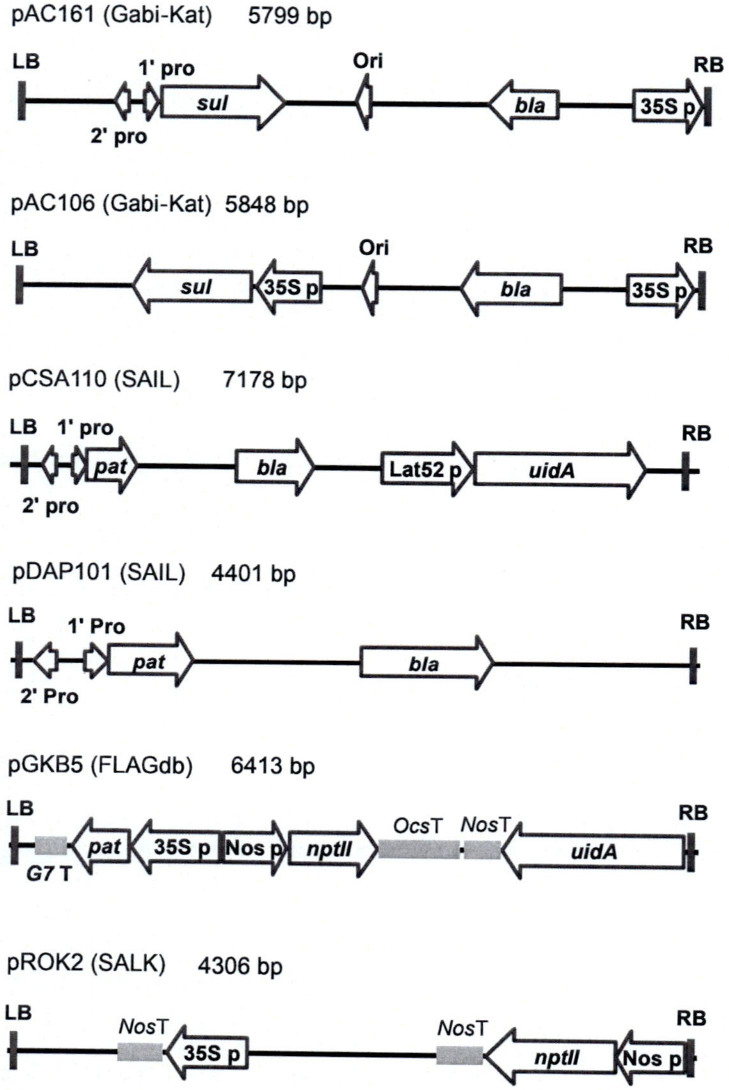

Fig. 19.5 T-DNA vectors used in generation of the Gabi-Kat, SAIL, FLAGdb, and SALK *lines*. pAC161 according to Rosso et al. (2003) and GenBank accession AJ537514, pCSA110 according to Mengiste et al. (1997), pGKB5 according to Bouchez et al. (1993b) and pROK2 according to Baulcombe et al. (1986). RB, right border; LB, left border; ori, origin of replication; *bla*, β-lactamase gene conferring resistance to ampicillin in bacteria; *sul*, resistance marker gene conferring resistance to antibiotic sulfadiazine in plants; *uidA* (GUS), coding region of the β-glucuronidase from *Escherichia coli*; *Nos*T, 3′ region (terminator) of the nopaline synthase gene from pTiC58; *Ocs*T, 3′ region (terminator) of the octopine synthase gene from pTiAch5; Lat52 p, a pollen-specific promoter from the tomato Lat52 gene; *nptII*, neomycin phosphotransferase II gene conferring resistance to kanamycin; Nos p, promoter region of the nopaline synthase gene; 35S p, promoter of the 35S transcript of cauliflower mosaic virus; *pat*, coding sequence of the basta resistance gene phosphinothricin acetyl transferase from *Streptomyces hygroscopicus*; *G7* T, 3′ region (terminator) of gene 7 from the T-DNA of pTi15955. Further information, links, and downloadable GenBank formatted files of these sequences are provided in Ülker et al. (2008b)

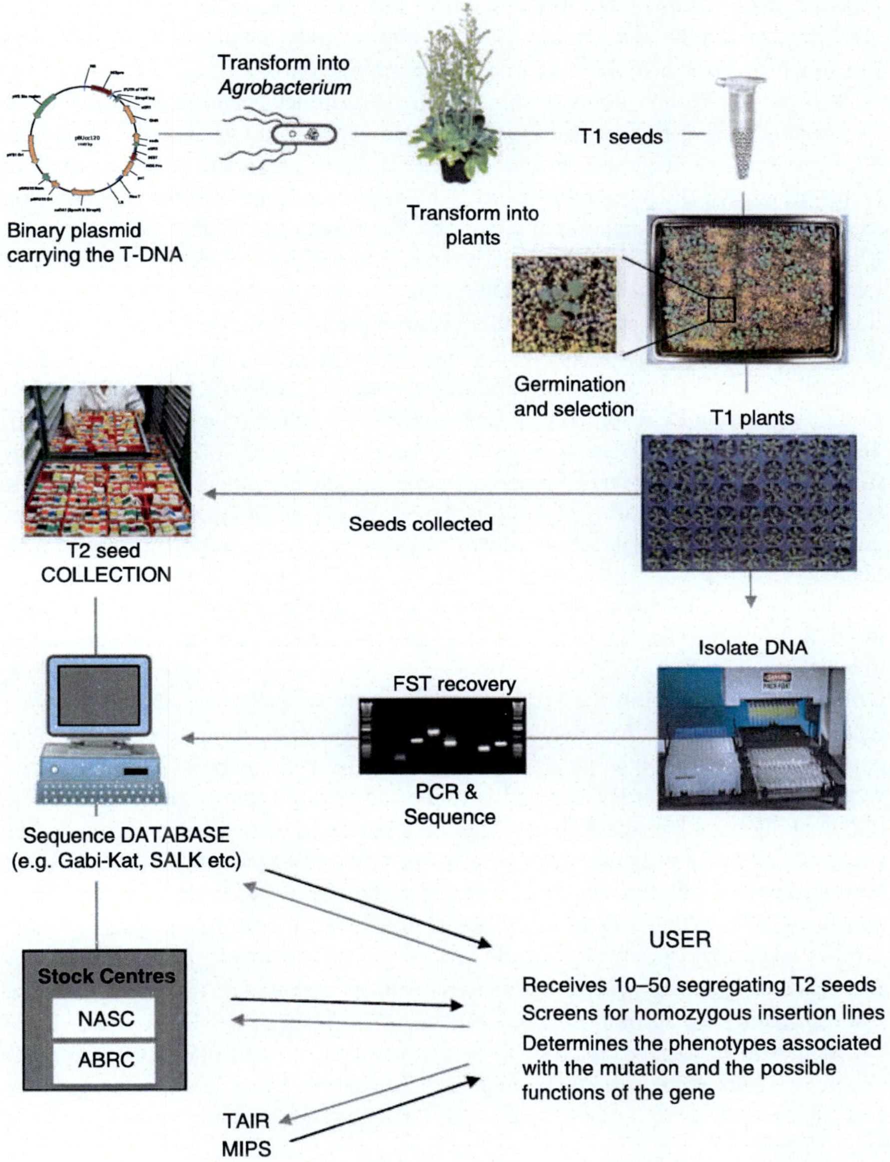

Fig. 19.6 Plant transformation and FST recovery procedure from T-DNA insertion populations and how the end user searches and obtains these lines; general procedure with a focus on GABI-Kat (for details please refer to the text)

Researchers who are interested in finding T-DNA insertion lines in their gene of interest can search using AGI (*Arabidopsis* Genome Initiative) gene code number or by comparing their sequence of interest with the T-DNA insertion sequence databases via BLAST. Once a promising T-DNA insertion event is found, the user requests seeds of this line to further confirm T-DNA insertion, to isolate a homozygous line, to amplify the amount of seeds, and to study phenotypes and functions that might be associated with the mutated gene (Fig. 19.6).

In order to obtain a homozygous line, the researcher needs to grow the T_2 or T_3 progeny of T-DNA insertion line and isolate DNA for PCR analysis. Typically, 8–16 progeny is sufficient to find at least one homozygous plant. DNAs from these plants are used in PCR reactions using gene-specific primers flanking the insertion site as well as the combination of gene-specific primers and T-DNA border primers (Fig. 19.7). As can be seen from the example of a T-DNA insertion into the *TT1* locus of *Arabidopsis* in Fig. 19.7, DNA samples from homozygous T-DNA insertion lines would fail to produce an amplification product with gene-specific primers but DNA from wild-type plants would amplify the fragment of expected size. Due to large sizes of T-DNA sequences, typically over 5 kb in length, under short extension times used in PCR screenings, fragments over 2 kb are typically not amplified. To make sure that a T-DNA is inserted in the locus of interest, in a separate reaction a combination of T-DNA border and gene-specific primers is used. Such a PCR amplification would result in amplification of the predicted fragment in heterozygous and homozygous plants but would fail to amplify any fragment in wild-type plants.

Ülker et al. 2008b have reported several pitfalls and untapped potentials of these publicly available T-DNA insertion lines that researchers should consider when using them (Ülker et al. 2008b). For example, regulatory sequences (promoters, terminators, etc.) used in the T-DNA vectors could influence the flanking regions in the genome where they are inserted. The constitutive promoters such as 35S could lead to activation of genes that are located near the insertion site. Similarly, weak terminators in T-DNAs could lead to read-through transcripts from T-DNA into flanking plant loci and influence gene expression by causing over-expression or antisense silencing of genes situated adjacent to the insertion site as well as genes with sequence similarity which are located somewhere else in the genome. They also demonstrated that T-DNA insertions in introns need to be used cautiously since T-DNA insertions in intronic regions can be spliced out together with the introns and hence fully functional full-length transcripts are generated. We also discovered that *Agrobacterium* chromosomal DNA (AchrDNA) other than T-DNA or the vector backbone can be transferred from *Agrobacterium* and may be integrated into plant genomes (Ülker et al. 2008a). We reported that one in 250 T-DNA insertion lines could have such AchrDNA associated with T-DNAs integrated in the genome (Ülker et al. 2008a).

Transposon-Mediated Insertional Mutations

Transposons are mobile genetic elements that can jump from one location in the genome to another as long as active transposase is present. This feature of transposons has been used by researchers to generate large mutant collection in plants for functional genomics analyses. These include Cold Spring Harbor Laboratory gene and enhancer trap population (Sundaresan et al. 1995, Martienssen 1998), the

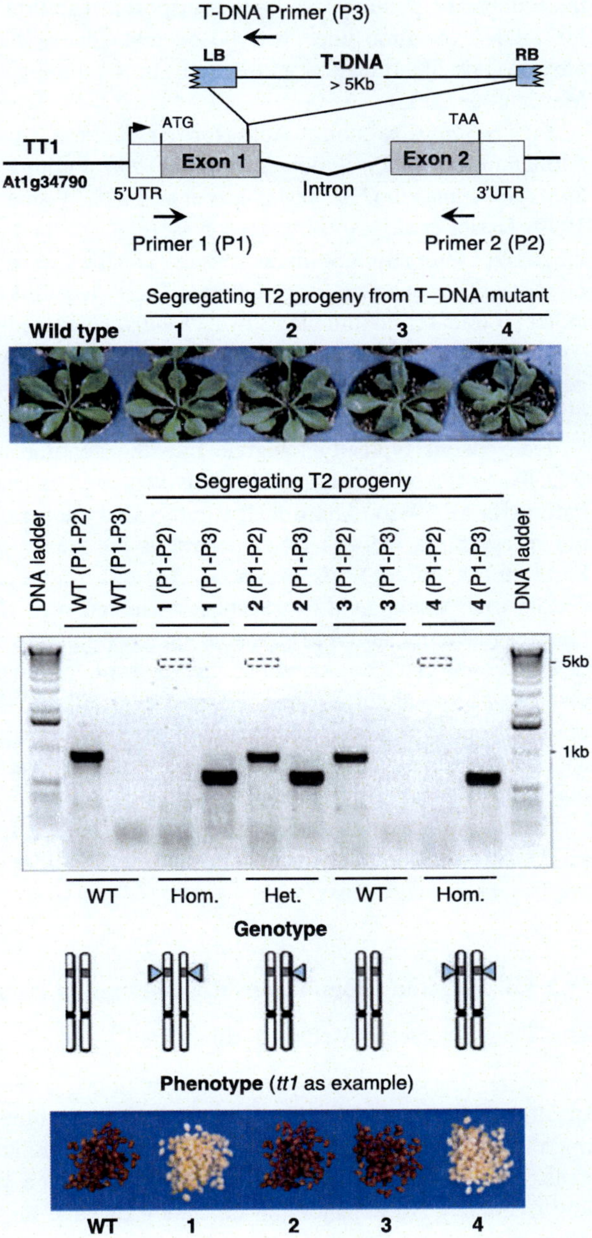

Fig. 19.7 PCR screening method for identification of homozygous T-DNA insertions in *Arabidopsis* genes/genome

Sainsbury Laboratory *A. thaliana* (SLAT) lines (Tissier et al. 1999), the Wageningen population (Aarts et al. 1995), AMAZE/ZIGIA Population (Wisman et al. 1998), and University of Cologne TAMARA (transposable element-mediated activation tagging mutagenesis in *Arabidopsis*) collection (Schneider et al. 2005). Although the number of publicly available transposon insertion lines is not as high as it is for T-DNA insertion lines, transposon mutagenesis has provided excellent examples of gene identification and cloning in *Arabidopsis* and other plants (May and Martienssen 2003).

In *Arabidopsis*, maize Activator/Dissociation (*Ac/Ds*) (Honma et al. 1993, Altmann et al. 1995, Sundaresan et al. 1995, Parinov et al. 1999) and *Enhancer/Supressor–mutator* (*En/Spm*) (Aarts et al. 1995, Wisman et al. 1998, Speulman et al. 1999, Tissier et al. 1999) transposon systems are the most frequently used ones. In the *Ac/Ds* system, immobilized *Ac* is used as the source of transposase and modified *Ds* elements as enhancer or gene traps. These two units of the transposon system are either co-transferred into plants by *Agrobacterium*-mediated transformation or a *Ds*-containing plant is crossed with a plant expressing transposase. The new mutations generated by mobilizing the transposon are then stabilized by crossing the active transposase out or selecting against it.

Transposons typically integrate into the genome as simple, intact insertions. T-DNA, on the other hand, tends to be integrated in arrays (Sundaresan et al. 1995, Parinov et al. 1999). Additionally, transposed elements can be remobilized resulting in reversion while T-DNA insertions are stably integrated into the genome. Furthermore, mutagenesis of genomic regions can be carried out by exploiting that certain transposons tend to transpose to linked sites (Muskett et al. 2003). These features make transposable elements uniquely suitable vehicles for a powerful form of insertional mutagen, known as the gene trap or enhancer trap. The drawback is that transposons are often present in multiple copies in the mutagenized lines, which are laborious to remove. And although the generation of high numbers of transposon insertion lines is straightforward, determining and selecting plants with stable insertions can be relatively cumbersome.

Transposon insertion lines can be screened by PCR for insertion in genes of interest. Isolation of homozygous transposon insertion lines can be done as described for the T-DNA insertion lines.

19.2.3.4 Targeted (Homologous Recombination Induced) Mutations

Zinc Finger Nuclease-Mediated Mutations

Gene targeting by homologous recombination is a method that allows to change specific DNA sequences at a defined locus in genomes. A DNA fragment carrying a desired sequence is introduced into a cell to replace the existing copy of the gene. As mentioned above, gene targeting takes place in higher plants only at quite low frequency. To enhance the efficiency of gene targeting in plants, a chromosome break is created at the target site of modification. An enzyme called a zinc finger nuclease (ZFN) is used to generate the chromosome break. ZFNs have two components: a DNA recognition domain (a zinc finger array) and a nuclease that cleaves the DNA. Several reports have demonstrated the feasibility of this technology in introducing desired mutations in plant genomes (Wright et al. 2005, Shukla et al. 2009, Townsend et al. 2009). Current research is directed at developing zinc finger nuclease-assisted gene targeting for widespread use, including establishing key parameters for high-frequency gene targeting and robust methods for the design of zinc finger arrays. The key step is the specificity of the Zinc finger array which must address ideally only one site in the (haploid) genome. Once these improvements are achieved, the gene-targeting platform will enable the engineering of crop plants with novel traits. Crops generated through gene targeting may be met with greater public acceptance than traditional genetically modified crops since gene targeting introduces changes in plant genomes in a highly specific and controlled manner.

19.2.4 Over-Expression/Activation-Mediated Functional Assays

Although usually the best way to unravel the function of a gene is to study a loss-of-function (NULL) allele, this strategy is not informative for every gene. For example, the functions of genes belonging to large families with redundant roles and genes whose loss-of-function mutants are lethal cannot be dissected using simple loss-of-function mutants. In these cases the use of other methodologies such as over-expression with a "constitutive" or inducible promoter, activation tagging, and gene silencing by engineered microRNAs may provide clues to gene function.

19.2.4.1 Collections of Transgenic Lines that Over-Express Plant Genes

Large-scale generation of over-expressors for *Arabidopsis* genes were developed by cloning the full length cDNAs under the control of 35S cauliflower mosaic virus promoter and transforming them into plants using T-DNA-mediated transformation (LeClere and Bartel 2001, Ichikawa et al. 2006). Due to frequently observed severe developmental deficiencies in transgenic plants over-expressing genes under a strong constitutive promoter, Papdi et al. (2008) have used an estradiol-inducible system for the identication of novel *Arabidopsis* loci involved in the control of abiotic stress responses, by generating transgenic lines expressing a library of *Arabidopsis* cDNAs (Papdi et al. 2008).

19.2.4.2 Activation Tagging Lines

Systematic gene activation tagging systems have been established in *Arabidopsis* using T-DNA vectors carrying four copies of 35S promoter enhancers at the left (Walden et al. 1994, Weigel et al. 2000, Nakazawa et al. 2003, Seki et al. 2005) or right T-DNA border region (Ichikawa et al. 2003). Other activation tagging systems also have been established in *Arabidopsis* using the transposon Ac/Ds system or En-I system, which carries CaMV 35S enhancers (Wilson et al. 1996, Marsch-Martinez et al. 2002, Schneider et al. 2005). Genes located up to 8.2 kb away from the enhancer sequence are activated in an *Arabidopsis* activation tagging population (Ichikawa et al. 2003).

Activation tagging has been successfully applied to uncover the function of novel genes in plant development such as *ADR2* (Aboul-Soud et al. 2009), *HD-START* (Yu et al. 2008), and *ADR1* conveying resistance to virulent pathogens (Grant et al. 2003), *BAK1* an *Arabidopsis* LRR receptor-like protein kinase gene modulating brassinosteroid signalling (Li et al. 2002a), *PAP1* encoding for a MYB transcription factor involved in phenylpropanoid biosynthesis (Borevitz et al. 2000), *LEAFY PETIOLE* affecting leaf petiole development (van der Graaff et al. 2000), *FLOWERING LOCUS T* (*FT*) which induces flowering of *Arabidopsis* (Kardailsky et al. 1999), *BAS1* regulating brassinosteroid levels and light responsiveness in *Arabidopsis* (Neff et al. 1999) and *CKI1* which is involved in cytokinin signal transduction (Kakimoto 1996).

Unlike FST indexed web-searchable T-DNA insertion line collections, most of these activation tagging populations require PCR screening of pooled DNAs of tagged mutant plants. Only the small collection of RIKEN Institute (Ichikawa et al. 2003) is sequence searchable and includes 500 lines that showed phenotype at the T1 generation out of 50,000 lines (http://rarge.gsc.riken.go.jp/activationtag/top.php). Another important source for activation tagging lines is the GABI-Kat T-DNA insertion collection. Although GABI-Kat lines were originally generated for activation tagging and hence contain 35S promoter at T-DNA right border, many researchers tend to use this resource for only T-DNA insertional mutagenesis (Rosso et al. 2003). Ülker et al. have demonstrated the use of this population for activation tagging as well (Ülker et al. 2008b). When using this resource for activation tagging one must be aware that the FSTs are recovered from the left border in GABI-Kat lines; hence presence of T-DNA in a locus of interest does not necessarily mean that there is 35S promoter on the other side of the T-DNA in the same locus due to typical complex T-DNA insertions in a locus. Therefore, researchers must experimentally determine if there is indeed a 35S promoter next to region of interest when using this resource for activation tagging. It is also important to establish if the phenotype is the result of activation or insertional mutation in these lines.

There are several advantages of using activation tagging lines over artificially generated transgenic over-expressors. Some *Arabidopsis* genes are very large, hence amplification and cloning of such genes can be problematic. Numerous international consortia were initiated to supply scientists with full length cDNAs of *Arabidopsis* (See MASC report 2008). Activation tagged lines, in contrast, have already been produced, saving researchers time, money, and effort. Over-expressing an endogenous gene in its native location with all the regulatory sequences at the 3′ end and intronic sequences may have additional advantages over artificially generated transgenic over-expressors. For some genes it has been shown that regions of importance for gene regulation lie at the 3′end or in introns, in such cases complementation experiments of mutants with over-expressed cDNAs may fail (Bolle et al. 1996, Deyholos and Sieburth 2000, Mun et al. 2002, Sheldon et al. 2002, Quesada et al. 2003).

19.2.5 Gene Silencing-Mediated Functional Analysis

Gene silencing lines are very important in elucidating gene function especially when no suitable (reduced/loss of function) T-DNA insertion alleles are available or for genes belonging to families where gene redundancy is a known issue.

Gene silencing, also called RNA interference (RNAi), co-suppression, or post-transcriptional gene silencing, is an innate gene silencing system that is triggered by double-stranded RNA (dsRNA) in plants and other eukaryotes. It is thought to have evolved as a defence mechanism against viruses and active endogenous transposons (Robertson 2004). Viral infection, inverted transgene repeats, or aberrant transcription products all lead to the production of dsRNA. This dsRNA is converted to small interfering RNA (siRNA) that cause gene silencing by various mechanisms. Plants exhibit a surprising diversity of small RNA types and the proteins that generate them. These small RNA classes include microRNAs (miRNAs), siRNAs, and transacting RNAs (taRNAs). These classes differ in their biogenesis, in modes of target regulation, and in the biological pathways they regulate (Ghildiyal and Zamore 2009).

Gene silencing occurs when cells respond to a dsRNA trigger by destroying a corresponding target, an mRNA sharing some or all of the sequence of the dsRNA (Zamore et al. 2000, Agrawal et al. 2003, Ghildiyal and Zamore 2009). RNA-silencing pathways also include the microRNA (miRNA) pathways of plants and animals, in which small RNAs derived from their own coding genes repress one or several target mRNAs to which they are partially complementary. Therefore gene silencing can be induced artificially by introducing an antisense RNA construct that is complementary to a sense mRNA in the genome, dsRNA-forming constructs that are aimed at targeting homologous mRNA sequences in the genome, or by engineering a miRNA gene such that the expression of the gene of interest is affected.

19.2.5.1 Antisense Lines

The feasibility of using antisense lines at large scale for functional genomics was demonstrated by expressing random antisense cDNAs in transgenic *Arabidopsis* lines (Jun et al. 2002). However, due to the higher efficiency of other gene silencing strategies that build on RNAi constructs of defined sequence or artificial miRNAs, simple antisense technology is not used as much anymore.

19.2.5.2 RNAi Lines

In order to silence a specific gene of interest, an expression construct in which full or partial length coding sequence of the target gene is cloned as inverted repeat mostly separated by an intron and transformed into plants. Transcription of the fragments in the construct generates hairpin-forming RNA molecules that are targets for the innate gene silencing/RNAi machinery. The resulting siRNAs subsequently assemble with proteins into an RNA-induced silencing complex (RISC). The RISC complex targets mRNAs with sequences related to siRNAs for degradation (Agrawal et al. 2003).

In *Arabidopsis*, a collection of RNAi lines covering 22,969 of *Arabidopsis* genes has been initiated in 2004 (The AGRIKOLA project, http://www.agrikola.org), (Hilson et al. 2004) and is becoming available. Currently, 3,592 transformed plant lines are available from NASC (Table 19.2).

Table 19.2 Various other resources for reverse genetics in *Arabidopsis*

Initiative	Number of lines	Ecotype	References
Stock centres			
NASC (Nottingham *Arabidopsis* stock centre) http://arabidopsis.info/			
ABRC (*Arabidopsis* biological resource center) http://www.biosci.ohio-state.edu/~plantbio/Facilities/abrc/abrchome.htm			
Transposon insertion collections			
Gene/enhancer trap collection at CSHL http://genetrap.cshl.org/	16,887	Landsberg	Martienssen (1998)
SLAT (Sainsbury laboratory *Arabidopsis thaliana* transposants) http://www.arabidopsis.info/info/slat_info1.html	1,000	Columbia-0	
TAMARA, University of Cologne, Germany http://arabidopsis.info/CollectionInfo?id=71 Available through the Nottingham Arabidopsis Stock Centre	9,471	Columbia-0	Schneider et al. (2005)
Enhancer/promoter trap and activation tagging collections			
SK Collection (Saskatoon Arabidopsis T-DNA mutant population) http://aafc-aac.usask.ca/FST	15,507 out of 50,000 are sequence indexed	Columbia	Robinson et al. (2009)
Weigel T-DNA activation lines http://arabidopsis.org/abrc/weigel.jsp Available from ABRC and NASC 22,600	22,600	Columbia-0	Weigel et al. (2000)
TAMARA, University of Cologne, Germany http://arabidopsis.info/CollectionInfo?id=71 Available through the Nottingham Arabidopsis stock centre	9,471	Columbia-0	Schneider et al. (2005)
GAL4-GFP enhancer trap lines http://www.plantsci.cam.ac.uk/Haseloff/	>8,000	C24	Kiegle et al. (2000)
GAL4-GFP enhancer trap lines http://enhancertraps.bio.upenn.edu/default.html		Columbia-0	Kiegle et al. (2000)
RIKEN Activation tagging line database http://rarge.gsc.riken.go.jp/activationtag/top.php Available from RIKEN BRC 4,000 transposon insertion lines which have the Ds transposon with visible phenotypes. Kuromori 2006 now is at http://rarge.gsc.riken.jp/phenome/	32,650	Columbia-0	Ichikawa et al. (2003), Nakazawa et al. (2003)

Table 19.2 (continued)

Initiative	Number of lines	Ecotype	References
JIC activation lines, John innes centre, UK http://arabidopsis.info/CollectionInfo?id=29 Available from ABRC and NASC	972	Landsberg	N/A
Gene/enhancer trap collection at CSHL http://genetrap.cshl.org/	14,204	Landsberg	Martienssen (1998)
Collections for gene-specific screening			
The *Arabidopsis* knockout facility (Alpha) http://www.biotech.wisc.edu/Arabidopsis ceased operations due to lack of funding	60,480	Wassilevskija	Sussman et al. (2000)
The *Arabidopsis* knockout facility (Basta) http://www.biotech.wisc.edu/Arabidopsis ceased operations due to lack of funding	72,960	Wassilevskija	Sussman et al. (2000)
TILLING			
Seattle *Arabidopsis* TILLING project http://tilling.fhcrc.org/ 3000 M2 plants harbouring EMS-induced mutations are screened for insertions in gene of interest		An erecta mutation in a Col-0 background	Till et al. (2003)
A single Col er105 plant provided seed for EMS mutagenesis. Coler105, donated by Keiko Torii, carries an induced erecta mutation in a Col-0 background			
TILLer http://www.cnb.csic.es/~tiller/Analysis of 2,300–3,700 mutant lines, ensuring that at least five independent mutations are identified		*Landsberg erecta glabrous1 (Ler gl1)* background	Colbert et al. (2001), Greene et al. (2003)
AGRICOLA (*Arabidopsis* genomic RNAi knockout line analysis) http://www.agrikola.org Available from NASC	3,592	Columbia	Hilson et al. (2004)

19.2.5.3 MicroRNAs and Targeted miRNA Lines

MicroRNAs (miRNAs) are encoded by the genome itself and the precursor transcripts form imperfect stem-loop structures. The mature miRNAs (20–25 nt long short RNAs) interact with their target mRNAs by sequence complementarity and lead to their degradation or inhibit their translation into proteins (Schwab et al. 2006, Ghildiyal and Zamore 2009). Plant miRNAs direct mRNA cleavage of a single or small number of targets with near-perfect complementarity. There are 1,638 distinct miRNAs in plants (miRBase, Release 12.0). miRNAs regulate diverse cellular pathways and are widely believed to regulate most biological processes, in both plants and animals, ranging from housekeeping functions to responses to environmental stress (Brodersen et al. 2008).

miRNA-based artificial miRNAs (amiRNAs) have been developed and successfully applied to the silencing of various plant endogenous genes as well as plant pathogens genes (Alvarez et al. 2006, Du et al. 2006, Niu et al. 2006, Schwab et al. 2006). This technology has great potential to be used in the functional characterization of genes with overlapping or partially redundant functions.

Silencing of multiple gene targets is to a degree possible using all of these silencing mechanisms. However, these techniques all require rational selection of target genes and follow-up studies to understand which subset of gene products has been inactivated. RNA interference does not entirely ensure that phenotypic traits will be inherited through subsequent generations, moreover silencing can vary, so results may be unpredictable and many independent transgenic lines might need to be analysed. A large collection of different siRNAs are generated once silencing is initiated, therefore it is also possible to have off-target gene silencing which is difficult to predict (Jackson et al. 2003, Tschuch et al. 2008). Furthermore, high throughput is somewhat limited since for each gene of interest a carefully designed silencing construct needs to be generated.

19.3 Outlook

Although T-DNA insertion-based reverse genetics methods are well established in *Arabidopsis* and have been extremely successful and productive in providing loss-of-function mutants to the community, there are still a few thousands of genes without a predicted and probably more without a confirmable T-DNA insertion allele. Additionally, detailed analysis of some of the mutants indicates that not every T-DNA insertion allele is useful. Therefore, these efforts should be continued until the genome is saturated with at least one and hopefully two or more useful T-DNA insertion alleles. For large phenotype screening analysis, homozygous T-DNA insertion mutants are necessary. Currently, the SALK institute has over 25,000 homozygous T-DNA insertion lines available and this number needs to be increased to cover all genes in *Arabidopsis*. Also the GABI-Kat project provides access to homozygous confirmed mutants via NASC.

Thanks to TILLING and other targeted screening methods, point mutations as well as INDELS are expected to be searched for and found in those genes without useful T-DNA insertion mutants. We predict that there will be further improvements in mutation detection and screening methods. For example, new developments in genome sequencing lead to drastic reductions in the cost of sequencing (Service 2006, Pettersson et al. 2009). Such price reductions are expected to continue and allow scientists to sequence whole genomes to detect polymorphisms and mutations in plants. Additionally, Zinc finger-mediated homologous recombination as well as similar strategies are expected to play important roles in further characterizing gene functions by allowing precise changes in the plant genome.

Only a small proportion of the knockout lines obtained to date display directly informative phenotypes at the whole-plant level. Functional redundancy within members of gene families appear to be one of the reasons for this. This problem can often be overcome by generating double and/or triple mutants or by silencing multiple genes. However, international community efforts might need to be initiated to generate genome-wide double, triple, or multiple T-DNA insertion mutant or gene silencing collections for members of gene families. Comprehensive phenotype analyses for every gene in *Arabidopsis* is expected to be developed. The main current bottleneck in gene function search is phenotyping, and automation as well as standardization in this area will allow to find more gene functions, including smaller contributions to a phenotype (or chemotype detected by metabolomics) that is difficult to score without statistical measures.

Arabidopsis research has been providing valuable insights into all aspects of modern biology. Significant advances have been achieved in applied research efforts in cloning disease resistance genes, genes involved in cold and drought tolerance, hormone biogenesis and flowering. Similarly, functions of thousands of genes were first determined in *Arabidopsis* and such discoveries started to be translated to crops. This tiny weed is expected to continue to be the best characterized plant model system and its community resources will continue to increase both in numbers and sophistication.

References

Aarts MG, Corzaan P, Stiekema WJ, Pereira A (1995) A two-element enhancer-inhibitor transposon system in Arabidopsis thaliana. Mol Gen Genet 247:555–564

Aboul-Soud MA, Chen X, Kang JG, Yun BW, Raja MU, Malik SI, Loake GJ (2009) Activation tagging of ADR2 conveys a spreading lesion phenotype and resistance to biotrophic pathogens. New Phytol 183(4):1163–1175

Agrawal N, Dasaradhi PV, Mohmmed A, Malhotra P, Bhatnagar RK, Mukherjee SK (2003) RNA interference: biology, mechanism, and applications. Microbiol Mol Biol Rev 67:657–685

Alonso JM, Ecker JR (2006) Moving forward in reverse: genetic technologies to enable genome-wide phenomic screens in Arabidopsis. Nat Rev Genet 7:524–536

Alonso JM, Stepanova AN, Leisse TJ, Kim CJ, Chen H, Shinn P, Stevenson DK, Zimmerman J, Barajas P, Cheuk R et al (2003) Genome-wide insertional mutagenesis of Arabidopsis thaliana. Science 301:653–657

Altmann T, Felix G, Jessop A, Kauschmann A, Uwer U, Pena-Cortes H, Willmitzer L (1995) Ac/Ds transposon mutagenesis in Arabidopsis thaliana: mutant spectrum and frequency of Ds insertion mutants. Mol Gen Genet 247:646–652

Alvarado MC, Zsigmond LM, Kovacs I, Cseplo A, Koncz C, Szabados LM (2004) Gene trapping with firefly luciferase in Arabidopsis. Tagging of stress-responsive genes. Plant Physiol 134: 18–27

Alvarez JP, Pekker I, Goldshmidt A, Blum E, Amsellem Z, Eshed Y (2006) Endogenous and synthetic microRNAs stimulate simultaneous, efficient, and localized regulation of multiple targets in diverse species. Plant Cell 18:1134–1151

Arabidiopsis Genome Initiative (2000) Analysis of the genome sequence of the flowering plant Arabidopsis thaliana. Nature 408:796–815

Balzergue S, Dubreucq B, Chauvin S, Le-Clainche I, Le Boulaire F, de Rose R, Samson F, Biaudet V, Lecharny A, Cruaud C et al (2001) Improved PCR-walking for large-scale isolation of plant T-DNA borders. Biotechniques 30:496–498, 502, 504

Barkley NA, Wang ML (2008) Application of TILLING and EcoTILLING as reverse genetic approaches to elucidate the function of genes in plants and animals. Curr Genomics 9:212–226

Baulcombe DC, Saunders GR, Bevan MW, Mayo MA, Harrison BD (1986) Expression of biologically-active viral satellite RNA from the nuclear genome of transformed plants. Nature 321:446–449

Bechtold N, Ellis J, Pelletier G (1993) In planta Agrobacterium mediated gene transfer by infiltration of adult Arabidopsis thaliana plants. CR Acad Sci Paris Life Sci 316:1194–1199

Birch RG (1997) PLANT TRANSFORMATION: problems and strategies for practical application. Annu Rev Plant Physiol Plant Mol Biol 48:297–326

Bolle C, Herrmann RG, Oelmuller R (1996) Intron sequences are involved in the plastid- and light-dependent expression of the spinach PsaD gene. Plant J 10:919–924

Borevitz JO, Xia Y, Blount J, Dixon RA, Lamb C (2000) Activation tagging identifies a conserved MYB regulator of phenylpropanoid biosynthesis. Plant Cell 12:2383–2394

Bouchez D, Camillieri C, Caboche M (1993a) A binary vector based on Basta resistance for in planta transformation of Arabidopsis thaliana. C R Acad Sci Paris 316:1188–1193

Bouchez D, Camillieri C, Caboche M (1993b) A binary vector based on Basta resistance for in planta transformation of *Arabidopsis thaliana*. CR Acad Sci Paris 316:1188–1193

Brodersen P, Sakvarelidze-Achard L, Bruun-Rasmussen M, Dunoyer P, Yamamoto YY, Sieburth L, Voinnet O (2008) Widespread translational inhibition by plant miRNAs and siRNAs. Science 320:1185–1190

Caldwell DG, McCallum N, Shaw P, Muehlbauer GJ, Marshall DF, Waugh R (2004) A structured mutant population for forward and reverse genetics in barley (*Hordeum vulgare* L.). Plant J 40:143-150

Campisi L, Yang Y, Yi Y, Heilig E, Herman B, Cassista AJ, Allen DW, Xiang H, Jack T (1999) Generation of enhancer trap lines in Arabidopsis and characterization of expression patterns in the inflorescence. Plant J 17:699–707

Casadaban MJ, Cohen SN (1979) Lactose genes fused to exogenous promoters in one step using a Mu-lac bacteriophage: in vivo probe for transcriptional control sequences. Proc Natl Acad Sci USA 76:4530–4533

Clough SJ, Bent AF (1998) Floral dip: a simplified method for agrobacterium-mediated transformation of Arabidopsis thaliana. Plant J 16:735–743

Colbert T, Till BJ, Tompa R, Reynolds S, Steine MN, Yeung AT, McCallum CM, Comai L, Henikoff S (2001) High-throughput screening for induced point mutations. Plant Physiol 126:480–484

Comai L, Young K, Till BJ, Reynolds SH, Greene EA, Codomo CA, Enns LC, Johnson JE, Burtner C, Odden AR, Henikoff S (2004) Efficient discovery of DNA polymorphisms in natural populations by ecotilling. Plant J 37:778–786

Cooper JL, Till BJ, Laport RG, Darlow MC, Kleffner JM, Jamai A, El-Mellouki T, Liu S, Ritchie R, Nielsen N, Bilyeu KD, Meksem K, Comai L, Henikoff S (2008) TILLING to detect induced mutations in soybean. BMC Plant Biol 8:9

Deyholos MK, Sieburth LE (2000) Separable whorl-specific expression and negative regulation by enhancer elements within the AGAMOUS second intron. Plant Cell 12:1799–1810

Dierking EC, Bilyeu KD (2009) New sources of soybean seed meal and oil composition traits identified through TILLING. BMC Plant Biol 9:89

Du G, Yonekubo J, Zeng Y, Osisami M, Frohman MA (2006) Design of expression vectors for RNA interference based on miRNAs and RNA splicing. FEBS J 273:5421–5427

Enns LC, Kanaoka MM, Torii KU, Comai L, Okada K, Cleland RE (2005) Two callose synthases, GSL1 and GSL5, play an essential and redundant role in plant and pollen development and in fertility. Plant Mol Biol 58:333–349

Eyal Y, Curie C, McCormick S (1995) Pollen specificity elements reside in 30 bp of the proximal promoters of two pollen-expressed genes. Plant Cell 7:373–384

Fobert PR, Miki BL, Iyer VN (1991) Detection of gene regulatory signals in plants revealed by T-DNA-mediated fusions. Plant Mol Biol 17:837–851

Gelvin SB (2003) Agrobacterium-mediated plant transformation: the biology behind the "gene-jockeying" tool. Microbiol Mol Biol Rev 67:16–37 table of contents

Gelvin SB, Kim SI (2007) Effect of chromatin upon agrobacterium T-DNA integration and transgene expression. Biochim Biophys Acta 1769:410–421

Ghildiyal M, Zamore PD (2009) Small silencing RNAs: an expanding universe. Nat Rev Genet 10:94–108

Gilchrist EJ, O'Neil NJ, Rose AM, Zetka MC, Haughn GW (2006) TILLING is an effective reverse genetics technique for Caenorhabditis elegans. BMC Genomics 7:262

Gong JM, Waner DA, Horie T, Li SL, Horie R, Abid KB, Schroeder JI (2004) Microarray-based rapid cloning of an ion accumulation deletion mutant in Arabidopsis thaliana. Proc Natl Acad Sci USA 101:15404–15409

Grant JJ, Chini A, Basu D, Loake GJ (2003) Targeted activation tagging of the Arabidopsis NBS-LRR gene, ADR1, conveys resistance to virulent pathogens. Mol Plant Microbe Interact 16:669–680

Greene EA, Codomo CA, Taylor NE, Henikoff JG, Till BJ, Reynolds SH, Enns LC, Burtner C, Johnson JE, Odden AR, Comai L, Henikoff S (2003) Spectrum of chemically induced mutations from a large-scale reverse-genetic screen in Arabidopsis. Genetics 164:731–740

Henikoff S, Till BJ, Comai L (2004) TILLING. Traditional mutagenesis meets functional genomics. Plant Physiol. 135:630–636

Hilson P, Allemeersch J, Altmann T, Aubourg S, Avon A, Beynon J, Bhalerao RP, Bitton F, Caboche M, Cannoot B et al (2004) Versatile gene-specific sequence tags for Arabidopsis functional genomics: transcript profiling and reverse genetics applications. Genome Res 14:2176–2189

Himelblau E, Gilchrist EJ, Buono K, Bizzell C, Mentzer L, Vogelzang R, Osborn T, Amasino RM, Parkin IA, Haughn GW (2009) Forward and reverse genetics of rapid-cycling Brassica oleracea. Theor Appl Genet 118:953–961

Honma MA, Baker BJ, Waddell CS (1993) High-frequency germinal transposition of DsALS in Arabidopsis. Proc Natl Acad Sci USA 90:6242–6246

Ichikawa T, Nakazawa M, Kawashima M, Iizumi H, Kuroda H, Kondou Y, Tsuhara Y, Suzuki K, Ishikawa A, Seki M, Fujita M, Motohashi R, Nagata N, Takagi T, Shinozaki K, Matsui M (2006) The FOX hunting system: an alternative gain-of-function gene hunting technique. Plant J 48:974–985

Ichikawa T, Nakazawa M, Kawashima M, Muto S, Gohda K, Suzuki K, Ishikawa A, Kobayashi H, Yoshizumi T, Tsumoto Y, Tsuhara Y, Iizumi H, Goto Y, Matsui M (2003) Sequence database of 1172 T-DNA insertion sites in Arabidopsis activation-tagging lines that showed phenotypes in T1 generation. Plant J 36:421–429

Ito T, Motohashi R, Kuromori T, Mizukado S, Sakurai T, Kanahara H, Seki M, Shinozaki K (2002) A new resource of locally transposed Dissociation elements for screening gene-knockout lines in silico on the Arabidopsis genome. Plant Physiol 129:1695–1699

Ito T, Motohashi R, Kuromori T, Noutoshi Y, Seki M, Kamiya A, Mizukado S, Sakurai T, Shinozaki K (2005) A resource of 5,814 dissociation transposon-tagged and sequence-indexed lines of Arabidopsis transposed from start loci on chromosome 5. Plant Cell Physiol 46:1149–1153

Jackson AL, Bartz SR, Schelter J, Kobayashi SV, Burchard J, Mao M, Li B, Cavet G, Linsley PS (2003) Expression profiling reveals off-target gene regulation by RNAi. Nat Biotechnol 21:635–637

Jun JH, Kim CS, Cho DS, Kwak JM, Ha CM, Park YS, Cho BH, Patton DA, Nam HG (2002) Random antisense cDNA mutagenesis as an efficient functional genomic approach in higher plants. Planta 214:668–674

Kakimoto T (1996) CKI1, a histidine kinase homolog implicated in cytokinin signal transduction. Science 274:982–985

Kardailsky I, Shukla VK, Ahn JH, Dagenais N, Christensen SK, Nguyen JT, Chory J, Harrison MJ, Weigel D (1999) Activation tagging of the floral inducer FT. Science 286:1962–1965

Kertbundit S, De Greve H, Deboeck F, Van Montagu M, Hernalsteens JP (1991) In vivo random beta-glucuronidase gene fusions in Arabidopsis thaliana. Proc Natl Acad Sci USA 88: 5212–5216

Kiegle E, Moore CA, Haseloff J, Tester MA, Knight MR (2000) Cell-type-specific calcium responses to drought, salt and cold in the Arabidopsis root. Plant J 23:267–278

Koncz C, Martini N, Mayerhofer R, Koncz-Kalman Z, Korber H, Redei GP, Schell J (1989) High-frequency T-DNA-mediated gene tagging in plants. Proc Natl Acad Sci USA 86:8467–8471

Koornneef M, Dellaert LW, van der Veen JH (1982) EMS- and radiation-induced mutation frequencies at individual loci in Arabidopsis thaliana (L.) Heynh. Mutat Res 93:109–123

Kuromori T, Hirayama T, Kiyosue Y, Takabe H, Mizukado S, Sakurai T, Akiyama K, Kamiya A, Ito T, Shinozaki K (2004) A collection of 11,800 single-copy *Ds* transposon insertion lines in *Arabidopsis*. Plant J 37:897–905

Lacroix B, Tzfira T, Vainstein A, Citovsky V (2006) A case of promiscuity: Agrobacterium's endless hunt for new partners. Trends Genet 22:29–37

Laubinger S, Zeller G, Henz SR, Sachsenberg T, Widmer CK, Naouar N, Vuylsteke M, Scholkopf B, Ratsch G, Weigel D (2008) At-TAX: a whole genome tiling array resource for developmental expression analysis and transcript identification in Arabidopsis thaliana. Genome Biol 9:R112

LeClere S, Bartel B (2001) A library of Arabidopsis 35S-cDNA lines for identifying novel mutants. Plant Mol Biol 46:695–703

Li X, Lassner M, Zhang Y (2002b) Deleteagene: a fast neutron deletion mutagenesis-based gene knockout system for plants. Comp Funct Genomics 3:158–160

Li Y, Rosso MG, Ulker B, Weisshaar B (2006) Analysis of T-DNA insertion site distribution patterns in Arabidopsis thaliana reveals special features of genes without insertions. Genomics 87:645–652

Li X, Song Y, Century K, Straight S, Ronald P, Dong X, Lassner M, Zhang Y (2001) A fast neutron deletion mutagenesis-based reverse genetics system for plants. Plant J 27:235–242

Li J, Wen J, Lease KA, Doke JT, Tax FE, Walker JC (2002a) BAK1, an Arabidopsis LRR receptor-like protein kinase, interacts with BRI1 and modulates brassinosteroid signaling. Cell 110: 213–222

Liu YG, Mitsukawa N, Oosumi T, Whittier RF (1995) Efficient isolation and mapping of Arabidopsis thaliana T-DNA insert junctions by thermal asymmetric interlaced PCR. Plant J 8:457–463

Marsch-Martinez N, Greco R, Van Arkel G, Herrera-Estrella L, Pereira A (2002) Activation tagging using the En-I maize transposon system in Arabidopsis. Plant Physiol 129:1544–1556

Martienssen RA (1998) Functional genomics: probing plant gene function and expression with transposons. Proc Natl Acad Sci USA 95:2021–2026

May BP, Martienssen RA (2003) Transposon mutagenesis in the study of plant development. CRC Crit Rev Plant Sci 22:1–35

McCallum CM, Comai L, Greene EA, Henikoff S (2000) Targeted screening for induced mutations. Nat Biotechnol 18:455–457

McElver J, Tzafrir I, Aux G, Rogers R, Ashby C, Smith K, Thomas C, Schetter A, Zhou Q, Cushman MA, Tossberg J, Nickle T, Levin JZ, Law M, Meinke D, Patton D (2001) Insertional mutagenesis of genes required for seed development in Arabidopsis thaliana. Genetics 159:1751–1763

Meinke DW, Cherry JM, Dean C, Rounsley SD, Koornneef M (1998) Arabidopsis thaliana: a model plant for genome analysis. Science 282(662):679–682

Meinke D, Scholl R (2003) The preservation of plant genetic resources. Experiences with Arabidopsis. Plant Physiol 133:1046–1050

Menand B, Desnos T, Nussaume L, Berger F, Bouchez D, Meyer C, Robaglia C (2002) Expression and disruption of the Arabidopsis TOR (target of rapamycin) gene. Proc Natl Acad Sci USA 99:6422–6427

Mengiste T, Amedeo P, Paszkowski J (1997) High-efficiency transformation of Arabidopsis thaliana with a selectable marker gene regulated by the T-DNA 1' promoter. Plant J 12:945–948

Mizoi J, Nakamura M, Nishida I (2006) Defects in CTP:PHOSPHORYLETHANOLAMINE CYTIDYLYLTRANSFERASE affect embryonic and postembryonic development in Arabidopsis. Plant Cell 18:3370–3385

Moens CB, Donn TM, Wolf-Saxon ER, Ma TP (2008) Reverse genetics in zebrafish by TILLING. Brief Funct Genomic Proteomic 7:454–459

Mun JH, Lee SY, Yu HJ, Jeong YM, Shin MY, Kim H, Lee I, Kim SG (2002) Petunia actin-depolymerizing factor is mainly accumulated in vascular tissue and its gene expression is enhanced by the first intron. Gene 292:233–243

Muskett PR, Clissold L, Marocco A, Springer PS, Martienssen R, Dean C (2003) A resource of mapped dissociation launch pads for targeted insertional mutagenesis in the Arabidopsis genome. Plant Physiol 132:506–516

Nakazawa M, Ichikawa T, Ishikawa A, Kobayashi H, Tsuhara Y, Kawashima M, Suzuki K, Muto S, Matsui M (2003) Activation tagging, a novel tool to dissect the functions of a gene family. Plant J 34:741–750

Neff MM, Nguyen SM, Malancharuvil EJ, Fujioka S, Noguchi T, Seto H, Tsubuki M, Honda T, Takatsuto S, Yoshida S, Chory J (1999) BAS1: a gene regulating brassinosteroid levels and light responsiveness in Arabidopsis. Proc Natl Acad Sci USA 96:15316–15323

Niu QW, Lin SS, Reyes JL, Chen KC, Wu HW, Yeh SD, Chua NH (2006) Expression of artificial micro RNAs in transgenic Arabidopsis thaliana confers virus resistance. Nat Biotechnol 24:1420–1428

Oleykowski CA, Bronson Mullins CR, Godwin AK, Yeung AT (1998) Mutation detection using a novel plant endonuclease. Nucleic Acids Res 26:4597–4602

Ostergaard L, Yanofsky MF (2004) Establishing gene function by mutagenesis in Arabidopsis thaliana. Plant J 39:682–696

Papdi C, Abraham E, Joseph MP, Popescu C, Koncz C, Szabados L (2008) Functional identification of Arabidopsis stress regulatory genes using the controlled cDNA overexpression system. Plant Physiol 147:528–542

Parinov S, Sevugan M, Ye D, Yang W-C, Kumaran M, Sundaresan V (1999). Analysis of flanking sequences from dissociation insertion lines: a database for reverse genetics in Arabidopsis. Plant Cell 11:2263–2270

Paterson AH, Freeling M, Sasaki T (2005) Grains of knowledge: genomics of model cereals. Genome Res 15:1643–1650

Perry JA, Wang TL, Welham TJ, Gardner S, Pike JM, Yoshida S, Parniske M (2003) A TILLING reverse genetics tool and a web-accessible collection of mutants of the legume lotus japonicus. Plant Physiol 131:866–871

Pettersson E, Lundeberg J, Ahmadian A (2009) Generations of sequencing technologies. Genomics 93:105–111

Preuss D, Rhee SY, Davis RW (1994) Tetrad analysis possible in Arabidopsis with mutation of the QUARTET (QRT) genes. Science 264:1458–1460

Puchta H, Hohn B (2005) Green light for gene targeting in plants. Proc Natl Acad Sci USA 102:11961–11962

Perry J, Brachmann A, Welham T, Binder A, Charpentier M, Groth M, Haage K, Markmann K, Wang TL, Parniske M (2009) TILLING in lotus japonicus identified large allelic series for symbiosis genes and revealed a bias in functionally defective EMS alleles towards glycine replacements. Plant Physiol 151(3)1281–1291

Quesada V, Macknight R, Dean C, Simpson GG (2003) Autoregulation of FCA pre-mRNA processing controls Arabidopsis flowering time. EMBO J 22:3142–3152

Rensing SA, Fritzowsky D, Lang D, Reski R (2005) Protein encoding genes in an ancient plant: analysis of codon usage, retained genes and splice sites in a moss, Physcomitrella patens. BMC Genomics 6:43

Robertson D (2004) VIGS vectors for gene silencing: many targets, many tools. Annu Rev Plant Biol 55:495–519

Robinson SJ, Tang LH, Mooney BA, McKay SJ, Clarke WE, Links MG, Karcz S, Regan S, Wu YY, Gruber MY, Cui D, Yu M, Parkin IA (2009) An archived activation tagged population of Arabidopsis thaliana to facilitate forward genetics approaches. BMC Plant Biol 9:101

Rosso MG, Li Y, Strizhov N, Reiss B, Dekker K, Weisshaar B (2003) An Arabidopsis thaliana T-DNA mutagenized population (GABI-Kat) for flanking sequence tag-based reverse genetics. Plant Mol Biol 53:247–259

Rubin GM, Yandell MD, Wortman JR, Gabor Miklos GL, Nelson CR, Hariharan IK, Fortini ME, Li PW, Apweiler R, Fleischmann W et al (2000) Comparative genomics of the eukaryotes. Science 287:2204–2215

Salmeron JM, Oldroyd GE, Rommens CM, Scofield SR, Kim HS, Lavelle DT, Dahlbeck D, Staskawicz BJ (1996) Tomato Prf is a member of the leucine-rich repeat class of plant disease resistance genes and lies embedded within the pto kinase gene cluster. Cell 86:123–133

Samson F, Brunaud V, Balzergue S, Dubreucq B, Lepiniec L, Pelletier G, Caboche M, Lecharny A (2002) FLAGdb/FST: a database of mapped flanking insertion sites (FSTs) of Arabidopsis thaliana T-DNA transformants. Nucleic Acids Res 30:94–97

Schneeberger RG, Zhang K, Tatarinova T, Troukhan M, Kwok SF, Drais J, Klinger K, Orejudos F, Macy K, Bhakta A, Burns J, Subramanian G, Donson J, Flavell R, Feldmann KA (2005) Agrobacterium T-DNA integration in Arabidopsis is correlated with DNA sequence compositions that occur frequently in gene promoter regions. Funct Integr Genomics 5:240–253

Schneider A, Kirch T, Gigolashvili T, Mock HP, Sonnewald U, Simon R, Flugge UI, Werr W (2005) A transposon-based activation-tagging population in Arabidopsis thaliana (TAMARA) and its application in the identification of dominant developmental and metabolic mutations. FEBS Lett 579:4622–4628

Schwab R, Ossowski S, Riester M, Warthmann N, Weigel D (2006) Highly specific gene silencing by artificial micro RNAs in Arabidopsis. Plant Cell 18:1121–1133

Seki H, Nishizawa T, Tanaka N, Niwa Y, Yoshida S, Muranaka T (2005) Hairy root-activation tagging: a high-throughput system for activation tagging in transformed hairy roots. Plant Mol Biol 59:793–807

Service RF (2006) Gene sequencing. The race for the $1000 genome. Science 311:1544–1546

Sessions A, Burke E, Presting G, Aux G, McElver J, Patton D, Dietrich B, Ho P, Bacwaden J, Ko C et al (2002) A high-throughput Arabidopsis reverse genetics system. Plant Cell 14:2985–2994

Sheldon CC, Conn AB, Dennis ES, Peacock WJ (2002) Different regulatory regions are required for the vernalization-induced repression of FLOWERING LOCUS C and for the epigenetic maintenance of repression. Plant Cell 14:2527–2537

Shukla VK, Doyon Y, Miller JC, DeKelver RC, Moehle EA, Worden SE, Mitchell JC, Arnold NL, Gopalan S, Meng X et al (2009) Precise genome modification in the crop species Zea mays using zinc-finger nucleases. Nature 459:437–441

Slade AJ, Fuerstenberg SI, Loeffler D, Steine MN, Facciotti D (2005) A reverse genetic, nontransgenic approach to wheat crop improvement by TILLING. Nat Biotechnol 23:75–81

Smits BM, Mudde J, Plasterk RH, Cuppen E (2004) Target-selected mutagenesis of the rat. Genomics 83:332–334

Speulman E, Metz PL, van Arkel G, te Lintel Hekkert B, Stiekema WJ, Pereira A (1999) A two-component enhancer-inhibitor transposon mutagenesis system for functional analysis of the Arabidopsis genome. Plant Cell 11:1853–1866

Steinmetz LM, Davis RW (2004) Maximizing the potential of functional genomics. Nat Rev Genet 5:190–201

Sun T, Goodman HM, Ausubel FM (1992) Cloning the Arabidopsis GA1 locus by genomic subtraction. Plant Cell 4:119–128
Sundaresan V, Springer P, Volpe T, Haward S, Jones JD, Dean C, Ma H, Martienssen R (1995) Patterns of gene action in plant development revealed by enhancer trap and gene trap transposable elements. Genes Dev 9:1797–1810
Sussman MR, Amasino RM, Young JC, Krysan PJ, Austin-Phillips S (2000) The Arabidopsis knockout facility at the University of Wisconsin-Madison. Plant Physiol 124:1465–1467
Suzuki T, Eiguchi M, Kumamaru T, Satoh H, Matsusaka H, Moriguchi K, Nagato Y, Kurata N (2008) MNU-induced mutant pools and high performance TILLING enable finding of any gene mutation in rice. Mol Genet Genomics 279:213–223
Talame V, Bovina R, Sanguineti MC, Tuberosa R, Lundqvist U, Salvi S (2008) TILLMore, a resource for the discovery of chemically induced mutants in barley. Plant Biotechnol J 6: 477–485
Teeri TH, Herrera-Estrella L, Depicker A, Van Montagu M, Palva ET (1986) Identification of plant promoters in situ by T-DNA-mediated transcriptional fusions to the npt-II gene. EMBO J 5:1755–1760
Thimm O, Blasing O, Gibon Y, Nagel A, Meyer S, Kruger P, Selbig J, Muller LA, Rhee SY, Stitt M (2004) MAPMAN: a user-driven tool to display genomics data sets onto diagrams of metabolic pathways and other biological processes. Plant J 37:914–939
Thomas CM, Jones DA, English JJ, Carroll BJ, Bennetzen JL, Harrison K, Burbidge A, Bishop GJ, Jones JD (1994) Analysis of the chromosomal distribution of transposon-carrying T-DNAs in tomato using the inverse polymerase chain reaction. Mol Gen Genet 242: 573–585
Till BJ, Colbert T, Tompa R, Enns LC, Codomo CA, Johnson JE, Reynolds SH, Henikoff JG, Greene EA, Steine MN, Comai L, Henikoff S (2003a) High-throughput TILLING for functional genomics. Methods Mol Biol 236:205–220
Till BJ, Cooper J, Tai TH, Colowit P, Greene EA, Henikoff S, Comai L (2007) Discovery of chemically induced mutations in rice by TILLING. BMC Plant Biol 7:19
Till BJ, Reynolds SH, Greene EA, Codomo CA, Enns LC, Johnson JE, Burtner C, Odden AR, Young K, Taylor NE, Henikoff JG, Comai L, Henikoff S (2003) Large-scale discovery of induced point mutations with high-throughput TILLING. Genome Res 13:524–530
Till BJ, Reynolds SH, Weil C, Springer N, Burtner C, Young K, Bowers E, Codomo CA, Enns LC, Odden AR, Greene EA, Comai L, Henikoff S (2004) Discovery of induced point mutations in maize genes by TILLING. BMC Plant Biol 4:12
Tissier AF, Marillonnet S, Klimyuk V, Patel K, Torres MA, Murphy G, Jones JD (1999) Multiple independent defective suppressor-mutator transposon insertions in Arabidopsis: a tool for functional genomics. Plant Cell 11:1841–1852
Topping JF, Agyeman F, Henricot B, Lindsey K (1994) Identification of molecular markers of embryogenesis in Arabidopsis thaliana by promoter trapping. Plant J 5:895–903
Topping JF, Wei W, Lindsey K (1991) Functional tagging of regulatory elements in the plant genome. Development 112:1009–1019
Townsend JA, Wright DA, Winfrey RJ, Fu F, Maeder ML, Joung JK, Voytas DF (2009) High-frequency modification of plant genes using engineered zinc-finger nucleases. Nature 459: 442–445
Tschuch C, Schulz A, Pscherer A, Werft W, Benner A, Hotz-Wagenblatt A, Barrionuevo LS, Lichter P, Mertens D (2008) Off-target effects of siRNA specific for GFP. BMC Mol Biol 9:60
Ülker B, Li Y, Rosso MG, Logemann E, Somssich IE, Weisshaar B (2008a) T-DNA-mediated transfer of Agrobacterium tumefaciens chromosomal DNA into plants. Nat Biotechnol 26:1015–1017
Ülker B, Peiter E, Dixon DP, Moffat C, Capper R, Bouche N, Edwards R, Sanders D, Knight H,, Knight MR (2008b) Getting the most out of publicly available T-DNA insertion lines. Plant J 56(4):665–677

van der Graaff E, Dulk-Ras AD, Hooykaas PJ, Keller B (2000) Activation tagging of the LEAFY PETIOLE gene affects leaf petiole development in Arabidopsis thaliana. Development 127:4971–4980

Van Montagu M (2003) Jeff schell (1935–2003): steering Agrobacterium-mediated plant gene engineering. Trends Plant Sci 8:353–354

Walden R, Fritze K, Hayashi H, Miklashevichs E, Harling H, Schell J (1994) Activation tagging: a means of isolating genes implicated as playing a role in plant growth and development. Plant Mol Biol 26:1521–1528

Wang S, Sim TB, Kim YS, Chang YT (2004) Tools for target identification and validation. Curr Opin Chem Biol 8:371–377

Weigel D, Ahn JH, Blazquez MA, Borevitz JO, Christensen SK, Fankhauser C, Ferrandiz C, Kardailsky I, Malancharuvil EJ, Neff MM et al (2000) Activation tagging in Arabidopsis. Plant Physiol 122:1003–1013

Wienholds E, van Eeden F, Kosters M, Mudde J, Plasterk RH, Cuppen E (2003) Efficient target-selected mutagenesis in zebrafish. Genome Res 13:2700–2707

Wilson K, Long D, Swinburne J, Coupland G (1996) A Dissociation insertion causes a semidominant mutation that increases expression of TINY, an Arabidopsis gene related to APETALA2. Plant Cell 8:659–671

Winkler S, Schwabedissen A, Backasch D, Bokel C, Seidel C, Bonisch S, Furthauer M, Kuhrs A, Cobreros L, Brand M, Gonzalez-Gaitan M (2005) Target-selected mutant screen by TILLING in Drosophila. Genome Res 15:718–723

Wisman E, Hartmann U, Sagasser M, Baumann E, Palme K, Hahlbrock K, Saedler H, Weisshaar B (1998) Knock-out mutants from an En-1 mutagenized Arabidopsis thaliana population generate phenylpropanoid biosynthesis phenotypes. Proc Natl Acad Sci USA 95:12432–12437

Wright DA, Townsend JA, Winfrey RJ Jr., Irwin PA, Rajagopal J, Lonosky PM, Hall BD, Jondle MD, Voytas DF (2005) High-frequency homologous recombination in plants mediated by zinc-finger nucleases. Plant J 44:693–705

Wu GZ, Shi QM, Niu Y, Xing MQ, Xue HW (2008) Shanghai RAPESEED Database: a resource for functional genomics studies of seed development and fatty acid metabolism of Brassica. Nucleic Acids Res 36:D1044–D1047

Xin Z, Wang ML, Barkley NA, Burow G, Franks C, Pederson G, Burke J (2008) Applying genotyping (TILLING) and phenotyping analyses to elucidate gene function in a chemically induced sorghum mutant population. BMC Plant Biol 8:103

Yamada K, Lim J, Dale JM, Chen H, Shinn P, Palm CJ, Southwick AM, Wu HC, Kim C, Nguyen M et al (2003) Empirical analysis of transcriptional activity in the Arabidopsis genome. Science 302:842–846

Yu H, Chen X, Hong YY, Wang Y, Xu P, Ke SD, Liu HY, Zhu JK, Oliver DJ, Xiang CB (2008) Activated expression of an Arabidopsis HD-START protein confers drought tolerance with improved root system and reduced stomatal density. Plant Cell 20:1134–1151

Zamore PD, Tuschl T, Sharp PA, Bartel DP (2000) RNAi: double-stranded RNA directs the ATP-dependent cleavage of mRNA at 21 to 23 nucleotide intervals. Cell 101:25–33

Zhang J, Guo D, Chang Y, You C, Li X, Dai X, Weng Q, Chen G, Liu H, Han B, Zhang Q, Wu C (2007) Non-random distribution of T-DNA insertions at various levels of the genome hierarchy as revealed by analyzing 13 804 T-DNA flanking sequences from an enhancer-trap mutant library. Plant J 49:947–959

Zhang S, Raina S, Li H, Li J, Dec E, Ma H, Huang H, Fedoroff NV (2003) Resources for targeted insertional and deletional mutagenesis in Arabidopsis. Plant Mol Biol 53:133–150

Zimmermann P, Hirsch-Hoffmann M, Hennig L, Gruissem W (2004) GENEVESTIGATOR. Arabidopsis microarray database and analysis toolbox. Plant Physiol 136:2621–2632

Zupan J, Muth TR, Draper O, Zambryski P (2000) The transfer of DNA from agrobacterium tumefaciens into plants: a feast of fundamental insights. Plant J 23:11–28

Chapter 20
Resources for Reverse Genetics Approaches in *Brassica* Species

Thomas Wood, Pauline Stephenson, and Lars Østergaard

Abstract Gene, genomic and genome sequences are being generated with incredible speed, thanks to the advent of cheaper and faster sequencing technologies. For plants, the *Arabidopsis thaliana* (Col-0) genome was sequenced in its entirety in 2000, and a number of whole-genome sequences in additional *A. thaliana* ecotypes have been completed since then, providing an amazing resource for functional genomics in this species. Although *Arabidopsis* is still the only genus of the Brassicaceae family to have its genome completely sequenced, a multinational effort is currently ongoing to obtain first the sequence of the A genome (*Brassica rapa*) and ultimately all the cultivated *Brassicas* of the "U triangle". The obvious challenge is therefore what to do with this massive amount of information. How does these data expand our knowledge of plant biology or aid in the development of tools for crop improvement? In this chapter we will describe the current status of the principal resources available for reverse genetics approaches in the *Brassica* genus, TILLING and RNAi, and discuss their advantages and disadvantages for the study of plant biology and the development of tools for crop improvement.

Keywords Reverse genetics · Gene silencing · TILLING · Gene function

Contents

20.1 Introduction 562
20.2 TILLING 562
  20.2.1 EMS 562
  20.2.2 EMS-Induced Mutations and the Genetic Code 563
  20.2.3 Mutation Load 564
  20.2.4 TILLING in *Brassica* Step by Step 565
20.3 RNA Interference 568
  20.3.1 Background 568

T. Wood (✉)
John Innes Centre, Norwich, UK
e-mail: tom.wood@bbsrc.ac.uk

R. Schmidt, I. Bancroft (eds.), *Genetics and Genomics of the Brassicaceae*, Plant Genetics and Genomics: Crops and Models 9, DOI 10.1007/978-1-4419-7118-0_20,

20.3.2 Classes of sRNA Associated with PTGS 569
20.3.3 RNAi/PTGS Mechanisms: Gene Silencing Approaches in Brassicaceae . . 570
20.3.4 Examples of RNAi in *Brassica* Species 577
20.4 Concluding Remarks 579
References 579

20.1 Introduction

The advent of high-throughput sequencing technologies, vast genomic databases and increasingly powerful genetic tools has had a gigantic impact on the development of our understanding of the biochemical networks regulating the multitude of genetic and physiological processes in plants. Insight from studies in the model species, *Arabidopsis thaliana,* is increasingly facilitating our ability to elucidate and beneficially exploit key regulatory processes in relevant crop species (Østergaard et al. 2006). The last decade has seen the development of a number of large-scale "reverse genetics" tools to study the effects of mutations in genes for which the sequence is known. These tools include T-DNA insertion (Alonso and Stepanova 2003); transposable elements (Sundaresan et al. 1995); TILLING (Targeting Induced Local Lesions in Genomes) (Till et al. 2003) and RNAi technologies (Horiguchi 2004), including hairpin RNA, co-suppression and antisense approaches and artificial microRNA (amiRNA) (Schwab et al. 2005, 2006). In this chapter we will describe reverse genetics technologies that were first validated in *Arabidopsis* to study gene function, but which are now being widely applied in *Brassica* species to both functionally validate target genes and to exploit resources for crop improvement. The focus will be on TILLING and RNAi technologies, since generation of near-saturation insertion mutant populations is too expensive and impractical to be feasible for *Brassica species* due to the larger size and extended generation time compared to *Arabidopsis*.

20.2 TILLING

TILLING (Targeting Induced Local Lesions in Genomes) is a reverse genetics tool, which was originally developed for *Arabidopsis* (McCallum et al. 2000) and has been successfully employed in other plant species as well as animal species (Henikoff et al. 2004, Winkler et al. 2005, Gilchrist et al. 2006). For plants, large mutant populations are generated by the treatment of seed or pollen with a chemical mutagen that can induce point mutations at a very high density, sufficient to allow for the possibility of establishing a series of allelic mutations in all genes. Amplified sequences are then screened using established high-throughput SNP discovery methods.

20.2.1 EMS

Ethyl methanesulfonate (EMS) is a mutagenic, teratogenic and possibly carcinogenic organic compound with the formula $C_3H_8O_3S$, and it is the mutagen of choice for the development of plant TILLING populations. It produces random mutations in genetic material by nucleotide substitution, primarily by alkylation on the O^6 position of guanine leading to GC→AT transition changes. The ethyl group of EMS reacts with guanine in DNA, forming the abnormal base O-6-ethylguanine. During DNA replication, DNA polymerases frequently place thymine, instead of cytosine, opposite O-6-ethylguanine. Following subsequent rounds of replication, the original GC base pair becomes an AT pair.

20.2.2 EMS-Induced Mutations and the Genetic Code

When initiating a TILLING screen in a particular gene of interest, it is worth analysing the coding region and taking a look at the genetic code (Fig. 20.1). Premature stop codon (nonsense) mutations within exons are often desirable as they are expected to truncate proteins and typically abolish function. However, out of the 96 positions that can be mutated by EMS (G>A or C>T) within the genetic code (GC→AT) there are only five ways in which a stop codon can be obtained. These comprise the two glutamine codons (CAA and CAG to TAA and TAG, respectively), one of the six arginine codons (CGA to TGA) and the tryptophan codon (TGG to TGA or TAG).

UUU	Phe		UCU	Ser	Phe	UAU	Tyr		UGU	Cys	Tyr
UUC	Phe		UCC	Ser	Phe	UAC	Tyr		UGC	Cys	Tyr
UUA	Leu		UCA	Ser	Leu	UAA	stop		UGA	stop	
UUG	Leu		UCG	Ser	Leu	UAG	stop		UGG	Trp	stop/stop
CUU	Leu	Phe	CCU	Pro	Ser/Leu	CAU	His	Tyr	CGU	Arg	Cys/His
CUC	Leu	Phe	CCC	Pro	Ser/Leu	CAC	His	Tyr	CGC	Arg	Cys/His
CUA	Leu		CCA	Pro	Ser/Leu	CAA	Gln	stop	CGA	Arg	stop/Gln
CUG	Leu		CCG	Pro	Ser/Leu	CAG	Gln	stop	CGG	Arg	Trp/Gln
AUU	Ile		ACU	Thr	Ile	AAU	Asn		AGU	Ser	Asn
AUC	Ile		ACC	Thr	Ile	AAC	Asn		AGC	Ser	Asn
AUA	Ile		ACA	Thr	Ile	AAA	Lys		AGA	Arg	Lys
AUG	Met	Ile	ACG	Thr	Met	AAG	Lys		AGG	Arg	Lys
GUU	Val	Ile	GCU	Ala	Thr/Val	GAU	Asp	Asn	GGU	Gly	Ser/Asp
GUC	Val	Ile	GCC	Ala	Thr/Val	GAC	Asp	Asn	GGC	Gly	Ser/Asp
GUA	Val	Ile	GCA	Ala	Thr/Val	GAA	Glu	Lys	GGA	Gly	Lys/Glu
GUG	Val	Met	GCG	Ala	Thr/Val	GAG	Glu	Lys	GGG	Gly	Lys/Glu

Fig. 20.1 Possible changes to the genetic code by EMS. The 64 codon triplets are organised in four columns with 16 in each. The amino acids normally encoded by the respective codons are shown in *black*. The consequences of nucleotide changes are shown by colour coding such that the resulting amino acid is shown in the same colour as the mutated nucleotide

As indicated in Fig. 20.1, far from all combinations of amino acid changes are possible when using EMS as a mutagen. First, 8 out of the 64 codons (12.5%) are unaffected by EMS-induced mutations because they do not contain guanine or cytosine. Second, out of the 96 mutatable positions, 33 (34%) would not lead to an amino acid sequence change (silent mutations). This leaves 58 positions of the genetic code at which mutations would result in the 26 possible amino acid substitutions indicated in Fig. 20.1 (missense mutations). Out of these, nine changes (corresponding to mutations at 21 out of 58 sites) result in incorporation of amino acids with similar chemical properties, such as, for example, exchange of one hydrophobic amino acid for another (Leu→Phe, Met→Ile, Val→Ile, Val→Met, Pro→Leu and Ala→Val). In many cases this would be less likely to significantly affect the function of the encoded protein.

Of the remaining 17 possible amino acid changes, some are more likely to have an effect on protein structure than others. Proline (Pro) and glycine (Gly) are so-called helix breakers and are often found at the end of α-helical structural elements. Mutations that changes Pro and Gly are thus very likely to disturb the three-dimensional structure and may dramatically effect protein activity. These include Pro→Ser, Pro→Leu, Gly→Ser, Gly→Asp, Gly→Lys and Gly→Glu. Depending on the function of the individual amino acids in a protein, it can also have devastating effects to change the charge at a given position. Especially, Arg→Trp and Glu→Lys would be expected to strongly reduce the native activity of a protein.

In conclusion, the genetic code appears to have considerable robustness in order to minimise the effect of point mutations. It may therefore be beneficial to analyse the sequence of interest to realise the potential of inducing mutations that may reduce/abolish the activity of the encoded protein.

20.2.3 Mutation Load

Under optimal mutagenesis conditions, individuals of an EMS mutant population carry a high mutation load but remain vigorous and fertile. It is important, therefore, to determine the level of mutagen treatment necessary to achieve the maximal mutation load. The possible mutation level is species dependent and relates to the ploidy/level of redundancy as well as size of the genome. Most plant species can be described as paleopolyploids, but such ancient genome multiplication events have frequently been followed by subfunctionalisation or loss of the duplicated genes. More recent polyploidisation events such as those seen in hexaploid wheat and *Brassica* species have resulted in genomes with several copies of seemingly the same gene. It is expected that this leads to a high degree of functional redundancy, and as a consequence, these species can cope with a very high level of mutation.

Brassica rapa is diploid and has a genome size of approximately 550 Mbp. It can be described as a paleopolyploid due to an ancient triplication event (Blanc and Wolfe 2004a, b, Lysak et al. 2005). A *B. rapa* TILLING population was recently developed at the John Innes Centre (UK). Based on TILLING in six individual genes, we estimate that each plant carries a mutation load in the region of 15,000

Table 20.1 Plant TILLING platforms including information regarding mutagen concentration used, genome size, mutation density and mutations per plant. The mutation load, which can be endured by a plant, correlates with genome size, ploidy status and the level of gene redundancy. *Medicago truncatula* link: http://www.gl-ttp.com/products_services/technical_services/genomic_resources_from_glip/functional_genomics/medicago_tilling_platform; *Brassica rapa* link: http://www.jic.ac.uk/staff/lars-ostergaard/tilling.htm

Organism	%EMS	Genome size (bp)	Mutation frequency (kb)	Mutations per plant	References
Arabidopsis	0.25–0.5	145×10^{-6}	1/310	468	Greene et al. (2003)
Rice	1.5	420×10^{-6}	1/300	1,400	Till et al. (2007)
Lotus	0.4–0.6	500×10^{-6}	1/380	1,300	Perry et al. (2003)
M. truncatula	0.2	500×10^{-6}	1/400	1,250	Web link in legend
B. rapa	0.3	550×10^{-6}	1/37	15,000	Web link in legend
Soybean (Glycine max)	0.5–0.625	$1,100 \times 10^{-6}$	1/320	34,000	Cooper et al. (2008)
B. napus	0.3	$1,200 \times 10^{-6}$	1/130	9,231	Wang et al. (2008)
B. napus	0.6	$1,200 \times 10^{-6}$	1/35	34,286	Wang et al. (2008)
Maize	Pollen	$2,500 \times 10^{-6}$	1/400	6,250	Till et al. (2004)
Pea (*Pisum sativum*)	0.25	$4,000 \times 10^{-6}$	1/200	20,000	Dalmais et al. (2008)
Barley	0.25–0.375	$4,900 \times 10^{-6}$	1/570	8,600	Caldwell et al. (2004)
Tetraploid wheat	0.75–1.0	$11,000 \times 10^{-6}$	1/41	270,000	Slade et al. (2005)
Hexaploid wheat	0.75–1.25	$16,000 \times 10^{-6}$	1/20	810,000	Slade et al. (2005)

mutations. This would seem to be more than the plant should be able to cope with; however, as only 11% of the genome is coding sequence, we expect no more than 1,600–1,700 point mutations within genes of which ~1,000 will cause amino acid changes and ~80 will introduce new stop codons.

We have included Table 20.1, which shows the level of mutations that has been achieved in a range of plant TILLING initiatives. It is clear that *B. rapa*, *Brassica napus* and hexaploid and tetraploid wheat are best able to cope with high mutation frequencies, which is likely to be a consequence of the high level of gene duplication in these species.

20.2.4 TILLING in Brassica *Step by Step*

20.2.4.1 Optimising Mutagen Dosage

The optimal conditions for EMS mutagenesis need to be established prior to treatment. Germination rate provides a suitable measure, and for *Brassica* species it is recommended to assay in the range of 0.1–1.0% EMS. For *B. rapa*, we have found that using a concentration at the point where germination starts to become compromised is optimal for obtaining a large mutation load while maintaining vigorous and fertile plants. This is in agreement with what has previously been reported for

Medicago truncatula mutagenesis (Penmetsa and Cook 2000). For the *B. rapa* population that we have developed, we observed high mutation loads using 0.3% EMS while still maintaining healthy-looking plants with high fertility. For a TILLING population in *B. napus*, 0.3 and 0.6% EMS concentrations were used (Wang et al. 2008) (Table 20.1). The 0.6% EMS treatment was considered optimal with respect to mutation load and seed set and does indeed make up the bulk of this *B. napus* population (7,100 M_2 lines for 0.6% versus 2,604 M_2 lines for 0.3%). Treatment with the lower level of EMS resulted in a mutation load of approximately one-fourth compared to the higher level. Although it will be less likely to identify the desired mutations, the ones that are found would have a lower amount of background mutations.

20.2.4.2 M_1 and M_2 Population Structure

When treating the seeds with the mutagen, it is only one or a few cells that will carry the mutation on to the next generation. These are so-called germ line or progenitor cells, and if embryos contain more than one, mutations in each will be independently generated. The number of germ line cells (the genetically effective cell number or GECN) varies even between seeds of the same species. For *Arabidopsis* the GECN ranges from one to four with an average of two (Koornneef 2002) and is likely to be similar in *Brassica* species. Theoretically, if there is only a single progenitor cell all the M_2 seeds produced by that plant, even though they are segregating, will have mutations derived from the genome of that cell. The mutation load of each of these M_2 seed will represent 75% of all the mutations in the initial M_1 progenitor cell, and the probability that two seeds will carry the same mutant allele is 50%. However, if there are four progenitor cells then each mutagenised M_1 seed will give rise to four independent sets of mutations. To select only one seed from this pool for progression from M_2 to M_3 would mean, in effect, discarding 75% of the mutation potential of that line even before any loss due to segregation is factored into the calculation.

This information can influence the strategy of population development as it is almost always desirable to minimise the number of plants necessary to build a useful resource. In cases of species with one to two germ line cells, it may be most efficient to grow a large number of M_1 plants and perhaps only grow one M_2 plant from each M_1 line in the next generation. Alternatively, if a higher number of germ line cells are present, one can grow a smaller number of M_1 plants and very likely obtain sufficient variation by growing several M_2 plants from each of these M_1 lines.

It has been estimated that effective treatments give mutation frequencies in M_1 plants with two progenitor cells (equivalent of four haploid genomes) of 2×10^{-3} per locus per M_1 plant and that mutation frequency is independent of genome size (Koornneef et al. 1982). Under such conditions, 1,500 M_1 plants would give a 95% chance of identifying a mutation in a specific locus.

For *B. rapa*, we decided to grow ~5,000 M_1 plants and subsequently grow 2 M_2 plants from each of those in the next generation. Growing such a high number of M_1 plants as a basis for the M_2 population should enhance the number of mutations in the population assuming that to a good approximation each M_1 sector is only represented once. In our *B. rapa* population, this does indeed seem to be the case, since analysing the mutation distribution in several genes by TILLING gave rise to very few cases where the two M_2 siblings have identical mutations (~2%).

20.2.4.3 Setting up the TILLING Platform

Mutation detection is carried out on pooled genomic DNA samples prepared from the M_2 plants. Leaf samples are taken from each M_2 seedling, DNA is isolated, the concentration accurately determined and stocks normalised to ensure that DNA pools are balanced. For efficient mutation detection it is important that all individual lines are equally represented within the pools.

The standard pooling strategy for screening large populations is a simple one-dimensional pooling system where each M_2 line is represented only once in a single pool, with each pool comprising four to eight M_2 lines. This is ideally suited to high-throughput mutation detection. Individual M_2 lines from pools identified as containing a mutant allele are subsequently sequenced in order to confirm the presence of the mutation and to identify the M_2 line carrying the mutation.

Two-dimensional pooling strategies are best suited for screening smaller populations. Each M_2 plant is included in two DNA pools and is therefore amplified twice during screening process, thereby reducing the occurrence of false mutation-calling errors.

20.2.4.4 Choosing the Amplicon

Mutations in genes of interests are most often detected by cell digestion at a mismatched base pair (Oleykowski et al. 1998, McCallum et al. 2000). Identification of digested fragments was originally done using dHPLC (McCallum et al. 2000); however, today sequencers such as ABI3730 or LI-COR are preferably used. In both cases, the optimal fragment length for analysis is ~1 kb (Till et al. 2003).

Choosing the precise amplicon for mutation detection is of crucial importance in order to ensure success, and several considerations need to be taken into account: (1) In the majority of cases, it will be advantageous to include as much coding sequence as possible and avoid intron or intragenic sequence. (2) Repetitive sequence such as microsatelites may cause "taq slippage" which could delete or insert extra repeats. This will lead to mismatches between intact and erroneous strands, which may become substrates for the cell nuclease. (3) As mentioned above, only a subset of codon changes is likely to have major effects on protein properties (see Section 20.2.2 above). It is therefore advisable to identify the region with most potential for generating stop codons and significant amino acid changes.

Finally, in the case of *Brassica* species, it is furthermore important to test for paralogue specificity. As the diploid *Brassica* genome is largely triplicated compared to *Arabidopsis*, one could find up to three copies (paralogues) of each single-copy *Arabidopsis* gene. When designing primers for the region of interest, it is therefore essential to verify that this primer set only amplifies the intended sequence before initiating the TILLING screen.

As mentioned above, TILLING platforms are often organised in pools with genomic DNA from four to eight individual M_2 plants mixed in one well to facilitate large-scale screenings. When a series of mutations have been identified by a cell digest of pooled DNA samples, it is therefore important to verify the mutations by sequencing PCR fragments amplified from the individual samples that were mixed in the wells where cell digestion products were identified.

Once the mutations of interest have been identified it is of extreme importance that one is able to locate the mutant lines and confirm the presence of the mutation. It is vital therefore that a robust sample tracking system is in place from the very beginning of the population development.

20.2.4.5 Linking Mutation to Phenotype

A classical backcrossing programme to remove the contaminating background mutation load is a prolonged procedure, which is expensive in both time and resources, especially for larger plants with relatively long generation times such as *Brassica*. Each backcross generation reduces the mutation load by 50%; therefore reducing the number of mutations from 1,000 to less than 10 will take seven generations of backcrossing ($1{,}000 \times 0.5^7 = 8$) along with genotyping to check for inheritance of the lesion. In many cases, it may be advantageous to cross to another ecotype/variety for introgression, which will make selection against non-target mutations more feasible and reduce the number of backcrosses.

Therefore, before embarking on such a lengthy programme it is advisable to verify the correlation between mutation and phenotype by comparing homozygous mutants to heterozygotes and homozygous wild-type sibling plants. A 100% correlation between homozygous mutant and observed phenotype would lend strong support to the hypothesis that the mutation in the "TILLed" gene is responsible for the phenotype. Moreover, if an allelic series of independent mutations in the same gene is obtained, similar phenotypes would strongly associate the phenotype with the gene and may often make a backcrossing scheme unnecessary.

20.3 RNA Interference

20.3.1 Background

The reverse genetics approach termed RNA interference (RNAi) has been utilised at fundamental and applied levels to both study and manipulate gene function in a highly specific manner. RNAi is a term used to describe the exploitation of a functionally diverse array of endogenous regulatory mechanisms, which mediate transcriptional and post-transcriptional gene silencing (TGS and PTGS, respectively) in plants and animals through short 20–26 nt RNAs (Fire et al. 1998, Baulcombe 2004, Vaucheret 2006). These short RNAs (sRNAs) target highly specific regions of mRNA or DNA, enabling their respective degradation or epigenetic modification, through action of specific protein complexes. Homology-dependent, RNA-directed downregulation of gene expression is evident at transcriptional, post-transcriptional and translational levels. RNA silencing appears to be part of an adaptive, genetic defense mechanism to protect against viral pathogens and to silence transposons (Waterhouse et al. 2001).

Endogenous sRNAs are generated from various types of precursor molecules including hairpin loops, imperfect fold-back precursors in the case of microRNAs (miRNAs) and even single-stranded RNA (ssRNA), requiring the presence of an RNA-dependent RNA polymerase (RDR). The activity of RDRs is necessary to generate double-stranded RNA molecules that will act as substrates from which siRNAs be processed (Dalmay et al. 2000). Precursors are processed by nucleases of the DICER family (Xie et al. 2004, Fusaro et al. 2006). This family of proteins mediates cleavage of precursors, producing different size dsRNA fragments (20–26 nt), according to the dicer/silencing pathway involved. sRNAs act as a complementary guide to target endogenous mRNAs for degradation or to induce epigenetic modifications, resulting in gene silencing. Targeted mRNAs are directed through RISC (RNA-induced silencing complex) protein complexes, whose core component is an Argonaute (AGO) protein. In *Arabidopsis*, the AGO proteins consist of a family of 10 members (Carmell et al. 2002). Certain members of this family have been demonstrated to exhibit an ability to "slice" RNA. Sequence specificity guides substrate selection, and the mRNA target is subsequently "sliced" by an AGO protein. AGO1 is reported to be responsible for cleavage of 21 nt siRNAs and miRNAs, while AGO4 is associated with the slicing of 24 nt siRNAs (Baumberger and Baulcombe 2005, Qi et al. 2005, 2006). Numerous endogenous RNA-directed silencing pathways have been identified in plants, including nasiRNA (natural antisense siRNA), tasiRNA (trans-acting siRNA), miRNA (microRNA), casiRNA (chromatin-targeted RNA silencing) and inverted repeat-mediated PTGS. These diverse gene silencing pathways, at least in *Arabidopsis,* vary in RNA precursor, processing complexes and the RNA-dependent RNA polymerases (RDRs) required to amplify single-stranded RNA (ssRNA) to active double-stranded RNA (dsRNA) (Shivaprasad et al. 2008 and for recent reviews on the mechanisms involved in RNAi in plants see, for example, Horiguchi (2004), Brodersen and Voinnet (2006), Vaucheret (2006), Small (2007), Eamens et al. (2008)).

The use of RNAi technology is well established as a powerful research tool for studying gene function especially in *Arabidopsis*, but such technology also has much potential for commercial application in important *Brassica* crop varieties to manipulate key traits (Byzova et al. 2004, Yu et al. 2008). The rate at which knowledge transfer is implemented using this technology is limited to the availability of sequence information. However, this is fast increasing with the advancement of *B. rapa* genome sequencing.

20.3.2 Classes of sRNA Associated with PTGS

Although other classes exist, two major types of sRNAs are associated with reverse genetics RNA silencing platforms: microRNAs (miRNAs) and short-interfering RNAs (siRNAs) (Hamilton et al. 2002).

miRNAs are generally 21–24 nt dsRNAs, produced from a single-stranded, non-coding transcript that includes imperfect fold backs processed by the DCL1 protein (Ossowski et al. 2008). miRNAs regulate endogenous genes by PTGS, and this can occur via translational inhibition or site-specific cleavage of mRNA (Brodersen et al. 2008). This class of sRNA is also documented to mediate epigenetic modification, with proposed effects on DNA methylation within the nucleus (Brodersen and Voinnet 2006).

siRNAs, although chemically similar to miRNAs, function in distinct silencing mechanisms. siRNA act as "guide molecules" in a range of endogenous silencing pathways to regulate a vast range of differing gene functions. These dsRNAs range in length from ~21 to 26 nt with their size varying according to which DCL protein they were processed by. siRNAs can induce both PTGS and TGS (Jones et al. 1999). PTGS is triggered by 21–22-nt fragments, whereas TGS is mediated by siRNAs of 24–26 nt, processed by DCL3. DCL1 and DCL4 each generates ~21 nt siRNAs, with DCL4 playing a central role in processing siRNAs involved in RNAi. DCL2 is reported to mediate 22 nt production and is associated with viral silencing pathways (Hamilton et al. 2002, Xie et al. 2004, Fusaro et al. 2006) (Fig. 20.2). siRNAs are processed from a range of precursors including transposons, retro elements and long dsRNA molecules arranged as complementary fold backs (Slotkin and Martienssen 2007, Kasschau et al. 2007, Meister and Tuschl 2004). Unlike miRNAs, processed siRNAs are not uniform in sequence and exhibit a diverse array of fragments, complementary to differing regions of the precursor molecule. This can lead to different silencing efficiencies of the various fragments directing RNAi.

20.3.3 RNAi/PTGS Mechanisms: Gene Silencing Approaches in Brassicaceae

Brassica species include a diverse range of natural, ornamental and agriculturally important species. A number of RNAi approaches have been developed for the targeted downregulation of genes in *Arabidopsis* and have been transferred to *Brassica*. These range from single transgenes antisense and co-suppression methods to more elaborate complementary hairpin RNAs and artificial microRNAs. Such applied techniques act to beneficially exploit and manipulate endogenous sRNA pathways, whose regulatory mechanisms are still not completely understood.

20.3.3.1 Virally Induced Gene Silencing

Using virally induced gene silencing (VIGS) to reduce the expression of target genes exploits the plant's natural, antiviral defense mechanism (Fig. 20.2). Viruses can be manipulated to silence endogenous genes, through expression of their cognate sequences, designating host mRNA for silencing (Lu et al. 2003). Modified viruses have been successfully used to introduce gene silencing constructs to *Brassica* and a wide range of other plant species (Jones et al. 1999, Pflieger et al. 2008, Zhang and Ghabrial 2006, 2009). VIGS vectors are derived from natural plant viruses, infecting specific plant species. A fragment of DNA, complementary to a desired target gene, is incorporated into the viral cassette and is then expressed under control of a viral promoter. An antisense strand, complementary to the ssRNA, is subsequently

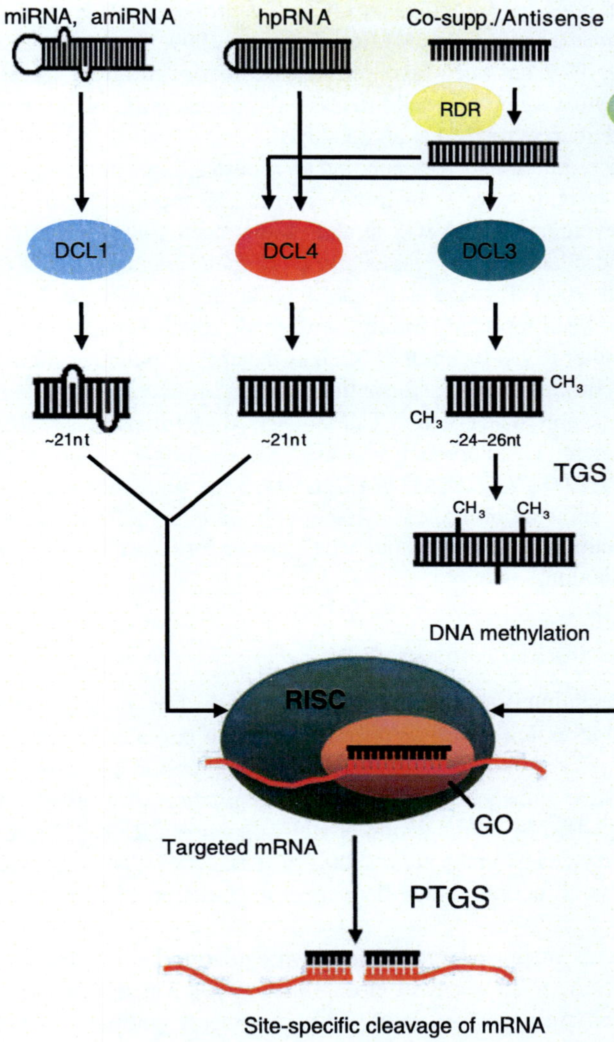

Fig. 20.2 The four main RNAi pathways that have been manipulated to study and beneficially exploit *Brassica*: microRNA/artificial miRNA (miRNA/amiRNA), hairpin (hpRNA), co-suppression/antisense and viral-induced gene silencing (VIGS) (RDR/vRDR, RNA-dependent polymerase/viral RDR, respectively)

produced by a virally encoded RDR to produce dsRNA which will trigger the silencing.

VIGS does not require *Agrobacterium*-mediated transformation and is therefore amenable to crop varieties that may be recalcitrant to this procedure (Burch-Smith et al. 2004). Viral siRNAs are reported to exhibit two main forms: the 21 nt class inducing PTGS and the 24–26 nt class eliciting TGS (Ossowski et al. 2008, Voinnet 2005). In *Arabidopsis,* VIGS strategies have exploited vectors based on a range of viruses including, *cabbage leaf curl virus* (CbLCV), *tobacco rattle virus* (TRV) and *turnip yellow mosaic virus* (TYMV) (Ratcliff et al. 1999, Turnage et al. 2002, Pflieger et al. 2008). TRV has been favoured in a number of different species due to a relatively wide host range and mild disease symptoms, minimising potential effects on a silenced phenotype (Liu et al. 2002a, b).

Due to increasing requirements for economical, accurate, high-throughput RNAi approaches, VIGS systems have seen relatively limited application in *Brassica* crops. Moreover, development of VIGS strategies has been somewhat limited by virus–host range, reproducibility of results and cost/labour associated with certain vectors (Watson et al. 2005, Turnage et al. 2002). Difficulty in construct delivery in *Arabidopsis* has also been reported (Burch-Smith et al. 2006). Even so, VIGS has some benefits over other RNAi techniques, such as the ability to induce silencing in lines recalcitrant to conventional transformation (Burch-Smith et al. 2004). Pflieger et al. (2008) recently reported an efficient TYMV-derived vector, pTY, containing an inverted repeat fragment. This approach was used successfully to induce silencing in the *Arabidopsis PHYTOENE DESATURASE* (*PDS*) gene, with only a small target fragment, by simple abrasion of the leaf surface with the intact pTY plasmid. This may represent a potential, high-throughput VIGS vector for future investigation of *Arabidopsis* gene function.

20.3.3.2 Co-suppression and Antisense RNA

A classical RNAi technique in plants is through introduction of a transgene in either a sense (co-suppression) or an antisense orientation into a recipient genome (Napoli et al. 1990, Ecker and Davis 1986). In co-suppression systems, expression of a sense-oriented transgene may lead to accumulation of a target ssRNA, which is converted to dsRNA by the action of RNA-dependent RNA polymerase 6 (RDR6) and other enzymes (Dalmay et al. 2000, Xu et al. 2006). Precursors are subsequently processed to direct homology-dependent degradation of complementary targets.

Antisense constructs mediate target gene downregulation by binding complementary nascent mRNAs (Van der Krol et al. 1988, 1990). These dsRNAs act as siRNA precursors and are processed into ~21-nt fragments guiding PTGS, as in other pathways. Both antisense and co-suppression have been utilised to repress the endogenous *FAD2* (*FATTY ACID DESATURASE2*) gene in *Brassica carinata* (Jadhav et al. 2005). Manipulation of this gene resulted in increases in oleic acid, with a concomitant increase in erucic acid, among other long-chain, monounsaturated fatty acids. This example demonstrates that such an approach is suitable for generating novel oil profiles in *Brassica* species.

Antisense technology has also been demonstrated to have application in improving storage quality of *Brassica* vegetables: An antisense construct was utilised to suppress an aleurain-like cysteine protease termed BoCP5, in broccoli, resulting in delayed yellowing and retarded chlorophyll degradation compared to wild-type plants (Eason et al. 2005). Both examples show how important such techniques could be for future *Brassica* crop improvement.

20.3.3.3 IR-PTGS: siRNA-Directed Gene Silencing Using hpRNA Constructs

Although co-suppression, antisense and VIGS have all been applied successfully in *Brassica* species, more efficient systems have since been developed through which endogenous genes can be silenced (Wesley et al. 2001, Helliwell and Waterhouse 2003). Inverted repeats (IR) or self-complementary hairpin RNAs (hpRNA) have seen the most widespread adoption for RNAi in *Brassica* and many other plant species. (Wesley et al. 2001, Liu et al. 2002). Techniques employing hpRNA exploit siRNA-mediated PTGS and TGS pathways in transformants, and hairpin stem-loop precursors are processed to generate 21 and 24 nt dsRNAs. In this system, precursors are diced to generate an array of siRNAs complementary to a target locus.

hpRNAi is amenable to silence single, multiple and even whole gene families in a highly specific, stable and heritable fashion. Thus, RNAi approaches of this nature clearly have an extremely important role in functional genomics and crop improvement efforts. Transgene constructs expressing inverted hairpin loop structures have been used to successfully induce gene silencing in a number of key metabolic and developmental genes in both model and crop *Brassica* species (Stoutjesdijk et al. 2002, Byzova et al. 2004, Yu et al. 2008).

20.3.3.4 hpRNA

The most basic hpRNA constructs contain an inverted repeat, homologous to a target locus, under control of a constitutive promoter, separated by a spacer sequence (Fig. 20.3). Examples of spacer sequences used in hpRNAi constructs include regions of the *GUS* (β-glucuronidase) and *GFP* (green fluorescent protein) reporter genes (Chuang and Meyerowitz 2000, Piccin et al. 2001, Hirai et al. 2007). Such constructs have been utilised to downregulate key developmental genes involved

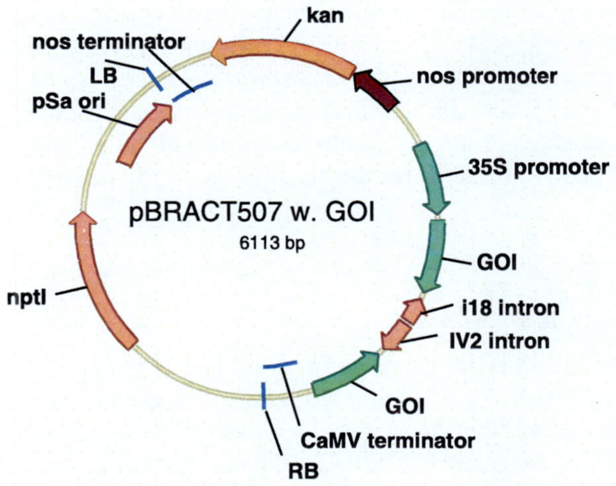

Fig. 20.3 Generalised pBRACT507 vector for delivery of *Brassica* RNAi constructs. GOI: gene of interest is expressed under control of 35S promoter

in floral patterning, such as *AGAMOUS* (*AG*) and *APETALA1* (*AP1*) (Chuang and Meyerowitz 2000). This presents an excellent proof of concept of how hpRNA constructs can be used to selectively downregulate genes in *Arabidopsis* but has the potential to modulate key regulatory genes in *Brassica* crop species. An extremely intriguing property of this approach is the variation in silencing efficiency in transformants, resulting in phenotypic series (weak, intermediate and strong). Such series could be of great use for studying dose-dependent gene function or for modulating beneficial traits in crop species.

ihpRNA

The strength of PTGS can be modulated depending on the structure of the RNAi construct (Hirai et al. 2007). This discovery has lead to the development of more efficient silencing constructs and effective vector systems for delivering RNAi cassettes into recipient genomes (Helliwell et al. 2002, Bartlett et al. 2008). Inclusion of an intron-derived spacer in constructs is reported to raise silencing efficiency from approximately 55% in hpRNA systems to approximately 90–100% in transformants expressing intron-spliced, hairpin RNA constructs (ihpRNA) (Wesley et al. 2001). This is a major benefit to the researcher and breeder alike, greatly reducing the space and resources required to produce RNAi lines.

ihpRNA constructs have been used to investigate and exploit a wide range of genes – hpRNA has even been used to investigate functional genomics in redundant gene pairs (Helliwell et al. 2002). Effective silencing can be achieved using fragments ranging from 50 bp to 1 kb, but a recommended length of 300–600 bp is suggested: The shorter the fragment, the less effective silencing will be, but long hairpins are reported to increase the chance of recombination in bacterial host strain and therefore complicate the cloning procedure. With respect to which region of the gene should be targeted, 5′UTR, 3′UTR and coding sequences have all been successfully used. The main consideration though is to avoid areas homologous to other loci, thereby increasing gene specificity. It is suggested to avoid the use of sequences containing blocks of more than 18 bp homology between non-targets and construct. The silencing efficiency of such constructs appears to be somewhat gene dependent and may reflect issues with accessibility of the siRNA and target mRNA or relative abundances of mRNA (Helliwell and Waterhouse 2003). A generalised example of a pBRACT vector for use in *Brassica* transformation is presented in Fig. 20.3.

Vectors for Delivering ihpRNA Contructs in *Arabidopsis* and *Brassica*

pHANNIBAL and pKANNIBAL

A major factor in achieving silencing of target loci is the generation of suitable transformation vectors. Such vectors are required to enable the transfer of constructs containing the silencing cassette. A series of intron-spliced, hairpin RNAi (ihpRNAi) vectors, termed pHANNIBAL and pKANNIBAL, have been developed to deliver efficient, accurate silencing of a target gene. These vectors have been designed so that constructs containing inverted repeats of a target gene, separated by an intron, can be cloned directionally using conventional restriction enzyme digestion and sequential DNA ligation (Wesley et al. 2001, Helliwell and Waterhouse 2003, 2005). The vectors differ in their selectable marker gene: pHANNIBAL contains bacterial ampicillin resistance, whereas pKANNIBAL contains bacterial kanamycin resistance. After construction of the silencing cassette in these bacterial vectors, a fragment containing the 35S promoter and the intron-containing hairpin sequence is transferred to the pART27 vector for plant transformation. The pH/KANNIBAL system is useful for silencing a small number of target genes (<10), but the cloning steps involved make it a relatively laborious procedure.

pHANNIBAL/KANNIBAL-based constructs have been utilised to downregulate genes involved in a range of developmental and metabolic functions in members of the Brassicaceae family. In *Arabidopsis* this has been successfully achieved for *CHALCONE SYNTHASE* (*CHS*), *ETHYLENE INSENSITIVE2* (*EIN2*) and *FLOWERING LOCUS C* (*FLC*) (Wesley et al. 2001). pH/KANNIBAL have also been used to perturb gene function to generate novel and beneficial properties in *Brassica* crop varieties. For example, manipulation of *FAD2* in *B. carinata,* using a pKANNIBAL vector to deliver a silencing construct, resulted in increases in oleic and erucic acid, as observed using co-suppression and antisense approaches (Mietkiewska et al. 2008, Jadhav et al. 2005).

pHELLSGATE Series

pHELLSGATE are a series of ihpRNA vectors designed around the Gateway™ recombination system, facilitating high-throughput cloning. Inclusion of a target sequence into pHELLSGATE vector by unidirectional recombination is mediated by incorporating *att*B1/B2 or *att*L1/L2 sites into the PCR primers. Silencing constructs are generated from entry clones in a single step, greatly increasing the efficiency of the cloning procedure. pHELLSGATE vectors allow directional cloning of target gene sequences and include two *ccdB*-negative selection markers at the sites where target genes will be recombined.

These types of vectors are designed to enable production of ihpRNA constructs for single genes up to large numbers of genes and gene families. Such vectors are amenable to high-throughput applications as the PCR; recombination and transformation stages required to introduce ihpRNA constructs can all be performed in 96-well plates (Helliwell et al. 2002, Helliwell and Waterhouse 2003, 2005).

Inducible and Tissue-Specific Gene Silencing

The use of inducible promoters, acting as a biochemical switch, enables gene silencing to be initiated at the cultivators desire. This "temporal control" mechanism has a multitude of applications, especially in silencing genes that would be lethal earlier in development.

Wielopolska et al. (2005) have engineered a high-throughput dexamethasone-inducible system, utilising a pHELLSGATE-derived vector, to transform *Arabidopsis*. This two-component system uses a pOp6 promoter to drive target gene expression under control of a synthetic transcription factor, LhGR. LhGR will only initiate transcription of the pOp6 promoter in the presence of dexamethasone (DEX) (Craft et al. 2005). Exposure resulted in the silencing of constructs targeted against *PHYTOENE DESATURASE* (*PDS*) and a luciferase transgene, respectively. RNAi in *PDS* lines was established within 24 h of the introduction of the inducer and maintained under continued exposure. Elegantly, on removal of the DEX, silencing of *PDS* acquiesced, not returning to levels in wild type until 10 days later. Inducible systems have application to *Brassica* crop species and could pose an extremely powerful strategy for dissecting gene function on a temporal scale.

Off-Targets and Transitive Silencing (Transitivity)

Although RNAi is capable of highly specific downregulation of target genes, the resolution at which siRNAs guide silencing creates an inherent flaw in the system. As only a small number of nucleotides in siRNAs are required to confer specificity to a target, transcripts from "off-targets" displaying close homology may also be subject to gene silencing (Lin et al. 2005). In efforts to minimise such effects, programmes have been designed, capable of predicting "off-targets" in 25 different species, minimising the potential for erroneous construct design (Xu et al. 2006). Another side-effect of RNAi is termed transitive silencing. This is where cleaved mRNAs cause silencing to spread into adjacent sequences through action of RDR, potentially leading to undesirable phenotypes that may mask the "true" effects of silencing on the target gene (Bleys et al. 2006a, b, Petersen and Albrechtsen 2005). However, these features may also be of benefit when trying to silence a number of conserved genes, i.e. a multi-gene family.

Novel Silencing Technologies

amiRNA – Artificial microRNAs

A recently developed approach termed artificial microRNAs (amiRNA) is now being adapted to manipulate endogenous miRNA silencing pathways (Schwab et al. 2005, 2006, Alvarez et al. 2006). Natural miRNAs encoded by endogenous genes are implicated in both TGS and PTGS and downregulate target genes by either translational inhibition or site-specific transcript cleavage (Schwab et al. 2005, Brodersen et al. 2008). The design of amiRNAs exploits natural miRNAs, which are used as a backbone to produce synthetic miRNAs with differing degrees of complementarity to a target locus. This approach is capable of silencing up to 10 genes successfully and can be regulated under the control of an inducible promoter (Schwab et al. 2006). amiRNA silencing is reported to be more accurate than hpRNAi, leading to greater specificity. The basis of this specificity is that only a single target miRNA is processed from the stem-loop precursor, as opposed to the diverse range of siRNA fragments derived from hpRNA constructs.

To produce 21 nt amiRNAs, individual miRNA is replaced with fragments designed to target a gene(s) of interest using overlapping PCR techniques. A number of *Arabidopsis* miRNAs have been developed as backbones in amiRNA applications, including miR319a, miR172a and miR156a. Using transcriptomics, parameters for amiRNA design has been suggested in *Arabidopsis* (Schwab et al.

2005) that are also likely to be applicable in *Brassica* species. This has been developed into a web-based design tool termed WMD (Web MicroRNA Designer: http//wmd2.weigelworld.org) which enables users to accurately design amiRNAs targeted towards genes of interest for around 30 different species (Ossowski et al. 2008). EST collections can be used if genome sequence data are sparse.

amiRNA-induced downregulation of target loci has been demonstrated to precisely phenocopy multiple mutant phenotypes in *Arabidopsis* and can be used to induce cross-species gene silencing (Alvarez et al. 2006). A molecular tool of this nature could potentially have extensive application in *Brassica* crop species. In *Arabidopsis,* amiRNA-mediated resistance to several plant viruses has been reported (Niu et al. 2006, Duan et al. 2008), and this could have significant implications on modulating disease resistance in *Brassica* crop varieties.

Target Mimicry

A relatively recent insight into the control of endogenous miRNA function is the development of the "target mimicry" approach. Franco-Zorrilla et al. (2007) demonstrated that a non-coding transcript can interact with complementary, natural miRNAs, thereby inhibiting the interaction with their normal targets. Using target mimicry, one can therefore produce transgenic lines to inhibit the activity of specific miRNAs and hence providing a promising additional tool for gene function studies and crop improvement.

20.3.4 Examples of RNAi in Brassica *Species*

As described above, RNA silencing technology has developed in a relatively short period of time. Constructs have been shown to have profound effects on gene function, allowing targets to be downregulated in a stable, heritable and highly specific manner. A variety of *Agrobacterium*-mediated vectors have been developed for wide-scale application in many different plant species (Helliwell et al. 2002). It is now practical to generate RNAi lines at high efficiency (close to 100% downregulation of gene activity), with reduced requirement for space, resources or funding. Although much stigma has been associated with the use of transgenics in some countries, benefits presented by RNAi technology are huge. Therefore, if RNAi is used to facilitate increased nutritional properties and more sustainable crop and food production, public opinion may eventually sway in favour of such technologies.

20.3.4.1 Metabolic Engineering and Manipulation of Biosynthetic Pathways Using RNAi

RNAi technology represents a powerful tool for functional genomics studies within members of the Brassicaceae family. Technologies also have potential application to manipulating a suite of important agronomic traits. Feasibly, any pathway could be investigated for any beneficial effects offered through targeted down-regulation of a candidate locus. Examples could be genes involved in biosynthetic and developmental pathways, disease resistance or even biotic/abiotic stress resistance. *Brassica* species, especially oil-producing varieties have a clear potential to be beneficially exploited using RNAi technology.

The targeted silencing of genes regulating metabolic pathways, using hpRNAi, is an excellent strategy for both fundamental genetic research and future development of crop varieties. We have already seen how basic hpRNA constructs have been used to modulate *Brassica* oil production to generate high erucic acid varieties (Mietkiewska et al. 2008). It is apparent that such technology can also be applied to improve the nutritional quality of crops. Yu et al. (2008) have used an RNAi approach to modulate carotenoid levels in *B. napus* using a hairpin construct containing a 300-bp GUS spacer. The construct induces PTGS in the epsilon cyclase gene, increasing beta-carotene, lutein, zeaxanthin and violaxanthin levels in the seed. RNAi-mediated downregulation of candidate genes is a promising tool for modulating the composition of a wide range of important *Brassica*-derived products. This proposes an extremely exciting strategy through which to tackle problems associated with food security.

20.3.4.2 Studying Gene Function Throughout *Brassica* Development

RNAi technology in the Brassicaceae family is not only limited to down regulating metabolic genes. A wide range of targets has been investigated, including loci with important roles in development and yield. One example is the reduction in seed loss in the dehiscent oilseed *Brassica* varieties by manipulation of a process known as pod shatter. Unsynchronised pod shatter is a serious economic issue in commercial varieties of *B. napus*, and a mechanism to reduce this would be of dramatic benefit to both the farmer and the environment. This was previously achieved in the oilseed *Brassica juncea* by expressing the *Arabidopsis FRUITFULL* (*FUL*) gene under control of the constitutive CaMV 35S promoter (Østergaard et al. 2006). Although very effective, the resulting phenotype was too severe, as pods were totally indehiscent, resulting in damaging the seed during harvest. Recently, pBRACT (Bartlett et al. 2008) and pHELLSGATE vectors were used to introduce constructs reducing pod shatter into *B. napus* by targeted downregulation of the *BnaX.IND.a* (*B. napus* orthologue of the *INDEHISCENT* (*IND*) gene in *Arabidopsis*). (For standardised *Brassica* gene nomenclature see Østergaard and King (2008).) In *Arabidopsis*, *IND* regulates development of the valve margin where fruit opening takes place upon maturity (Liljegren et al. 2004). *B. napus IND* RNAi lines exhibited a complete lack of separation layer, resulting in valves that were much harder to shed, but not completely indehiscent as in the *35S::FUL* lines (unpublished data and Østergaard et al. 2006). In addition to the obtained fine-tuning of the pod shatter trait, this example also illustrates how knowledge gained in the model *Arabidopsis* can be utilised to improve *Brassica* crop varieties.

20.3.4.3 Conferring Tissue Specificity in RNAi Approaches in *Brassica* Species

RNAi has been utilised to silence developmental genes in *Brassica* in a tissue-specific manner. Byzova et al. (2004) utilised an hpRNAi approach to modify floral development in *Arabidopsis* and *B. napus*. Silencing of the *Arabidopsis*

PISTILLATA gene and the orthologous gene family in *B. napus* (*BPI*) resulted in male fertile flowers, where petals were converted to sepals (*Arabidopsis*) or into sepaloid petals (*B. napus*). Phenotypes were observed to be both stable and heritable. This apetalous phenotype is of potential interest to breeders of oilseed rape as it results in greater penetrance of photosynthetically active radiation (PAR) to the canopy (up to 60%). In the case of *B. napus,* a chimeric promoter consisting of the *APETELA1* (*AP1*) promoter from *Arabidopsis*, adjacent to a modified fragment from the *Arabidopsis AP3* promoter, was utilised to direct tissue-specific expression of the silencing construct. The development of constructs, which confer tissue-specific PTGS, vastly increases the power of RNAi strategies by improving the spatial precision in gene silencing.

Mietkiewska et al. (2008) have also demonstrated tissue-specific RNAi constructs in *Brassica* crops: Utilising a pKANNIBAL vector, a silencing construct targeting the endogenous *FAD2* gene, under control of a seed-specific napin promoter resulted in an increase in erucic acid proportions in *B. carinata* compared to wild-type plants and transformants expressing transgenes targeting the *FAD2* gene alone.

20.4 Concluding Remarks

Reverse genetics platforms are proving to be invaluable tools for functional genomics and the improvement of crop species belonging to the Brassicaceae family. In this chapter, we have aimed to provide an overview of the current status of resources and technologies, which we believe will develop further and remain crucial for both fundamental science and crop improvement initiatives in many years to come. In order to take full advantage of these approaches, it is necessary to have access to whole-genome sequence, which will reveal important information of, for example, gene copy numbers. For *Brassica*, reverse genetics is therefore expected to reach full bloom as complete genome sequences become available.

References

Alonso JM, Stepanova AN (2003) T-DNA mutagenesis in Arabidopsis. Methods Mol Biol 236:177–188

Alvarez JP, Pekker I, Goldshmidt A et al (2006) Endogenous and synthetic microRNAs stimulate simultaneous, efficient, and localized regulation of multiple targets in diverse species. Plant Cell 18:1134–1151

Bartlett JG, Alves SC, Smedley M et al (2008) High-throughput Agrobacterium-mediated barley transformation. Plant Methods 4:22

Baulcombe D (2004) RNA silencing in plants. Nature 431:356–363

Baumberger N, Baulcombe DC (2005) Arabidopsis ARGONAUTE 1 is an RNA slicer that selectively recruits microRNAs and short interfering RNAs. Proc Natl Acad Sci USA 102:11928–11933

Blanc G, Wolfe KH (2004a) Functional divergence of duplicated genes formed by polyploidy during Arabidopsis evolution. Plant Cell 16:1679–1691

Blanc G, Wolfe KH (2004b) Widespread paleopolyploidy in model plant species inferred from age distributions of duplicate genes. Plant Cell 16:1667–1678

Bleys A, Van Houdt H, Depicker A (2006a) Down-regulation of endogenes mediated by a transitive silencing signal. Rna-a Publication of the Rna Society 12:1633–1639
Bleys A, Vermeersch L, Van Houdt H et al (2006b) The frequency and efficiency of endogene suppression by transitive silencing signals is influenced by the length of sequence homology. Plant Physiol 142:788–796
Brodersen P, Sakvarelidze-Achard L, Bruun-Rasmussen M et al (2008) Widespread translational inhibition by plant miRNAs and siRNAs. Science 320:1185–1190
Brodersen P, Voinnet O (2006) The diversity of RNA silencing pathways in plants. Trends Genet 22:268–280
Burch-Smith TM, Anderson JC, Martin GB et al (2004) Applications and advantages of virus-induced gene silencing for gene function studies in plants. Plant J 39:734–746
Burch-Smith TM, Schiff M, Liu Y et al (2006) Efficient virus-induced gene silencing in Arabidopsis. Plant Physiol 142:21–27
Byzova M, Verduyn C, De Brouwer D et al (2004) Transforming petals into sepaloid organs in Arabidopsis and oilseed rape: implementation of the hairpin RNA-mediated gene silencing technology in an organ-specific manner. Planta 218:379–387
Caldwell DG, McCallum N, Shaw P et al (2004) A structured mutant population for forward and reverse genetics in Barley (*Hordeum vulgare* L.). Plant J 40:143–150
Carmell MA, Xuan Z, Zhang MQ et al (2002) The Argonaute family:tentacles that reach into RNAi, developmental control, stem cell maintainance, and tumorigenesis. Genes Dev 16: 2733–2742
Chuang CF, Meyerowitz EM (2000) Specific and heritable genetic interference by double-stranded RNA in Arabidopsis thaliana. Proc Natl Acad Sci USA 97:4985–4990
Cooper JL, Till BJ, Laport RG et al (2008) TILLING to detect induced mutations in soybean. BMC Plant Biol 8:9
Craft J, Samalova M, Baroux C et al (2005) New pOp/LhG4 vectors for stringent glucocorticoid-dependent transgene expression in Arabidopsis. Plant J 41:899–918
Dalmais M, Schmidt J, Le Signor C et al (2008) UTILLdb, a Pisum sativum in silico forward and reverse genetics tool. Genome Biol 9:R43
Dalmay T, Hamilton A, Rudd S et al (2000) An RNA-dependent RNA polymerase gene in Arabidopsis is required for posttranscriptional gene silencing mediated by a transgene but not by a virus. Cell 101:543–553
Duan CG, Wang CH, Fang RX et al (2008) Artificial MicroRNAs highly accessible to targets confer efficient virus resistance in plants. J Virol 82:11084–11095
Eamens A, Wang MB, Smith NA et al (2008) RNA silencing in plants: yesterday, today, and tomorrow. Plant Physiol 147:456–468
Eason JR, Ryan DJ, Watson LM et al (2005) Suppression of the cysteine protease, aleurain, delays floret senescence in Brassica oleracea. Plant Mol Biol 57:645–657
Ecker JR, Davis RW (1986) Inhibition of gene expression in plant cells by expression of antisense RNA. Proc Natl Acad Sci USA 83:5372–5376
Fire A, Xu S, Montgomery MK et al (1998) Potent and specific genetic interference by double-stranded RNA in Caenorhabditis elegans. Nature 391:806–811
Franco-Zorrilla JM, Valli A, Todesco M et al (2007) Target mimicry provides a new mechanism for regulation of microRNA activity. Nat Genet 39:1033–1037
Fusaro AF, Matthew L, Smith NA et al (2006) RNA interference-inducing hairpin RNAs in plants act through the viral defence pathway. EMBO Rep 7:1168–1175
Gilchrist EJ, O'Neil NJ, Rose AM et al (2006) TILLING is an effective reverse genetics technique for Caenorhabditis elegans. BMC Genomics 7:262
Greene EA, Codomo CA, Taylor NE et al (2003) Spectrum of chemically induced mutations from a large-scale reverse-genetic screen in Arabidopsis. Genetics 164:731–740
Hamilton A, Voinnet O, Chappell L et al (2002) Two classes of short interfering RNA in RNA silencing. Embo J 21:4671–4679
Hammond SM, Caudy AA, Hannon GJ (2001) Post-transcriptional gene silencing by double-stranded RNA. Nat Rev Genet 2:110–119

Helliwell C, Waterhouse P (2003) Constructs and methods for high-throughput gene silencing in plants. Methods 30:289–295

Helliwell CA, Waterhouse PM (2005) Constructs and methods for hairpin RNA-mediated gene silencing in plants. Methods Enzymol 392:24–35

Helliwell CA, Wesley SV, Wielopolska AJ et al (2002) High-throughput vectors for efficient gene silencing in plants. Funct Plant Biol 29:1217–1225

Henikoff S, Till BJ, Comai L (2004) TILLING. Traditional mutagenesis meets functional genomics. Plant Physiol 135:630–636

Hirai S, Oka S, Adachi E et al (2007) The effects of spacer sequences on silencing efficiency of plant RNAi vectors. Plant Cell Rep 26:651–659

Horiguchi G (2004) RNA silencing in plants: a shortcut to functional analysis. Differentiation 72:65–73

Jadhav A, Katavic V, Marillia EF et al (2005) Increased levels of erucic acid in Brassica carinata by co-suppression and antisense repression of the endogenous FAD2 gene. Metabolic Eng 7:215–220

Jones L, Hamilton AJ, Voinnet O et al (1999) RNA-DNA interactions and DNA methylation in post-transcriptional gene silencing. Plant Cell 11:2291–2301

Kasschau KD, Fahlgren N, Chapman EJ et al (2007) Genome-wide profiling and analysis of Arabidopsis siRNAs. PLoS Biol 5:e57

Koornneef M (2002) Classical mutagenesis in higher plants. In: Gilmartin PM, Bowler C (eds) Molecular plant biology, vol 1, pp 1–11. Oxford, GB, Oxford University Press

Koornneef M, Dellaert LMW, van der Veen JH (1982) EMS-induced and radiation-induced mutation frequencies at individual loci in *Arabidopsis-thaliana* (L) Heynh. Mutat Res 93:109–123

Liljegren SJ, Roeder AH, Kempin SA et al (2004) Control of fruit patterning in Arabidopsis by INDEHISCENT. Cell 116:843–853

Lin X, Ruan X, Anderson MG et al (2005) siRNA-mediated off-target gene silencing triggered by a 7 nt complementation. Nucleic Acids Res 33:4527–4535

Liu Y, Schiff M, Dinesh-Kumar SP (2002a) Virus-induced gene silencing in tomato. Plant J 31:777–786

Liu Y, Schiff M, Marathe R et al (2002b) Tobacco Rar1, EDS1 and NPR1/NIM1 like genes are required for N-mediated resistance to tobacco mosaic virus. Plant J 30:415–429

Liu Q, Singh SP, Green AG (2002) High-stearic and High-oleic cottonseed oils produced by hairpin RNA-mediated post-transcriptional gene silencing. Plant Physiol 129:1732–1743

Lu R, Martin-Hernandez AM, Peart JR et al (2003) Virus-induced gene silencing in plants. Methods 30:296–303

Lysak MA, Koch MA, Pecinka A et al (2005) Chromosome triplication found across the tribe Brassiceae. Genome Res 15:516–525

McCallum CM, Comai L, Greene EA et al (2000) Targeting induced local lesions IN genomes (TILLING) for plant functional genomics. Plant Physiol 123:439–442

Meister G, Tuschl T (2004) Mechanisms of gene silencing by double-stranded RNA. Nature 431:343–349

Mietkiewska E, Hoffman TL, Brost JM et al (2008) Hairpin-RNA mediated silencing of endogenous FAD2 gene combined with heterologous expression of crambe abyssinica FAE gene causes an increase in the level of erucic acid in transgenic Brassica carinata seeds. Mol Breed 22:619–627

Napoli C, Lemieux C, Jorgensen R (1990) Introduction of a chimeric chalcone syntahse gene into petunia results in revesible co-suppression of homologous genes in trans. Plant Cell 2: 279–289

Niu QW, Lin SS, Reyes JL et al (2006) Expression of artificial microRNAs in transgenic Arabidopsis thaliana confers virus resistance. Nat Biotechnol 24:1420–1428

Oleykowski CA, Bronson Mullins CR, Godwin AK et al (1998) Mutation detection using a novel plant endonuclease. Nucleic Acids Res 26:4597–4602

Ossowski S, Schwab R, Weigel D (2008) Gene silencing in plants using artificial microRNAs and other small RNAs. Plant J 53:674–690

Østergaard L, Kempin SA, Bies D et al (2006) Pod shatter-resistant Brassica fruit produced by ectopic expression of the FRUITFULL gene. Plant Biotechnol J 4:45–51

Østergaard L, King GJ (2008) Standardized gene nomenclature for the Brassica genus. Plant Methods 4:10

Penmetsa RV, Cook DR (2000) Production and characterization of diverse developmental mutants of Medicago truncatula. Plant Physiol 123:1387–1398

Perry JA, Wang TL, Welham TJ et al (2003) A TILLING reverse genetics tool and a web-accessible collection of mutants of the legume Lotus japonicus. Plant Physiol 131:866–871

Petersen BO, Albrechtsen M (2005) Evidence implying only unprimed RdRP activity during transitive gene silencing in plants. Plant Mol Biol 58:575–583

Pflieger S, Blanchet S, Camborde L et al (2008) Efficient virus-induced gene silencing in Arabidopsis using a "one-step" TYMV-derived vector. Plant J 56:678–690

Piccin A, Salameh A, Benna C et al (2001) Efficient and heritable functional knock-out of an adult phenotype in Drosophila using a GAL4-driven hairpin RNA incorporating a heterologous spacer. Nucleic Acids Res 29:E55–55

Qi Y, Denli AM, Hannon GJ (2005) Biochemical specialization within Arabidopsis RNA silencing pathways. Mol Cell 19:421–428

Qi Y, He X, Wang XJ, Kohany O, Jurka J, Hannon GJ (2006) Distinct catalytic and non-catalytic roles of ARGONAUTE4 in RNA-directed DNA methylation. Nature 443:1008–1012

Ratcliff FG, MacFarlane SA, Baulcombe DC (1999) Gene silencing without DNA. RNA-mediated cross-protection between viruses. Plant Cell 11:1207–1216

Schwab R, Ossowski S, Riester M et al (2006) Highly specific gene silencing by artificial microRNAs in Arabidopsis. Plant Cell 18:1121–1133

Schwab R, Palatnik JF, Riester M et al (2005) Specific effects of MicroRNAs on the plant transcriptome. Dev Cell 8:517–527

Shivaprasad PV, Rajeswaran R, Blevins T, Schoelz J, Meins F Jr, Hohn T, Pooggin MM (2008) The CaMV transactivator/viroplasmin interferes with RDR6-dependent trans-acting and secondary siRNA pathways in Arabidopsis. Nucleic Acids Res 36:5896–5909

Slade AJ, Fuerstenberg SI, Loeffler D et al (2005) A reverse genetic, nontransgenic approach to wheat crop improvement by TILLING. Nat Biotechnol 23:75–81

Slotkin RK, Martienssen R (2007) Transposable elements and the epigenetic regulation of the genome. Nat Rev Genet 8:272–285

Small I (2007) RNAi for revealing and engineering plant gene functions. Curr Opin Biotechnol 18:148–153

Stoutjesdijk PA, Singh SP, Liu Q et al (2002) hpRNA-mediated targeting of the Arabidopsis FAD2 gene gives highly efficient and stable silencing. Plant Physiol 129:1723–1731

Sundaresan V, Springer P, Volpe T et al (1995) Patterns of gene action in plant development revealed by enhancer trap and gene trap transposable elements. Genes Dev 15:1797–1810

Till BJ, Cooper J, Tai TH et al (2007) Discovery of chemically induced mutations in rice by TILLING. BMC Plant Biol 7:19

Till BJ, Reynolds SH, Greene EA et al (2003) Large-scale discovery of induced point mutations with high-throughput TILLING. Genome Res 13:524–530

Till BJ, Reynolds SH, Weil C et al (2004) Discovery of induced point mutations in maize genes by TILLING. BMC Plant Biol 4:12

Turnage MA, Muangsan N, Peele CG et al (2002) Geminivirus-based vectors for gene silencing in Arabidopsis. Plant J 30:107–114

van der Krol AR, Lenting PE, Veenstra J et al (1988) An antisense chalcone synthase gene in transgenic plants inhibits flower pigmentation. Nature 333:866–869

van der Krol AR, Mur LA, Beld M et al (1990) Flavonoid genes in petunia: addition of a limited number of gene copies may lead to suppression of gene expression. Plant Cell 2:291–299

Vaucheret H (2006) Post-transcriptional small RNA pathways in plants: mechanisms and regulations. Genes Dev 20:759–771

Voinnet O (2005) Induction and suppression of RNA silencing: insights from viral infections. Nat Rev Genet 6:206–220

Wang N, Wang Y, Tian F et al (2008) A functional genomics resource for Brassica napus: development of an EMS mutagenized population and discovery of FAE1 point mutations by TILLING. New Phytol 180:751–765

Waterhouse PM, Wang MB, Lough T (2001) Gene silencing as an adaptive defence against viruses. Nature 411:834–842

Watson JM, Fusaro AF, Wang M et al (2005) RNA silencing platforms in plants. FEBS Lett 579:5982–5987

Wesley SV, Helliwell CA, Smith NA et al (2001) Construct design for efficient, effective and high-throughput gene silencing in plants. Plant J 27:581–590

Wielopolska A, Townley H, Moore I et al (2005) A high-throughput inducible RNAi vector for plants. Plant Biotechnol J 3:583–590

Winkler S, Schwabedissen A, Backasch D et al (2005) Target-selected mutant screen by TILLING in Drosophila. Genome Res 15:718–723

Xie Z, Johansen LK, Gustafson AM et al (2004) Genetic and functional diversification of small RNA pathways in plants. PLoS Biol 2:E104

Xu P, Zhang Y, Kang L et al (2006) Computational estimation and experimental verification of off-target silencing during posttranscriptional gene silencing in plants. Plant Physiol 142:429–440

Yu B, Lydiate DJ, Young LW et al (2008) Enhancing the carotenoid content of Brassica napus seeds by downregulating lycopene epsilon cyclase. Transgenic Res 17:573–585

Zhang C, Ghabrial SA (2006) Development of Bean pod mottle virus-based vectors for stable protein expression and sequence-specific virus-induced gene silencing in soybean. Virology 344:401–411

Zhang C, Yang C, Whitham SA et al (2009) Development and Use of an Efficient DNA-Based Viral Gene Silencing Vector for Soybean. Mol Plant Microbe Interact 22:123–131

Chapter 21
Bioinformatics Resources for *Arabidopsis thaliana*

Neil Graham and Sean May

Abstract *Arabidopsis thaliana* was the first plant to be sequenced in 2000 and is widely used as a model organism for flowering plants. This has led to the generation of large data sets, such as sequences (DNA, RNA and proteins), gene predictions, protein information and microarray data. A number of different databases have been developed to store, distribute and analyse these data. These range from large repositories that contain a wide range of data (e.g. sequences, gene models, germplasm data) and small specialised databases containing one type of data (e.g. protein localisation). In addition tools have been developed to enable users to access, download and analyse the data. These include tools to analyse sequences, gene function, metabolomic pathways and microarray data. A number of these databases and tools will be described.

Keywords *Arabidopsis* · Transcriptomics · Genomics · Proteomics · Bioinformatics

Contents

21.1 Background to *Arabidopsis thaliana* 586
21.2 Genome Browsers 586
21.3 Transcriptomics Data 588
21.4 Gene and Protein Analysis Resources 590
21.5 Gene Interactions and Pathways 592
21.6 Small RNA Databases 593
21.7 Metabolomic Data 594
21.8 Integration of Data 594
References 595

N. Graham (✉)
Plant and Crop Sciences Division, Nottingham Arabidopsis Stock Centre, University of Nottingham, Loughborough, LE12 5RD UK
e-mail: neil@arabidopsis.org.uk

R. Schmidt, I. Bancroft (eds.), *Genetics and Genomics of the Brassicaceae*, Plant Genetics and Genomics: Crops and Models 9, DOI 10.1007/978-1-4419-7118-0_21,

21.1 Background to *Arabidopsis thaliana*

A. thaliana (*Arabidopsis*) is a widely adopted model organism for flowering plants and is used to study many processes, including development and response to biotic and abiotic stresses. This is due to its small size, fast life cycle, small genome size and large collections of germplasm, data and other resources. *Arabidopsis* has one of the smallest plant genomes and was the first plant to be sequenced, in 2000, by the *Arabidopsis* Genome Initiative (2000). This has enabled large data sets to be produced, including gene predictions, Gene Ontology, protein function prediction and whole genome microarrays. A wide range of bioinformatics resources and databases has subsequently been developed to visualise and analyse these data. Data repositories include large, long-established databases containing information from a range of sources and small specialised databases developed by a single group. These databases and associated web pages greatly aid the interpretation of expression; proteomic and metabolomic experiments and the development of techniques to combine data from different sources also greatly aids this process.

21.2 Genome Browsers

The availability of a full genome sequence has enabled a number of *Arabidopsis* genome browsers to be developed, in order to visualise this and associated data sets, such as gene models, peptide information, polymorphisms and germplasm information. These browsers allow users a graphical interface in order to browse or search the genome and have many tools to download data sets. Examples of *Arabidopsis* genome browsers include AtEnsembl (James et al. 2007, http://atensembl.arabidopsis.info), The *Arabidopsis* Information Resource (TAIR; http://arabidopsis.org/), Gramene (Liang et al. 2008, http://www.gramene.org/), and Salk Institute Genomic Analysis Laboratory browsers (http://signal.salk.edu/). The Nottingham *Arabidopsis* Stock Centre (NASC; http://arabidopsis.info/) has developed the AtEnsembl genome browser based on the Ensembl browser developed at the EBI (European Bioinformatics Institute) and Sanger Institute. This browser displays a wide range of data sets including gene models, protein information, single nucleotide polymorphisms (SNPs), Affymetrix array information, other plant species data and position of germplasm insert lines (Fig. 21.1). The data are taken from many sources, for example, the gene models are taken from both The *Arabidopsis* Information Resource (TAIR) based at the Carnegie Institution of Washington, USA, and the Munich Information Centre for protein sequences (MIPS; http://mips.gsf.de/) in Germany. In addition, peptide information and SNP data are taken from the large public repositories such as Genbank, EMBL and Swissprot and there are numerous links from the pages in AtEnsembl to these external databases. The data in AtEnsembl can be accessed and downloaded using a web interface or alternatively the whole database can be downloaded and installed locally. Users can also display their own data in AtEnsembl along with the other data

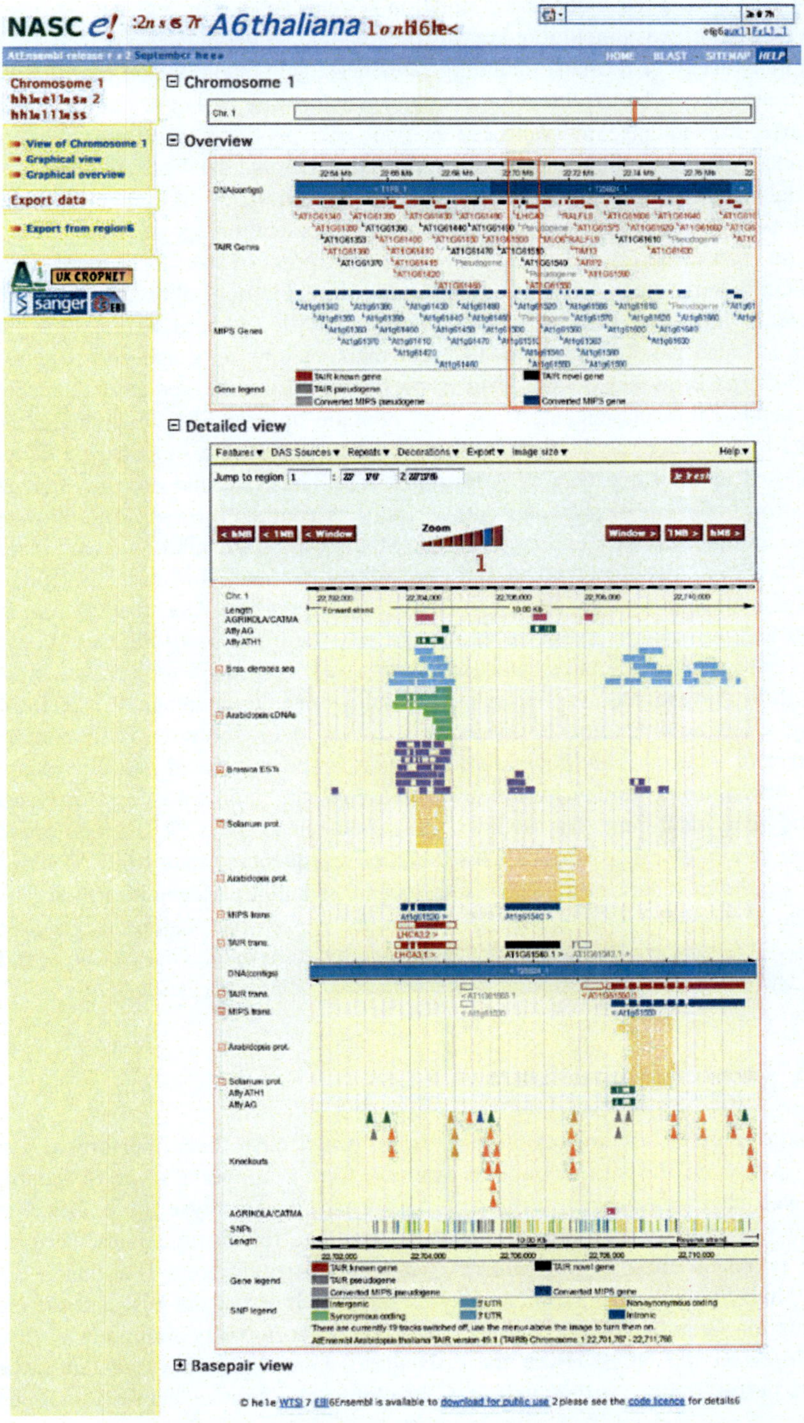

Fig. 21.1 Screenshot of the contig view from the AtEnsembl genome browser displaying information about sequences, gene models, proteins, inserts, etc.

via DAS (distributed sequences annotation system) lanes. The data can be searched in a number of ways including keyword search and by using BLAST. Detailed data sets in a variety of formats, including sequences, gene, peptide and repeat features information, can be exported for a single gene/peptide or larger data sets can be exported for sequence intervals between two genes/peptides, markers or by physical distance. The Ensembl databases use a MySQL relational database structure to store the data and a set of Perl Application Programme Interfaces (APIs) act as a middle layer between the databases and more specific application programmes. This API can be used to perform detailed searches and exports of data from the databases. The Gramene project developed at the Cold Spring Harbor Laboratory and Cornell University also uses the Ensembl database. In addition to an *Arabidopsis* browser, there are also genome browsers for rice, maize, sorghum, grape and poplar. The *Arabidopsis* browser contains data taken from AtEnsembl (sequence, gene models, peptide information, etc.) and also contains information about homologous genes in the other species present in Gramene. As this is also based on Ensembl it is used in the same way as AtEnsembl. The TAIR website contains numerous tools for analysing *Arabidopsis* data including a genome browser. This browser is based on the Generic Genome Browser (Stein et al. 2002, Gbrowse) and displays similar data to that displayed on AtEnsembl and Gramene. This includes gene models (generated in-house), peptide information, expression data, methylation data, SNPs and information about orthologs and gene families. TAIR also has an extensive bibliographic reference database that can be accessed via the genome browser and allows investigators to find relevant publications on the genes or proteins they study. Users data can also be loaded and displayed in Gbrowse. TAIR contains a number of tools for bulk downloads of genomic and associated data, for example, gene descriptions, sequences and GO ontologies can be downloaded from a user's list of gene identifiers. The SIGnAl site developed at the Salk Institute provides access to a wide range of data. This includes resequencing data from 20 ecotypes of *A. thaliana* (Clark et al. 2007, http://signal.salk.edu/perlegen.html), which are displayed using the *Arabidopsis* SNP sequence viewer, the position of germplasm inserts, *Arabidopsis* Tiling array data and methylation data. This can be searched and data downloaded in a similar way to the other genome browsers.

21.3 Transcriptomics Data

A large number of transcriptomic experiments have been performed using *Arabidopsis* and the data are publicly available from a number of sources. Databases and web pages have been developed to distribute and analyse these data. There are two large array data repositories that store data for all organisms, performed on a range of microarray platforms. These are ArrayExpress (Parkinson et al. 2009, http://www.ebi.ac.uk/microarray-as/ae/) developed at the EBI and the Gene Expression Omnibus (GEO; Barrett et al. 2009, http://www.ncbi.nlm.nih.gov/geo/) developed at the NCBI. Both these databases contain a wide range of experiments performed on *Arabidopsis* (e.g. developmental series, chemical treatments, mutant line analysis) using a range of platforms including commercial arrays (e.g. Affymetrix, Agilent) and spotted arrays. Detailed information on the experimental design, samples (germplasm details, growth conditions, etc.) and processing details (RNA extraction, labelling protocol) can be accessed. Both raw and normalised data can be downloaded from both databases, which can then be analysed locally. ArrayExpress has also developed a tool for analysing the data; this allows a user to visualise the expression profiles of particular genes or find genes that change under specific conditions.

NASC has also developed NASCArrays (Craigon et al. 2004, http://affymetrix.arabidopsis.info/narrays/experimentbrowse.pl), an *Arabidopsis*-specific array database. This houses data generated using Affymetrix GeneChips (http://affymetrix.com/), either from experiments performed using the NASC transcriptomics service or from data donations. As with GEO and ArrayExpress, both raw and normalised data can be downloaded from the database, as well as detailed experimental and sample information. NASCArrays also has a number of associated tools to analyse the data it holds. These include spot history (Fig. 21.2), which displays the expression level of a particular gene across all the experiments in the database as a histogram. Other tools include a two-gene

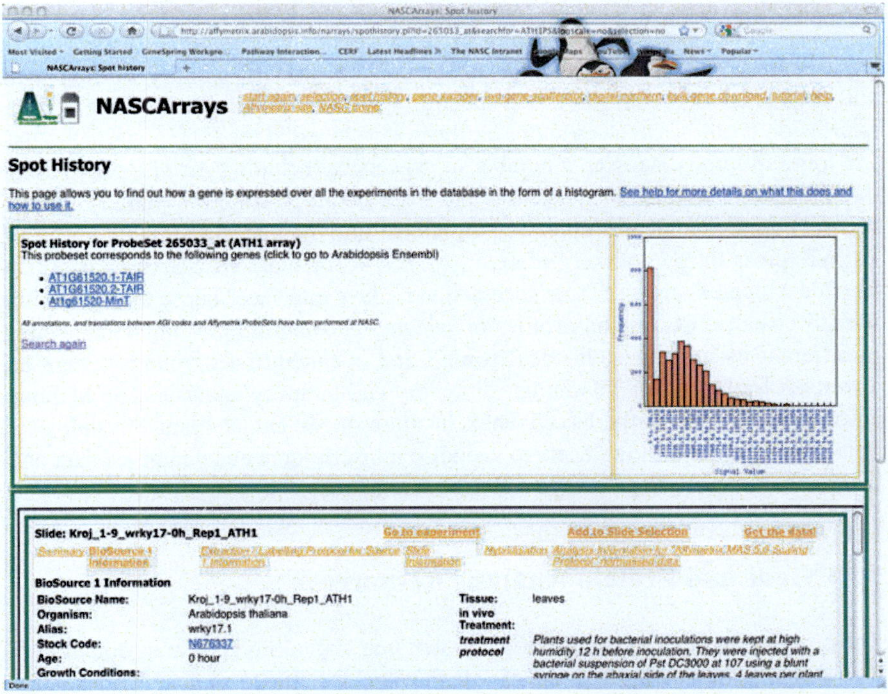

Fig. 21.2 Screenshot taken from the NASCArrays spot history tool showing a histogram of signal intensity (*horizontal axis*) against frequency (*vertical axis*) for a selected gene. Experimental details are shown for a selected signal intensity

scatter plot, to compare expression of two genes, and a facility to download the expression values for up to 300 genes. NASC has also implemented a Genespring workgroup (http://affy.arabidopsis.info/workgroup.html/), which is a database version of the Genespring GX analysis software developed by Agilent Technologies (http://www.chem.agilent.com/). The workgroup allows users to access and analyse the data using the Genespring software, which performs detailed statistical analysis (e.g. volcano plots, ANOVA tests, clustering and pathway analysis) of the data. The AREX database (Birnbaum et al. 2003, Brady et al. 2007, http://arexdb.org/) developed at Duke University is a database of *Arabidopsis* gene expression data and includes both genome-wide microarray and gene specific (in situ, GFP-reporter, etc.) data. This can be used to identify the expression profiles of selected genes or find genes that have a specific expression profile within a tissue.

Due to the large amount of *Arabidopsis* transcriptome data available, a number of websites have been developed to analyse the data. These include the Bio-Array Resource (BAR; Winter et al. 2007, http://bar.utoronto.ca/), Genevestigator (Hruz et al. 2008, https://www.genevestigator.ethz.ch/) and Mapmann (Thimm et al. 2004, http://gabi.rzpd.de/projects/MapMan/). The Bio-Array Resource for *Arabidopsis* functional genomics has been developed at the University of Toronto, Canada, to provide tools for the analysis of transcriptome and other genomic data. The tools available include the *Arabidopsis* eFP browser (Fig. 21.3), which can display a gene expression profile as a pictorial heat map either in a developmental series, in specific tissues or after abiotic and biotic treatments. The expression angler and the sample angler can be used to identify co-regulated, anti-correlated or condition/tissue-specific genes or samples exhibiting similar expression profiles. Genevestigator has been developed at the ETH Zurich, to analyse transcriptome data from a number of organisms including *Arabidopsis*. All the data in the Genevestigator databases has been obtained from the public repositories such as GEO (Gene Expression Omnibus; http://www.ncbi.nlm.nih.gov/geo/), ArrayExpress (http://www.ebi.ac.uk/microarray-as/ae/) and NASCArrays and they have developed a range of tools accessed via a Java interface. These tools allow the user to visualise expression profiles of selected gene across developmental series, in different tissues, after chemical stimuli and in mutant backgrounds. There are also tools for clustering, biomarker discovery and pathway analysis. The Mapman software developed at the Max Planck Institute of Molecular Plant Physiology in Potsdam, Germany, allows users to visualise transcriptome and genomics data onto diagrams of metabolic pathways and gene interactions.

21.4 Gene and Protein Analysis Resources

The ongoing analysis of the *Arabidopsis* genome, including gene and protein function, has generated large amounts of data that are available to analyse. For example, genes can be classified by their function, cellular localisation and processes involved in, using controlled vocabularies developed by the Gene Ontology (GO) consortium (Ashburner et al. 2000, http://www.geneontology.org/). These data can be

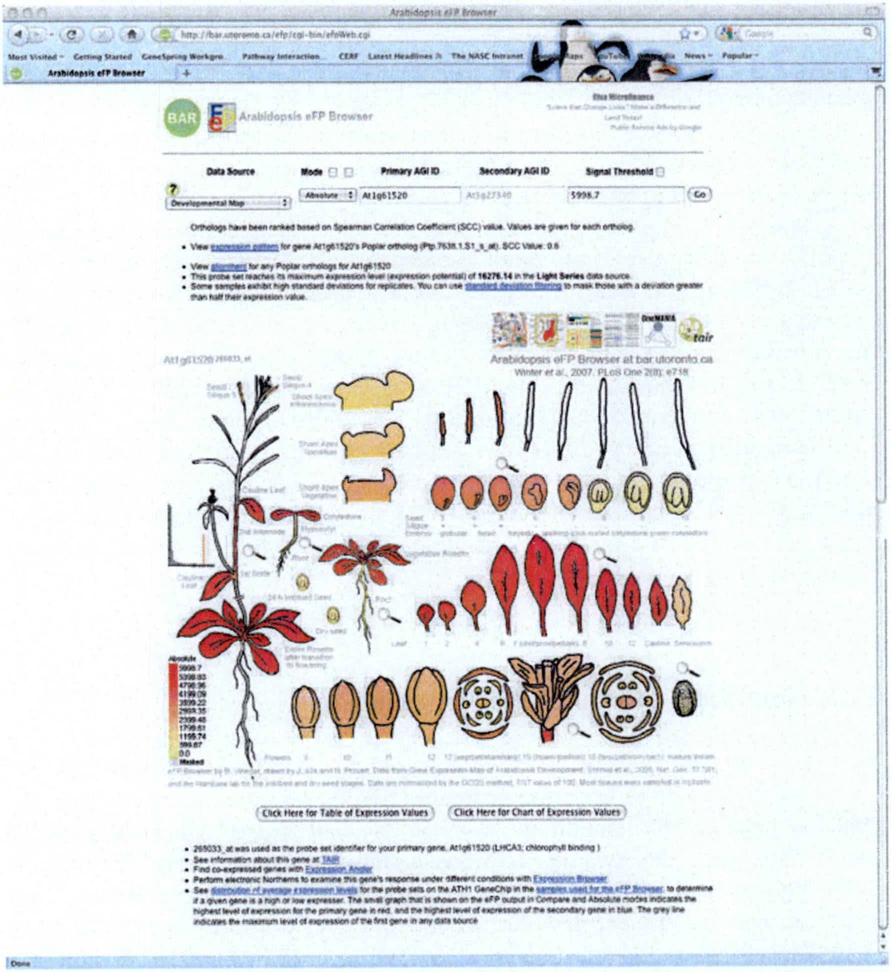

Fig. 21.3 Screenshot taken from the *Arabidopsis* eFP browser at BAR. Schematic diagrams of *Arabidopsis* developmental stages are coloured dependent on the expression level of a selected gene

accessed and used in a number of ways, for example, the classification superviewer at BAR and the GO annotation search at TAIR; both group a list of genes based on their GO ontologies. This can be useful when interpreting results of transcriptomic experiments.

As well as analysing coding sequences, tools are available to analyse promoter sequences. For example, the pronmomer tool available at BAR identifies overrepresented *n*-mer words in the promoters of a single gene or a group of co-regulated genes. The motif analysis tool (http://www.arabidopsis.org/tools/bulk/motiffinder/index.jsp) available at the TAIR site can also identify motifs in a set of sequences or a list of gene identifiers. These tools can be used to identify potential transcription factor-binding sites within a group of genes identified by array analysis. The PlantCARE database (Rombauts et al. 1999, http://bioinformatics.psb.ugent.be/webtools/plantcare/html/) developed at Gent University contains information about known transcription factors and binding sites. The data can be searched for information on a transcription factor and binding site or a list of genes can be searched for known binding sites.

A range of *Arabidopsis* protein data are also available. Sequences can be downloaded from the large multi-species repositories such as the EBI and NCBI and also more detailed information (e.g. protein motifs) from the genome browsers at NASC, TAIR and Gramene. Tools are also available to visualise the protein localisation of proteins within a cell. The SUB-cellular location database for *Arabidopsis* proteins (SUBA; Heazlewood et al. 2007, http://www.plantenergy.uwa.edu.au/suba2/) has been developed at the ARC Centre of Excellence in Plant Energy Biology, Australia, by bringing together protein localisation data sets and developing a web interface. This allows users to visualise the cellular localisation of proteins of interest. The same data can be accessed and viewed using the Cell eFP browser (http://bar.utoronto.ca/cell_efp/cgi-bin/cell_efp.cgi) at BAR (Fig. 21.4).

21.5 Gene Interactions and Pathways

Data are available on gene/protein interactions and metabolomic pathways. The Kyoto encyclopaedia of gene and genomes (KEGG; http://www.genome.jp/kegg/) contains detailed information on a range of *Arabidopsis* pathways including metabolism (e.g. carbohydrate, lipid and amino acid metabolism), transcription, DNA replication, signal transduction and cell cycle. *Arabidopsis* metabolic pathways can also be viewed, downloaded and analysed using the AraCyc database (http://arabidopsis.org/biocyc/index.jsp) hosted at TAIR. The OMICs viewer of AraCyc allows a range of data (from expression, proteomic or metabolomic experiments) to be overlaid onto the pathways. A similar database has been developed at the John Innes Centre, UK, called the *Arabidopsis* Reactome (Tsesmetzis et al. 2008, http://arabidopsisreactome.org/). Like AraCyc and KEGG this contains core pathways and reactions in *Arabidopsis* and can be searched for particular genes, proteins or pathways. *Arabidopsis* protein–protein interaction networks, domain architecture, ortholog information and GO annotations have also been integrated in the *A. thaliana* protein interactome database (AtPID; Cui et al. 2008, http://atpid.biosino.org/) developed at the Northeast Forestry University, Harbin, China. In addition to known protein interactions, predicted interactions can be viewed using the *Arabidopsis* Interactions Viewer at BAR which contains predicted protein interactions from interacting orthologs in yeast nematodes, fruit fly and humans (Geisler-Lee et al. 2007). All these databases can be used to view the pathways and reactions and also the databases can be searched for specific genes or proteins to visualise which pathways they are present in. One tool that has been developed to use these pathways is the Mapman software (http://gabi.rzpd.de/projects/MapMan/), which can be used to overlay expression data on these pathways.

21.6 Small RNA Databases

Recent work has demonstrated the role that small non-coding RNAs play in transcriptional and post-transcriptional regulation. An increasing number of small RNAs and their targets have been identified and databases have been developed to store and exploit these data. Examples of this type of database are the *Arabidopsis thaliana* Small RNA project (Gustafason et al. 2005, http://asrp.cgrb.oregonstate.edu/) developed at the Oregon State University and *Arabidopsis* MPSS database (http://mpss.udel.edu/at/) developed at the University of Delaware. Both these databases contain detailed information about small RNAs including sequences, expression level and target genes. Both databases have genome browsers

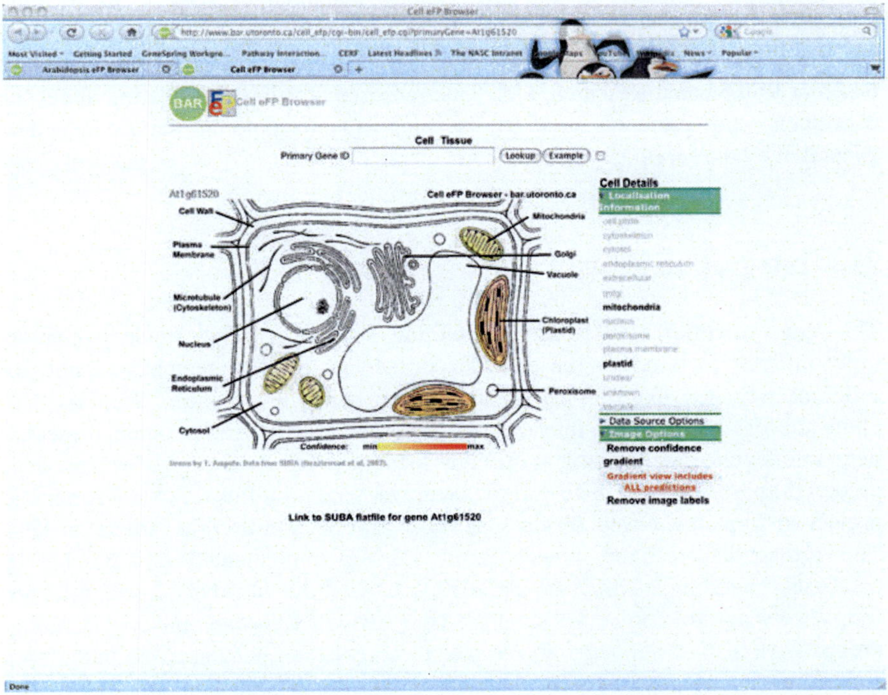

Fig. 21.4 Screenshot taken from the *Arabidopsis* cell eFP browser at BAR. A schematic diagram of an *Arabidopsis* cell is coloured dependent on the localisation of a selected *Arabidopsis* protein

to visualise the genomic position of the small RNAs and which genes they target. The data can be searched by RNA name, gene name or genomic location, etc. Alternatively all the data can be downloaded and mined locally.

21.7 Metabolomic Data

An increasing number of metabolomic studies are being conducted in *Arabidopsis* and databases have been developed to store and distribute these data. These resources include the NSF2010 metabolomics database (http://lab.bcb.iastate.edu/projects/plantmetabolomics/index.php), tools available at TAIR (http://www.arabidopsis.org/biocyc/index.jsp) and the MeT-RO service (http://www.metabolomics.bbsrc.ac.uk/MeT-RO.htm). The NSF2010 metabolomics database is a NSF-funded project developed at the Iowa State University and contains data from a number of experiments including environmental effects and mutant analysis. The data are available to download and there are also links to experimental protocols and tools for analysing metabolomic data. The TAIR site contains a number of tools for investigating metabolomic pathways, which have developed with the Plant Metabolic Network project (http://www.plantcyc.org). These include the Metabolic Map which enables users to view, provides a "bird's eye" view of *Arabidopsis* metabolism and also enables metabolomic data to be overlaid onto the maps. The MeT-RO resource has been developed at the National Centre for Plant and Microbial Metabolomics (http://www.metabolomics.bbsrc.ac.uk/) based at Rothamsted Research, UK. This resource contains information about the instruments and protocols used by the MeT-RO metabolomic service and data generated by the service.

21.8 Integration of Data

The range of different databases containing data formats has made it increasingly difficult to integrate the data. Experimenters have to search a number of different websites in order to obtain all the data they require. Web services allow the databases to be interrogated programmatically and provide a mechanism to integrate data residing at different locations. The *Arabidopsis* web services project (http://bioinfo.mpiz-koeln.mpg.de/araws/) is a joint European and American project to implement web services at eight European and eight American labs. This project has resulted in web services being implemented at a number of databases including NASC, the AGRIKOLA database of RNAi knockout lines (http://www.agrikola.org/), MPSS at the University of Delaware and AREX. Other groups have also developed web services to access their databases including both general and *Arabidopsis*-specific databases such as NCBI, EBI and BAR. The Taverna project (Oinn et al. 2004, http://taverna.sourceforge.net/) has developed software that can utilise these web services and combine them into workflows. These workflows can be used with a single or multiple search terms (e.g. gene/protein name, accession number) to obtain data from a database (e.g. sequence data) and use this to search one or more databases (e.g. BLAST search or expression database) to obtain more results. The results can then be combined to provide detailed information from a range of sources without the user having to visit each website individually.

References

Arabidopsis Genome Initiative (2000) Analysis of the genome sequence of the flowering plant Arabidopsis thaliana. Nature 408:796–815

Ashburner M, Ball CA, Blake JA, Botstein D, Butler H, Cherry JM, Davis AP, Dolinski K, Dwight SS, Eppig JT et al (2000) Gene ontology: tool for unification of biology. The gene ontology consortium. Nat Genet 25:25–29

Barrett T, Troup DB, Wilhite SE, Ledoux P, Rudnev D, Evanglista C, Kim IF, Sobleva A, Tomashevsky M, Marshall KA et al (2009) NCBI GEO: archive for high-throughput functional genomic data. Nucleic Acids Res 37:D885–D890

Birnbaum K, Shasha DE, Wang JY, Jung JW, Lambert GM, Galbraith DW, Benfey PN (2003) A gene expression map of the Arabidopsis root. Science 302:1956–1960

Brady SM, Orlando DA, Lee Ji Y, Wang JY, Koch J, Dinneny JR, Mace D, Ohler U, Benfey PN (2007) A high-resolution root spatiotemporal map reveals dominant expression patterns. Science 318:801–806

Clark RM, Schweikert G, Toomajian C, Ossowski S, Zeller G, Shinn P, Warthmann N, Hu TT, Fu G, Hinds DA et al (2007) Common sequence polymorphisms shaping genetic diversity in *Arabidopsis thaliana*. Science 317:338–342

Craigon DJ, James N, Okyere J, Higgins J, Jotham J, May S (2004) NASCArrays: a repository for microarray data generated by NASC's transcriptomics service. Nucleic Acids Res 32: D575–D577

Cui J, Li P, Li G, Xu F, Zhao C, Li Y, Yang Z, Wang G, Yu Q, Li Y, Shi T (2008) AtPID: Arabidopsis thaliana protein interactome database and integrative platform for systems biology. Nucleic Acids Res 36:D999–D1008

Geisler-Lee J, O'Toole N, Ammar R, Provart NJ, Millar AH, Geisler M (2007) A predicted interactome for Arabidopsis. Plant Physiol 145:317–329

Heazlewood JL, Verboom RE, Tonti-Filippinin J, Small I, Millar AH (2007) SUBA: the Arabidopsis subcellular database. Nucleic Acids Res 35:D213–D218

Hruz T, Laule O, Szabo G, Wessendorp F, Bleuler S, Oertle L, Widmayer P, Gruissem W, Zimmerman P (2008) Genevestigator V3: a reference expression database for meta-analysis of transcriptomes. Adv Bioinformatics 2008:420747

James N, Graham N, Clemects D, Schildknecht B, May S (2007) AtEnsembl – a post genomic resources browser for Arabidopsis. In: Edwards D (ed) Plant bioinformatics – methods and protocols, Humana Press. ISBN10:1-58829-653-9

Liang C, Jaiswal P, Hebbard C, Avraham S, Buckler ES, Casstevens T, Hurwitz B, McCouch S, Ni J, Pujar A et al (2008) Gramene: a growing plant comparative genomics resource. Nucleic Acids Res 36:D947–D953

Oinn T, Addis M, Ferris J, Marvin D, Senger M, Greenwood M, Carver T, Glover K, Pocock MR, Wipat A, Li P (2004) Taverna: a tool for the composition and enactment of bioinformatics workflows. Bioinformatics 20:3045–3054

Parkinson H, Kapushesky M, Kolenikov N, Rustici G, Shojatalab M, Abeygunawardena N, Berube H, Dylag M, Emam I, Farne A et al (2009) ArrayExpress update-from an archive of functional genomics experiments to the atlas of gene expression. Nucleic Acids Res 37:D868–D872

Rombauts S, dehais P, Van Montagu M, Rouze P (1999) PlantCARE, a plant cis-acting regulatory element database. Nucleic Acids Res 27:295–296

Stein LD, Mungall C, Shu S, Caudy M, Mangone M, Day A, Nickerson E, Stajich JE, Harris TW, Arva A, Lewis S (2002) The generic genome browser: a building block for model organism system database. Genome Res 12:1599–1610

Thimm O, Blaesing O, Gibon Y, Nagel A, Meyer S, Kruger P, Selbig J, Muller LA, Rhee SY, Stitt M (2004) MAPMAN: a user-driven tool to display genomics data sets on diagrams of metabolic pathways and other biological processes. Plant J 37:914–939

Tsesmetzis N, Couchman M, Higgins J, Smith A, Doonan JH, Seifert GJ, Schmidt EE, Vastrik I, Birney E, Wu GM et al (2008) Arabidopsis reactome: a foundation knowledgebase for plant systems biology. Plant Cell 20:1426–1436

Winter D, Vinegar B, Nahal H, Ammar R, Wilson GV, Provart NJ (2007) An "electronic fluorescent pictograph" browser for exploring and analyzing large-scale biological data sets. PLoS ONE 2:e718

Chapter 22
Bioinformatics Resources for the *Brassica* Species

Martin Trick

Abstract The science of *Brassica* genomics is now at an exciting stage with capillary-based BAC-by-BAC sequencing of the *Brassica rapa* gene-space, combined with the application of next generation sequencing technologies, opening up new possibilities to relate genome-wide variation to traits with unprecedented resolution and cost-effectiveness. We are at the real beginnings of the translation of the fundamental discoveries being made every day in the *Arabidopsis* model system to a group of closely related crop genomes of global economic importance. It is imperative that the appropriate bioinformatics resources to allow navigation between these genomes and also an integration of the emerging *B. rapa* genome sequence with the genetic maps that underpin trait analysis are put in place. In this chapter the current status and future prospects for these bioinformatics infrastructures will be discussed.

Keywords Bioinformatics · Computational tools · Gene annotation · Next generation sequencing · SNP discovery

Contents

22.1 Introduction 598
22.2 First Steps in *Brassica* Bioinformatics 598
22.3 A Directory of Current Web Resources 599
22.4 EST Resources, Transcript Assemblies, and Microarrays 599
22.5 The *B. rapa* Genome Sequencing Project 603
 22.5.1 Methodology 603
 22.5.2 BAC End Sequencing 603
 22.5.3 Physical Maps and Informatics 603
 22.5.4 Bioinformatic Selection of Seed BACs 605

M. Trick (✉)
Computational and Systems Biology Department, John Innes Centre, Norwich Research Park, Norwich NR4 7UH, UK
e-mail: martin.trick@bbsrc.ac.uk

R. Schmidt, I. Bancroft (eds.), *Genetics and Genomics of the Brassicaceae*, Plant Genetics and Genomics: Crops and Models 9, DOI 10.1007/978-1-4419-7118-0_22,

22.5.5 Coordination of Sequencing Programme 607
22.5.6 Automated Annotation 608
22.6 Next Generation Sequencing and the Re-sequencing of *Brassica* Genomes 611
22.7 Future Developments 613
References 614

22.1 Introduction

The field of *Brassica* crop genetics and genomics is being revolutionized by the production of genome sequences from traditional BAC-by-BAC approaches, recently enhanced by whole genome de novo assemblies obtained from next generation sequencing technologies. These latter platforms and methods hold the key to relatively cheap re-sequencing experiments that could allow genome-wide association genetics approaches to the identification of variation controlling traits of agronomic importance. At the heart of this still remains the intimate phylogenetic relationship between the various *Brassica* genomes and that of the *Arabidopsis* model plant, allowing the translation of a large body of accumulated functional genomics data to this group of important crops.

To facilitate this translation, bioinformatics resources need to be available that allow easy navigation between the model and the crop genomes and also an integration of the crop genome sequences with the genetic maps on which trait analyses are built.

The challenges represented by the paleopolyploid genome structures of the diploid *Brassica* species, intensified by the amphidiploid structures of important species such as *B. napus*, should not be underestimated. These challenges start first with the assembly of a *Brassica* genome sequence but continue with ensuring its correct interpretation and integration with the genetic map. We are well used to the rich bioinformatics infrastructure that has sprung up around the *Arabidopsis* sequence (see Chapter 21) and must now seek to achieve something similar for the emerging *Brassica* genome sequence. This chapter summarizes and describes the available resources.

22.2 First Steps in *Brassica* Bioinformatics

With the publication of the complete genome sequence of *Arabidopsis thaliana* (The Arabidopsis Genome Initiative 2000) there came the first prospects of exploiting this model plant system for crop improvement in the closely related *Brassica* species. A clear objective was to establish relationships between *Brassica* genetic maps and physical segments of *Arabidopsis* DNA, with candidate genes from the sequence of the model genome emerging from the newly completed annotation. The BrassicaDB database, part of the UK CropNet programme that was initiated in 1996 (Anderson et al. 2000), was one of the first attempts both to systematize *Brassica* sequence information and to store and represent genetic maps. Using ACEDB (Stein and Thierry-Mieg 1999) as the database management system in combination with the

Aceserver interface for web access (Thierry-Mieg and Stein 1998), it initially contained all publicly available *Brassica* sequences from *Brassica oleracea*, *Brassica rapa*, and *Brassica napus*, bibliographic data associated with the sequence information and two related genetic maps of *B. napus* (Parkin et al. 1995, Sharpe et al. 1995). Visualization of the genetic maps and the underlying genotyping data is enhanced through graphical tools supplied by Java applets. Later, publicly available microsatellite marker data and genetic maps were added (Lowe et al. 2004). Most recently, computational analysis using BLASTN (Gish 1996) of similarity between sequenced *Brassica* RFLP markers comprising the *B. napus* genetic map and the *Arabidopsis* genome sequence finally allowed the *in silico* alignments long sought after. This same methodology has allowed a thorough description of the *B. napus* genome in terms of a segmental structure derived from the *Arabidopsis* sequence (Parkin et al. 2005). Although there is now no further manual curation or software development of BrassicaDB, the database is still supported and available (http://brassica.bbsrc.ac.uk) and is automatically updated daily with newly deposited *Brassica* sequences obtained directly from the EMBL databank.

In 2002, the first true genomics-scale resource became available when the *B. oleracea* genome database was put online (http://brassica.jcvi.org/cgi-bin/brassica/index.cgi). This database contains nearly 600,000 genome survey sequences (GSSs) from a *B. oleracea* whole genome shotgun library ($\sim$0.5 $\times$ coverage). BLASTN analysis using an E-value cutoff of 1×10^{-10} identified *Arabidopsis* homologues for around 53% of these sequences. This has been exploited for *Arabidopsis* gene discovery and re-annotation (Ayele et al. 2005). A BLAST query interface and a text search facility against the annotation for this dataset are still available.

22.3 A Directory of Current Web Resources

Now there are a number of web resources available that are dedicated to *Brassica* genomics and genetics and which facilitate remote querying of databases, BLAST searches, and other forms of bioinformatic manipulation. Table 22.1 lists these and many of them will be referred to in detail later in this chapter.

22.4 EST Resources, Transcript Assemblies, and Microarrays

An important contribution to the provision of an infrastructure for *Brassica* genomics has been the deposition into the nucleotide databanks of very large numbers of EST sequences derived from all the major *Brassica* species. To date these have been obtained through traditional Sanger sequencing and capillary electrophoresis (CE) methodology. Although next generation sequencing technologies are likely to revolutionize such studies, with the steady improvements over the last few years in the CE instrumentation, the associated base-calling software, and the arrival of high-throughput methods for library preparation and processing, these data nonetheless represent an enormous resource. At the time of writing there were

Table 22.1 A list of web resources providing bioinformatic support and analysis for *Brassica* genetics and genomics

Name of resource	Affiliation	URL	Key contents
AAFC/PBI	AAFC Saskatoon Research Center and Plant Biotechnology Institute, Saskatoon, Canada	http://aafc-aac.usask.ca/cgi-bin/gbrowse/gbrowse/BAGI2/	Numerous *Brassica* EST alignments to *Arabidopsis* genome sequence
ACPFG Applied Bioinformatics Group	Australian Centre for Plant Functional Genomics, University of Queensland, Australia	http://acpfg.imb.uq.edu.au	(Under development after relocation)
AtEnsembl	Nottingham Arabidopsis Stock Centre, UK	http://atensembl.arabidopsis.info	*Arabidopsis* genome viewer with *Brassica* tracks
Brassica Genome Gateway	John Innes Centre, UK	http://brassica.bbsrc.ac.uk	All *B. rapa* BAC annotations; 95k *Brassica* microarray; Physical maps; BrassicaDB
Brassica.info	Rothamsted Research, UK	http://www.brassica.info	Genetic maps; AdVab/Oregin sites; BAC registry for BrGSP
BrGP	National Institute of Agricultural Botany and National Genome Information Center, Korea	http://www.brassica-rapa.org	Some annotation for selected seeds and A3/A9 BACs; *B. rapa* FPC DB; ESTs and 24 k *B. rapa* microarray
TIGR *Brassica oleracea* genome project	JC Venter Institute, USA	http://blast.jcvi.org/er-blast/index.cgi?project=bog1	BLAST/annotation searches of *B. oleracea* WGS

approximately 840,000 ESTs in EMBL/GenBank/DDBJ originating from 12 different *Brassica* species and from numerous, yet significantly overlapping, tissue types or developmental stages and, occasionally, defined experimental treatments such as pathogen infection or abiotic stress.

Clearly, this dataset is characterized by a very high level of redundancy and in this form is unwieldy for most purposes. Some initial work on clustering and assembling a smaller dataset into tentative consensus sequences had been done (Love et al. 2005). More recently, as a first step in the design of a Brassica microarray resource, reduction through clustering and assembly into a non-redundant set of transcript assemblies (or "unigenes") was undertaken (Trick et al. 2009a). In this study an expanded set of *Brassica* species ESTs available in 2007 were obtained, comprising three principal sets corresponding to the AC, A, and C genomes: *B. napus* (567,240), *B. rapa* (180,611), and *B. oleracea* (59,696) making a total of 803,326 ESTs after vector cleaning and removal of low-quality and short (<100 bp) sequences. The ESTs were assembled together using the TGICL software package (Pertea at al. 2003) using the default parameters of minimum identity of 94% and minimum coverage of 90%. The aim was to co-assemble ESTs of orthologous genes (including homoeologous pairs of genes in *B. napus* from each of the A and C genomes), but resolve into separate assemblies of paralogous or what were termed paleo-homoeologous genes (i.e., the genes related by the ancestral genome triplication observed in diploid *Brassica* species). This resulted in 90,864 distinct sequences of mean length 677 bp comprising 42,642 assemblies and 48,222 singletons (Fig. 22.1).

In the TGICL pipeline, sequences were oriented 5′–3′ either based on their alignment with a known protein in the UniProt database (using a BLASTX E-value threshold of 1×10^{-5}) or, failing this, by the presence of a polyA (polyT) tail.

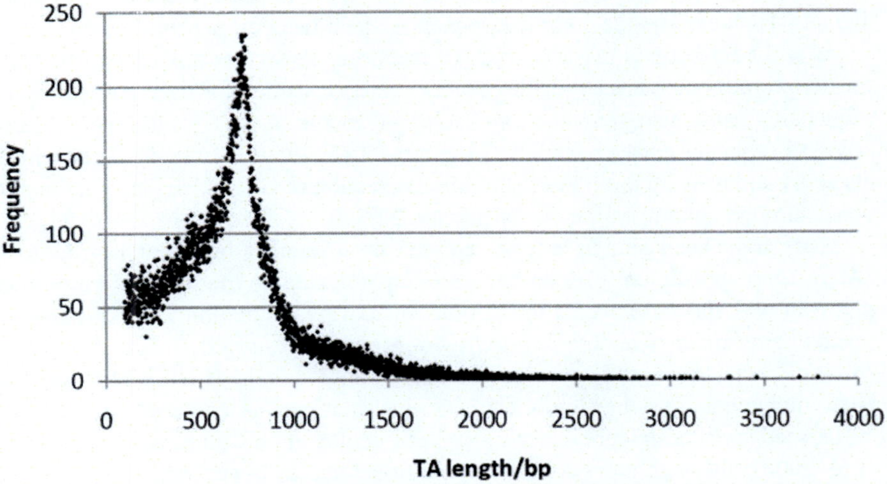

Fig. 22.1 Length distribution of the 90,864 unigenes assembled from ~800,000 *Brassica* ESTs. Raw ESTs were clustered and assembled by the TGICL pipeline (Pertea et al. 2003)

Protein similarities were annotated for the assemblies and singletons with significant BLASTX alignments. Just 4% of the 90,864 sequences (3,694 – comprising 330 assemblies and 3,364 singletons) could not be characterized in this way and were thus left represented in both orientations. After the complete process, the raw EST dataset was thus reduced to a unigene set of 94,558 sequences (64,044,420 bp) comprising 42,972 assemblies and 51,586 singletons, with a total of 72,148 sequences annotated with protein similarities.

The assembled *Brassica* sequences were submitted to Agilent Technologies' eArray system for probe design (https://earray.chem.agilent.com/earray/). It was found that unique 60-mer probes could be designed for a total of 91,854 of the 94,558 unigenes (including 6,989 derived from oppositely oriented pairs of sequences). Gene expression experiments can be conducted using this microarray by entering into a standard commercial agreement with Cogenics (Morrisville, NC 27560, USA). However, since the unigene dataset is made publicly available, researchers are free to use the data to design array resources on a platform of their choice.

Details of the composition of the unigene set together with the annotation have been loaded into a web-searchable database (http://brassica.bbsrc.ac.uk/). This database may be searched with text terms or fragments (which are wild carded) for matches on a number of fields, including assembly or singleton identifier, the identifier, gene name, description or source organism of the best UniProt BLASTX hit, and, where appropriate, the identifiers, tissue sources, and source *Brassica* species of the ESTs contributing to an assembly. Search results are returned in HTML tabular form and, where appropriate, are marked up with hyperlinks to GBrowse views, EBI sequence and InterPro descriptions, and NCBI dbEST records. The sequence of the unigene is also returned and, if it appears on the array, the 60-mer Agilent probe designed is rendered in lower case. The same web portal contains a link for similarity matching with BLASTN (with results hyper-linked to the above database) and its FTP site offers the complete sequence file in fasta format for download.

As will be described in Section 22.6 it has been shown that this virtual *Brassica* transcriptome constitutes a useable pseudo-reference sequence for the purpose of re-sequencing using next generation technology (Trick et al. 2009b). Retention of the raw EST sequence data is nonetheless important to facilitate correction of gene models generated from ab initio predictions and detection of alternative instantiations of genes through splicing variants (see Section 22.5.6).

There have been similar efforts reported by a number of research groups. A 24k *B. rapa*-specific unigene set has been implemented on the NimbleGen 60-mer platform with database access to both the microarray annotation and experimental results being made available at http://www.brassica-rapa.org/BrEMD/. A 94k all-*Brassica* array, based on a very similar raw EST source to the JIC/JCVI 95k set, has been implemented on the Combimatrix 35-mer platform (Xiang et al. 2008) and a ~10k seed-specific unigene set has also been deployed on Combinatrix (Bekkaoui et al. 2008), although there is currently no direct access to either of these datasets offered.

22.5 The *B. rapa* Genome Sequencing Project

22.5.1 Methodology

The *B. rapa* genome sequencing project (BrGSP) is an international, coordinated initiative to sequence ~330 Mb of euchromatin (gene-space) using a BAC-by-BAC approach (see Chapter 15). This involves extension from seed BACs both genetically anchored to given *B. rapa* chromosomes and physically anchored to *Arabidopsis* genome sequence. The project thus presents both significant challenges and exciting opportunities in terms of the bioinformatics resources it consumes and offers to the wider research community, respectively.

22.5.2 BAC End Sequencing

One of the first facilitating steps in the BrGSP programme was the end sequencing of the clones in the *B. rapa* BAC libraries that were to form the source material. Various members of the initiative undertook the sequencing of subsets of the total of 110,592 clones in the three KBrH, –B and, –S BAC libraries (Park et al. 2005). Ninety percent of the ends were successfully sequenced and 200,031 are now deposited in the EMBL/GenBank/DDBJ sequence databanks. This database is one of several that are available for BLAST searches at http://brassica.bbsrc.ac.uk.

These end sequence data can, in isolation, be used simply like the *B. oleracea* GSSs (Section 22.2), but in pairs they also impart important synteny information and, as such, are now crucial to seed BAC selection for the BrGSP (Section 22.5.4). Several groups have implemented similar strategies to align the BACs with the *Arabidopsis* genome. Assuming sequence similarities for each of the paired ends reflects microsynteny with the *Arabidopsis* genome sequence, algorithms can be written to comparatively map the BACs *in silico*. A number of criteria need to be employed: location to the same *Arabidopsis* pseudo-molecule, separation by a physical distance adjusted for ~threefold inter-genome inflation/deflation and consistency of strands of the BLASTN matches for each end. This approach has been adopted in the ATIDB database where, by satisfying these criteria, 14% of the complete set of BAC clones was mapped, the results displayed graphically (http://atidb.org/cgi-perl/gbrowse/atibrowse/).

A similar strategy using data generated by Love et al. (2006) has been adopted by the AtEnsembl database (http://atensembl.arabidopsis.info).

22.5.3 Physical Maps and Informatics

Critically important to the success of a BAC-based sequencing strategy is the availability of a high-quality physical map of the target genome. Many projects have constructed such physical maps by fingerprinting of BAC clone libraries with restriction enzymes and then assembling into contigs using the FingerPrinted

Contigs programme FPC (Soderlund et al. 2000). The use of physical maps is not restricted to construction of tiling paths for sequencing, however, especially when the contigs are anchored genetically or physically. In these circumstances the contigs represent very useful resources for comparative studies of short-range genome structure and also for candidate gene isolation.

The first reported draft physical maps for the diploid *Brassica* A and C genomes were integrated with the *Arabidopsis* genome sequence through nucleic acid hybridization by selected gene sequence tag (GST) probes. The GSTs were designed through a computational approach in which every exon of every annotated gene model was tested *in silico* against the entire *Arabidopsis* genome for sequence uniqueness. Candidates were then selected to represent an even chromosomal distribution of about one GST per 100 kb. In these studies, conducted between 2001 and 2004, approximately 30,000 clones each from *B. rapa* (JBr) and *B. oleracea* (JBo and BoB) BAC libraries were fingerprinted, band profiles captured from either agarose gel or capillary electrophoresis by Image v3.10 (Wobus et al. 2001), and then contigs assembled with FPC v6.4 using parameters (Sulston cutoff score of 1×10^{-10}) expected to resolve paralogous genomic regions. In the case of the *B. rapa* map, the FPC contigs were then re-ordered with CORAL (Flibotte et al. 2004). In parallel with the fingerprinting, the entire BAC libraries were colony hybridized with around 1,300 *Arabidopsis* GSTs and these data integrated into the FPC contigs as physical markers.

FPC is essentially a standalone desktop application so remote querying of the project database itself via web browsers was not possible at that time. Originally, therefore, the integrated data, periodically cycled through the separate FPC project databases to power incremental builds, were maintained in an ACEDB database (Stein and Thierry-Mieg 1999) that could be accessed through the AceBrowser web interface to the Aceserver (Thierry-Mieg and Stein 1998). Latterly, direct searches for contigs, clones, and markers were implemented through the WebFPC Java applet interface (Pampanwar et al. 2005). Finally, all the project data were migrated to a MySQL relational database with users' searches entered in HTML forms being brokered through CGI scripts to form SQL queries and graphical displays of contigs (with functionality emulating that of WebFPC) directly rendered through a Perl `Bio::Graphics` layer. The legacy data are still fully searchable using this system at http://brassica.bbsrc.ac.uk/IGF/. A little over 2,000 contigs were assembled for *B. rapa* (with an estimated coverage of 462 Mb) and almost 1,400 contigs for *B. oleracea*. Comparative genomics studies utilizing the *B. rapa* map have confirmed that paralogous A genome regions were indeed resolved by the FPC build process and could be aligned with their counterparts within the A genome component of *B. napus* (Rana et al. 2004.)

More recently, a *B. rapa* physical map has been assembled from the BrGSP BAC libraries (Mun et al. 2008). High-Information Content Fingerprinting (HICF) was carried out using SNaPshot™ technology (Luo et al. 2003, Nelson et al. 2007) for fluorescence-labeled capillary electrophoresis of the products. GenoProfiler (You et al. 2007) was used to convert the capillary data to band sizes and then FPC v8.5.3 used to assemble contigs in an iterative procedure, starting with an initial cutoff of 1×10^{-45} followed by successive merges at lower cutoffs ending at 1×10^{-15}. This automated phase was followed by some manual intervention to remove artifacts, excessively deep contigs and falsely joined contigs. The final contig assembly, named Build 2, consists of 1,428 contigs, estimated to cover 717 Mb. In principle, this dataset should represent a very informative resource by providing supporting evidence for the selection of extension clones selected on the basis of BLASTN similarities of BAC end sequences to existing scaffold sequence. The FPC database is made available by the investigators through the WebFPC interface at http://www.brassica-rapa.org/BRGP/physicalMap.jsp. Example screenshots of a clone query, followed by rendering of the returned contig through the WebFPC applet, are shown in Fig. 22.2.

There are, however, limitations to the interpretation possible through this web-based query. For example, in Fig. 22.2b, the three highlighted BAC clones KBrH113G14, KBrH009D02, and KBrH108A17 within Contig 11 would appear from the WebFPC display not to directly overlap. These clones have now been sequenced and both end-sequence analysis and gene annotation strongly indicate that they do indeed physically overlap.

22.5.4 Bioinformatic Selection of Seed BACs

The original selection of seed BACs for the BrGSP was based on analysis of cognate BAC end-sequence data (Section 22.5.2). Essentially, those BACs whose end-sequences mapped *in silico* to regions of the *Arabidopsis* genome that would be consistent with microsynteny over the ~100–200 kb range were retained and further selections made from this set. This strategy has, ipso facto, restricted the source material to clones whose paired end-reads (~400 bp read-length) each contains exon sequence(s). Either casual inspection of the annotated gene density of sequenced BACs or database interrogation (predicted exons occupy about 12% of the total sequence) indicates that this will exclude a significant proportion. Also, depending on the sophistication of the algorithm employed for selection, micro-rearrangements where genes may be transposed over short distances, thus interrupting microsynteny (Town et al. 2006), might constitute a confounding factor. The consequence is that the selected seed BACs tend to be clustered, with a greater than expected number of overlaps occurring from the outset, a rapid merging of scaffolds built from them, and a concomitant paucity of "free ends" from which to extend.

Novel strategies are therefore being pursued in order to identify new seed BACs. These include directed searches of the end-sequence data for similarity to regions of the *Arabidopsis* genome currently "missing" and pattern discovery within the data to identify candidate clones that span breakpoints in synteny. With regard to the latter strategy, we estimate that there must be around 50 such major breakpoints over which there could be corresponding physical segments of *B. rapa* DNA represented within the BAC libraries. As this estimate is based on the results of previous studies (e.g., Parkin et al. 2005), which used methods unlikely to detect relatively small

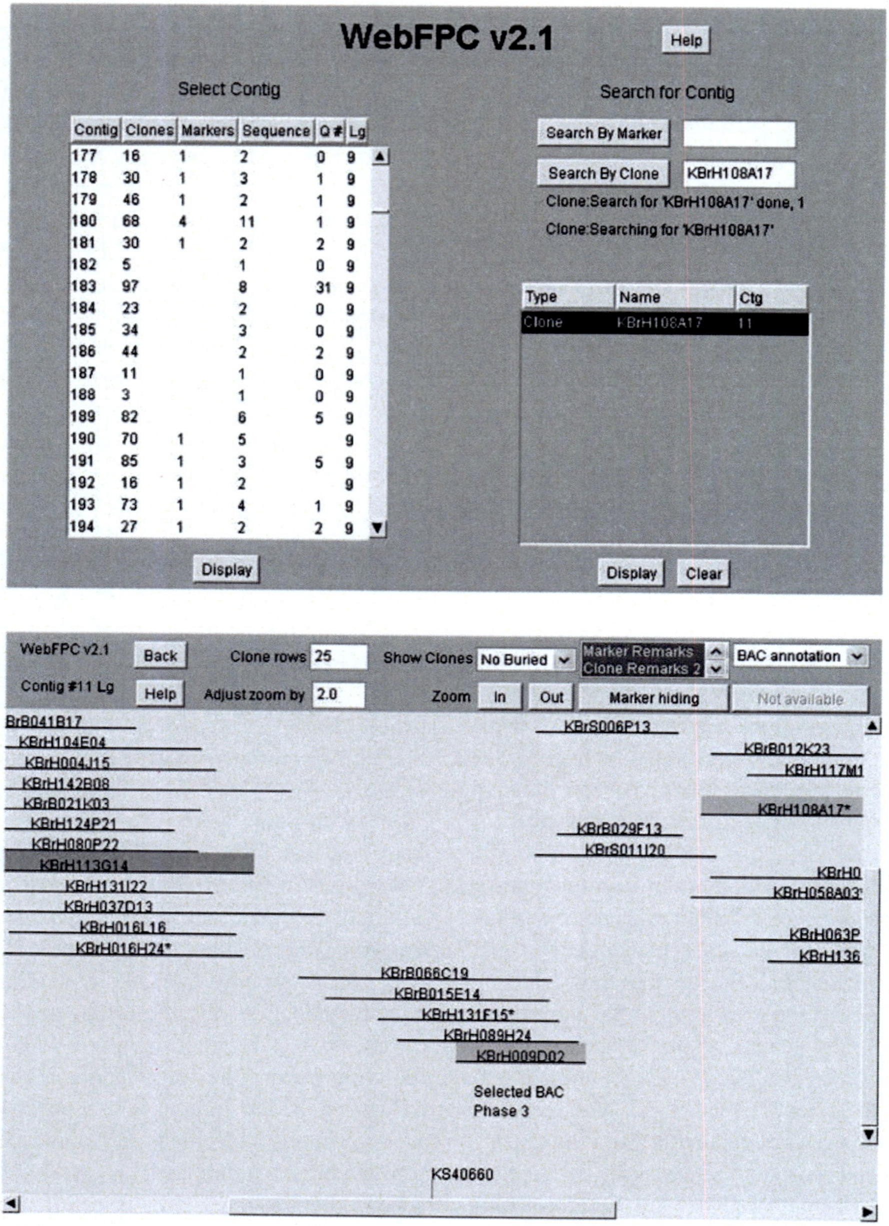

Fig. 22.2 Screenshots from a WebFPC interface to the *Brassica rapa* Build 2 physical map (Mun et al. 2008). (**a**) The query screen and (**b**) Graphical view of the contig returned by the query

collinear genome segments or segmental inversions within collinear genome segments, we anticipate that there will be many more minor breakpoints. One approach to address this issue has been a programmatic analysis of the data (Trick et al. 2009c) to identify pairs of non-contiguous bins of *Arabidopsis* genome sequence that are

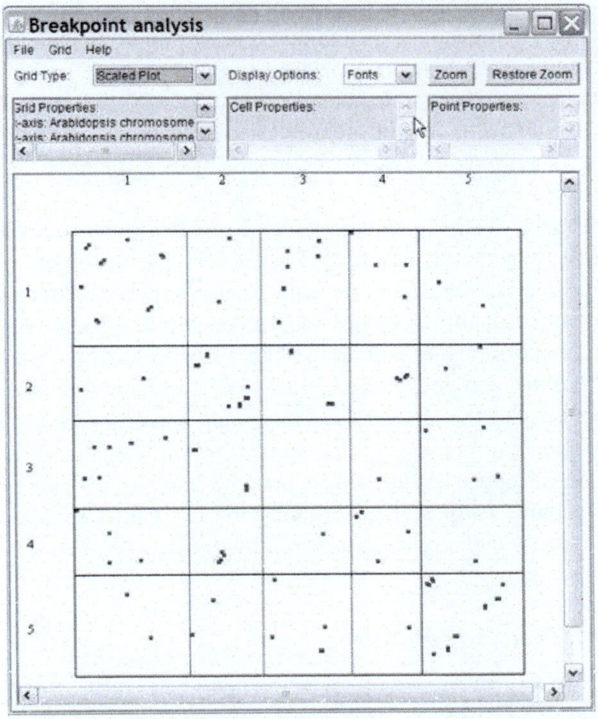

Fig. 22.3 Screenshot of the Grid Map display applet rendering the results of a synteny breakpoint analysis. Each Arabidopsis chromosome pseudo-molecule (Chr 1–5) was divided into 500 kb bins and each bin then compared against all other non-contiguous bins for connections through *B. rapa* BAC end mappings. A dot signifies an instance of at least three such connections between bins so, for example, there were five candidate chromosome 1/chromosome 3 breakpoints identified

"connected" by significant numbers of pairs of end mappings derived from different BAC clones. A local copy of the ATIDB database containing these *in silico* mapping data was interrogated in order to compile a list of such connections. Figure 22.3 illustrates an example set of connections rendered through the Grid Map display (Priestly et al. 2002) which essentially produces a dot-plot of similarity data. At the time of writing, about half of the candidate BAC clones identified in this way and then submitted for sequencing have been validated as containing major synteny breakpoints.

22.5.5 Coordination of Sequencing Programme

Given the distributed nature of the BrGSP, with various national groups building sequence scaffolds by extension from seed BACs genetically mapped to allocated chromosomes, it is inevitable that there will be occasions when it is discovered either that a seed BAC had been incorrectly mapped or that a selected extension clone actually corresponds to a paralogous region and hence maps elsewhere within the genome. Such sequence data will remain nonetheless important to the progress of the project as a whole, as indeed is the declared intention of a group to sequence a given clone, so that effort is not needlessly duplicated. It is thus imperative that there is a rapid and centralized means of disseminating this coordinating information.

A comprehensive Website detailing the Korean groups' activities and some data from their sequencing of the A3 and A9 chromosomes is published at http://www.brassica-rapa.org. The status pages available here retrospectively list the BACs completed for all chromosomes and, in the case of A3 and A9, report genetic mapping data supporting the anchoring. The Website at http://www.brassica.info also hosts similar data but goes further to act as a comprehensive reference point for all the collaborators in the BrGSP. Downloadable spreadsheets indicate all stages of the BAC sequencing process: "in progress," "completed," and "submitted" and hence reference to these should help to prevent "collisions" between sequencing groups. This resource is updated monthly by the UK/China collaborators and it is the intention to include all collaborators joining the BrGSP.

22.5.6 Automated Annotation

A major agreed objective of the BrGSP is the rapid dissemination of assembled and functionally annotated sequence for the *B. rapa* gene space. As a significant fraction of the BACs to be sequenced is being completed to Phase 2 standard only, such annotation cannot be accepted by the sequence databanks within the primary submissions. The community must therefore devise its own systems until such time as a complete Phase 3 sequence becomes available, along with a standard annotation, for deposition in the databanks.

The first automated annotation pipeline for completed BACs sequenced under the auspices of the BrGSP (http://brassica.bbsrc.ac.uk) was initiated in 2005 as part of the preparatory work for the UK/China collaboration to sequence chromosomes A1 and A8. During the first phase all publicly available seed BACs, mapping throughout the genome, were annotated. Subsequently all BACs deposited in the EMBL/GenBank/DDBJ databases have been annotated and the results immediately published to the public domain. In November 2008, after the implementation and extensive trials of a development version, the pipeline was significantly upgraded to a production version which will be automatically applied to all completed BACs from the BrGSP. At the time of writing, 887 BAC annotations have been published on this site describing some 106 Mb of *B. rapa* genome sequence. Future enhancements will include BLASTX searches against all of UniProt in order to give information on gene predictions for proteins not present in the annotated *Arabidopsis* proteome. Also in 2008, both seed BACs and BACs completed for chromosomes A3 and A9 were independently annotated using the commercial

Pedant-Pro system by groups in Korea and some of these results are now available at http://www.brassica-rapa.org.

The UK's annotation pipeline consists of a Perl script which takes as its input the completed BAC sequence (to a minimum of Phase 2) in fasta format. In the case of Phase 2 sequences, assembly gaps and unresolved bases are marked up for later display. As a first step the entire BAC sequence is deconstructed into overlapping 1 kb segments and these used for BLASTN searches against the annotated *Arabidopsis* gene complement (TAIR 6). The gene model codes and annotation for the best hits, together with the E-value and the coordinates for the BLAST HSP alignments, are recorded and summarized in tab-delimited text format which is made available for download as a so-called synteny report. This simplistic yet effective analysis is used for initial quality control checks by the UK's clone coordination group.

The next step is a BLASTN analysis of the completed BAC sequence with all 200,031 available BAC end sequences from the three libraries. An empirical 94% identity threshold is used to perform a secondary filter on hits with E-values of less than 1×10^{-30}. This is to reduce noise from end sequences containing repetitive elements and also attempts to discriminate between cognate alignments and false alignments with paralogous genomic segments. Again, the E-values and coordinates for the BLAST HSP alignments are recorded. The principal use of this analysis is to facilitate the informed selection of extension clones from seed BACs or from growing scaffolds and is thus of critical importance to the coordination of the project.

There follows two other BLASTN analyses, one using the 1,300 *Arabidopsis* gene sequence tags that anchor a previous *B. rapa* physical mapping project (http://brassica.bbsrc.ac.uk/IGF/) and the other using 175 sequenced RFLP markers that both underpin a number of published *Brassica* genetic maps (Parkin et al. 1995, Rusholme et al. 2007) and a detailed comparative analysis with *Arabidopsis* (Parkin et al. 2005).

Lastly in this *in silico* marker analysis, the *Brassica* BAC sequence is searched for microsatellites with the msatfinder program (Thurston and Field 2005) and Primer3 (Rozen and Skaletsky 2000) is then used to generate PCR primers for candidate amplicons containing SSR tracts. The primer oligonucleotides can be directly searched against the entire reference sequence database using GBrowse's OligoFinder plugin, and hence screened for potential cross-hybridization.

The final stages in the annotation process involve the application of a number of ab initio gene prediction programs, SNAP (Korf 2004), Augustus (Stanke and Waack 2003), GlimmerHMM (Majoros et al. 2004), Genezilla (Majoros et al. 2004, 2005), FGENESH (Softberry, Inc.), and Genscan (Burge and Karlin 1997) to generate independent gene models. The majority of the open source programs have been used with parameter matrices developed on *Brassica* (F Cheung and CD Town, personal communication). The best gene models, in the opinion of the authors, are based on the SNAP ab initio gene predictions trained on *Brassica* and then post-corrected with PASA (Haas et al. 2003, Campbell et al. 2006) using the ~800 k available *Brassica* raw ESTs as evidence. PASA refines the SNAP gene model using

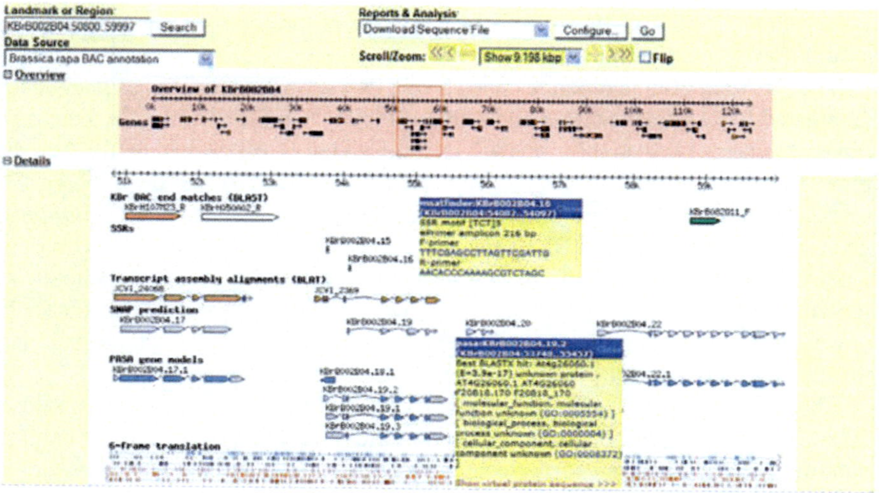

Fig. 22.4 Screenshot from a GBrowse display of an example annotated *B. rapa* BAC, KBrB002B04. The BAC end sequence match, SSR, Transcript assembly alignment, SNAP gene prediction, and PASA gene model tracks have been enabled. Mouseovers for two example features have been included, an SSR (msatfinder: KBrB002B04.16) and a PASA gene model (pasa: KBrB002B04.19.2). In this latter instance PASA has corrected the ab initio SNAP gene model by including an exon for which there is EST evidence

the EST data and can identify and mark up both 5′ and 3′ UTRs and also multiple instantiations of the gene based on alternative splicing variants.

As supporting evidence to aid interpretation of these various ab initio gene predictions, the BLAT program (Kent 2002) is then used to attempt a direct alignment between the *Arabidopsis* CDS corresponding to the best BLASTX hit for the PASA-corrected SNAP gene model and the *Brassica* genomic DNA sequence. Finally, the BAC sequence is searched by BLASTN with the 95 k unigene set developed (see Sect. 22.4). The BLAST hits recovered from this are then re-aligned to genomic sequence using BLAT. Clicking on a Transcript assembly feature allows the user to launch a real-time ClustalW alignment (Chenna et al. 2003) and to inspect it graphically with the Jalview applet (Clamp et al. 2004).

The Genes track in the Overview panel and the PASA gene models track in the Details panel now use SNAP ab initio gene predictions trained on Brassica and corrected with PASA using ~800k Brassica raw ESTs. These represent our best models.

The annotation script parses the results of all of these various analyses to the GFF1 format (http://www.sanger.ac.uk/Software/formats/GFF/GFF_Spec.shtml) and loads these records into a MySQL database driving the Generic Model Organism Database project's GBrowse genome viewer (Stein et al. 2002). GBrowse is an open source, highly configurable, and extensible software system supporting a plug-in architecture based on Perl modules. The `Bio::DB::GFF` schema used for the underlying MySQL database is highly optimized for querying on feature names or on genome locations in relative or absolute coordinates. End-users can customize their displays, perform a variety of queries and analyses, and even securely upload and overlay their own annotation (or that of a third party) onto the publicly provided dataset for comparison. The GBrowse system is also used by the Korean group to deliver the results of their annotation.

An example screen shot of an annotated BAC clone is shown in Fig. 22.4. The graphical rendering of the BAC annotations available from http://brassica.bbsrc.ac.uk largely uses GBrowse's inbuilt methods, but also a few extensions employing client-side Javascript, real-time interrogation of the database using `Bio::DB::GFF` methods, and delivery of content to popup windows. For instance, the BACend BLAST track shows glyphs that are color-coded using computed options available through the GBrowse configuration (e.g., red signifies a potential overlap with another sequenced clone, orange signifies candidate clones that may bridge between two sequenced clones, and green indicates candidate extending clones). Clicking on a BACend feature produces a popup with links to GBrowse views of these cross-referenced BACs and to a ClustalW/Jalview view of the alignment. Similarly, the PASA gene model track is enhanced, with mouseovers giving (currently) the best *Arabidopsis* BLASTX hit, links to GO terms for that gene, the virtual protein translation, and a link to real-time alignment between these proteins and with ClustalW and visualization with Jalview.

The Brassica Genome Gateway portal (http://brassica.bbsrc.ac.uk) contains a number of convenient search interfaces to the `Bio::DB::GFF` database that serve to insulate the casual user from the internals but will deliver her or him straight to the appropriate GBrowse views via constructed URLs. For instance, a complete listing of all completed BACs and all BACs annotated within the last 31 days are provided as links. In addition, wild-carded name searches and free-text feature searches on annotated features (e.g., *Arabidopsis* gene models, annotations, GO terms, *Brassica* marker names) are also provided. A "do-it-yourself" version of the annotation pipeline, currently with reduced functionality with respect to the full production version, is also made available over the web at http://brassica.bbsrc.ac.uk/annotate.html for processing of user-submitted BAC-scale sequence data. A similar facility is due to be offered by the bioinformatics group at the Australian Centre for Plant Functional Genomics group (http://acpfg.imb.uq.edu.au/index.php).

22.6 Next Generation Sequencing and the Re-sequencing of *Brassica* Genomes

The genomes of the *Brassica* species are relatively large for analysis by Sanger/Capillary Electrophoresis (CE) sequencing. Given that the advent of next generation sequencing (NGS) technologies has somewhat set expectations with funding agencies, it is unlikely that the ~1.2 Gb genome of *B. napus*, for example, will now ever be sequenced by exclusively CE methodology. Three commercial NGS systems are currently available: GSFLX (from Roche), Solexa (from

Illumina), and SOLiD (from Applied Biosystems). All are based on massively parallel, fluorescence-tagged pyrosequencing. NGS typically produces very much shorter high-quality read-lengths than the >800 bases produced by CE sequencing, but at very much lower unit cost per nucleotide. However, the computational challenges associated with the de novo assembly of NGS data, especially Solexa and SOLiD platforms with read-lengths of 35–70 bases, represent a major difficulty. Some studies have demonstrated the synergistic benefit obtained by combining NGS with CE methodology (e.g., Goldberg et al. 2006).

Nonetheless, the NGS systems are very well-suited to re-sequencing applications, where a reference genome sequence is available for computational alignment, for example, as applied in *Caenorhabditis elegans* using Solexa (LaDeana et al. 2008). However, there are no complete reference sequences available for the larger genomes of most crop species. We are thus left with a somewhat recursive problem which amply demonstrates the critical importance of the BrGSP in developing a *Brassica* reference genome sequence. Ultimately an NGS/CE hybrid strategy may well be implemented to achieve this. It is presently unclear whether an A genome reference sequence would suffice for alignment with C genome reads.

Although a reference genome per se is not yet available, an extensive EST dataset is, as was previously described (see Section 22.4). The unigene set developed there may be viewed as an approximation to the transcriptome components of the *Brassica* A and C genomes. Such a genome compartment can theoretically be used as a pseudo-reference sequence in NGS alignments of reads obtained from mRNA. The feasibility of this approach has recently been demonstrated in a study directed at single nucleotide polymorphism (SNP) discovery in *B. napus* (Trick et al. 2009b). About 20 million single-ended reads from juvenile leaf mRNA of each of two oilseed rape cultivars, Tapidor and Ningyou 7, were obtained using the Solexa platform and aligned to the *Brassica* unigene pseudo-reference sequence using the MAQ software suite (Li et al. 2008). It was found that the reads could be aligned to about 26 Mb of the 64 Mb reference sequence with an average read-depth of 4–5. SNPs called by MAQ relative to the reference were then compared between the cultivars in order to refine a list of robust candidate SNPs. Between 23,330 and 41,593 putative SNPs, depending on the minimum read-depth threshold applied, were identified in this way, the inferred polymorphism rate (0.047–0.084%) being comparable with that previously observed between these cultivars. The vast majority of the polymorphisms (87.5–91.2%) were of a type, termed in the study "hemi-SNPs," that are indicative of transcription from homoeologous genes from the two parental genomes within oilseed rape. Analysis of a small number of segregants from a mapping population constructed from the parental cultivars over a specimen region of the genome showed that segregation of SNP alleles (almost all hemi-SNPs) largely, though not entirely, followed the pattern expected for genomic markers. This is encouraging for the future deployment of this kind of technology for genome-wide association genetics studies in *B. napus*.

These results are promising in that even the polyploid structure of the *B. napus* genome appears to be computationally tractable for NGS short read alignment and moreover the use of these methods provides a rich source of identifiable markers.

With respect to the problem of de novo assembly of a diploid *Brassica* genome, we might expect to encounter difficulties arising from its general paleopolyploid structure and, specifically, false joins being made across paralogous gene segments or members of middle-repetitive repeat sequence families. These issues are currently being addressed with the use of mate-pair reads from combinations of size-fractionated libraries designed to defeat this kind of conflation. However, even with the single-end, 35 base reads from the reduced complexity RNA-seq dataset from the Ningyou cultivar, some progress was still possible. When analyzed with the Velvet short read assembler software suite (Zerbino and Birney 2008) using a k-mer value of 23, some 25,320 contigs could be assembled, with half of these contigs being shorter than 78 bases (N_{50}=78) and the longest contig (N_{max}) being 1,763 bases. There appeared to be the expected correlation between assembled contig length and probable transcript abundance in un-normalized leaf mRNA (and, hence, read depth). So, for instance, the N_{max} contig had a coverage metric of 106 and was found to encode the beta subunit of ATP synthase. The results from this pilot experiment not only are certainly interesting but also amply demonstrate the deficiencies that we currently face.

22.7 Future Developments

It seems clear that the most pressing requirement for the anticipated post-genomic era for *Brassica* will be an integration of the *B. rapa* gene space sequence (and its annotation) with genetic maps and QTL for traits inferred using those maps. It is expected that the A genome data generated from the BrGSP will serve just as well for the exploration of traits mapped to the A genome component of *B. napus*.

The methodology for achieving this is quite straightforward. There needs to be a careful integration of the existing *Brassica* A genome genetic maps, based on a variety of legacy markers but relative to which a large number of important agronomic traits have been positioned, with the new sequence-based markers that are being used to anchor the emerging *B. rapa* sequence scaffolds. Once this is achieved, there is a choice of database/browser systems available to render the integration and *Brassica* chromosome/linkage groups can then be displayed either in pure centiMorgan coordinates or some physical transform based on genetic distance. Sequence scaffolds, which can be made to link through to the BAC-based annotations, can then be added to decorate this integrated map. The UK sequencing programme is funded to implement this objective and so we look forward to seeing the first results in the near future (to be published from http://www.brassica.info).

In the medium term it is expected that the *B. rapa* genome annotation will remain BAC-based. It remains to be seen whether funding and/or technology will be available for the finishing of the genome to permit the construction of *Brassica* pseudo-molecules and thus a genome-wide re-annotation. Indeed, the impact of NGS re-sequencing technologies on *Brassica* genome science may well completely reshape the way we leverage these new data.

Acknowledgments The author would like to acknowledge support from the UK Biotechnology and Biological Sciences Research Council (Grant BB/E017363/1 and Competitive Strategic Grant to the John Innes Centre).

References

Anderson ML, Cardle L, Cartinhour S, et al (2000) UK cropnet: a collection of databases and bioinformatics resources for crop plant genomics. Nucl Acids Res 28:104–107

Ayele M, Haas BJ, Kumar N et al (2005) Whole genome shotgun sequencing of *Brassica oleracea* and its application to gene discovery and annotation in Arabidopsis. Genome Res 15: 487–495

Bekkaoui F, Xiang D, Datla R et al. (2008) Development of Brassica seed cDNA microarray. http://www.intl-pag.org/14/abstracts/PAG14_P740.html

Burge C, Karlin S (1997) Prediction of complete gene structures in humangenomic DNA. J Mol Biol 268:78–94

Campbell MA, Haas BJ, Hamilton JP et al (2006) Comprehensive analysis of alternative splicing in rice and comparative analyses with Arabidopsis. BMC Genomics 7:327–344

Chenna R, Sugawara H, Koike T et al (2003) Multiple sequence alignment with the Clustal series of programs. Nucl Acids Res 31:3497–3500

Clamp M, Cuff J, Searle SM et al (2004) The Jalview Java alignment editor. Bioinformatics 20:426–427

Flibotte S, Chiu R, Fjell C et al (2004) Automated ordering of fingerprinted clones. Bioinformatics 20:1264—1271

Gish W (1996–2008) WU-BLAST: http://blast.wustl.edu

Goldberg SMD, Johnson J, Busam D, et al (2006) A Sanger/pyrosequencing hybrid approach for the generation of high-quality draft assemblies of marine microbial genomes. Proc Natl Acad Sci USA 103:11240–11245

Haas BJ, Delcher AL, Mount SM, et al (2003) Improving the Arabidopsis genome annotation using maximal transcript alignment assemblies. Nucl Acids Res 31:5654–5666

Kent WJ (2002) BLAT – the BLAST-like alignment tool. Genome Res 4:656–664

Korf I (2004) Gene finding in novel genomes. BMC Bioinformatics 5:59

LaDeana WH, Marth GT, Quinlan AR, et al (2008) Whole-genome sequencing and variant discovery in *C. elegans*. Nat Methods 5:183–188

Li H, Ruan J, Durbin R (2008) Mapping short DNA reads and calling variants using mapping quality scores. Genome Res 18:1851–1858

Love C, Logan E, Erwin T, et al (2006) Integrating and interrogating diverse Brassica data within an Ensembl structured database. Acta Hort (ISHS) 706:77–82

Love CG, Robinson AJ, Lim GAC, et al (2005) Brassica ASTRA: an integrated database for Brassica genomic research. Nucl Acids Res 33:D656–D659

Lowe AJ, Moule C, Trick M, et al (2004) Efficient large-scale development of microsatellites for marker and mapping applications in Brassica crop species. Theor Appl Genet 108:1103–1112

Luo MC, Thomas C, You FM, et al (2003) High-throughput fingerprinting of bacterial artificial chromosomes using the SNaPshot labeling kit and sizing of restriction fragments by capillary electrophoresis. Genomics 82:378–389

Majoros WH, Pertea M, Delcher AL, et al (2005) Efficient decoding algorithms for generalized hidden Markov model gene finders. BMC Bioinformatics 5:616

Majoros WH, Pertea M, Salzberg SL (2004) TigrScan and GlimmerHMM: two open source ab initio eukaryotic gene-finders. Bioinformatics 20:2878–2879

Mun J-H, Kwon S-J, Yang T-J, et al (2008) The first generation of a BAC-based physical map of *Brassica rapa*. BMC Genomics 9:280

Nelson WM, Dvorak J, Luo MC, et al (2007) Efficacy of clone fingerprinting methodologies. Genomics 89:160–165

Pampanwar V, Engler F, Hatfield J, et al (2005) FPC web tools for rice, maize and distribution. Plant Physiol 138:116–126

Park J, Koo DH, Hong CP, et al (2005) Physical mapping and microsynteny of *Brassica rapa* ssp. *pekinensis* genome corresponding to a 222 kbp gene-rich region of Arabidopsis chromosome 4 and partially duplicated on chromosome 5. Mol Genet Genomics 274:579–588

Parkin IAP, Gulden SM, Sharpe AG, et al (2005) Segmental structure of the *Brassica napus* genome based on comparative analysis with *Arabidopsis thaliana*. Genetics 171: 765–781

Parkin IA, Sharpe AG, Keith DJ, et al (1995) Identification of the A and C genomes of amphidiploid *Brassica napus* (oilseed rape). Genome 38:1122–1131

Pertea G, Huang X, Liang F, et al (2003) TIGR Gene Indices clustering tools (TGICL): a software system for fast clustering of large EST datasets. Bioinformatics 19:651–652

Priestly M, et al. (2002) Grid Map. http://cbr.jic.ac.uk/dicks/software/Grid_Map/

Rana D, van den Boogaart T, O'Neill CM, et al (2004) Conservation of the microstructure of genome segments in *Brassica napus* and its diploid relatives. Plant J 40:725–733

Rozen S, Skaletsky H (2000) Primer3 on the WWW for general users and for biologist programmers. In: Krawetz S, Misener S (eds) Bioinformatics methods and protocols: methods in molecular biology. Humana Press, Totowa, NJ

Rusholme RL, Higgins EA, Walsh JA, et al (2007) Genetic control of broad-spectrum resistance to turnip mosaic virus in *Brassica rapa* (Chinese cabbage). J Gen Virol 2007(88):3177–3186

Sharpe AG, Parkin IAP, Keith DJ, et al (1995) Frequent nonreciprocal translocations in the amphidiploid genome of oilseed rape (*Brassica napus*). Genome 38:1112–1121

Soderlund C, Humphray S, Dunham I, et al (2000) Contigs built with fingerprints, markers, and FPC V4.7. Genome Res 11:934–941

Stanke M, Waack S (2003) Gene prediction with a hidden-Markov model and a new intron submodel. Bioinformatics 19(2):ii215–ii225

Stein LD, Mungall C, Shu S, et al (2002) The generic genome browser: a building block for a model organism system database. Genome Res 12:1599–1610

Stein LD, Thierry-Mieg J (1999) AceDB: a genome database management system. Comput Sci Eng 1:44–52

The *Arabidopsis* Genome Initiative (2000) Analysis of the genome sequence of the flowering plant *Arabidopsis thaliana*. Nature 408:796–815

Thierry-Mieg J, Stein LD (1998) Scriptable access to the *Caenorhabditis elegans* genome sequence and other ACEDB databases. Genome Res 8:1308–1315

Thurston MI, Field D (2005) Msatfinder: detection and characterisation of microsatellites. Distributed by the authors at http://www.genomics.ceh.ac.uk/msatfinder/ CEH Oxford, Mansfield Road, Oxford OX1 3SR

Town CD, Cheung F, Maiti R, et al (2006) Comparative genomics of *Brassica oleracea* and *Arabidopsis thaliana* reveals gene loss, fragmentation and dispersal following polyploidy. Plant Cell 18:1348–1359

Trick M, Cheung F, Drou N, et al (2009a) A newly-developed community microarray resource for transcriptome profiling in Brassica species enables the confirmation of Brassica-specific expressed sequences. BMC Plant Biol 9:50

Trick M, Kwon S-J, Choi SR, et al (2009c) Complexity of genome evolution by segmental rearrangement in *Brassica rapa* revealed by sequence-level analysis. BMC Genomics [*in press*]

Trick M, Long Y, Meng J, et al (2009b) SNP discovery in the polyploid *Brassica napus* using Solexa transcriptome sequencing. Plant Biotechnol J 7:334–346

Wobus F, et al (2001) Image. http://www.sanger.ac.uk/Software/Image/

Xiang D, Yongguo C, Schwab D, et al. (2008) High density Combimatrix Brassica oligo microarray. http://www.intlpag.org/16/abstracts/PAG16_P07a_708.html

You FM, Luo MC, Gu YQ, et al (2007) GenoProfiler: batch processing of high-throughput capillary fingerprinting data. Bioinformatics 23:240–242

Zerbino DR, Birney E (2008) Velvet: Algorithms for de novo short read assembly using de Bruijn graphs. Genome Res 18:821–829

Chapter 23
Perspectives on Genetics and Genomics of the Brassicaceae

Renate Schmidt and Ian Bancroft

Abstract The *Arabidopsis thaliana* genome project laid the foundation for the advancement of structural and functional genomics in this species which resulted in an ever-increasing understanding of a multitude of processes at the molecular level. Ongoing progress in high-throughput genome sequencing technologies will now allow for a boost of genome sequencing activities in various Brassicaceae species, accessions, and even populations. Such studies will provide unique insights into the evolution of plant genomes and may ultimately advance breeding of the *Brassica* crops. The development of genome-wide transcriptome analyses in Brassicaceae species other than *A. thaliana* will also heavily depend on the rapid advancement of cost-effective high-throughput sequencing technologies. Important contributions to fields as diverse as developmental biology, evolutionary biology, population genetics, plant physiology, and ecology can be expected since genetically tractable Brassicaceae species that are particularly suitable for the study of a specific trait and/or adaptation are currently developed as additional model systems.

Keywords Brassicaceae · Diploidisation · Duplication · Genome evolution · Genome-wide transcriptome analysis · Model system · Next generation sequencing · Polyploidy

Abbreviations

BAC	Bacterial artificial chromosome
cDNA	Complementary DNA
dbEST	Expressed sequence tags database
EST	Expressed sequence tag
Gb	Giga base
GS FLX	Genome sequencer FLX
GS 20	Genome sequencer 20

R. Schmidt (✉)
Leibniz Institute of Plant Genetics and Crop Plant Research (IPK), 06466 Gatersleben, Germany
e-mail: schmidtr@ipk-gatersleben.de

R. Schmidt, I. Bancroft (eds.), *Genetics and Genomics of the Brassicaceae*, Plant Genetics and Genomics: Crops and Models 9, DOI 10.1007/978-1-4419-7118-0_23,

MPSS	Massively parallel signature sequencing
RNAi	RNA interference
SAGE	Serial analysis of gene expression
siRNA	Small interfering RNA
SOLiD	Sequencing by oligonucleotide ligation and detection
SNP	Single nucleotide polymorphism

Contents

23.1 Brassicaceae Species as Models for Studying Genome Evolution Following Polyploidy 618
23.2 The Impact of Advances in Genome Sequencing Technology 621
23.3 Prospects for Transcriptome Analysis in the Brassicaceae 623
23.4 Upcoming Model Systems in the Brassicaceae 626
References 628

23.1 Brassicaceae Species as Models for Studying Genome Evolution Following Polyploidy

Genome sequencing in *Arabidopsis thaliana*, interpreted in the context of corresponding data from other species, led to the elucidation of ancient polyploidy events: a hexaploidy event (the γ whole genome duplication), which was followed by successive tetraploidy events (the β and α whole genome duplications) (Jaillon et al. 2007). The two tetraploidy events are specific for the core Brassicales, including the Brassicaceae (Hall et al. 2004). The last of these, the α whole genome duplication, post-dates the Cleomaceae–Brassicaceae split (Schranz and Mitchell-Olds 2006). Genes duplicated by the α event, which occurred around 35 million years ago, retain sufficient nucleotide identity for cross-hybridization to have complicated early physical mapping experiments in the *Brassica* species (e.g., O'Neill and Bancroft 2000). Thus, although the γ and β events, which are much more ancient, have little impact on nucleotide-level genome analyses, relationships across the α event need to be taken into account in comparative studies.

Polyploidy subsequent to the α whole genome duplication event is frequently observed in the Brassicaceae. The *Brassica* species have provided a particular focus for studies aiming to understand the evolution of genomes following polyploidy, despite the lack of complete genome sequences. This is because each polyploidy event is followed by a process commonly termed "diploidization," whereby the genome stabilizes in terms of both gene content and genome organization, and the *Brassica* species provide the opportunity to study the impacts of these processes over a wide range of timescales, as summarized in Fig. 23.1. Physical mapping studies (e.g., O'Neill and Bancroft 2000, Rana et al. 2004) clarified earlier conclusions based upon comparative linkage mapping (e.g., Lagercrantz 1998) that the genomes of the *Brassica* species had a fundamentally triplicated structure, indicative of hexaploid ancestry. These studies showed that the triplicated genome regions

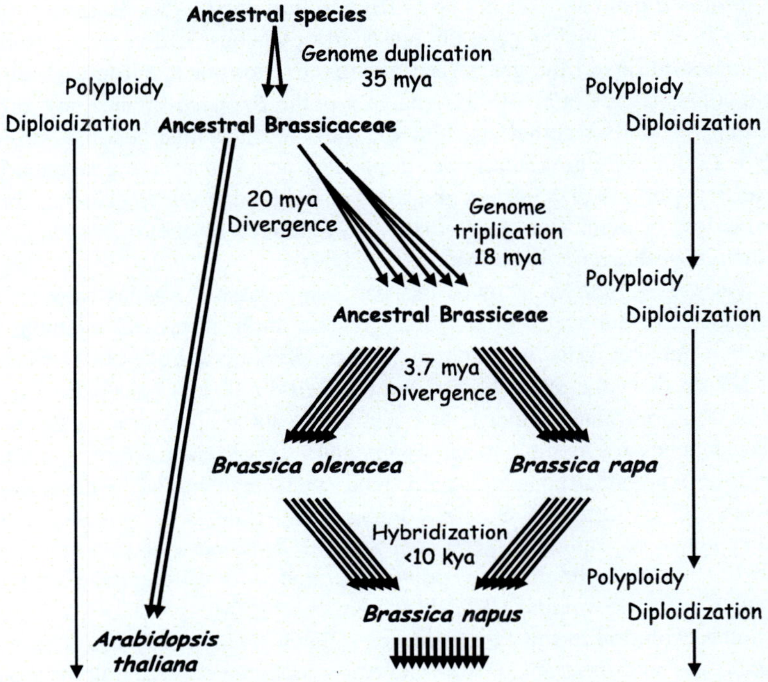

Fig. 23.1 Contrasting genome evolution histories in *A. thaliana* and *B. napus*. The genome of *A. thaliana* has undergone a continuous period of diploidization since the last whole genome duplication that is common to the Brassiceae. In contrast, the genomes of *B. rapa* and *B. oleracea* underwent a triplication that is common to the Brassicaceae, re-launching the diploidization process. Furthermore, *B. napus* has undergone another round of polyploidy, launching a third round of diploidization. Thus the Brassicaceae in general, and the *Brassica* species in particular, presents an excellent opportunity to study the impacts of polyploidy and diploidization over a wide range of time periods

remained largely collinear with each other, and the corresponding regions of the genome of *A. thaliana*, but there was extensive interspersed gene loss. More recent sequence-level studies confirmed these conclusions and showed that the triplicate genomes observed in the *Brassica* species diverged from a common ancestor shortly after the divergence of lineages leading to *Arabidopsis* and *Brassica*, i.e., around 18 million years ago (Town et al. 2006, Yang et al. 2006). It has also been demonstrated that relatively recent segmental duplications have occurred, for example, a segmental duplication that occurred around 0.8 million years ago led to the presence of additional copies of numerous genes, including the flowering time control gene *FLC*, in a cultivar of Chinese cabbage (Yang et al. 2006).

Duplicated genes await different fates during evolution. In most cases gene duplicates are lost rather than preserved. Degenerative mutations may render a gene copy non-functional and thus lead to pseudogene formation and ultimately gene loss. The average half-life of a gene duplicate has been estimated at approximately 4 million years (Lynch and Conery 2000). Nonetheless, a substantial proportion of duplicated genes that were generated by the α event are still present in duplicate in the *A. thaliana* genome (Arabidopsis Genome Initiative 2000). Retention of both copies is likely if a higher gene dosage is advantageous, if divergence leads to a new function in one of the genes (neofunctionalization) or to subfunctionalization of both copies (Force et al. 1999). Analyses in the *Brassica* species now offer the opportunity to study whether and how the more recent whole genome duplication events may influence the retention of duplicated genes that were generated by the more ancient polyploidizations. A first study of this kind provided evidence for more extensive loss of duplicated genes that were generated by the α event in *Brassica rapa* than in *Brassica oleracea* (Mun et al. 2009).

Analyses in *A. thaliana* showed that particular classes of genes were preferentially retained as duplicates after whole-genome duplications, for example, those involved in signal transduction and transcription (Blanc and Wolfe 2004). Moreover, it was shown that a large fraction of the retained duplicated genes showed divergence of transcriptional profiles (Blanc and Wolfe 2004, Haberer et al. 2004), with duplicated genes responding to environmental stress showing more pronounced expression divergence than duplicated genes that are involved in developmental processes (Ha et al. 2007). So far, comparatively little is known about expression divergence of duplicated gene copies in the *Brassica* species. Only recently, a study of ribosomal protein genes provided evidence for differential expression of paralogues in *Brassica napus* (Whittle and Krochko 2009).

Genome evolution in the earliest stages following polyploidy has been studied in *Arabidopsis* and *Brassica* species. Several studies have focused upon *B. napus*, an allotetraploid that occurred naturally by the hybridization of *B. rapa* and *B. oleracea*, probably within the last 10,000 years, and which can be artificially "resynthesized" in the laboratory from the same progenitor species. Elegant studies based upon comparative linkage mapping (i.e., involving markers which could be anchored to the genome of *A. thaliana* based upon sequence similarity) demonstrated a segmental pattern of collinearity between the genome of *B. napus* and that of *A. thaliana*, indicative of numerous chromosome rearrangements having occurred (Parkin et al. 2005). A broader study across the Brassicaceae subsequently identified 24 conserved chromosomal blocks, relating them to a proposed ancestral karyotype ($n = 8$) (Schranz et al. 2006b). A more recent study based upon the sequencing of BAC (Bacterial Artificial Chromosome) clones that contain discontinuities in collinearity between the genomes of *B. rapa* and *A. thaliana* showed that there exist numerous additional small collinearity blocks that were not identified in the initial linkage mapping-based study, resulting in greater complexity (Trick et al. 2009b). These results suggest that, in the Brassicaceae, chromosome rearrangement is likely to have played an important role in the stabilization of genomes following polyploidy. Studies involving resynthesized *B. napus* have shown that their genomes undergo a high rate of change (Song et al. 1995, Udall et al. 2005, Lukens et al. 2004), which continues for many generations following polyploidy, involves homoeologous non-reciprocal transpositions and leads to qualitative changes in gene expression (Gaeta et al. 2007). In contrast, mapping studies in natural *B. napus* have detected such events at a much lower frequency (Parkin et al. 1995, Sharpe et al. 1995, Udall et al. 2005). Moreover, sequence-level studies of the genome of naturally formed *B. napus*, in the form of an oilseed rape cultivar, showed remarkably high conservation of genome structure (Cheung et al. 2009). The difference may be due to naturally formed *B. napus* having inherited a locus controlling homoeologous recombination (Jenczewski et al. 2003) or the genome changes being observed in the resynthesized *B. napus* being an artifact of the polyploidy induction methods used. It is also conceivable that natural lines with few rearrangements were selected for in nature and/or during domestication because resynthesized lines with extensive rearrangements often exhibit low fertility (Gaeta and Pires 2010).

Analyses of artificially synthesized *Arabidopsis* and *Brassica* hybrids have yielded important insights into the genetic and epigenetic processes that confer changes in genome organization and gene expression in plant polyploids (Chapter 7 by Pires and Gaeta, this book). Such studies may also yield insights into heterosis (Ni et al. 2009). Work on *Arabidopsis* hybrids has been focused on the synthetic allotetraploids that were generated by crossing tetraploid *A. thaliana* with tetraploid *Arabidopsis arenosa*. Recently, the generation of fertile hybrids between *A. thaliana* and *Arabidopsis lyrata* has also been reported (Beaulieu et al. 2009). The availability of genome sequences for both parents may offer interesting new perspectives for the study of synthesized *Arabidopsis* hybrids.

The study of nucleolar dominance in *Arabidopsis* hybrids illustrates particularly well on how the power of reverse genetic approaches is currently used to elucidate the molecular details of epigenetic switches in hybrids. Nucleolar dominance describes the phenomenon that the ribosomal RNA genes derived from one of the parents are expressed whereas the genes originating from the other parent are silenced in interspecific hybrids. The important role of DNA methylation and histone modifications in this process has been described more than 10 years ago (Chen and Pikaard 1997) but RNAi (RNA interference) knockdown screens studies in *Arabidopsis* hybrids were necessary to identify genes that are required for this process. It was shown that erasure of histone acetylation is required for Megabase-scale gene silencing in nucleolar dominance (Lawrence et al. 2004, Earley et al. 2006). Furthermore, the important role of the siRNA (small interfering RNA)-directed DNA methylation pathway in this process has been disclosed (Preuss et al. 2008). Despite these advances the mechanism that chooses only one parental set of rRNA genes for complete inactivation remains to be elucidated.

23.2 The Impact of Advances in Genome Sequencing Technology

At the time of writing, despite their economic importance, under 1 Gb (Giga bases) of sequences from *Brassica* species have been placed in the public databases (~1,000 BAC clones, ~880,000 ESTs (expressed sequence tags) and ~1× genome equivalent as genome survey sequences, including BAC end sequences). Indeed, apart from *Arabidopsis*, there are relatively few sequence data for any of the Brassicaceae, although genome projects have been initiated for some species in this family (http://www.jgi.doe.gov/genome-projects/). However, advances in sequencing technology are already beginning to impact genome analysis in these species.

In particular, three commercial systems for massively parallel sequencing (typically referred to as Next Generation Sequencing) are available and in widespread use: GS FLX (Genome Sequencer FLX from Roche), Solexa (from Illumina), and SOLiD (Sequencing by Oligonucleotide Ligation and Detection from Applied Biosystems). These systems implement differing approaches to massively parallel sequencing, but all rely on spatial resolution and amplification of single DNA molecules, followed by sequencing based upon imaging of fluorescent signals. Previously, genome sequencing had been based upon capillary electrophoresis using, for example, Applied Biosystems 3730 $\times$ 1 instruments, which typically provide high-quality reads of >800 bases. In contrast, the typical high-quality read lengths for the GS FLX are currently ca. 400 bases, for Solexa ca. 75 bases, and for the SOLiD ca. 50 bases. However, all have very much lower costs per nucleotide than capillary electrophoresis sequencing.

One of the key resources required to underpin the acceleration of crop improvement by breeding is the availability of molecular markers that can be used to select for the presence of desired alleles. Crop species with polyploid genomes, such as oilseed rape (*B. napus*), present particular challenges for both marker discovery and marker assay. In one of the first publications of the application of Next Generation Sequencing in *Brassica* (Trick et al. 2009c), the Solexa platform was used for the discovery of single nucleotide polymorphisms (SNPs) between two cultivars of oilseed rape. Using established sequence alignment tools and novel bioinformatics processing of alignment data, allelic sequence polymorphisms could be distinguished from inter-homoeologue polymorphisms (i.e., those arising from sequence differences between corresponding genes in the two genomes of *B. napus*), enabling the discovery of ~40,000 SNP markers from a data set of ~1.4 Gb of transcriptome sequences.

Advances in genome sequencing have already led to the launch of projects aiming to characterize the sequences of 1000+ different strains of *A. thaliana* (Weigel and Mott 2009, http://1001genomes.org/). Such work is underpinned by the availability of a high-quality "reference" genome sequence, which has been available for *A. thaliana*, from accession Columbia, since 2000 (Arabidopsis Genome Initiative 2000). Most of the new accessions will be used for "re-sequencing," whereby sequence variation is inferred by alignment of sequence reads to the reference genome and analysis of variation. However, for most of the Brassicaceae, high-quality reference sequences are not available. The relatively short sequence reads produced by Next Generation Sequencing platforms are not ideally suited to the sequencing of paleopolyploid genomes. This is because identical short sequences can occur at multiple locations in the genome, confounding sequence assembly. However, the progressively increasing sequence read lengths produced by current instruments, and improving assembly algorithms and packages, have led to the launch of de novo sequencing of the genomes of numerous members of the Brassicaceae, including two *Brassica* species (*B. oleracea* and *B. rapa*). Although these are unlikely to reach the very high standard of the fully "finished" genome sequence of *A. thaliana*, they should represent key resources to underpin future genome analysis and molecular marker development.

Comparisons of *A. thaliana* accessions suggest that the genomes of Brassicaceae species are evolving and diverging sufficiently quickly that the genome of the species is not fully represented by the genome sequence of any single line (Santuari et al. 2010). Therefore we should perhaps expand our perspective to consider their "pan-genomes." The pan-genome is a concept that has been put forward for some bacterial species (Tettelin et al. 2005) and comprises a core shared genome and a variable fraction partially shared between lines. Ongoing advances in genome sequencing technologies are likely to make the sequencing of genomes across genetic diversity collections feasible within a few years, even for the relatively difficult species such as *B. napus*. The remarkable power of high-throughput sequencing technologies to the study of adaptive variation in natural populations was recently documented (Turner et al. 2010). Variation in *A. lyrata* was detected on a genomic scale by high-throughput sequencing of DNA pooled from 25 plants from four different populations that were either adapted to soil with a high heavy metal content and poor nutrient content or not. Intriguingly, among the 96 polymorphisms that were most strongly associated with soil type many mapped to genes coding for genes involved in heavy metal detoxification and metal ion transport.

In addition to the power of such technologies to identify allelic variation that may also be of benefit for crop improvement, their application across the Brassicaceae will lead to an ever-greater understanding of the evolution of plant genomes following polyploidy, and provide insights of value to understanding the bases of genome evolution across higher plants.

23.3 Prospects for Transcriptome Analysis in the Brassicaceae

The development of various transcript profiling techniques allows the analysis of gene expression of many genes simultaneously and has become an important tool for the discovery of gene function. Over the years different array platforms were developed ranging from spotted cDNA (complementary DNA) arrays through spotted gene-specific tag and long oligonucleotide arrays to on-slide synthesized arrays (Rensink and Buell 2005). For *A. thaliana* all these techniques have been used and a wealth of data can be accessed through various databases (Brady and Provart 2009).

Genome and/or transcript sequence information is a key prerequisite to develop such arrays, but as detailed in the previous section, for most species of the Brassicaceae very limited genome sequence information is available to date. Alternatively, large-scale collections of EST and/or cDNA sequences can be used to develop arrays. However, at the time of writing less than 2,800,000 million ESTs were available in dbEST (expressed sequence tags database; http://www.ncbi.nlm.nih.gov/dbEST/) for species of the Brassicaceae. Collections encompassing more than 40,000 ESTs are currently limited to *A. thaliana* (1,500,000), *B. napus* (640,000), *B. rapa* (190,000), *Raphanus raphanistrum* (162,000), *Raphanus sativus* (120,000), *B. oleracea* (60,000), and *Thellungiella halophila* (44,000). Adopting next generation sequencing technologies for the generation of EST sequences will very likely increase the depth of EST coverage in the coming years for species of the Brassicaceae. EST sequencing of *A. thaliana* seedling tissue already demonstrated the power of massively parallel pyrosequencing. With two sequencing runs on a GS 20 (Genome Sequencer 20) instrument more than 500,000 ESTs were generated that corresponded to transcripts of over 17,000 loci (Weber et al. 2007). Similarly, the data produced from a single Solexa instrument run more than doubled the EST data available for *Brassica* species (Trick et al. 2009c).

EST microarrays that were used for transcript profiling in Brassicaceae species other than *A. thaliana* were more or less limited to *Brassica* species (e.g., Xiang et al. 2008) and *T. haplophila* (Wong et al. 2006), reflecting the rather restricted EST resources in the Brassicaceae. More recently, the wealth of EST information for the *Brassica* species has been used to develop oligonucleotide microarrays, one such array encompasses, for example, probes for 24,000 *B. rapa* genes (Lee et al. 2008). Furthermore, the EST information for *B. napus*, *B. rapa*, and *B. oleracea* was combined in order to assemble a large unigene set of orthologous genes. Based on this information a 60-mer oligonucleotide microarray comprising almost 95,000 probes was developed for transcript profiling in *Brassica* species (Trick et al. 2009a). The latter two resources promise to advance transcript profiling in *Brassica* species greatly. Nevertheless, a caveat remains, long-oligonucleotide arrays are not well-suited to discriminate invariably the expression patterns of homoeologous genes in *B. napus* since these genes usually show sequence identities between 94 and 97% (Trick et al. 2009a). In contrast, next generation sequencing technologies for transcriptome analyses have the potential to differentiate between the expression of closely related members of gene families, such as that occur in polyploids, thus, they may complement or even supplant microarray analyses in the *Brassica* species in future.

Researchers have also adopted the resources which were generated for transcript profiling of *A. thaliana* for the analysis of related species. Species in this family show a considerable average sequence identity in protein-coding genes, a value of 87% has been for example reported for *B. napus* and *A. thaliana* (Cavell et al. 1998). When *A. thaliana* EST microarrays were hybridized to *A. thaliana* and *B. napus* probes, respectively a similar set of genes were classified as seed specific. However, it was also noted that heterologous probes very likely compromise the detection of weakly expressed genes (Girke et al. 2000). Even the very extensive EST collections of *A. thaliana* do not cover the entire transcriptome of this species, moreover transcript profiles of members of multigene families may not be discernable by EST array experiments. Due to these shortcomings the annotated *A. thaliana* genome has been exploited to develop short and long synthetic oligonucleotide arrays as well as gene-specific tag arrays for transcript profiling. The sensitivity for detection of expression changes was found to be similar for 70-mer oligonucleotides and cDNA features when *A. thaliana* arrays were hybridized with probes from *A. thaliana* and its related species *A. arenosa* and *B. oleracea* (Lee et al. 2004). Based on these results, the 70-mer oligonucleotides were used to study gene regulation in *Arabidopsis* and *Brassica* allopolyploids

(Wang et al. 2006, Gaeta et al. 2009). Long-oligonucleotide and gene-specific tag arrays were also used to study gene expression in *T. halophila* (Gong et al. 2005), *Thlaspi caerulescens* (Van de Mortel et al. 2006), *Thlaspi arvense* (Sharma et al. 2007) and *Capsella bursa-pastoris* (Slotte et al. 2007).

Affymetrix GeneChips® have been exceptionally useful to study gene expression in *A. thaliana*. Extensive data sets have been compiled with ATH1 arrays; gene expression during development, in response to stress and hormones were, for example, studied in this way (Schmid et al. 2005, Kilian et al. 2007, Goda et al. 2008). *A. thaliana* GeneChips® were also successfully used for comparative transcript profiling of *A. thaliana* and *Arabidopsis halleri* (Becher et al. 2004, Weber et al. 2004). Hammond et al. (2005) improved the sensitivity of the GeneChips® by limiting the cross-species transcriptome analysis to those oligonucleotide probes that efficiently hybridized to the genomic DNA of the species of interest. A large fraction of the array probes were removed from the analysis by adopting this strategy for *B. oleracea*, *T. caerulescens*, and *T. arvense*, nonetheless due to the fact that genes are represented by multiple probes on the ATH1 array it was possible to evaluate the majority of the genes on the array with the heterologous probes (Hammond et al. 2005, 2006).

All cross-species transcript profiling studies described above suffer from the same important deficiency; the transcript profiling experiments are limited to those genes that are present on the *A. thaliana* arrays. Comparisons of EST and cDNA sequences from different species of the Brassicaceae with the *A. thaliana* reference genome have, however, shown that considerable fractions of *T. halophila*, *T. caerulescens*, and *Brassica* unigene sets lacked counterparts in the *A. thaliana* genome (Wong et al. 2005, Rigola et al. 2006, Lee et al. 2008, Trick et al. 2009a). Moreover, using *A. thaliana* arrays it is not possible to discriminate the expression patterns of homoeologous and/or paralogous genes in polyploid species such as *C. bursa-pastoris*, *Arabidopsis suecica*, or *B. napus*.

In contrast to array-based techniques, tag-based profiling technologies do not depend on prior availability of sequence information. In procedures such as Massively Parallel Signature Sequencing (MPSS, Brenner et al. 2000) and Serial Analysis of Gene Expression (SAGE, Velculescu et al. 1995) a small specific sequence tag is isolated from each transcript and sequencing of the tags reveals not only the identity of the different tags but also their abundance. Transcript profiling by MPSS yielded important new insights into the transcriptome of *A. thaliana* due to the very high sampling depth that this technology provides (Meyers et al. 2004a–c). Due to the short length of MPSS and SAGE tags not all tags match unique positions in the genome. Only 78% of the 14 bp-long SAGE-tags mapped for example to a single locus in the *A. thaliana* genome (Robinson et al. 2004). This disadvantage of SAGE has been ameliorated by introducing modifications that result in longer sequence tags (LongSAGE, Saha et al. 2002, SuperSAGE, Matsumura et al. 2003). LongSAGE has recently been used for transcript profiling of *B. napus* (Wu et al. 2008, Obermeier et al. 2009). A comprehensive analysis of transcripts requires a high sampling depth, thus it is of prime importance that massively parallel sequencing technologies now allow the sequencing of SuperSAGE tags at justifiable costs. Recently, more than 400,000 SuperSAGE tags were generated in this manner for *Boechera* accessions (Sharbel et al. 2009). A disadvantage of the tag-based technologies is that only a specific part of the transcript is assessed at sequence level, but this drawback can be overcome by sampling the transcriptome of a given species in parallel with massively parallel pyrosequencing, preferably by analyzing normalized cDNA libraries. Thus, combining SuperSAGE experiments with the power of massively parallel sequencing technologies hold promise for transcript profiling experiments in species without extensive sequence information such as many species of the Brassicaceae to date.

23.4 Upcoming Model Systems in the Brassicaceae

A. thaliana has been extremely useful to study numerous processes in plant biology. Despite its unique position as a model plant it is clear that this species does not cover the vast diversity found in the plant kingdom at the whole plant, physiological, biochemical, genetic, or molecular level. Moreover, *A. thaliana* is not suitable to study certain biological phenomena that are of great importance in plant biology. However, some aspects for plant research for which *A. thaliana* is not ideally suited can be effectively studied in other species of the Brassicaceae. A notable example in this context is the study of self-incompatibility (Chapter 14 by Nasrallah, this book). In addition to crop species that are actively being studied due to their agronomical importance (Chapter 2 by Warwick, this book) more recently also other species of the Brassicaceae gain more and more importance such as *A. lyrata* (Chapter 12 by Savolainen and Kuittinen, this book) and *Capsella* (Chapter 13 by Theißen, this book).

Species that thrive in extreme environments may represent important genetic reservoirs that may reveal unique insights into stress tolerance. The Brassicaceae as a whole have an extremely wide distribution and have colonized virtually all types of environments, thus members of this family display many of the environmental stress adaptations occurring in plants and hence may represent well-suited models for the study of particular traits and/or adaptations. However, in order to qualify as a model for molecular-genetic studies such species should ideally fulfill a number of criteria. The potential model species should be a self-pollinator and produce many seeds. Its growth habit should be compact and the plant should have a reasonable generation time. Moreover, the model should possess a diploid, compact genome that is closely related to *A. thaliana* and it should be easily amenable to *Agrobacterium*-mediated transformation (Bressan et al. 2001, Peer et al. 2006, Schranz et al. 2007).

In contrast to *A. thaliana*, *Thellungiella salsuginea* (*halophila*) thrives in extreme conditions of cold, salt, and drought and possesses all attributes noted above for efficient molecular-genetic studies (Bressan et al. 2001, Zhu 2001, Inan et al. 2004). The tool-box for gene-function studies is steadily growing and a genome project has been launched for this species (Amtmann 2009). *T. salsuginea* and *A. thaliana* are closely related and morphologically similar and both species use similar components to cope with stress. However, the regulation appears to be more specific for the different stress situations in *Thellungiella* and even under non-stressful conditions the plant is constitutively prepared for stress (Amtmann 2009).

Metal hyperaccumulators are plants that are able to grow and reproduce in heavy metal-polluted soils and accumulate metals to very high concentrations in their aerial tissues. It is of high interest to study how plants manage to adapt to these adverse conditions, furthermore, the knowledge gained by studying such processes can be used to develop plants suited for phytoremediation. Approximately 20% of the plants which are classified as metal hyperaccumulators belong to the Brassicaceae (Milner and Kochian 2008). Peer et al. (2006) assessed species and accessions from the Brassicaceae in order to identify a species which is best suited as a model for the study of nickel and zinc hyperaccumulation. An accession of *T. caerulescens* turned out to be the best candidate overall and the molecular basis of heavy metal hyperaccumulation is currently elucidated by combining molecular, genetic, and physiological studies in this species (Milner and Kochian 2008). *A. halleri* also proved to be a powerful system in order to study zinc tolerance and hyperaccumulation (Roosens et al. 2008, Pauwels et al. 2008). Candidate genes for metal hyperaccumulation were identified via transcript profiling studies and some of these co-segregated with zinc and cadmium hypertolerance. However, *in planta* loss-of-function studies were necessary to demonstrate that one of the candidates, *HMA4*, is required for the process of zinc hyperaccumulation and hypertolerance to zinc and cadmium in *A. halleri*. Gene triplication together with *cis*-regulatory changes lead to enhanced expression of *HMA4* in *A. halleri* when compared to *A. thaliana* (Hanikenne et al. 2008).

The Brassicaceae also offers ample opportunity to investigate the molecular basis of phenotypic and reproductive traits and their evolution (Bowman 2006, Mummenhoff et al. 2009, Chapter 13 by Theißen, this book). The *Boechera holboellii* complex promises, for example, to deliver insight into the important process of apomixis (asexual reproduction through seeds; Schranz et al. 2006a, Sharbel et al. 2009). Recent studies have highlighted how genetically tractable species of the Brassicaceae can be adopted to study the genetic basis for diversification of plant form and life history traits. For example, the genetic basis for differences in leaf form was investigated by comparing *A. thaliana* and *Cardamine hirsuta* (Hay and Tsiantis 2006, Barkoulas et al. 2008). *A. thaliana* is an annual plant, whereas many of its close relatives are perennials. Mutant studies showed that perennial flowering in *Arabis alpina* is regulated by *PEP1*, an orthologue of the *A. thaliana* gene *FLC*. Most notably, a study of the regulation of these orthologous loci revealed distinct expression patterns that correlated with chromatin modifications. Based on these findings chromatin regulation was proposed as a mechanism of how diversification of traits may be generated (Wang et al. 2009).

These few examples document the potential which the comparative study of *A. thaliana* and its relatives holds for the discovery of the underlying molecular mechanisms and/or physiological processes of a wide range of traits. Hence, such comparative studies of cruciferous species promise to deliver important contributions to the fields of evolutionary genomics, genetics, and ecology.

References

Amtmann A (2009) Learning from evolution: *Thellungiella* generates new knowledge on essential and critical components of abiotic stress tolerance in plants. Mol Plant 2:3–12

Arabidopsis Genome Initiative (2000) Analysis of the genome of the flowering plant *Arabidopsis thaliana*. Nature 408:796–815

Barkoulas M, Hay A, Kougioumoutzi E, Tsiantis M (2008) A developmental framework for dissected leaf formation in the *Arabidopsis* relative *Cardamine hirsuta*. Nat Genet 40: 1136–1141

Beaulieu J, Jean M, Belzile F (2009) The allotetraploid *Arabidopsis thaliana-Arabidopsis lyrata* subsp. *petraea* as an alternative model system for the study of polyploidy in plants. Mol Genet Genomics 281:421–435

Becher M, Talke IN, Krall L, Krämer U (2004) Cross-species microarray transcript profiling reveals high constitutive expression of metal homeostasis genes in shoots of the zinc hyperaccumulator *Arabidopsis halleri*. Plant J 37:251–268

Blanc G, Wolfe KH (2004) Functional divergence of duplicated genes formed by polyploidy during Arabidopsis evolution. Plant Cell 16:1679–1691

Bowman JL (2006) Molecules and morphology: comparative developmental genetics of the Brassicaceae. Plant Syst Evol 259:199–215

Brady SM, Provart NJ (2009) Web-queryable large-scale data sets for hypothesis generation in plant biology. Plant Cell 21:1034–1051

Brenner S, Johnson M, Bridgham J et al (2000) Gene expression analysis by massively parallel signature sequencing (MPSS) on microbead arrays. Nat Biotechnol 18:630–634

Bressan RA, Zhang C, Zhang H et al (2001) Learning from the Arabidopsis experience. The next gene search paradigm. Plant Physiol 127:1354–1360

Cavell AC, Lydiate DJ, Parkin IA et al (1998) Collinearity between a 30-centimorgan segment of *Arabidopsis thaliana* chromosome 4 and duplicated regions within the *Brassica napus* genome. Genome 41:62–69

Chen ZJ, Pikaard CS (1997) Epigenetic silencing of RNA polymerase I transcription: a role for DNA methylation and histone modification in nucleolar dominance. Genes Dev 11: 2124–2136

Cheung F, Trick M, Drou N et al (2009) Comparative analysis between homoeologous genome segments of *Brassica napus* and its progenitor species reveals extensive sequence-level divergence. Plant Cell 21:1912–1928

Earley K, Lawrence RJ, Pontes O et al (2006) Erasure of histone acetylation by *Arabidopsis HDA6* mediates large-scale gene silencing in nucleolar dominance. Genes Dev 20:1283–1293

Force A, Lynch M, Pickett FB et al (1999) Preservation of duplicate genes by complementary, degenerative mutations. Genetics 151:1531–1545

Gaeta RT, Pires JC (2010). Homoeologous recombination in allopolyploids: the polyploid ratchet. New Phytol 186:18–28

Gaeta RT, Pires JC, Iniguez-Luy F et al (2007) Genomic changes in resynthesized *Brassica napus* and their effect on gene expression and phenotype. Plant Cell 19:3403–3417

Gaeta RT, Yoo SY, Pires JC et al (2009) Analysis of gene expression in resynthesized *Brassica napus* allopolyploids using Arabidopsis 70mer oligo microarrays. PLoS One 4:e4760

Girke T, Todd J, Ruuska S et al (2000) Microarray analysis of developing Arabidopsis seeds. Plant Physiol 124:1570–1581

Goda H, Sasaki E, Akiyama K et al (2008) The AtGenExpress hormone and chemical treatment data set: experimental design, data evaluation, model data analysis and data access. Plant J 55:526–542

Gong Q, Li P, Ma S et al (2005) Salinity stress adaptation competence in the extremophile *Thellungiella halophila* in comparison with its relative *Arabidopsis thaliana*. Plant J 44: 826–839

Ha M, Li WH, Chen ZJ (2007) External factors accelerate expression divergence between duplicate genes. Trends Genet 23:162–166

Haberer G, Hindemitt T, Meyers BC, Mayer KF (2004) Transcriptional similarities, dissimilarities, and conservation of cis-elements in duplicated genes of Arabidopsis. Plant Physiol 136: 3009–3022

Hall JC, Iltis HH, Sytsma KJ (2004) Molecular phylogenetics of core brassicales, placement of orphan genera *Emblingia, Forchhammeria, Tirania*, and character evolution. Syst Bot 29: 654–669

Hammond JP, Bowen HC, White PJ et al (2006) A comparison of the *Thlaspi caerulescens* and *Thlaspi arvense* shoot transcriptomes. New Phytol 170:239–260

Hammond JP, Broadley MR, Craigon DJ et al (2005) Using genomic DNA-based probe-selection to improve the sensitivity of high-density oligonucleotide arrays when applied to heterologous species. Plant Methods 1:10

Hanikenne M, Talke IN, Haydon MJ et al (2008) Evolution of metal hyperaccumulation required cis-regulatory changes and triplication of *HMA4*. Nature 453:391–395

Hay A, Tsiantis M (2006) The genetic basis for differences in leaf form between *Arabidopsis thaliana* and its wild relative *Cardamine hirsuta*. Nat Genet 38:942–947

Inan G, Zhang Q, Li P et al (2004) Salt cress. A halophyte and cryophyte Arabidopsis relative model system and its applicability to molecular genetic analyses of growth and development of extremophiles. Plant Physiol 135:1718–1737

Jaillon O, Aury JM, Noel B et al (2007) The grapevine genome sequence suggests ancestral hexaploidization in major angiosperm phyla. Nature 449:463–467

Jenczewski E, Eber F, Grimaud A et al (2003) *PrBn*, a major gene controlling homoeologous pairing in oilseed rape (*Brassica napus*) haploids. Genetics 164:645–653

Kilian J, Whitehead D, Horak J et al (2007) The AtGenExpress global stress expression data set: protocols, evaluation and model data analysis of UV-B light, drought and cold stress responses. Plant J 50:347–363

Lagercrantz U (1998) Comparative mapping between *Arabidopsis thaliana* and *Brassica nigra* indicates that *Brassica* genomes have evolved through extensive genome replication accompanied by chromosome fusions and frequent rearrangements. Genetics 150:1217–1228

Lawrence RJ, Earley K, Pontes O et al (2004) A concerted DNA methylation/histone methylation switch regulates rRNA gene dosage control and nucleolar dominance. Mol Cell 13:599–609

Lee SC, Lim MH, Kim JA et al (2008) Transcriptome analysis in *Brassica rapa* under the abiotic stresses using *Brassica* 24K oligo microarray. Mol Cell 26:595–605

Lee HS, Wang J, Tian L et al (2004) Sensitivity of 70-mer oligonucleotides and cDNAs for microarray analysis of gene expression in *Arabidopsis* and its related species. Plant Biotechnol J 2:45–57

Lukens LN, Quijada PA, Udall J et al (2004) Genome redundancy and plasticity within ancient and recent *Brassica* crop species. Biol J Linnean Soc 82:665–674

Lynch M, Conery JS (2000) The evolutionary fate and consequences of duplicate genes. Science 290:1151–1155

Matsumura H, Reich S, Ito A et al (2003) Gene expression analysis of plant host-pathogen interactions by SuperSAGE. Proc Natl Acad Sci USA 100:15718–15723

Meyers BC, Lee DK, Vu TH et al (2004a) Arabidopsis MPSS. An online resource for quantitative expression analysis. Plant Physiol 135:801–813

Meyers BC, Tej SS, Vu TH et al (2004b) The use of MPSS for whole-genome transcriptional analysis in *Arabidopsis*. Genome Res 14:1641–1653

Meyers BC, Vu TH, Tej SS et al (2004c) Analysis of the transcriptional complexity of *Arabidopsis thaliana* by massively parallel signature sequencing. Nat Biotechnol 22:1006–1011

Milner MJ, Kochian LV (2008) Investigating heavy-metal hyperaccumulation using *Thlaspi caerulescens* as a model system. Ann Bot 102:3–13

Mummenhoff K, Polster A, Mühlhausen A, Theißen G (2009) *Lepidium* as a model system for studying the evolution of fruit development in Brassicaceae. J Exp Bot 60:1503–1513

Mun JH, Kwon SJ, Yang TJ et al (2009) Genome-wide comparative analysis of the *Brassica rapa* gene space reveals genome shrinkage and differential loss of duplicated genes after whole genome triplication. Genome Biol 10:R111

Ni Z, Kim ED, Ha M et al (2009) Altered circadian rhythms regulate growth vigour in hybrids and allopolyploids. Nature 457:327–331

Obermeier C, Hosseini B, Friedt W, Snowdon R (2009) Gene expression profiling via LongSAGE in a non-model plant species: a case study in seeds of *Brassica napus*. BMC Genomics 10:29

O'Neill CM, Bancroft I (2000) Comparative physical mapping of segments of the genome of *Brassica oleracea* var. *alboglabra* that are homoeologous to sequenced regions of chromosomes 4 and 5 of *Arabidopsis thaliana*. Plant J 23:233–243

Parkin IA, Gulden SM, Sharpe AG et al (2005) Segmental structure of the *Brassica napus* genome based on comparative analysis with *Arabidopsis thaliana*. Genetics 171:765–781

Parkin IA, Sharpe AG, Keith DJ, Lydiate DJ (1995) Identification of the A and C genomes of amphidiploid *Brassica napus* (oilseed rape). Genome 38:1122–1131

Pauwels M, Roosens N, Frérot H, Saumitou-Laprade P (2008) When population genetics serves genomics: putting adaptation back in a spatial and historical context. Curr Opin Plant Biol 11:129–134

Peer WA, Mahmoudian M, Freeman JL et al (2006) Assessment of plants from the Brassicaceae family as genetic models for the study of nickel and zinc hyperaccumulation. New Phytol 172:248–260

Preuss SB, Costa-Nunes P, Tucker S et al (2008) Multimegabase silencing in nucleolar dominance involves siRNA-directed DNA methylation and specific methylcytosine-binding proteins. Mol Cell 32:673–684

Rana D, van den Boogaart T, O'Neill CM et al (2004) Conservation of the microstructure of genome segments in *Brassica napus* and its diploid relatives. Plant J 40:725–733

Rensink WA, Buell CR (2005) Microarray expression profiling resources for plant genomics. Trends Plant Sci 10:603–609

Rigola D, Fiers M, Vurro E, Aarts MG (2006) The heavy metal hyperaccumulator *Thlaspi caerulescens* expresses many species-specific genes, as identified by comparative expressed sequence tag analysis. New Phytol 170:753–765

Robinson SJ, Cram DJ, Lewis CT, Parkin IA (2004) Maximizing the efficacy of SAGE analysis identifies novel transcripts in Arabidopsis. Plant Physiol 136:3223–3233

Roosens NH, Willems G, Saumitou-Laprade P (2008) Using *Arabidopsis* to explore zinc tolerance and hyperaccumulation. Trends Plant Sci 13:208–215

Saha S, Sparks AB, Rago C et al (2002) Using the transcriptome to annotate the genome. Nat Biotechnol 20:508–512

Santuari L, Pradervand S, Amiguet-Vercher AM et al (2010) Substantial deletion overlap among divergent Arabidopsis genomes revealed by intersection of short reads and tiling arrays. Genome Biol 11:R4

Schmid M, Davison TS, Henz SR et al (2005) A gene expression map of *Arabidopsis thaliana* development. Nat Genet 37:501–506

Schranz ME, Kantama L, de Jong H, Mitchell-Olds T (2006a) Asexual reproduction in a close relative of *Arabidopsis*: a genetic investigation of apomixis in *Boechera* (Brassicaceae). New Phytol 171:425–438

Schranz ME, Lysak MA, Mitchell-Olds T (2006b) The ABC's of comparative genomics in the Brassicaceae: building blocks of crucifer genomes. Trends Plant Sci 11:535–542

Schranz ME, Mitchell-Olds T (2006) Independent ancient polyploidy events in the sister families Brassicaceae and Cleomaceae. Plant Cell 18:1152–1165

Schranz ME, Song BH, Windsor AJ, Mitchell-Olds T (2007) Comparative genomics in the Brassicaceae: a family-wide perspective. Curr Opin Plant Biol 10:168–175

Sharbel TF, Voigt ML, Corral JM et al (2009) Molecular signatures of apomictic and sexual ovules in the *Boechera holboellii* complex. Plant J 58:870–882

Sharma N, Cram D, Huebert T et al (2007) Exploiting the wild crucifer *Thlaspi arvense* to identify conserved and novel genes expressed during a plant's response to cold stress. Plant Mol Biol 63:171–184

Sharpe AG, Parkin IA, Keith DJ, Lydiate DJ (1995) Frequent nonreciprocal translocations in the amphidiploid genome of oilseed rape (*Brassica napus*). Genome 38:1112–1121

Slotte T, Holm K, McIntyre LM et al (2007) Differential expression of genes important for adaptation in *Capsella bursa-pastoris* (Brassicaceae). Plant Physiol 145:160–173

Song K, Lu P, Tang K, Osborn TC (1995) Rapid genome change in synthetic polyploids of *Brassica* and its implications for polyploid evolution. Proc Natl Acad Sci USA 92:7719–7723

Tettelin H, Masignani V, Cieslewicz MJ et al (2005) Genome analysis of multiple pathogenic isolates of *Streptococcus agalactiae*: implications for the microbial "pan-genome". Proc Natl Acad Sci USA 102:13950–13955

Town CD, Cheung F, Maiti R et al (2006) Comparative genomics of *Brassica oleracea* and *Arabidopsis thaliana* reveals gene loss, fragmentation and dispersal following polyploidy. Plant Cell 18:1348–1359

Trick M, Cheung F, Drou N et al (2009a) A newly-developed community microarray resource for transcriptome profiling in *Brassica* species enables the confirmation of *Brassica*-specific expressed sequences. BMC Plant Biol 9:50

Trick M, Kwon S-J, Choi SR et al (2009b) Complexity of genome evolution by segmental rearrangement in *Brassica rapa* revealed by sequence-level analysis. BMC Genomics 10:539

Trick M, Long Y, Meng J, Bancroft I (2009c) Single nucleotide polymorphism (SNP) discovery in the polyploid *Brassica napus* using Solexa transcriptome sequencing. Plant Biotechnol J 7:334–346

Turner TL, Bourne EC, Von Wettberg EJ et al (2010) Population resequencing reveals local adaptation of *Arabidopsis lyrata* to serpentine soils. Nat Genet 42:260–263

Udall JA, Quijada PA, Osborn TC (2005) Detection of chromosomal rearrangements derived from homologous recombination in four mapping populations of *Brassica napus* L. Genetics 169:967–979

van de Mortel JE, Almar Villanueva L, Schat H et al (2006) Large expression differences in genes for iron and zinc homeostasis, stress response, and lignin biosynthesis distinguish roots of *Arabidopsis thaliana* and the related metal hyperaccumulator *Thlaspi caerulescens*. Plant Physiol 142:1127–1147

Velculescu VE, Zhang L, Vogelstein B, Kinzler KW (1995) Serial analysis of gene expression. Science 270:484–487

Wang R, Farrona S, Vincent C et al (2009) *PEP1* regulates perennial flowering in *Arabis alpina*. Nature 459:423–427

Wang J, Tian L, Lee HS et al (2006) Genomewide nonadditive gene regulation in Arabidopsis allotetraploids. Genetics 172:507–517

Weber M, Harada E, Vess C et al (2004) Comparative microarray analysis of *Arabidopsis thaliana* and *Arabidopsis halleri* roots identifies nicotianamine synthase, a ZIP transporter and other genes as potential metal hyperaccumulation factors. Plant J 37:269–281

Weber AP, Weber KL, Carr K et al (2007) Sampling the Arabidopsis transcriptome with massively parallel pyrosequencing. Plant Physiol 144:32–42

Weigel D, Mott R (2009) The 1001 genomes project for *Arabidopsis thaliana*. Genome Biol 10:107

Whittle CA, Krochko JE (2009) Transcript profiling provides evidence of functional divergence and expression networks among ribosomal protein gene paralogs in *Brassica napus*. Plant Cell 21:2203–2219

Wong CE, Li Y, Labbe A et al (2006) Transcriptional profiling implicates novel interactions between abiotic stress and hormonal responses in Thellungiella, a close relative of Arabidopsis. Plant Physiol 140:1437–1450

Wong CE, Li Y, Whitty BR et al (2005) Expressed sequence tags from the Yukon ecotype of *Thellungiella* reveal that gene expression in response to cold, drought and salinity shows little overlap. Plant Mol Biol 58:561–574

Wu GZ, Shi QM, Niu Y et al (2008) Shanghai RAPESEED Database: a resource for functional genomics studies of seed development and fatty acid metabolism of *Brassica*. Nucleic Acids Res 36(Database issue):D1044–D1047

Xiang D, Datla R, Li F et al (2008) Development of a *Brassica* seed cDNA microarray. Genome 51:236–242

Yang TJ, Kim JS, Kwon SJ et al (2006) Sequence-level analysis of the diploidization process in the triplicated *FLC* region of *Brassica rapa*. Plant Cell 18:1339–1347

Zhu JK (2001) Plant salt tolerance. Trends Plant Sci 6:66–71

Index

A

Aarts, M. G., 545–546
Abdel-Ghany, S. E., 106
Abel, C., 352
Abel, S., 200
Abiotic environment adaptation
  A. lyrata genetics, 364–365
Abiotic stress tolerance
  B. rapa, 243–244
  See also Tolerances
Aboul-Soud, M. A., 547
Abraham, V., 237
Abrahams, S., 492
Abscisic acid (ABA), 79
Acarkan, A., 177–178, 377
Activation tagging, 547–548
Adam, A., 352, 355
Adams, K. L., 197, 201, 203–204
Addis, M., 441
Adenot, X., 110
Advanced intercross line (AIL), 128
Aethionema (Aethionemeae)
  genome and chromosomal evolution, 10
  polyploidy and hybridization, 15, 18
Agerbirk, N., 44
Agnihotri, A., 42, 48
Agrawal, N., 549
Ågren, J., 348, 361–362, 364
Agriculture, *see Brassicaceae*, crop species
Agrobacterium rhizogenes
  transformation method, 511
Agrobacterium T-DNA mediated insertional mutations
  Arabidopsis reverse genetics, 540–544
Agrobacterium tumefaciens
  transformation method, 508–510
Agronomic/yield traits
  B. juncea, 337–338
  B. rapa, 236–237
Aguinagalde, I., 44
Aharoni, A., 471, 479
Ahmed, S. U., 237
Ahuja, I., 43
Ajisaka, H., 220–221, 224, 240
Albert, T. J., 297
Albertin, W., 199, 201–202, 205, 277
Albrechtsen, M., 576
Albugo candida, 244, 248, 338
Alcázar, R., 135, 138
Ali, H., 46
Ali, H. B. M., 18
Alix, K., 158, 202
Ali, Z., 46
Allan, H. H., 3
Allard, R. W., 304
Allen, E., 94, 101, 104–105
Allen, R. S., 100
Allender, C., 294
Allgayer, H., 267
Alliaria petiolata
  chemical traits, 43
  See also Brassicaceae
Allopolyploidy, 15, 196
  B. juncea allopolyploid, 324–326
  See also Polyploidy
Allotetraploids, Tragopogon, 202
Alonso-Blanco, C., 123–141
Al-Shammary, K. I., 43
Al-Shebaz, I. A., 124, 126
Al-Shehbaz, I. A., 3–9, 12, 14–15, 19–20, 23, 35, 42–45, 348
Altenbach, S. B., 309
Alternaria blight
  B. juncea disease resistance, 341
Altmann, T., 508, 546
Altuvia, Y., 91
Alvarado, M. C., 533
Alvarez, J., 552, 576–577

R. Schmidt, I. Bancroft (eds.), *Genetics and Genomics of the Brassicaceae*, Plant Genetics and Genomics: Crops and Models 9, DOI 10.1007/978-1-4419-7118-0,

Alysseae tribe, 6
Alyssum borzaeanum, 45
Alyssum murale, 45
Amar, S., 140, 300, 310–311
Amasino, R., 129, 131, 137, 198, 305, 447
Amplicon
  selection, 567–568
  See also TILLING
Amplified Fragment Length Polymorphism (AFLP), 218–220, 230
  B. juncea mapping, 330
  B. napus genomic tools, 297
  molecular markers in diversity studies, 219
Amtmann, A., 626–627
Ananiev, E. V., 416
Anastatica hierochuntica
  morphological traits, 42
Ancestral crucifer karyotype (ACK), 20–22
  ancestral Proto-Calepineae Karyotype (PCK), 23
  n=8, 20
  See also Karyotype
Anchonieae, 8
  chromosomes and chromosome number variation, 13
  genome size (GS) variation, 12–13
Anderson, M. D., 49
Anderson, M. L., 598
Anderson, R. C., 44
Aneuploidy
  B. oleracea genetics, 278–279
  defined, 196
  See also Polyploidy
An, G., 508
Angelini, L., 43
Angelini, L. G., 41
Anssour, S., 201, 205
Anthocyanin pigmentation
  in *B. rapa*, 238–239
Antisense lines
  gene silencing mediated functional analysis
    Arabidopsis reverse genetics, 549
    RNAi/PTGS mechanisms in *Brassicaceae*, 572
  See also Genomics
Aphragmeae, 8
Apel, P., 44
Apomixis
  Brassicaceae crop species, 45
Appel, O., 3–4, 6, 8, 12
Applequist, L. A., 294
Arabideae tribe, 6
Arabidopsis
  BAC, 20, 22
  chromosome homeology revealing through CCP, 18, 22
  FAD2 gene, 242
  genome and chromosomal
    collinearity, 19
    evolution, 10–11
  karyotype evolution in aspects, 24
  molecular resources, 451–452
    AGI BAC and P1 libraries, filters and clones sources, 457
    AGI, resources utilized by, 456
    bacterial artificial chromosome and P1 libraries and clones used to generate genome sequence, 456–457
    expressed sequence TAGs And cDNA clones, 460
    gene specific TAG and RNA interference clones, 452
    genomic library/clone resources for *A. thaliana*, 456–458
    large insert libraries from various *Arabidopsis* accessions, 458
    large insert libraries utilization, 458
    multifunctional vectors, 463
    sequenced full length cDNA and ORF clones in entry vectors, 462
    transformable BAC libraries and clones, 458
    yeast artificial chromosome libraries and clones, 457–458
  plant regeneration and transformation, 45
  polyploids, 199
  resynthesized polyploids, 200
    epigenetic changes, 204
    future research, 205
    genetic changes effects, 204–205
    phenotypic effects, 200
  reverse genetics (gene function analysis), 529
    expression analyses, 531–533
    mutation analyses, 533, 536–547
    similarity to other known proteins, 530
  reverse genetics (gene silencing mediated functional analysis), 548
    antisense lines, 549
    microRNAs and targeted miRNA lines, 552
    RNAi gene silencing approaches, 576–577
    RNAi lines, 549

reverse genetics (over-expression/activation mediated functional assays), 547–548
secondary metabolites in
LC/API-MS-based profiling approaches, 488–493
transformation technology in
Agrobacterium rhizogenes, 510
as research tool, 517
bacterial strains and plasmids, use of, 511–512
direct uptake transformation methods, 511
in vitro shoot regeneration, 512
vectors for delivering ihpRNA contructs in, 574–575
See also Brassica; Brassicaceae
Arabidopsis arenosa
polyploidy and hybridization, 15–18
sequence comparisons between *A. thaliana* and, 176, 178
Arabidopsis Genome Initiative (AGI)
resources utilized by, 456
See also Brassica rapa genome sequencing project (BrGSP)
Arabidopsis halleri, 2, 45
ancestral crucifer karyotype (ACK), 20
polyploidy and hybridization, 15
sequence comparisons between *A. thaliana* and, 174–176
Arabidopsis kamchatica
polyploidy and hybridization, 15
Arabidopsis lyrata, 2
A. thaliana comparative mapping extension to related species, 166–167
ancestral crucifer karyotype (ACK), 20–22
genetics, 345
A. lyrata genome, 350–351
abiotic environment adaptation, 364–365
disease resistance, 363–364
flowering time, 364
functional and population genomics, 365–366
glucosinolates, 362–363
highly differentiated disjunct populations, 358–361
local adaptation genetics, 361–365
mating system and genome evolution, 353–354
population genetic diversity in individual populations, 354–358
self-incompatibility and inbreeding depression aspects, 351–353
genome and chromosome collinearity, 18
polyploidy and hybridization, 15
sequence comparisons between *A. thaliana* and, 177
systematics and distribution, 348–350
Arabidopsis suecica
polyploidy and hybridization, 15–18
Arabidopsis thaliana, 2, 5, 45
ACK, 20–22
as model genome for *Brassicaceae*, 160–162
A. thaliana reference genome, 173–174
across A, B and C genomes, 162
comparative mapping for trait analysis, 165–166
conserved chromosome landmarks, 164
rearrangement hotspots, 164–165
sequence comparative analysis, 172–173
bioinformatics resources, 585
A. thaliana background, 586
gene and protein analysis resources, 590–592
gene interactions and pathways, 592
genome browsers, 586–588
integration of data, 594
metabolomic data, 594
small RNA databases, 593
transcriptomics data, 588–590
chromosomes and chromosome number variation, 13, 15
comparative mapping
and candidate genes identification for *B. rapa* traits, 249
of *B. oleracea* to *A. thaliana*, 279–280
comparative mapping extension to related species, 165
Arabidopsis lyrata, 166–167
Capsella rubella, 166–167
genome and chromosomal
collinearity, 18
evolution, 12
genome size (GS)
analysis, 172–173
variation, 12
genomics, 67, 68
cis elements, 69–71
Cis-regulatory elements detection, 76–87
core promoter, 71–76
long ncRNAs, 90

Arabidopsis thaliana (*cont.*)
non-coding RNA landscape, 88–110
noncoding transcripts, 69
small RNAs (sRNAs), 91–110
proximal and distal promoter, 76
germplasm resources, 440–451
glucosinolates (GSL) pathways, 272
karyotype evolution in aspects, 23
molecular resources
expressed sequence TAGs And cDNA clones, 460
gene specific TAG and RNA interference clones, 462
multifunctional vectors, 463
sequenced full length cDNA and ORF clones in entry vectors, 462
molecular resources (genomic library/clone resources for *A. thaliana*), 456
AGI, 456–457
bacterial artificial chromosome and P1 libraries and clones used to generate genome sequence, 456–457
large insert libraries from various *Arabidopsis* accessions, 458
transformable BAC libraries and clones, 458
yeast artificial chromosome libraries and clones, 457–458
natural variation in, 124–126
A. thaliana genetic information use in *Brassica*, 139–141
genetic and molecular analysis, 127–130
genetic bases of adaptation (QTL underlying aspects), 128, 135–137
geographical distribution and demographical history of *A. thaliana*, 126–127
molecular bases of adaptation (QTL underlying aspects), 137–139
polyploidy and hybridization, 15–18
self-fertility in, 404–405
sequence comparative analysis
A. thaliana accessions, 174–176
sequence comparisons between *A. thaliana* and *Brassica*, 181
B. napus, 185–187
B. oleracea, 184–185
B. rapa, 182–184
sequence comparisons between *A. thaliana* and near relatives
A. arenosa, 176, 178
A. halleri, 177–178
A. lyrata, 179
Boechera drummondii, 178
Boechera stricta, 178–179
Capsella rubella, 176–178
OlimArabidopsis pumila, 178, 180
transformation technology in, 506–507
Agrobacterium tumefaciens, 508–510
chloroplast transformation, 511
See also Brassica
Arabis, 2
Arabis alpina, 3
Arabis gunnisoniana
breeding systems and apomixis, 45
Arabis holboellii, 45
polyploidy and hybridization, 15
Aranzana, M. J., 130
Aravin, A., 108
Aravin, A. A., 108
Armoracia rusticana, 2
crop species
chemical traits, 43
morphological traits, 42
under-utilized crucifer crops, 41
Armstrong, K. C., 278, 513
Armstrong, M. E., 275
Arnison, P. G., 278
Array related expression data microarrays
Arabidopsis reverse genetics, 531
Artificial microRNAs
gene silencing approaches in *Brassicaceae*, 576–577
Arus, P., 279
Ashburner, M., 590
Ashraf, M., 46
Association mapping (AM), 130
Assunção, A. G. L., 46
Atibalentja, N., 49
Atmospheric pressure ionization (API)
chemical ionization (APCI), 477
LC/API-MS-based profiling approaches, 488–492
photoionization (APPI), 492
Aubrieta, 2
Aukerman, M. J., 104, 137
Auld, D. L., 302
Aung, K., 107
Autopolyploidy, 15, 196
See also Polyploidy
Autotetraploid, 16
Autran, D., 75
Auxin response transcription factor (ARF), 100
Awadalla, P., 351

Axelsson, T., 158, 159, 199–200, 205, 224, 233, 328, 330–331
Axtell, M. J., 94, 105
Ayele, M., 185, 428, 459, 599

B

Babula, D., 279, 296
Backcrossed inbred lines (BILs), 128
Backstrom, S., 74
Bacterial artificial chromosome (BAC), 19
- *Arabidopsis*, 18–23
- *Brassica rapa* genome sequencing
  - BAC libraries and BAC-end sequences for *B. rapa*, 420
  - seed BAC selection, 428–430
  - seed BAC sequences characteristics, 430–432
- *Brassica rapa* genome sequencing project (BrGSP)
  - BAC end sequencing, 603
  - bioinformatic selection of seed BACs, 605–607
- genomic library/clone resources for *A. thaliana*, 456
  - AGI BAC and P1 libraries, filters and clones sources, 457
  - transformable BAC libraries and clones, 458

Bacterial strains and plasmids
- use of, 511–512
- *See also* Transformation technology

Badenes-Perez, F. R., 49
Baer, J., 391
Baggett, J. R., 267, 271
Bailey, C. D., 6, 8, 35
Bailey, M. A., 509
Bailey, T. L., 83
Bain, D. C., 276
Bajaj, Y. P. S., 199
Baker, B. J., 101
Bakhetia, D. R. C., 49
Bakker, E. G., 175–176, 356, 363
Balana-Alcaide, D., 356
Balasubramanian, S., 129, 131, 137
Balding, D. J., 360
Baltisberger, M., 17
Bancroft, I., 171–191, 246, 250, 419, 6, 617–627
Banga, S. K., 45
Banga, S. S., 45, 331, 337
Bansal, V. K., 48
Barbarea verna
- chemical traits, 43

Barbour, M. G., 46
Barciszewski, J., 88
Barfield, D. G., 509
Bari, R., 107
Barker, D. H., 238
Barker, G. C., 298, 303
Barkley, N. A., 538
Barkoulas, M., 378, 627
Barret, P., 302
Barrett, S. C. H., 374, 382, 403
Barrett, T., 588
Barry, S. J., 476
Barsby, T. L., 513
Bartel, B., 94, 106, 547
Bartel, D. P., 94, 106–107
Bartelheim, S., 378–379
Bartholmes, C., 384–385, 508
Bateman, A. J., 391, 403
Bateman, R. M., 380
Baulcombe, D. C., 68, 108, 52, 568–569
Baumberger, N., 569
Bauwe, H., 44
Baxter, C., 132, 136
Beaulieu, J., 205, 35, 621
Beaumont, M. A., 360
Becher, M., 625
Bechsgaard, J., 391, 401, 404
Bechtold, N., 508, 515, 539–541
Beck, J. B., 126–127, 139
Becker, H. C., 43, 200, 219, 307
Beckmann, J. S., 219
Beilstein, M. A., 6–8, 11, 414, 486
Bekkaoui, F., 602
Bell, C. J., 129
Bell, J. M., 310
Bellostas, N., 41
Benfey, P. N., 125, 137
Ben-Ghnaya, A., 46
Bennett, M. D., 197
Bennett, R. N., 43
Bennetzen, J. L., 178, 185, 418
Bent, A. F., 508, 515, 540–541
Bent, A., 507, 512
Bento, M., 201, 205
Benton, H. P., 482
Bentsink, L., 129, 131, 136–138, 240
Berezikov, E., 92
Berg, O. G., 71, 80
Bergelson, J., 133
Berr, A., 351
Berteroa, 4
Bertrand, C., 74
Besnard, G., 16

Best, M. M., 311
Bett, K. E., 166
Bevan, M., 512
Bhalla, P. L., 509, 514
Bhat, M. A., 332
Bhat, S. R., 43, 45, 47–49, 332
Bhatia, C. R., 237
Bhinu, V. S., 310
Bhowmik, T. P., 247
Bigfoot genes, 87
Bilyeu, K. D., 538
Biogenesis
  microRNAs (miRNAs), 91–92
Bioinformatics resources
  A. thaliana, 585
    gene and protein analysis resources, 590–592
    gene interactions and pathways, 592
    genome browsers, 586–588
    integration of data, 594
    metabolomic data, 594
    small RNA databases, 593
    transcriptomics data, 588–590
  Brassica species, 597–599
    B. rapa genome sequencing project (BrGSP), 603–611
    directory of current web resources, 599
    EST resources, transcript assemblies and microarrays, 599–602
    future developments, 613
    next generation sequencing (NGS), 611–613
  See also Germplasm resources; Molecular resources
Biologically generated insertional mutants
  Arabidopsis reverse genetics, 540
    Agrobacterium T-DNA mediated insertional mutations, 540–544
    transposon mediated insertional mutations, 544, 546
Biomarkers
  B. oleracea, 279
  See also Molecular markers
Biosynthetic pathways manipulation using RNAi
  RNAi in *Brassica* species, 578
Birnbaum, K., 590
Birney, E., 613
Biscutella, 16
Biscutella laevigata
  polyploidy and hybridization, 15
Biscutelleae, 8
Bisht, N. C., 165, 243, 250, 328, 335
Bjorkman, T., 270
Black, L. D., 36, 201, 294
Black rot
  B. oleracea disease, 276
  genetics and mapping disease resistance in *B. rapa*, 248–249
Blanc, G., 10, 162, 174, 184, 198, 564, 620
Blanchette, M., 83
BLASTN, 599
Block, G., 272
Bloor, S. J., 492
Bodnaryk, R. P., 49
Boechera, 45
  Boechera stricta
    karyotype evolution in aspects, 22
    polyploidy and hybridization, 15–18
    sequence comparisons between *A. thaliana* and, 181–184
  chromosomes and chromosome number variation, 13
  karyotype evolution in aspects, 22
  polyploidy and hybridization, 15
    Boechera divaricarpa, 17–18
    Boechera holboellii, 17–18
Boechera drummondii
  sequence comparisons between *A. thaliana* and, 174
Boffelli, D., 86
Boggs, N. A., 402, 404–405
Bohuon, E. J., 155, 158
Boivin, K., 18, 166, 177–178, 377
Bolle, C., 548
Bolting
  B. oleracea genetics, 271
  B. rapa, 233–234
Bomblies, K., 132, 135
Bonaventure, G., 494–495
Booth, E. J., 310
Borchert, G. M., 91
Borevitz, J. O., 129, 136, 445
Böttcher, C., 469–496
Bouchard, C. J., 47
Bouchez, D., 533, 542
Boukema, I., 440, 447
Bowers, J. E., 10, 12, 162, 174, 198
Bowers, W. S., 44
Bowman, J. L., 384, 627
Boyd, R. S., 46
Boyes, D. C., 400
Brachycarpea, 3
Bradshaw, J. E., 294
Brandle, J. E., 306
Brar, D. S., 340

Brassica
A. *thaliana* natural variation genetic information use in, 137–139
bioinformatics resources, 597–611
B. rapa genome sequencing project (BrGSP), 603–611
directory of current web resources, 599
EST resources, transcript assemblies and microarrays, 599–602
future developments, 613
next generation sequencing (NGS), 611–613
Brassica A genomes
B. rapa as A genome reference, 414–415
sequence relationships between *B. rapa* and *B. napus*, 188
Brassica C genomes
sequence relationships between *B. oleracea* and *B. napus*, 188–189
chromosomes and chromosome number variation, 13, 15
crops, 5
cytoplasmic male sterility, 45
disease resistance, 48–49
drought tolerance, 47
genome and chromosomal
collinearity, 18
evolution, 11
genome evolution and organisation within, 154
genome sequencing technology, 621–623
genome size analysis, 172–173
germplasm resources, 448–449
DFFS, 450
educational resources, 451
emerging resources, 451
mapping populations, 450–451
herbicide resistance, 47–48
insect and nematode resistance, 49–50
molecular resources, 451–455, 460
plant regeneration and transformation, 45
polyploids, 196–197
polyploidy contribution to genome evolution, 156
B. juncea, 158
B. napus, 158, 159
B. oleracea, 158
resynthesized polyploids, 199–200
epigenetic changes, 204
future research, 205–206
genetic changes, 201–203
phenotypic effects, 200
proteome changes, 204–205
reverse genetics (RNAi), 561
background, 568, 569
conferring tissue specificity in RNAi approaches in *Brassica* species, 578–579
gene function studying throughout *Brassica* development, 578
gene silencing approaches in *Brassicaceae*, 570–577
metabolic engineering and manipulation of biosynthetic pathways using RNAi, 577
RNAi in *Brassica* species, 577–579
RNAi/PTGS mechanisms, 570–577
sRNA classes associated with PTGS, 569–570
reverse genetics (TILLING), 562
amplicon selection, 567–568
EMS-induced mutations and genetic code, 563–564
M_1 and M_2 population structure, 566
mutagen dosage optimization, 565
mutation linking to phenotype, 568
mutation load, 564–565
TILLING platform setting, 567
salt and heavy metal tolerances, 46
transcriptome analysis prospects, 623–626
See also Arabidopsis; Brassicaceae
Brassica carinata, 45
crop species, 38
polyploidy and hybridization, 15–18
resynthesized polyploids, 199–200
future research, 205
phenotypic effects, 200
Brassica cretica
chemical traits, 43
coenospecies and cytodemes, 264
Brassica elongata
chemical traits, 43
Brassica fruticulosa
under-utilized crucifer crops, 41
Brassica hilarionis
morphological traits, 42
Brassica incana
chemical traits, 43
Brassica juncea, 45
allopolyploid, 324–326
available variability, 326
east European origins, 326–327
Indian origins, 326–327

Brassica juncea (*cont.*)
crop species, 36
morphological traits, 42
genetics, 324
future prospects/inheritance pattern, 340
genome mapping, 327–331
genetics (traits genetics and mapping), 331
agronomic and yield traits, 337–338
disease resistance, 338–339
erucic acid content and oil content, 332–334
glucosinolates, 334–336
seed coat colour, 336–337
polyploidy
and hybridization, 15–18
contribution to genome evolution, 158
resynthesized polyploids, 199
future research, 205
genetic changes, 201
phenotypic effects, 200
Brassica macrocarpa
morphological traits, 42
Brassica napus, 2, 45
A. thaliana as model genome for, 160
comparative mapping for trait analysis, 165–166
ancestral crucifer karyotype (ACK), 22
comparative mapping extension to related species, 166
crop species, 37
genetics, 291, 312
flowering time variation (winter vs. spring), 304–306
hybrids, population development and seed yield improvement, 306–309
modified FA and specialty oil and meal profiles, 302–303
oil content, 303–304
oil quality related traits, 309–311
genomic tools
genetic linkage maps, 297–302
molecular marker technology, 295–297
oilseed crop, 295
origin and domestication, 293–294
polyploidy
and hybridization, 15–18
contribution to genome evolution, 158–160
resynthesized polyploids, 200
epigenetic changes, 204
future research, 205–206
genetic changes, 201–203
phenotypic effects, 200
proteome changes, 204–205
sequence comparative analysis
B. napus accessions, 189
sequence comparative analysis between *Brassica* genomes
B. oleracea and *B. napus* (*Brassica* C genomes), 188–189
B. rapa and *B. napus* (*Brassica* A genomes), 187–188
sequence comparisons between *A. thaliana* and, 188–189
Brassica nigra
chemical traits, 43
crop species, 37
polyploidy
and hybridization, 15–18
resynthesized polyploids, 200–201
Brassica oleracea, 2, 45
coenospecies and cytodemes, 264–265
crop species, 36–37
morphological traits, 42
crops importance, 263
broccoli, 263
Brussels' sprouts, 263
cabbage, 263
cauliflower, 263
kale, 263
kohlrabi, 263
genetics, 263, 266, 282
aneuploidy, 278
broccoli traits, 269–270
brussel sprouts traits, 269
cabbage traits, 267
cauliflower traits, 269–270
chromosome number variation, 277–278
disease and insect resistance, 275–277
flower color and bolting, 271
genomic tools (markers, genetic and physical maps), 279
kale traits, 268
kohlrabi traits, 268–269
map development in, 279–280
monoploids/microspore culture, 278
polyploidy, 277
secondary metabolites (glucosinolates and carotenoids), 272–275
synteny maps, 280–281
genomics, 281
interspecific and intergeneric hybridizations, 266

origin, distribution and domestication, 263–264
polyploidy, 277
and hybridization, 15–18
contribution to genome evolution, 158
resynthesized polyploids, 199–200
genetic changes, 201, 203
phenotypic effects, 200
sequence comparative analysis between *Brassica* genomes
B. oleracea and *B. napus* (*Brassica* C genomes), 188–189
sequence comparisons between *A. thaliana* and, 182–184
Brassica rapa, 46
Brassica A genome reference, 414–415
crop species, 36, 37
disease resistance mapping
black rot resistance genes, 248–248
clubroot resistance genes, 244–246
turnip mosaic virus genes, 249
white rust resistance genes, 247–248
genetics, 216–217, 251
breeding and trait genetics, 217–218
comparative mapping and identification of candidate genes for important traits, 249–250
disease resistance mapping, 244–249
molecular markers, 218–233
genome sequencing, 414
BAC libraries and BAC-end sequences, 420
euchromatic regions sequencing by clone-by-clone strategy, 427–428
expressed sequence tags and transcriptome analysis, 424–427
genetic map, 422
information resources, 427
perspective, 433
physical map, 422–424
process, 432–433
repetitive sequences of *B. rapa*, 416–419
seed BAC selection, 428–430
seed BAC sequences characteristics, 430–432
triplicated nature of genome, 419–420
genome structure
cytogenetic study, 415–416
repetitive sequences, 416–419
triplicated nature of genome, 419–420
genomic resources
BAC libraries and BAC-end sequences, 420
genetic map, 422
information resources, 427
physical map, 422–424
mapping
abiotic stress tolerance, 243–244
agronomic and morphological traits, 236–237
anthocyanin pigmentation, 238–239
bolting, flowering and vernalizaton requirement, 233–234
disease resistance, 244–249
embryogenic ability in microspore culture, 239–240
fatty acid composition, 241
glucosinolates traits, 242–243
mineral accumulation, 240–241
plant height, 234–235
root traits, 235–236
seed coat color, 237–238
self incompatibility, 239
molecular markers, 218
and molecular genetics maps development in *B. rapa*, 219–232
and trait genetics, 230, 233
in diversity studies, 219
polyploidy and hybridization, 15–18
resynthesized polyploids, 199
future research, 205
genetic changes, 201
phenotypic effects, 200
sequence comparative analysis between *Brassica* genomes
B. rapa and *B. napus* (*Brassica* A genomes), 186–187
sequence comparisons between *A. thaliana* and, 181–187
Brassica rapa genome sequencing project (BrGSP)
automated annotation, 608–611
BAC end sequencing, 603
bioinformatic selection of seed BACs, 605–607
methodology, 603
physical maps and informatics, 603–605
sequencing programme coordination, 607–608
See also Arabidopsis Genome Initiative (AGI)
Brassica rupestris
chemical traits, 43

Brassica villosa
chemical traits, 43
morphological traits, 42
Brassicaceae, 1, 2, 5
A. thaliana as model genome for, 160
A. thaliana reference genome, 173–174
across A, B and C genomes, 162
comparative mapping for trait analysis, 165–166
conserved chromosome landmarks, 164
rearrangement hotspots, 164–165
sequence comparative analysis, 172–173
cold tolerance, 46–47
comparative analysis at sequence level, *see* Sequence comparative analysis
comparative mapping extension to related species, 166
B. napus, 166
Raphanus sativa, 166
comparative mapping in, 153–167
sequenced genomes analysis promise, 167
trait analysis, 165–166
crop species, 34–35
as sources of agronomic and economic traits, 42–45
B. carinata, 38
B. juncea, 39
B. napus, 38
B. nigra, 38
B. oleracea, 36–37
B. rapa, 37–38
breeding systems and apomixis, 45
C_3–C_4 photosynthesis, 44
Camelina, 39
chemical traits, 43–44
Crambe, 39–40
cytoplasmic male sterility, 45
Eruca, 40
morphological traits, 42
Raphanus, 40
Sinapis, 41
taxonomy and genetic relationships, 35–36
under-utilized crucifer crops, 41
disease resistance, 48–49
drought tolerance, 47
family, 2–5
genome and chromosomal evolution, 12
ancestral crucifer karyotype, 20–22
chromosome homeology revealing through CCP, 18–20
chromosomes and chromosome number variation, 13–15
genome and chromosome collinearity, 18
genome size (GS) variation, 12–13
karyotype evolution, 22–24
polyploidy and hybridization, 15–18
whole-genome duplications and age of family, 10–12
herbicide resistance, 47–48
insect and nematode resistance, 49–50
mating-type dimorphism, 403–405
molecular resources
expressed sequence TAGs And cDNA clones, 462
genomic library/clone resources for *Brassicaceae*, 459
resources associated with *Brassica* sequencing projects, 459
phylogenetic position and recognition of infrafamiliar taxa, 5–10
Alysseae, 6
Anchonieae, 8
Aphragmeae, 8
Arabideae, 6
Biscutelleae, 8
Brassiceae, 6, 10
Buniadeae, 8
Calepineae, 8
Camelineae, 8
Conringieae, 8
Dontostemoneae, 8
Erysimeae, 8
Euclidieae, 6, 8
Heliophileae, 6
Hesperideae, 6
Lepidieae, 6
Malcolmieae, 8
multigene phylogenies, 6
Schizopetaleae, 6, 10
Sisymbrieae, 6
plant regeneration and transformation, 45–46
salt and heavy metal tolerances, 46
self-incompatibility in, 390–391
future prospects, 405–406
loss of SI and self-fertile mode of mating, 403–405
self pollen recognition/inhibition mechanism (cytological responses), 392

self pollen recognition/inhibition mechanism (molecular studies), 393–400
S-haplotype structure, suppressed recombination, and diversification, 400–403
sequence comparative analysis
B. napus accessions, 188
B. oleracea and *B. napus* (*Brassica* C genomes), 188–189
B. rapa and *B. napus* (*Brassica* A genomes), 188–189
sequence comparisons between *A. thaliana* and, 181
B. napus, 185–187
B. oleracea, 184–185
B. rapa, 182–184
taxonomy and genetic relationships
Brassicinae, 35–36
Moricandiinae, 35
Raphaninae, 35
transformation technology in, 506–508
Agrobacterium rhizogenes, 510
Agrobacterium tumefaciens, 508–510
as research tool, 517
bacterial strains and plasmids, use of, 511–512
chloroplast transformation, 511
direct uptake transformation methods, 511
floral dipping/micro-injection, 515–516
hyperhydricity and tissue necrosis, 514–515
shoot regeneration
transgenics selection, 516
whole genome duplications, 197–199
See also Arabidopsis; Brassica
Brassicales, 10
Brassiceae, 10
karyotype evolution in aspects, 24
Brassiceae tribe, 6
Brassicinae
Brassica crop species, 33
Braun, D. M., 398
Braun, R. H., 518
Braya
polyploidy and hybridization, 15
Breeding
B. rapa genetics, 217–218
Brassicaceae crop species, 45
Bringmann, G., 489
Brismar, K., 278, 281
Broadhurst, D. I., 483
Broadley, M. R., 240
Broccoli, 262
traits, 269–271
See also Brassica oleracea
Brodersen, P., 93–94, 107, 552, 569–570, 576
Brown, A. H. D., 356
Brown, G., 298
Brown, G. G., 45, 179
Brown, P. D., 489
Brun, H., 48
Brussel sprouts, 263
traits, 269
Buczacki, S. T., 244
Budin, J. T., 43
Buell, C. R., 623
Buente, R., 50
Buggs, R. J. A., 202
Bulow, L., 77, 80
Buniadeae, 8
genome size (GS) variation, 12, 13
Bunias
Bunias orientalis, 12
chromosomes and chromosome number variation, 13
genome size (GS) variation, 12–13
Buratowski, S., 74
Burch-Smith, T. M., 571–572
Burdzinski, C., 228, 230, 238
Burge, C., 609
Burgess, B., 279
Burnett, L., 515
Burns, J., 303, 333–334, 339
Burton, W., 326
Busch, M. A., 86
Butler, J. E., 71, 74
Buzza, G. C., 295, 306
Byzova, M., 517, 569, 573, 578

C

CAAT-box, 75
Cabbage, 262
traits, 267
See also Brassica oleracea
Cabrillac, D., 398
C_3–C_4 photosynthesis
Brassicaceae crop species, 44
Cai, C. C., 300, 306
Caicedo, A. L., 139, 379
Cakile
morphological traits, 42
Calepina irregularis
polyploidy and hybridization, 15–16

Calepineae, 8
  karyotype evolution in aspects, 22
Camargo, L. E. A., 276
Camelina sativa, 39, 45
  chemical traits, 43–44
Camelineae, 8
  ancestral crucifer karyotype (ACK), 20–22
  chromosomes and chromosome number variation, 14
  genome and chromosome collinearity, 18
  genome size (GS) variation, 12
  karyotype evolution in aspects, 22, 23
Campbell, M. A., 609
Campisi, L., 533
Campos, H., 264, 272
Candidate genes
  identification for *B. rapa* traits, 249–251
CanOLA, *see* Rapeseed *B. napus*
Cao, J., 507, 516
Capparaceae, 5
CAPS markers, 230
Capsella
  genetics, 374–375, 382
    genome and chromosome evolution, 377–378
    traits, 378–384
  traits genetics
    floral size, 379
    flowering time, 378–379
    fruit structure, 383–384
    leaf development, 378
    saltational change in floral architecture, 380–381
    self-incompatibility, 382
  origin
    C. bursa-pastoris, 376–377
    C. grandiflora, 375
    C. rubella, 376
Capsella bursa-pastoris, 376–377
  chemical traits, 43
  origin, 376, 377
  traits genetics, 378–384
  under-utilized crucifer crops, 41
Capsella grandiflora, 375–376
  ancestry, 375
  genome and chromosome evolution, 377
  traits genetics (floral structure and function)
    floral size, 379
    self-incompatibility, 382
Capsella rubella, 2, 376–377
  A. thaliana comparative mapping extension to related species, 16
  genome and chromosome
    collinearity, 18
    evolution, 377–378
  origin, 376
  sequence comparisons between *A. thaliana* and, 176–177
  traits genetics
    floral size, 379
    fruit structure, 383–384
    self-incompatibility, 382
Cap-trapper transcript isolation followed by massively parallel signature sequencing (CT-MPSS), 73
Carcamo, H., 50
Cardamine, 2, 10
  chromosomes and chromosome number variation
    Cardamine angustata, 14
    Cardamine concatenata, 14
    Cardamine diphylla, 14
    Cardamine dissecta, 14
    Cardamine maxima, 14
  karyotype evolution
    Cardamine hirsuta, 23
    Cardamineae, 23
  morphological traits, 42
    Cardamine chenopodiifolia, 42
  polyploidy and hybridization, 15–18
    Cardamine amara, 15–16
    Cardamine digitata, 15
    Cardamine insueta, 17
    Cardamine majovskii, 16
    Cardamine matthioli, 16
    Cardamine rivularis, 17
    Cardamine schulzii, 17
Cardamine kokaiensis
  breeding systems and apomixis, 45
Cardaminopsis petraea, 348
Cardoza, V., 507, 509, 513–515
Carica papaya
  genome and chromosomal evolution, 10
Caricaceae
  genome and chromosomal evolution, 10
Carlson, D. G., 272
Carlson, E. E., 476
Carlson, K. D., 40
Carmell, M. A., 569
Carotenoids
  B. oleracea, 272, 274
Casadaban, M. J., 532
Casselman, A. L., 400
Castric, V., 350–352
Cataldi, T. R. I., 489

Catchpole, G. S., 472
Cauliflower, 262
  traits, 269–271
  See also Brassica oleracea
Cavell, A., 250
Cavell, A. C., 197, 624
Cavin Perier, R., 77
CCA1–binding site (CBS), 84
cDNA clones
  A. thaliana molecular resources, 460
    sequenced full length cDNA, 461
  Brassicaceae molecular resources, 463
Ceplitis, A., 377
Chai, Y-R., 140
Chamira, 3
Chander, H., 49
Chand, J., 338
Chandler, J., 518
Chang, W. C., 80
Chapman, V., 202
Chapple, C. C. S., 488
Charlesworth, B., 352, 360
Charlesworth, D., 348, 351–354, 357, 365, 391, 400–401
Chemically generated mutants, 533–539
  Arabidopsis reverse genetics, 533, 536
  See also TILLING
Chen, F., 492
Chen, B., 299, 308
Chen, B. Y., 223, 237, 307
Chen, C. Y., 48
Cheng, X., 300
Chen, H., 49, 219, 235
Chen, H. F., 206
Chen, H. G., 205–206
Chen, J. L., 333
Chen, L., 49, 299, 308
Chenna, R., 610
Chen, S., 136, 272–273
Chen, W., 299, 308
Chen, X., 104, 205–206
Chen, Y. P., 219
Chen, Z., 492
Chen, Z. J., 197, 201, 203–206, 621
Chernushevich, I. V., 479, 481
Cheung, F., 186–190, 198–199, 203, 298–299, 328, 332–334, 339, 609, 621
Cheung, W. Y., 158
Chèvre, A. M., 17, 218, 277
Chiang, C. M., 71, 74
Chiang, M. S., 263, 266, 275, 278
Chiang, S. F., 71, 74
Chi, G. L., 514–515
Chiou, T. J., 107
ChIP-chip technology, 77–78, 88
  See also Cis-regulatory elements (CREs)
Chi, S. W., 94
Chiu, L., 270
Chloroplast transformation, 511
  See also Transformation technology
Cho, H., 507
Choi, S., 129, 141
Choi, S. R., 219–220, 227, 230, 232, 251, 296, 419, 421–422
Cho, K., 186, 188
Cho, K. S., 244
Chookajorn, T., 396, 398, 402–403
Chopra, V. L., 42, 45, 266, 324, 512
Choudhary, B. R., 336
Christey, M. C., 510, 513, 518
Christianson, J. A., 329–330
Christie, R. G., 249
Christoffers, M. J., 47
Chromosome evolution
  Capsella, 377–378
  genome and chromosomal evolution (*Brassicaceae*), 12
    ancestral crucifer karyotype, 20–22
    chromosome homeology revealing through CCP, 18–20
    chromosomes and chromosome number variation, 13–15
    genome and chromosome collinearity, 18
    genome size (GS) variation, 12–13
    karyotype evolution, 22–24
    polyploidy and hybridization, 15–18
    whole-genome duplications and age of family, 10–12
Chromosome number variation
  B. oleracea genetics
    aneuploidy, 278
    polyploidy, 277
  See also Chromosome evolution
Chrungu, B., 48
Chua, N. H., 100
Chuang, C. F., 573–574
Chuong, P. V., 240
CID (non-targeted metabolites profiling) molecular structures, 483–488
Cifuentes, M., 199, 202, 205
Cipollini, D., 44
Cis elements
  A. thaliana genomics, 68–70

Cis-regulatory elements (CREs), 69
computational approaches (*A. thaliana* genomics), 79–87
known sites detection, 79
unknown sites detection, 79–80
experimental approaches (*A. thaliana* genomics), 76–77
See also ChIP-chip technology
Cis-regulatory modules (CRMs), 70
Claeys, M., 492
Clamp, M., 610
Clark, R. M., 124, 128, 175–176, 588
Clarke, J. D., 272
Clauss, M. J., 43, 129, 348, 352, 353–363
Cleomaceae, 5, 10, 12
Clones
cDNA
A. thaliana molecular resources, 460
Brassicaceae molecular resources, 463
clone-by-clone strategy, 427–428
GST
A. thaliana molecular resources, 462
gene specific TAG clones, 462
ORF
A. thaliana molecular resources, 462
sequenced full length cDNA and ORF clones in entry vectors, 462
P1, 456–457
P1 libraries, filters and clones sources, 457
RNAi, 462
transformable BAC libraries and clones, 458
yeast artificial chromosome libraries and, 457
Clerkx, E. J., 128, 135
Clerkx, E. J. M., 128, 135
Clough, S. J., 508, 515, 539–540
Cloutier, S., 299
Clubroot disease
B. oleracea disease, 275
genetics and mapping disease resistance in *B. rapa*, 244–246
Cochlearia
chromosomes and chromosome number variation, 13
polyploidy and hybridization
Cochlearia anglica, 15
Cochlearia officinalis, 15
under-utilized crucifer crops
Cochlearia arctica, 41
Cochlearia officinalis, 41
Coelho, P. S., 276
Cogan, N. O. I., 509
Cohen, S. N., 532
Colbert, T., 536, 538
Cold tolerance
Brassicaceae, 44, 46
Collier, R., 510
Collinearity
defined, 156
See also Comparative mapping
Collins, F. S., 296
Color
flower (*B. oleracea* genetics), 271
seed coat (*B. juncea* traits genetics), 336–337
Comai, L., 20, 196–201, 203–204, 538
Comes, H. P., 355
Comparative chromosome painting (CCP), 11
chromosome homeology revealing through, 18, 22
Comparative genenome analysis, 172
See also Sequence comparative analysis
Comparative mapping
and candidate genes identification for *B. rapa* traits, 249–251
B. oleracea to *A. thaliana*, 279–280
basics, 156–157
RFLP markers, 154
definitions
collinearity, 156
homoeology, 156
homology, 156
orthologues and paralogues, 156
polyploidy, 156
synteny, 156
in *Brassicaceae*, 154–159
A. thaliana as model genome for *Brassicaceae*, 160–164
ancestral hexaploid genome, 160
extension to related species, 166–167
for trait analysis, 165–166
polyploidy (inter-specific hybridisation) contribution to *Brassica* genome evolution, 158–160
sequenced genomes analysis promise, 167
See also Genomics
Compton, D. L., 44
Conery, J. S., 157, 619
Conner, J. A., 396
Conringia orientalis
morphological traits, 42
under-utilized crucifer crops, 41

Conringieae, 8
  karyotype evolution in aspects, 22
Constantinescu, O., 276
Cook, D. R., 566
Cooper, J. L., 538, 565
Copenhaver, G. P., 174, 416
Cordewener, J. H. G., 278
Core promoter
  A. thaliana genomics, 71–76
Co-suppression RNA
  RNAi/PTGS mechanisms
    gene silencing approaches in *Brassicaceae*, 572
Coulhart, M., 199
Couvreur, T. L. P., 7, 12
CpG islands, 75
Craft, J., 576
Craig, B. M., 46, 302, 456, 464
Craigon, D. J., 84–86, 589
Crambe
  chemical traits, 43
  morphological traits, 42
  polyploidy and hybridization, 15
Crambe abyssinica, 39–40, 46
  chemical traits, 43–44
Crambe hispanica, 40, 46
  chemical traits, 43
Cravatt, B. F., 476
Cremolobus peruvianus
  crop species
    morphological traits, 42
Crepet, W. L., 380
Crete, R., 275
Crist, D., 437–466
Crop species, *see under Brassicaceae*
Crucifers, *see Brassicaceae*
Crucihimalaya wallichii, 23
Crute, I. A., 49
Crute, I. R., 244–245
Cui, W., 44
Curto, G., 44
Cuyckens, F., 492
Cytodeme
  defined, 264
Cytogenetic study
  B. rapa genome, 415–416
  self-incompatibility in *Brassicaceae*, 392
Cytoplasmic male sterility
  Brassicaceae crop species, 44–45

D
Dal-Corso, G., 45
Dalin, P., 362
Dalmais, M., 565
Dalmay, T., 569, 572
Damania, A. B., 294
Damm, B., 508, 511
Daniell, H., 511
Das, S., 219
Dass, H., 199
Data analysis
  LC/MS-based non-targeted metabolites profiling
    data acquisition, 476–481
    data extraction, 481–483
D'Auria, J. C., 479, 488–489
Daun, J., 43
Davey, M. P., 363
David, C., 510
Davis, A. R., 42
Davis, C. D., 275
Davis, R. W., 531, 572
Davuluri, R. V., 77
Daxenbichler, M. E., 43
Dean, C., 128
Dearnaley, J., 392
De Block, M., 509, 512, 515, 517
De Bodt, S., 10, 12
DeBodt, S., 196–198
De Candolle, A. P., 4
DeCook, R., 136
Deinzer, M. L., 492
Delourme, R., 298, 303–304, 333
Delwiche, P. A., 247
Demeke, T., 219
Denby, K. J., 136
Deng, Z., 179
De novo motif discovery, 80–81
Dentaria
  chromosomes and chromosome number variation, 14
Deol, J. S., 45
De Quiros, C., 272
Derivatization
  non-targeted profiling of semi-polar plant metabolites, 481
Descurainia sophia
  crop species chemical traits, 43–44
  under-utilized crucifer crops, 41
Descurainieae
  ancestral crucifer karyotype (ACK), 20–22
  karyotype evolution in aspects, 23
Desfeux, C., 515
Deslandes, L., 128, 134
Desloire, S., 179
Devaiah, S. P., 493

De Vos, C. H., 473
Devos, K., 13, 154
Deyholos, M. K., 548
Dhawan, A. K., 45
Dhinrga, A., 511
Dickinson, H., 392
Dickinson, H. G., 392
Dickson, M. D., 48–49, 266–267, 270, 276–277
Didiano, D., 94
Diederichsen, A., 41, 264–265, 271, 298
Dierig, D. A., 41, 45–46
Dierking, E. C., 538
Diers, B. W., 306–307
Dietrich, M. R., 380
DiMichele, W. A., 380
Dinesh-Kumar, S. P., 464
Ding, Y., 423
Diploidization, 11, 16
  defined, 618
  See also Polyploidy
Diploidized paleopolyploids, 196
Diplotaxis
  chromosomes and chromosome number variation, 14
  crop species
    morphological traits, 42
    chemical traits, 43
  under-utilized crucifer crops, 41
Diplotaxis berthautii, 45
Diplotaxis erucoides, 45
  crop species chemical traits, 44
Diplotaxis harra, 45
Diplotaxis muralis, 45–46
  polyploidy and hybridization, 16
Diplotaxis siifolia, 45
Diplotaxis tenuifolia, 43
  crop species chemical traits, 43–44
  polyploidy and hybridization, 16
Diplotaxis viminea
  crop species chemical traits, 41
  polyploidy and hybridization, 16
Dipping
  floral, 515–516
  See also Transformation technology
Direct uptake transformation methods, 510
Disease resistance
  A. lyrata genetics, 362–363
  B. juncea traits genetics, 338–339
  B. oleracea, 275–276
    type A, 275
    type B, 275
  B. rapa
    black rot resistance genes, 248–249
    clubroot resistance genes, 244–246
    turnip mosaic virus resistance genes, 249
    white rust resistance genes, 247–248
  Brassicaeae, 48–49
Distal promoter, *see* Proximal and distal promoter
Diversity Fixed Foundation Sets (DFFS), 450
  See also Germplasm resources
Dixelius, C., 340
Dixit, R., 398
Dixon, G. R., 275
Dixon, R. A., 492
DNA binding domains (DBD), 69
Dobeš, C., 17
Dobigny, G., 24
Dodd, I. C., 238
Dontostemoneae, 8
Dosdall, L. M., 50
Doubled haploid (DH) production
  B. oleracea genetics, 278
Double-stranded RNA (dsRNAs), 108
Downey, R. K., 237, 302
Downstream core element (DCE), 71
Downstream promoter (DPE), 71
Downy mildew
  B. oleracea disease, 275–276
Doyle, J. J., 131, 138, 197, 201
Draba, 2, 10
  polyploidy and hybridization, 15–18
    Draba aizoides, 17
    Draba ladina, 17
    Draba tomentosa, 17
Dressiantha
  genome and chromosomal evolution, 12
Drought tolerance
  Brassicaceae, 47
Duan, C. G., 577
Dubcovsky, J., 197
Du, G., 552
Dugas, D. V., 94, 107
Duijs, J. G., 278
Dunn, W. B., 479
Dutta, I., 46
Duplications, 197
  See also Whole genome duplications
Dvorak, J., 197

E

Eamens, A., 569
Eapen, S., 43, 515

Earle, E. D., 513
Earley, K., 621
Early Fuji resistance
  B. oleracea disease, 276
Eason, J. R., 517, 572
Easow, G., 94
Eastburn, D. M., 49
Easterly, N. W., 14
Ebrahimi, A. G., 247
Eckenrode, C. J., 277
Ecker, J. R., 129, 506, 529, 540, 572
Ecker, R., 41, 43
Ecke, W., 303–304, 333–334
EcoTILLING
  advantages and disadvantages, 538
  Arabidopsis mutation analyses, 539
Edger, P. P., 197–198
Edwards, D., 75, 327, 339
Edwards, M. D., 247
Edwardson, J. R., 249
Eilam, T., 201, 205
Eimert, K., 511
El-Assal, S. E-D., 129, 131, 137
Electrospray ionization (ESI)
  ESI-MS based profiling approaches (lipids), 494
  non-targeted metabolites profiling
    data acquisition aspects, 476–481
    data extraction aspects, 481
Electrophoresis, proteins
  stigma/pollen determinants of SI, 393
Elemento, O., 86
El-Khatib, A. A., 44
Elleman, C. J., 392
Ellestrom, S., 275
Ellis, P. R., 49–50
El-Lithy, M. E., 128, 135
Elsey, K. D., 277
Embryogenic ability in microspore culture
  Brassica rapa, 239–240
Emerson, B. C., 361, 363
Enarthrocarpus lyratus, 45
Endoreduplication, 196
Enns, L. C., 538
Epigenetic changes
  in resynthesized polyploids, 204
Erickson, D. B., 40
Erickson, L. R., 324
Ericson, L., 349, 361
Eruca sativa, 40
Eruca vesicaria, 43–46
  crop species, 40
    chemical traits, 43–44
Erucastrum canariense, 45
Erucastrum gallicum, 45
Erucastrum
  crop species chemical traits, 43
    Erucastrum cardaminoides, 43
Erucic acid content
  B. juncea traits genetics, 332–333
  erucic acid content and oil content, 332–334
Erysimeae, 8
Erysimum, 10
  chromosomes and chromosome number variation, 14
  Erysimum cheiri chemical traits, 43
Ethylene inhibitors
  hyperhydricity and tissue necrosis transformation method and, 514–516
Ethylmethanesulfonate (EMS)
  mutagenesis with, 533, 536
  TILLING and *Brassica* species reverse genetics, 562
    EMS-induced mutations and genetic code, 563–564
    mutation load, 564
Euchromatic regions sequencing (*Brassica*)
  clone-by-clone strategy for, 427–428
Euclidieae, 8
Euclidieae tribe, 6
Eutrema wasabi, 41
Eutremeae, 24
Evening Element (EE), 84
Excoffier, L., 358
Expressed sequence tags (ESTs)
  A. thaliana molecular resources, 460
  Brassicaceae
    B. rapa, 424–427
    molecular resources, 463
  EST resources, transcript assemblies and microarrays
    Brassica species bioinformatics resources, 599–602
Expression analyses
  Arabidopsis reverse genetics
    array related expression data microarrays, 531
    reporter gene fusions and promoter, gene and enhancer trap lines, 532–533
Extraction
  non-targeted profiling of semi-polar plant metabolites, 473
Eyal, Y., 535

Eysimum cheiri
crop species chemical traits, 44
under-utilized crucifer crops, 41
Eysimum diffusum, 41

F
Fabijanski, S. F., 278
Facciola, S., 39
FAD2 gene
Arabidopsis, 242
Fagbenro, O. A., 40
Fahey, J. W., 43, 243, 272
Fahlgren, N., 105, 110
Falentin, C., 302
Falk, K. C., 34, 39
Fang, Y., 93
Farinho, M., 276
Farmer, E. E., 493
Farnham, M. W., 261–282
Farnham, M., 50
Farrell, J. A., 49
Farsetia somalensis, 42
Fatehi, J., 276
Fatty acid (FA)
modified FA and specialty oil *B. napus* profiles, 302–303
composition and *Brassica rapa*, 241
Federico, M. L., 291–312
Feldman, K. A., 508
Feldman, M., 201, 205
Feldmann, K. A., 447
Felippes, F. F., 105
Fernie, A. R., 470–471, 487
Ferrándiz, C., 384, 518
Ferreira, M. E., 154, 298–299, 305–306
Fiber and tannin
B. napus, 310
Fiebig, A., 178–179
Fiehn, O., 470, 473, 481, 486
Field, D., 609
Figdore, S. S., 275
Filiault, D. L., 131, 137
Finley, J. W., 275
Fitz Gerald, J. N., 135
Flannery, M. L., 199
Flibotte, S., 604
Floral dipping
transformation method, 515–516
Floral structure and function
Capsella traits genetics
floral size, 379
saltational change in floral architecture, 380–381
self-incompatibility, 382
Flower color
B. oleracea genetics, 271
Flowering
Brassica rapa, 233–234
plants, 196
polyploidy phenomenon in, 196–197
time
A. lyrata genetics, 364
B. napus, 304–306
Capsella traits genetics, 378–379
See also Polyploidy
Fluorescence in situ hybridization (FISH)
B. oleracea synteny maps, 281
for *B. rapa* cytogenetic study, 416
Fobert, P. R., 533
Fobis-Loisy, I., 382–383, 508
Foisset, N., 296, 299
Force, A., 198, 418, 620
Ford, C. S., 325
Fourmann, M., 140, 302, 332
Foxe, J. P., 353, 376, 380, 382, 403
Fractionation
non-targeted profiling of semi-polar plant metabolites, 474
Francis, A., 43
François, O., 127
Franco-Zorrilla, J. M., 106, 506, 577
Franklin-Tong, V. E., 391, 393
Fransz, P., 14
Franzke, A., 3, 6–8, 11–12, 374
Fray, M. J. P., 299
Freeling, M., 85, 197–198
Frenkel, M., 135
Friedt, W., 43, 218, 296
Frost, L. N., 131
Fruit structure
Capsella traits genetics, 383–384
Fu, F., 471, 496
Fu, F.-Y., 140, 300
Fu, J., 471, 496
Fujii, H., 107
Fujimoto, R., 396, 400
Fujisawa, M., 311
Fukui, K., 281
Fusaro, A. F., 569–570
Fu, T. D., 306
Fu, Y., 102

G
Gaeta, R. T., 195–206, 294, 296, 620–621, 625
Gaj, M. D., 45
Gale, M. D., 154
Galil, J., 45
Gandolfo, M. A., 12
Gan, Y., 309
Gao, H. L., 46
Gao, M., 181, 270, 274, 276, 280, 296
Gao, X., 74
Garcia, D., 110
Gardner, P. P., 88
Gas chromatography (GC), 471
  GC-MS based profiling approaches and lipids, 493
Gasic, K., 509, 513–514
Gassmann, W., 133
Gaudeul, M., 355, 359
Gazzani, S., 138, 364
Gehringer, A., 39
Geisler-Lee, J., 592
Gelvin, S. B., 540
Gene silencing mediated functional analysis
  Arabidopsis reverse genetics, 548–549
    antisense lines, 549
    microRNAs and targeted miRNA lines, 552
    RNAi lines, 549
  See also Post-transcriptional gene silencing (PTGS)
Gene specific TAG (GST) clones
  A. thaliana molecular resources, 462
General transcription factors (GTFs), 71, 74
Genetic analysis
  A. thaliana natural variation, 127–137
    eQTL/mQTL/pQTL, 136–137
    genetic information use in *Brassica*, 139–141
  genetic changes in resynthesized *Brassica* polyploids, 201–202
  See also Genome sequencing; Reverse genetics
Genetic maps
  B. juncea, 327–331
  B. napus genomic tools, 297–302
  B. oleracea, 279–280
  B. rapa, 422
Genetically modified (GM) technology, 507
  See also Transformation technology
Genome evolution
  A. lyrata, 353–354
  Brassica species as models for studying genome evolution following polyploidy, 618–621
Genome sequencing
  B. rapa, 414
    BAC libraries and BAC-end sequences, 420
    euchromatic regions sequencing by clone-by-clone strategy, 475–428
    expressed sequence tags and transcriptome analysis, 424–427
    genetic map, 422
    information resources, 427
    physical map, 422–424
    repetitive sequences, 416–419
    seed BAC selection, 428–430
    seed BAC sequences characteristics, 430–432
    sequencing perspective, 433
    sequencing process, 432–433
    triplicated nature of genome, 419–420
  Brassica species, 621, 622–623
  See also Brassica rapa genome sequencing project (BrGSP); Sequence comparative analysis
Genome size (GS) variation
  Arabidopsis thaliana, 172–174
  Brassicaceae genome and chromosomal evolution, 12–13
  Brassica spp., 172–174
Genomic in situ hybridization (GISH), 17–18
Genomics
  A. thaliana, 67–87, 617
    cis elements, 69–71
    core promoter, 71–72
    non-coding RNA (ncRNA), 88–110
    noncoding transcripts, 68
  B. napus
    genetic linkage maps, 297–302
    molecular marker technology, 295–297
  B. oleracea, 281
  B. rapa
    as *Brassica* A genome reference, 414–415
    cytogenetic study, 415–416
    genetic maps, 422
    genomic resources, 420
    physical map, 422–424
    repetitive sequences, 416–419
    triplicated nature of genome, 419–420

Genomics (*cont.*)
Brassica species, 618
as models for studying genome evolution following polyploidy, 618–621
genome sequencing technology, 621–623
transcriptome analysis prospects, 623–626
upcoming model systems, 626–627
See also Comparative mapping
Geococcus pusillus
crop species morphological traits, 42
George, L., 515
German, D. A., 8–9, 35
German, D., 374
Germplasm resources, 440–451
A. thaliana, 440–447
Brassica species, 447–451
Diversity Fixed Foundation Sets (DFFS), 450
educational resources, 451
emerging resources, 451
mapping populations, 450–451
See also Bioinformatics resources; Molecular resources
Gershenzon, J., 243, 335, 479, 488–489
Ge, X. H., 205–206
Ghabrial, S. A., 570
Ghaderian, S. M., 46
Ghildiyal, M., 89, 93, 549, 552
Giamoustaris, A., 272
Giavalisco, P., 473–474, 486
Gilchrist, E. J., 538, 562
Gillott, C., 49
Gilman, J. C., 275
Giranton, J., 398
Girke, T., 624
Gladis, T. h, 264
Gladis, T., 325
Gland, A., 327, 334
Glauser, G., 473, 488
Glimelius, K., 513
Glucosinolates (GSL)
A. thaliana genetics, 272
A. lyrata genetics, 362–363
Brassica traits genetics
B. juncea, 334–336
B. napus, 310
B. oleracea, 272–274
B. rapa, 242–243
Goda, H., 625
Goffman, F., 43
Goldbachia laevigata, 16
Goldberg, S. M. D., 612
Goldsack, C. M. P., 505–518
Göllner, K., 136
Gómez-Campo, C., 35, 42, 44, 199, 217–218, 294, 325
Gong, J. M., 539
Gong, Q., 46, 625
Goring, D. R., 394, 403
Gos, G., 363
Gottlieb, L. D., 196
Graef, G. L., 303
Graham, N., 585–595
Grange, A. H., 486
Grant, I., 306
Grant, J. J., 547
Grant, M. R., 133, 182
Grant, V, 200
Grata, E., 474, 478
Gray, A. R., 264, 271
Greene, E. A., 129, 533, 551, 565
Greenhalgh, J. G., 48
Gregory, S. G., 423
Grewal, H. S., 240
Griffin, J. L., 41
Griffin, T. S., 41
Griffith, M., 46
Griffiths, D. W., 294
Griffiths, P. D., 279
Griffiths, S., 202
Grotewold, E., 72–73, 80, 437–466
Grundy, S. M., 303
Gruner, B., 44
Guan, C. Y., 49
Guan, R., 41
Guan, Z. Q., 46
Gugel, R. K., 34, 39, 48–49
Gugel, R., 34, 40
Guo, H. S., 101
Guo, J. X., 219
Guo, M., 197
Guo, Y.-L., 376, 403
Gupta, A. K., 236
Gupta, S. K., 293
Gupta, V., 241, 328, 509
Gustafson, J. P., 201, 205
Gustafson, P., 201, 205
Gu, T., 398
Gutierrez, L., 101–102

H
Haas, B. J., 174, 609
Haberer, G., 67–111, 174, 184, 620

Ha, C. M., 199, 205–206, 620
Hagemeier, J., 489
Hagenblad, J., 138, 400
Haig, D., 354
Hairpin RNAs (hpRNAs)
ihpRNA contructs, vectors for delivering
pHANNIBAL and pKANNIBAL, 574–575
pHELLSGATE, 575
RNAi/PTGS mechanisms
gene silencing approaches in *Brassicaceae*, 573–574
siRNA-directed gene silencing using RNAi/PTGS mechanisms, 573
Halfhill, M., 507
Halimolobeae
genome size (GS) variation, 12
Halkier, B. A., 243, 335
Hall, A. E., 164, 174, 179, 377
Hall, J. C., 5, 10, 48, 274, 618
Hall, S. E., 179
Hameister, S., 381
Hamilton, A., 569–570
Hamilton, A. J., 68, 108
Hammell, M., 89, 94
Hammer, K., 264, 325
Hammond, J. P., 625
Han, H. P., 249
Hanikenne, M., 178, 627
Han, J., 92
Hansen, L. B., 206
Hanson, B. D., 47
Hanson, S. L., 44
Hansson, B., 350
Han, T. H., 249
Harada, H., 136, 240
Harberd, D. J., 35, 264
Harmer, S. L., 84
Harrabi, S., 495
Harrison, G. E., 417
Harvey, B. L., 302
Hasan, V., 310
Haseloff, J., 550
Hasterok, R., 17, 281
Hauser, M. -T., 131, 138
Hausmann, N., 135
Hausmann, N. J., 135
Hawkins, D. J., 303
Hay, A., 378, 627
Hayek, A., 4
Heap, I. M., 45
Heavy metal tolerances
Brassicaceae, 46
Heazlewood, J. L., 592
Hedge, I. C., 3
Heegeri variant
fruit structure, 383–384
See also Capsella bursa-pastoris
He, F., 126, 139
Hegarty, M. J., 15, 197, 201
Hegeman, A. D., 486
Heidel, A. J., 351–352, 363
Height, plant
Brassica rapa, 234–235
Hein, J., 353, 357
He, J., 298
Heliophila, 3, 6, 10
polyploidy and hybridization, 18
Heliophila scandens
crop species morphological traits, 42
Hellens, R., 512
Helliwell, C., 573–575
Helliwell, C. A., 574–575, 577
Henderson, A. E., 49
Henderson, I. R., 204
Heneen, W. K., 278, 281
Henikoff, S., 538, 562
Hennion, F., 47
Henry, Y., 7, 10–11, 162
Herbicide resistance
Brassicaceae, 47–48
Herron, C., 42, 48
Herve, Y., 263–264
Heslop-Harrison, J., 392
Heslop-Harrison, J. S., 417
Hesperideae, 6
Hesperis matronalis
crop species chemical traits, 43
under-utilized crucifer crops, 41
He, W., 298
Hewson, H. J., 3
Hexaploid genome, ancestral
Brassicaceae, 158
He, Y. T., 219
High-Information Content Fingerprinting (HICF) methods, 423
Hill, C. B., 218
Hill, D. W., 487
Hilson, P., 462, 549, 551
Himelblau, E., 538
Hinata, H., 264
Hinata, K., 36, 38, 45, 198–199, 216–217, 278, 293, 393
Hinds, D. A., 357
Hirai, M., 226, 275
Hirai, M. Y., 140, 471

Hirai, S., 573
Hirschfeldia incana, 42
crop species morphological traits, 42
Hirsch, P., 101
Hobbs, D. H., 136
Hobert, O., 94
Hochheimer, A., 74
Hoekema, A., 511
Hoekstra, H. E., 362
Hoffman, M. H., 349–350, 355
Hoffmann, M. H., 124, 126
Hohn, B., 539
Holland, W. H., 24
Holmes, M. C., 73–74
Homoeologous non-reciprocal transposition (HNRT), 202
B. napus genomic tools, 301
See also Polyploidy
Homoeologous reciprocal transpositions (HRT)
B. napus genomic tools, 301
Homoeology
defined, 156
See also Comparative mapping
Homologous recombination induced mutations
Arabidopsis reverse genetics, 546–547
Homology
defined, 156
See also Homoeology
Hong, C. P., 167, 182
Hong, R. L., 86
Hong, X., 94
Honma, M. A., 546
Honma, S., 271, 275
Hood, E. E., 508, 512
Hooker, W. J., 275
Hopley, C., 487
Hornungia
ancestral crucifer karyotype (ACK), 22
Hornungia alpina
ancestral crucifer karyotype (ACK), 21
Hosaka, K., 265
Hou, B., 511
Hougen, F. W., 294
Howell, B. C., 279, 281
Howell, E. C., 17, 164, 279, 281
Howell, P. M., 164, 298–299, 303, 310
Hruz, T., 590
Huang, B. Q., 41
Huang, H., 78
Hudson, R. R., 357
Hughes, J. D., 83
Hughes, S. L., 249
Huhman, D. V., 492
Hu, J., 298, 302
Hunter, J. E., 276
Hu, Q., 46
Hurka, H., 16, 374, 376, 379, 382
Hüsken, A., 310
Hussiney, H. A., 44
Hu, X., 140, 302–303
Hybridization
Brassicaceae
B. napus, 306–309
B. oleracea, 266–267
genome and chromosomal evolution, 14–24
FISH, 281, 416
intergeneric, 266–267
inter-specific, *see* Polyploidy
GISH, 17–18
Hyperhydricity
transformation method, 517–518

I

Iapichino, G., 41
Iberis, 2, 4
Iberis amara
crop species chemical traits, 44
Ichikawa, T., 547–548, 550
Ignatov, A., 49, 277
Ignatov, A. N., 248
Immunogenetics
stigma/pollen determinants of SI, 393
Inaba, R., 186, 190
Inbreeding depression
A. lyrata, 351–353
Inducible and tissue-specific gene silencing, 575–576
See also Post-transcriptional gene silencing (PTGS)
Inheritance pattern
B. juncea genetics, 340
Iniguez-Luy, F. L., 291–312
Inomata, N., 199
Insect and nematode resistance
B. oleracea, 275–277
Brassicaceae, 49–50
See also Disease resistance
Introgression lines (ILs), 128
Inverted-repeats (IR) -PTGS
siRNA-directed gene silencing in *Brassicaceae*, 573
See also Hairpin RNAs (hpRNAs)
Isaac, G., 493, 495

Isatideae
karyotype evolution in aspects, 24
Isatis tinctoria
polyploidy and hybridization, 15
Isothiocyanate (ITC), 272–274
Ito, T., 383
Iwano, M., 392

J
Jackson, A. L., 552
Jackson, S. A., 20, 281
Jacobsen, S. E., 204
Jagadeeswaran, G., 94, 107
Jagannath, A., 340
Jaillon, O., 10–11, 618
Jain, A., 326–327
Jalas, J., 349, 355
James, C., 507
James, N., 586
James, R. V., 244–245
Janchen, E., 5
Jander, G., 129, 175
Janeja, H. S., 45
Janssen, A. M. P., 244
Jenczewski, E., 158, 202, 277, 621
Jensen, B. D., 276
Jiang, J. J., 298
Jin, H., 90, 110
Jin, R. G., 46, 507, 515
Johanson, U., 137
Johnston, J. S., 172–173, 186, 324, 414
Jones, J. D. G., 512
Jones, L., 570
Jones, L. R., 275
Jones-Rhoades, M. W., 106
Jonsell, B., 355, 359
Jordon-Thaden, I., 15
Jørgensen, M. H., 15
Jorgensen, S., 126, 139
Jorgensen, T. H., 361, 363
Jourdan, P. S., 200
Jourdren, C., 298–299, 302
Jubault, M., 136, 250
Judd, W. S., 5
Juenger, T., 135
Juenger, T. E., 135
Jugulam, M., 48
Jung, J. H., 102, 104
Jun, J. H., 549
Jyoti, J. L., 50

K
Kachlicki, P., 492
Kachroo, A., 396, 398
Kai, K., 491
Kakimoto, T., 548
Kakizaki, T., 396
Kale
traits, 266, 268
See also Brassica oleracea
Kamau, E., 400
Kamchatica, 349–350
Kanada, I., 45
Kandasamy, M. K., 393
Kang, J., 46
Kanne, H. J., 275
Kantama, L., 13, 17–18
Kapranov, P., 70, 90
Kärkkäinen, K., 352, 361–362
Karling, J. S., 244
Karlin, S., 609
Karpechenko, G. D., 266
Karyotype
evolution in *Brassicaceae*, 22–24
See also Ancestral crucifer karyotype (ACK)
Kashkush, K., 202, 204
Katajamaa, M., 481–482
Katari, M. S., 185, 459
Kaushal, P., 205
Kawabe, A., 351, 354
Kawasakiana, 349
Kawashima, C. G., 106
Kawecki, T. J., 361
Kay, S. A., 85
Kell, D. B., 483
Keller, W. A., 278
Kelly, A. L., 299
Kempin, S. A., 131
Kendziorski, C., 472
Kennard, W. C., 267, 271
Kent, W. J., 610
Kenton, A., 201
Kerhoas, L., 491
Kertbundit, S., 533
Keurentjes, J. J. B., 128, 130, 136–138, 496
Khan, N., 205
Kho, Y. O., 391
Khush, G. S., 235, 340
Kickhoefer, V. A., 88
Kiefer, C., 3
Kiefer, M., 14, 18, 20, 22
Kiegle, E., 533, 550
Kikuchi, M., 245
Kilian, J., 625
Kimber, D. S., 293
Kim, D. W., 80

Kim, H. R., 219, 230, 251
Kim, J., 102
Kim, J. H., 102
Kim, J. S., 130, 227, 230, 296, 415–416, 419, 421–422
Kim, S., 130, 357
Kim, S. I., 540
Kimura, R., 401
Kim, V. N., 91
Kim, Y. K., 91
Kind, T., 481, 486
King, G., 268–269
King, G. J., 269, 506
Kivimäki, M., 348, 361–362
Klaper, R., 238
Klewer, A., 48
Kliebenstein, D., 133, 138
Kliebenstein, D. J., 133, 136, 138, 242, 470
Klimazewska, K., 513
Kmec, P., 49
Knee, E. M., 437–466
Knight, C. A., 366
Knutzon, D. S., 303
Kochian, L. V., 627
Koch, M., 127, 172–173, 180–181, 348
Koch, M. A., 1–24, 34, 160, 172, 176, 180, 348, 350, 352, 355, 361, 377
Kohlrabi, 263
  traits, 268
  See also Brassica oleracea
Kohno-Murase, J., 309
Koike, S. T., 275–276
Kojima, M., 515
Kole, C., 220, 222–224, 234, 244, 247–248, 250, 298
Kole, P., 220, 222–223, 250
Kolte, S. J., 48
Koncz, C., 511, 533
Konishi, Y., 486
Konno, N., 45
Koornneef, M., 13–14, 123–141, 305, 533, 539, 566
Kopsell, D. A., 240, 274
Kopsell, D. E., 240, 274
Korban, S. S., 509, 514
Korf, I., 609
Korves, T. M., 135
Kotova, V. V., 44
Kozlowska, H., 310
Kozma, P., 509
Kramer, C. C., 307
Krens, F. A., 50
Kresovich, S., 219
Kridl, J. C., 303
Krochko, J. E., 620
Kruckeberg, A. R., 46
Kubo, H., 238
Ku, H. M., 174
Kuittinen, H., 18, 128–129, 140–141, 166, 177, 347–366, 626
Kulikova, O., 416
Kumar, A., 418
Kumar, P. R., 43
Kumar, R., 311
Kunin, W. E., 355, 361
Kusterer, B., 135
Kuvshinov, V., 510, 515
Kwan, C. C., 267
Kwon, S. J., 418

L

Labate, J., 269
Lacroix, B., 540
Lagercrantz, U., 11, 18, 155, 159–160, 162, 173, 182, 184, 198, 246, 298, 305, 330–331, 419, 618
Lago, C., 74
Lahiri, A., 80
Lall, S., 136, 513
Lamb, R. J., 49
Lambrix, V., 133
Lambrix, V. M., 133
Lamos, S. M., 476
Landry, B. S., 154, 275, 299
Lange, W., 41
Lan, T. H., 155, 165, 270, 279
Larkin, R. P., 41
Laroche, A., 46
Laserna, M. P., 135
Last, R. L., 471
Lau, J. A., 136
Laubinger, S., 531
Law, M., 311
Lawrence, R. J., 621
Lazo, G. R., 512
Leaf development
  Capsella traits genetics, 378
Leaf lamina length (LAL), 236
Leaf length (LL), 236
LEAFY transcription factor, 86
Leavens, W. J., 476
LeClere, S., 447, 547
LeComber, S. C., 196
Le corre, V., 139, 364
Lee, H. S., 204, 624
Lee, J., 78

Lee, J. H., 518
Lee, J. S., 518
Lee, J. Y., 87
Lee, M. K., 516
Lee, R. C., 68, 91
Lee, S. C., 244, 421, 427
Lee, T. I., 77
Lee, Y., 91
Leflon, M., 202
Lefol, C., 49
Lehtila, K., 49
Leinonen, P., 358, 362
Leitch, A. R., 197, 200–201, 203
Leitch, I. J., 12, 197, 200–201, 203
Leitch, L. J., 197
Lelivelt, C. L., 50
Lelivelt, C. L. C., 50
Lempe, J., 139
Lenhard, M., 380
Leonard, E. C., 39
Lepidieae, 4, 6
Lepidium, 2, 4, 10, 43
  crop species morphological traits, 42
  polyploidy and hybridization, 16
Lepidium draba, 43
Lepidium meyenii
  under-utilized crucifer crops, 41
Lepidium perfoliatum, 44
Lepidium peruvianum, 43
Lepidium sativumm
  crop species chemical traits, 44
  under-utilized crucifer crops, 41
Lepidium scandens, 42
Lescot, M., 77
Lesquerella fendleri, 43–46
  crop species chemical traits, 43–44
  under-utilized crucifer crops, 41
Levin, D. A., 196–197, 200
Levy, A. A., 201, 205
Lewis, D., 392
Lexer, C., 13, 17–18, 22
Leyman, B., 516
Liang, C., 586
Li, C., 297, 305
Lichter, R., 278
Li, C. X., 339
Li, D., 299, 307
Li, D. R., 243
Li, F., 612
Li, G., 43, 140, 243, 273–274, 279, 296
Li, H., 48, 612
Li, H. J., 492
Lihová, J., 15–16, 18
Li, H. Q., 46
Li, J., 90, 299, 307, 548
Li, L., 90, 274
Liljegren, S. J., 578
Li, M., 299, 307
Lim, K. B., 415–418, 420, 432
Lim, K. Y., 201, 203, 205
Li, M. R., 46
Li, M. Y., 494–495
Lim, Y. P., 215–251
Linkage disequilibrium (LD), 130
Linkage maps
  Brassicaceae, 154
  See also Comparative mapping
Lin, S. I., 107
Lin, W. Y., 107
Lin, X., 576
Lionneton, E., 328, 332–334, 337–338
Li, P., 46
Lipids
  LC/API-MS-based profiling approaches, 493–495
Lippman, Z. B., 340
Liquid chromatography (LC), 471
  LC-MS-based profiling approaches, 488–495
    lipids, 493
    secondary metabolites in *Arabidopsis*, 488–492
  non-targeted metabolites profiling
    data acquisition aspects, 476–481
    data extraction aspects, 481–483
    molecular structures, 483–488
Li, R., 297, 305
Li, S., 100
Litcher, R., 513
Liu, F., 515
Liu, B., 201, 204–205
Liu, C., 511
Liu, D., 106
Liu, N., 106
Liu, P., 405
Liu, P. P., 100
Liu, Q., 48, 573
Liu, Y., 572
Liu, Y. G., 541
Liu, Z. Q., 202
Li, W. X., 102
Li, X., 205, 539
Li, X. F., 31
Li, Y., 48, 135, 299, 307, 494–495, 540
Li, Z., 46
Li, Z. Y., 205–206

Lloyd, A. M., 508
Lobularia maritima
crop species chemical traits, 43–44
under-utilized crucifer crops, 41
Local distribution of short sequences (LDSS), 72–73
Lockton, S., 348, 354
Loe, G., 361–362
Lohmann, J. U., 86
Lombi, E., 366
Long, M. L., 49
Long non-coding RNAs (ncRNAs)
A. thaliana genomics, 89–90
natural antisense transcripts, 89–90
transcripts generated by RNA polymerase V, 90
See also Small RNAs (sRNAs)
Long, Y., 140, 165, 306
Lou, P., 140, 229–230, 234–236, 243
Loudet, O., 128–129, 131–132, 135–136, 138
Love, C., 459, 603
Love, C. G., 601
Love, H. R., 334
Lowe, A. J., 219, 599
Lu, C., 46, 108
Lu, G., 220, 236
Lu, J., 49
Lunaria, 2
Lunaria annua
crop species chemical traits, 44
under-utilized crucifer crops, 41
Lukens, L., 155, 157, 160, 296, 304–305
Lukens, L. N., 199, 204–205, 414, 620
Lundemo, S., 355
Luo, J., 491
Luo, M. C., 423, 604
Luo, P., 41, 43
Lu, R., 570
Lu, S., 270
Lydiate, D., 18, 166
Lydiate, D. J., 155, 159–161, 166, 173, 182, 298, 310, 419
Lynch, M., 157, 198, 619
Lynch, V. J., 197–198
Lysak, M., 11–14, 16–20, 22–24
Lysak, M. A., 1–24
Lysak, M. L., 11

M

Mable, B. K., 196, 351–352, 355, 365, 374, 391, 403
MacKay, T., 308
Madlung, A., 199, 204–205
MADS-box genes, 86
Maere, S., 10, 12
Magallon, S., 5, 12
Magliano, T. M. A., 128, 135
Magrath, R., 48, 272
Magruder, R., 267
Mahajan, V., 276
Maher, C., 94
Mahmood, T., 329, 332–334, 337–338
Majoros, W. H., 609
Malabat, C., 310
Malcolmieae, 8
Maleck, K., 84
Maliga, P., 511
Malik, M., 45
Malitsky, S., 488
Mallet, J., 15
Mallory, A. C., 100–102
Malmberg, R. L., 135
Maloof, J. N., 131, 135, 137
Maltais, B., 47
Maluszynska, J., 17
Mamula, D., 49
Mancoa mexicana
crop species morphological traits, 42
Mandáková, T., 11, 16, 19, 20, 23–24, 163, 166
Mandal, P., 46
Manton, I., 13, 16
Mapping populations, 450–451
See also Genomics; Germplasm resources
Margulies, M., 297
Marks, M. D., 508
Marra, M., 423
Marsch-Martinez, N., 547
Martens, S. N., 46
Marthe, F., 49
Martienssen, R. A., 532–534, 544, 546, 550–551
Martienssen, R., 570
Marvin, H. J. P., 41
Mascher, J. W., 349, 361
Masle, J., 132
Mass spectrometry (MS), 471
See also Liquid chromatography (LC); Time-of-flight mass spectrometry (TOF-MS)
Mastebroek, H. D., 41, 43
Masterson, J., 196
Mathews, S., 41, 44
Mating
A. lyrata genome evolution and, 353–354
type dimorphism

self-fertility in *A. thaliana*, 404–405
self-fertility in non-model members of *Brassicaceae*, 403–404
See also Self-incompatibility (SI)
Matschinger, M., 16, 20, 348, 350
Matsuda, F., 484, 491
Matsui, A., 83
Matsumoto, E., 220, 223, 245
Matsumura, H., 625
Matsuoka, M., 296
Matsuzawa, Y., 45
Matthäus, B., 43
Matthiola
chromosomes and chromosome number variation, 13–15
genome size (GS) variation, 12–13
Matthiola incana
crop species chemical traits, 43
under-utilized crucifer crops, 41
Matzke, A. J. M., 197
Mauricio, R., 126, 139
Mauro, A. O., 509
Ma, X. F., 201, 205
May, B. P., 546
May, S., 585–595
Mayer, K., 67–111
Mayerhofer, M., 296, 298–299, 305
Mayerhofer, R., 166, 296, 298–299, 305
Mazzeo, J. R., 478
McCallum, C. M., 536, 538, 562, 567
McCarthy, B. C., 44
McCarthy, M. I., 296
McClenaghan, R., 278
McClintock, B., 204
McDowell, J. M., 134
McGrath, J. M., 220–221, 278
McGregor, W. G., 293
McNaughton, I. H., 277, 294
McVetty, P. B. E., 45, 47, 306
Megacarpaea gigantea
crop species morphological traits, 42
Megacarpaea megalocarpa
breeding systems and apomixis, 45
Megacarpaea polyandra
crop species morphological traits, 42
Megdiche, W., 46
Megraw, M., 93
Mei, D. S., 298, 300, 306
Meister, G., 570
Melchinger, A. E., 135
Mellon, F. A., 489
Mengiste, T., 542
Meng, J., 298, 310, 433
Meng, J. L., 46
Menonvillea
chromosomes and chromosome number variation, 13
Mensuali-Sodi, A., 46
Mercer, T. R., 88
Mesopolyploid, 15
Mesopolyploid WGDs, 11
Messenger RNA (mRNA), 88
Metabolic engineering and manipulation of biosynthetic pathways using RNAi
RNAi in *Brassica* species, 578
Metabolites
LC/API-MS-based profiling approaches
lipids, 493–495
secondary metabolites in *Arabidopsis*, 488–492
non-targeted profiling of semi-polar plant (sample preparation)
derivatization, 476
extraction, 473
fractionation, 474–475
non-targeted profiling of semi-polar plant (using UPLC/ESI-QTOF-MS)
data acquisition aspects, 476–481
data extraction, 481–483
experimental design and sampling (general considerations), 472–473
molecular structures, 483–488
Metabolomics
defined, 470
metabolomic data (*Arabidopsis* bioinformatics resources), 594
Metz, T., 507
Meyer, R. C., 135–137, 496
Meyers, B. C., 91, 625
Meyers, L. A., 196
Meyer-Walf, M., 379
Michael, T. P., 85
Michaels, S. D., 129, 131, 137–138, 198, 305
Microarray-based genomic selection (MGS) method
B. napus genomic tools, 297
Microarrays
Arabidopsis reverse genetics, 231
Micro-injection
transformation method, 515–516
MicroRNAs (miRNAs)
A. thaliana genomics, 91–107
and target interaction, 93–94
Arabidopsis miRNA families, 94–107
miR156, 100
miR157, 100

MicroRNAs (miRNAs) (*cont.*)
miR159, 100
miR160, 100–101
miR161, 101
miR162, 101
miR163, 101
miR164, 101–102
miR165, 102
miR166, 102
miR167, 102
miR168, 102
miR169, 104
miR170, 104
miR171, 104
miR172, 104
miR173, 104–105
miR319, 105
miR390, 105
miR391, 105
miR393, 105–106
miR395, 106
miR396, 106
miR397, 106–107
miR398, 107
miR399, 107
biogenesis, 91–93
lines (gene silencing mediated functional analysis)
Arabidopsis reverse genetics, 552
Microspore culture
B. oleracea genetics, 278–279
embryogenic ability in *B. rapa*, 239–240
Microthlaspi perfoliatum
polyploidy and hybridization, 17
Mietkiewska, E., 506, 575, 578–579
Miki, B., 516
Millam, S., 46
Millenaar, F. F., 136
Mimicry, *see* Target mimicry
Mindrinos, M., 134
Mineral accumulation
B. rapa, 240–241
Ming, R., 10, 11
Mishima, M., 396
Mishra, R. S., 47
Mitchell, J. P., 41
Mitchell, N. D., 48–49
Mitchell-Olds, T., 5–7, 10–12, 125–126, 137, 174, 348, 355–360, 363, 379, 618
Mithen, R. F., 42–43, 48, 243, 272–273
Mithen, R., 140, 242, 266, 272
Mithila, J., 48
Miyake, T., 354
Mizoi, J., 538
Mock, H. P., 491
Moco, S., 489
Mohapatra, T., 45
Mohiuddin, S., 44
Molecular analysis
A. thaliana natural variation, 127–137
Molecular cloning
stigma/pollen determinants of SI, 393
Molecular markers
B. napus genomic tools, 295–297
B. rapa, 218
abiotic stress tolerance mapping, 243–244
agronomic and morphological traits mapping, 236–237
anthocyanin pigmentation mapping, 238–239
bolting, flowering and vernalizaton requirement mapping, 233–234
disease resistance mapping, 244–249
embryogenic ability in microspore culture mapping, 239–240
fatty acid composition mapping, 241–242
glucosinolates traits mapping, 242–243
mineral accumulation mapping, 240–241
plant height mapping, 234–235
root traits mapping, 235–236
seed coat color mapping, 237–238
self incompatibility mapping, 239
trait genetics, 233–249
development aspects, 219–220
in diversity studies, 219
See also Genomics
Molecular resources, 437
Arabidopsis species, 452–453
expressed sequence TAGs And cDNA clones, 460
gene specific TAG and RNA interference clones, 462
multifunctional vectors, 463
sequenced full length cDNA and ORF clones in entry vectors, 462
Brassica species, 451–452, 455
expressed sequence TAGs And cDNA clones, 460–461
genomic library/clone resources for, 459
resources associated with *Brassica* sequencing projects, 459–460

genomic library/clone resources
(*A. thaliana*)
AGI BAC and P1 libraries, filters and clones sources, 457
Arabidopsis Genome Initiative (AGI), resources utilized by, 456
bacterial artificial chromosome and P1 libraries and clones used to generate genome sequence, 456–457
large insert libraries from various *Arabidopsis* accessions, 458
large insert libraries utilization, 458
transformable BAC libraries and clones, 458
yeast artificial chromosome libraries and clones, 457–458
See also Bioinformatics resources; Germplasm resources
Molecular structures
LC/MS-based non-targeted metabolites profiling, 483–488
Molecular studies
self-incompatibility in *Brassicaceae*
pollen determinant of SI specificity, 394–396
S haplotype and control of recognition specificity, 396–398
signal transduction, 398–400
stigma determinant of SI specificity, 393–394
See also Genomics
Molina, C., 72–73, 80
Monoploids, 278–279
Monteiro, A. A., 276
Montgomery, F. H., 14
Montgomery, T. A., 105
Moore, G., 154, 164
Moore, R., 198
Moricandia
Brassica crop species, 35
Moricandia arvensis, 44–45
Moricandia nitens, 44
Moricandia sinaica, 44
Moricandia spinosa, 44
Moricandia suffruticosa, 44
polyploidy and hybridization, 15
Morikawa, T., 205
Morinaga, S., 45
Morisia monanthos, 42
Morrison, I. N., 48
Mosaic virus, turnip
genetics and mapping disease resistance in *B. rapa*, 249
Mosher, R. A., 108–110
Moskal, W. A. Jr, 185
Motif discovery, 79, 85
known sites detection, 79–80
phylogenetic footprinting, 80
unknown sites detection, 80–81
Mouchel, C. l. F., 128, 132, 135, 137
Mouradov, A., 379
Mozo, T., 456
Muangprom, A., 165–166, 225
Mukherjee, A. K., 328, 339
Mukhopadhyay, A., 511
Mulder, J. H., 43
Muller, M. H., 355–357, 359–361
Multifunctional vectors
A. thaliana molecular resources, 463
Mummenhoff, K., 3, 6, 16, 376, 627
Munir, M., 46
Mun, J. -H., 183–184, 188, 413–433, 459, 548, 604, 606, 620
Munshi, M. K., 46
Murase, K., 399, 403
Murayama, K., 45
Muroi, A., 491
Murphy, A. S., 352
Muskett, P. R., 546
Mustard family, *see Brassicaceae*
Mutation analyses
Arabidopsis reverse genetics, 533
biologically generated insertional mutants, 539–546
chemically generated mutants, 533–536
physical agents generated mutants, 539
targeted (homologous recombination induced) mutations, 546–547
Brassica species reverse genetics
amplicon selection, 567–568
EMS-induced mutations and genetic code, 563–564
M_1 and M_2 population structure, 566–567
mutagen dosage optimization, 565–566
mutation linking to phenotype, 568
mutation load, 564–565
TILLING platform setting, 567
See also TILLING
Mu, X., 44
MYB transcription factors, 100
Myers, C. H., 267

N

Naczk, M., 310
nad4 intron, 6, 8

Nah, G., 179
Nakazawa, M., 547, 550
Nam, J., 509
Napoli, C., 572
Narasimhulu, S. B., 509, 512
Nasrallah, J. B., 350–351, 382, 389–406, 626
Nasrallah, M. E., 352, 382
Nasturtium officinale
  crop species chemical traits, 44
  under-utilized crucifer crops, 41
Natural antisense transcripts
  long non-coding RNA (ncRNA), 90–91
  siRNA (nat-siRNA), 108
Naumova, T. N., 45
Navarro, L., 106
ndhF gene, 6
Necrosis, *see* Tissue necrosis
Neff, M. M., 548
Nelson, M. N., 166
Nelson, W. M., 423, 604
Neobeckia lacustris, 42
Nesi, N., 309, 311
Neslia
  ancestral crucifer karyotype (ACK), 22
  Neslia paniculata, 22
Newman, C. S., 460–461
Next Generation Sequencing (NGS), 312
  B. napus genomic tools, 297
  Brassica species bioinformatics resources, 611–613
NF-Y binding sites, 75
Nichols, R. A., 360
Niklas, K. J., 380
Nishio, T., 186, 190, 239, 393
Nissinen, M., 311
Niu, Q. W., 552, 577
Ni, Z., 197, 201, 203, 621
Noccaeeae
  karyotype evolution in aspects, 24
Non-coding RNAs (ncRNAs)
  Arabidopsis genomics, 88–111
  long
    natural antisense transcripts, 90
    transcripts generated by RNA polymerase V, 90
  small RNAs (sRNAs), 91
    microRNAs, 91–107
    natural antisense transcript siRNA (nat-siRNA), 110
    repeat associated siRNAs (rasiRNAs), 108–110
    small/short interfering RNAs (siRNAs), 107–108
    trans-acting siRNAs (tasiRNAs), 110
Noncoding transcripts
  A. thaliana genomics, 68
Non-reciprocal translocations (NRTs)
  B. napus genomic tools, 301
Non-targeted profiling
  semi-polar plant metabolites using UPLC/ESI-QTOF-MS
    data acquisition aspects, 476–481
    data extraction, 481–483
    experimental design and sampling (general considerations), 472–473
    molecular structures, 483–488
  sample preparation aspects
    derivatization, 476
    extraction, 473
    fractionation, 474–475
  See also Metabolomics
Noor, R., 46
Nordborg, M., 127–128, 130, 175, 353–354, 357–359
Nordstrom, A., 479
Nou, I. S., 225, 239, 391
Nozaki, T., 220, 223, 239
Nuclear Factor Y (NF-Y), 75
Nuclear magnetic resonance (NMR), 471
  non-targeted metabolites profiling
    molecular structures, 488
  See also Liquid chromatography (LC)
Nucleolar organizing regions (NORs), 14
Nucleotide binding leucine-rich repeat (NB-LRR) genes, 176
Nutt, P., 375, 380–382
Nybom, H., 199

O

Obermeier, C., 625
Ockendon, D., 278, 391
Off targets
  and transitive silencing (transitivity), 576
  See also Post-transcriptional gene silencing (PTGS)
Oh, E., 78
Ohlson, R., 294
Ohta, D., 471
Oil content
  B. juncea traits genetics
    erucic acid content and oil content, 332–334
  rapeseed, *see* Rapeseed *B. napus*
Oilseed crop (*B. napus*), 295

Oinn, T., 594
Ojanpera, S., 486
Okamoto, M., 202
Okamoto, S., 403
O'Kane, S. L., 348
ÓKane, S. L., 124, 126
Okazaki, K., 140
Okou, D. T., 297
Oksman-Caldentey, K. M., 471
Oleracea crops
taxonomy (coenospecies and cytodemes)
B. cretica, 265
B. oleracea, 264–265
See also Brassica oleracea
Oleykowski, C. A., 536, 567
Olimarabidopsis pumila
sequence comparisons between *A. thaliana* and, 178–180
Olsen, K. M., 139
Olson, M., 277, 297
O'Neill, C. M., 173, 181, 184, 186, 246, 250, 419, 618
Ono, Y., 512–513
Onyilagha, J., 44
Opabode, J. T., 508
Oresic, M., 481
ORF clones
in entry vectors, 462
See also Molecular resources
Orr, H. A., 362
Orthologues
defined, 156
See also Comparative mapping
Orton, T., 279
Orton, T. J., 512
Orychophragmus violaceus
crop species chemical traits, 43
under-utilized crucifer crops, 41
Osborn, T. C., 38, 165, 197–204, 220–222, 225, 233–235, 237, 241, 250, 293, 296, 298, 301, 304–308, 331
Ossowski, S., 129, 176, 569, 571, 577
Østergaard, L., 505–518, 561–579
Ostrowski, M. F., 127
Otto, S. P., 196–198
Over-expression/activation mediated functional assays
Arabidopsis reverse genetics
activation tagging lines, 547–548
transgenic lines collections, 547
See also Genomics
Oyama, R., 172, 180–181
Oyama, R. K., 13
Özeker, E., 44
Ozkan, H., 201, 205

P

Pachagounder, P., 49
Pachycladon exilis
karyotype evolution in aspects, 23
Padmaja, K. L., 328, 337
Pahwa, R. S., 45
Palatnik, J. F., 105
Paleopolyploids
diploidized, 196
events, 10
WGDs, 11
Paleopolyploidy, 174
Pampanwar, V., 604
Papdi, C., 547
Paralogues
defined, 156
See also Comparative mapping
Parinov, S., 535, 546
Parisod, C., 16
Parizotto, E. A., 104
Park, B. S., 183, 188, 413–434
Park, C. M., 102
Parker, J. E., 134, 141
Park, J., 603
Parkin, I., 153–167, 296, 298
Parkin, I. A., 173, 184, 599, 609, 620
Parkin, I. A. P., 11, 20, 22, 34, 141, 197–198, 202, 296, 298–299, 301, 305, 330–331, 599, 605, 609
Parkinson, H., 588
Paterson, A. H., 154–155, 165, 270, 279–280, 419, 428, 531
Pathania, A., 45
Pattison, A. B., 50
Paunescu, A., 45
Pauwels, M., 491, 627
Payne, K. A., 240
Pecinka, A., 14, 20
Pedras, M. S., 489
Pedras, M. S. C., 48
Peer, W. A., 626–627
Peisker, M., 44
Peláez, F., 44
Pellegrin, V., 486
Pelletier, G., 266
Pelofske, P. J., 267, 271
Penmetsa, R. V., 566
Pental, D., 323–341
Perchepied, L., 136
Perez-Perez, J. M., 135

Perez-Vich, B., 303
Peronospora parasitica, 250
Perry, J., 538
Perry, J. A., 538, 565
Perry, S. E., 78
Pertea, G., 601
Petersen, B. L., 489
Petersen, B. O., 576
Peterson, C. J., 44
Petrie, G., 339
Petrie, G. A., 247, 339
Pettersson, E., 553
Pfalz, M., 128, 136
Pflieger, S., 570, 572
pHANNIBAL vector, 574–575
pHELLSGATE Series, 575
pHELLSGATE vector, 575
Phenotypic effects
  resynthesized *Brassica* polyploids, 200
Photosynthesis (C_3–C_4), 44
Phylogenetic footprinting, 80, 82–83
Phylogeny
  Brassicaceae, 5–10
Physaria
  chromosomes and chromosome number variation, 13–15
  genome size (GS) variation, 12–13
Physical agents generated mutants
  Arabidopsis reverse genetics, 539
Physical maps
  B. oleracea, 279
  B. rapa, 422–424
  Brassica rapa genome sequencing project (BrGSP), 603–605
Phytochrome and Flowering Time 1 (PFT1) phenotype, 75
Piao, Z. Y., 226, 245–247
PIC, 70, 72
Piccin, A., 573
Picó, F. X., 124, 126–127, 139
Pigmentation
  anthocyanin, 238–239
  See also Brassica rapa
Pignatta, D., 197, 201, 203
Pikaard, C. S., 204–205, 621
Pilet, M. L., 298–299
Pilon, M., 106
Pink, D. A. C., 451
Pires, C. J., 621
Pires, J. C., 195–206, 331
pKANNIBAL vector, 574–575
Plamodiophora brassica, 244
Plant height
  B. rapa, 234–235
Plant regeneration and transformation
  Brassicaceae, 45–46
Plasmids, 511–512
  See also Transformation technology
Plasmodiophora brassicae, 244, 275
Plastid transformation, 511
Plessers, A. G., 39
P1 libraries, 456–457
Pollen determinant of SI specificity, 394–396
Polyctenium fremontii
  karyotype evolution in aspects, 23
  polyploidy and hybridization, 16
Polyploidy
  Brassica species
    as models for studying genome evolution following polyploidy, 618–621
    B. oleracea genetics, 277
    genome and chromosomal evolution, 17
    genome evolution comparative mapping, 158–160
  defined, 156, 196
  phenomenon in flowering plants, 196–197
  polyploids
    Arabidopsis, 199
    Brassica species, 195, 199
  resynthesized polyploids (*Brassica* and *Arabidopsis*), 199–200
    future research, 205–206
    genetic changes, 201–203
    phenotypic effects, 200
    proteome changes, 204–205
  See also Allopolyploidy; Aneuploidy; Autopolyploidy; Comparative mapping; Diploidization
Pontes, O., 199, 201, 205, 366
Potter, D., 272
Population development
  B. napus, 306–309
Position-specific scoring matrix (PSSM), 80–81
Post-transcriptional gene silencing (PTGS)
  RNAi/PTGS mechanisms, 570
    artificial microRNAs, 576–577
    co-suppression and anti-sense RNA, 572
    hpRNA, 573–574
    inducible and tissue-specific gene silencing, 575–576
    off targets and transitive silencing (transitivity), 576

siRNA-directed gene silencing using hpRNA constructs, 573
target mimicry, 577
virally-induced gene silencing, 570–572
sRNA classes associated with, 569–570
Potyvirus, 249
Prabhu, K. V., 339
Pradhan, A. K., 45, 323–341
Prakash, S., 36, 38, 42–43, 45, 47–50, 83, 198–200, 216–218, 266, 293, 324–326, 340
Prantl, K., 4
Prasad, M. N. V., 46
Prasanth, K. V., 88
Prem, D., 46
Preuss, D., 535
Preuss, S. B., 621
Price, J. S., 518
Priestly, M., 607
Primard-Brisset, C., 166
Promoter
Arabidopsis (reverse genetics) expression analyses using, 532–533
proximal and distal, 76
Protein content
B. napus, 309
Proteome changes
in resynthesized polyploids, 204–205
Proto-Calepineae Karyotype (PCK), 23
Provart, N. J., 623
Proximal and distal promoter
A. thaliana genomics, 76
Przedpelska, E., 46
Pseudomonas syringae, 250
Pua, E. C., 509, 514–515
Puddephat, I. J., 510–511
Pumphrey, M., 205
Punt, P. J., 72
Purity, R. S., 506
Putnam, D. H., 39
Putterill, J., 305

Q
Quadrapole MS (QTOF-MS), *see* Time-of-flight mass spectrometry (TOF-MS)
Quantitative Trait Locus (QTL) analyses
A. thaliana genetic information use in *Brassica*, 139–141
genetic and molecular analysis of *A. thaliana*, 127–130
genetic bases of adaptation, 130–137
molecular bases of adaptation, 137–139
natural variation in *A. thaliana*, 124–125
Qi, L., 164, 202
Qing, C. M., 510, 515
Qiu, D., 165, 274, 281, 299, 303, 333–334
Qi, Y., 109
Quesada, V., 136, 548
Quijada, P. A., 298, 301, 306–308
Quiniou, S. M. A., 423
Quiros, C. F., 140, 154, 181, 199, 219–221, 243, 261–282, 296

R
Radke, S. E., 509–510, 515
Radoev, M., 298, 300, 306–307
Rafalski, A., 296
Rahman, M. H., 228, 230, 241, 298, 311
Rajagopalan, R., 92, 108, 109
Rajcan, I., 302
Rakow, G., 302
Ramachandran, S., 49, 507
Ramos-Onsins, S. E., 350, 352, 404
Ramsay, L. D., 294
Ramsey, A. D., 49
Ramsey, J., 196–197, 200
Random Amplified Polymorphic DNA (RAPD), 219–220, 230
molecular markers in diversity studies, 219
Rao, G. U., 45
Rapeseed *B. napus*
genomic tools, 295–297
genetic linkage maps, 297–302
molecular marker technology, 295–297
oilseed crop importance, 295
specific traits genetics
flowering time variation (winter vs. spring), 304–306
hybrids, population development and seed yield improvement, 306–309
modified FA and specialty oil and meal profiles, 302–303
oil content, 303–304
oil quality related traits, 309–311
Rapeseed oil, 294–295
Raphaninae
Brassica crop species, 35
Raphanus
morphological traits, 42
R. radicula, 40
R. raphanistrum, 40

Raphanus (*cont.*)
R. sativa, 45
comparative mapping extension to related species, 165
crop species, 40
Rapid stress response element (RSRE), 84
Rapistrum rugosum, 44
Rare genomic changes (RGCs), 24
Ratcliff, F. G., 572
Rauh, B. L., 135
Raut, R. N., 200, 202
Ravetta, D. A., 47
Ray, N., 358
Razmjoo, K., 44
Rea, A. C., 382
Reciprocal translocation (RT)
B. napus genomic tools, 301
Recombinant inbred lines (RILs), 128
Recombination induced mutations
targeted homologous, 546–547
See also Reverse genetics
Reed, J., 516
Rehn, F., 49
Reichelt, M., 489
Reinhart, B. J., 91
Renard, M., 306
Ren, B., 77
Ren, S., 87
Rensing, S. A., 530
Rensink, W. A., 623
Renwick, J. A. A., 49
Repeat associated siRNAs (rasiRNAs)
A. thaliana genomics, 108–109
Reporter gene fusions and promoter, gene and enhancer trap lines
Arabidopsis reverse genetics, 532–533
Resequencing
Brassica genomes, 611
See also Bioinformatics resources
Resistance, *see* Disease resistance
Resources, *see* Bioinformatics resources; Germplasm resources; Molecular resources
Restriction Fragment Length Polymorphism (RFLP), 154, 156, 218–220
molecular markers and trait genetics, 233
B. juncea mapping, 330
B. napus genomic tools, 297, 302
B. oleracea biomarkers, 279
molecular markers
and trait genetics, 234
in diversity studies, 219
PCR-RFLP, 239
PCVR-RFLP, 239
Resynthesized polyploids
Brassica and *Arabidopsis*, 199, 200
epigenetic changes, 204
future research, 205–206
genetic changes, 201–203
genetic changes effects on gene expression and phenotypes, 201–203
proteome changes, 204–205
phenotypic effects, 200
Reverse genetics
gene function analysis, 529–553
expression analyses, 531–533
mutation analyses, 533, 536–547
similarity to other known proteins, 530
gene silencing mediated functional analysis, 548
antisense lines, 549
microRNAs and targeted miRNA lines, 552
RNAi gene silencing approaches, 576–577
RNAi lines, 549
over-expression/activation mediated functional assays, 547–548
RNAi, 562
background, 568–570
conferring tissue specificity in RNAi approaches in *Brassica* species, 578–579
gene function studying throughout *Brassica* development, 578
gene silencing approaches in *Brassicaceae*, 570–577
metabolic engineering and manipulation of biosynthetic pathways using RNAi, 577–578
RNAi in *Brassica* species, 577–579
RNAi/PTGS mechanisms, 569–577
sRNA classes associated with PTGS, 569–570
TILLING, 562
amplicon selection, 567–568
EMS-induced mutations and genetic code, 563–564
M_1 and M_2 population structure, 566
mutagen dosage optimization, 565
mutation linking to phenotype, 568
mutation load, 564–565
TILLING platform setting, 567
See also Genomics

Reymond, M., 123–141
Richards, E. J., 138
Riddle, N. C., 138
Riechmann, J. L., 69
Rieseberg, L. H., 15, 45, 196–197
Rigola, D., 625
Riihimäki, M., 361, 364
Riley, H. P., 375
Riley, R., 202
Rimmer, S., 339
Rimmer, S. R., 48, 247
Riungu, T. C., 45
Rivero, L., 437–465
RNA
 induced silencing complex (RISC), 92
 interference clones
  A. thaliana molecular resources, 462
  See also RNAi
 non-coding, *see* Non-coding RNAs (ncRNAs)
 polymerase V/VI, 90
 sequencing (*A. thaliana* genomics), 68–71
RNAi
 in *Brassica* species, 577
  conferring tissue specificity in RNAi, 578–579
  gene function studying throughout development, 578
  metabolic engineering and manipulation of biosynthetic pathways, 577–578
 lines (gene silencing mediated functional analysis
  Arabidopsis reverse genetics, 549
 RNAi/PTGS mechanisms (gene silencing approaches), 570
  artificial microRNAs, 576–577
  co-suppression and anti-sense RNA, 572
  hpRNA, 573–574
  inducible and tissue-specific gene silencing, 575–576
  off targets and transitive silencing (transitivity), 576
  siRNA-directed gene silencing using hpRNA constructs, 573
  sRNA classes associated with PTGS, 569–570
  target mimicry, 577
  virally-induced gene silencing, 570–571
Röbbelen, G., 201, 7, 336
Robert, L. S., 305
Roberts, K. J., 44
Roberts, L. D., 493
Robertson, D., 549
Robinson, S. J., 625
Rochfort, S. J., 489
Rodman, J. E., 43
Rohde, A., 491
Rokas, A., 24
Rombauts, S., 75, 592
Roosens, N., 627
Roosens, N. H., 627
Roosens, N. H. C. J., 20
Root traits
 Brassica rapa, 235–236
Rooting, 514
 See also Shoot regeneration
Rorippa, 10
 polyploidy and hybridization, 16
Rorippa indica
 under-utilized crucifer crops, 31
Rosa, E. A. S., 242, 272
Roscoe, T. J., 165
Rose, L. E., 134
Ross-Ibarra, J., 348, 356–358, 360, 363
Rouhier, H., 236
Roulund, N., 278
Routaboul, J. M., 492
Roux, F., 378
Rowe, H. C., 136
Roy, B. A., 45
Rozen, S., 609
Rubin, G. M., 530
Rudloff, E., 46
Rus, A., 132, 136
Ruvkun, G., 68
Ruwandi, A., 49
Ryder, C. D., 305

S

Sabhyata, B., 159
Sachan, G. C., 50
Sachan, J., 339
Sadowski, J., 281
Saha, P., 625
Saharan, G., 338
Saharan, G. S., 338
Sainudiin, R., 402
Saito, K., 471–472
Saito, M., 226, 245–246, 250
Sakai, H., 104
Sakamoto, K., 228, 230, 245–246, 250
Salazar, E., 200
Salisbury, P. A., 42, 47–48
Salmeron, J. M., 539
Salmon, A., 201

Salt and heavy metal tolerances
Brassicaceae, 46
Saltational change in floral architecture
Capsella traits genetics, 380–381
Salt, D. E., 136
Salywon, A. M., 41, 43
Sampson, D. R., 267, 391
Samson, F., 533–534, 541
Samuel, M. A., 391, 399
Sánchez-Yélamo, M. D., 44
Sanderson, M. J., 12
Sandring, S., 348, 364
Santos, J. L., 16
Santuari, L., 623
Sarla, N., 200, 202
Sato, K., 401–402
Sauder, C., 17, 35–36
Savolainen, O., 347–366
Sax, K., 218
Scarcelli, N., 128
Scarecrow-like (SCL) transcription factors, 104
Schaad, N. W., 248
Schauer, S. E., 86
Scheel, D., 469–496
Schein, M., 136, 178
Scheiner, S. M., 136
Schemske, D. W., 196–197, 200
Schiantarelli, E., 513
Schierup, M. H., 351, 355, 361, 391, 401
Schilling, A., 220
Schivereckia doerfleri
crop species chemical traits, 43
Schizopetaleae, 6, 10
Schläppi, M. R., 131
Schlechteria, 3
Schmickl, R., 16, 349, 355, 360
Schmid, K. J., 126–127, 139, 356
Schmid, M., 84, 625
Schmidt, A., 479
Schmidt, R., 171–190
Schmitt, J., 126, 379
Schmuths, H., 174
Schneeberger, R. G., 540
Schneider, A., 545–547
Schoen, D. J., 356
Scholl, R., 437–466, 529
Schommer, C., 105
Schopfer, C. R., 393, 395–396, 400–401, 403
Schrader, O., 281
Schranz, E., 18
Schranz, M. E., 5, 7, 10–12, 18, 20, 22–24, 34–35, 45, 140–141, 161, 165–166, 173–174, 177, 184, 197–200, 225, 233–234, 250, 304–305, 324, 330, 350, 366, 378, 439, 618, 620, 626–627
Schrijver, B., 278
Schubert, I., 23
Schulman, A., 225, 242
Schulz, O. E., 4–6
Schwab, R., 104, 506, 552, 560, 576
Schwab, W., 462
Sciaky, D., 512
Sclerotinia stem rot
B. juncea disease resistance, 338–339
SCR diversification, 400–403
See also Self-incompatibility (SI)
Sears, E. R., 202
Sebastian, R. L., 269, 279, 296
Secondary metabolites in *Arabidopsis*
LC/API-MS-based profiling approaches, 488–495
Seed
coat color
B. juncea traits genetics, 331–332
B. rapa, 235–236
yield (*B. napus*), 306–309
Seed BACs
Brassica genome sequencing
BAC selection, 428–430
BAC sequences characteristics, 430–433
Brassica rapa genome sequencing project (BrGSP)
bioinformatic selection of seed BACs, 605–607
Seki, H., 547
Seki, M., 72, 83
Self-fertility
in model plant *A. thaliana*, 404–405
in non-model members of *Brassicaceae*, 403–404
See also Self-incompatibility (SI)
Self-incompatibility (SI)
A. lyrata, 351–352
B. rapa, 239
Capsella traits genetics, 384
cytological responses, 392
genetics, 391–393
in *Brassicaceae*, 389–391
future prospects, 405–406
loss of SI and self-fertile mode of mating, 403–405
self pollen recognition/inhibition mechanism, 391–398

S-haplotype structure, suppressed recombination, and diversification, 400–342
SRK and SCR diversification, 400–402
molecular studies
pollen determinant of SI specificity, 394–396
S haplotype and control of recognition specificity, 396–398
signal transduction, 398–399
stigma determinant of SI specificity, 393–394
See also Mating
Sequence comparative analysis, 163, 174–176
A. thaliana
accessions, 174–176
reference genome, 173–174
B. napus accessions, 189
between *Brassica* genomes
B. oleracea and *B. napus* (*Brassica* C genomes), 188–189
B. rapa and *B. napus* (*Brassica* A genomes), 187–188
sequence comparisons between *A. thaliana* and *Brassica* species, 176–181
B. napus, 185–187
B. oleracea, 184–185
B. rapa, 178–180
sequence comparisons between *A. thaliana* and near relatives, 176–181
See also Comparative mapping; Genome sequencing
Sequence related amplified polymorphism (SRAP)
molecular markers and trait genetics, 241
Sequenced full length cDNA and ORF clones in entry vectors
A. thaliana molecular resources, 460
Sergeeva, L. I., 132, 136–137
Sernyk, J. L., 306
Service, R. F., 553
Sessions, A., 534, 540–541
Shahidi, F., 294–295, 310
Shahmuradov, I. A., 72, 75
Shaked, H., 201, 205
S haplotype and control of recognition specificity, 396–398
S-haplotype structure, suppressed recombination, and diversification, 400
SRK and SCR diversification, 400–402
See also Self-incompatibility (SI)
Sharbel, T. F., 122, 626–627
Sharma, G., 46, 328, 333, 339
Sharma, N., 44, 625
Sharma, R., 328, 333, 339
Sharma, R. P., 328, 333, 339
Sharma, T. R., 46
Sharpe, A. G., 202, 298–299, 301, 310, 599, 620
Shattuck, V. I., 249
Shcherban, A. B., 205
Sheldon, C. C., 135, 548
Shepherd, T., 472–473
Sherman-Broyles, S., 382, 391, 400, 404–405
Shewmaker, C., 311
Shi, J., 308
Shiba, H., 396
Shimizu, K., 374
Shimizu, K. K., 16, 349, 351–352, 404–405
Shimizu-Inatsugi, R., 349, 351
Shimosato, H., 398
Shindo, C., 135, 138–139, 304
Shirzadegan, M., 336
Shivpuri, A., 338
Shoot regeneration
choice of explant and tissue culture media, 513–514
in vitro
genetics, 512
intolerance to, 513
shoot elongation and rooting, 514
See also Transformation technology
Short interfering RNA (siRNA), 569–570
directed gene silencing, 573
Short RNA (sRNA)
classes associated with PTGS, 569–570
See also Long non-coding RNAs (ncRNAs); Small RNAs (sRNAs)
Shukla, V. K., 546
Shukla, V. K. S., 43
Shull, G. H., 383
Sicard, O., 136
Siddharthan, R., 83
Sieber, P., 101
Sieburth, L. E., 86, 548
Siefers, N., 75
Siegemund, F., 511
Siemens, J., 46, 48–49
Signal transduction
self-incompatibility in *Brassicaceae*, 398–399
Signal-sensing domains (SSD), 69
Sikdar, S. R., 46
Sikorski, T. W., 74
Silicularia, 3

Simillion, C., 10, 12, 198
Simon, M., 126, 133
Simple Sequence Repeats (SSRs), 218–220, 230
  molecular markers in diversity studies, 219
  B. oleracea biomarkers, 279
Sinapate ester, 310
Sinapidendron angustifolia, 43
Sinapis, 41, 44
Sinapis alba, 34, 43–44
Sinapis arvensis, 43–44
Sinclair, B. K., 510
Singh, A., 392
Singh, J., 48
Singh, M., 509, 514
Singh, R., 49
Singh, R. R., 514
Singh, S. J., 48
Singh, V. P., 50
Single nucleotide Polymorphism (SNPs), 218
  B. napus genomic tools, 297–298
Sinha, S., 83
Sinskaia, E. N., 217
Sisymbrieae, 6
  karyotype evolution in aspects, 24
Skaletsky, H., 609
Skarjinskaia, M., 46
Slade, A. J., 538, 565
Slatkin, M., 358
Slominski, B. A., 336
Small, I., 569
Small RNAs (sRNAs)
  A. thaliana genomics, 90
    MicroRNAs, 91–108
    natural antisense transcript siRNA (nat-siRNA), 110
    repeat associated siRNAs (rasiRNAs), 108–109
    small/short interfering RNAs (siRNAs), 107
    trans-acting siRNAs (tasiRNAs), 110
  databases (*Arabidopsis* bioinformatics resources), 593
  See also Long non-coding RNAs (ncRNAs)
Small/short interfering RNAs (siRNAs), 107
Smisek, A., 48
Smith, C., 196
Smith, C. A., 482
Smith, L. B., 269
Smith, N., 509
Smith, R. H., 508
Smits, B. M., 538
Snowdon, R. J., 17–18, 218, 281, 296
Soengas, P., 227, 230, 248
Solanki, Z. S., 336
Soller, M., 219
Soltis, D. E., 15–16, 196–197, 199, 201, 203
Soltis, K., 197, 201, 203
Soltis, P. S., 15–16, 196–197, 199, 201, 203
Somers, D. J., 298, 339
Somerville, C., 508
Somerville, C. R., 494
Song, K., 37–38, 199–201, 205, 222, 233–235, 236, 331, 620
Song, K. M., 199, 217, 221, 293–294, 324–325
Song, L., 93
Sonntag, K., 46
Soriano, A., 47
Soroka, J., 49–50
Sparrow, P. A. C., 505–518
Spe, 371
Spector, D. L., 88, 93
Speulman, E., 546
Sphaerocardamum
  genome size (GS) variation, 12
Spillane, C., 354
Squamosa promoter binding-like (SPL) transcription factors, 100
Srivastava, A., 326
Srivastava, P. S., 200
Srivastava, S., 200
SRK
  diversification, 400–402
  See also Self-incompatibility (SI)
Staal, J., 134, 141
Stahl, E. A., 139
Stamenoid petals (*Spe*), 380
Stanke, M., 609
Starmer, K. P., 306
Stebbins, G. L., 200, 374
Stefansson, B. R., 294, 306
Stegemann, S., 511
Stein, J., 397
Stein, J. C., 394, 401
Stein, L. D., 588, 598, 604, 610
Steinmetz, K. A., 272
Steinmetz, L. M., 531
Stenøien, H. K., 126–127, 139, 356, 359
Stenopetalum
  chromosomes and chromosome number variation, 14
Stepanova, A. N., 562
Stephenson, P., 561–579
Steuer, R., 483
Stevens, P. F., 5
Stewart, C. jr, 507

Stewart, N., 507, 509, 513–514
Stiewe, G., 306
Stigma determinant
SI specificity, 393–394
Stinchcombe, J. R., 139
Stirling, A., 49
Stobbs, L. W., 49
Stobiecki, M., 492
Stolc, V., 90
Stoll, N., 486
Stone, S. L., 399
Stoner, K., 42, 49
Stoner, K. A., 277
Stormo, G. D., 71, 81, 83
Storz, G., 88
Stout, A. B., 391
Stoutjesdijk, P., 573
Strains (bacterial)
use of, 511–512
See also Transformation technology
Stransfeld, L., 380
Strauss, S. Y., 49
Stress tolerance
abiotic, 231–244
See also Brassica rapa
Stringam, G. R., 237, 295, 334, 336
STRUWWELPETER (SWP) phenotype, 75
Sturm, M., 482
Subularia aquatica, 45
Sumner, L. W., 487, 492
Sundaresan, V., 533–534, 544–545, 562
Sung, S., 305
Sunkar, R., 94, 107
Sun, T., 539
Sun, Z., 129, 300
Suominen, J., 349, 355
Suwabe, K., 140–141, 155, 158, 219–220, 225, 227, 230, 245–246, 250, 296
Suzuki, G., 396
Suzuki, T., 394, 538
Svistoonoff, S., 129, 132, 138
Symonds, V. V., 135
Synteny
defined, 156
maps, 280–281
See also Comparative mapping
Sysi-Aho, M., 483
Systemic acquired resistance (SAR), 78
Szadkowski, E., 199, 202, 205
Szymanski, M., 88

T
Tadage, M., 305
Taji, T., 46
Takahata, Y., 199, 240
Takasaki, T., 394
Takayama, S., 391, 396, 398
Takeda, S., 296
Talalay, P., 243
Talame, V., 538
Tamaoki, M., 136
Tanaka, C., 72–73, 83
Tanaka, T., 72–73, 83
Tang, C., 383, 404–405
Tang, C. L., 352, 365
Tang, G. X., 515
Tang, H., 154, 173–174
Tang, H. B., 10–11
Tang, J., 298
Tang, W., 78, 298
Tanhuanpää, P., 248, 302
Tanhuanpää, P. K., 222, 225, 241–242, 248, 333
Tan, J., 489
Tannin content
B. napus, 310–311
Tan, X., 136
Tapsell, C. R., 264–265, 271
Target interaction
microRNAs (miRNAs), 93–100
Target mimicry
gene silencing approaches in *Brassicaceae*, 570–577
Targeted (homologous recombination induced) mutations
Arabidopsis reverse genetics, 546
zinc finger nuclease mediated mutations, 546–547
Targeting Induced Local Lesions In Genomes, *see* TILLING
TATA-box, 73
containing core promoters, 74
TATA-box binding protein (TBP), 71, 73–74
associated factors (TAFs), 71–72, 74
related factor (TRF), 73
Tate, J. A., 197, 199, 202, 205
Tatsuzawa, F., 44
Tautenhahn, R., 482
Taylor, C. G., 510
Taylor, J. D., 248
Taylor, J. P., 391, 510
T-DNA mediated insertional mutations
Agrobacterium, 540–544
See also Reverse genetics

Teesdalia, 4
Telomerase Activator 1 (TAC1), 87
Tempe, J., 510
Teng, S., 133, 138
Tetraploidization, 10
Tetraploids
 B. oleracea genetics, 277
Tettelin, H., 623
Teutonico, R. A., 220, 237, 241, 296, 305–306
Teutonico, R., 220–222, 233, 250
Tewari, I., 48
Thaveeschai, N., 248
Theißen, G., 373–384, 626–627
Thellungiella halophila, 2
Thellungiella salsuginea
 cold tolerance, 46
 drought tolerance, 47
 salt and heavy metal tolerances, 46
Thiagarajah, M. R., 334, 336
Thibaud-Nissen, F., 70, 78, 82
Thierfelder, A., 50
Thierry-Mieg, J., 598–599, 604
Thijs, G., 83
Thiocone, A., 479
Thlaspi arvense
 chemical traits, 43
Tholl, D., 492
Thomas, B. C., 87
Thomas, C. M., 541
Thomas-Chollier, M., 83
Thomas, G., 275
Thomas, M. C., 71, 74
Thompson, K. F., 49, 391
Thormann, C. E., 219, 302–303
Thow, R. F., 294
Thurston, M. I., 609
Tian, D., 139
Tian, Q. G., 489
Tian, Z. H., 46
Till, B. J., 536, 538, 551, 562, 565, 567
TILLING, 451
 advantages and disadvantages, 538
 Arabidopsis mutation analyses, 536–538
 Brassica species reverse genetics, 562
  amplicon selection, 567–568
  EMS for, 562–564
  M_1 and M_2 population structure, 566
  mutagen dosage optimization, 565
  mutation linking to phenotype, 568
  mutation load, 564–565
  TILLING platform setting, 567
 EMS and
  induced mutations and genetic code, 563–564
  use, 562–563
Time-of-flight mass spectrometry (TOF-MS)
 lipids, 495
 non-targeted metabolites profiling (QTOF-MS)
  data acquisition aspects, 476–481
  data extraction aspects, 481
  molecular structures, 483–488
Tisné, S., 135
Tissier, A. F., 545–546
Tissue necrosis
 transformation method, 514–515
Tissue specificity
 in RNAi approaches in *Brassica* species, 578–579
Tiwari, A., 247, 339
Tiwari, A. S., 247
Tiwari, J. P., 339
Tjallingii, F., 244
Tjian, R., 73, 74
Tohge, T., 471, 487, 492
Tolerances
 abiotic stress, 243–244
 cold tolerance, 46–47
 disease resistance, 48–49
 drought tolerance, 47
 herbicide resistance, 47–48
 insect and nematode resistance, 49–50
 salt and heavy metal, 46
 See also Disease resistance
Tomer, K. B., 477
Tompa, M., 83
Tonguc, M., 277, 279
Tonsor, S. J., 136
Tookey, H. L., 40
Topping, J. F., 533
Torjek, O., 128
Tör, M., 134
Town, C., 171–190, 282
Town, C. D., 141, 160, 166, 198, 415, 419, 431, 605, 619
Townsend, J., 546
Trachystoma ballii, 45
Tragopogon, 203
 allotetraploids, 202
Trait genetics
 B. juncea, 331
  agronomic and yield traits, 337–338
  disease resistance, 338–339

erucic acid content and oil content, 332–334
glucosinolates, 334–36
seed coat color, 336–337
B. napus
flowering time variation (winter vs. spring), 304–306
hybrids, population development and seed yield improvement, 306–309
modified FA and specialty oil and meal profiles, 302–303
oil content, 303–304
oil quality related traits, 309–311
B. oleracea
broccoli, 269–271
brussel sprouts, 269
cabbage, 267
cauliflower, 269–271
kale, 268
kohlrabi, 268
B. rapa, 215–216
abiotic stress tolerance, 243–244
agronomic and morphological traits, 236–237
anthocyanin pigmentation, 238–239
bolting, flowering and vernalizaton requirement, 233–234
comparative mapping and identification of candidate genes for important traits, 249–251
disease resistance, 244–249
embryogenic ability in microspore culture, 239–240
fatty acid composition, 241–242
glucosinolates traits, 242–243
molecular markers and trait genetics, 233–249
mineral accumulation, 240–241
plant height, 234–235
root traits, 235–236
seed coat color, 237–238
self incompatibility, 239
Capsella
floral structure and function, 379–383
flowering time, 378–379
fruit structure, 383–384
leaf development, 378
comparative mapping in *Brassicaceae*, 163–164
Trans-acting siRNAs (tasiRNAs)
A. thaliana genomics, 110
Trans-activating domain (TAD), 69
Transberingia bursifolia, 23
transcription factors (TFs), 69
GTFs, 71
Transcription initiation site (TIS), 69
Transcription start site (TSS), 69, 71
Transcriptome analysis
Arabidopsis species, 588–590
B. rapa, 424–427
Brassica species, 623–627
Transcripts
generated by RNA polymerase V, 90
natural antisense, 89–90
See also Long non-coding RNAs (ncRNAs)
Transcripts of Unknown Functions (TUFs), 111
Transformation technology
Agrobacterium rhizogenes, 510
Agrobacterium tumefaciens, 508–510
as research tool (*Arabidopsis*/*Brassica* relationship), 517
bacterial strains and plasmids, use of, 511–512
chloroplast transformation, 511
direct uptake transformation methods, 511
floral dipping/micro-injection, 515–516
hyperhydricity, 514–515
in *Arabidopsis*, 506–507
Agrobacterium rhizogenes, 510
Agrobacterium tumefaciens, 508–509
as research tool, 517
bacterial strains and plasmids, use of, 511–512
chloroplast transformation, 511
direct uptake transformation methods, 511
shoot regeneration (in vitro), 512
in *Brassicaceae*, 505–507
Agrobacterium rhizogenes, 510
Agrobacterium tumefaciens, 508–510
as research tool, 517
bacterial strains and plasmids, use of, 511–512
chloroplast transformation, 511
direct uptake transformation methods, 511
floral dipping/micro-injection, 515–516
hyperhydricity and tissue necrosis, 514–515
shoot regeneration (in vitro), 512–514
transgenics selection, 516
shoot regeneration
choice of explant and tissue culture media, 513–514
in vitro, 512–513

Transformation technology (*cont.*)
shoot elongation and rooting in vitro, 514
tissue necrosis, 514–515
transgenics selection, 516
See also Genetically modified (GM) technology
Transgenic lines collections, 547
Transgenics selection, 516
Transitive silencing (transitivity), 576
See also Post-transcriptional gene silencing (PTGS)
Transposon mediated insertional mutations
Arabidopsis reverse genetics, 544
See also Reverse genetics
Trautwein, E. A., 311
Tremetsberger, K., 15–16
Tribodet, M., 48
Trick, M., 183–184, 190, 297, 597–614, 620, 622, 624–625
Triplicated nature
of *B. rapa* genome, 419–420
trnF pseudogene, 6
Tschuch, C., 552
Tsesmetzis, N., 592
Tsiantis, M., 378, 627
Turnage, M. A., 572
Turner, T. L., 364, 623
Turnip mosaic virus (TuMV)
genetics and mapping disease resistance in *B. rapa*, 249
Turritis, 22
Turritis glabra, 21

U
Uchimiya, H., 324
Udall, J. A., 202–203, 296, 298–299, 301, 307, 620
Ueno, M., 205
Ueno, O., 44
Ueno, Y., 102
Ulker, B., 527–553
Ulmer, B. J., 50
U, N., 36, 158, 173, 198–199, 217, 293, 324, 414
UPLC (non-targeted metabolites profiling)
data acquisition aspects, 481–483
data extraction aspects, 481
Uprety, D. C., 44
Urbanska, K. M., 17
Usuda, H., 236
Uzunova, M., 155, 299, 310

V
Valvekens, D., 508
Van de Mortel, J. E., 625
Van den Berg, R. A., 48, 483
Van de Peer, Y., 197
Vandepoele, K., 77, 85
Van der Graaff, E., 548
Van de Weg, M., 244
Van Hintum, J., 440
Van Leeuwen, H., 136
Van Montagu, M., 540
Van Treuren, R., 355, 359
Varshney, A., 328, 339
Vaughan, J. G., 5, 43, 199
Vaughn, M. W., 138
Vaughn, S. F., 43
Veena, V., 510
Veit, M., 492
Vegetable oil, 295
Velasco, L., 43
Velculescu, V. E., 625
Vella
polyploidy and hybridization, 15
Vera, C. L., 336
Vergara-Silva, F., 380
Vergeer, P., 361, 365
Verma, R., 339, 514
Verma, S. S., 45, 514
Vernalizaton
Brassica rapa, 233
Vicente, J. G., 277, 464
Vigeolas, H., 304
Vignal, A., 296
Vinterhalter, B., 45
Vioque, J., 43
Virally-induced gene silencing (VIGS)
RNAi/PTGS mechanisms
gene silencing approaches in *Brassicaceae*, 570–572
Vision, T. J., 197
Vitis vinifera
genome and chromosomal evolution, 10
Vitrification, *see* Hyperhydricity
Vlieghe, D., 77
Vogel, J., 74
Voinnet, O., 93, 569–571
Von Hippel, P. H., 71, 80
Von Roepenack-Lahaye, E., 469–496
Vreugdenhil, D., 136
Vriesenga, J. D., 275
Vuylsteke, M., 136

W

Wagner, G. P., 197
Wahlroos, T., 510
Walden, R., 547
Walker, J. C., 275
Walker, K. C., 310
Walker, R. L., 41
Wallace, D. H., 266–267, 393
Walley, J. W., 84
Wallis, J. G., 493
Walsh, J. A., 249
Wan, Z., 45
Wang, C. H., 565–566
Wang, C., 46
Wang, H. Z., 225
Wang, J., 43, 131, 199–200, 205, 309, 623
Wang, J. W., 100–101, 104
Wang, L. J., 101
Wang, M., 218, 225, 276–277
Wang, M. B., 565–566
Wang, M. L., 538
Wang, N., 140, 565–566
Wang, Q., 131
Wang, R., 179, 309, 627
Wang, R. H., 366
Wang, S., 539
Wang, T., 83
Wang, W., 46
Wang, W. C., 515
Wang, X., 140, 493, 495
Wang, X. J., 100
Wang, Y., 67–109, 140, 538, 565–566
Wang, Y. P., 43
Wang, Z., 127, 309
Ward, J. L., 471
Warren, R. F., 133
Warthmann, N., 129, 179
Warwick, S. I., 2, 7–9, 14–15, 17, 20, 33–50, 201, 294, 449, 455, 459–460, 463, 626
Watanabe, M., 396, 401
Waterhouse, P., 573–574
Waterhouse, P. M., 568, 575
Watson, J. M., 572
Watson, L. M., 572
Weber, A. P., 624
Weber, K. L., 624
Weber, M., 625
Wechter, W. P., 49
Weier, D., 310
Weigel, D., 105, 129, 176, 354, 447, 547, 550, 622
Weinig, C., 135
Weisshaar, B., 527–553
Welti, R., 493, 495
Wendell, D. L., 228, 230, 238
Werner, E., 128, 131, 138, 483, 487
Wesley, S. V., 573–575
West, M. A. L., 136, 138
Westermeier, P., 297
Wheeler, D. A., 297
White, P. J., 460
White rust
    B. juncea disease resistance, 338–339
    B. rapa disease resistance, 247–248
Whittle, C. A., 620
Whitton, J., 196
Whole genome duplications
    Brassicaceae genome and chromosomal evolution, 10–12
    in *Brassicaceae*, 197–198
Whole genome shotgun (WGS) sequencing, 185
Widmer, A., 17
Wielopolska, A., 575
Wienholds, E., 538
Wierzbicka, M., 46
Wierzbicki, A. T., 90
Wightman, B., 68
Wikström, N., 5, 7
Wildman, S. G., 324
Willems, G., 350, 364, 366
Williams, P. H., 48, 218, 247–248, 276
Willis, J. H., 15, 196–197
Willmitzer, L., 508
Wilson, I. W., 128
Windsor, A. J., 178, 180, 362
Wingender, E., 77, 80
Winter, D., 590
Winter, H., 48
Wisconsin Ballhead
    B. oleracea disease, 275
Wisconsin Hollander
    B. oleracea genetics, 275
Wisman, E., 545–546
Wittkop, B., 302, 309, 311
Wobus, F., 604
Wolfe, K. H., 184, 196, 564, 620
Wolfender, J. L., 492
Wong, C. E., 46–47, 624–625
Won, S. K., 79
Wood, T. E., 196–197
Wood, T., 561–579
Woods, D. L., 336
Wright, D. A., 546
Wright, K. M., 205

Wright, S., 348, 355–356, 359
Wright, S. I., 348, 353–354, 357–358, 360, 363, 365
Wu, C. C., 100
Wu, G., 100, 102, 302
Wu, G. M., 538
Wu, G. Z., 538, 625
Wu, H., 100
Wu, J., 229–230, 240
Wu, J. S., 205–206
Wu, M. F., 100
WUSCHEL transcription factor, 86
Wu, Y., 302

X

Xanthomonas campestris, 244, 248
  B. oleracea disease, 274
Xiang, D., 602, 624
Xiang, H., 507
Xiang, Y., 515
Xie, D. Y., 93
Xie, Q., 93
Xie, Z., 93, 108, 569–570
Xin, Z., 538
Xu, C., 205
Xue, Y., 393
Xu, P., 572, 576
Xu, T., 46
Xu, Y., 205
Xu, Z., 423
Xu, Z. H., 513

Y

Yamagishi, H., 45
Yamamoto, R., 77
Yamamoto, Y. Y., 72–75, 77, 83, 86
Yamasaki, H., 107
Yang, F., 24
Yang, C., 44
Yang, M. Z., 513
Yang, Q., 278
Yang, T. J., 167, 182–183, 185, 246, 415, 418–420, 428–429, 431, 619
Yang, W. C., 514
Yang, Y. W., 7, 172, 181, 198
Yang, Z. H., 510, 513–514
Yaniv, Z., 40–41, 43
Yan, J. Y., 510, 516
Yan, M., 337
Yanofsky, M. F., 536
Yan, X., 136
Yarnell, S. H., 218, 266–268
Yeast artificial chromosome libraries and clones, 457
Ye, G. N., 515
Yellows (*B. oleracea* disease), 275
Yield traits
  B. juncea traits genetics, 337–338
Yoshikawa, H., 244–245
Yoshikawa, M., 104
You, F. M., 604
Yu, B., 517, 569, 573, 578
Yu, H., 547
Yu, J. M., 354, 358, 360
Yu, S., 303
Yu, S. C., 220, 225, 236, 243
Yu, Z., 205

Z

Zarhloul, M. K., 303
Zeller, G., 176
Zeng, C., 136
Zerbino, D. R., 613
Zhai, J., 138
Zhandong, Y., 515
Zhang, F. L., 236, 510
Zhang, C., 570
Zhang, J., 238, 540
Zhang, L., 46, 136
Zhang, L. G., 220, 223
Zhang, Q., 540
Zhang, S., 539
Zhang, S. F., 238
Zhang, T., 46
Zhang, W., 46, 79, 93
Zhang, X., 129, 136, 185, 238, 428, 432
Zhang, Y., 79, 93, 272, 393, 416
Zhang, Y. X., 46
Zhang, Z., 133
Zhang, Z. W., 238
Zhao, H., 298, 309
Zhao, J., 130, 135, 203, 298–299, 303–304, 310
Zhao, J. J., 219
Zhao, J. Y., 219
Zhao, K., 130, 135, 358
Zhao, X. Y., 217
Zheng, H. G., 48
Zhou, G. K., 102
Zhou, N., 179
Zhu, C., 354
Zhu, J. K., 626
Ziermann, J., 380–381
Zilberman, D., 138

Zilla spinosa
chemical traits, 43
Zimmermann, P., 531
Zinc finger nuclease mediated mutations
Arabidopsis reverse genetics, 546–547
Ziolkowski, P. A., 11, 175, 281
Ziv, M., 514
Zuberi, M. I., 237
Zubr, J., 39, 43
Zum Felde, T., 310
Zunk, K., 6, 374